Handbook of Research on Emerging Developments and Environmental Impacts of Ecological Chemistry

Gheorghe Duca
Academy of Sciences of Moldova, Moldova

Ashok Vaseashta
International Clean Water Institute, USA & New Jersey City University, USA & Academy of Sciences of Moldova, Moldova

A volume in the Advances in Environmental Engineering and Green Technologies (AEEGT) Book Series

IGI Global
DISSEMINATOR OF KNOWLEDGE

Published in the United States of America by
IGI Global
Engineering Science Reference (an imprint of IGI Global)
701 E. Chocolate Avenue
Hershey PA, USA 17033
Tel: 717-533-8845
Fax: 717-533-8661
E-mail: cust@igi-global.com
Web site: http://www.igi-global.com

Library of Congress Cataloging-in-Publication Data

Names: Duca, Gheorghe, 1952- editor. | Vaseashta, A. (Ashok) editor.
Title: Handbook of research on emerging developments and environmental impacts of ecological chemistry / Gheorghe Duca, Ashok Vaseashta, editors.
Description: Hershey : Engineering Science Reference, 2019. | Includes bibliographical references and index. | Summary: "This book examines the environmental impacts of ecological chemistry"-- Provided by publisher.
Identifiers: LCCN 2019026566 (print) | LCCN 2019026567 (ebook) | ISBN 9781799812418 (hardcover) | ISBN 9781799812432 (ebook)
Subjects: LCSH: Green chemistry.
Classification: LCC TP155.2.E58 .E63 2019 (print) | LCC TP155.2.E58 (ebook) | DDC 660.028/6--dc23
LC record available at https://lccn.loc.gov/2019026566
LC ebook record available at https://lccn.loc.gov/2019026567

This book is published in the IGI Global book series Advances in Environmental Engineering and Green Technologies (AEEGT) (ISSN: 2326-9162; eISSN: 2326-9170)

British Cataloguing in Publication Data
A Cataloguing in Publication record for this book is available from the British Library.

Advances in Environmental Engineering and Green Technologies (AEEGT) Book Series

Sang-Bing Tsai
Zhongshan Institute, University of Electronic Science and Technology of China, China & Wuyi University, China
Ming-Lang Tseng
Lunghwa University of Science and Technology, Taiwan
Yuchi Wang
University of Electronic Science and Technology of China Zhongshan Institute, China

ISSN:2326-9162
EISSN:2326-9170

Mission

Growing awareness and an increased focus on environmental issues such as climate change, energy use, and loss of non-renewable resources have brought about a greater need for research that provides potential solutions to these problems. Research in environmental science and engineering continues to play a vital role in uncovering new opportunities for a "green" future.

The **Advances in Environmental Engineering and Green Technologies (AEEGT)** book series is a mouthpiece for research in all aspects of environmental science, earth science, and green initiatives. This series supports the ongoing research in this field through publishing books that discuss topics within environmental engineering or that deal with the interdisciplinary field of green technologies.

Coverage

- Industrial Waste Management and Minimization
- Contaminated Site Remediation
- Biofilters and Biofiltration
- Policies Involving Green Technologies and Environmental Engineering
- Alternative Power Sources
- Radioactive Waste Treatment
- Cleantech
- Sustainable Communities
- Pollution Management
- Air Quality

The Advances in Environmental Engineering and Green Technologies (AEEGT) Book Series (ISSN 2326-9162) is published by IGI Global, 701 E. Chocolate Avenue, Hershey, PA 17033-1240, USA, www.igi-global.com. This series is composed of titles available for purchase individually; each title is edited to be contextually exclusive from any other title within the series. For pricing and ordering information please visit http://www.igi-global.com/book-series/advances-environmental-engineering-green-technologies/73679. Postmaster: Send all address changes to above address.

Titles in this Series

For a list of additional titles in this series, please visit: http://www.igi-global.com/book-series/advances-environmental-engineering-green-technologies/73679

Analyzing the Physics of Radio Telescopes and Radio Astronomy
Kim Ho Yeap (Universiti Tunku Abdul Rahman, Malaysia) and Kazuhiro Hirasawa (University of Tsukuba, Japan)
Engineering Science Reference • © 2020 • 327pp • H/C (ISBN: 9781799823810) • US $225.00

Climate Change and Anthropogenic Impacts on Health in Tropical and Subtropical Regions
Ahmed Karmaoui (Southern Center for Culture and Sciences, Morocco) Abdelkrim Ben Salem (Southern Center for Culture and Sciences, Morocco & Research Center of Plant and Microbial Biotechnologies, Biodiversity, and Environment (BioBio), Faculty of Sciences, Mohammed V University, Rabat, Morocco) and Ashfaq Ahmad Shah (National University of Sciences and Technology, Pakistan)
Engineering Science Reference • © 2020 • 177pp • H/C (ISBN: 9781799821977) • US $225.00

Physiological Aspects of Imprinting and Homing Migration in Salmon Emerging Research and Opportunities
Hiroshi Ueda (Hokkaido University, Japan)
Engineering Science Reference • © 2020 • 200pp • H/C (ISBN: 9781799820543) • US $175.00

Artificial Intelligence and IoT-Based Technologies for Sustainable Farming and Smart Agriculture
Pradeep Tomar (Gautam Buddha University, India) and Gurjit Kaur (Delhi Technological University, India)
Engineering Science Reference • © 2020 • 350pp • H/C (ISBN: 9781799817222) • US $215.00

Predicting, Monitoring, and Assessing Forest Fire Dangers and Risks
Nikolay Viktorovich Baranovskiy (National Research Tomsk Polytechnic University, Russia)
Engineering Science Reference • © 2020 • 417pp • H/C (ISBN: 9781799818670) • US $235.00

Effects of Emerging Chemical Contaminants on Water Resources and Environmental Health
Victor Shikuku (Kaimosi Friends University College, Kenya)
Engineering Science Reference • © 2020 • 339pp • H/C (ISBN: 9781799818717) • US $225.00

Waste Management Techniques for Improved Environmental and Public Health Emerging Research and Opportunities
Sang-Bing Tsai (Zhongshan Institute, University of Electronic Science and Technology of China, China & Wuyi University, China) Zhengxi Yuan (Zhongshan Institute, University of Electronic Science and Technology of China, China) Jian Yu (Research Center for Environment and Sustainable Development of China Civil Aviation, Civil Aviation University of China, China) and Xuexin Liu (Capital University of Economics and Business, China)
Engineering Science Reference • © 2020 • 229pp • H/C (ISBN: 9781799819660) • US $190.00

IGI Global
DISSEMINATOR OF KNOWLEDGE

701 East Chocolate Avenue, Hershey, PA 17033, USA
Tel: 717-533-8845 x100 • Fax: 717-533-8661
E-Mail: cust@igi-global.com • www.igi-global.com

Editorial Advisory Board

List of Contributors

Table of Contents

Section 3
Interaction of Nanoparticles

Detailed Table of Contents

environment and health. The need for contemporary technological societies to develop mechanisms like education, environmental and health governance, and public accountability for environmental health equity and justice is highlighted. The chapter concludes by proposing a multidimensional framework, based on both natural and social sciences, for the transition to healthy and sustainable societies and for improving the welfare of all people, as well as future research directions for environmental health sciences.

The ecological processes have complicated structure. The equations, describing their behavior, also have a complicated structure; hence, to solve them faces with some difficulties. One of the effective ways is to collect data and make statistical data analysis of these data and afterward to make corresponding decisions. As ecological processes depend on a lot of parameters, and moreover, the number of these parameters can increase in time. In the chapter, the conception of regression models with increasing number of unknown parameters is introduced. In the frame of regression models with increasing number of unknown parameters, the approach for estimating of the main parameters of the ecological processes is suggested. Using estimates of unknown parameters, the approach allows to construct a confidence band for unknown functions in these models. Numerical examples demonstrating these results are given.

Long-term prediction of the complex chemical processes and their practical consequences is only feasible by using simulation models. The chapter presents a thermodynamic approach to the study of complex chemical equilibria in natural aqueous multicomponent systems. Its quintessence consists in the thermodynamic analysis of the conditions of various processes on the basis of derived overall thermodynamic characteristics. Ways to shift from chemical-analytical, descriptive characteristics in heterogeneous multicomponent systems "mineral – natural aqueous systems" to the quantitative study of involved chemical equilibria by thermodynamic and computer modeling are extensively described. The analyzed chemical processes include the heterogeneous interaction of minerals and aqueous solution, homogeneous reactions of hydrolysis, complex formation, and acid-base equilibria.

Significant efforts of mankind and huge funds were spent to study the mechanisms of environmental processes. Recent decades have been marked by exponential growth of computer power and the accompanying decrease in the cost of computing. With regard to the mathematical modeling of physical and chemical processes that determine the quality of natural waters, atmosphere, and soil, this has led to the development of an extensive approach based on an increase in the number of components and reactions between them taken into consideration. In this chapter, the authors describe features of ecochemical systems and discuss the moments that complicates their prediction. Using the method of numerical experiment, they investigate the behavior of periodic systems with multiple stationary states. One conclusion is that the actual manifestation cannot be used to determine at what stage the impact occurred and to what stage of the food chain it relates. Another conclusion is that systems involving multiple stationary states are prone to bifurcations and chaotic jumps from one limit cycle to another.

The general model of global carbon circulation (in particular 14C) have been known for half a century and includes both natural and anthropogenic sources and sinks. The main problem was the lack of information about rates of migration from one form to another. The authors made fitting of model data concerning both 12C and 14C dynamics in atmosphere and in sea water. The 70-times discrepancy between generation and decay rates for radiocarbon was explained. The main channel of 14C disappearance is not decay but dissolution in sea waters. By this reason, they deny the Suesse hypothesis because atmospheric carbon dioxide is diluted not by 14C-free substance (after burning of fossil fuels) but from the source of oceans. Multiple numeric experiments modeling 14C pulse or permanent input to the environment had demonstrated the exclusive stability of natural system. It obeys homeostatic principle and resists any attempts to shift steady state concentration. So, all speculations about extraordinary threat from technogenic 14C from nuclear industry do not have any scientific base.

Redox processes involving hydrogen peroxide, as oxidizing agent, and compounds of sulfur as carrier of reducing equivalents form the quality of natural waters. The inflow of reductants interacting with H2O2 can lead to a toxic quasi-reductive state. Dynamic change of redox in a natural aqueous medium is

Water pollution represents the direct or indirect change of its normal composition as a result of human activity in such extent that is affecting all other possibilities to use the water in its natural state. This study was carried out to investigate the seasonal variability and distribution of heavy metals in the waters of Somesul Mic (Romania), Bic (Republic of Moldova), and Prut (Romanian-Moldovan natural border) rivers. Water samples were collected from nine sampling sites in Autumn 2016, Spring and Summer 2017. Dissolved metals concentrations were measured using inductively coupled plasma mass spectrometry (ICP-MS) and flame atomic absorption spectrometry (AAS). The metal index (MI) was calculated to assess the contaminations of the Somesul Mic, Bic, and Prut rivers. MI values revealed that Somesul Mic and Prut rivers are pure to moderately affected by metal pollution, while Bic river is slightly to strongly affected by metal pollution.

The rapid increase in population and urbanization leads to the scarcity of water resources in the present era. Therefore, effective wastewater treatment is a prerequisite for a growing economy. Development and implementing the advanced treatment technologies of wastewater with high efficiency and low capital is difficult. In the recent advancements among various treatment processes, nanomaterial science has been attracting the attention of researchers. However, limited collective knowledge is available in this context. The chapter reviews the potential of nano catalysis's process, mechanism, and current drawbacks in treatment technologies. It explains the different nano catalysts that are widely utilized for the treatment and removal of organic and inorganic pollutants in water and wastewater and discusses the nano-based photocatalytic, nano-based electrocatalysis, nano-based Fenton catalysis and their efficiency in various removal of pollutants from wastewater.

Nanotechnology is one of the most reliable techniques for the remediation of heavy metals. As nanoparticles have a higher surface area to volume, ratio, and high surface energies, so nano-based absorbents are very efficient. Adsorption technique is the most preferred for the remediation of wastewater pollutants. In the current study, a comparative study was done between bio sorbents, nanosorbents and bio nanosorbents. The chapter studies with the synthesis and characterization of the bio sorbents, bionanosorbents, their mechanism of sorption, their synthesis, in addition, application for the remediation of heavy metals from wastewater. The fly ash is an industrial byproduct. Biosorbents have immense applications in the field of bioremediation of heavy metals. Further, their components have also enhanced removal efficiency from the wastewater.

Anaerobic digestion of organic substrates in wastewaters from agro-industrial sector can be efficient method to reduce the pollutant contents, and also a source of biogas. The research is focused on developing of intensified biogas technology and integrated equipment, based on a series of original approaches. New types of new phyto-catalysts of methanogenic digestion were proposed: aescinum, squalene, and betulinol, introduced in the digested biomass in micro-concentrations of 10-3-10-5%. The proposed substances promote about 1.5-2 times acceleration of methanogenic process and increase biomethane contents in biogas close to the natural gas level – up to 93-97%. As a feedstock for biochemical digestion, post-distillery vinasse in mixture with farmyard manure was used. By combining of biochemical treatment with microfiltration an advanced treatment degree was reached, reducing COD from 28530 mg O2/L to 187 mg O2/L. In bioreactor proposed, biomass using coefficient reaches 85-93%, as compared to 60% in the conventional biogas processes.

Uninterrupted supply of adequate and clean drinking water is a fundamental human necessity, and essential to maintain the healthy surroundings. It also influences the development and economic growth, specifically in developing countries. It provides an excellent opportunity for treatment of surface water, groundwater, and wastewater contaminated by toxic metal ions, organic and inorganic solutes, and microorganisms. It facilitates an opportunity to optimize the current conventional techniques and to offer novel and innovative methods of water purification. Various nanomaterials on which purification of contaminated drinking water is based includes nanostructured catalytic membranes, nanosorbents, nanocatalysts, and bioactive nanoparticles. Current and potential applications of nanotechnology in water purification are adsorption, membrane separation processes, photocatalysis, disinfection, and microbial control. This chapter reviews the recent research efforts accomplished in the area of nanotechnology for drinking water purification.

Remote sensing is an efficient way to detect environmentally vulnerable terrain, where waste chemicals and excess fertilizers must result to considerable pollution of soil and groundwater. The indicators of such vulnerable formations were analyzed and applied for investigation of ground microdepressions. First indicator was the humidity level that was well recognized from the satellite images by calculation of normalized water index (NWI). Another active indicator used for the detection was the relief. The method of relief plastics joint with morphodynamic analysis was suggested for the data processing. The scheme of water basins and the map of environmentally fragile areas were created by combination of the indicators of soil humidity and topography. On the basis of the created maps, the restrictions for used chemicals can be developed in order to preserve considerable environmental impact of the pollution.

Halogenated hydrophobic organic compounds (HOCs) have been used in various industrial applications and are present in many commercial products. Due to their emissions during manufacturing and discharges as wastes, halogenated HOCs such as polychlorinated biphenyls and polybrominated diphenyl ethers are ubiquitously found in the environment and create contaminated sites. To remove the contamination from these sites, various remediation techniques have been useful. The purpose of this chapter is to investigate the impacts of traditional and emerging remediation techniques on ecosystem. One of the traditional remediation techniques is dredging and the mostly studied emerging remediation techniques are bioaugmentation and biostimulation. The efficiency of these techniques is also evaluated regarding reduction in contaminant mass. Overall, this chapter presents the efficiency and possible impacts of dredging, bioaugmentation and biostimulation of soils and sediments, and the implications include the evaluation of most feasible remediation techniques by using life cycle assessment.

Section 3
Interaction of Nanoparticles

In this chapter, the authors discussed the recent as well as significant development in the synthesis of ionic liquid-mediated nanoparticles (considered a less toxic way to make nanoparticles). Apart from that, they also disused a small background of ionic liquid considering types, properties, and effect of ionic liquid's counterparts on nanoparticles synthesis. There are promising consequences proposing that the ionic liquid route can lead the design and fabrication of nanoparticles through the variety of size, nature, configuration, and functionality. All discussed examples in this chapter clearly depicted that ionic liquids add great value to the area of nanoparticle synthesis.

Most countries worldwide face the problem related to clean and safe water. The major source of contamination of water bodies includes the discharges of wastes from various domestic and industrial sources (heavy metal ions, dyes, and pharmaceuticals). To overcome those problems, various water purification methods such as coagulation, flocculation, adsorption, membrane separation, biological, and electrochemical methods have been engaged. Adsorption using nano-biosorbents based on chitin, chitosan and modified chitosan, cellulose, alginate, micro algal has emerged as a better alternative for the removal of contaminants from drinking and wastewater treatment. The main advantages of these nano-biosorbents include its biodegradability and eco-friendliness, which have attracted researchers to this field.

Zinc oxide (ZnO) nanoparticles prepared using simple co-precipitation method are characterized and photocatalytic activity is tested on the degradation of methylene blue and rhodamine B organic pollutants. Morphological and structural properties of synthesized nanomaterial have been characterized using FESEM, EDAX spectroscopy, and XRD, while UV-visible DRS spectroscopy and photoluminescence have been used to understand their optical properties. The photocatalytic behaviour of synthesized nanoparticles was evaluated on the degradation of methylene blue (MB) and rhodamine B (RhB) organic pollutants under solar light irradiation. The highest degradation was achieved for MB (100%) over RhB (96%). Preliminary investigation shows the effective degradation of organic pollutants by ZnO nanoparticles.

Foreword

Pollution of the environment produced by human activity, along with climate warming, becomes one of the most threatening and challenging problems of the human society on our planet. In this respect, the publication of the present book devoted to environmental impacts on ecology and the latest developments in ecological chemistry in fighting pollution problems should be most welcome.

The book *Handbook of Research on Emerging Developments and Environmental Impacts of Ecological Chemistry* is a collection of chapters with up-to-date information, original research, reviews, and discussions by authors from different parts of the world, in entirety presenting a picture of what is presently being done in the field of Ecological and Environmental Chemistry. It reflects also the current trends in the specific areas of research, aimed at proposing solutions to emerging ecological, man-caused problems by applying an arsenal of methods and approaches from chemistry, biology, physics, mathematics, and geography. The more specific problems considered in this book include health issues related to toxic products and carcinogens, agriculture related food safety, atmospheric air quality, soil systems and the effects of pollutants on their quality, methods of assessment of water, air and soil quality, and several chapter devoted to clean water-related problems. The latter include experimental and theoretical models of both polluted and natural water bodies, and improvement of biogas production in the treatment of strongly polluted (with organics) wastewaters, as well as employing advanced and efficient oxidation systems and photocatalysts based on nanomaterials.

It goes beyond doubt that this book is important to a wide, interdisciplinary variety of students and researchers involved in ecological chemistry and related problems, as well as to the people in health care, industry, and agriculture, engaged in the pollution and human safety issues discussed in this book. The work done by its Editors, Professors Ashok Vaseashta and Gheorghe. Duca, should be highly appreciated.

Isaac B. Bersuker
Department of Chemistry, The University of Texas at Austin, USA

Preface

It is well known that industrial waste discharge modifies the chemical composition of the environment and may render negative impacts on the biosphere. Furthermore, these pollutants enter the water, the food supply-chain and the ambient air, impacting living organisms and human health. As of this writing, major cities worldwide are in a state of continuous expansion. However, booming global economic growth is a double-edged sword, while enhancing the standard of living, it also presents enormous challenges for environmental ecology and associated infrastructures, which need to sustain long-term well-being for the habitat. This dichotomy may be humanity's greatest challenge ever: sustaining high living standards for a large fraction of humanity while simultaneously preserving the ecological integrity of the environment. Hence, current human civilization will continue to face a direct threat from anthropogenic pollution, which will most likely increase with future industrial development.

On a global scale, industrial activity has reached significantly high levels leading to changes in the heat and energy balance, impacting weather conditions in some regions, and, perhaps even the normal climate. As scientists, the most often asked questions are: is humanity responsible for increasing the water level in the oceans and seas, is there a slight rise in global temperature, and are variations in weather patterns observable? It is therefore, essential to examine how the chemicals produced and new and improved materials released into the environment affect, for example, the quality of water, and how negatively this in turn affects the health and well-being of humans, plants and animals. The chemical aspect of these interactions is related to the qualitative and quantitative composition of the chemical impurities and to their transformations in the environment.

Emerging developments from research activities and their topical directions are strongly linked with the most stringent needs and requirements of society - first and foremost, those related to human health, the environment, food chain safety and quality, efficient and environmentally friendly agriculture, and well-grounded environmental policies and education. These challenges, although seemingly different at first glance, nevertheless relate to one and the same problem: how to ensure a safe and sustainable existence and the continued development of human civilization on our planet.

Ecological chemistry is in emerging scientific field that is unraveling the processes that determine the composition and chemical properties of the environment and the interaction and impact of biological diversity and habitation. Ecological chemistry is closely linked to other scientific disciplines, the most important being the field of physical chemistry that includes the study of mass transfer processes in the environment; chemical kinetics, such as pollutant transformation mechanisms; analytical chemistry, such as physico-chemical methods of analysis; and biochemistry illustrating molecular mechanisms regarding the influence of pollutants on vital processes. Thus, the environmental impacts of innovation technologies, new materials and substances, and modern methods of energy conversion open a Pandora's box of questions for scientists, ecological chemists, policy- and decisions-makers, social and mass media reporters, and concerned citizens. Ecological chemistry includes the study of anthropogenic and natural contaminants; the extent to which pollutants are harmful to human beings, animals and plants; the rates of degradation and the pathways of migration of hazardous materials and ways to neutralize them; and the scientific assessment of risks and uncertainties - all focused on what we can do to protect all life on our unique planet. As this ecological chemistry continues to unfold, this subject demands immediate attention by the scientific community to ensure our continued survival going forward.

Our natural surroundings and their components, including the atmosphere, surface and oceanic waters, lithosphere, soil, flora, and fauna, continue to be subjected to an unprecedented level of anthropogenic impact. Polluting chemicals and industrial wastes entering the environment disrupt the steady-state positions, homeostatic cycles, characteristic times and other chemical-physical parameters of the circulation of substances in the biosphere and render negative impacts on habitats; chemical composition, and the structure of biocenosis, interfering with the natural biota and the natural self-cleaning processes. Pollutants penetrate water and intrude in the food supply-chain and ambient air, impacting living organisms and human health. Anthropogenic activities have reached astronomic levels globally, as is evident by the shifts in the heat-energy balance in certain urban settings, aerosol smog, ozone depletion in the stratosphere and, quite the contrary, active generation near the ground surface. Dozens of biological species are disappearing, and many of the remaining species are changing their usual habitats.

To investigate the reasons for stressed water quality, it is critical to study how ecological chemistry and other environmental sciences can help the human race. It is vital to note that changes in the environment are occurring faster than the development of methods to counter them and that decision makers can and should implement policies to slow the progression of these changes. Thus, we witness the progression of unfavorable ecological phenomena without having the means and ability to prevent them. It is hoped that this book will help to enhance awareness and understanding – as one of the most important steps to mitigate such ecological impacts. In addition to testing physico-chemical methods to assess pollution levels, it is crucial to implement principles of green chemistry along with practices to reduce stress in water and green-house gas emission in the environment.

The thematic objective of this book, *Handbook of Research on Emerging Developments and Environmental Impacts of Ecological Chemistry*, is to present several research viewpoints and to identify and formulate new approaches to the processes of chemicals and their behaviors within the environment. While the notions of Ecological Chemistry and Environmental Chemistry emerged in the mid-1950s, the formation of these scientific areas is rapidly becoming one of the most essential directions in research. This area of scientific studies seems to be among the most significant ones, playing a key role in modern chemistry and revealing the relations between the living world and the surrounding environment. Compared to other fields of chemistry and chemical technology, this domain demands a comprehensive investigation.

This book consists of 25 interlinked chapters that cover topics ranging from pollutants in the environment and their health impacts to toxicity and potential health hazards due to incidental and engineered nanoparticles. The chapter begins with a brief overview of environmental health challenges, exposure and research trends and how such challenges provide pathways to moving towards a healthy and sustainable future (Papadopolou, Marouli). A holistic approach to ecological chemistry is introduced through several chapters on statistical methods in ecological processes and kinetic considerations of natural systems. Several chapters provide the theoretical basis and history of the evolution of Ecological Chemistry vs. Environmental Chemistry; since both play a significant role in modern chemistry, the chapters also cover the associations between the living world and the surrounding environment. In another chapter, food safety is discussed with its links to Ecological Chemistry (Sturza). As there is great interest among both researchers and readers, one of the chapters focuses on studies describing the fate, transformations and health effects of potential carcinogens (Gonta).

It should be noted that most of this book's chapters are connected to water-related issues, which is not surprising, especially when considering that humanity faces ever-growing water pollution every single day and that the quality of life and health are directly dependent on water's availability and quality. The water-related studies featured in this book demonstrate the theoretical approaches (Povar) or show field and laboratory research results (Scurlatov, Gladchi, Duca). These studies have been conducted using a model of polluted waters or natural water bodies. Special attention was paid to the improvement of biogas production systems (Covaliova) during the treatment of wastewaters that were strongly polluted with organic compounds. Another approach to water treatment covered in this book uses advanced and efficient oxidation systems, along with the elaboration of photocatalysts based on nanomaterials (Karnena, Srivastava, Dutta, Sidorenco).

Several other natural environmental aspects were also studied by the authors who contributed to this book, including the soil systems and the effects of pollutants on their quality (Demirtepe) and atmospheric air quality. Also covered in this book is research on kinetics and the mechanisms of chemical processes which occur in environmental systems that make it possible to predict changes in air quality and elaborate on preventive measures (Travin). In this regard, the methods to assess water, air and soil quality receive specific importance (Zincovskaia, Romanciuc). The well-known fact is that without proven analytical and assessment methods, research results may be incorrect and may trigger unreliable conclusions.

Finally, a teaching methodology is essentially the way in which a teacher chooses to explain or teach so that students learn the material. There are many different methodologies that can be utilized by a teacher, and the methods chosen often depend on the teacher's educational philosophy and preferences. In one chapter, an interesting approach to introduce Ecological Chemistry through popular scientific articles is presented (Kupatadze). Several thematically related chapters on Ecological Chemistry and Environmental Chemistry as well as related topics are also featured in this book. It is anticipated that due to ongoing awareness, this book may need to be updated every 2-3 years in the form of a series. The subsequent editions will consist of chapters on societal implications and network-to-network knowledge transfer to promote this ecologically relevant topic.

The book is oriented towards professional scientists and students, who have at least preliminary knowledge in math, physics and chemistry. Nevertheless, a much wider audience can find interesting details in the text between formulas and equations. The editors acknowledge the contributions of several authors and reviewers who assisted us by providing their independent assessment of chapters. Such feedback was critical to the quality of this book. The editors also acknowledge the publisher IGI for its guidance and help with the book. We certainly hope that may researchers, students and scientists benefit from these chapters and look forward to the possibility of becoming contributors to our subsequent editions.

Disclaimer: The opinions, recommendations, and propositions presented in various chapters represent the findings, views, and opinions of their authors and may not coincide with the position(s) of the editors. The editors do not promote, endorse, or express their dissent against any product mentioned in the chapters. Although due precautions are exercised before publication, the editors, however, do not take any responsibility for infringement of any copyrighted material of either their own or copied from elsewhere.

Gheorghe Duca
Academy of Sciences of Moldova, Moldova

Ashok Vaseashta
International Clean Water Institute, USA & New Jersey City University, USA & Academy of Sciences of Moldova, Moldova

Section 1
Environment, Ecology, and Processes

Chapter 1
Current Environmental Health Challenges: Part I – Exposures and Research Trends

Paraskevi Papadopoulou
https://orcid.org/0000-0002-8194-7495
Deree – The American College of Greece, Greece

Anastasia Misseyanni
Deree – The American College of Greece, Greece

Christina Marouli
Deree – The American College of Greece, Greece

ABSTRACT

This is the first of two overview chapters of important contemporary environmental health challenges. The exciting developments in the environmental health fields are approached in an interdisciplinary manner covering cutting-edge scientific developments and research. In the first chapter, environmental exposures to a variety of toxins, diseases, and stressors that challenge the individual and affect public health are examined. The handling, storage, big data management related to medical and health-informatics are discussed. Issues such as single gene polymorphisms, gene expression, transcriptomics, epigenetics, metabolomics, exposure to carcinogens, endocrine disruptors, heavy metals, physical hazards, airborne particulates, quality of food and water, toxin metabolism, bioinformatics, and exposome analysis are considered. Important recommendations and solutions are provided emphasizing the collaboration between researchers/scientists and the community.

DOI: 10.4018/978-1-7998-1241-8.ch001

INTRODUCTION

Environmental concerns in contemporary societies continue to be high. They are becoming more challenging today because they are determined by multiple interrelated factors leading to a cascade of effects, with multiple impacts on human health and other environmental components. Scientists, researchers and people from all around the world have become aware of various environmental health challenges and the increased need to improve health and healthcare in an efficient and effective manner. Many improvements and technological innovations have been implemented in relation to environmental issues (e.g., increased use of renewable energy sources, cleaner vehicles, air pollution abatement and improvements in sanitation and water quality, biodegradable plastics, etc.); however, environmental problems persist and some have become worse. Climate change is surely a major concern both for the environment and for human health. More recently, in May 2019, the chair of the Intergovernmental Science-Policy Platform on Biodiversity and Ecosystem Services (IBPES) Robert Watson stated, "The health of the ecosystems on which we and other species depend is deteriorating more rapidly than ever. We are eroding the very foundations of economies, livelihoods, food security, health and quality of life worldwide" (IBPES, 2019). More than 450 scientists and diplomats after a three-year study reveal in their United Nations (UN) Global Assessment Report that 1 million species are at risk of extinction. This type of ecosystem assessment goes much further by looking not just at an inventory of species, but at the web of interactions between biodiversity, climate and human wellbeing. According to the report, the current global response is insufficient; *'Transformative changes'* are needed to restore and protect nature. Most scientists are in agreement with Robert Watson's statement, "We have lost time. We must act now." For more information, see the reports and publications of the Intergovernmental Science-Policy Platform on Biodiversity and Ecosystem Services (IPBES, 2019) and of the UN Office for Disaster Risk Reduction (UN, 2019).

Although we have experienced some significant health achievements in the 20th century, the quality of life and the goal of having "healthy people" continue to be challenging, especially as environmental problems continue to be significant. There is an increasing need to effectively connect environmental factors and human health challenges today. Environmental health aims, among other things, to prevent disease and to-promote health supportive environments.

This chapter (Part I) presents some major contemporary environmental health hazards/risks and their impacts. As these challenges are many and complex, there is a general understanding that, unless people decide in a collaborative and interdisciplinary way to commit themselves to address them, the future of humanity is at risk. The authors of this chapter aim to add a positive contribution toward raising further awareness of contemporary environmental exposures/hazards, risks and health impacts and also offer some insights for environmental and health policy development.

This chapter's (Part I) aims and scope are to:

- provide an overview of the most important contemporary environmental health challenges,
- review the exciting developments in the environmental health field drawing from diverse bodies of knowledge, including chemistry, biology, physics and informatics, and to identify recent trends in the field. More specifically, Part I presents current issues of concern in the relevant sciences, such as single gene polymorphisms, genome sequences, gene expression data, transcriptomics, metabolomics, exposure to carcinogens, endocrine disruptors, heavy metals, physical hazards such as airborne particulates, quality of food and drinking water and other toxins, their metabolism and health impacts, bioinformatics and exposome analysis.

- provide cues and a multidimensional framework regarding the direction towards which environmental health research and health care should move. Acknowledging the contribution of machine learning methodologies and big data mining and analytics to contemporary clinical science and practice, we propose that turning 'big" and "smart" data to "wise" data becomes a main challenge that needs to be addressed today (Papadopoulou, Lytras and Marouli, 2018) on this level. The socio-political aspects of our framework will be analyzed in our second chapter (Part II).

Figure 1. Environmental exposures and human health impacts
Source: Our own

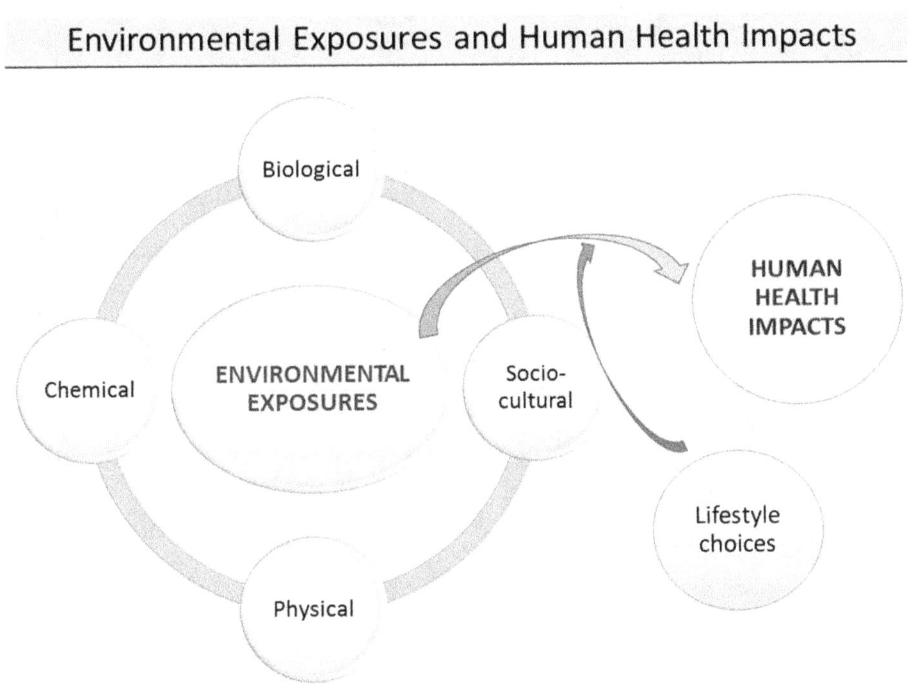

CONTEMPORARY ENVIRONMENTAL CHALLENGES: AN OVERVIEW

Environmental exposures related to hazards (see Fig. 1, Fig. 2 and Table 1) can be described as anything in our environment that can cause injury, disease, death, economic loss, damage to human property or damage to environmental components. Hazards can be of a different nature: either imposed by the external environment such as biological, chemical, physical or cultural hazards, or associated with lifestyle choices (e.g., smoking or alcohol abuse). Risk expresses the probability of suffering harm because of exposure to a hazard. Risk analysis is an essential process in estimating the severity of the hazard and the probability of suffering harm from it, as well as for taking preventive measures to manage risk and protect human health.

Table 1. Types of hazards and risks with major examples

Biological hazards	Physical hazards	Chemical hazards	Cultural & lifestyle-related Hazards
• Allergies • Anthrax • Arbovirus • Avian influenza • Bovine spongiform encephalopathy (BSE) • Cholera • Ebola • Epidemics • Food poisoning • Malaria • Molds • Onchocerciasis (river blindness) • Pandemics • Pathogens • Pollen for allergic people • Rabies • Severe acute respiratory syndrome (SARS) • Sick building syndrome	• Cosmic rays • Desertification and drought • Earthquakes • Electromagnetic fields • E-waste • Floods • Fires • Fog • Light pollution • Lighting • Noise pollution • Quicksand • Ultraviolet radiation • Vibrations • X-rays • Soil degradation (erosion and loss of fertility) • Urban conditions (e.g. overcrowding, housing conditions) • Climate change (extreme temperatures, extreme weather phenomena, etc.)	• Antibiotics in animals destined for human consumption • Airborne oxides of nitrogen • Airborne sulphur dioxide and sulphates • Asbestos • Bisphenol A • Carbon monoxide • Carcinogens • Dioxins • DDT • Endocrine disruptors • Explosive material • Fungicides • Furans • Haloalkanes • Heavy metals, especially mercury, lead, cadmium, chromium and arsenic • Herbicides • Hormones in animals destined for human consumption • Marine debris • Mutagens • Ozone • Pesticides • Photochemical oxidants and reactive hydrocarbons • Polycyclic aromatic hydrocarbons • Polychlorinated biphenyls (PCBs) • Phthalates • Radon and other natural sources of radioactivity • Soil pollution • Suspended particulate matter • Toxic waste • Toxins of biological origin (from algae, fungi, and bacteria)	• Alcohol addiction • Cultural • Drug Use • Lifestyle choices • Poor food choices • Stress • Violence • School stressors • Sedentary lifestyle • Tobacco smoking • Workplace stressors

As mentioned above, environmental hazards can be classified into biological, chemical, physical, cultural and lifestyle choices.

Biological hazards refer to disease-causing organisms such as bacteria, viruses and parasites. Communicable diseases transmitted through contaminated water, food, air or human contact are a major cause of death in less developed countries. **Chemical hazards** include all harmful chemicals to which people get exposed in their environment (air, water, soil, food or human made products). Major pathways of exposure include ingestion (oral pathway), inhalation (respiratory pathway), and absorption through the skin. **Physical hazards** can be natural phenomena such as fires, earthquakes, hurricanes and volcanic eruptions, but also physical agents such as electromagnetic radiation (ultraviolet, radio waves), noise, nanoparticles and climate change. **Cultural hazards** are associated with the social environment of an individual; e.g., conditions such as an unhealthy workplace, poverty and criminality. **Lifestyle choices** refer to those hazards that are a matter of individual choice, or at least are influenced by choice. Examples include smoking, overeating, alcohol or drug abuse and engaging in risky activities such as extreme sports or driving too fast.

A short discussion of the main contemporary environmental health picture focusing on the different categories of hazards follows:

Biological Hazards

Major biological hazards of today are listed in table I. Respiratory infections, diarrheal diseases, intestinal nematode infections and malaria account for millions of deaths every year (Pruss-Ustun, Wolf, Corvalán, Bos and Neira, 2016). **Respiratory infections** are the most important cause of mortality in children; they account for 18% of deaths in children under five; for over 900,000 deaths every year (Pruss-Ustun et al., 2016). Pneumonia, an infection of the lungs, is caused by exposure to a variety of infectious agents (bacteria, viruses, fungi; Pruss-Ustun et al., 2016). Household air pollution, ambient air pollution and second-hand tobacco smoke are important environmental risk factors for respiratory infections. Lower respiratory infections show sensitivity to climate change, as weather patterns affect their incidence. Upper respiratory infections (e.g. pharyngitis, laryngitis) seem also to be associated with housing-related risks, such as exposure to ventilators or molds, and crowding.

Diarrheal diseases account for 20% of all deaths in children under five years and are caused to a large extent by fecal-oral pathogens. Inadequate drinking water, sanitation and hygiene are the most important environmental risk factors for such diseases (Pruss-Ustun et al., 2016). It has been estimated that environmental factors account for 58% of diarrheal cases in low- and middle-income countries and for 842,000 deaths annually.

Nematode infections, such as ascariasis, trichuriasis and ancylostomiasis, are infections of the intestinal system caused by soil-transmitted helminths, affecting more than 2 million people worldwide (WHO, 2012). Transmission of such diseases mainly occurs from soil contaminated with human excreta that may contain infectious eggs or larvae. Improved sanitation and hygiene, as well as health education, are measures that will reduce morbidity from such infectious diseases.

Malaria, an infectious disease caused by the parasite *Plasmodium* and transmitted via the *Anopheles* mosquito, is the most important vector-borne disease around the globe, accounting for over 500,000 deaths annually (Pruss-Ustun et al., 2016), mostly among children in Africa. Environmental vector management seems to be the most efficient way of addressing this environmental threat.

HIV/AIDS affects mostly Sub-Saharan Africa. The HIV virus is transmitted through sexual contact mainly, but also through drug injection in some regions; 35 million people were infected in 2013. It also constitutes an occupational risk mainly for commercial sex workers and health care workers (cultural hazard). Behavioral interventions would be important to reduce the incidence of HIV infections (Wariki et al., 2012) and other sexually-transmitted diseases.

Tuberculosis, caused by the bacterium *Mycobacterium tuberculosis,* is the cause of death for 1.5 million people per year, 95% of which live in low and middle-income countries (WHO, 2014). Risk factors include crowding in houses (Baker, Venugopal & Howden-Chapman, 2011), malnutrition (Jaganath & Mupere, 2012), exposure to indoor smoke from solid fuels (Sumpter & Chandramohan, 2013) and second-hand smoke from tobacco (Leung et al., 2010). Occupational groups that are more likely to suffer from increased risk of tuberculosis are miners who are exposed to silica dust, and health-care workers who come in contact with tuberculosis patients and detained populations.

Other diseases are listed in Table 1.

Chemical Hazards

Examples of major chemical hazards are included in Table 1.

While global chemical production is on the rise, with the increase taking place mainly in non-OECD countries (BRIICS: Brazil, India, Indonesia, China, South Africa), concerns about the impact on human health rise, especially in the countries where this increase in the production takes place. A need for sound management of chemicals becomes imperative (OECD, 2012). Humans are exposed to a wide and increasing range of chemicals and the effects of long-term and low dose exposure to mixtures of chemicals are not well understood. Carcinogenic and mutagenic chemicals and endocrine disruptors are among the groups of chemicals that are of particular concern. It is also known that early human life stages are more sensitive to chemicals; a growing number of studies show a link between early life or even prenatal exposures and adult diseases such as cancer (Grandjean & Landrigan, 2014; WHO, 2012).

Heavy Metals

Heavy metals such as mercury, cadmium, arsenic, chromium and lead, known for their toxicity, have decreased in terms of the amount of emissions because of regulations, but remain a problem today. Their effects on human health have been studied and are better understood today; however, there are new metals to which humans are exposed (e.g. tungsten, thorium, platinum, palladium, rhodium, etc.), the health effects of which are still unknown.

Lead is known for its neurotoxic effects and its negative effect on the neurodevelopment of children, leading to a loss in IQ points. Unfortunately, no threshold level for this effect on the IQ has been identified; so, although lead levels have experienced significant declines following source reduction (e.g., phasing out of lead in gasoline mixtures), further measures are needed to prevent exposures to low levels in the environment (EEA & JRC, 2013).

Cadmium, a heavy metal that is toxic to the bones and kidneys and is linked to lung cancer, has also seen a decline in ambient air concentrations, at least in Europe (EEA & JRC, 2013); smokers are exposed to high levels of cadmium through tobacco smoke, while non-smokers mainly get it through their diets. Despite the overall decline in its levels in the air, it seems that its levels in the bodies of non-smokers have not decreased in the past decades (WHO, 2007) and that some populations, especially vegetarians, children, people living in contaminated areas and smokers may be exposed to levels that exceed the tolerable weekly intake (TWI) (EFSA, 2009). This means that measures should be taken to reduce exposure to it at the level of the population (EFSA, 2011).

Mercury and its compounds remain a concern for public health. All forms of mercury are toxic, but the one that is of particular concern is methylmercury, known for its neurotoxic effects, especially in the prenatal stage (fetuses) and in young children. Mercury is emitted from a variety of sources (including fluorescent light bulbs, batteries, silver and gold extraction and power plants) and can be extensively transported through air to remote places such as the Arctic. Studies have shown that people consuming fish and sea mammals have higher concentration of mercury in their body (WHO, 2007; Castano et al., 2012). International policy on mercury has existed since the 1970s and the most recent one is the "Minamata Convention on Mercury" (UNEP, 2013).

Arsenic has been linked by a good number of studies to various health problems like skin diseases, anemia, bronchitis, gastrointestinal problems, hormonal imbalance and cancer. A major source of arsenic is drinking water, as in some parts of the world it is present in high levels in the groundwater that is used for drinking purposes. Arsenic compounds were widely used as pesticides in the past; today, arsenic is also released from mining, metal processing and combustion processes. Studies have shown that both prenatal and postnatal exposure to inorganic arsenic (iAs) is detrimental to the health of the individual (Chikkanna, Mehan & Ghosh, 2019); (Kumar & Ghosh, (2019). As an example, Rojas et al. (2015) examined a set of 16 genes which were correlated with iAs-associated changes in DNA methylation and mRNA expression and all were highly enriched for binding sites of the early growth response (EGR) and CCCTC-binding factor (CTCF) transcription factors. This complex interplay between DNA methylation, functional changes in gene expression and health outcomes seems to increase the risk of disease development later in life.

Unfortunately, consumer products are also sources of heavy metals. Waste electrical and electronic equipment (WEEE) contains, among others, heavy metals and rare earth metals, and needs to be properly disposed of. Of particular concern is the unsafe disposal of this waste in less developed countries that exposes people to these chemical hazards (Waeger et al., 2012).

Persistent Chemicals

We continuously get exposed to a wide variety of persistent chemicals that have the ability to bioaccumulate in living organisms or can be biomagnified through the food chain. Synthetic organic chemicals such as those present in pesticides, plastics, textiles, cosmetics, electronic goods, are used in large quantities, and humans get exposed to them through consumer products. Additionally, when WEEE is discarded, its hazardous components, namely brominated flame-retardants and heavy metals, easily migrate to the environment.

Some of the **persistent organic pollutants (POPs)** include chlorinated hydrocarbons such as DDT and other first-generation pesticides, phthalates, PCBs, bisphenol A, perfluorinated chemicals and brominated flame-retardants. Studies link them with adverse health effects (WHO, 2012); biomonitoring studies indicate their presence in the environment and in humans (Koch & Angerer, 2012). Pharmaceutical products released to the environment from production processes, household waste and sewage treatment plants are an additional source of concern. Epidemiological studies on the effects of postnatal exposure to neurotoxic POPs through breast-feeding have been conducted. Gascon et al. (2013b) in their study showed that breastfeeding increases children's blood persistent organic pollutants (POPs) levels during postnatal life; deleterious effects of polychlorinated biphenyl-153 (PCB-153) on neuropsychological development are mainly attributable to prenatal exposure.

Phthalates (Diethyl phthalate and di(2-ethylhexyl) phthalate) are substances used as plasticizers, to improve the properties of plastics (flexibility, durability, etc.) such as PVC, so they can be found in a variety of consumer products, including food packaging, children's toys, baby care products, chemical stabilizers in cosmetics, vinyl floors, electronics, pharmaceuticals and textiles; they can easily leach to the environment because they are not chemically bound to PVC. There is concern for the effects of these substances on testes, kidneys, and fertility, as well as for their developmental toxicity for people exposed to them (JRC, 2008). Preau, Wong, Silva, Needham & Calafat, (2010) report that phthalates are metabolized and eliminated in urine within hours after exposure and when collecting multiple spot urine samples, changing the time of collection may provide the most complete approach to assess exposure to diverse phthalates.

Bisphenol A (BPA) is a substance that is mainly used in the production of epoxy resins and polycarbonate plastics, so it is present in many consumer products. Studies have measured its leaching in baby bottles and food containers, thus humans are constantly exposed to it. Due to its structural similarities to the female sex hormone estrogen, BPA was shown to have weak estrogen-like properties in animal studies and cell-based test systems; this indicates it may act as an endocrine disruptor and may account for a number of endocrine-related diseases that are increasingly being observed among European populations (e.g., decrease in sperm counts, increase in breast cancer in women and in testicular cancer in men, as well as developmental malformations in male newborn babies) (EEA & JRC, 2013).

Polychlorinated biphenyls (PCBs) are a group of industrial organochlorine chemicals that are persistent in the environment and can bioaccumulate in living systems. Because of their widespread use as coolant fluids in power transformers and capacitors or as plasticizers, and because of careless disposal practices, they became common environmental contaminants. They are suspected to cause adverse health effects in humans. Ulaszewska et al. (2012) measured breast milk concentrations of PCBs and utilized modeling approaches based on environmental data to predict the human dietary intake, hoping that their approach will help to reconstruct accurately the exposure and to fill in data gaps on such exposure;

Other categories of synthetic organic chemicals are the brominated flame retardants like **polybrominated diphenyl ethers (PBDE),** the effects of which are still uncertain, and the **perfluorinated compounds (PFCs)** which due to their unique properties (resistance to heat and acid, hydrophobic and lipophobic at the same time), are found in a large number of consumer products such as non-stick coatings, textile coatings, food packaging, and fire-fighting foams.

A number of scientists have looked into exposure assessment, epidemiology, risk assessment and management of various by-products in water in relationship to various cancers, small for gestational age (SGA)/intrauterine growth retardation (IUGR) and preterm delivery, low birth weight (LBW), stillbirth, congenital anomalies and semen quality among other risks. Nieuwenhuijsen et al. (2009), for example, have examined the relationship between chlorination by-products in drinking water, as measured by trihalomethanes (THMs), and various health problems. The overall aim of the HIWATE study is to investigate potential human health risks associated with long-term exposure to low levels of disinfectants (such as chlorine) and DBPs occurring in water for human consumption and use in the food industry. Recent organic diet intervention studies such as that of Bradman et al. (2013) suggest that diet is a significant source of pesticide exposure in young children. These studies have focused on children living in suburban communities.

To improve understanding of chemical hazards, risk assessment approaches need to be revised. Environmental and human monitoring should also be improved to provide early signals of health impacts.

Air Pollution

Outdoor Air

Urban environments impose a stress on humans and wildlife through air pollution, traffic accidents, and sedentary lifestyles. Urban air pollution, mainly caused by motor vehicles, industry and power plants, is estimated to kill 1.2 million people every year (Kenworthy and Laube, 2002). More developed countries have taken measures to reduce pollution, and the levels of some pollutants have declined; the problem is getting more severe in less developed countries (Health Effects Institute (HEI), 2004; Peden et al., eds., 2004).

Air pollutants of concern are particulate matter, ground-level ozone, sulfur dioxide, nitrogen dioxide, and polycyclic hydrocarbons. The health impacts of air pollutants have been extensively studied; effects can be acute (leading to premature death), chronic or less severe.

According to epidemiologic studies, **particulate matter (PM)** is the pollutant with the most severe health effects and for which no threshold levels exist, below which no effects would be observed (WHO, 2006a; WHO, 2006b). Exposure to PM was shown to increase mortality from cardiovascular and respiratory diseases (Colais et al., 2012), with more specific links to myocardial infarction (Mustafic, 2012) and to ischemic stroke (Mateen and Brook, 2011). Average life expectancy in the EU is reduced 8.6 months as a result of exposure to $PM_{2.5}$ (EEA, 2012a). The WHO air quality guidelines for $PM_{2.5}$ may need to be revised, in light of new findings on the health effects of this pollutant (WHO, 2013a).

Ground level ozone is a pollutant that remains at high levels in urban centers (photochemical pollutant) and is associated with respiratory problems and premature mortality. Children, elderly people, people with chronic lung and heart disease, but also healthy people who exercise outdoors are more threatened by ozone pollution. **Nitrogen dioxide** is another pollutant that affects the respiratory system (both in short-term and long-term exposures). It is linked also with increase in symptoms of bronchitis in asthmatic children.

There is evidence of prenatal effects and effects of exposure to air pollutants during childhood. Reduced fetal growth, pre-term birth and spontaneous abortions are some of the effects observed in cases of such exposure during pregnancy (WHO, 2005). The effects of exposure to air pollution during pregnancy on birth weight are of the same order of magnitude as in the case of passive smoking (Pearce et al., 2012). Air pollution may also affect the nervous system and epidemiological studies have shown positive associations between air pollution and impaired cognitive function (Van Kempen et al., 2012).

Groups particularly vulnerable to air pollution are those who are already ill or have pre-existing medical problems, children - as they spend more time outdoors, and socially-deprived populations, such as those belonging to lower social classes.

Pollution prevention approaches in more developed countries have been successful at reducing the levels of primary pollutants. Sulfur dioxide is an example of a success story. Carbon monoxide and other health-relevant pollutants such as benzene and heavy metals have also been significantly reduced in the EU (EEA & JRC, 2013). Policies for reducing pollution have mainly targeted individual pollutants. Some examples can be seen in the following papers: Kenworthy and Laube, 2002; HEI, 2004 and Peden et al., 2004.

Furthermore, Huck, Whyatt, Coulton, Davison & Gradinar, (2017) in yet another study, which represents a step forward in the application of portable sensor technologies and in the monitoring of personal exposure pollutants in the air and the frequency and volume of that air which enters the lungs, utilize high-resolution spatio-temporal data drawn from sensors for NO_2, nasal airflow and location (GPS) making possible the empirical analysis into pollution exposure. This study demonstrates both the potential and limitations of sensor-based approaches and the relevance of such technologies for the advancement of the citizens' well-being.

Nitrogen dioxide (NO_2) and nitrogen oxides (NO_x) were measured by Land Use Regression (LUR) models which are able to explain small-scale within-city variations. Beelen et al. (2013) developed LUR models in a standardized way in 36 study areas in Europe for the ESCAPE (European Study of Cohorts for Air Pollution Effects) project. Within the ESCAPE project it was possible to develop LUR models that explained a large fraction of the spatial variance in measured annual average NO_2 and NO_x concentrations.

Indoor Air

Indoor chemicals and particulate matter (also molds and other biological agents) are also important hazards. The main effects include increased prevalence of respiratory symptoms, allergies and asthma; and immunological ones. In the EU, there is no legislation on indoor air quality except for tobacco smoke and some ventilation standards for buildings.

As an example, Marques & Pitarma (2016) in their paper state that the ambient assisted living (AAL) residencies are designed to meet the needs of the aging population and in order to maintain their independence in indoor environments, high indoor air quality must be ensured by utilizing advances in networking, sensors, and embedded devices. The collected data can be accessed by doctors in order to support medical diagnostics.

Radon is a radioactive gas that is released from uranium-containing soils and rocks and can be found at high concentrations in buildings. It is an indoor air pollutant of concern and has become the second largest cause of lung cancer after smoking in Europe. There seems to be a synergism between exposure to radon and smoking (EEA & JRC, 2013). Its levels vary, depending on region, and the most efficient way to reduce indoor exposure to it is by increasing ventilation in buildings.

Physical Hazards

Noise-affects people in psychological and physiological ways; traffic noise and airport noise seem to be the most disturbing. According to WHO (WHO/JRC, 2011), Europeans lose more than one million healthy life years due to traffic noise.

Electromagnetic fields are an additional physical hazard. Radio-frequency EM fields have been characterized as possibly carcinogenic to humans by WHO's International Agency for Research on Cancer (IARC). Mobile phones risks and the possible sensitivity of children are to be further examined.

Ultraviolet radiation is a form of radiation to which we get increasingly exposed. It has been classified as a category 1 carcinogen to humans by IARC. It is related to different types of skin and eye diseases, with the most aggressive form being malignant melanoma. The increase in skin cancer in Europe and larger parts of the world observed in the last decades can be attributed to changing lifestyles and the thinning ozone layer that acts as a protection against UV radiation.

Climate Change, this change in weather patterns and extreme events such as heat waves, floods, fires and windstorms, is expected to adversely affect human health and lead to the spread of diseases; more than 50 infectious diseases reportable in the EU can be affected by climate change (directly or indirectly). Some regions are expected to be affected more than others.

Nanotechnology which is concerned with the impact of very small particles (nanoparticles, 1-100 nm) on human health is an area of high concern. Such nanoparticles are released to the environment from a variety of sources (e.g., electronics, textiles, cosmetics, pharmaceuticals) and can enter the human body via inhalation, oral intake or absorption through the skin. Little is known about their effects on human health, as the risk assessment processes for these nanoparticles are still not well developed.

Cultural Hazards and Lifestyle Choices

Psychosocial factors such as poverty, stress, violence, wars, school, family, workplace, etc. contribute to the well-being or not of the individual. Some of them are related to biological or chemical hazards described in previous paragraphs. The **poverty/malnutrition/disease** cycle has been considered as the major cause of death worldwide, with only one third of the world's population having access to sanitation and less than one billion people having access to safe drinking water and to enough food for good health. Impacts of **occupational factors** such as chemicals, physical and biological exposures, as well as stress in the workplace, injuries and illnesses can also pose significant health hazards. Such impacts account for over one million deaths per year worldwide. **Lifestyle choices** such as smoking, alcohol abuse, obesity, and unsafe sex are important hazards as well with smoking killing over 6 million people per year worldwide. Automobile accidents account for over 1.2 million deaths annually.

Access to Green Spaces seems to offer multiple benefits, especially to people living in urban centers. Studies have shown an increase in longevity of urban senior citizens and improvement in social contacts, particularly for the poor (Takano et al., 2002; HCN, 2004). Contacts with nature also seem to improve psychological well-being and social cohesion. (Velarde et al., 2007; HCN, 2004). The mechanisms on how green spaces improve physical, mental and social health need to be explored further.

Figure 2. Traditional analytical approaches
Source: Our own

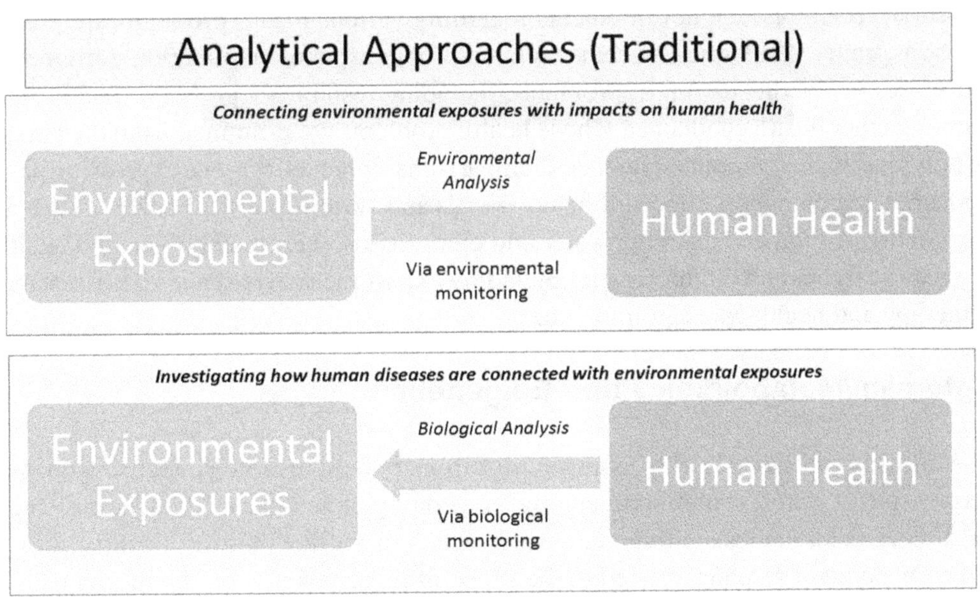

Figure 3. New analytical approaches
Source: Our own

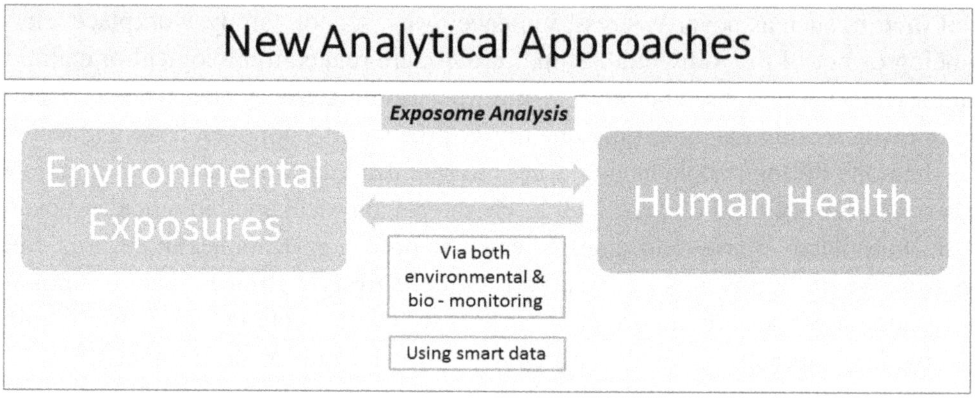

ENVIRONMENTAL EXPOSURES AND HUMAN HEALTH: ANALYTICAL APPROACHES AND HEALTHCARE

Traditionally, researchers have used various analytical approaches of environmental analysis in their effort to connect environmental exposures with human health impacts via environmental monitoring as well as biological monitoring (see Fig. 2). More recently, new analytical approaches, especially Exposome Analytics, have been developed, which are supported by new biological and environmental analysis methods that utilize sophisticated ICT tools as depicted in Fig. 3.

In this section, recent reseach developments regarding various health problems are presented such as inflammations, cancers, allergies, vaccine- related problems, autism, attention deficit disorder, autoimmune diseases, endocrine disruptor problems, including respiratory problems and nervous system disorders, among other health challenges. Those health problems are examined from the perspectives of single gene polymorphisms, genome sequences comparisons, epigenetics, gene expression data, metabolomics, exposure to carcinogens, effects of heavy metals and various physical hazards such as airborne particulates, quality of food and drinking water and other toxins, their metabolism and health impacts. Through big data analytics and bioinformatics including exposome analysis, and risk assessment, finally prognosis, therapy and health care can improve.

Transcriptomics/Metabolomics and Epigenetics

Genomics has already had a long history as an interdisciplinary field of biology, focusing on the structure, function, expression of genes, evolution, mapping, and editing of genomes. As such, it is still a growing field in which new technologies continue to emerge which fast affect the way scientists consider solving problems. One type of research related to the scope of this chapter is on the mechanism of cellular response to genotoxicity and the unique ways of *epigenetic modifications* and *gene expression*. The role of *microRNAs* (miRNAs) in the modulation of toxic responses, in addition to serving as biomarkers, has helped scientists to establish better ways of addressing risk assessment. Thus, the elucidation of the effects of chemical exposure on microRNAs expression has become a high priority. As an example, Cavill et al., (2011) have used *transcriptomic* and *metabolomic* measurements from the NCI60 cell line panel, where they integrated molecular profile data and showed that the biochemical pathways associated with tumor cell chemosensitivity to platinum-based drugs describe a consensus phenotype. They concluded that metabolic phenotyping could be used for both biomarker discovery and in the prediction of the complex relationships between biological pathways and drug response.

MicroRNAs are short, single-stranded, non-coding molecules with unique functions mainly acting as negative regulators of gene expression to either silence or suppress genes. There is growing evidence linking miRNAs and environmental chemical exposure; when subjected to environmental stimuli, such as air pollution, cigarette smoking, heavy metals and bisphenol A, miRNAs can exhibit differential expression (Hou, Wang & Baccarelli, 2011; Hou, Zhang, Wang & Baccarelli, 2012). As critical regulators of gene expression, this type of differential miRNA expression has been implicated in several pathogenic pathways and numerous nervous system disorders. The unique role of miRNAs and the ways environmental exposures and epigenetic modifications are coupled to their regulatory role make miRNAs a potential biomarker of environmental diseases and a model for studying induced alterations of miRNA expression in inflammatory and oxidative stress pathways with the goal of improving prevention and treatment of environmental diseases (Hou, Wang & Baccarelli, 2011; Hou, Zhang, Wang & Baccarelli, 2012).

In a recent study, for example, Lin, Liu, Hu, Cottrell & Kass (2018) examined the effect of a general anesthetic, as an environmental stimulus, on miRNA expression of the developing brain. It appears that general anesthetics have potential long-lasting neurotoxic effects on the developing brain, resulting in behavioral changes in adulthood.

Furthermore, the unique single gene polymorphisms together with their expression and possible epigenetic modifications have pushed scientists to the development of applications where genomic or genetic changes are interpreted in the context of biochemical and physiological profiles. Smart data mining and big data analysis are now achieved by systems biology and bioinformatics approaches so that the large "omics" data is reduced to interpretable data. That in turn can help scientists better understand the mechanisms of action as well as be able to reach wiser decisions (Lytras & Papadopoulou, 2018; Papadopoulou, Lytras & Marouli, 2018). Already, there are applications of genomics in risk assessment in terms of genotoxic and carcinogenic potential (Bonnefoi et al., 2010).

It is interesting to point out that adverse drug reactions and hypersensitivity issues constitute a common problem we need to address. Examples include but are not limited to human leukocyte antigen (HLA) variants, genetic polymorphisms in susceptibility genes, socioeconomic factors, gene-environment interaction and carcinogenic risks as a genetic predisposition. All those examples and more are well examined in the Bonnefoi et al. (2010) article. In the long run, they will help us define sensitive populations to certain drugs, stratify individuals into high/medium/low metabolizers using SNP variants to predict their biological/pharmacological response but also consider the socioeconomic factors that are affected by the economic status, the maternal/paternal status and education, the micronutrient status together with exposures to heavy metals, airborne particulates, quality of food and drinking water which together with the genetic influences can affect the "sensitivity" of the individual in terms of susceptibilities/vulnerabilities. According to Bonnefoi et al. (2010), the US Food and Drug Administration (FDA) and the European Medicines Agency (EMA) along with HESI support the preclinical trials of novel biomarkers international initiatives and the efforts for translational marker evaluation which seem to be a significant promise for the future. Another challenge area includes the development of improved test systems such as use of stem cells that will help reduce the use of animals and come up with more accurate prediction of human risk, and improve risk assessment models, ultimately guiding intervention to prevent diseases or their progression.

Gene Expression and the Epigenome in Relation to Environmental Exposures and Lifestyles

In this section, obesity and diabetes, allergies and respiratory problems, autism, and attention deficit disorders, stress and cancer, brain/body development and brain/body disorders are discussed. The epigenome is thought to be at the interface between the genome and the environment. Environmental and lifestyle factors elicit changes in the epigenome leading to a variety of cancers. Studies have led to discovery of biomarkers suitable for exposure and risk assessment and better understanding of causal pathways of childhood cancer providing evidence for better diagnosis, prognosis and prevention (Ghantous, Hernandez-Vargas, Byrnes, Dwyer & Herceg, 2015). Advances in epigenetics and our recent progress in our understanding of the epigenetic mechanisms and the way environmental stressors and endogenous factors determine gene expression in health and disease, such as metabolic syndrome and cancer have well been summarized in the study of Herceg (2016). Interestingly enough, persistent organic pollutants (POPs) are found in Greenlandic Inuit who have some of the highest reported POP levels worldwide. POPs may influence epigenetic mechanisms; therefore, they could affect chromosomal stability and gene expression associated with cancer initiation and progression. (Rusiecki et al., 2008).

Epigenetic mechanisms such as DNA methylation, histone modifications, chromatin remodeling and/or regulatory feedback by microRNAs, all modulators of gene expression and promoters of the metabolic syndrome phenotype, link early environmental exposures during pregnancy with programmed changes in gene expression, growth and development. There is irrefutable evidence that this type of 'gestational programming' in terms of tissue-specific transcriptome patterns and phenotypes continues in subsequent progeny and generations even in the absence of continued adverse environmental exposures, yet propagating the metabolic syndrome. If indeed such epigenetic gene modifications can be reversed, then a good understanding of how such transgenerational transmission of obesity and metabolic syndrome results may help toward better detection and prevention of obesity and the metabolic syndrome. (Desai, Jellyman & Ross, 2015). For example, organochlorine chemicals and peroxisome proliferator-activated receptor agonists are identified as possible risk factors for obesity (La Merrill & Birnbaum, 2011). Heindel (2019) points to the obesogen field developed from two separate scientific research areas, endocrine disruptors and the Developmental Origins of Health and Disease (DOHaD). This paper makes reference to exposure studies to endocrine-disrupting chemicals (EDCs) during early development (DOHaD) which has been shown to increase susceptibility to a variety of diseases such as obesity, infertility, asthma, cancers, puberty onset, and susceptibility to infections, heart disease, autoimmune disease, including attention deficit hyperactivity and disorder/learning disability.

In the review of Trasande, Cronk, Durkin, Weiss, Schoeller, Gall, et al. (2009), the authors describe the approach taken by the National Children's Study (NCS), a 21-year prospective study of 100,000 American children, to understanding the role of environmental factors in the development of obesity with the goal to identify risk factors for childhood obesity, and for their subsequent elimination through prevention.

Van Dijk et al. (2015) in their article provide evidence on how environmental exposures during early life may lead to an increased risk of obesity at later stages of life. They also introduce EpiSCOPE, a multidisciplinary research program aimed at the understanding of epigenetic changes and obesity. They suggest that by changing lifestyles, and the exposure in utero, the potential for interventions in postnatal life with the aim to modify unfavourable epigenomic profiles is possible.

A very recent and interesting study (Michelet et al., 2018) showed that up to 49% of certain types of cancer are attributed to *obesity*. The authors showed that "obesity induces robust peroxisome proliferator-activated receptor (PPAR)-driven lipid accumulation in Natural Killer cells, causing complete 'paralysis' of their cellular metabolism and trafficking, and potential mechanisms include overproduction of hormones, adipokines, and insulin." Their results demonstrate that the lipotoxic obese environment impairs immunosurveillance and suggest that metabolic reprogramming of NK cells may improve cancer outcomes in obesity.

New research could help explain "why stress early in life can create vulnerabilities to mood and anxiety disorders later on" (Ohio State University, 2018). Brain signature of depressed mood is unveiled in a new study via direct recordings of human brain activity which "seem to link memory, emotion, and anxiety during bouts of low mood" (University of California - San Francisco, 2018).

It is also well known that the brain's immune cells, such as mast cells, as well as microglia and astrocytes are implicated in *psychiatric disorders,* as brain sex differences are sex biased in terms of prevalence or symptomology as they are programmed by early life exposure to hormones and sex chromosome genes. The neuroimmune cells are targeted by sex hormones leading to different neuroimmune response and actively contribute to sex-specific brain development, plasticity and behavioral development (Nelson, Saulsbery & Lenz, 2018; Theoharides, Tsilioni, Patel & Doyle, 2016).

When it comes to *allergies,* Lozano et al. (2018) demonstrated what they call sex-dependent impact of Roundup on the rat gut microbiome which could have a role in the development of metabolic disorders. They examined the long-term effects of 3 doses of the Roundup (R) herbicide (made of glyphosate and formulants) on the gut microbiota in male and female Sprague-Dawley rats. They observed a sex-dependent toxicity where only female rats showed "dysbiosis" and that their microbiome disturbances overlapped with liver dysfunction as previously indicated by previous studies as well. The authors raise their concerns on the alteration of the Firmicutes to Bacteroides ratio due to the exposure to the Roundup(R) herbicide and to other pesticides in general and their possible role in the rise of intestinal disorders thus warranting further investigation.

Bousquet et al. (2011) in their effort to improve the health of European citizens while at the same time addressing global health issues but also ethical issues, report in their paper that an FP7 European Union project (No. 264357) abbreviated as MeDALL (Mechanisms of the Development of ALLergy) aims to generate novel knowledge on the mechanisms of initiation of allergy and to propose early diagnosis, prevention, and targets for therapy. This novel phenotype definition and the integrative translational approach used already seems to have helped toward better understanding of allergic phenotypes and has contributed to the development of novel diagnostic tools for the early diagnosis of allergy, targets for the development of novel treatment modalities, and prevention of allergic diseases. It has become obvious that "disruption of the developing immune and respiratory systems by early-life exposure to persistent organic pollutants (POPs) could result into reduced capacity to fight infections and increased risk to develop allergic manifestations later in life" (Gascon, Morales, Sunyer & Vrijheid, 2013a); (Theoharides, Tsilioni, Patel & Doyle, 2016).

In an interview with Linda Birnbaum, Director of the US National Institute of Environmental Health Services on "Exposure, effects, and public health challenges" in an EMBO report, Dr. Birnbaum stated "We're beginning to understand that even very low levels of chemicals may have adverse effects during vulnerable stages, such as development" and continues further by saying "It may sound a little too deterministic, but I think what happens to you early in life sets your trajectory for the rest of your life." At the end of her interview, she also warns us of the fact that their mission as researchers is to provide the best science that can be used by others in decision-making. However, she thinks that there is a conflict of interest among scientists, as well as among regulators and clearly among legislators (Birnbaum, 2014).

Vrijheid et al., (2012) provide an overview of existing pregnancy and birth cohort studies such as the European birth cohorts for environmental health research, which investigate the health effects of early-life environmental contaminant exposure so as to improve collaboration, harmonization, and future project planning. (Molitor, Papathomas, Jerrett & Richardson, 2010) have used a biostatistical method that addresses problems for categorical covariates in epidemiological studies and has been implemented using the standard Bayesian modeling software, WinBUGS, National Survey of Children's Health.

There is evidence that the "developmental origins of disease" paradigm is the result of interactions between the fetal environment and genome, which in turn can determine the risk of postnatal disease such as metabolic and cardiovascular disease, as well as the individual's capacity to cope with the postnatal changing environment (Gluckman & Hanson, 2004). Barouki, Gluckman, Grandjean, Hanson & Heindel (2012) refer to the developmental period as a plastic phase. Plasticity is more prominent prenatally and during early postnatal life and it is greatly influenced by nutritional imbalance or exposures to environmental chemicals with potentially adverse consequences for health later in life.

In relation to prenatal exposures to environmental chemicals and children's neurodevelopment, Braun et al. (2011) as well as Braun et al. (2014) examined if endocrine-disrupting chemicals (EDCs) are involved in the etiology of *autism spectrum disorder (ASP)*. They measured the concentrations of 8 phthalate metabolites, bisphenol A, 25 polychlorinated biphenyls (PCBs), 6 organochlorine pesticides, 8 brominated flame retardants, and 4 perfluoroalkyl substances in blood or urine samples from 175 pregnant women in the HOME (Health Outcomes and Measures of the Environment) Study (Cincinnati, OH). Mothers of 4-5- year-old children completed the Social Responsiveness Scale (SRS), a measure of autistic behaviors. While the researcher concluded that some EDCs were associated with autistic behaviors in this modest sample, size cohort precludes them from dismissing chemicals with null associations. However, PFOA, β-hexachlorocyclohexane, PCB-178, PBDE-28, PBDE-85, and trans-nonachlor deserve additional scrutiny as factors that may be associated with childhood autistic behaviors.

While for some chemicals, higher prenatal exposures have conclusively been linked to postnatal expressions of neurotoxicity, autism and cancers, it is exceedingly challenging to characterize exposure as various study findings are, at this stage, difficult to reconcile or integrate. Bellinger (2013) in his review focuses primarily on chemicals of recent concern, including phthalates, bisphenol-A, polybrominated diphenyl ethers, and perfluorinated compounds, but also addresses chemicals with longer histories of investigation, including air pollutants, lead, methylmercury, manganese, arsenic, and organophosphate pesticides. The author emphasizes the value of coordinating studies in a way that meaningful comparisons can be made in order to discern coherent signals in the evidence available for a chemical or class of chemicals, the type of exposure measures, as biomarkers or concentration in an environmental medium or whether it is best to measure the concentration of the parent chemical or of a metabolite plus to consider the stage of development and the various assessment tools used.

According to Theoharides, Sant, & Giota (2019) "Autism spectrum disorder (ASD) is a developmental condition characterized by impaired social interactions and communication, as well as by stereotypic movements, which affect 1 in 59 children," perhaps 1 in about 40 children by 2020. These authors (Theoharides, Sant, & Giota, 2019) propose that environmental stimuli spur the unique immune cells, known as mast cells (MC), which then trigger microglia, leading to dysfunctional neuronal connectivity in the amygdala. The corticotropin-releasing hormone (CRH) could have a synergistic effect with environmental stimuli especially mycotoxins, affecting the "fear response" and the "fight-or-flight" reaction. Recognition of this association and prevention of stimulation of mast cells/microglia could lead to effective treatment of ASD, which remains without distinct pathogenesis, and effective treatment (Theoharides, Tsilioni, Patel & Doyle, 2016). Bagasra & Bagasra (2019) summarize in their chapter the prevailing theories regarding the aetiologies of autism and present scientific data supporting the thesis that some of the most commonly used environmental agents that we have so easily accepted into our social fabric of life, such as fragrances, may be major contributors in the development of autism spectrum disorders (ASD).

Grazuleviciene, Danileviciute, Nadisauskiene & Vencloviene (2009), when they examined *maternal smoking,* glutathione S-transferases (GSTs), noted that "The GSTs are a polymorph super-gene family of detoxification enzymes that are involved in the metabolism of numerous toxins and provide critical defense against xenobiotics." More specifically, studies on GSTM1 and GSTT1 polymorphism and susceptibility to adverse pregnancy outcomes concluded that smoking, even at a low-level, ought to be considered a potential risk factor for adverse birth outcomes and that genetic polymorphism may contribute to individual variation in tobacco smoke response. Another study comes from Larsen et al. (2013) "on pregnancy and birth cohort resources in Europe: a large opportunity for aetiological child health research." Philippat et al. (2013) used maternal urinary biomarkers to assess fetal exposure to phenols and their precursors and concluded that "Their effectiveness as a measure of exposure in epidemiological studies depends on their variability during pregnancy and their ability to accurately predict fetal exposure."

Furthermore, Gallagher et al. (2011) designed the Mechanistic Indicators of Childhood Asthma (MICA) study to pilot an integrative approach in children's health research that incorporates "exposure metrics, internal dose measures, clinical indicators and blood gene expression measures to decipher the biological complexity inherent in diseases with etiology related to complex gene-environment interactions". MICA was a nested study within the EPA's Detroit Children's Health Study (DCHS) that identified 1,157 children who completed health questionnaires as well as clinical examinations of lung function and exhaled breath. The authors claim that while MICA is an asthma study, relationships between early indicators of cardiovascular disease, obesity, and asthma can also be explored leading to a more holistic analytical approach in the interpretation of biomarker data toxicology studies and risk assessments of chemical exposures.

As for *cancer,* Herceg & Vaissière (2011), in their review paper, summarized how epigenetic mechanisms may function as an interface between environmental factors and the genome. At the cellular level, environmental stressors seem to disrupt different cellular processes and contribute to cancer risk. Their paper focuses on human lung cancer. Presently, the progress in our understanding of the epigenetic mechanisms through which environmental factors may promote tumor development, has increased as also seen in more recent studies. One such example is the effort to understand the mechanism of epigenetic reprogramming during the occurrence of pulmonary hypertension (PH) (Wang et al., 2018). In that article the discovery of novel epigenetic mechanisms associated with PH including DNA methylation, histone modification, and noncoding RNA interference is reviewed, including how the undesirable perinatal environments affect adult Pulmonary Arterial Hypertension (PAH) as in the case of extrauterine growth restriction (EUGR) and intrauterine growth retardation (IUGR). Moreover, Kaur, Shorey, Ho, Dashwood & Williams (2013) in their review paper related to the epigenome as a potential mediator of cancer and disease prevention in prenatal development discuss on the one hand, the complex interplay of epigenetic events and various environmental factors and on the other, the impact of various dietary phytochemicals on the epigenetic programs during prenatal exposure. A number of studies point to maternal dietary modifications both during gestation and lactation, capable of modulating the epigenome, leading to various health problems including cancers, and the efforts that exist to identify chemo-preventive agents as dietary interventions which could be beneficial in the treatment of such health problems by also developing methods of epigenetic targeted therapy.

GMOs/Pesticides and Food Safety

The past few years, consumers, researchers, activists, and environmental Non-Governmental Organizations (NGO) have suggested that all Genetically Modified (GM) foods/plants should be subjected to long-term free and independent animal feeding studies and only then be approved for human consumption (Domingo, Gine Bordonaba, 2011). This suggestion came out of the concern that until then most of these studies had been conducted by biotechnology companies responsible for producing and commercializing these GM plants. In order to protect GM plants, glyphosate-based herbicides (GBHs) were developed to replace the previous herbicides, which caused well-documented problems for both plants and human health. The use of the herbicide glyphosate (common trade name "Roundup"), in particular, which is used to support many GM crops, has increased 100-fold from the time it was first sold to farmers in 1974. Glyphosate was assumed to be a low hazard to vertebrates because its herbicidal mechanism of action targeted a plant enzyme which is not present in vertebrates.

The initial industry toxicity testing suggested that GBHs posed relatively low risks to non-target species and that as a consequence lead the regulatory authorities to set high exposure limits until epidemiological studies evaluated by the World Health Organization's International Agency for Research on Cancer (IARC) lead to the classification of glyphosate as "probably carcinogenic to humans." In a report on "Glyphosate cancer determination challenged by world consensus" IARC (2016), one reads: "IARC classified glyphosate as 'probably carcinogenic to humans' (Group 2A). This was based on 'limited' evidence of cancer in humans (from real-world exposures that actually occurred) and 'sufficient' evidence of cancer in experimental animals (from studies of 'pure' glyphosate). IARC also concluded that there was 'strong' evidence for genotoxicity, both for 'pure' glyphosate and for glyphosate formulations". Krimsky (2019) in his chapter explores the glaring scientific differences in the human health assessment of glyphosate between European and American institutions. The International Agency for Research on Cancer (IARC) classified glyphosate as a probable human carcinogen, while the U.S. Environmental Protection Agency (EPA) concluded that glyphosate is not likely to be carcinogenic to humans. Both IARC's and the EPA's carcinogenic risk assessment processes are discussed in Krimsky's chapter revealing the uncertainties in the sciences of toxicology and epidemiology related to the assumptions made in their applications for evaluating glyphosate and the political context of chemical risk assessment which could explain the divergent findings on the carcinogenic risks of glyphosate.

After that major concern, in the past decade, a number of researchers have examined the effects of glyphosate (G) and more recently its commercial formulations. Their results point to identifying the glyphosate-based herbicides (GBH) as substances that have teratogenic, tumorigenic and hepatorenal effects due to endocrine disruption and oxidative stress. Those impacts seem to affect the gut microbiome and carcinogenicity in general, but also bring about multigenerational effects on reproductive capability and the frequency of birth defects. (Seralini et al., 2011; Séralini et al., 2014; Mesnage, Defarge, Spiroux de Vendomois, Seralini, 2015; Myers et al., 2016). Some effects were in fact detected in the range of the recommended acceptable daily intake. Townsend et al. (2017) reported significant DNA damage and cellular death to take place when Raji cells are exposed to glyphosate concentrations at or above 10 mM while G concentrations below 10 μM do not induce significant DNA damage and cells maintain full long-term viability. In the case of intermediate glyphosate concentrations, the induced initial damage could be repaired. All researchers seemed to agree that the current evidence raises serious concerns and they point to the need for further studies. They suggest that neurodevelopmental, reproductive, and transgenerational effects of GBH must be revisited. (Mesnage, Defarge, Spiroux de Vendomois & Seralini, 2015; Townsend, et al., 2017; Lozano et al., 2018)

Mayers et al. (2016) concluded that: (1) GBHs are the most heavily applied herbicides in the world and usage continues to rise; (2) Worldwide, GBHs often contaminate drinking water sources, precipitation, and air, especially in agricultural regions; (3) The half-life of glyphosate in water and soil is longer than previously recognized; (4) Glyphosate and its metabolites are widely present in the global soybean supply; (5) Human exposures to GBHs are rising; (6) Glyphosate is now authoritatively classified as a probable human carcinogen; (7) Regulatory estimates of tolerable daily intakes for glyphosate in the United States and European Union are based on outdated science.

As a result, the aforementioned researchers together with many others have begun to offer a series of recommendations related to the need for more toxicological studies but also stress the need for new investments in epidemiological studies and biomonitoring by major health centers which aim at control and prevention worldwide.

Furthermore, Defarge, Spiroux de Vendômois & Séralini (2018) in their recent mass spectrometry study report that glyphosate-based herbicides (GBH) and 14 of its chemical formulants are mainly families of petroleum-based oxidized molecules, such as POEA, and other contaminants. When they exposed plants and human cells either to glyphosate alone or to the formulants alone and measured their toxicity including the possible human cellular endocrine disruption, they found that contrary to the wide held belief that glyphosate is the active ingredient, the POEA formulant family exhibited stronger herbicidal and toxic properties than glyphosate. Heavy metals such as arsenic, chromium cobalt, lead and nickel known as endocrine disruptors were also identified by mass spectrometry and these heavy metals are found as contaminants in 22 pesticides and among them 11 Glyphosate-based ones. It becomes obvious that new studies must be done to shed more light on the effects of pesticides but more importantly to inform the public of the active ingredients of all pesticides because other active ingredients could be in formulants.

Exposome Analysis

The term *exposome* was coined in 2005 in an effort to underscore the importance of the environment to human health and to align human genome research with environmental exposures research. The ability to characterize environmental exposures through biomonitoring is key to exposome research efforts (Dennis et al., 2017; Wild, 2005; Wild, 2012; Lioy & Rappaport, 2011). "The exposome, defined as the totality of environmental exposures from conception onward, may advance our understanding of environmental contributors to disease by more fully assessing the multitude of human exposures across the life course" (Stingone et al., 2017).

Patel, Bhattacharya & Butte (2010) in reference to *enviromics*, "the study of a wide array of environmental factors in relation to health and biology," state that it is imperative that patients and physicians be informed (and not only epidemiologists and geneticists) on how multiple environmental factors may influence disease in a *systematic fashion* and that we must conduct our analyses in *a non-selective fashion* as evidenced by the "Human Toxome Project".

The disease of type 2 diabetes (T2D) is a good example of both environmental and genetic factor triggering, and with variable risk in the population. Patel, Chen, Kodama, Ioannidis & Butte (2013) in their study provide a systematic identification of interaction effects between genome- and environment-wide associations in type 2 diabetes mellitus. "Through genome-wide association studies (GWAS), the investigators have identified over 90 single nucleotide polymorphisms (SNPs) associated to T2D." They used a method analogous to GWAS [environment-wide association study (EWAS)] and found five environmental factors associated with the disease. They concluded that "unbiased consideration of environmental and genetic factors may help identify larger and more relevant effect sizes for disease associations."

In their review, Wild, Scalbert & Herceg (2013) considered the areas of genomics, transcriptomics, epigenomics and metabolomics and the evidence existing up to date that people exposed to well-defined factors (for example, tobacco, diet, occupational exposures, environmental pollutants) have specific omics profiles. They propose that through exposure assessment and an interdisciplinary approach, the opportunity to elucidate the underlying mechanisms of carcinogenesis, earlier detection and more refined molecular classification of disease increases.

Vrijheid et al. (2014) describe the general design of the collaborative project named the Human Early-Life Exposome (HELIX) funded by European Union FP7 that "aims to implement novel exposure assessment and biomarker methods to characterize early-life exposure to multiple environmental factors and associate these with omics biomarkers and child health outcomes, thus characterizing the "early-life exposome". This project is focused on studying the early life exposome by combining six mother-child cohorts in Europe: BiB (Born in Bradford, UK), EDEN (Étude des Déterminants pré-et postnatals du développement et de la santé de l'Enfant, France), INMA (INfancia y Medio Ambiente, Spain), KANC (Kaunus Cohort, Lithuania), MoBa (Norwegian Mother and Child Cohort Study, Norway), and the Rhea Mother-Child Cohort Study (Rhea, Greece).

Robinson & Vrijheid (2015) concluded in their report that a holistic approach is needed to incorporate knowledge of the exposome and the new framework which will integrate disease. There is an etiology from the pregnancy period to adult years and this can ultimately lead to better prevention strategies. There is a need therefore to develop technologies for longitudinal, internal and external exposure monitoring, and to use bioinformatics to integrate and analyze the datasets which are generated.

Stingone et al. (2017), in their review, point to many challenges on exposome research due to mainly limitations in infrastructure and the need for coordinated support from the new technologies. Investigators have nevertheless initiated efforts, such as the HELIX project (Vrijheid et al., 2014), to design epidemiologic research that capitalizes on new technologies in order to characterize the early-life exposome and relate it to specific health end points. The infrastructure available to support these types of studies has also expanded. Their recommendations focus on the collective efforts to promote the environmental health and well-being of all populations and the need for greater data sharing and coordination, methods development, acknowledgment and minimization of multiple types of measurement error, thus encouraging researchers to embark on exposome research. Limitations are in part due to theoretical and practical challenges, lack of appropriate study designs and analytic methods, lack of infrastructure for exposure assessment, difficulty in differentiating genetic variation from environmentally-induced changes, plus the challenge of variation in susceptibility.

As an example, Buck Louis, Smarr & Patel (2017), in their paper, present an overview of the exposome on human reproduction and development and its implications for health across a lifespan.

Leaders in the field of exposomics met at a recent Symposium on "Lifetime Exposures and Human Health: The Exposome," held at Yale School of Public Health. "The aim was to highlight the most recent technological advancements for measurement of the exposome, bioinformatics development, current limitations, and future needs in environmental health." Advancements in management of big data, standardization for exposure and data analyses, and data sharing, metabolomics and bioinformatics are the driving force in biomonitoring and emphasis is placed toward monitoring the totality of exposures over the lifetime of individuals than the one-chemical one-health outcome (Johnson et al., 2017).

More recently, Sarigiannis et al. (2018a) and Sarigiannis (2019) provide a comprehensive overview of the modelling and data analytics needed towards refined external and internal exposure assessment, for elucidating the human personal exposome starting from more accurate methods of measurement such as use of wearable sensors and for extrapolation to larger population groups as in the case of agent-based modelling. Exposure reconstruction algorithms, coupled with biokinetics, support biomonitoring data interpretation and assimilation. In their chapter, Sarigiannis, Gotti, Handakas & Karakitsios (2018b) outline the field of computational exposure biology and demonstrate how the bioinformatics techniques and algorithms are coupled with big data analytics to perform studies of exposome-wide associations with putative adverse health outcomes and help in the deciphering of the molecular and metabolic pathways of induced toxicity related to environmental chemical stressors such as airborne chemical mixtures, organic compounds found in consumer goods, and mixtures of organic chemicals and metals through multiple exposure pathways.

Exposures, Biomonitoring and Types of Sensors

Clewell, Tan, Campbell & Andersen (2008) while they define biomonitoring as the measurement of chemicals in human tissues and fluids, they raise concerns due to low concentrations of the chemicals detected in human samples typically in the parts per billion (ppb) or parts per trillion (ppt) range and that it is often difficult to relate animal toxicity studies with a measured concentration of a chemical in a human tissue or fluid. They also point to challenges associated with interpretation of biomonitoring data and their contribution is a physiologically based pharmacokinetic (PBPK) modeling, a general approach, referred to as reverse dosimetry that could give rise to measured biomarker concentrations so as to support the interpretation of human biomonitoring data from the perspective of exposure reconstruction and risk characterization. Interestingly enough, Hebels et al. (2013) examined the suitability for omic analysis of biosamples collected in previous decades and currently stored in biobanks and found out that most samples are amenable to meaningful omics analysis, provided that they satisfy collection and storage criteria

According to Bonnefoi et al. (2010), biomonitoring and epidemiology are tools for reliable exposure metrics necessary for risk assessment and regulatory decision-making. Even though the interpretation of epidemiological studies is often difficult due to uncertainty associated with exposure estimates (e.g., if exposure is based on residues) or if exposure estimates are based on recall, Bonnefoi et al. (2010) propose data collection using reliable methodologies for assessing exposure and its effects. Similarly, Sobus, Tan, Pleil & Sheldon (2011) report that biomonitoring is used in exposure and risk assessments to reduce uncertainties along the source-to-outcome continuum. They developed a biomonitoring framework which can be used as a guide for understanding biomarker data and developing new risk analysis studies in an integrative manner.

Gruden et al. (2012) in their article presented the first *crossomic* study performed measuring variability of different components in peripheral blood of healthy Caucasoid individuals. They present a pilot study on 16 healthy individuals, applying a combination of transcriptomics, metabolomics and cell biology tools in order to evaluate the physiological variability of different peripheral blood measurables. Despite the fact that the distribution of measured values in their dataset was in agreement with previously published data, their results emphasize the importance of the use of personalized medicine approaches.

Turner et al. (2017) have reviewed categories of exposures that can be assessed externally, the current state of the science in external exposure assessment, current tools available for external exposure assessment, and priority research needs. Examples include use of geographic information systems; remote sensing; global positioning system and geolocation technologies; portable and personal sensing, including smartphone-based sensors and assessments; and self-reported questionnaire assessment.

Turner et al. (2017) also discuss priority research needs related to "methodological and technological improvement, data analysis and interpretation, data sharing, and other practical considerations, including improved assessment of exposure variability as well as exposure in multiple, critical life stages," and as such this is a good resource paper.

Dennis et al. (2017) in their paper described why both traditional and nontraditional (exposomic) biomonitoring are critical in studies aiming to capture the exposome. They suggest hybrid approaches and the integration of emerging techniques, biomonitoring strategies including exposomic approaches which differ from traditional biomonitoring methods in that they can include all endogenous or exogenous exposures. Nevertheless, issues of "sample availability and quality, identification of unknown analytes, capture of nonpersistent chemicals, integration of methods, and statistical assessment of increasingly complex data sets" remain as challenges.

Larkin & Hystad (2017) present a review of emerging technologies such as air pollution sensors, smartphones, and air pollution models and how these can transform personal air pollution exposure assessment and health research. They recommend a multi-disciplinary approach for epidemiological research and prevention measures while taking into consideration ethical, privacy, logistical, and data science challenges

Steckling et al. (2018) examined and summarized the availability of specific biomarkers of exposure (BoE) for a broad range of environmental stressors and exposure determinants, as well as the corresponding reference and exposure limit values and biomonitoring equivalents as part of the Human Biomonitoring (HBM) tool which greatly contributes to the understanding of human exposures. The authors tried to refine the methodology as BoE are limited yet useful for unraveling the exposome by using the framework of environment-wide association studies (EWAS) under the European Union's 7th Framework Programme (EU's FP7) project HEALS - Health and Environment-wide Associations based on Large Population Surveys.

De Nazelle et al. (2013) track people's movement and physical activity level using smart phone technology. They illustrate the use of technology to assess air pollution exposure and inhalation and manage to integrate data from smart phones with air quality models. This paper demonstrates a promising use of smart phones for exposure assessment, and highlights some further challenges with GPS data processing and automated activity recognition algorithms.

Using personal sensors to assess the exposome and acute health effects had also been addressed successfully by Nieuwenhuijsen, Donaire-Gonzalez, Foraster, Martinez, & Cisneros (2014). Nieuwenhuijsen et al. (2015) used smart phones and sensors as a means to provide insights regarding personal exposure to air pollution. They examined the correlation between modeled home estimates and overall personal black carbon levels which seemed to vary substantially during the day, especially during commuting times.

Loh et al. (2017) in their study lay out important criteria for using mobile technology, various sensors and the "internet of things," smart apps and wireless devices for external exposome studies. They also address various limitations and challenges and discuss ways of combining "traditional" measurement data with gathering personal exposure-factor data, such as location and activities.

Big Data Analytics, Bioinformatics and Health-informatics

Lytras & Papadopoulou (2018) in their book 'Applying Big Data Analytics in Bioinformatics and Medicine´point to the current need that academics, commercial groups, and industry in general must come together and through good teamwork decide on what is best for an individual's well-being, health and psychological and social needs. They state that improved synergies and good teamwork of talented and well-trained people could help improve personalized medicine and possibly precision medicine. Talented people need to be given the space and also be allowed to creatively form the conditions and manifestations of intelligent and wise delivery.

In the same book, Papadopoulou, Lytras & Marouli (2018), in their chapter, point to the challenge that exists on how to apply 'smart data' in ways that lead to wise decisions concerning individual wellness and social welfare, and not focusing on market imperatives alone. Aside from the focus of smart

Figure 4. Moving from conventional medicine to personalized and precision medicine
Source: Modified from Papadopoulou, Lytras & Marouli, 2018

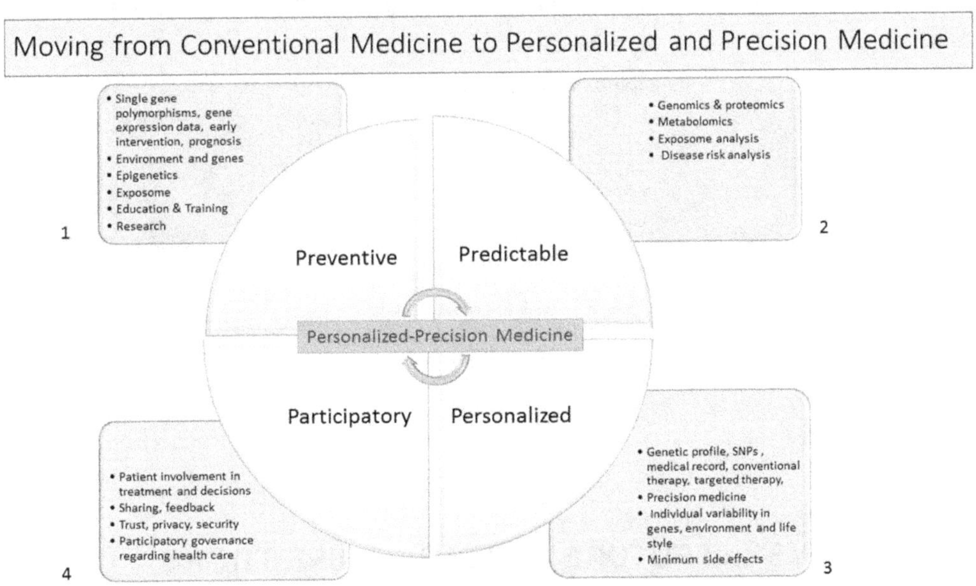

data on veracity and value, 'wise data' should also reflect the distinction between individual health and wellness on the one hand, and social improvement, cohesion and longevity on the other. That in turn would lead to more efficient medical systems, healthier individuals and more socially cohesive societies. In this chapter, the authors communicate and disseminate the importance of establishing the biological relevance of genomic and proteomic discoveries and related specific gene expression to realizing the clinical potential to make possible targeted therapy, personalized medicine and the enhancement of human wellness and social cohesion (see Fig. 4). They recommend that the recent advances, therefore, in bioinformatics and in healthcare informatics for next generation medical research should be examined with the goal to improve medical practices and the management of medical systems both at the national and global level.

Spruit & Lammertink (2018) provide an excellent example of research that focuses on the design process of an effective and efficient dashboard which displays management information for an Electronic Health Record (EHR) in Dutch long-term and chronic healthcare. This research shows how effective and efficient dashboard design can benefit from theoretical insights related to human perception and computer interaction.

Manrai et al. (2017) in their review describe the extant and novel analytic methods needed to associate the exposome with critical health outcomes and also provide recommendations on how to inform the public health community and how to train scientists who can bridge *public health, genomics, and biomedicine in informatics and statistics.*

Figure 5. Proposed approach to analysis of exposures
Source: Our own

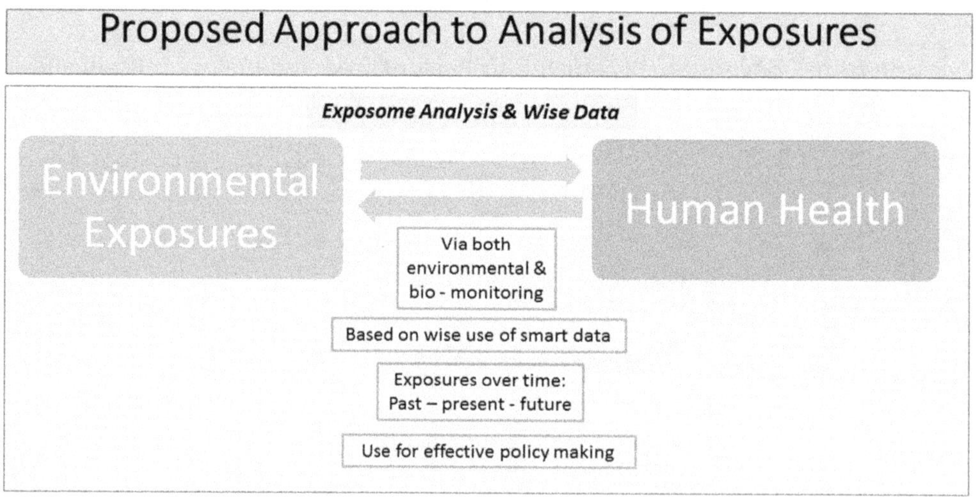

RECOMMENDATIONS and FUTURE RESEARCH DIRECTIONS

This chapter proposes a novel approach to analysis of exposures (see Fig. 5).

Environmental exposures should be examined both via environmental analysis and bio-monitoring stressing their connection. Exposures should be recorded over time and the long-term data should be collected and analyzed by making wise use of the "smart data" with the help of sophisticated ICT tools.

Research on the environmental positive exposures enhancing health and their positive impacts should be strengthened.

A "Green Chemistry" approach is important; such an approach would include not only considerations of the toxicity and eco-toxicity of chemicals, but would address all the steps in the life cycle of a product, from its production to its final disposal, aiming to minimize resource use, waste and pollution. The adoption of such green, sustainable chemistry approaches has shown to provide financial benefits, according to the OECD report (2012). To further encourage such an approach, a combination of policy tools, economic incentives and information-based instruments would be required.

This, of course, requires high investment in research and in the healthcare industry. It also requires effective policy-making, a dimension which will be further discussed in Part II. Both chapters together provide innovative future recommendations and promotion of sustainable health-related solutions

Additionally, this chapter points to three main needs:

1. Reconceptualization of health towards wellness
    a. To adopt the best practices available to move from conventional medicine to personalized and precision medicine.
    b. When there is sufficient evidence and well-established relationships between the individual's health and environmental exposures, to ensure that policies and efforts to minimize exposure are well implemented and that biomonitoring remains effective. Periodic surveys and cohort studies with measurements of contaminant levels in blood, urine and other samples and tests must be ensured.
    c. When there is not sufficient evidence or correlations, proactive actions are necessary to ensure targeted research and the development of better policies on the basis of the precautionary principle. Evaluation and assessment strategies are required, including population-based, multidisciplinary and collaborative research on the many relationships supported by inadequate evidence, as these represent major knowledge gaps, especially in relation to risk assessment.
2. **Policy and research implications – future directions** (more detailed coverage in Part II)
    - To connect research with policies and real needs
        a. To examine how research is affected by funding in the new research scene (very non-ear-marked funding but funding for a specific project on selected policy priorities) and recognition of social problems
        b. To carefully consider how we understand real needs (hearing the voices of the silenced). There is a need for different procedures, but also input from scientists
        c. To improve ways on how to evaluate epidemiologic evidence of potential causal relationships by expanding synergies of scientists from different disciplines, including better and more effective use of ICTs
        d. To invest in research infrastructure, human resources and research methods and better use of analytic laboratories

3. **New culture: understanding and incorporating logic of interconnectedness; sustainability**
    a. To learn from alternative approaches to health
    b. To follow a more holistic approach to solving problems in a sustainable way
    c. To help increase awareness of interconnectedness and to raise awareness that through 'transformative change,' nature can still be conserved, restored and used sustainably. Humans should help maintain the essential, interconnected web of life on Earth because current losses and species extinctions are a direct result of human activity and constitutes a direct threat to human well-being throughout the world.

CONCLUSION

This chapter provides an overview of the most important contemporary environmental health challenges. It then covers the exciting developments in the environmental health field drawing from diverse bodies of knowledge, including chemistry, biology, physics and informatics by summarizing the level of epidemiologic evidence for relationships between prenatal and/or early life exposure to environmental chemical contaminants and fetal, child, and adult health. It presents exposures to a variety of environmental hazards (chemical, biological, physical and psychosocial influenced by cultural, as well as lifestyle choices) which challenge the individual and affect public health. The contributions to envioronmental health from social sciences and social justice, as well as policy studies are covered in our second chapter (Part II). The extensive literature review includes original epidemiologic studies conducted on environmental exposures which include chemical toxicants in air, water, soil/house dust and foods, including milk, cosmetics and consumer products.

This chapter (Part I) presents current issues on the relevant sciences such as single gene polymorphisms, genome sequences, gene expression data, transcriptomics, metabolomics, exposure to carcinogens, endocrine disruptors, heavy metals, physical hazards such as airborne particulates, quality of food and drinking water and other toxins, their metabolism and health impacts, bioinformatics and exposome analysis. Discussion focuses on preterm birth and birth defects, respiratory and other childhood diseases, neuropsychological deficits, hormone disruptor problems, and cancers linked to fetal or childhood exposures.

The chapter concludes by proposing a multidimensional framework for more effective environmental health analysis, recommendations and future research directions. The importance of prognosis with the aim of wellness is underlined, while the need for flexible and effective personalized therapy and medicine is proposed. Lastly, acknowledging the contribution of machine learning methodologies and big data mining and analytics to contemporary clinical science and practice, the proposition that turning 'big" and "smart" data to "wise" data becomes the main challenge that needs to be addressed today at this level. An integrated and participatory approach to environmental and health policy making is advocated in Part II.

REFERENCES

Bagasra, O., & Bagasra, A. (2019). Exposure to Toxic Environmental Agents in the Form of Fragrances and Development of Autism Spectrum Disorders (ASD). In P. Papadopoulou, C. Marouli, & A. Misseyanni (Eds.), *Environmental Exposures and Human Health Challenges* (pp. 52–85). Hershey, PA: IGI Global. doi:10.4018/978-1-5225-7635-8.ch003

Baker, M. G., Venugopal, K., & Howden-Chapman, P. (2011). Household crowding and tuberculosis. In M. Braubach, D. Jacobs, & D. Ormandy (Eds.), *Environmental burden of disease associated with inadequate housing*. Copenhagen: WHO Regional Office for Europe.

Barouki, R., Gluckman, P. D., Grandjean, P., Hanson, M., & Heindel, J. J. (2012). Developmental origins of non-communicable disease: Implications for research and public health. *Environmental Health*, *11*(42), 42. doi:10.1186/1476-069X-11-42 PMID:22715989

Beelen, R., Hoek, R., Vienneau, D., Eeftens, M., Dimakopoulou, K., Pedeli, X., ... de Hoogh, K. (2013). Development of NO_2 and NO_x land use regression models for estimating air pollution exposure in 36 study areas in Europe—The ESCAPE project. *Atmospheric Environment*, *72*, 10–23. doi:10.1016/j.atmosenv.2013.02.037

Bellinger, D. C. (2013). Prenatal Exposures to Environmental Chemicals and Children's Neurodevelopment: An Update. *Safety and Health at Work*, *4*(1), 1–11. doi:10.5491/SHAW.2013.4.1.1 PMID:23515885

Birnbaum, L. (2014). Exposure, effects, and public health challenges, Interview. *EMBO Reports*, *15*(4). doi:10.1002/embr.201438564 PMID:24623404

Bollati, V., & Baccarelli, A. (2010). Environmental epigenetics. *Heredity*, *105*(1), 105–112. doi:10.1038/hdy.2010.2 PMID:20179736

Bonnefoi, M. S., Belanger, S. E., Devlin, D. J., Doerrer, N. G., Embry, M. R., Fukushima, S., ... van der Laan, J. W. (2010). Human and environmental health challenges for the next decade (2010–2020). *Critical Reviews in Toxicology*, *40*(10), 893–911. doi:10.3109/10408444.2010.506640 PMID:20854192

Bousquet, J., Anto, J., Auffray, C., Akdis, M., Cambon-Thomsen, A., Keil, T., ... Zuberbier, T. (2011). MeDALL (Mechanisms of the Development of ALLergy): An integrated approach from phenotypes to systems medicine. *Allergy*, *66*(5), 596–604. doi:10.1111/j.1398-9995.2010.02534.x PMID:21261657

Bradman, A., Kogut, K., Eisen, E. A., Jewell, N. P., Quiros-Alcala, L., Castorina, R., ... Eskenazi, B. (2013). Variability of organophosphorous pesticide metabolite levels in spot and 24-hr urine samples collected from young children during 1 week. *Environmental Health Perspectives*, *121*(1), 118–124. doi:10.1289/ehp.1104808 PMID:23052012

Braun, J. M., Kalkbrenner, A. E., Calafat, A. M., Bernert, J. T., Ye, X., Silva, M. J., ... Lanphear, B. P. (2011). Variability and predictors of urinary bisphenol A concentrations during pregnancy. *Environmental Health Perspectives*, *119*(1), 131–137. doi:10.1289/ehp.1002366 PMID:21205581

Braun, J. M., Kalkbrenner, A. E., Just, A. C., Yolton, K., Calafat, A. M., Sjödin, A., ... Lanphear, B. P. (2014). Gestational exposure to endocrine-disrupting chemicals and reciprocal social, repetitive, and stereotypic behaviors in 4- and 5-year-old children: The HOME study. *Environmental Health Perspectives*, *122*(5), 513–520. Epub2014Mar12. doi:10.1289/ehp.1307261 PMID:24622245

Buck Louis, G. M., Smarr, M. M., & Patel, C. J. (2017). The Exposome Research Paradigm: An Opportunity to Understand the Environmental Basis for Human Health and Disease. *Current Environmental Health Reports*, *4*(1), 89–98. doi:10.100740572-017-0126-3 PMID:28194614

Castano, A., Sanchez-Rodriguez, J.E., Canas, A., Esteban, M., Navarro, C., Rodrνguez-Garcνa, A.C., Arribas, M., Dνaz, G., & Jimenez-Guerrero, J.A. (2012). Mercury, lead and cadmium levels in the urine of 170 Spanish adults: A pilot human biomonitoring study. *International Journal of Hygiene and Environmental Health, 215*(2), 191–195.

Cavill, R., Kamburov, A., Ellis, J. K., Athersuch, T. J., Blagrove, M. S., Herwig, R., ... Keun, H. C. (2011). Consensus-phenotype integration of transcriptomic and metabolomic data implies a role for metabolism in the chemosensitivity of tumour cells. *PLoS Computational Biology*, *7*(3), e1001113. doi:10.1371/journal.pcbi.1001113 PMID:21483477

Chikkanna, A., Mehan, L. P. K. S., & Ghosh, D. (2019). Arsenic Exposures, Poisoning, and Threat to Human Health: Arsenic Affecting Human Health. In P. Papadopoulou, C. Marouli, & A. Misseyanni (Eds.), Environmental Exposures and Human Health Challenges (pp. 86-105). Hershey, PA: IGI Global. doi:10.4018/978-1-5225-7635-8.ch004

Clewell, H. J., Tan, Y. M., Campbell, J. L., & Andersen, M. E. (2008). Quantitative interpretation of human biomonitoring data. *Toxicology and Applied Pharmacology*, *231*(1), 122–133. doi:10.1016/j.taap.2008.04.021 PMID:18589468

Colais, P., Faustini, A., Stafoggia, M., Berti, G., Bisanti, L., Cadum, E., Cernigliaro, A., Mallone, S., Pacelli, B., Serinelli, M., Simonato, L., Vigotti, M.A., & Forastiere, F. (2012). Particulate Air Pollution and Hospital Admissions for Cardiac Diseases in Potentially Sensitive Subgroups. *Epidemiology, 23*(3), 473–481.

de Nazelle, A., Seto, E., Donaire-Gonzalez, D., Mendez, M., Matamala, J., Nieuwenhuijsen, M. J., & Jerrett, M. (2013). Improving estimates of air pollution exposure through ubiquitous sensing technologies. *Environ Pollut*, *176*, 2–99. doi:10.1016/j.envpol.2012.12.032 PMID:23416743

Defarge, N., Spirouxde Vendômois, J., & Séralini, G. E. (2018). Toxicity of formulants and heavy metals in glyphosate-based herbicides and other pesticides. *Toxicology Reports*, *5*, 156–163. doi:10.1016/j.toxrep.2017.12.025 PMID:29321978

Dennis, K. K., Marder, E., Balshaw, D. M., Cui, Y., Lynes, M. A., Patti, G. J., ... Barr, D. B. (2017). Biomonitoring in the Era of the Exposome. *Environmental Health Perspectives*, *125*(4), 502–510. doi:10.1289/EHP474 PMID:27385067

Desai, M., Jellyman, J. K., & Ross, M. G. (2015). Epigenomics, gestational programming and risk of metabolic syndrome. *International Journal of Obesity*, *39*(4), 633–641. doi:10.1038/ijo.2015.13 PMID:25640766

Domingo, J. L., & Gine Bordonaba, J. (2011). A literature review on the safety assessment of genetically modified plants. *Environment International*, *37*(4), 734–742. doi:10.1016/j.envint.2011.01.003 PMID:21296423

EFSA. (2009). Cadmium in food. Scientific Opinion of the Panel on Contaminants in the Food Chain (Question No EFSA-Q-2007-138). *EFSA Journal*, *980*, 1–139.

EFSA. (2011). *Panel on Contaminants in the Food Chain (CONTAM). Statement on tolerable weekly intake for cadmium. EFSA Journal, 9(2), 1975.*

European Environmental Agency (EEA). (2012). *Air quality in Europe — 2012 report*. EEA Report No 4/2012, European Environment Agency, Copenhagen, Denmark.

European Environmental Agency (EEA) & Joint Research Center (JRC). (2013). *Environment and human health — Joint EEA-JRC report* (EEA Report No 5/2013), 1/112.

Gallagher, J., Hudgens, E., Williams, A., Inmon, J., Rhoney, S., Andrews, G., ... Hubal, E. C. (2011). Mechanistic indicators of childhood asthma (MICA) study: Piloting an integrative design for evaluating environmental health. *BMC Public Health*, *11*(1), 344. doi:10.1186/1471-2458-11-344 PMID:21595901

Gascon, M., Morales, E., Sunyer, J., & Vrijheid, M. (2013a). Effects of persistent organic pollutants on the developing respiratory and immune systems: A systematic review. *Environment International*, *52*, 51–65. doi:10.1016/j.envint.2012.11.005 PMID:23291098

Gascon, M., Verner, M. A., Guxens, M., Grimalt, J. O., Forns, J., Ibarluzea, J., ... Vrijheid, M. (2013b). Evaluating the neurotoxic effects of lactational exposure to persistent organic pollutants (POPs) in Spanish children. *Neurotoxicology*, *34*, 9–15. doi:10.1016/j.neuro.2012.10.006 PMID:23085522

Ghantous, A., Hernandez-Vargas, H., Byrnes, G., Dwyer, T., & Herceg, Z. (2015). Characterising the epigenome as a key component of the fetal exposome in evaluating *in utero* exposures and childhood cancer risk. *Mutagenesis*, *30*(6), 733–742. doi:10.1093/mutage/gev010 PMID:25724893

Gluckman, P. D., & Hanson, M. A. (2004). Living with the past: Evolution, development, and patterns of disease. *Science*, *305*(5691), 1733–1736. doi:10.1126cience.1095292 PMID:15375258

Grandjean, P., & Landrigan, P. J. (2014). Neurobehavioural effects of developmental toxicity. *Lancet Neurology*, *13*(3), 330–338. doi:10.1016/S1474-4422(13)70278-3 PMID:24556010

Grazuleviciene, R., Danileviciute, A., Nadisauskiene, R., & Vencloviene, J. (2009). Maternal smoking, GSTM1 and GSTT1 polymorphism and susceptibility to adverse pregnancy outcomes. *International Journal of Environmental Research and Public Health*, *6*(3), 1282–1297. doi:10.3390/ijerph6031282 PMID:19440446

Gruden, K., Hren, M., Herman, A., Blejec, A., Albrecht, T., Selbig, J., ... Jeras, M. (2012). A "crossomics" study analysing variability of different components in peripheral blood of healthy Caucasoid individuals. *PLoS One*, *7*(1), e28761. doi:10.1371/journal.pone.0028761 PMID:22253695

HCN. (2004). Nature and Health: The influence of nature on social, psychological and physical well. being, 2004/09E. Health Council of the Netherlands and Dutch Advisory Council for Research on Spatial Planning, Nature and the Environment (RMNO), The Hague, The Netherlands.

Health Effects Institute (HEI). (2004). *Health Effects of Outdoor Air Pollution in Developing Countries of Asia: A Literature Review*. Retrieved from https://www.healtheffects.org/publication/health-effects-outdoor-air-pollution-developing-countries-asia-literature-review

Heindel, J. J. (2019). History of the Obesogen Field: Looking Back to Look Forward. *Frontiers in Endocrinology (Lausanne)*, *10*, 14. doi:10.3389/fendo.2019.00014 PMID:30761083

Herceg, Z. (2016). Epigenetic Mechanisms as an Interface Between the Environment and Genome. *Advances in Experimental Medicine and Biology*, *903*, 3–15. doi:10.1007/978-1-4899-7678-9_1 PMID:27343085

Herceg, Z., & Vaissière, T. (2011). Epigenetic mechanisms and cancer: An interface between the environment and the genome. *Epigenetics*, *6*(7), 804–819. doi:10.4161/epi.6.7.16262 PMID:21758002

Hou, L., Wang, D., & Baccarelli, A. (2011). Environmental chemicals and microRNAs. *Mutation Research*, *714*(1–2), 105–112. doi:10.1016/j.mrfmmm.2011.05.004 PMID:21609724

Hou, L., Zhang, X., Wang, D., & Baccarelli, A. (2012). Environmental chemical exposures and human epigenetics. *International Journal of Epidemiology*, *41*(1), 79–105. doi:10.1093/ije/dyr154 PMID:22253299

Huck, J. J., Whyatt, J. D., Coulton, P., Davison, B., & Gradinar, A. (2017). Combining physiological, environmental and locational sensors for citizen-oriented health applications. *Environmental Monitoring and Assessment*, *189*(3), 114. doi:10.100710661-017-5817-6 PMID:28210895

IARC. (2016). *Glyphosate cancer determination challenged by world consensus*. Retrieved from https://geneticliteracyproject.org/glp-facts/iarc-international-agency-research-cancer-glyphosate-determination-world-consensus/

IPBES. (2019). *Assessment-reports*. Retrieved from https://www.ipbes.net/assessment-reports

IPBES. (2019). *Global Assessment*. Retrieved from https://www.ipbes.net/news/Media-Release-Global-Assessment

Jaganath, D., & Mupere, E. (2012). Childhood tuberculosis and malnutrition. *The Journal of Infectious Diseases*, *206*(12), 1809–1815. doi:10.1093/infdis/jis608 PMID:23033147

Johnson, C. H., Athersuch, T. J., Collman, G. W., Dhungana, S., Grant, D. F., Jones, D. P., & Vasiliou, V. (2017). Yale school of public health symposium on lifetime exposures and human health: The exposome; summary and future reflections. *Human Genomics*, *11*(1), 32. doi:10.118640246-017-0128-0 PMID:29221465

JRC. (2008). *European Union Risk Assessment Report: Bis(2-ethylhexyl) phthalate (DEHP), Existing Substances Second Priority List* (Vol. 80). Brussels: European Commission Joint Research Centre.

Kaur, P., Shorey, L. E., Ho, E., Dashwood, R. H., & Williams, D. E. (2013). The epigenome as a potential mediator of cancer and disease prevention in prenatal development. *Nutrition Reviews*, *71*(7), 441–457. doi:10.1111/nure.12030 PMID:23815143

Kenworthy & Laube. (2002). Urban Transport Patterns in a Global Sample of Citiesand their Linkages to Transport Infrastructure, Land Use, Economics and Environment. *World Transport Policy and Practice*, *3*(3).

Koch, H., & Angerer, J. (2012). Phthalates: Biomarkers and Human Biomonitoring. In Biomarkers and Human Biomonitoring, Volume 1: Ongoing Programs and Exposures. Royal Society of Chemistry.

Krimsky, S. (2019). Glyphosate Toxicology: What We Can Learn From the Current Controversy. In P. Papadopoulou, C. Marouli, & A. Misseyanni (Eds.), *Environmental Exposures and Human Health Challenges* (pp. 343–356). Hershey, PA: IGI Global. doi:10.4018/978-1-5225-7635-8.ch015

Kumar, A., & Ghosh, A. K. (2019). Arsenic and Cancer. In P. Papadopoulou, C. Marouli, & A. Misseyanni (Eds.), *Environmental Exposures and Human Health Challenges* (pp. 106–132). Hershey, PA: IGI Global. doi:10.4018/978-1-5225-7635-8.ch005

La Merrill, M., & Birnbaum, L. S. (2011). Childhood obesity and environmental chemicals. *The Mount Sinai Journal of Medicine, New York, 78*(1), 22–48. doi:10.1002/msj.20229 PMID:21259261

Larkin, A., & Hystad, P. (2017). Towards Personal Exposures: How Technology Is Changing Air Pollution and Health Research. *Current Environmental Health Reports, 4*(4), 463–471. doi:10.100740572-017-0163-y PMID:28983874

Larsen, P. S., Kamper-Jørgensen, M., Adamson, A., Barros, H., Bonde, J. P., Brescianini, S., ... Andersen, A.-M. N. (2013). Pregnancy and birth cohort resources in Europe: A large opportunity for aetiological child health research. *Paediatric and Perinatal Epidemiology, 27*(4), 393–414. doi:10.1111/ppe.12060 PMID:23772942

Leung, C. C., Lam, T. H., Ho, K. S., Yew, W. W., Tam, C. M., Chan, W. M., ... Au, K. F. (2010). Passive smoking and tuberculosis. *Archives of Internal Medicine, 170*(3), 287–292. doi:10.1001/archinternmed.2009.506 PMID:20142576

Lin, D., Liu, J., Hu, Z., Cottrell, J. E., & Kass, I. S. (2018). Neonatal anesthesia exposure impacts brain microRNAs and their associated neurodevelopmental processes. *Scientific Reports, 8*(1), 10656. doi:10.103841598-018-28874-0 PMID:30006558

Lioy, P. J., & Rappaport, S. M. (2011). Exposure science and the exposome: An opportunity for coherence in the environmental health sciences [Editorial]. *Environmental Health Perspectives, 119*(11), A466–A467. doi:10.1289/ehp.1104387 PMID:22171373

Loh, M., Sarigiannis, D., Gotti, A., Karakitsios, S., Pronk, A., Kuijpers, E., ... Cherrie, J. W. (2017). How Sensors Might Help Define the External Exposome. *International Journal of Environmental Research and Public Health, 14*(4), 434. doi:10.3390/ijerph14040434 PMID:28420222

Lozano, V. L., Defarge, N., Rocque, L. M., Mesnage, R., Hennequin, D., Cassier, R., ... Amiel, C. (2018). Sex-dependent impact of Roundup on the rat gut microbiome. *Toxicology Reports, 5*, 96–107. doi:10.1016/j.toxrep.2017.12.005 PMID:29854581

Lytras, M. D., & Papadopoulou, P. (2018). *Applying Big Data Analytics in Bioinformatics and Medicine*. Hershey, PA: IGI Global. doi:10.4018/978-1-5225-2607-0

Manrai, A. K., Cui, Y., Bushel, P. R., Hall, M., Karakitsios, S., Mattingly, C., ... Patel, C. J. (2017). Informatics and Data Analytics to Support Exposome-Based Discovery for Public Health. *Annual Review of Public Health, 38*(1), 279–294. doi:10.1146/annurev-publhealth-082516-012737 PMID:28068484

Marques, G., & Pitarma, R. (2016). An Indoor Monitoring System for Ambient Assisted Living Based on Internet of Things Architecture. *International Journal of Environmental Research and Public Health*, *13*(11), 1152. doi:10.3390/ijerph13111152 PMID:27869682

Mateen, F. J. and Brook, R. D. (2011). Air pollution as an emerging global risk factor for stroke. *JAMA: the Journal of the American Medical Association*, *305*(12), 1 240–1 241.

Mesnage, R., Defarge, N., Spiroux de Vendomois, J., & Seralini, G. E. (2015). Potential toxic effects of glyphosate and its commercial formulations below regulatory limits. *Food and Chemical Toxicology*, *84*, 133–153. doi:10.1016/j.fct.2015.08.012 PMID:26282372

Michelet, X., Dyck, L., Hogan, A., Loftus, R. M., Duquette, D., Wei, K., ... Lynch, L. (2018). Metabolic reprogramming of natural killer cells in obesity limits antitumor responses. *Nature Immunology*, *19*(12), 1330–1340. doi:10.103841590-018-0251-7 PMID:30420624

Molitor, J., Papathomas, M., Jerrett, M., & Richardson, S. (2010). Bayesian profile regression with an application to the National Survey of Children's Health. *Biostatistics (Oxford, England)*, *11*(3), 484–498. doi:10.1093/biostatistics/kxq013 PMID:20350957

Mustafic, H. (2012). Main Air Pollutants andMyocardial Infarction: A Systematic Review and Meta-analysis. *Journal of the American Medical Association*, *307*(7), 713. doi:10.1001/jama.2012.126 PMID:22337682

Myers, P. J., Antoniou, M. N., Blumberg, B., Carroll, L., Colborn, T., Everett, L. G., ... Benbrook, C. M. (2016). Concerns over use of glyphosate-based herbicides and risks associated with exposures: A consensus statement. *Environmental Health*, *15*, 19. doi:10.118612940-016-0117-0

Nelson, L. H., Saulsbery, A. I., & Lenz, K. M. (2018). Small cells with big implications: Microglia and sex differences in brain development, plasticity and behavioral health. *Progress in Neurobiology*. doi:10.1016/j.pneurobio.2018.09.002 PMID:30193820

Nieuwenhuijsen, M. J., Donaire-Gonzalez, D., Foraster, M., Martinez, D., & Cisneros, A. (2014). Using Personal Sensors to Assess the Exposome and Acute Health Effects. *International Journal of Environmental Research and Public Health*, *11*(8), 7805–7819. doi:10.3390/ijerph110807805 PMID:25101766

Nieuwenhuijsen, M. J., Donaire-Gonzalez, D., Rivas, I., de Castro, M., Cirach, M., Hoek, G., ... Sunyer, J. (2015). Variability in and agreement between modeled and personal continuously measured black carbon levels using novel smartphone and sensor technologies. *Environmental Science & Technology*, *49*(5), 2977–2982. doi:10.1021/es505362x PMID:25621420

Nieuwenhuijsen, M. J., Smith, R., Golfinopoulos, S., Best, N., Bennett, J., Aggazzotti, G., ... Kogevinas, M. (2009). Health impacts of long-term exposure to disinfection by-products in drinking water in Europe: HIWATE. *Journal of Water and Health*, *7*(2), 185–207. doi:10.2166/wh.2009.073 PMID:19240347

OECD. (2012). *OECD Environmental Outlook to 2050*. Paris, France: Organisation for Economic Co-operation and Development.

Ohio State University. (2018). How childhood stress contributes to anxiety, depression: Research in animals illuminates important role of mast cells. *ScienceDaily*. Retrieved November 18, 2018 from www.sciencedaily.com/releases/2018/11/181105160833.htm

Papadopoulou, P., Lytras, M., & Marouli, C. (2018). Bioinformatics as Applied to Medicine: Challenges Faced Moving from Big Data to Smart Data to Wise Data. In M. Lytras & P. Papadopoulou (Eds.), *Applying Big Data Analytics in Bioinformatics and Medicine* (pp. 1–25). Hershey, PA: IGI Global. doi:10.4018/978-1-5225-2607-0.ch001

Patel, C. J., Bhattacharya, J., & Butte, A. J. (2010). An environment-wide association study (EWAS) on type 2 diabetes mellitus. *PLoS One*, *5*(5), e10746. doi:10.1371/journal.pone.0010746 PMID:20505766

Patel, C. J., Chen, R., Kodama, K., Ioannidis, J. P., & Butte, A. J. (2013). Systematic identification of interaction effects between genome- and environment-wide associations in type 2 diabetes mellitus. *Human Genetics*, *132*(5), 495–508. doi:10.100700439-012-1258-z PMID:23334806

Pearce, M.S., Glinianaia, S.V., Ghosh, R., Rankin, J., Rushton, S., Charlton, M., Parker, L., & Pless-Mulloli, T. (2012). Particulate matter exposure during pregnancy is associated with birth weight, but not gestational age, 1962–1992: a cohort study. *Environmental Health*, *11*(1), 13.

Peden, M. M. & World Health Organization. (2004). *World report on road traffic injury prevention*. Geneva: World Health Organization.

Philippat, C., Wolff, M. S., Calafat, A. M., Ye, X., Bausell, R., Meadows, M., ... Engel, S. M. (2013). Prenatal exposure to environmental phenols: Concentrations in amniotic fluid and variability in urinary concentrations during pregnancy. *Environmental Health Perspectives*, *121*(10), 1225–1231. doi:10.1289/ehp.1206335 PMID:23942273

Preau, J. L. Jr, Wong, L. Y., Silva, M. J., Needham, L. L., & Calafat, A. M. (2010). Variability over 1 week in the urinary concentrations of metabolites of diethyl phthalate and di(2-ethylhexyl) phthalate among eight adults: An observational study. *Environmental Health Perspectives*, *118*(12), 1748–1754. doi:10.1289/ehp.1002231 PMID:20797930

Pruss-Ustun, A., Wolf, J., Corvalán, C., Bos, R., & Neira, M. (2016). *Preventing disease through healthy environments: A global assessment of the burden of disease from environmental risks*. Geneva: WHO.

Rappaport, S. M. (2011). Implications of the exposome for exposure science. *Journal of Exposure Science & Environmental Epidemiology*, *21*(1), 5–9. doi:10.1038/jes.2010.50 PMID:21081972

Rappaport, S. M., & Smith, M. T. (2010). Environment and disease risks. *Science*, *330*(6003), 460–461. doi:10.1126cience.1192603 PMID:20966241

Robinson, O., & Vrijheid, M. (2015). The Pregnancy Exposome. *Current Environmental Health Reports*, *2*(2), 204–213. doi:10.100740572-015-0043-2 PMID:26231368

Rojas, D., Rager, J. E., Smeester, L., Bailey, K. A., Drobná, Z., Rubio-Andrade, M., ... Fryet, R. C. (2015). Prenatal arsenic exposure and the epigenome: Identifying sites of 5-methylcytosine alterations that predict functional changes in gene expression in newborn cord blood and subsequent birth outcomes. *Toxicological Sciences*, *143*(1), 97–106. doi:10.1093/toxsci/kfu210 PMID:25304211

Rusiecki, J. A., Baccarelli, A., Bollati, V., Tarantini, L., Moore, L. E., & Bonefeld-Jorgensen, E. C. (2008). Global DNA hypomethylation is associated with high serum-persistent organic pollutants in Greenlandic Inuit. *Environmental Health Perspectives*, *116*(11), 1547–1552. doi:10.1289/ehp.11338 PMID:19057709

Sarigiannis, D. A. (2017). Assessing the impact of hazardous waste on children's health: The exposome paradigm. *Environmental Research*, *158*, 531–541. Epub2017Jul14. doi:10.1016/j.envres.2017.06.031 PMID:28711809

Sarigiannis, D. A. (2019). The Exposome Paradigm in Environmental Health. In P. Papadopoulou, C. Marouli, & A. Misseyanni (Eds.), *Environmental Exposures and Human Health Challenges* (pp. 1–29). Hershey, PA: IGI Global. doi:10.4018/978-1-5225-7635-8.ch001

Sarigiannis, D. A., Gotti, A., Handakas, E., & Karakitsios, S. P. (2018). Informatics and Data Analytics to Support Exposome-Based Discovery: Part 2 - Computational Exposure Biology. In M. Lytras & P. Papadopoulou (Eds.), *Applying Big Data Analytics in Bioinformatics and Medicine* (pp. 145–187). Hershey, PA: IGI Global. doi:10.4018/978-1-5225-2607-0.ch007

Sarigiannis, D. A., Karakitsios, S. P., Handakas, E., Papadaki, K., Chapizanis, D., & Gotti, A. (2018). Informatics and Data Analytics to Support Exposome-Based Discovery: Part 1 - Assessment of External and Internal Exposure. In M. Lytras & P. Papadopoulou (Eds.), *Applying Big Data Analytics in Bioinformatics and Medicine* (pp. 115–144). Hershey, PA: IGI Global. doi:10.4018/978-1-5225-2607-0.ch006

Séralini, G.-E., Clair, E., Mesnage, R., Gress, S., Defarge, N., Malatesta, M., ... Spiroux de Vendômois, J. (2014). Republished study: Long-term toxicity of a Roundup herbicide and a Roundup-tolerantgenetically modified maize. *Environmental Sciences Europe*, *26*(1), 14. doi:10.118612302-014-0014-5

Seralini, G.-E., Mesnage, R., Clair, E., Gress, S., de Vendomois, J., & Cellier, D. (2011). Genetically modified crops safety assessments: Present limits and possibleimprovements. *Environmental Sciences Europe*, *23*(1), 10. doi:10.1186/2190-4715-23-10

Shaffer, R. M., Smith, M. N., & Faustman, E. M. (2017). Developing the Regulatory Utility of the Exposome: Mapping Exposures for Risk Assessment through Lifestage Exposome Snapshots (LEnS). *Environmental Health Perspectives*, *125*(8), 085003. doi:10.1289/EHP1250 PMID:28796633

Sobus, J. R., Tan, Y. M., Pleil, J. D., & Sheldon, L. S. (2011). A biomonitoring framework to support exposure and risk assessments. *The Science of the Total Environment*, *409*(22), 4875–4884. doi:10.1016/j.scitotenv.2011.07.046 PMID:21906784

Spruit, M., & Lammertink, M. (2018). Effective and Efficient Business Intelligence Dashboard Design: Gestalt Theory in Dutch Long-Term and Chronic Healthcare. In M. Lytras & P. Papadopoulou (Eds.), *Applying Big Data Analytics in Bioinformatics and Medicine* (pp. 243–271). Hershey, PA: IGI Global. doi:10.4018/978-1-5225-2607-0.ch010

Steckling, N., Gotti, A., Bose-O'Reilly, S., Chapizanis, D., Costopoulou, D., De Vocht, F., ... Sarigiannis, D. A. (2018). Biomarkers of exposure in environment-wide association studies - Opportunities to decode the exposome using human biomonitoring data. *Environmental Research*, *164*, 597–624. Epub2018. doi:10.1016/j.envres.2018.02.041 PMID:29626821

Stingone, J. A., Buck Louis, G. M., Nakayama, S. F., Vermeulen, R. C., Kwok, R. K., Cui, Y., ... Teitelbaum, S. L. (2017). Toward Greater Implementation of the Exposome Research Paradigm within Environmental Epidemiology. *Annual Review of Public Health*, *20*(38), 315–327. Epub2017Jan6. doi:10.1146/annurev-publhealth-082516-012750 PMID:28125387

Sumpter, C., & Chandramohan, D. (2013). Systematic review and metaanalysis of the associations between indoor air pollution and tuberculosis. *Tropical Medicine & International Health*, *18*(1), 101–108. doi:10.1111/tmi.12013 PMID:23130953

Takano, T., Nakamura, K., & Watanabe, M. (2002). Urban residential environments and senior citizens' longevity in megacity areas: the importance of walkable green spaces. *Journal of Epidemiology and Community Health*, *56*(12), 913–918.

Theoharides, T. C., Sant, J., & Giota, M. (2019). Autism Spectrum Disorder, Fear Response, and Environmental Exposures. In P. Papadopoulou, C. Marouli, & A. Misseyanni (Eds.), *Environmental Exposures and Human Health Challenges* (pp. 30–51). Hershey, PA: IGI Global. doi:10.4018/978-1-5225-7635-8.ch002

Theoharides, T. C., Tsilioni, I., Patel, A. B., & Doyle, R. (2016). Atopic diseases and inflammation of the brain in the pathogenesis of autism spectrum disorders. *Translational Psychiatry*, *6*(6), e844. doi:10.1038/tp.2016.77 PMID:27351598

Townsend, M., Peck, C., Meng, W., Heaton, M., Robison, R., & O'Neill, K. (2017). Evaluation of various glyphosate concentrations on DNA damage in human Raji cells and its impact on cytotoxicity. *Regulatory Toxicology and Pharmacology*, *85*, 79–85. doi:10.1016/j.yrtph.2017.02.002

Trasande, L., Cronk, C., Durkin, M., Weiss, M., Schoeller, D. A., Gall, E. A., ... Gillman, M. W. (2009). Environment and obesity in the National Children's Study. *Environmental Health Perspectives*, *117*(2), 159–166. doi:10.1289/ehp.11839 PMID:19270782

Turner, M. C., Nieuwenhuijsen, M., Anderson, K., Balshaw, D., Cui, Y., Dunton, G., ... Jerrett, M. (2017). Assessing the Exposome with External Measures: Commentary on the State of the Science and Research Recommendations. *Annual Review of Public Health*, *20*(38), 215–239. doi:10.1146/annurev-publhealth-082516-012802 PMID:28384083

Ulaszewska, M. M., Ciffroy, P., Tahraoui, F., Zeman, F. A., Capri, E., & Brochot, C. (2012). Interpreting PCB levels in breast milk using a physiologically based pharmacokinetic model to reconstruct the dynamic exposure of Italian women. *Journal of Exposure Science & Environmental Epidemiology*, *22*(6), 601–609. doi:10.1038/jes.2012.36 PMID:22760444

UN. (2019). *Office for Distaster Risk Reduction*. Retrieved from https://www.unisdr.org/we/inform/publications

UNEP. (2013). *Minamata Convention Agreed by Nations*. Retrieved from https://www.unenvironment.org/news-and-stories/press-release/minamata-convention-agreed-nations

University of California - San Francisco. (2018). Brain signature of depressed mood unveiled in new study: Direct recordings of human brain activity link memory, emotion, and anxiety during bouts of low mood. *ScienceDaily*. Retrieved from www.sciencedaily.com/releases/2018/11/181108142410.htm

Van Dijk, S. J., Molloy, P. L., Varinli, H., Morrison, J. L., & Muhlhausler, B. S. (2015). Recent developments on the role of epigenetics in obesity and metabolic disease. *Clinical Epigenetics*, *7*(66), 66. doi:10.118613148-015-0101-5 PMID:27408648

Van Kempen, E., Fischer, P., Janssen, N., Houthuijs, D., Van Kamp, I., Stansfeld, S., & Cassee, F. (2012). Neurobehavioral effects of exposure to traffic-related air pollution and transportation noise in primary schoolchildren. *Environmental Research*, *115*, 18–25. doi:10.1016/j.envres.2012.03.002 PMID:22483436

Velarde, M.D., Fry, G., & Tveit, M. (2007). Health effects of viewing landscapes — Landscape types in environmental psychology. *Urban Forestry & Urban Greening*, *6*(4), 199–212.

Vrijheid, M. (2014). The exposome: A new paradigm to study the impact of environment on health. *Thorax*, *69*(9), 876–878. doi:10.1136/thoraxjnl-2013-204949 PMID:24906490

Vrijheid, M., Casas, M., Bergstrom, A., Carmichael, A., Cordier, S., Eggesbo, M., & Nieuwenhuijsen, M. (2012). European birth cohorts for environmental health research. *Environmental Health Perspectives*, *120*(1), 29–37. doi:10.1289/ehp.1103823 PMID:21878421

Vrijheid, M., Slama, R., Robinson, O., Chatzi, L., Coen, M., van den Hazel, P., ... Nieuwenhuijsen, M. J. (2014). The Human Early-Life Exposome (HELIX): Project rationale and design. *Environmental Health Perspectives*, *122*(6), 535–544. doi:10.1289/ehp.1307204 PMID:24610234

Waeger, P.A., Schluep, M., Móller, E., & Gloor, R. (2012). RoHS regulated Substances in Mixed Plastics from Waste Electrical and Electronic Equipment. *Environmental Science & Technology, 46*(2), 628–635.

Wang, Y., Yan, L., Zhang, Z., Prado, E., Fu, L., Xu, X., & Du, L. (2018). Epigenetic regulation and its therapeutic potential in pulmonary hypertension. *Frontiers in Pharmacology*, *9*, 241. doi:10.3389/fphar.2018.00241 PMID:29615911

Wariki, W. M., Ota, E., Mori, R., Koyanagi, A., Hori, N., & Shibuya, K. (2012). Behavioral interventions to reduce the transmission of HIV infection among sex workers and their clients in low- and middle-income countries. *Cochrane Database of Systematic Reviews (Online)*, *2*. doi:10.1002/14651858.CD005272.pub3 PMID:22336811

WHO. (2006a). *Air quality guidelines. Global update 2005. Particulate matter, ozone, nitrogen dioxide and sulfur dioxide*. Copenhagen, Denmark: World Health Organization, Regional Office for Europe.

WHO. (2006b). *Health risks of particulate matter from long-range transboundary air pollution*. Copenhagen, Denmark: World Health Organization, Regional Office for Europe.

WHO. (2007). *Health risks of heavy metals from long-range transboundary air pollution. Joint WHO/Convention Task Force on the Health Aspects of Air Pollution*. Copenhagen, Denmark: World Health Organization, Regional Office for Europe.

WHO. (2012). *Soil-transmitted helminthiases: eliminating soil-transmitted helminthiases as a public health problem in children: progress report 2001-2010 and strategic plan 2011-2020*. Geneva: World Health Organization.

WHO. (2012b). *Endocrine disrupters and child health - Possible developmental early effects of endocrine disrupters on child health*. Geneva, Switzerland: World Health Organization.

WHO. (2013a). *Review of evidence on health aspects of air pollution — REVIHAAP*. Copenhagen, Denmark: World Health Organization, Regional Office for Europe.

WHO. (2014). *Global Tuberculosis Report 2014*. Geneva: World Health Organization.

WHO/JRC. (2011). Burden of disease from environmental noise. In L. Fritschi, A. L. Brown, R. Kim, D. Schwela, & S. Kephalopoulos (Eds.), *World Health Organization, Regional Office for Europe*. Copenhagen, Denmark: Author.

Wild, C. P. (2005). Complementing the genome with an "exposome": The outstanding challenge of environmental exposure measurement in molecular epidemiology. *Cancer Epidemiology, Biomarkers & Prevention, 14*(8), 1847–1850. doi:10.1158/1055-9965.EPI-05-0456 PMID:16103423

Wild, C. P., Scalbert, A., & Herceg, Z. (2013). Measuring the exposome: A powerful basis for evaluating environmental exposures and cancer risk. *Environmental and Molecular Mutagenesis, 54*(7), 480–499. doi:10.1002/em.21777 PMID:23681765

KEY TERMS AND DEFINITIONS

Bioinformatics: Bioinformatics uses computation to understand, organize, analyze, and interpret biological data.

Environmental Hazard: Is a substance, a state or an event, which has the potential to threaten the health of people and the natural environment.

Environmental Risks: The probability or chance of suffering an actual or potential threat, danger or hazard of adverse effects on living organisms and environment.

Epigenetics: The study of heritable phenotype changes in organisms caused by modification of gene expression rather than alteration in the DNA sequence itself.

Exposome: The totality of environmental exposures from conception onward, assessing the multitude of human exposures across the life.

Health-Informatics: Is the practice of acquiring, studying, designing, developing, applying and managing health and medical data in conjunction with health information technology systems to help improve healthcare.

Metabolomics: Refers to systematic identification and quantification of small molecules and their interactions within a biological system, commonly known as metabolites, within cells, biofluids, tissues or organisms.

Next Generation Sequencing: Massively parallel or deep sequencing of DNA technology allowing one to determine in a single experiment the sequence of a DNA molecule(s) with total size significantly larger than 1million base pairs (1millionbp or 1Mb).

Personalized Medicine: Is a therapeutic approach involving the use of an individual's genetic and epigenetic information where the medical decisions, practices, interventions and/or products are being tailored to the individual's drug therapy scheme and/or preventive care.

Sustainability: Avoidance of the depletion of natural resources in order to maintain an ecological balance, all in harmony and enhancement both for current and future generations potential that meet the human needs and aspirations without compromising the survival of other species.

Chapter 2
Current Environmental Health Challenges:
Part 2 – Moving Toward a Healthy and Sustainable Future

Christina Marouli
Deree – The American College of Greece, Greece

Paraskevi Papadopoulou
https://orcid.org/0000-0002-8194-7495
Deree – The American College of Greece, Greece

Anastasia Misseyanni
Deree – The American College of Greece, Greece

ABSTRACT

This is part two of two overview chapters of the most important contemporary environmental health challenges. This second chapter discusses environmental health as a socio-political and ethical issue. It argues that effectively moving towards healthier and sustainable societies requires not only sound scientific knowledge but also policies, medical practices, healthcare systems, and health-related attitudes and behaviors that are informed by a deep socio-political understanding and that reflect a new integrated approach to environment and health. The need for contemporary technological societies to develop mechanisms like education, environmental and health governance, and public accountability for environmental health equity and justice is highlighted. The chapter concludes by proposing a multidimensional framework, based on both natural and social sciences, for the transition to healthy and sustainable societies and for improving the welfare of all people, as well as future research directions for environmental health sciences.

DOI: 10.4018/978-1-7998-1241-8.ch002

INTRODUCTION

With the increasing awareness of and concern for environmental issues and impacts on human health in the last century or so, the field of environmental health has burgeoned. Environmental exposures and related human health impacts or health problems and their environmental causes have extensively been studied in recent decades (World Health Organization (WHO), 2006A; WHO, 2006B; WHO, 2007, WHO, 2012, WHO, 2013A; WHO, 2014; Bonnefoi, et al., 2010; Bollati & Baccarelli, 2010; Barouki, Gluckman, Grandjean, Hanson, & Heindel, 2012; Pruess-Ustuen et al., 2014; Herceg &Vaissière, 2011; Herceg, 2016; Seralini, Mesnage, Clair, Gress, de Vendomois & Cellier, 2011; Séralini, Clair, Mesnage, Gress, Defarge & Spiroux de Vendômois, 2014; Larkin & Hystad 2017; Lytras & Papadopoulou, 2018; Papadopoulou, Lytras & Marouli, 2018; Bagasra & Bagasra, 2019). Environmental health is surely a very significant field of study nowadays since escalating environmental and health problems are interdependent; climate change is further enhancing environmental impacts on human health (Charron, 2012). Natural scientists have also investigated appropriate methods to document and evaluate such relations (Wild, 2005; Wild, Scalbert & Herceg, 2013; Rappaport & Smith, 2010; Lioy & Rappaport, 2011; Vrijheid, Casas, Bergstrom, Carmichael, Cordier, Eggesbo & Nieuwenhuijsen, 2012;Vrijheid, Slama, Robinson, Chatzi, Coen, van den Hazel, & Nieuwenhuijsen 2014;Vrijheid, 2014; Buck Louis, Smarr & Patel, 2017; Jeanette et al., 2017; Sarigiannis, 2017; Sarigiannis, Karakitsios, Handakas, Papadaki, Chapizanis, & Gotti, 2018; Sarigiannis, Gotti, Handakas, & Karakitsios, 2018; Sarigiannis, 2019) as well as to assess related risks (Hines et al., 2010; Domingo & Gine Bordonaba, 2011; Elliott, Briggs, Morris, et al. 2001; Myers et al., 2016; Dennis et al., 2017; Turner et al., 2017). On the basis of this vastly expanded knowledge, many significant successes have been achieved in the field of health and medicine, with new cures for previously deadly diseases and with the subsequent lengthening of the average human life.

Despite these developments, it cannot be asserted that we live in healthy societies. Instead, many diseases and human health threats plague contemporary societies. Many of these threats have been associated with environmental and social exposures, while for others the causes are still not known. The World Health Organization (WHO) highlights the significance of environmental factors – which it defines as "factors external to a person, and all related behaviors, but excluding those natural environments that cannot be reasonably modified" (Pruss-Usten, Wolf, Corvalán, Bos & Neira, 2016) – for the state of human health.

According to the same report, 23% of all deaths and 22% of the disease burden measured in Disability Adjusted Life Years (DALYs) were attributable to modifiable environmental risks in 2012. The diseases with the highest preventable disease burden attributable to environmental risks include diarrheal diseases (57%), malaria (42%), respiratory problems (lower respiratory infections - 35%, asthma – 44%, chronic obstructive pulmonary disease – 35%), cancer (20%), cardiovascular diseases (31%), musculoskeletal diseases (22%), injuries (road traffic related – 39%, other unintentional ones – 50%) and unipolar depressive disorder (11%). The main environmental exposures associated with these diseases include air pollution, water sanitation, management of chemicals, agricultural practices, occupational parameters and stressors, smoking, quality of housing (e.g. molds), land use planning, road/household/workplace design, and work-life imbalances. WHO has also estimated that air pollution is the biggest environmental risk to human health in the European Union, causing about 400,000 premature deaths each year as well as significant health costs amounting to billions of euro (European Court of Auditors, 2018). The main culprits for these early deaths are particulate matter, nitrogen dioxide and ground level ozone.

Figure 1. Human health in environment & society
Source: Amended from Papadopoulou, Marouli, Misseyanni, 2019

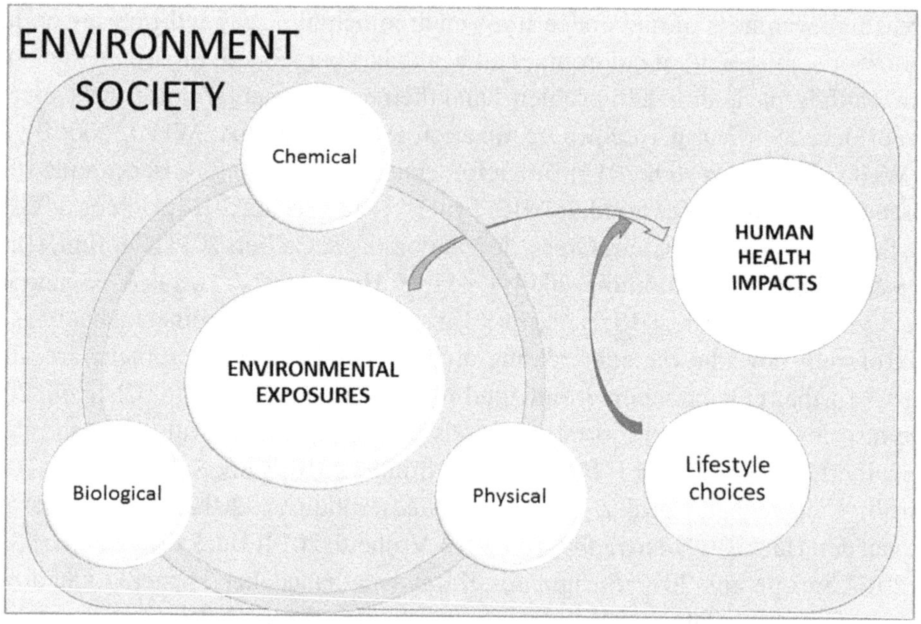

At the same time, we now acknowledge that environmental hazards and degradation are to a large degree human-induced and a consequence of socio-economic activities and choices. The United Nations Sustainable Development Goals for 2030 clearly demonstrate the understanding that the environmental, economic and social aspects of life are closely interlinked (United Nations, 2015). In contemporary societies, the overall trend is increasing prevalence and severity of environmental hazards, exacerbating overall health inequities. There is thus a growing need to understand and effectively connect environmental exposures and human health challenges on the one hand, with social organization, and environmental parameters on the other in the pursuit of healthier individuals and societies. The extensive introduction of machine learning methodologies and big data mining and analytics to contemporary clinical science and practice opens immense possibilities and poses significant new challenges requiring socially wise use of big data (Lytras & Papadopoulou, 2018; Papadopoulou, Marouli & Lytras, 2018). This development further highlights the need for a deep socio-political understanding of health and environment in order to render it an effective tool towards the goal of universal health and wellbeing.

After the previous chapter, Part I, which documented the biological and chemical basis of major environmental health challenges today and discussed relevant scientific research, this chapter discusses the socio-political and cultural dimensions of environmental health issues. It advocates that environmental health issues and their assessment as risks constitute a socio-political process and their root causes lie in the socio-political arena. It concludes by proposing a strategic framework for the health policy making and medical practice that goes closer to the structural sources of the problems. The main questions that will be addressed include: What is an environmental risk? Are environmental risks the same for all? What is health? What are the characteristics of an effective health promotion and health care system? How can environmental health science contribute in this direction?

ENVIRONMENTAL EXPOSURES AND ENVIRONMENTAL RISKS: THE SIGNIFICANCE OF THE SOCIAL CONTEXT

Risk Analysis: A Valuation Process or the Social Construction of Risk

The defining mission of environmental health sciences has been to shed light on how the environment affects human health. According to Kumwenda, Samanyika, Chingaipe, Mamba, Lungu, Mwendera and Kalulu (2014), environmental health also aims to prevent diseases and promote health-supportive environments. In these fields, emphasis has been placed on environmental factors or exposures that constitute risks to human health.

Risk expresses the probability of suffering some harm (including injury, disease, death, economic loss, or damage to environmental components, etc.) because of an exposure to a hazard. Risk analysis is the effort to estimate the chance that a hazard may cause harm; thus, it explores and tries to manage uncertainties and the significance of potential impacts. Beck (1992), who introduced the term of "risk society," referred to it as "a systematic way of dealing with hazards and insecurities induced and introduced by modernisation itself" (p.21), thus hinting to the social construction of risk (Cooper & Thorogood, 2013). Giddens (1998) on the other hand, highlighted modern societies' increasing preoccupation with risk and safety. Thus, risk assessment is a valuation process, which, below the scientific quantification techniques used, generally reflects prevailing value systems and paradigms or specific worldviews. It is socially constructed. Shostak (2013) writes:

It is time for an open dialogue among scientists, risk assessors, regulators, community groups, activists, and regulated industries about what certainty is or could be in the context of environmental health controversies. As suggested by scientists whom I interviewed, this conversation should include explicit consideration of differences in what certainty means across institutional domains within the environmental health arena:

The basic scientist who deals with certainty—something has to be close to 100%. The toxicologists say, "Well that's not true, maybe 90% is good for me." And the policymaker might say, "Well, maybe 80% is good for me, because I have to make a decision. I have to make a decision on what's safe, I can't wait until all the information is in, so I want to use your system now to give me 80% confidence of what my decision is —for [a] safe exposure level (Interview S81). (p.84)

A good example is the case of Genetically Modified Organisms (GMOs). Although they were introduced as a very promising tool against malnutrition, recent scientific evidence point to their negative health impacts (Seralini, Mesnage, Clair, Gress, de Vendomois & Cellier, 2011; Séralini, Clair, Mesnage, Gress, Defarge & Spiroux de Vendômois, 2014). Despite that, they continue to be extensively used. Strong economic interests influencing policy makers emphasize the lack of certainty of scientific results in order to keep GMOs and the required chemicals for their cultivation in the market (Krimsky, 2019a; Krimsky, 2019b; Domingo & Gine Bordonaba, 2011). In some societies like the European Union, precaution has been accepted as a basic principle in policy making requiring that action is taken for significant environmental risks before scientific data is considered conclusive.

Thus, it can be understood that the perception of the significance of potential health impacts depends considerably on the positioning of each actor, expert or policy maker or member of an exposed community, in relation to the risk exposure or to the impacted people. Harrington & Elliott (2015) state:

There is an inherent spatiality to the risk construct that emphasizes distance, direction, position, and co-location of hazards with other social or physical phenomena... The concept of place, however, extends the notion of spatiality to include the characteristics and meanings of these spaces. (p. 286)

Thus, it is not only physical dimensions of positioning that influence potential risk impacts and relevant perceptions, but also the meanings and power dynamics that are associated with this positioning.

The conception of an environmental exposure as a risk or not worthy of attention in society (and to a certain degree in science too) is a process that reflects the specific social, economic, cultural and political context. Harrington and Elliott (2015) highlight the difference between the "objective" quantifiable risk assessment experts seek to perform and the intuitive and experiential understanding of risk by the public; they highlight that these two conceptions of risk "deviate from one another in a systematic fashion" (Harrington & Elliott, 2015). In our societies where science and technology have a predominant role, experts' quantifiable risk assessment is conceived as the norm against which we compare other risk conceptions. However, they also correspond to different bodies of knowledge, which may be useful for effective environmental and health interventions.

Environmental Exposures: Risks or Benefits or "Non-Issues"

Environmental health sciences have been particularly concerned with environmental exposures that have hazardous impacts on human health. However, it is also widely recognized that the environment provides very significant benefits and services for human health and healthy communities and it is the fundamental support of all life forms (Hanna & Coussens, 2001). Given the import of positive environmental – external influences on human health and resilience to attacks of pathogenic organisms or chemical factors (e.g. through the quality of food we eat, or the quality of water we use), it is essential for environmental health sciences to consider environmental benefits to health as well. In the environmental health sciences, it is important to discuss environmental risks and benefits together, since multiple exposures constitute the health exposure reality of an individual and a community. Masuda, Poland and Baxter (2010) state: "Human-produced environmental risks result from both exposures to hazards and limitations on access to environmental opportunities." (p. 453).

But as environmental risks are socially constructed, so are environmental benefits. Green spaces in cities have been repeatedly associated with good health as they clean the air, provide a place for athletic activities and control the local climate among other benefits. However, there are cases in which they are considered a problem and a threat to their wellbeing by local communities, e.g. when they are not well maintained and they attract criminal activities (Walker, 2012; Marouli, 1995). Such cases reveal that health is multidimensional and includes not only tangible parameters (like physical ones) but also political-economic (e.g. distribution of funds among different areas or municipalities) and socio-psychological ones too (e.g. sense of safety), the significance of which is well recognized in social scientific literature as well (Walker, 2012; Spencer, 2016). Thus, the evaluation of environmental exposures as hazardous or beneficial is not a uniquely scientific process. Lay knowledge is also relevant and useful.

Risk Assessment and Multiple Exposures

Risk assessment continues to be a useful tool for identifying some potentially hazardous environmental exposures, as well as concentrations that are likely to negatively impact people's health. Risk models are useful for evaluating future possible impacts with current knowledge (Goerlandt & Reniers, 2018).

Figure 2. Valuation and categorization of environmental exposures
Source: Our Own

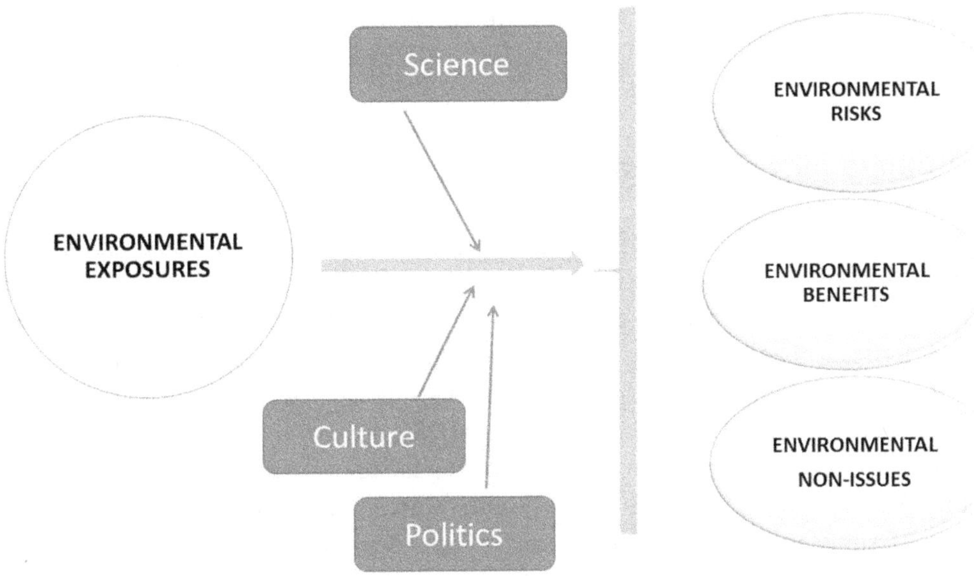

Their predictive capacity is questionable in a changing context. The increased prevalence of risks at all levels of life in contemporary "risk societies," including environmental risks, and the preoccupation with safety in a world of uncertainty constitute a new context for risk analysis. According to Giddens (1998), it is the modern way of life that has generated the multiplicity of risks. Global chemical production is on the rise and yet understanding the impact of chemicals on human health is limited. Many chemicals have been introduced in the market and are presently in use although they have not been sufficiently studied and their health impacts are not known. Furthermore, chemicals such as heavy metals, hormone disruptors, carcinogens, food preservatives, which are well studied and their negative health impacts have been identified (Bonnefoi et al., 2010; Myers et al., 2016; Nieuwenhuijsen et al., 2009; Seralini, Mesnage, Clair, Gress, de Vendomois & Cellier, 2011; Séralini, Clair, Mesnage, Gress, Defarge & Spiroux de Vendômois, 2014; Chikkanna, Mehan, & Ghosh, 2019) are extensively used in most parts of the world, despite efforts to reduce their use. A new major risk to human societies, human health and ecosystems alike is climate change, with a complex array of potential hazardous impacts on all aspects of societies. This new challenge further enlarges the complexity of the risk matrix that affects health.

However, risk assessment is not designed for quantifying multiple exposures, with possible synergistic effects, and which change over time (Kristie et al., 2016). Furthermore, environmental exposures begin from the uterus and childhood exposures are very significant sources of influence on adult health (Syme & Wang, 2012; Evans & Kim, 2007; Cecil, Brubaker, Adler, Dietrich, Altaye, Egelhoff, et al., 2008). Linda Birnbaum, Director of the US National Institute of Environmental Health Services, in her interview talks about "windows of higher susceptibility" in a person's lifetime (Birnbaum, 2014). The exposome analysis, a new promising approach, tries to address this multiplicity of hazardous exposures to health throughout an individual's lifetime. However, as it continues focusing on the individual body it cannot account for structural parameters affecting health. Shostak (2013) writes:

With its focus on human bodies living in complex social and physical environments, exposomics offers a new way of studying the effects of environmental exposures. However, like other molecular techniques, it locates the scientific gaze wholly within the human body. This raises questions about the capacity of exposomics to encompass the social, political, and economic conditions that put people at risk of risk. (p. 81)

Risk and Vulnerability: The Same Environment is not the Same for all

As scientific risks, "safe" thresholds and significant impacts are defined based on probabilities, there is a good possibility that exposures below the indicated thresholds will affect some people's health negatively. The categorization of an environmental exposure as a risk (or not) also depends on the specific conditions and characteristics of an individual, which change over time. Children, elders and pregnant women are considered particularly vulnerable groups. For other specific environmental exposures, other socio-spatial characteristics – like sexual preferences (e.g. AIDS), poverty and malnutrition (associated with many diseases), living close to a main street (e.g. asthma due to exposure to air pollutants) – contribute to vulnerability. Vulnerability – most often related with socio-economic parameters such as poverty, age, and general health status – plays an important role in the determination of the risk level (from non-risk to different levels of risk) for the individual and the community. Public health research has shown that vulnerabilities may be more significant determinants of the possible risk level for a population and an individual. Syme and Wang (2012) indicate that poverty and low socioeconomic status has been consistently correlated with higher rates of morbidity and mortality in public health literature. For climate change, Kristie et al. (2016) state that "Vulnerabilities, such as poverty or failing infrastructure, may be more important determinants of the magnitude and pattern of climate-sensitive health outcomes" (p.5).

ENVIRONMENT AND HEALTH: SAME FOR ALL?

But are environments the same for all? Do all people live in environments of similar quality, or even sufficient quality to support their health? Or, even in the same environment, are all people affected – in terms either of negative impacts on their health or their access to environmental amenities – the same way?

Different Environments: Inequalities in Exposure

A study of the World Health Organization (Pruss-Usten et al., 2016) indicates that there are differences between countries depending on their economic status. As the study reports, infectious diseases decreased at a global level, but they are significantly higher in low and middle-income countries while they predominate as a cause of death in Sub-Saharan Africa. Infectious diseases have been found to be attributable to environmental factors to a significant degree. For example, main risk factors contributing to susceptibility to respiratory infections include malnutrition and a compromised immune system, smoke from heating or cooking with biomass, or living in crowded homes. Diarrheal diseases, which are a main contributor to global child mortality mainly prevalent in poor countries, relate to poor sanitation and lack of access to clean water for good hygiene, all issues of concern in these poor countries. On the other hand, non-communicable diseases (e.g. ischemic heart disease (IHD), a leading cause of mortality and disability worldwide) have increased, but they constitute overwhelmingly the main cause of deaths in Europe and other wealthy nations. Main risk factors for non-communicable diseases include air pollution, noise, poor quality water, chemicals, radiation, biological agents and otherwise degraded environments (European Environmental Agency (EEA) & Joint Research Center (JRC), 2013), along with high blood pressure, limited physical activity and diet. Although some of these factors may appear to be individual choices, they correlate with an urban lifestyle and sedentary work. Due to the increasing urbanization, more than half of the global population, more than 70% of Europeans and 80% of people in North America live in urban areas.

Cities have the potential to concentrate several modern risks, such as ambient air pollution, low physical activity due to poor urban planning and sedentary work. (Pruss-Usten et al., 2016)

Air pollution in cities remains a major environmental problem that significantly affects health and well-being. Furthermore, climate change is another upcoming threat, with noteworthy direct and indirect impacts on health.

Climate change is likely to have significant effects on crop productivity of major cereals, and thus potentially increase the risk of undernutrition (IPCC, 2013; WHO, 2014z) as agriculture is climate sensitive. Furthermore, increases in water scarcity, population growth, meat consumption and land degradation are likely to additionally reduce food security and possibly to aggravate malnutrition. (Pruss-Usten et al., 2016)

Climate change is not expected to affect all geographic regions in a similar way. There are some regions that are particularly vulnerable to global warming effects, like the Arctic and the Mediterranean region. Furthermore, the health impacts of global warming depend on the vulnerability of the population and its adaptability, which is closely linked to ecological, social, economic and cultural parameters (EEA & JRC, 2013). The lack of resources for appropriate preparedness measures for protection from and mitigation of climate change impacts contributes to the creation of risk-prone areas.

Syme and Wang (2012) highlight that environments are different and unequal for different groups of people. They note: "…socioeconomic status is often treated as if it were a quality of individuals. Poverty, wealth and social status are qualities of environments as well as of people" (p. 414). Some places have better infrastructures, more greenery, more public and private services, better housing, more well-developed networks of social support, while others have less or fewer of all or some of these resources. Such structural/ infrastructural elements of a place constitute the inscription of social inequalities in the physical environment and render it more or less conducive to good health (Marouli, 1995). Health and social scientists have begun to investigate these environmental differences, rather than simply focusing on individuals or population characteristics. They explore the built environment including walkability, availability of healthy food, availability of resources and services and quality of social networks and other such parameters, which influence the ease and possibility for individuals to adopt healthy behaviors (Syme and Wang, 2012).

Environmental justice scientists began their research on the differential distribution of environmental hazards back in the 1990s, and they started revealing the existence of different environments for different groups of people and a systemic connection between social inequalities and proximity to environmental risk exposures. As an example, many studies have documented that waste management facilities are overwhelmingly located in people-of-color and poor communities not only in the United States of America (Faber & Krieg, 2002; Goldman & Fitton, 1994; Bullard, Mohai, Saha & Wright, 2007; Kearney & Kiros, 2009; Brown, 1995) but also in other parts of the world, like Europe (Elliott, Briggs, Morris, et al., 2001; Laurian, 2008; Martuzzi, Mitis, Bianchi, et al., 2009; Varga, Kiss & Ember, 2002). Martuzzi, Mitis & Forastiere (2010) indicate that environmental and social factors (like poverty, less education, ethnic minorities, etc.) work together to produce marked inequalities in the health pressures people experience. Proximity to such facilities unquestionably poses serious challenges to the health of members of nearby communities, but also lowers the economic value of property and housing; these are undesirable environments. Blowers (2003), in his text about nuclear oases – places where nuclear waste is disposed / stored – used the term of "peripheral communities" to refer to "places which experience the disadvantage of risk and environmental degradation. …They are the outcome of a process of 'peripheralization' whereby a coalition of economic interests (of labour and capital) reinforces the presence of an activity that is resisted by a coalition of interests (environmental and economic) defending the intrusion of the unwanted industry elsewhere." These places then have become degraded.

Peripheral communities like these appear to condescend to the presence of nuclear waste or other hazardous activities because of limited resources and a relative powerlessness in front of the external forces that influence their lives; they compromise their health for survival (Walker, 2012, p.98). Bullard has called this "environmental blackmail." Toxic waste – like waste of electric and electronic equipment – has traditionally been moved to other countries, away from its production site, to communities that have no responsibility for the generation of this waste but are also desperate for jobs and income for survival. Nevertheless, they bear the environmental health burden from the toxic materials that wealthy societies generate and use. The selection of the location for these facilities is a political process and reveals the import of social inequalities in the quality of the physical environments as well as inequities in resources and access to decision making.

Environmental Health: The Significance of Vulnerability and Access

But even in the same environment, not all people are affected by environmental burdens the same way, nor do they share environmental benefits equally. The social characteristics of individuals or specific groups influence their interaction with the environment.

Climate change affects all aspects of societies, including food production, forests, infrastructures, vector diseases, and health. It leads to intense weather phenomena that intensify the risk of floods, droughts or wildfires. Some communities are more vulnerable than others due to where they live. For example both Venice and Bangladesh are flood prone areas, but the impacts of floods there are quite different. But what determines whether a risk translates into disaster lies in the socio-economic sphere (i.e. resources and technology to protect from and adapt to the occurrence of floods and its negative impacts). Walker (2012) indicates that two important parameters that influence the level of vulnerability of a community are: "'resistance', the capacity of an individual or a group to withstand the impact of a hazard, and 'resilience', the ability to cope with or adapt to hazard stress, encompassing preparedness and processes of relief, rescue and recovery." The case of Hurricane Katrina is enlightening. Elders that could not be evacuated during the event, people with pre-existing health problems or a disability, being female or a single parent with children were the groups that suffered the most from the impacts of the hurricane. Most of the evacuation was done by private cars; thus, African-Americans that overwhelmingly did not own a vehicle were more at risk. Some of them had particular difficulty in coping or recovering from the disaster. For example:

People with disabilities (6 per cent of the population) were widely excluded from food relief and rehabilitation programmes because of accessibility problems and the difficulties in providing for their particular care needs. (Walker, 2012; p. 139)

So, vulnerability, arising from different social characteristics, is a fundamental parameter that differentiates environmental health risks even within the same environment and the same environmental risk. Exposure differs depending on a person's or a group's level of vulnerability.

Social roles and characteristics are also important parameters influencing vulnerability and exposure levels, as they influence human daily activity patterns. Gaffron (2012), in a study on urban transport in the city of Hamburg, Germany and the UK, looked into time use surveys to identify different exposures to air pollutants near major motorways, since relevant environmental risks (like noise and air pollution) affect people when both people and moving vehicles are present simultaneously. He found that gender was a very important factor differentiating time spent at home, and thus, people's level of exposure to air pollutants. Women and girls spent more time at home in almost all age, education, employment and income groups. Elders also spent a lot of time at home, while young people between 15 and 24 years of age the least. Thus, he concludes:

Even in areas, where the legal emissions limits are not exceeded, members of certain socioeconomic population groups may suffer excessive cumulative exposure to transport emissions simply because of the amount of time they spend at home. (Gaffron, 2012; p.125)

This discussion also points to the importance of social characteristics in determining individual issues, like health problems. As C. Wright Mills (1959) indicated, private troubles are social problems; and so is health.

But specific social characteristics also influence access to environmental amenities, i.e. physical environments and their services that are beneficial to human health. Urban green areas are generally valued as a positive health influence as they clean the air and provide a space for physical exercise and social relations. However, they are unequally distributed in cities. Access to urban green spaces is an issue in lower class areas. However, accessibility is not uniquely related with proximity (Walker, 2012). The quality and structure of green spaces affect their accessibility by local communities. A green space that does not have ramps or flat entrance points for people with special mobility needs renders this place inaccessible to this group. Parks that do not have benches are unfriendly to elders. Or green spaces without adequate lighting deter women from their use as they feel unsafe after dark. Parks that are poorly maintained and guarded may become unwelcoming places for different groups of people. Women in Aigaleo, a lower class area in Athens Greece were concerned about the existence of a park near their houses as they felt unsafe in and around it due to the illegal activities (e.g. drug use) that took place there, thus limiting their mobility (Marouli, 1995). Consequently, environmental health goods also are not experienced the same way by all; social characteristics of individuals influence their beneficial health exposure.

Thus, the previous sections have demonstrated that the characterization of environmental exposures but also environmental health impacts are to a large extent socially constructed, influenced by social, economic and political parameters.

ENVIRONMENTAL HEALTH: THEORETICAL APPROACHES

Previous sections discussed the socio-economic and political dimensions of environmental health and of the evaluation of environmental risks and their health impacts. Although these are generally acknowledged, they have been approached in different ways.

Figure 3. Environmental exposures and human health: A sociopolitical understanding
Source: Our own

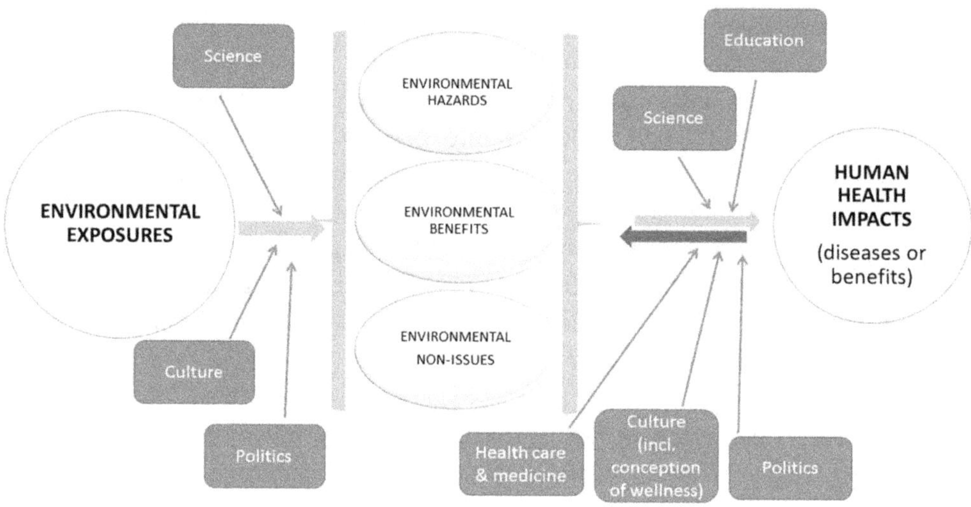

Biomedical Sciences

A very prominent theoretical approach is that of the biomedical sciences: a scientific theory that looks primarily to chemistry, biology and genetics, along with scientific risk assessment as the main explanatory frameworks. Traditionally, scientific studies investigated the impacts of different chemical and physical exposures on human health, through animal experiments or epidemiologic studies (Wigle, 1977; Delfino, Sioutas & Malik, 2005; Hubbs-Tait, Nation, Krebs & Bellinger, 2005). Furthermore, biological studies investigate how the human organism works, the significance of genetic inheritance, different biochemical processes and the biological pathways for different environmental exposures. An extensive coverage is provided in our first chapter (Part I) and references within. Thus, a significant body of very valuable knowledge has been collected regarding the ways human organisms react to different environmental exposures.

Environmental health sciences have built on biomedical knowledge and approaches, but the focus is on preventing disease by offering an empirical basis for health policy. Thus, they also have a strong connection with the science of risk assessment in their effort to handle issues of certainty. This combination leaves environmental health sciences vulnerable to contestation from strong stakeholders, like industry and multinationals in the fields of production of chemicals, agribusiness, etc.

In the last decades, new methods have been proposed in order to provide useful information for public policy, and they aim to consider the influence of environmental exposures on human health. Epigenetics, the exposome and other omics analyses offer very helpful and promising insights contributing to a deeper understanding of the biological mechanisms that connect environment, genetic inheritance and health. An extensive coverage is provided in our first chapter (Part I) and references within. These natural scientific approaches focus on the individual and even more specifically at the molecular level, advocating that medical and health care practices based on this approach will lead to more precise and more effective therapy and potentially health protection policy approaches.

Public Health and Health Promotion

Public health started in the nineteenth century in response to the prevalence of different epidemics and focused on improving environmental conditions such as sanitation for better community health. Industrialization led to populous working communities and rapidly expanding cities with poor housing and sanitation conditions. These environments contributed to the increase of epidemics of infectious diseases. As epidemics were recognized as a threat not only to a nation's health but also to the socio-economic-political system, health policies with an emphasis on improving the environment of these communities especially in terms of housing conditions and sanitation were introduced. But public health has changed focus and approach over time (Mold & Berridge, 2013). Significant developments in laboratory and medical sciences (e.g. development of different medicines) that were achieved in this same period, contributed to retracting the public health focus on the individual and away from the environment. Vaccinations were also imposed in the same period in order to control these epidemics. In the early to mid twentieth century, public health shifted its focus to families and individuals, while in the 1950s faith in high-tech medicine increased. Social medicine introduced the notion of health as a positive state of health and not simply absence of disease. Since the 1970s, the concepts of risk and prevention have become central to the new public health approaches.

Notions of risk, safety, prevention, and individual behaviour – as both a cause of disease and a way to combat it – were central to what was called the new public health. (Mold & Berridge, 2013: 12)

In the last decades of the 20th century, the health promotion approach started becoming more popular. According to the Ottawa Charter for Public Health (WHO, 1986), health promotion is "the process of enabling people to increase control over, and to improve, their health." This approach differs from the more medicalized new public health as it emphasizes social – environmental influences upon health, both of communities and of individuals. This was an effort to reintroduce the significance of environment in determining health. But now, environment was conceived in a new way, reflecting the increased awareness of environmental problems and their connection with social practices. The Ottawa Charter for Public Health states that improvement in health requires social and environmental changes, such as sanitation and good quality housing (still an issue in some parts of the world), control of environmental pollution and resource depletion, peace and social justice.

Environmental Justice and Political Economy

Environmental justice approaches have highlighted the political dimensions of health inequalities; they analyze the way social inequalities predispose and even determine a community health situation and an individual's health status (e.g. Bullard, 2007; Walker, 2012; Agyeman, 2003). Sciences like the political economy or political ecology inform their analyses. Walker (2012) demonstrates and analyzes how environmental burdens are unequally distributed in marginalized communities; how new sets of evidence are needed to document the nexus of environment, social inequalities, and health outcomes; and how the organization of societies and social structures (e.g. decision-making or health care and rehabilitation systems) contributes to the creation and maintenance of this intricate relation.

Several social scientists have tried to theorize the creation of environmental inequalities and injustices, and different theories have been proposed. Faber Blowers (2003), in his discussion of nuclear oases, connects the theory of the risk society, with globalization, imperialism and the political economy. He starts by indicating that the framework of the risk society does not recognize that risks are unequally distributed at the global but even regional level. Globalization has contributed to a new reality, where local problems are often generated by geographically remote producers. Consequently, at the local level, only adaptation measures are possible; root causes lie far away and cannot be effectively controlled by local communities. Imperialism and inequalities in the economic and political spheres influence where environmental risks are concentrated. Political processes and decision making, in which large corporations and other big economic players have a strong presence and influence, favor those with many resources and render the socially marginalized groups silenced.

Rixecher and Tipene-Matua (2003), while discussing the experience of Maori people in relation to genetic engineering and scientific research for medicines, points to yet another inequality: the extraction of knowledge and natural resources of remote communities without their approval and without any respect to their cultures or needs. They use the terms "biopiracy" and "bioprospecting" to refer to contemporary genetic engineering and pharmaceutical/ medical scientific practices, informed by the theory of imperialism and neocolonialism. They state:

...[S]cientists require access to two things: knowledge and natural resources. The natural resources, or life forms, usually come from a non-local geographic location, and rarely are the items attained with permission from the local peoples who are the original stewards of the organism. When researchers do seek out local or indigenous knowledge, it is usually with the intent of determining which local flora and fauna have the greatest chance for generating medicinal or pharmaceutical properties which can be marketed. Thus, these processes of natural resource and knowledge extraction are the most recent form of colonization and imperialism and may be referred to as biocolonialism. (Rixecher & Tipene-Matua, 2003; p. 259)

The aforementioned social scientists have highlighted the significance of power and economic inequalities at the international level. Faber & McCarthy (2003) reinforce the previous scientists' observations and argue that this "ecological imperialism" in their terminology is supported by the same mechanisms that operate within the United States national context. They identify the main culprit in neoliberalism (i.e. less state intervention in the market and individualistic ethic) and globalization, which have also led to the commodification of pollution (e.g. emissions rights and emissions trade schemes). Corporations seek to increase their profits / decrease their costs through increasingly efficient use of resources and of labor, as well as low-cost and politically expedient (least resistance) ways of handling waste (e.g. international movement of waste for low-cost and politically feasible ways of treatment, incineration or disposal). They state that these political and economic forces generate and/or exacerbate environmental and social disparities.

In order to bolster profits and competitiveness, companies typically adopt strategies for the exploitation of nature that are not only economically 'efficient' but politically 'expedient' (that offer the path of least social resistance). The less political power a community of people commands, the fewer resources a community possesses to defend itself; the lower the level of community awareness and mobilization against potential ecological threats, the more likely they are to experience arduous environmental and human health problems at the hands of capital and the state.... In the US (as elsewhere in the world), it is the most politically oppressed segments of the population, or the subaltern [sic] – dispossessed peoples of colour, industrial labourers, the underemployed and the working poor (especially women), rural farmers and farm workers, and undocumented immigrants – who are being selectively victimized [sic] to the greatest extent by corporate practices. (Faber & McCarthy, 2003; p.39)

Neoliberalism and ecological imperialism have provided the context for companies to seek "to extract more value from labour power and nature in less time and at less cost" (Faber & McCarthy, 2003; p. 42). Science and technology focusing on increasing efficiency, with the pretense of environmental protection, provide the tools to increase efficiency but also the concomitant negative effects on nature and people's health, with the environmental burdens - considered externalities in economics, but in reality amounting to less favorable health situations - being overwhelmingly born by marginalized groups at a global scale, and the environmental benefits shared by select few that can avoid degraded environments and protect their health more effectively.

A Critical Discussion of Theories for Environmental Health

The theoretical approach scientists presuppose, or subconsciously adopt, has noteworthy implications for the evidence they seek to collect and the policy suggestions that can arise from the results of their studies. Shostak (2013) indicates that the shift of environmental health studies towards analyses focusing on the molecular level "has consequences not only for how scientists do their work, but also for how we, as a society, understand and seek to control the potentially harmful effects of environmental exposures" (p. 78). Biomedical approaches focus on individual characteristics and genetic information that is valuable for understanding biological mechanisms, but not a good basis for solutions that target the structural causes of bad health or the structural supports of good health. Molecular or individual-centered health care approaches cannot effectively account for the social context and the social characteristics of both physical environments and individuals that significantly influence an individual's health and even his/her genetic makeup (e.g. mutagens). Individual-centered public health approaches may acknowledge individuals' social characteristics, but as they shun away from discussions of power, they also cannot correct root causes of health problems or provide a context for health promotion. Thus, these approaches point to clinical approaches and personalized medicine, while ignoring the root causes that lie at the socio-political context and the physical environment.

Simply put, molecular measurements—which locate genetic susceptibilities, environmental exposures, and their effects inside the human body—individualize the risks of exposures most often experienced in settings shared by multiple individuals, such as neighborhoods, workplaces, schools, and playgrounds. Individualization, in turn, supports clinical approaches to problems heretofore addressed via public policy. (Shostak, 2013; p. 79)

Resorting back to Mills (1959), individual troubles (like health problems or an individual's desire but inability to exercise more) are public issues. This implies that their causes lie in the socio-political organization and the culture of the societies in which the individuals find themselves; their individual issues are incidences of social patterns. Therefore, effective solutions lie at the social rather than the individual level. Individuals can only make minor adjustments (i.e. healthier choices) or manage the risks as best as they can.

Elsey, Manandah, Sah, Khanal, MacGuire, King, et al. (2016) very usefully distinguished environmental exposures leading to health problems to *intermediate and structural determinants* of health. Intermediate determinants relate with individual choices and behaviors (e.g. in their study of urban slum women in Nepal, cooking in open fires, poor ventilation, over-crowding and insufficient child supervision), while structural ones refer to socio-economic status (including poverty and lack of access to clean water and sanitation) and environmental parameters (such as insufficient and inappropriate waste management and slum conditions). In our view, this distinction provides a valuable handle for the transition to healthy societies.

Understanding the intermediate and structural determinants allows us to identify which areas are likely to be the most effective to focus on in interventions. For example, only focusing on the behavioural aspects -through education or behaviour change techniques- may be of limited value when material or structural determinants are the stronger influences in driving risk. For example, women were well aware of the risks of smoke and burns from chulas[1], but are unable to switch to using gas due to their low income. No amount of education will change this….

Identifying which determinants operate at household level and which at community level is also key to effective intervention development. (Elsey et al., 2016; p. 14)

In ecology and conservation sciences, protecting a species can be done in two ways: adopting an ecosystems approach – with measures that aim to protect the ecosystem as a whole and as a consequence a specific species as well, or a species approach – which specifically targets the endangered or threatened species. It is generally believed that the ecosystems approach is more effective, but the species approach also has benefits in specific contexts. This analysis is useful for our case as well, since humanity constitutes one of the many species in the intricate webs of natural ecosystems and thus, it is subjected to the same natural laws.

Thus, a combination of structural and intermediate measures to eliminate root causes, minimize potential negative effects and maximize positive impacts on health for the community and the individual is called for, but one that is fit for the particular context. In all cases, we advocate that the ultimate goal of environmental health policies should be to create social environments that support good health for all and eliminate the root causes of health and environmental degradation. At the same time, we should supplement our efforts with health reform initiatives targeting the individuals, with full understanding that these make individuals more adaptive and resilient to a situation that should soon change.

HEALTH: WHAT IS IT?

For any policy framework and any effort to move towards healthy/healthier societies, it is important to have a clear vision and a common understanding of the main goal. Thus, in the case of environmental health policy, it is significant to be clear about the meaning of health.

In contemporary risk-sensitive societies, emphasis is placed on health risks and their prevention or enhanced ability to deal with their consequences. As discussed in previous sections and demonstrated in the first chapter of this series, relevant sciences have focused on health risks, while the mission of environmental health sciences is their prevention. A similar trend is observed in existing health policies and health care systems, which aim at disease control and at best, health promotion by prevention from health risks (Mold & Berridge, 2013). Reducing ill health for vulnerable groups has become an additional major aim (e.g. European Union).

In the case of the European Union (EU), better health has been a major concern and a driving force for environmental action. In 2003, the first EU environment and health strategy was adopted, using public health principles that clearly connect environment and health. Its objectives were:

- "To reduce the disease burden caused by environmental factors in the EU;
- To identify and to prevent new health threats caused by environmental factors, and
- To strengthen EU capacity for policymaking in this area" (Health and Environment Alliance, 2010; p. 7)

A reading of these objectives reveals that the strategy's focus was on disease control and prevention from new health threats; health was perceived as control of ill-health and health problems. The EU Environmental Health Action Plan 2004-2010 (European Commission, COM(2004)), in an effort to materialize the objectives of the EU Environment and Health Strategy, included a set of actions aiming to build a body of integrated environment and health information, to strengthen research on environment and health, and to enhance communication and awareness raising so that individuals could make better health choices and health professionals would be alert to environment and health interactions. In addition, the Directorate-General's (DG) Strategic Plan for Health and Food Safety indicates that they seek to develop health systems and support actions "that help prevent and reduce the impact of ill-health on individuals and economies" (DG Health & Food Safety, 2016; p.4). Finally, the European Union has incorporated the precautionary principle in environmental and health policy, as well as a concern for protecting the health of vulnerable groups from exposures to environmental hazards.

So, the EU's conception of health is more akin to disease control and individual health prevention. In contrast, the World Health Organization had defined health as "a state of complete physical, mental and social well-being and not merely the absence of disease or infirmity" as early as in the late 1940s, when it adopted its constitution (WHO, 1946). However, this approach does not seem to be effectively implemented as a policy guiding principle in the world at any significant degree.

Health can be conceived in an even more positive and inclusive manner, including not only well-being but also people's ability to develop to their full potential (Corburn, 2009). Further still, the Healing Cities approach calls for holistic health and environments that support people and communities being "healthier and more whole" (Healing cities, 2011). This approach proposes that the definition / understanding of health should not only include physical health, social and mental well-being, with the ability for all people to develop to their full potential, but also spiritual aspects. According to this conception of health, "whole communities" – i.e. places where people live, work, learn and play – support one's "aliveness", social interactions and networks of support; the built environment makes users feel good and at peace; open and green spaces are available nearby and provide a connection with the earth and other living beings; the society in which one lives makes them feel they are connected with and accepted in the local community, while s/he knows of the available resources in case one needs them. All these (e.g. fulfilling and healthy work, learn and play; individual "aliveness", social interactions and connectedness with the community, connection with nature, feeling of peace, sufficient available resources and services for different life situations) are conceived as important elements of health and wellness. We advocate that although such a holistic conception of health may be difficult to implement, it is very useful as a main guiding principle of policies, social planning and environmental/urban design in order to promote healthy societies.

Figure 4. Transitioning towards healthy and sustainable societies: Important points for interventions
Source: Our own

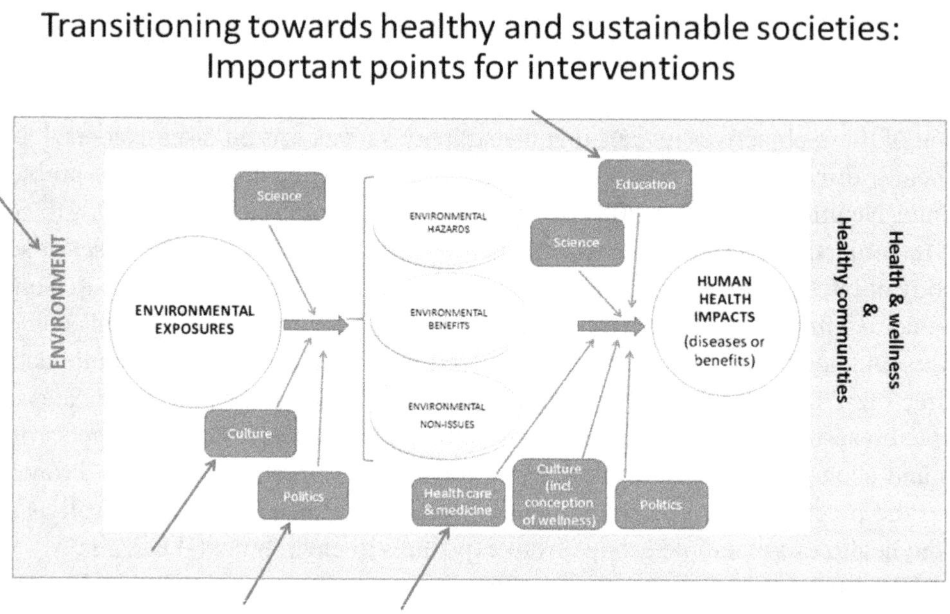

FUTURE POLICY DIRECTIONS FOR ENVIRONMENTAL HEALTH: A FRAMEWORK PROPOSITION

Acknowledging the present environmental situation with the many old and new challenges (e.g. chemical pollution, overexploitation of natural resources, environmental degradation, climate change) and understanding the import of the economic and political organization of our societies (e.g. unhealthy work places, stressful lifestyles especially in cities, unequal distribution of environmental burdens and benefits, social inequalities) to human health, we propose a framework for the protection of human health and the transition to healthy societies.

A framework for the transition to healthy societies and the promotion of health should identify the key points in the environment – health system and target actions appropriately. It actually promotes both health and environmental sustainability at the same time. The key points where interventions are needed in order to promote health in contemporary societies are indicated in Figure 4 below and include culture and reconceptualization of health, politics, science, education, and health care systems.

Main Characteristics of a Framework for the Transition to Healthy Societies

The framework for the transition to healthy societies and individuals should:

1. Address different levels of planning and implementation: vision, strategic goals, means of implementation, monitoring and evaluation.
2. Combine a structural / community-centered (similar to the ecosystems approach in conservation sciences) and an individual-centered medicine approach.

Figure 5. Framework for healthy societies: Priorities and strategies per priority
Source: Our own

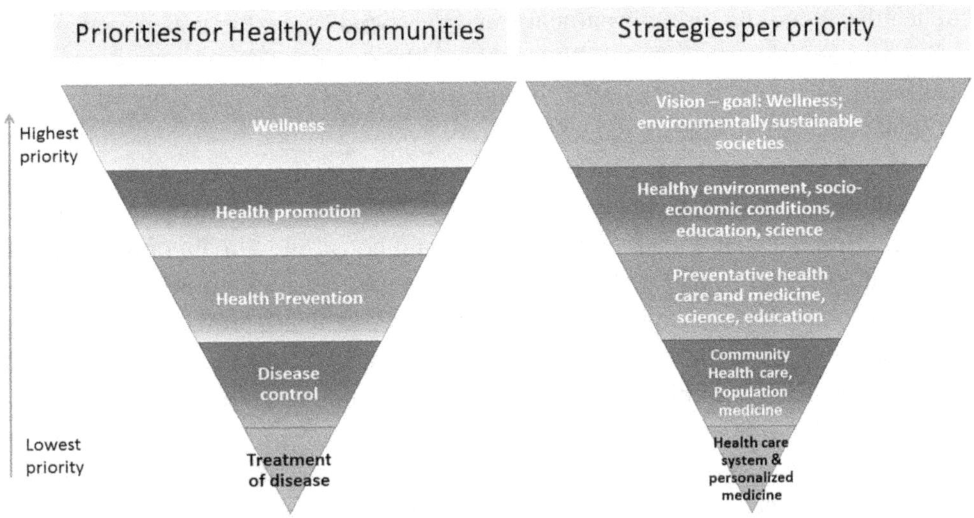

3. Target the health and wellness of all people (i.e. favoring an environment-centered approach but supplemented with appropriate health care institutions for vulnerable groups and already unhealthy individuals).
4. Incorporate recent technological developments (e.g. bioinformatics, biomonitoring, omics approaches) at the health care level.

Environmental Health Vision, Strategy and Institutional Framework

An effort towards healthy societies with healthy individuals is an intricate and demanding task and requires a clear vision, a well-conceived strategy, good governance with participatory processes, and effective implementation, monitoring and evaluation with proper feedback mechanisms.

We advocate that the priorities of all health policies and systems should be, presented from the most desirable to the least desirable, wellness, health promotion, health prevention, disease control and disease treatment.

These priorities also require appropriate strategies. The vision and overall goals should be decided and shared by all social actors (so participatory governance is significant) and they should connect wellness with healthy ecosystems. Health promotion requires a healthy environment and socioeconomic conditions that promote health for all and do not undermine individuals' health. Such an approach would also require the adoption of the precautionary principle in relevant (both environmental and health related) policy making, not waiting for conclusive scientific evidence regarding a certain risk or the need for a valuable environmental "good" for action to be taken. Furthermore, science and education should cultivate individuals, scientists, policymakers, corporate managers that value and are concerned about health promotion. Health care systems and medicine – in terms of science, education and practice – should aim to health promotion first and foremost, while they should prioritize community over individual health approaches.

Structural Versus Individual Health Approaches

As individual health depends on the environmental and the social context in which someone lives, structural approaches are needed to promote wellness and healthy communities and thus enhance individual health for all as well (Kristie et al., 2016). Well-designed structural approaches that take into account the knowledge of how ecosystems work (e.g. closely interwoven networks, interconnectedness of all parts in ecosystems, etc.) and the understanding that humans are a member of ecosystems equally affected by the environment as other species may lead to effective health and wellness promotion. Thus, strategies for healthy environments and health promotion should constitute a main component of the effort to transition to healthy societies and sustainable environments. Systems-thinking is required. For the needed shift towards more structural approaches for wellness and health promotion and the concomitant environmental protection, environmental health sciences should also shift their focus from the genetics – molecular level to the social, political and economic factors that affect human health (e.g. Masuda et al., 2010; Shostak, 2013).

However, each individual is unique at the micro level, with a particular genetic predisposition and a specific array of environmental exposures; thus, the health care system – aiming to promote his/her health and effectively treat his/her diseases – should aim to a holistic understanding of his/her health context (e.g. exposome approach). Furthermore, when a person's health is compromised it should be addressed with the goal of establishing his/her healthy status. The two approaches – health promotion strategies and community health and population medicine on the one hand, and individual disease treatment and precision medicine on the other – should be implemented concurrently, without losing sight of the ultimate goal: health and wellness for all. As Elsey et al. (2016) indicate, health policy makers and practitioners should be clear about the relative significance of intermediate and structural determinants of health so that they can draft policies and choose courses of action that can effectively promote, prevent and enhance human health. This is especially true for socially marginalized groups.

Health and Wellness for All

Health promotion as well as disease control systems should aim to secure and enhance all people's health. The protection of vulnerable groups' health has already been adopted as a goal in some health policy frameworks (e.g. the European Union, WHO). This has arisen from the recognition of unavoidable or little controlled connections among individuals, different groups of people, different places. As history has repeatedly demonstrated, the impacts of diseases – bad health conditions are very rarely contained within the limits of a certain environment or context (e.g. AIDS, air pollution, greenhouse gases from burning fossil fuels). Thus, environmental health approaches should adopt multi-scalar analyses (i.e. at molecular – individual level, at local – community level, at regional, national and international levels) and integrated socio-environmental health approaches in order to effectively target the health and wellness of all people. At the implementation level, this implies that adopted strategies and actions should favor an environment-centered approach, but they should also be supplemented with appropriate health care institutions for vulnerable groups and already unhealthy individuals.

Wise Health Care Systems

The amazing developments in the fields of bioinformatics, biomonitoring, omics etc. have come up with some very useful tools especially for the health care and medicine levels. Recent Information and Communication Technologies (ICTs) and the huge body of smart health/disease data that exist can contribute to more targeted disease treatments and more effective medical approaches for individual health challenges. If these are also complemented with the huge amount of data on environmental exposures, they can lead to further developments that can contribute to health promotion. However, this should be done with full recognition that individual approaches can only lead to partial solutions to enhance the health status of an individual that experiences diverse health pressures but cannot eliminate the health pressures themselves that s/he (and so many others along with him/her) experiences. As environmental justice proponents have indicated "interventions 'after the fact' of exposure are insufficient; rather, the goal must be preventing harmful exposures in the first place" (Shostak, 2013).

These new technologies also require careful and wise use as they are both a blessing and a challenge (Papadopoulou, Lytras & Marouli, 2018). Technological developments themselves can easily be identified as main culprits for several contemporary diseases (e.g. vehicles and heating systems in cities related with air pollution and subsequent prevalence of asthma; technologies to support a more comfortable urban lifestyle lead to increased heart problems; introduction of computers in many work settings leading to a sedentary lifestyle and subsequent musculoskeletal and eye problems) and probably many more that we are still not aware of. ICTs and the availability of huge amounts of health-related big data, promising as they may be for identifying connections and potentially a greater potential for precision medicine approaches, also pose significant ethical (e.g. use of health data for exclusion from certain jobs) and social challenges (e.g. unequal access to new technologies and their effective use) that contemporary societies have to tackle.

Our Framework Proposition for the Transition to Healthy Societies

With the above considerations in mind, healthy societies should be characterized by social cohesion and solidarity, participatory governance, integrated environment and health policies, well-funded multidisciplinary science, health promotion education, well-funded preventative health care structures, smart health care institutions and personalized medicine, and they should exist in healthy and sustainable ecosystems. These components scan the whole range from system, structural, institutional and individual levels.

We propose the following framework for the transition to healthy societies, with individuals living in healthy environments.

For the effective implementation of the healthy societies vision and strategies we proposed above, an open and participatory governance approach is required for shared visioning and drafting of strategies. For such a governance approach, appropriate, independent and closely collaborating institutions are needed. As Figure 7 shows, our proposition incorporates already existing institutions such as community health centers, health care systems and health education and research institutions, but these need to be re-designed and re-directed to serve the goals of wellness and healthy societies, as well as to effectively incorporate new technologies. Additionally, we propose three interconnected Hubs – an Environmental Protection Hub, a Hub for Healthy Communities for Wellness, and a Bio-informatics Hub.

Figure 6. Components of healthy societies
Source: Our own

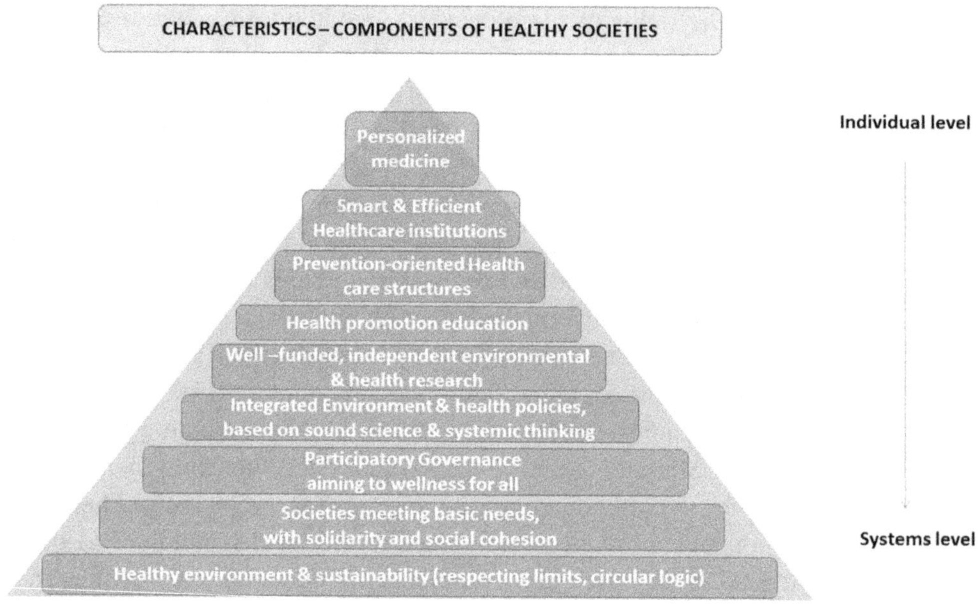

Figure 7. Framework for the transition to healthy societies: Strategies and institutions
Source: Our own

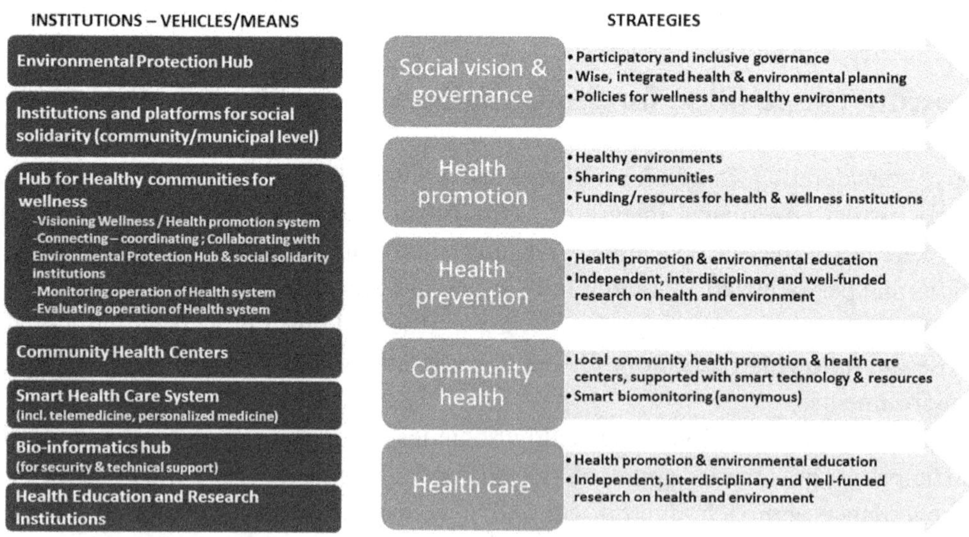

- The Environmental Protection Hub can be developed in each country with a multiscale structure – national, regional and local – with the aim to ensure that all social activities are promoting or at least not compromising healthy environments. This Hub should include representatives of the scientific community, policy makers, corporate world and the civic society. Common and equal information sharing should be a main goal, along with sharing in decision-making. A main goal of this Hub should also be the promotion of the culture for sustainable and healthy societies and environments.

- The Hub for Healthy Communities for Wellness should have a similar structure but focus on health issues. It should aim to a shared visioning of healthy communities and wellness, the formulation of relevant strategies in collaboration with the Environmental Protection Hub, and monitoring for their effective implementation. The two Hubs should be in close collaboration. Furthermore, this Hub should seek knowledge not only from the mainstream health sciences but also from alternative (not mainstream) health approaches which are often represented by small, marginalized "scientific" groups (such as homeopathy, herbal medicines, Ayurveda, indigenous medicines, etc.); these should be considered for effective approaches to health promotion and wellness with an open mind.
- The Bio-informatics Hub is more technical and should have the responsibility to holistically ponder and decide what type of health information should be collected, shared, used (or not used) in which contexts, with two primary aims: health promotion for all and the protection of individuals from potential harmful use of such information. This is a Hub that should involve health, Information Technology (IT) and legal experts, as well as a sufficient representation from lay people (so that their real needs are expressed).

In addition, these institutions should be completed by social institutions and local-regional platforms that have the explicit goal to support social solidarity; not with the aim to "structuralize" such grassroots efforts, but to enhance them and support synergies among citizen initiatives where these exist, and to stimulate the creation of new ones where they are needed.

CONCLUSION

Environmental health points to the connection between environment and health. Although this connection has been recognized for some decades now, health policies, health care systems and practices continue to focus on individuals and promote individual-centered approaches that cannot effectively address environmental determinants of health, structural or intermediate according to Elsey's et al. (2016) insightful categorization. Structural interventions are required so that contemporary societies can effectively promote health and wellness for communities and individuals together, rather than simply trying to mitigate health risks.

Furthermore, environmental and health policies continue to develop independently of each other. Environmental and health policy making should become more integrally connected and should develop integrated strategies for both environmental protection and the promotion of healthy and sustainable societies and healthy individuals.

For the transition to healthy societies, we have proposed a new framework. Moving towards healthy societies requires a new health visioning process with more inclusive and participatory governance. Health policies and strategies should be guided by a new set of priorities (informed by the developments in the environmental field) with the highest priority being wellness, followed by health promotion, health prevention, and disease control and treatment. Furthermore, new strategies and appropriate structures and institutions are required to support this transition. Health care and medicine should target community health in order to support individual health as well, while personalized medicine should wisely use biomonitoring and environmental monitoring big data for effective individual cures and health enhancement.

In this transitioning to healthy societies, environmental health sciences have an important role to play; they are called to provide an appropriate scientific basis for the needed reorientation in health policy making and strategies. For this new role, they should redirect themselves from the molecular level and environmental risk analysis to a systemic understanding of environmental (social, economic and physical environmental) influences on health, as well as to analyses of the significance of positive environmental influences on health (such as frequent exposure to healthy green areas).

REFERENCES

Agyeman, J., Bullard, R. D., & Evans, B. (Eds.). (2003). *Just Sustainabilities: Development in an Unequal World*. London: Earthscan.

Barouki, R., Gluckman, P. D., Grandjean, P., Hanson, M., & Heindel, J. J. (2012). Developmental origins of non-communicable disease: Implications for research and public health. *Environmental Health*, *11*(42), 42. doi:10.1186/1476-069X-11-42 PMID:22715989

Beck, U. (1992). *Risk Society: Towards a New Modernity*. New Delhi: Sage.

Birnbaum, L. (2014). Exposure, effects, and public health challenges, Interview. *EMBO Reports*, *15*(4), 338–341. doi:10.1002/embr.201438564 PMID:24623404

Blowers, A. (2003). Inequality and community and the Challenge to Modernization: Evidence from the Nuclear Oases. In J. Agyeman, R. Evan, & R. D. Bullard (Eds.), *Just Sustainabilities: Development in an Unequal World*. Earthscan Publications Ltd.

Bollati, V., & Baccarelli, A. (2010). Environmental epigenetics. *Heredity*, *105*(1), 105–112. doi:10.1038/hdy.2010.2 PMID:20179736

Bonnefoi, M. S., Belanger, S. E., Devlin, D. J., Doerrer, N. G., Embry, M. R., Fukushima, S., ... van der Laan, J. W. (2010). Human and environmental health challenges for the next decade (2010–2020). *Critical Reviews in Toxicology*, *40*(10), 893–911. doi:10.3109/10408444.2010.506640 PMID:20854192

Brown, P. (1995). Race, class, and environmental health: A review and systematization of the literature. *Environmental Research*, *69*(1), 15–30. doi:10.1006/enrs.1995.1021 PMID:7588491

Buck Louis, G. M., Smarr, M. M., & Patel, C. J. (2017). The Exposome Research Paradigm: An Opportunity to Understand the Environmental Basis for Human Health and Disease. *Current Environmental Health Reports*, *4*(1), 89–98. doi:10.100740572-017-0126-3 PMID:28194614

Bullard, R., Mohai, P., Saha, R., & Wright, B. (2007). *Toxic wastes and race at twenty: 1987–2007. Grassroots struggles to dismantle environmental racism in the United States*. A report prepared for the United Church of Christ Justice and Witness Ministries: United Church of Christ. Retrieved from https://www.ucc.org/assets/pdfs/toxic20.pdf

Cecil, K. M., Brubaker, C. J., Adler, C. M., Dietrich, K. N., Altaye, M., Egelhoff, J. C., ... Lanphear, B. P. (2008). Brain Volume in Adults with Childhood Lead Exposure. *PLoS Medicine*, *5*(5), e112. doi:10.1371/journal.pmed.0050112 PMID:18507499

Charron, D. F. (Ed.). (2012). *Ecohealth Research in Practice: Innovative Applications of an Ecosystem Approach to Health, Insight and Innovation in International Development 1*. Ottawa, Ontario, Canada: International Development Research Centre. doi:10.1007/978-1-4614-0517-7

Chikkanna, A., Mehan, L. P. K. S., & Ghosh, D. (2019). Arsenic Exposures, Poisoning, and Threat to Human Health: Arsenic Affecting Human Health. In P. Papadopoulou, C. Marouli, & A. Misseyanni (Eds.), Environmental Exposures and Human Health Challenges (pp. 86-105). Hershey, PA: IGI Global. doi:10.4018/978-1-5225-7635-8.ch004

Cooper, S., & Thorogood, N. (2013). Social construction of health and health promotion. In M. Davies, W. Macdowall, & L. Cragg (Eds.), *Health Promotion Theory* (2nd ed.). Maidenhead, UK: McGraw-Hill Education.

Corburn, J. (2009). *Toward the Healthy City: People, Places, and the Politics of Urban Planning*. Cambridge, MA: The MIT Press. doi:10.7551/mitpress/7583.001.0001

Delfino, R. J., Sioutas, C., & Malik, S. (2005). Potential Role of Ultrafine Particles in Associations between Airborne Particle Mass and Cardiovascular Health. *Environmental Health Perspectives*, *113*(8), 934–946. doi:10.1289/ehp.7938 PMID:16079061

Dennis, K. K., Marder, E., Balshaw, D. M., Cui, Y., Lynes, M. A., Patti, G. J., ... Barr, D. B. (2017). Biomonitoring in the Era of the Exposome. *Environmental Health Perspectives*, *125*(4), 502–510. Epub-2016Jul6. doi:10.1289/EHP474 PMID:27385067

DG Health and Food Safety. (2016). Strategic plan 2016-2020. *European Commission*. Retrieved from https://ec.europa.eu/info/sites/info/files/strategic-plan-2016-2020-dg-sante_en_0.pdf

Domingo, J. L., & Gine Bordonaba, J. (2011). A literature review on the safety assessment of genetically modified plants. *Environment International*, *37*(4), 734–742. doi:10.1016/j.envint.2011.01.003 PMID:21296423

Elliott, P., Briggs, D., & Morris, S. (2001). Risk of adverse birth outcomes in populations living near landfill sites. *BMJ (Clinical Research Ed.)*, *323*(7309), 363–368. doi:10.1136/bmj.323.7309.363 PMID:11509424

Elsey, H., Manandah, S., Sah, D., Khanal, S., MacGuire, F., King, R., ... Baral, S. C. (2016). Public Health Risks in Urban Slums: Findings of the Qualitative 'Healthy Kitchens Healthy Cities' Study in Kathmandu, Nepal. *PLoS One*, *29*(9), 1–17. doi:10.1371/journal.pone.0163798 PMID:27685999

European Commission. (2004, June 9). Communication of 9 June 2004: "The European Environment & Health Action Plan 2004-2010" [COM(2004) 416 - Official Journal C 49, 28.02.2006].

European Court of Auditors. (2018). *Air pollution: Our health still insufficiently protected*. Special report 23. Retrieved from https://www.eca.europa.eu/Lists/ECADocuments/SR18_23/SR_AIR_QUALITY_EL.pdf

European Environmental Agency (EEA) & Joint Research Center (JRC). (2013). *Environment and human health.* Joint EEA-JRC report (EEA Report No 5/2013), 1/112.

Evans, G. W., & Kim, P. (2007). Childhood Poverty and Health: Cumulative Risk Exposure and Stress Dysregulation. *Psychological Science, 18*(11), 953–957. doi:10.1111/j.1467-9280.2007.02008.x PMID:17958708

Faber, D. R., & Krieg, E. J. (2002). Unequal exposure to ecological hazards: Environmental injustices in the Commonwealth of Massachusetts. *Environmental Health Perspectives, 110*(Suppl 2), 277–288. doi:10.1289/ehp.02110s2277 PMID:11929739

Faber, D. R., & McCarthy, D. (2003). Neo-liberalism, Globalization and the Struggle for Ecological Democracy: Linking Sustainability and Environmental Justice. In J. Agyeman, R. Evan, & R. D. Bullard (Eds.), *Just Sustainabilities: Development in an Unequal World.* Earthscan Publications Ltd.

Gaffron, P. (2012). Urban transport, environmental justice and human daily activity patterns. *Transport Policy, 20*, 114–127. doi:10.1016/j.tranpol.2012.01.011

Giddens, A. (1998). *The Third Way: The Renewal of Social Democracy.* Cambridge: Polity.

Goerlandt, F., & Reniers, G. (2018). Prediction in a risk analysis context: Implications for selecting a risk perspective in practical applications. *Safety Science, 101*, 344–351. doi:10.1016/j.ssci.2017.09.007

Hanna, K. E., & Coussens, C.Institute of Medicine. (2001). *Rebuilding the Unity of Health and the Environment: A New Vision of Environmental Health for the 21st Century.* Washington, DC: The National Academies Press. doi:10.17226/10044

Harrington, D. W., & Elliott, S. J. (2015). Understanding emerging environmental health risks: A framework for responding to the unknown. *The Canadian Geographer. Geographe Canadien, 59*(3), 283–296. doi:10.1111/cag.12198

Healing Cities. (2011). Retrieved from http://healingcities.org/

Health and Environment Alliance. (2010). *The EU Environment and Health Action Plan (EHAP): Assessment and Outlook for future action.* Retrieved from https://www.env-health.org/IMG/pdf/ehap_final_report_final.pdf

Herceg, Z. (2016). Epigenetic mechanisms as an interface between the environment and genome. *Advances in Experimental Medicine and Biology, 903*, 3–15. doi:10.1007/978-1-4899-7678-9_1 PMID:27343085

Herceg, Z., & Vaissière, T. (2011). Epigenetic mechanisms and cancer: An interface between the environment and the genome. *Epigenetics, 6*(7), 804–819. doi:10.4161/epi.6.7.16262 PMID:21758002

Hines, R. N., Sargent, D., Autrup, H., Birnbaum, L. S., Brent, R. L., Doerrer, N. G., ... Slikker, W. (2010). Approaches for assessing risks to sensitive populations: Lessons learned from evaluating risks in the pediatric population. *Toxicological Sciences, 113*(1), 4–26. doi:10.1093/toxsci/kfp217 PMID:19770482

Hubbs-Tait, L., Nation, J. R., Krebs, N. F., & Bellinger, D. C. (2005). Neurotoxicants, Micronutrients, and Social Environments: Individual and Combined Effects on Children's Development. *Psychological Science in the Public Interest, 6*(3), 57–121. doi:10.1111/j.1529-1006.2005.00024.x PMID:26158603

Kearney, G., & Kiros, G. E. (2009). A spatial evaluation of socio demographics surrounding National Priorities List sites in Florida using a distance-based approach. *International Journal of Health Geographics*, *8*(1), 33. doi:10.1186/1476-072X-8-33 PMID:19531266

Krimsky, S. (2019a). *GMOs Decoded: A Skeptic's View of Genetically Modified Foods*. Cambridge, MA: The MIT Press. doi:10.7551/mitpress/10824.001.0001

Krimsky, S. (2019b). Glyphosate Toxicology: What We Can Learn From the Current Controversy. In P. Papadopoulou, C. Marouli, & A. Misseyanni (Eds.), *Environmental Exposures and Human Health Challenges* (pp. 343–356). Hershey, PA: IGI Global. doi:10.4018/978-1-5225-7635-8.ch015

Kristie, L. E., Semenza, J. C., & Rocklöv, J. (2016). Current medical research funding and frameworks are insufficient to address the health risks of global environmental change. *Environmental Health*, *15*(1), 108. doi:10.118612940-016-0183-3 PMID:27835959

Kumwenda, S., Samanyika, Y., Chingaipe, E., Mamba, K., Lungu, K., Mwendera, C., & Kalulu, K. (2014). The Emerging Environmental Health Risks and Challenges for Tomorrow: Prospects for Malawi. *Malawi Journal of Applied Sciences and Innovation*, *1*(1), 21–26.

Larkin, A., & Hystad, P. (2017). Towards Personal Exposures: How Technology Is Changing Air Pollution and Health Research. *Current Environmental Health Reports*, *4*(4), 463–471. doi:10.100740572-017-0163-y PMID:28983874

Laurian, L. (2008). Environmental Injustice in France. *Journal of Environmental Planning and Management*, *51*(1), 55–79. doi:10.1080/09640560701712267

Lioy, P. J., & Rappaport, S. M. (2011). Exposure science and the exposome: An opportunity for coherence in the environmental health sciences [Editorial]. *Environmental Health Perspectives*, *119*(11), A466–A467. doi:10.1289/ehp.1104387 PMID:22171373

Lytras, M. D., & Papadopoulou, P. (2018). *Applying Big Data Analytics in Bioinformatics and Medicine*. Hershey, PA: IGI Global. doi:10.4018/978-1-5225-2607-0

Marouli, C. (1995). Women Resisting (in) the City: Struggles, Gender, Class and Space in Athens. *International Journal of Urban and Regional Research*, *19*(4), 534–548. doi:10.1111/j.1468-2427.1995.tb00527.x

Martuzzi, M., Mitis, F., Bianchi, F., Minichilli, F., Comba, P., & Fazzo, L. (2009). Cancer mortality and congenital anomalies in a region of Italy with intense environmental pressure due to waste. *Occupational and Environmental Medicine*, *66*(11), 725–732. doi:10.1136/oem.2008.044115 PMID:19416805

Martuzzi, M., Mitis, F., & Fiorastere, F. (2010). Inequalities, inequities, environmental justice in waste management and health. *European Journal of Public Health*, *20*(1), 21–26. doi:10.1093/eurpub/ckp216 PMID:20061348

Masuda, J. R., Poland, B., & Baxter, J. (2010). Reaching for environmental health justice: Canadian experiences for a comprehensive research, policy and advocacy agenda in health promotion. *Health Promotion International*, *25*(4), 453–463. doi:10.1093/heapro/daq041 PMID:20615911

Mills, C. W. (1959). *The Sociological Imagination*. Oxford University Press.

Mold, A., & Berridge, V. (2013). The history of health promotion. In Health Promotion Theory (2nd ed.). Maidenhead, UK: McGraw-Hill Education.

Myers, P. J., Antoniou, M. N., Blumberg, B., Carroll, L., Colborn, T., Everett, L. G., ... Benbrook, C. M. (2016). Concerns over use of glyphosate-based herbicides and risks associated with exposures: A consensus statement. *Environmental Health*, *15*, 19. doi:10.118612940-016-0117-0

National Research Council. (2012). *Exposure Science in the 21st Century: A Vision and Strategy*. Washington, DC: National Academies Press.

Nieuwenhuijsen, M. J., Smith, R., Golfinopoulos, S., Best, N., Bennett, J., Aggazzotti, G., ... Kogevinas, M. (2009). Health impacts of long-term exposure to disinfection by-products in drinking water in Europe: HIWATE. *Journal of Water and Health*, *7*(2), 185–207. doi:10.2166/wh.2009.073 PMID:19240347

Papadopoulou, P., Lytras, M., & Marouli, C. (2018). Bioinformatics as Applied to Medicine: Challenges Faced Moving from Big Data to Smart Data to Wise Data. In M. Lytras & P. Papadopoulou (Eds.), *Applying Big Data Analytics in Bioinformatics and Medicine* (pp. 1–25). Hershey, PA: IGI Global. doi:10.4018/978-1-5225-2607-0.ch001

Pruss-Usten, A., Wolf, J., Corvalán, C., Bos, R., & Neira, M. (2016). *Preventing disease through healthy environments. A global assessment of the burden of disease from environmental risks.* World Health Organization (WHO).

Prüss-Ustün, A., Bartram, J., Clasen, T., Colford, J. M. Jr, Cumming, O., Curtis, V., ... Cairncross, S. (2014). Burden of disease from inadequate water, sanitation and hygiene in low- and middle-income settings: A retrospective analysis of data from 145 countries. *Tropical medicine & international health. TM & IH*, *19*(8), 894–905. doi:10.1111/tmi.12329 PMID:24779548

Rappaport, S. M., & Smith, M. T. (2010). Environment and disease risks. *Science*, *330*(6003), 460–461. doi:10.1126cience.1192603 PMID:20966241

Rixecher, S. S., & Tipene-Matua, B. (2003). Maori Kaupapa and the Inseparability of Social and Environmental Justice: An Analysis of Bioprospecting and a People's Resistance to (Bio)cultural Assimilation. In J. Agyeman, R. Evan, & R. D. Bullard (Eds.), *Just Sustainabilities: Development in an Unequal World*. Earthscan Publications Ltd.

Sarigiannis, D. A. (2017). Assessing the impact of hazardous waste on children's health: The exposome paradigm. *Environmental Research*, *158*, 531–541. doi:10.1016/j.envres.2017.06.031 PMID:28711809

Sarigiannis, D. A. (2019). The Exposome Paradigm in Environmental Health. In P. Papadopoulou, C. Marouli, & A. Misseyanni (Eds.), *Environmental Exposures and Human Health Challenges* (pp. 1–29). Hershey, PA: IGI Global. doi:10.4018/978-1-5225-7635-8.ch001

Sarigiannis, D. A., Gotti, A., Handakas, E., & Karakitsios, S. P. (2018). Informatics and Data Analytics to Support Exposome-Based Discovery: Part 2 - Computational Exposure Biology. In M. Lytras & P. Papadopoulou (Eds.), *Applying Big Data Analytics in Bioinformatics and Medicine* (pp. 145–187). Hershey, PA: IGI Global. doi:10.4018/978-1-5225-2607-0.ch007

Sarigiannis, D. A., Karakitsios, S. P., Handakas, E., Papadaki, K., Chapizanis, D., & Gotti, A. (2018). Informatics and Data Analytics to Support Exposome-Based Discovery: Part 1 - Assessment of External and Internal Exposure. In M. Lytras & P. Papadopoulou (Eds.), *Applying Big Data Analytics in Bioinformatics and Medicine* (pp. 115–144). Hershey, PA: IGI Global. doi:10.4018/978-1-5225-2607-0.ch006

Séralini, G.-E., Clair, E., Mesnage, R., Gress, S., Defarge, N., Malatesta, M., ... Spiroux de Vendômois, J. (2014). Republished study: Long-term toxicity of a Roundup herbicide and a Roundup-tolerant genetically modified maize. *Environmental Sciences Europe, 26*(1), 14. doi:10.118612302-014-0014-5 PMID:27752412

Seralini, G.-E., Mesnage, R., Clair, E., Gress, S., de Vendomois, J., & Cellier, D. (2011). Genetically modified crops safety assessments: Present limits and possible improvements. *Environmental Sciences Europe, 23*(1), 10. doi:10.1186/2190-4715-23-10

Shostak, S. (2013). *Exposed Science: Genes, the Environment, and the Politics of Population Health.* University of California Press. doi:10.1525/california/9780520275171.001.0001

Spencer, T. (2016). *Risk Perception: Theories and Approaches.* Hauppauge, NY: Nova Science Publishers, Inc.

Stingone, J. A., Buck Louis, G. M., Nakayama, S. F., Vermeulen, R. C., Kwok, R. K., Cui, Y., ... Teitelbaum, S. L. (2017). Toward Greater Implementation of the Exposome Research Paradigm within Environmental Epidemiology. *Annual Review of Public Health, 38*(1), 315–327. doi:10.1146/annurev-publhealth-082516-012750 PMID:28125387

Syme, S., & Wang, C. (2012). The social Environment and Health. In *The praeger handbook of environmental health.* ABC-CLIO, LLC. Retrieved from http://ebookcentral.proquest.com

Turner, M. C., Nieuwenhuijsen, M., Anderson, K., Balshaw, D., Cui, Y., Dunton, G., ... Jerrett, M. (2017). Assessing the Exposome with External Measures: Commentary on the State of the Science and Research Recommendations. *Annual Review of Public Health, 20*(38), 215–239. doi:10.1146/annurev-publhealth-082516-012802 PMID:28384083

United Nations. (2015). *Transforming our world: the 2030 Agenda for Sustainable Development.* Retrieved from https://www.un.org/ga/search/view_doc.asp?symbol=A/RES/70/1&Lang=E

Varga, C., Kiss, I., & Ember, I. (2002). The lack of environmental justice in Central and Eastern Europe. *Environmental Health Perspectives, 110*(11), A662–A1. doi:10.1289/ehp.110-a662 PMID:12417491

Vrijheid, M. (2014). The exposome: A new paradigm to study the impact of environment on health. *Thorax, 69*(9), 876–878. doi:10.1136/thoraxjnl-2013-204949 PMID:24906490

Vrijheid, M., Casas, M., Bergstrom, A., Carmichael, A., Cordier, S., Eggesbo, M., & Nieuwenhuijsen, M. (2012). European birth cohorts for environmental health research. *Environmental Health Perspectives, 120*(1), 29–37. doi:10.1289/ehp.1103823 PMID:21878421

Vrijheid, M., Slama, R., Robinson, O., Chatzi, L., Coen, M., van den Hazel, P., ... Nieuwenhuijsen, M. J. (2014). The Human Early-Life Exposome (HELIX): Project rationale and design. *Environmental Health Perspectives, 122*(6), 535–544. doi:10.1289/ehp.1307204 PMID:24610234

Walker, G. (2012). *Environmental Justice: Concepts, Evidence and Politics*. London: Routledge. doi:10.4324/9780203610671

WHO. (1986). *The Ottawa Charter for Health Promotion*. Retrieved from: https://www.who.int/health-promotion/conferences/previous/ottawa/en/

WHO. (2006a). *Air quality guidelines. Global update 2005. Particulate matter, ozone, nitrogen dioxide and sulfur dioxide*. Copenhagen, Denmark: World Health Organization, Regional Office for Europe.

WHO. (2006b). *Health risks of particulate matter from long-range transboundary air pollution*. Copenhagen, Denmark: World Health Organization, Regional Office for Europe.

WHO. (2007). *Health risks of heavy metals from long-range transboundary air pollution. Joint WHO/ Convention Task Force on the Health Aspects of Air Pollution*. Copenhagen, Denmark: World Health Organization, Regional Office for Europe.

WHO. (2012). *Soil-transmitted helminthiases: eliminating soil-transmitted helminthiases as a public health problem in children: progress report 2001-2010 and strategic plan 2011-2020*. Geneva: World Health Organization.

WHO. (2012b). *Endocrine disrupters and child health - Possible developmental early effects of endocrine disrupters on child health*. Geneva, Switzerland: World Health Organization.

WHO. (2013a). *Review of evidence on health aspects of air pollution — REVIHAAP*. Copenhagen, Denmark: World Health Organization, Regional Office for Europe.

WHO. (2014). *Global Tuberculosis Report 2014*. Geneva: World Health Organization.

Wigle, D. T. (1977). Cancer mortality in relation to asbestos in municipal water supplies. *Archives of Environmental Health, 32*(July/August), 185–190. doi:10.1080/00039896.1977.10667278 PMID:889357

Wild, C. P. (2005). Complementing the genome with an "exposome": The outstanding challenge of environmental exposure measurement in molecular epidemiology. *Cancer Epidemiology, Biomarkers & Prevention, 14*(8), 1847–1850. doi:10.1158/1055-9965.EPI-05-0456 PMID:16103423

Wild, C. P., Scalbert, A., & Herceg, Z. (2013). Measuring the exposome: A powerful basis for evaluating environmental exposures and cancer risk. *Environmental and Molecular Mutagenesis, 54*(7), 480–499. doi:10.1002/em.21777 PMID:23681765

World Health Organization (WHO). (1946). *Constitution of the World Health Organization*. Retrieved from http://apps.who.int/gb/bd/PDF/bd47/EN/constitution-en.pdf?ua=1

KEY TERMS AND DEFINITIONS

Environmental Hazard: Is a substance, a state, or an event that has the potential to cause harm on people's health and/or the environment

Environmental Risk: Expresses the probability of suffering some harm (including injury, disease, death, economic loss, or damage to environmental components, etc.) because of an exposure to an environmental hazard. It can also mean exposure to an environmental peril.

Environmental Risk Analysis: Is the effort to estimate the chance that an environmental hazard may cause harm. Thus, it explores and tries to manage uncertainties and the significance of potential impacts, while it includes risk assessment but also risk perception.

Exposome: The totality of environmental exposures from conception onward, assessing the multitude of human exposures across the life.

Personalized Medicine: is a therapeutic approach in which the medical decisions, practices, interventions and/or products are tailored to the individual's therapy scheme and/or preventive care, with the use of an individual's genetic and epigenetic information.

Political Ecology: Studies the interactions between environmental issues and the socio-political context and looks to social, political and economic factors for the causes and the solutions to environmental problems.

Political Economy: Is a branch of economics that studies on the one hand the organization of the economic activities in human societies, and on the other the relation and interaction of political and economic institutions and processes.

Sustainability: Refers to a way of life and social organization that allow people to meet their needs but within the limits of natural ecosystems, ensuring healthy ecosystems and leading to viable economies and socially cohesive and healthy communities. Integration, connectedness and social responsibility are important elements of sustainability.

Wellness: Is a state of complete physical, mental, spiritual and social well-being.

ENDNOTE

[1] A chula is an open burn pit.

Chapter 3
Statistical Methods in Ecological Processes

Asaf Hajiyev
Azerbaijan National Academy of Sciences, Azerbaijan

ABSTRACT

The ecological processes have complicated structure. The equations, describing their behavior, also have a complicated structure; hence, to solve them faces with some difficulties. One of the effective ways is to collect data and make statistical data analysis of these data and afterward to make corresponding decisions. As ecological processes depend on a lot of parameters, and moreover, the number of these parameters can increase in time. In the chapter, the conception of regression models with increasing number of unknown parameters is introduced. In the frame of regression models with increasing number of unknown parameters, the approach for estimating of the main parameters of the ecological processes is suggested. Using estimates of unknown parameters, the approach allows to construct a confidence band for unknown functions in these models. Numerical examples demonstrating these results are given.

INTRODUCTION

The investigation of ecological processes leads to creating the new scientific approaches and methods for their studying. These processes have a complicated stochastic structure, because depend on many parameters. The differential equation describing a behavior of these processes so complicated that there is no chance to find analytical solutions. Even using computers for modelling of ecological processes or for finding direct solution of differential equations, describing behavior of ecological processes is faced with some difficulties. One of the effective approaches for investigation of such processes are statistical methods, which allow to collect data and estimate the various unknown parameters in these processes. As the number of parameters in these problems can increase when number of observations goes up, hence their investigation is outside of the framework of methods and approaches of classic methods of statistics. Thus, an investigation of such processes leads to the appearance of new direction in statistics, which can be called *statistical models with increasing number of unknown parameters*. Although in world scientific literature there are a lot of publications in the field of statistical data analysis but (Craqmile and Haran,

DOI: 10.4018/978-1-7998-1241-8.ch003

2014; Chatterjee and Hadi, 2012; Harrel, 2011; Dreiper and Smith, 2016) but models with increasing number of unknown parameters not widely investigated. In (Belyaev and Hajiyev, 1979; Hajiyev, 2011) some universal regression models are partially investigated but for statistical data analysis of ecological problems it is necessary to create new approaches and methods, which must take into account specify of these problems.

Mathematical Models

Let us assume that we observe some ecological process, for instant climate change or changing of temperature in the forest. At the each point of observation as a result of measurement, we observe some quantity y_i, (response) which depends on many factors (parameters) as some function $f(\theta)$, θ is a vector with many components. Making measurement we always face with some random errors, which are denoted ε_i, $i=1,2,...,N$. Hence, our observation has the form

$$y(x_i) = f(x_i, \theta) + \varepsilon_i \tag{1.1}$$

where x_i is the point of observation. The function $f(.,.)$ is unknown and describes a behavior of ecological process. Below, for simplicity, instead of $y(x_i)$ we will use y_i.

The parameter θ is a vector, which has a lot of components and moreover the number of these components can increase, when number of observations go up, i.e. $\theta^T = (\theta_1, \theta_2,..., \theta_{m(N)})$, m is a number of unknown parameters, which depends on N and N is a number of observations.

Remark 1.1 In the practice, for more exact studying a behavior of ecological process during an observation, depending of number of observations new parameters can be introduced.

Our aim to estimate the parameter θ and try to estimate unknown function $f(.)$.Hajiyev (2004) proposed the approach for finding least square estimates for linear models with fixed number of unknown parameters . As ecological process has a complicated structure then function $f(.)$ can have non-linear structure, i.e. in this paper it is assumed that $f(.)$ is non-linear function. We also assume that parameter θ has true value θ^* but we do not know it.

About a random error $\varepsilon^T = (\varepsilon_1, \varepsilon_2,..., \varepsilon_N)$ we assume that $\varepsilon_1, \varepsilon_2,..., \varepsilon_N$ is the sequence of independent and identically distributed random variables $E\varepsilon_i = 0$, $Var(\varepsilon_i)=\sigma_i^2$, $i=1,2,...,N$. The variances σ_i^2 are different and unknown. The approach of investigation such models and some results were given in -(Wu Jacknife, 1986; Demidenko, 1989).

Remark 1.2 Such proposition reflects typical problems of the practice, because the measurements are doing independently, and their results can have the same distribution.

Consider Taylor's expansion of the function $f(.)$ at the neighborhood of the point θ_0 .

$$f(x_i, \theta)= f(x_i, \theta_0)+ (\partial f(x_i, \theta_0)/\partial x)(\theta-\theta_0)+o(\theta-\theta_0)$$

This expansion allows to present (1) in linear form, i.e.

$$y(x_i) = f(x_i, \theta_0)+ (\partial f'(x_i, \theta_0)/\partial x)(\theta-\theta_0)+o(\theta-\theta_0) + \varepsilon_i \tag{1.2}$$

Although we do not know the value of θ_0 but we will assume that we can take some value from the neighborhood of $\theta_{0.}$ It means that we can neglect by the value of $(\theta-\theta_0)$. Then equation (1.2) can be rewritten in the following form

$$y(x_i) - f(x_i, \theta_0) = (\partial f'(x_i, \theta_0)/\partial x)(\theta-\theta_0)+ \varepsilon_i \quad (1.3)$$

Expression (1.3) has already a linear form but with increasing number of unknown parameters and can be rewritten in the following vector form

$$Y = F\theta + \varepsilon \quad (1.4)$$

where $Y=(y_1\text{-}f(x_1,\theta_0), y_N\text{-}f(x_2,\theta_0), \ldots\ldots, y_N\text{-}f(x_N,\theta_0))^T$,

$\varepsilon = (\varepsilon_1, \varepsilon_2,\ldots,\varepsilon_N)^T$

F =//fij//, i=1,2,…,m; j=1,2,…,N; $f_{ij} = \dfrac{\partial f(x_j,\theta)}{\partial \theta_i}$

Thus, we have linear regression model with increasing number of unknown parameters.

Then we can apply approach suggested in (Belyaev and Hajiyev, 1979; Hajiyev, 2011) and for calculation of least square estimates of the vector θ_0 we will get iteration process. Specify of this problem is that a number of unknown parameters can increase, and hence it can create some difficulties for convergence of iterated process. For construction of iteration process let us introduce the following notations, where s means a number of iterations

$$f_{ij}(s) = \frac{\partial f(x_j,\theta(s))}{\partial \theta_i(s)}, \quad f_{ij}^* = \frac{\partial f(x_j,\theta^*)}{\partial \theta^*}$$

where $\theta*$ is a some estimate of unknown parameter θ. Below θ^* means least square estimate and for its calculation we introduce the following conceptions.F is the matrix with elements f_{ij}, $i=1,2,\ldots,N;\ j=1,2,\ldots m(N)$.

$$\Delta_{ij}^{k,l} = \delta_{ij}\frac{\partial f(x_i,\theta^*)}{\partial \theta^*_k}\frac{\partial f(x_j,\theta^*)}{\partial \theta^*_l}$$

Consider Fisher's matrix (F^TF /N) and denote $0 < \lambda_1(\theta)$ £ $\lambda_2(\theta)$ £ … £ $\lambda_m(\theta)$ its eigenvalues. As the elements of the matrix (F^TF /N) depend on θ, hence eigenvalues

From (1.4) formally we can find l.e.s.

$$\theta^* = (F^TF)^{-1}F^TY \quad (1.5)$$

Then we have

$$\theta^* - \theta = (F^TF)^{-1}F^TY - \theta = (F^TF)^{-1}F^TF\theta + (F^TF)^{-1}F^T\varepsilon - \theta = (F^TF)^{-1}F^T\varepsilon.$$

and for covariance matrix of the vector of deviation $N^{1/2}(\theta^* - \theta)$ (denote it as C_N) we have

$$C_N = EN(\theta^* - \theta)(\theta^* - \theta)^T = (F^TF/N)^{-1}F^TE\varepsilon\varepsilon^T F(F^TF/N)^{-1} \quad (1.6)$$

Our aim to find l.s.e. and afterward to estimate the elements of the matrix C_N. In (1/6) the elements of the matrix $(F^TF/N)^{-1}$ and F^T are known. Hence, for estimating of the elements of the matrix C_N it is necessary to estimate the elements of the matrix $E\varepsilon\varepsilon^T$. The elements of the matrix $E\varepsilon\varepsilon^T$ depend on σ_i^2, which are unknown and moreover can be different. Hence, it is necessary to create new approach for estimating of the elements of this matrix. Some approach can be found in (Belyaev and Hajiyev, 1979; Wu Jacknife, 1986).

Consider the sphere $B(r)$ of the radius $r>0$ and assume that we take some value θ_0 which belongs to this sphere.

Remark 1.3 At the practice it means that we do not known the value of θ but we have information about approximately location of its, i.e. where it can be located.

Least Square Estimates (l.s.e.)

Using Gauss-Markov approach (Demidenko, 1989) for fixed N we can formally construct the following well known iterated process

$$\theta(s+1) = \theta(s) + (F^T(s)F(s))^{-1}F^T(s)(y - f(x,\theta(s))) \quad (2.1)$$

which can be used for finding the least square estimates. First of all it is necessary to find the conditions when this iterated process converges. As the parameter θ is a random variable below everywhere convergence means convergence in probability.

Let us take some $\theta_0(r)$ from the sphere $B(r)$. We put $\theta_0(r)$ into the iteration process (2.1). Consider the matrix (F^TF/N) and denote $\lambda_1(\theta) \leq \lambda_2(\theta) \leq \ldots \leq \lambda_{m(N)}(\theta)$ its eigen values. Introduce the condition

$$m(N)/N\lambda_1(\theta(s)) \to 0, \text{ when } N \to \infty,\ s \to \infty \quad (2.2)$$

Statement. If the condition (2.2) is held, then there exists some $\theta(r)$ such that $\theta(s) \to \theta(r)$.

The proof of convergence of the iterated process (2.1) is quite complicated and very long, but the main idea and some way of proof can be found in [9].

Thus, if number of iterations is increasing then the iteration process converges to some value $\theta(r)$, which depends on a radius of sphere $B(r)$. It is a clear that $\theta(r)$ also depends on a zero iteration.

Some steps of proof of convergence of the iterated process (2.1) will be presented below.

1. It is necessary to proof convergence of the iterated process (2.1) for fixed N, i.e existence of some θ_N, for which $\theta_N(s) \to \theta_N$, when $s \to \infty$.
2. Afterward taking some $\theta_0(r) \in B(r)$ it is necessary to find conditions, when there exists some θ^*, for which we have $\theta_N(r) \to \theta^*(r)$.

3. Last step is to find the conditions when $\theta^*(r) \to \theta^*$, $r \to 0$, i.e. existence of θ^*, which is taken in the capacity of l.s.e.

Thus, for finding of l.s.e. of unknown parameter θ we have to follow to the following procedure:

1. It is necessary to take θ_0 (zero iteration), which is located in some small neighborhood of the true value θ, i.e. belongs to the sphere B(r). We can do it using some information about ecological process, which is studied.
2. Put this value into the iterated procedure (1.4) and find value of Afterward, using the iteration process (2.1) we can find the value, where iteration process converges.
3. As soon as we will take the sphere $B(r)$ with radius $r \to 0$ then can be uses θ^* as the l.s.e.

Example 2.1 Consider example for finding of number of unknown parameters (depending on number of observations), when we can use consistent l.s.e. Let us assume that we have some process, which is described by behavior of unknown function f(x). We will try to approximate this function by Chebyshe's polynomials [11] of the first order (these polynomials are least deviating from zero). We put $\varphi_j(p) = p_i(p)$ $[h(p)]^{1/2}$ where $p_1(p), p_2(p),.., p_j(p),...,$ is the system of orthogonal polynomials with weighted function $h(p)=(1-p^2)^{-1/2}$, p $(-1,1)$. In the capacity of orthogonal functions we take Chebyshev's polynomials of the first order which have the following form

$p_{n+1}(p)=1$, $p_1(p)=p$, $p_{n+1}(p)=2pp_n(p)-p_{n-1}(p)$, p [-1,1].

which are convenient for approximation, because as it was mentioned above they have a minimum deviation from zero. Then the elements of (X^TX/N) have the following form

$f_{ij}(p_k) = (1/N)_i(p_k)p_j(p_k)h(p_k)$.

Let us take

$p_1 = -1$, $p_i = -1+2(i-1)/N$, i=2,3,…,N.

Then for N we have

0, if ij

$f_{ij}(p)$ $_i(x)p_j(x)h(x)dx = \pi$ if i=j>0

0, if i=j=0

Thus, for N the matrix (X^TX/N) has the following diagonal form

$f_{ii} = \pi/2N$, i =1,2,..,m(N)-1; $f_{mm} = \pi/N$ and $f_{ij}=0$, if ij.

Hence, for eigen values of the matrix (X^TX/N) we have

$\lambda_1 = \lambda_2 = ... = \lambda_{m-1} = \pi/2N$ and $\lambda_m = \pi/N$.

Let us take $m = N^{1/2}$. Then the conditions (2-4) are held and hence, for unknown parameters $m = N^{1/2}$ the *l.s.e.,* defined by (6) will be consistent and unbiased estimator.

Construction of a Confident Band

Now using l.s.e. we will try to construct a confident band for unknown function *f(.)*. Consider the matrix C_N, which is a covariance matrix of the deviation vector $N^{1/2}(\theta - \theta^*)$ and has the following form

$$C_N = NE((\theta - \theta^*)(\theta - \theta^*)^T = (X^TX)^{-1}X^TI_{k,l}(\sigma)X((X^TX)^{-1}$$

where $I_{k,l}(\sigma) = E\varepsilon\varepsilon^T$, with elements $\Delta_{ij}^{k,l}$. The matrix $I_{k,l}(\sigma)$ has diagonal formi.e.

$\Delta_{ii}^{k,l} = \delta_i^2 \dfrac{\partial f(x_i,\theta)}{\partial\theta_k}\dfrac{\partial f(x_i,\theta)}{\partial\theta_l}$, if $i=j$ and $\Delta_{ij}^{k,l} = 0$, if $i \neq j$.

Introduce the matrix $I_{k,l}(\theta^*) = (\Delta_{ij}^{k,l})$, $k=1,2,..,m;\ i=1,2,...,N;\ j=1,2,...,m(N)$ and denote

$$y^* = F \cdot \theta^*,\ y^*_i = y_i - f(x_i,\theta^*),\ y^* = (y^*_1,...,y^*_N)^T.$$

Let us introduce the following notations:

$$a_{kl} = \frac{1}{N}(y^* - y)^T I_{k,l}(\theta^*)(y^* - y) \qquad (3.1)$$

In the capacity of estimator of the elements of the matrix $I_{k,l}(\sigma)$ we will take the elements a_{kl}. The following statement is true.

If $[m(N)]^{3/2}/(N\lambda_1(N) \to 0$, when $N \to \infty$, $r \to 0$ then

$E(a_{kl} - \Delta_{ij}^{k,l}) \to 0$ and $(a_{kl} - \Delta_{ij}^{k,l}) \to 0$ (in probability).

Now having asymptotically unbiased and consistent estimator of the elements a covariance matrix C_N we can construct a confident band for unknown function *f(.)*.

Let us consider the quadratic form

$$N(\theta^* - \theta)^T C_N^{-1}(\theta^* - \theta) \qquad (3.2)$$

For fixed *m(N)* it has asymptotically hi-square distribution with a degree of freedom *m(N)*. But if $N \to \infty$, according to central limit theorem normed form (3.2) has normal distribution, i.e. the following quadratic form

$$N(\theta^*-\theta)^T C_N^{-1}(\theta^*-\theta)/m(N)$$

has normal distribution. Then the inequality

$$N(\theta^*-\theta)^T C_N^{-1}(\theta^*-\theta)/m(N) \leq \chi_\gamma^2(m(N)/N$$

with probability not less than $\gamma>0$, where $\chi_\gamma^2(m(N)$ is quantile of the level $\gamma>0$ of standard normal distribution. Consider

$$\varepsilon_\gamma(\theta) = \{\theta: (\theta^* - \theta)^T C_N^{-1} (\theta^* - \theta) \leq \chi_\gamma^2(m(N)/N \quad (3.3)$$

which is a confident ellipsoid. For construction of confident band for unknown function *f(.)* it is necessary to find

$$\inf f(x, \theta) \text{ and } \sup f(x, \theta) \quad (3.4)$$

which are lower and upper border of confident band and θ belongs to the ellipsoid (3.4).

CONCLUSION

Suggested approach allows to make statistical analysis of data, when the number of unknown parameters is unknown and can increase, moreover we do not have information about variances of errors of observations. Moreover, by thus approach we can estimate the number of unknown parameters, which is necessary to take for consistent l.s.e. (see, example 2.1). This situation is typical for wide practical processes, like ecological problems, biology, medicine and other. Although we do not know the value of unknown parameter but from practice or additional information we can define some neighborhood and take first iteration for process (2.1) from this region. Convergence of iterated process (2.1) depends how we are close to true value of θ_0. For numerical calculation in the capacity of θ_0 we can take some value from the sphere *B(r)*. This value is called first or zero iteration value and a speed of convergence of the iterated process (2.1) depends how close we are to θ_0. As soon as we establish convergence and solution (θ^*) of (2.1) then the value of θ^* can be used as l.s.e. Afterward using θ^* we can estimate the elements of covariance matrix C_N (see, 3.1)and construct confidence ellipsoid (see, 3.3). Calculations show that for some practical cases usage of this approach gives more slim confidence band than usage of standard approach, when it is necessary to have information about variances of errors of observations. This approach has also advantage because it is not necessary to have information about variances of errors.

REFERENCES

Belyaev, Y. K., & Hajiyev, A. H. (1979). Statistical Data Analysis [Статистический анализ данных]. Известия АН СССР, Техническая кибернетика, 4, 79-83.

Chatterjee, S., & Hadi, A. S. (2012). *Regression Analysis by Examples*. Springer.

Craqmile, P., & Haran, M. (2014). Statistical Methodology. In *Modern Statistical Methods in Ecology*. Elsevier.

Demidenko, E. (1989). *Optimization and Regression [Оптимизация и регрессия]*. Наука.

Dreiper, N., & Smith, G. (2016). *Applied Regression Analysis /Прикладной регрессионный анализ*. Диалектика.

Hajiyev, A. H. (2004). *Mathematical Theory of Regression Models with Increasing Number of Unknown Parameters. Nonlinear Models*. Research Report. Department of Computer Science. *Lincoln University*.

Hajiyev, A. H. (2011). Regression Models with Increasing Numbers of Unknown Parameters. In International Encyclopedia. Statist. Sciences. Springer.

Hajiyev, A. H., & Hajiyev, V. S. (2009). Nonlinear Regression Models with Increasing Numbers of Unknown Parameters. *Russian Acad. Sci. Mathem.*, *79*(3), 339–341.

Harrel, F. (2001). *Regression Modeling Strategies*. Springer. doi:10.1007/978-1-4757-3462-1

Ryabenkii, V., & Tsynkov, S. (2019). *A Theoretical Introduction to Numerical Analysis*. Chapman and Hall/CRC.

Wu Jacknife, C. F. (1986). Bootstrap and other Resampling Methods in Regression Analysis. *Annals of Statistics*, *14*(4), 1261–1350. doi:10.1214/aos/1176350142

Chapter 4
Thermodynamic Stability of Natural Aqueous Systems

Igor Povar
Institute of Chemistry, Moldova

Oxana Spinu
https://orcid.org/0000-0002-3166-1931
Institute of Chemistry, Moldova

Tudor Lupascu
Institute of Chemistry, Moldova

Gheorghe Duca
Institute of Chemistry, Moldova

ABSTRACT

Long-term prediction of the complex chemical processes and their practical consequences is only feasible by using simulation models. The chapter presents a thermodynamic approach to the study of complex chemical equilibria in natural aqueous multicomponent systems. Its quintessence consists in the thermodynamic analysis of the conditions of various processes on the basis of derived overall thermodynamic characteristics. Ways to shift from chemical-analytical, descriptive characteristics in heterogeneous multicomponent systems "mineral – natural aqueous systems" to the quantitative study of involved chemical equilibria by thermodynamic and computer modeling are extensively described. The analyzed chemical processes include the heterogeneous interaction of minerals and aqueous solution, homogeneous reactions of hydrolysis, complex formation, and acid-base equilibria.

INTRODUCTION

The chapter presents a new thermodynamic approach to the study of complex chemical equilibria in real conditions, taking into account the complex formation reactions in multicomponent heterogeneous systems. Its quintessence consists in the thermodynamic analysis of the conditions of various processes

DOI: 10.4018/978-1-7998-1241-8.ch004

on the basis of their total thermodynamic characteristics. For reactions of complex formation, hydrolysis and protonation in heterogeneous multicomponent systems, the total change of Gibbs energy cannot be presented as a sum of contributions of several separate reactions. The quintessence of the elaborated formal thermodynamic approach consists in the analysis of the conditions of different processes based on the overall thermodynamic characteristics. The calculation of the global Gibbs energy variation of complex chemical processes allows determination of the thermodynamic stability domains of solid phases, the direction and optimum conditions for chemical processes under conditions close to the real ones.

BACKGROUND

As a result of thermodynamic data accumulation for separate reactions, the calculation of the overall thermodynamic characteristics of the previously unstudied multicomponent processes becomes available. The multi-component heterogeneous system is formulated herein as biphasic system "precipitate (solid phase) - solution (liquid phase)", in which a series of simultaneous reactions occur and the contribution of each depends on the chemical composition of the system, that is the ratio of the concentrations of the constituent components. In this regard, an increased interest acquires the direct and indirect problems of the chemical thermodynamics in the heterogeneous multi-component systems. In this context, the main problem consists in the calculation of the thermodynamic functions of the overall process based on the thermodynamic characteristics of the separate reactions. The secondary problem is to calculate the thermodynamic characteristics of separate reactions based from the overall functions. Moreover, the practical importance of buffer systems, as well as the need for focused search and the creation of new perspective buffer systems, have created the necessity to develop quantitative principles of the theory of buffering properties of multicomponent heterogeneous systems. This issue is examined in a separate chapter of this book.

Thermodynamic parameters of the system are important not only for a prediction of the dependence of the system properties on the aforementioned ones, but also for the more profound characteristic and understanding, of many complicated processes, taking into account the contributions of different reactions. Also, they are applied by geologists and geochemists for studying the transfer of elements of geothermal solutions, the formation of ores under endogenous conditions and the hydrothermal synthesis of crystals and minerals (Dutova & Nikitenko, 2010; Li et al., 2020; Seravkin et al., 2017; Vilor & Kuzmin, 2007; Dutta & Lodhari, 2018; Yakymchuk, 2017).

MAIN FOCUS OF THE CHAPTER

In this chapter the thermodynamics of the precipitation-dissolution processes of the sparingly soluble salts of different nature and the arbitrary composition under the conditions of the various side reactions between the precipitated ions (solid phase) and the saturated solution components are examined. As side reactions, protonation reactions of the ligand, hydrolysis of the metal ion, formation of neutral, protonated and hydroxocomplex are considered. For this purpose, an original method of analyzing complex chemical equilibria in the presence of solid phases is used. The notion of generalized reaction equation is introduced in order to take into account all conjugated side reactions. This equation is the generalization of a set of common chemical equations, taking into account the simultaneous formation of several different chemical species. The necessity of examining a global process derives from the fact that the nature and ratio of the concentrations of the complex species resulting from the reactions depend on the ratio of the concentrations of the initial substances (metal ion and ligand), temperature and other factors (thermodynamic parameters).

In the first paragraph, based on the generalized equation, the thermodynamic functions are derived under conditions close to real ones. The thermodynamic analysis of the equilibrium conditions exposed here allow the examination of a variety of chemical interactions in aqueous solutions (such as hydrolysis of metal ions, ligand protonation, complex formation) from unitary positions. A particular attention is paid to the influence of pH on variations in thermodynamic functions, taking into account the participation of hydrogen ions in many chemical reactions in aqueous solutions. The solution to the problem addressed in this chapter can be characterized as a transition from simple to complex and further to something even more complex. At the end of the paragraph, as far as possible, all the factors that may potentially affect the equilibrium state of the system are taken into account and the most generalized equation is derived, considering numerous equilibrium parameters.

Original methods of graphical representation of chemical transformations in multicomponent heterogeneous systems under real conditions, by constructing the diagrams of heterogeneous chemical equilibria (DHCE) in the coordinates "the partial molar fraction of the insoluble or soluble chemical species - the initial concentration of the components of the mixture (or solution pH)" have been developed. On the basis of DHCE for a particular composition of the heterogeneous mixture, the amount of formed precipitate can be appreciated.

Finally, in the second paragraph, the thermodynamics of equilibria of different hydroxy aluminium sulfate (HAS) minerals in soils, the conditions under which solids involving common constituents can coexist at equilibrium, the acid-base, complex formation and mineral equilibria have been investigated. It has been proved that the presence of sulfate ion dramatically alters the HAS mineral solubility under acidic conditions. The obtained data regarding the factors influencing *Al* speciation, based on the constructed diagrams of heterogeneous chemical equilibria, are in accordance with the current experimental and theoretical evidence. The used approach reveals the dominant processes that are responsible for the Al^{3+} concentration levels in acid soils by HAS as well. The results presented here suggest that the formation of HAS minerals plays a major role in sulfate retention in *S*-impacted soils. Based on our calculations of the soil solution composition, alunite, basaluminite and jurbanite are predicted to form in acidic soils and control the aluminium ion concentration level, for various soil environments.

Thermodynamics of the Overall Process Precipitation - Dissolution of Sparingly Soluble Compounds

At the beginning, from the diversity of possible chemical changes in the solution of the cation *M* and anion *A* of the sparingly soluble salt of a simple composition $MA_{(S)}$ (*here the symbol "S" denotes the solid phase*), the hydrolysis of the cation and the protonation of the anion will be analyzed according to the following set of reactions, named here as *Scheme A*:

$$MA_{(S)} = M + A\,,\ \ K_S^0, \Delta H_S^0, \Delta c_{p(S)}^0 \tag{1}$$

$$M + iH_2O = M(OH)_i + iH\ ,\ K_i^0, \Delta H_i^0, \Delta c_{p(i)}^0 \tag{2}$$

$$A + jH = H_jA\ ,\ K_j^0, \Delta H_j^0, \Delta c_{p(j)}^0 \tag{3}$$

Here K_S^0, K_i^0, K_j^0 are the equilibrium thermodynamic constants; $\Delta H_S^0, \Delta H_i^0, \Delta H_j^0$ denote the variations of standard enthalpies and $\Delta c_{p(s)}^0, \Delta c_{p(i)}^0, \Delta c_{p(j)}^0$ symbolize the variation of heat capacities of respective reactions[1]:

$$K_S^0 = a_M a_A \tag{4}$$

$$K_i^0 = \frac{a_{M(OH)_i} a_H^i}{a_M} \tag{5}$$

$$K_j^0 = \frac{a_{H_jA}}{a_A a_H^j} \tag{6}$$

Further trough the indices *s, i* and *j* are all the quantities that refer to the species $MA_{(S)}$, $M(OH)_i$ or H_jA or for reactions (1) - (3) are noted respectively. The variations of the other thermodynamic functions are calculated by means of well-known relationships in thermodynamics:

The variations of the other thermodynamic functions, in particular, ΔG^0 and ΔS^0 are calculated by means of well-known relationships in thermodynamics:

$$\Delta G_S^0 = -RT \ln K_S^0 = \Delta H_S^0 - T\Delta S_S^0 \tag{7}$$

$$\Delta G_i^0 = -RT \ln K_i^0 = \Delta H_i^0 - T\Delta S_i^0 \tag{8}$$

$$\Delta G_j^0 = -RT \ln K_j^0 = \Delta H_j^0 - T\Delta S_j^0 \tag{9}$$

The *"0"* index for equilibrium constants (4) - (6) indicates that these constants are thermodynamic. Concentration constants that are correlated with the thermodynamic ones are also used:

$$K_S^0 = K_S \gamma_M \gamma_A \tag{10}$$

$$K_i^0 = K_i \frac{\gamma_i \gamma_H^i}{\gamma_M} \tag{11}$$

$$K_j^0 = K_j \frac{\gamma_j \gamma_H^i}{\gamma_A \gamma_H^j} \tag{12}$$

where by γ is noted the activity coefficient of the respective species.

The following task is formulated: to calculate the variations of the thermodynamic functions of the overall process (1) - (3) under conditions different from the standard ones.

To solve it, the overall process (1) - (3) is described by a single equation. In conditions where the metal ion is represented by a single hydroxyl complex $M(OH)_i$ and the anion exists in the form of a single protonated *HjA* species, the overall reaction equation is a simple combination of equations (1) - (3):

$$MA_{(S)} + (j-i)H + iH_2O = M(OH)_i + H_jA \tag{13}$$

In this case, the task is reduced only to the calculation of the thermodynamic characteristics of the overall reaction (13). The thermodynamic relationships of this process are governed by the principles of the classical thermodynamics. In particular, the variation of the thermodynamic functions of the overall reaction (13) is expressed linearly by the thermodynamic characteristics of reactions (1) - (3).

Under real conditions, the metal ion and anion are usually present in the saturated solution in the form of several hydroxocomplexes $M(OH)_i$ and protonated species *HjA*; therefore, generalization of equation (13) is required. For the correct mass balance (MB) evidence, it is recognized that the hydroxocomplex and the protonated species are formed within the reactions proportional to their partial molar fractions. The last is formulated in the following way:

$$f_i = \frac{[M(OH)_i]}{C_M} = \frac{[M(OH)_i]}{\sum_{i=0}^{m}[M(OH)_i]} \tag{14}$$

$$f_j = \frac{[H_jA]}{C_A} = \frac{[H_jA]}{\sum_{j=0}^{n}[H_jA]} \tag{15}$$

Thus, the mass balance after the metal ion and anion is considered correct because $\sum_{i=0}^{m} f_i = 1$ and $\sum_{j=0}^{n} f_i = 1$. Next, by simple calculation, the amount of hydrogen ions and water molecules are added. As a result, the generalized equation of reactions takes the form:

$$MA_{(S)} + \left(\sum_{j=1}^{n} jf_j - \sum_{i=1}^{m} if_i\right)H + \left(\sum_{i=1}^{m} if_i\right)H_2O = \sum_{i=0}^{m} f_i M(OH)_i + \sum_{j=0}^{n} f_j H_j A \tag{16}$$

The specificity of this equation is that the stoichiometric coefficients are variable quantities depending on the composition of the system. For integer values of the partial molar fractions f_i and f_j the equation (16) takes the form of a trivial equation (13). The description of the equilibria in the "*sparingly soluble salt (solid phase) - aqueous solution*" system with the generalized interaction equation has a number of advantages. First of all, such an approach makes it possible to determine in an explicit form both the contribution of certain chemical species in the equilibrium process and the contribution of certain chemical equations describing the complex chemical equilibria. In other words, the stoichiometric aspects of the chemical interactions in the system are revealed. Second, as it will be demonstrated below, the thermodynamic analysis of the generalized equation (16) allows the determination of any thermodynamic functions of the system.

The calculation of the Gibbs energy variation of the reaction (1) under the conditions of the simultaneous run of the reactions (2) and (3) is examined, therefore for the whole process described by the generalized equation (16). First, the standard Gibbs energy variation $\log\alpha_M = \frac{1}{p}\log p\beta_{pq} + \frac{p-1}{p}\log C_M + \frac{q}{p}\log[L]$ is computed:

$$\Delta G_s^0 = \sum_{i=0}^{m} f_i \Delta G_f^0(M(OH)_i) + \sum_{j=0}^{n} f_j \Delta G_f^0(H_jA) - \bar{m}\Delta G_f^0(H_2O) - (\bar{n}-\bar{m})\Delta G_f^0(H) - \Delta G_f^0(MA_{(S)}) \qquad (17)$$

where $\Delta G_f^0(i)$ presents the standard Gibbs energy for the formation of the "i" species. In equation (17) $\bar{m}$ and $\bar{n}$ symbolize the Bjerrum formation functions:

$$\bar{m} = \sum_{i=1}^{m} if_i = \frac{\sum_{i=1}^{m} iK_i[H]^{-i}}{1+\sum_{i=1}^{m} K_i[H]^{-i}} \qquad \bar{n} = \sum_{j=1}^{n} jf_j = \frac{\sum_{j=1}^{n} jK_j[H]^{j}}{1+\sum_{j=1}^{n} K_j[H]^{j}}$$

Quantities (values) $\Delta G_f^0(i)$ can be expressed through the respective equilibrium constants:

$$\Delta G_f^0(M) + \Delta G_f^0(A) - \Delta G_f^0(MA_{(S)}) = -RT\ln K_S^0$$

$$\Delta G_f^0(M(OH)_i) + i\Delta G_f^0(H) - i\Delta G_f^0(H_2O) - \Delta G_f^0(M) = -RT\ln K_i^0$$

$$\Delta G_f^0(H_jA) + i\Delta G_f^0(H) - \Delta G_f^0(A) = -RT\ln K_j^0$$

Considering these equations, after some transformations, the expression (17) takes the form:

$$\Delta G_S^0 = -RT\ln K_S^0 \prod_{i=1}^{m} (K_i^0)^{f_i} \prod_{j=1}^{n} (K_j^0)^{f_j} \qquad (18)$$

Analogously, under conditions different from the standard, the Gibbs energy variation of the reaction (16) is calculated with the equation of the reaction isotherm in the following way:

$$\Delta G_S = \Delta G_S^0 + RT\ln \frac{\prod_{i=0}^{m} a_{M(OH)_i}^{f_i} \prod_{j=0}^{n} a_{H_jA}^{f_j}}{a_H^{n-m}} \qquad (19)$$

where ΔG_S^0 is determined by the expression (18). The activities of hydroxocomplexes $a_{M(OH)_i}$ and protonated species a_{H_jA} are thus determined:

$$a_{M(OH)_i} = \gamma_i [M(OH)_i] = \gamma_i f_i C_M \tag{20}$$

$$a_{H_jA} = \gamma_j [H_jA] = \gamma_j f_j C_A \tag{21}$$

For hydrogen ion activity the expression is valid:

$$a_H = \gamma_H [H] \tag{22}$$

Substituting (18) in (19) and considering the equations (17), (20)-(22), the following expression is obtained:

$$\Delta G_S = -RT \ln K_S^0 \frac{\prod_{i=1}^{m} (K_i^0)^{f_i} [H]^{-m} \prod_{j=1}^{n} (K_j^0)^{f_j} [H]^n}{\prod_{i=0}^{m} f_i^{f_i} \prod_{j=0}^{n} f_j^{f_j}} + RT \ln a_M a_A$$

The following equality can be demonstrated:

$$\frac{\prod_{i=1}^{m} (K_i^0)^{f_i} [H]^{-m}}{\prod_{i=0}^{m} f_i^{f_i}} = 1 + \sum_{i=1}^{m} K_i [H]^{-i} \equiv \alpha_M \tag{23}$$

$$\frac{\prod_{j=1}^{n} (K_j^0)^{f_j} [H]^{-n}}{\prod_{j=0}^{n} f_j^{f_j}} = 1 + \sum_{j=1}^{n} K_j [H]^{j} \equiv \alpha_A \tag{24}$$

For example, for the first equality (23), one obtains:

$$\frac{\prod_{i=1}^{m} (K_i^0)^{f_i} [H]^{-m}}{\prod_{i=0}^{m} f_i^{f_i}} = \frac{1}{f_m^{f_m}} \prod_{i=1}^{m} \left(\frac{K_i}{f_i} \right)^{f^i} [H]^{-m} = \frac{1}{\left(\frac{1}{\alpha_M} \right)^{f_M}} \prod_{i=1}^{m} \left(\frac{K_i}{\frac{K_i [H]^{-i}}{\alpha_M}} \right)^{f_i} [H]^{-m} =$$

$$= \frac{1}{\left(\frac{1}{\alpha_M} \right)^{\sum_{i=0}^{m} f_i}} \prod_{i=1}^{m} \left(\frac{1}{[H]^{-if_i}} \right) [H]^{-m} = \alpha_M \frac{[H]^{-\bar{m}}}{[H]^{-\sum_{i=1}^{m} if_i}} = \alpha_M$$

Thus, the variation of Gibbs energy of the overall process (1) - (3) under conditions different from the standard ones is described by the expression:

$$\Delta G_S = -RT\ln K_S\alpha_M\alpha_A + RT\ln C_M^0 C_A^0 = \Delta G_S^{sum} = -mRT\ln\frac{C_M^r}{C_M^0} - nRT\ln\frac{C_A^r}{C_A^0}, \quad (25)$$

where

$$\alpha_M = C^r/[M] = 1 + \sum_{i=1}^{m} K_i[H]^{-i} \quad (26)$$

$$\alpha_A = C_A^r = 1 + \sum_{j=1}^{n} K_j[H]^j \quad (27)$$

Here through C_M^0, C_A^0, C_M^r and C_A^r are denoted respectively the total (initial) concentrations of metal ion and anion in heterogeneous mixture and their residual concentration in aqueous solution. Between the total concentration of the metal ion C_M^0 and its residual concentration C_M^r in the solution there is a reciprocal relationship

$$C_M^0 = C_M^r + \Delta C_M,$$

$$C_A^0 = C_A^r + \Delta C_A,$$

where ΔC_M and ΔC_A is, respectively, the quantity of metal ion and anion precipitated in a unit volume of the solution.

By the sign and value of ΔG_S^{sum} the possibility of the formation of sparingly soluble salt and the precipitation completeness under specified conditions can be determined. Thus, for $\Delta G_S^{sum} > 0$ the precipitation takes place, otherwise, if $\Delta G_S^{sum} < 0$, then the solid phase is not formed. Equality $\Delta G_S^{sum} = 0$ reflects the beginning of precipitation-dissolution of the sparingly soluble salt.

As a result, the contribution of reactions (2) and (3) in the Gibbs energy variation of process (1) is expressed by concentration coefficients α_M and α_A. In literature, these concentration functions are well known and are called coefficients of side reactions. Note that these functions include concentration constants instead of thermodynamic ones. The relation (25) can be regarded as the sum of three components:

$$\Delta G_S = \Delta G(1) + \Delta G(2) + \Delta G(3)$$

where

$$\Delta G(1) = -RT\ln K_S^0 + RT\ln a_M a_A \quad (28)$$

$$\Delta G(2) = -RT \ln a_M \tag{29}$$

$$\Delta G(3) = -RT \ln a_A \tag{30}$$

Similarly, the generalized equation (16) can also be decomposed into three simple equations:

$$MA_{(s)} = M + A \tag{31}$$

$$M + \bar{m}H_2O = \sum_{i=0}^{m} f_i M(OH)_i + \bar{m}H \tag{32}$$

$$A + \bar{n}H = \sum_{j=0}^{n} f_j H_j A \tag{33}$$

If $C_M^0 = C_A^0 = 1$ mol/L, the term under the sign of the logarithm is equal to the product of conditional solubility K_S^{cond}:

$$\Delta G_r = -RT \ln K_S \alpha_M \alpha_A = -RT \ln K_S^{cond} \tag{34}$$

Therefore, the postulated conditional solubility product (Karak et al., 2016: Stumm & Morgan, 1996) has a rigorous thermodynamic sense, representing the measure of the Gibbs energy variation of the whole process under conditions different from the standard:

$$MA_{(s)} + (\bar{n} - \bar{m})H + \bar{m}H_2O = M(OH)_{\bar{m}} + H_{\bar{n}}A,$$

where $M(OH)_{\bar{m}}$ and $H_{\bar{n}}A$ are the average formulas of the species.

The equations (18) and (25) remain valid also in the presence of polynuclear hydroxocomplexes, only in this system the coefficient α_M is determined by the expression:

$$\alpha_M = \sum_i \sum_j i K_{M_i(OH)_j} [M]^{i-1} [H]^{-j}$$

In addition, the ligand L, which forms with the metal complexes of the ML composition, is additionally introduced into the "*sparingly soluble salt-saturated aqueous solution*" system. At the same time it is considered that these complexes participate in the secondary reactions with the formation of mixed species of the composition $M(OH)_iL$ $(i=0,1,...,p)$ and MH_iL $(i=1,2,...,q)$. Often the ligand L is a Brønsted base and, depending on the conditions, forms H_iL species $(i = 0.1, ..., k)$. The total constants of the respective species are denoted by $K_{M(OH)_i}, K_{H_jA}, \beta_{M(OH)_iL}, \beta_{MH_iL}$ and β_{H_iL}. The equilibria in this system are described by the following generalized equation:

$$MA_{(S)} + \bar{s}\sum_{i=0}^{k} f_{H_iL}H_iL + (\bar{n} + \bar{q} - \bar{m} - \bar{p} - \bar{ks})H + (\bar{m} + \bar{p})H_2O = \\ = \sum_{i=0}^{m} f_{M(OH)_i}M(OH)_i + \sum_{i=1}^{p} f_{M(OH)_iL}M(OH)_iL + \sum_{i=1}^{q} f_{H_iA}H_iA \tag{35}$$

An analogous analysis of the previous system "*sparingly soluble salt - saturated aqueous solution*" in the absence of ligand L shows that the standard Gibbs energy variation of the process, expressed by equation (35), is equal to:

$$\Delta G_S^0 = -RT\ln K_S \frac{\prod_{i=1}^{m} K_{M(OH)_i}^{f_{M(OH)_i}} \prod_{i=1}^{p} \beta_{M(OH)_iL}^{f_{M(OH)_iL}} \prod_{i=1}^{q} \beta_{MH_iL}^{f_{MH_iL}} \prod_{i=1}^{n} K_{H_iA}^{f_{H_iA}}}{\prod_{i=1}^{k} \beta_{H_iL}^{f_{H_iL\bar{s}}}}$$

For the variation of Gibbs energy under conditions different from the standard, the analogical equation with (25) is valid, but in this case α_M is determined by the expression:

$$\alpha_M = 1 + \sum_{i=1}^{m} K_{M(OH)_i}[H]^{-i} + [L]\left(\sum_{i=0}^{p} \beta_{M(OH)_iL}[H]^{-i} + \sum_{i=1}^{q} \beta_{MH_iL}[H]^{i}\right)$$

In this expression $[L]$ is determined from the mass balance condition if the ligand is in excess of the metal ion:

$$[L] = \frac{C_L}{1 + \sum_{i=1}^{K} \beta_{H_iL}[H]^i}$$

Analogous consideration is also given to the formation of different polynuclear complexes, having a more complex composition $M_i(OH)_pH_qL_j$:

$$\alpha_M = 1 + \sum_i\sum_p\sum_q\sum_j i\beta_{M_i(OH)_pH_qL_j}[M]^{i-1}[H]^{q-p}[L]^j$$

Here $[M]$ is the equilibrium concentration of the metal ion and is also determined from the mass balance conditions:

$$C_M = \sum_i\sum_p\sum_q\sum_j i\beta_{M_i(OH)_pH_qL_j}[M]^i[H]^{q-p}[L]^j$$

$$C_L = \sum_i\sum_p\sum_q\sum_j j\beta_{M_i(OH)_pH_qL_j}[M]^i[H]^{q-p}[L]^j$$

In the general case, for the salt deposition in precipitate of the composition M_pA_q an analogical analysis leads to the expression:

$$\Delta G_r = -RT \ln K_S \alpha_M^p \alpha_A^q + RT \ln C_M^p C_A^q$$

Therefore, within the generalized interaction equation, the description of chemical equilibria in the "*sparingly soluble salt - aqueous solution*" system, taking into account the contribution of various side reactions, has a number of advantages. Firstly, total (overall) thermodynamic functions can be introduced and therefore both the possibility and the optimum conditions of the precipitation-dissolution process can be determined. Secondly, generalized interaction equations allow explicitly taking into account the contribution of all species present in the system at equilibrium. It has been shown that the concept of conditional solubility product, previously postulated, has a rigorous thermodynamic sense. The importance of the treatment described above is of particular interest in the study of complex chemical equilibria when the use of traditional calculation methods, based on the law of mass action and mass balance conditions is a difficult issue because of the extremely large number of species that are present (Butler, 1998).

The evidence of formation in solution of polynuclear complex species leads to certain peculiarities in the behavior of the system, therefore the influence of the formation of polynuclear and small complexes on variations of thermodynamic functions will be examined here separately. The most general case of possible chemical processes in the system "$MA_{(S)}$ - saturated aqueous solution - complexing agent" will be analyzed. Possible chemical processes can be described by the equations:

$$MA_{(S)} = M + S,\ K_S, \Delta H_S^0 \tag{36}$$

$$M + mH + nL + jA = M_iH_mL_nA_j,\ \beta_{imnj}, \Delta H_{imnj}^0 \tag{37}$$

$$L + pH = H_pL\ ,\ \beta_p, \Delta H_p^0 \tag{38}$$

$$A + qH = H_qL\ ,\ \beta_q, \Delta H_q^0 \tag{39}$$

Here, in addition to the basic properties of the anion and ligand, the possibility of forming complex micronized and protonated polynuclear compounds is also taken into account. The calculation of the thermodynamic functions of the process (36) under the simultaneous performance of the reactions (37) - (39) can be reduced to the calculation of the thermodynamic functions of an overall reaction:

$$MA_{(S)} + \frac{n}{i}H_pL = \frac{1}{i}M_iH_mL_nA_j + \left(1-\frac{j}{i}\right)H_qA + \left[\frac{np}{i} + \left(\frac{j}{i}-1\right)q - \frac{m}{i}\right]H \tag{40}$$

By the indexes *i, m, n, j* and *k*, will be marked all the quantities, which refer to equations (37) - (39) or to species $M_iH_mL_nA_j, H_qL$ and H_qA respectively, with the indices *i, m, n, j* and *k*. The partial molar fractions of these species are determined as shown below:

$$f_{imnj} = \frac{i\beta_{imnj}[M]^{i-1}[H]^m[L]^n[A]^j}{1+\sum_{i=1}\sum_{m=1}\sum_{n=1}\sum_{j=1} i\beta_{imnj}[M]^{i-1}[H]^m[L]^n[A]^j} \tag{41}$$

$$f_q = \frac{\beta_q[H]^q}{1+\sum_{q=1}\beta_q[H]^q} \tag{42}$$

$$f_p = \frac{\beta_p[H]^p}{1+\sum_{p=1}\beta_p[H]^p} \tag{43}$$

When the metal ion, anion and ligand are represented by an arbitrary number of chemical species, equation (40) must be generalized. In supposition, that species react in proportions corresponding to their molar fractions, the whole process scheme takes the form:

$$MA_{(S)} + \sum_{p=0} f_p H_p L \rightarrow \sum_{i=1}\sum_{m=0}\sum_{n=0}\sum_{j=0} \frac{1}{i} f_{imnj} M_i H_m L_n A_j + \sum_{q=0} f_q H_q A$$

The mass balance will be taken into account. In relation to the metal ion, the mass balance is considered because the identity occurs:

$\sum_{i=1}\sum_{m=0}\sum_{n=0}\sum_{j=0} f_{imnj} = 1$, because $\sum_{p=0} f_p = 1$, on the left side of the equation there is one particle L, whereas on the right the number of moles of L is $\sum_{i=1}\sum_{m=0}\sum_{n=0}\sum_{j=0} \frac{n}{i} f_{imnj}$.

Analogously the number of anions and hydrogen ions is equal. Finally, the following generalized reaction equation is obtained:

$$\begin{aligned} MA_{(S)} &+ \sum_{p=0} f_p H_p L\left(\sum_{i=1}\sum_{m=1}\sum_{n=1}\sum_{j=1} \frac{n}{i} f_{imnj}\right) = \sum_{i=1}\sum_{m=0}\sum_{n=0}\sum_{j=0} \frac{1}{i} f_{imnj} M_i H_m L_n A_j + \\ &+ \sum_{q=0} f_q H_q A\left(1 - \sum_{i=1}\sum_{m=1}\sum_{n=1}\sum_{j=1} \frac{j}{i} f_{imnj}\right) + \left[\left(\sum_{i=1}\sum_{m=1}\sum_{n=1}\sum_{j=1} \frac{n}{j} f_{imnj}\right)\left(\sum_{p=1} p f_p\right) - \sum_{i=1}\sum_{m=1}\sum_{n=1}\sum_{j=1} \frac{m}{i} f_{imnj}\right] H - \\ &- \left[\left(1 - \sum_{i=1}\sum_{m=1}\sum_{n=1}\sum_{j=1} \frac{j}{i} f_{imnj}\right)\left(\sum_{q=1} q f_q\right)\right] H_2O \end{aligned} \tag{44}$$

It is easy to see that under the conditions $f_{imnj} = f_p = f_q = 1$ the equation (44) takes form (43).

By analyzing the generalized process equation obtained by the above-mentioned method, we will observe that in this case mathematical analysis is complicated. Only basic results will be shown here.

It can be demonstrated that the variation of Gibbs energy of the overall process in the non-stanadrd (real) conditions is governed by the relationship:

$$\Delta G_S = -RT\ln K_S\alpha_M\alpha_A + RT\ln C_M^0 C_A^0 \tag{45}$$

where the coefficient of the secondary reactions of the metal ion α_M has the form:

$$\alpha_M = 1 + \sum_{i=1}\sum_{m=1}\sum_{n=1}\sum_{j=1} i\beta_{mnj}[M]^{i-1}[H]^m[L]^n[A]^j \tag{46}$$

It should be noted that the evidence of polynuclear complex compounds results in the coefficient of the secondary metal ion reactions exhibiting a function of the equilibrium concentrations of the metal ion *M*, the anion *A* and the ligand *L*, which can be calculated from the mass balance conditions and, therefore, are functions of temperature.

The generalized equation in the case of formation-dissolution of the sparingly soluble hydroxide is:

$$M(OH)_{n(S)} = \sum_{i=0}^{m} f_{M(OH)_i} M(OH)_i + (n-\bar{m})OH\ , \tag{47}$$

where $f_{M(OH)_i}$ and $\bar{m}$ is the partial molar fraction of the hydroxocomplex $M(OH)_I$ and Bjerrum function respectively:

$$f_{M(OH)_i} = \frac{[M(OH)_i]}{\sum_{i=0}^{m}[M(OH)_i]} = \frac{\beta_{M(OH)_i}[OH]^i}{\alpha_M} \tag{48}$$

$$\bar{m} = \sum_{i=1}^{m} i f_{M(OH)_i} = \frac{\sum_{i=1}^{m} i\beta_{M(OH)_i}[OH]^i}{1+\sum_{i=1}^{m}\beta_{M(OH)_i}[OH]^i} \tag{49}$$

In this case, the expression for computing the overall Gibbs energy takes the form

$$\Delta G_S = -RT\ln K_S\alpha_M + RT\ln C_M[OH]^n = -RT\ln\frac{C_M^r}{C_M^0}\ , \tag{50}$$

where

$$\alpha_M = 1 + \sum_{i=1}^{m}\beta_{M(OH)_i}[OH]^i \tag{51}$$

It is easy to prove that the relation (51) has the same form in the cases of polynuclear hydrolysis, complex formation of the metal ion with the complexing agent etc.

The thermodynamic functions of the formation-dissolution reactions of the oxide precipitate with the composition $M_nO_{m(S)}$ are described with analogous expressions, except that the equilibrium constant K_S refers to the reaction:

$$\frac{1}{n}M_nO_{m(S)} + \frac{2m}{n}H = M + \frac{m}{n}H_2O,\ K_S\ .$$

The Gibbs energy variation of this reaction, taking into account the side reactions of the metal ion, is:

$$\Delta G_S = -RT\ln K_S\alpha_M + RT\ln\frac{C_M}{[H]^{2m/n}}\ ,$$

where α_M is determined by the above expression.

In the case of the sparingly soluble $H_nA_{(S)}$ acid, the following set of equilibria is considered:

$$H_nA_{(S)} = nH + A,\ K_{Sa}$$

$$jH + A = H_jA,\ K_j \tag{52}$$

The expression for the Gibbs energy variation of the system (52) takes the form:

$$\Delta G_S = -RT\ln K_{Sa}\alpha_A + RT\ln C_A^0[H]^n\ , \tag{53}$$

where

$$\alpha_A = 1 + \sum_{j=1} K_j[H]^j\ .$$

Using equation (53), the conditions of formation of sparingly soluble acid precipitates can be calculated: $\Delta G_S > 0$, which correspond to inequality $C_A^0[H]^n > K_{Sa}\alpha_A$.

Thermodynamic Stability of Aluminium-Containing Minerals

It is well-known that soil minerals are the key source of aluminium (Al) to aqueous and environments. For example, in the lithosphere, Al contains mostly in highly crystalline aluminosilicate minerals. A potential cycle of Al in natural waters involves aluminum minerals, dissolved monomeric species, dissolved and aggregated polymers. The hydrolysis and complex formation reactions in aqueous solution generate a large variety of organic and inorganic aluminum species, influencing the precipitation of solid aluminum-containing phases (Bouza et al., 2019). Models founded on equilibrium thermodynamics have been mainly generally used approach to predicting Al quantities in aqueous solutions. The degree of errors in predicting Al has important implications for plant growth. For example, in nutrient solutions, the activities of Al as low as 2 μM are toxic to barley (Cameron et al., 1986). The most poisonous aluminium species is widely considered to be the Al^{3+} ion and, to a lesser level, monomeric hydroxy-aluminium species; in contrast, stable AlF_x complexes are not considered phytotoxic (Chen et al., 2018). On the other hand, the acid deposition presently affects large soil areas. Long-term forecast of the complex chemical processes and their realistic consequences is only possible by using simulation models. Hence, it is of interest to calculate the change in Gibbs energy under actual conditions, in order to detect how much this change differs from the quantity under standard conditions. Such sort of problems were widely analyzed in (Fishtik & Povar, 1987; Povar et al., 1989; Povar, 1994; Povar, 1997; Povar & Rusu, 2012).

In aqueous media aluminum ion Al^{3+} contains in a range of soluble and insoluble species. The high concentrations of aluminum may influence the buffer action and transport of elements in aquatic ecosystems, this element being poisonous for aquatic organisms. Although comprehension of the aluminum chemistry and its effects in aqueous environment has substantially increased in last years, some supplementary researches are compulsory to identify the sources of aluminum and mechanisms of its transport from the soil solution to ground surface waters in order to evaluate the effects of aluminium on trace metals and organic carbon cycle, phosphorus compounds, in addition to the action of different aluminum containing species on aquatic organisms. Since many plant species are susceptible to micromolar concentrations of *Al,* the potential for soils to be *Al* poisonous is significant. Speciation of aluminium *Al* is an important question when assessing the effects of *Al* in soil solutions since not all chemical species of *Al* are similarly poisonous. The soluble aluminium species in soil solutions is found as the Al^{3+} ions together with chemical combinations with OH^-, F^-, SO_4^{2-}, PO_4^{3-} and organic compounds. While the current investigations have often ignored the presence of aluminum non – hydroxo- complexes, it can be assumed that the complex organic species and fluorides are the main forms of aluminum in dilute acidic surface waters (with a low ionic strength). The concentration of inorganic aluminum forms rises exponentially with diminishing the solution pH. This phenomenon is similar to the theoretical dependence of the solubility of mineral phases on pH solution values. At the same time, the concentration of organic aluminium species correlates better with the variations in the organic carbon concentration from surface waters than with the environmental pH values. Opportunely, most of the *Al* is bound by ligands or occurs in other non-toxic forms such as aluminosilicates and precipitates. Nevertheless, the solubilization of *Al* is enhanced by low pH and *Al* toxicity is a major factor restricting plant growing on acid soils; the toxic effect being dependent on the plant species and the concentrations of individual *Al* species in soil solution.

Ecologically considerable concentrations of aluminum are identified in surface waters due to acidic deposition, which represents the chief way by which the atmospheric pollutants influence the soil complex equilibria. Mineral acids consequent from rainfalls mobilize aluminium from minerals formed as a result of the soil development, being transported to the adjacent surface waters. Therefore, the discharge of toxic *Al* is one of the gravest consequences of soil acidification. Introduction of acidic rainwater in the soils results in a substantial change in soil solution composition and especially in the concentration. In the soil solution, *Al* is one of the major anions, generally ranging from $1.5 \cdot 10^{-4}$ to $1.25 \cdot 10^{-3}$ mol•L^{-1} in the diverse natural waters. The sulfate activity of surface waters and ground waters is relatively constant and is set at $2.5 \cdot 10^{-3}$ mol•L^{-1}. In some soils, the concentrations of inorganic monomeric aluminum are typically between $1 \cdot 10^{-4}$ and $2 \cdot 10^{-4}$ mol•L^{-1} (Larssen et al., 1998). Under usual soil conditions, the *Al* activity is generally controlled by the solubility of gibbsite or kaolinite (Stumm & Morgan, 1996). However, in areas of sulfide oxidation, high concentrations of sulfate modify the aqueous *Al* geochemistry. The acid sulfate soil waters contain commonly high concentrations of dissolved aluminum. In these acid soils, the *Al* solubility appears to be controlled by a variety of basic *Al* sulfate phases (as alunogen, alunite, jurbanite and basaluminite (Driscoll & Schecher, 1990; Jones et al., 2011)).

The detection of mineral species that control the *Al* solubility in acid soil solutions, reached in sulfate ion, is a complicated process. Firstly, as it was showed by X-ray diffraction, the solid phases can be frequently amorphous. Secondly, the discriminating of adsorbed sulfate from these mineral structures presents an additional complexity. In several studies theoretical calculations have been performed for determination of the Al mineral species, formed in acidic sulfate soils (Jones et al., 2011). Authors (Adams & Rawajfih, 1997) stated that the retention of sulfate ions by acid soils is due to the solubility of basaluminite and/or alunite. Basaluminite solubility and its control of *Al* concentration in aqueous sulfate acidic systems have been recently investigated for a set of natural acidic waters (Jones et al., 2011; Sánchez-España et al., 2011). Using the *Al*, *S* and *O*, K-XANES spectroscopy along with elemental analysis of groundwater precipitates, it has been established that for the range of pH values 4.0-5.2 the *Al* concentration is controlled by poor crystallinity mineral - hydrobasaluminite (with the empirical formula $Al_4(OH)_{10}SO_4{\cdot}15H_2O$ that under conditions of comparatively low humidity is not stable and predisposes to dehydrate with formation of basaluminite. The same authors have proved that the concentration of soluble *Al* in the range of pH values 3.5-7 in acidic sulfate soils is controlled by a combination of *Al* amorphous hydroxide and basaluminite. Besides, at low pH values and high activity, alunite can be the most stable hydroxy sulfate mineral (Stoffregen & Cydan, 1990).

Conversely, the majority of soil water samples considered by authors (Larssen et al., 1998) with pH below 5 were oversaturated relating to jurbanite. Their best estimations were achieved by supposing a jurbanite nature of equilibrium for pH<5 and a gibbsite nature of equilibrium for higher pH values, but how they in fact reproduced the authentic processes was uncertain.

Additionally, authors (Appelo & Postma, 2005) found that in groundwater the gibbsite model was realistically correct when pH > 4.5 and a jurbanite model was superior for pH < 4.5. Singh and others (Singh et al., 2017) supposed that the change occurring between pH 4.5–5 was the boundary for control between basic aluminum sulfates and aluminosilicates. The activity in soil solutions with pH < 5 seems to be controlled by jurbanite, alunite, and gibbsite. In solutions with pH > 5 an amorphous phase controls the *Al* activity (Karathanasis et al., 1988).

High concentrations of aluminum in dilute acidic waters are of interest from a number of points of view: aluminium is a significant buffer of pH; aluminum can control the cycle of important elements, such as phosphorus, organic carbon and metal ions; aluminum is potentially poisonous to aquatic organisms. Authors (Delfosse et al., 2005) found that the formation of basic aluminium sulfate minerals $(K,Na)_nAl_x(OH)_y(SO_4)_z(H_2O)_m$, also named hydroxyl - aluminium sulfate (*HAS*), may contribute to sulfate maintenance in soils. The detection of the mineral species controlling the solubility of *Al* in acidic waters rich in sulfate has represented several challenges. It is noticeable that there are disagreements in literature on the process of adjusting aqueous aluminum concentration in dilute acidic waters. If one means to refine the models developed to evaluate the extent and effects of surface water acidification, it is crucial to develop a better comprehension of the mechanisms of aluminum mobilization.

In order to better understand the effects of acid precipitation on soil and forecast the poisonous concentration of *Al* in soil solutions, it is crucial to have a method to predict how the diverse species of *Al* will react to changes in the composition of soil solutions containing *HAS* minerals. Chemical modelling represents an exclusive method for the forecast of aluminium behavior in soil solutions (Povar & Rusu, 2012). The main objective of this study consists in the thermodynamic analysis and graphical depiction of the distribution of different species of *Al*, both in saturated aqueous solutions and *HAS* solid phases. Diagrams of heterogeneous complex equilibria, constructed by using graphical and numerical computerized methods, which quantitatively describe the repartition of different soluble and insoluble, inorganic and organic, monomeric and polymeric aluminium species in heterogeneous multicomponent systems "*HAS mineral – saturated aqueous solution*", have been used for identification of the dominant processes that control the aluminium ion concentration levels in acidic soils.

Computational Part

As aluminium insoluble species, a range of minerals with general chemical formula $(K,Na)_nAl_x(OH)_y(SO_4)_z(H_2O)_m$, have been considered. Their solubility is controlled by the following heterogeneous equilibrium:

$$(K,Na)_n Al_x(OH)_y(SO_4)_z(H_2O)_m + yH^+ \leftrightarrow n(K^+,Na^+) + xAl^{3+} + zSO_4^{2-} + (m+y)H_2O,$$
$$K_S = [K,Na]^n[Al]^x[SO_4]^z / [H]^y \quad (54)$$

where K_s is the solubility product of aluminium minerals. The ionic charges and influence of ionic strength are disregarded for convenience.

The Al^{3+} concentration is computed from the K_s expression (54):

$$[Al] = K_S^{1/x}[K,Na]^{-n/x}[SO_4]^{-z/x}[H]^{y/x} \quad (55)$$

In this paragraph the following minerals have been examined: 1) jurbanite $Al(OH)(SO_4)\cdot 5H_2O$; 2) basaluminite $Al_4(OH)_{10}SO_4\cdot 15H_2O$; 3) alunite $KAl_3(OH)_6(SO_4)_2$; 4) alunogen $Al_2(SO_4)_3\cdot 17H_2O$ and 5) gibbsite $Al(OH)_3$. Along with the process (54) a large set of potential equilibria in the system „*mineral – soil solution*" has been taken into account. In the soil solution Al contains in different soluble species, including Al^{3+} and its hydroxo, fluoride, sulfate and organic complexes. Phosphate complexes and polymeric species of aluminium can also exist. Aluminium (III) composes with the Org^{3-} ligand two complexes. As polynuclear species of aluminium, the dimmer $Al_2(OH)_2^{4+}$ and trimmer $Al_3(OH)_4^{5+}$ have been considered. The polymeric species with the high molecular masses $Al_i(OH)_j^{3i-j}$, where i = 6, 8 and 13 are not considered because they almost are not formed in the soil solution as a result of the low polymerization kinetics. All together, the organic acids in soil solution block their formation (Povar & Rusu, 2012). The phosphate complexes of aluminum can exist as $AlH_2PO_4^{2+}$. The silicate complexes of aluminium are disregarded because of their weak complex formation degree. The mass balance (MB) conditions in the analyzed systems have been expressed within the method of residual concentrations (RC) (Povar, 2000; Povar, 2001):

$$C_{Al}^0 = C_{Al}^r + \Delta C_{Al} = \sum_{i=1}\sum_{j=0} i[Al_i(OH)_j^{3i-j}] + \sum_{j=1}[AlF_j^{3-j}] + [AlSO_4^+] + [Al(SO_4)_2^-] + [AlOrg^0] +$$
$$+[AlHOrg^+] + [AlH_2PO_4^{2+}] + \Delta C_{Al} \quad (56)$$

$$C_F^0 = [F^-] + [HF^0] + \sum_{j=1} j[AlF_j^{3-j}] = [F^-] + K_{HF}[H^+][F^-] + K_S[H^+]^3 \sum_{j=1} jK_{f,j}[F^-]^j \quad (57)$$

$$C_{SO_4}^0 = [SO_4^{2-}] + [AlSO_4^+] + 2[Al(SO_4)_2^-] + \Delta C_{SO_4} \quad (58)$$

$$C^0_{Org} = \sum_{m=0} [H_m Org^{m-3}] + [AlHOrg^+] + [AlOrg^0] \tag{59}$$

$$C^0_{PO_4} = \sum_{l=0} [H_l PO_4^{l-3}] + [AlH_2PO_4^{2+}] \tag{60}$$

$$C^0_K = [K^+] + \Delta C_K \tag{61}$$

$$C^0_H = C^r_H + \Delta C_H = [H^+] - [OH^-] - \sum_{i=1}\sum_{j=1} j[Al_i(OH)_j^{3i-j}] + [AlHOrg^+] + 2[AlH_2PO_4^{2+}] + [HF^0] + \\ + \sum_{m=1} m[H_m Org^{m-3}] + [HCO_3^-] + 2[H_2CO_3^0] + \sum_{l=1} l[H_l PO_4^{l-3}] + \Delta C_H \tag{62}$$

The quantities C_i^0 and C_i^r (mol•L^{-1}) in equations (56)-(61) symbolize correspondingly the analytical and residual concentrations of component *i* in the heterogeneous system, whereas ΔC_i denotes its quantity in the solid phase in one liter of solution (Povar, 1997). Within the RC method, the relations ΔC_{OH}=-ΔC_H and $C^r_{OH} = -C^r_H$ are valid. In the equation (62) C^0_H represents the excess of H^+ ions versus hydroxyl ions in biphasic mixtures, $C^0_H = -C^0_{OH}$ (Fishtik & Povar, 1987). From the mineral stoichiometric composition of type $(K,Na)_nAl_x(OH)_y(SO_4)_z(H_2O)_m$ the following relations are resulting:

$$\frac{1}{y}\Delta C_{OH} = \frac{1}{z}\Delta C_{SO_4} = \frac{1}{n}\Delta C_k = \frac{1}{x}\Delta C_{Al} \tag{63}$$

The equilibrium concentration of ion Al^{3+} is established from the relation (60). From the MB equations (56) - (62) for a definite pH value, considering the relations (63) and the equations of the law of mass action (LMA), one can get a system of five nonlinear equations (or 6 equations, in the case of alunite $KAl_3(OH)_6(SO_4)_2$ with 5 or 6 variables, respectively [*F*], [*Org*], [SO_4], [PO_4], ΔC_i and [*K*], that can be easily found by computerized numerical methods (Fishtik et al., 1986; Fishtik et al., 1987; Povar, 1997). After that, the residual concentrations of the component *i* are computed C_i^r, knowing the initial composition of heterogeneous mixture, e.g. the quantities C_i^0. Using the LMA and MB joint equations, it is possible to compute the diagrams of distribution of aluminium ion both in solid and liquid phases (species in aqueous solution), i.e. the diagrams of heterogeneous chemical equilibrium (DHCE) (Povar & Rusu, 2012). The following equations for computation of the species partial molar fractions γ_i in the heterogeneous system, where the considered reactions proceed concurrently, can be written:

$$\gamma_S = \frac{\Delta C_{Al}}{C^0_{Al}}; \quad \gamma_{Al} = \frac{[Al]}{C^0_{Al}}; \quad \gamma_{ij} = \frac{i[Al_i(OH)_j]}{C^0_{Al}};$$

$$\gamma_{PO_4} = \frac{[AlH_2PO_4]}{C^0_{Al}}; \quad \gamma_{f,j} = \frac{[AlF_j]}{C^0_{Al}}; \quad \gamma_{O1} = \frac{[AlOrg]}{C^0_{Al}};$$
$$\gamma_{O2} = \frac{[AlHOrg]}{C^0_{Al}}; \quad \gamma_{S1} = \frac{[AlSO_4]}{C^0_{Al}}; \quad \gamma_{S2} = \frac{[Al(SO_4)_2]}{C^0_{Al}}; \quad \gamma_{tot} = \frac{C^r_{Al}}{C^0_{Al}}. \tag{64}$$

The subscript index "tot" denotes the sum of partial molar fractions of all the soluble species. From the expressions (64) the following identity derives:

$$\gamma_S + \gamma_{tot} = \gamma_S + \gamma_{Al} + \sum_{i=1}\sum_{j=1}\gamma_{ij} + \sum_{j=1}\gamma_{f,j} + \gamma_{O1} + \gamma_{O2} + \gamma_{S1} + \gamma_{S2} + \gamma_{PO_4} = 1$$

The *DHCE*, apparently, are analogous to the diagrams of species distribution in the homogeneous systems. The last usually are built-in in coordinates $\gamma_i(pH)$, since in absence of polynuclear complexes the partial molar fractions of species γ_i in aqueous solution (more often, symbolized as f or α) are functions just of the solution pH and do not depend on initial concentrations of components. In the case of heterogeneous equilibria, the partial molar fractions of species also depend on the initial composition of mixture and, for a certain pH, are functions of five variables $\gamma_i = f(C^0_{Al}, C^0_F, C^0_{SO_4}, C^0_{Org}, C^0_{PO_4})$. In this case it is rationally to construct diagrams in coordinates (γ_i, pH) whilst values of other quantities are kept invariable.

Within the thermodynamic approach, earlier developed (Fishtik et al., 1987; Povar, 1997; Povar & Rusu, 2012) and extended in this paper for the solid phases with a more complex composition, it has been established that the values of the total Gibbs energy $\Delta G_{S,tot}$ for the analyzed complex processes, where a large set of concurrent reactions occur, can be calculated by the equation:

$$\Delta G_{S,tot} = -xRT\ln\frac{C^r_{Al}}{C^0_{Al}} - zRT\ln\frac{C^r_{SO_4}}{C^0_{SO_4}} - nRT\ln\frac{C^r_K}{C^0_K} \tag{65}$$

In order to compare the $\Delta G_{S,tot}$ values for different minerals, the Gibbs energy, attributed to one mole of aluminium, $\Delta G'_{S,tot}$, has been computed:

$$\Delta G'_{S,tot} = -RT\ln\frac{C^r_{Al}}{C^0_{Al}} - \frac{z}{x}RT\ln\frac{C^r_{SO_4}}{C^0_{SO_4}} - \frac{n}{x}RT\ln\frac{C^r_K}{C^0_K} \tag{66}$$

The solid phase $(K,Na)_nAl_x(OH)_y(SO_4)_z(H_2O)_m$ is stable if $\Delta G_{S,tot}>0$. The condition $\Delta G_{S,tot}=0$ matches the start of mineral dissolution/precipitation.

The procedure of the *DHCE* construction includes the following essential steps (Povar & Rusu, 2012):

1. Thermodynamic calculation of the mineral stability area. Consequently, from equations (65) or (66), the values of the Gibbs energy for the analyzed process are calculated;
2. The partial molar fractions of all the species γ_i, containing the aluminium ion, for the thermodynamic stability area of mineral found in the previous step, are calculated by equations (64);

3. In order to finish the diagram, in the case of aqueous homogeneous solution, the partial molar fractions are computed from the equations used for the typical diagrams of species repartition. In this case, under the conditions of the mononuclear species formation, the γ values do not depend on the initial concentrations of components.

Thermodynamic Stability of Hydroxyl-Aluminium Sulfate Minerals

The constructed diagrams $\Delta G_{S,tot}(pH)$ easily permit to calculate the stability area of minerals in function of the initial composition of multicomponent systems. Fig. 1 and 2 shows the variation of the Gibbs energy versus pH in the "aqueous saturated solution – mineral" systems for diverse concentrations of individual components. These dependencies show that the aluminium residual concentration can vary within the great range for the same pH value. Thus, the pH value is not the single factor controlling the soluble aluminium concentration. The analysis of achieved data demonstrates that the thermodynamic stability areas of examined soil minerals become significantly larger at growing total (analytical) concentrations. Calculations in addition show that significant amounts of inorganic ligand F^-, as well as organic ligand Org^{3-} affect appreciably the solubility of minerals. Under the considered conditions alunogel is not thermodynamically stable. For elevated concentrations of the aluminum and sulfate ions $C^0_{Al} = C^0_{SO_4} = 1 \cdot 10^{-3}$ mol•L^{-1} (Fig. 1) and $C^0_{Al} = C^0_{SO_4} = 1 \cdot 10^{-2}$ mol•L^{-1} (Fig. 2) at different pH values, the condition $\Delta G_{S,tot} > 0$ happens for all the minerals, excluding for alunogel. One can distinguish the pH areas in which the aluminium minerals are thermodynamically stable, for example in Fig. 1: alunite (3.2 ÷ 8.6), basaluminite (4.0 ÷ 9.0), gibbsite (4.2 ÷ 10.6) and jurbanite (3.0 ÷ 6.8). Besides, the maximal thermodynamic stability of minerals varies as a function of the pH value for heterogeneous mixtures. In the pH range of 3.0 ÷ 5.0 the most stable mineral is jurbanite, next for the pH values of 5.2 ÷ 5.8 is basaluminite, and for the pH values bigger than 6.0 is gibbsite.

The analysis of data depicted on Fig. 1 and 2 exposes that thermodynamic stability area for the *Al* hydroxyl-sulfate minerals becomes significantly larger by raising the total (analytical) concentration of aluminium ion. Sulfate ion modifies as well considerably the *Al* solubility under acidic conditions, where less soluble minerals than gibbsite, control the aqueous geochemistry of aluminum. Our analysis proves that a considerable influence on the mineral solubility plays additionally the presence of large quantities of inorganic ligand $F^-$and organic ligand Org^{3-}. These minerals are alunite $KAl_3(OH)_6(SO_4)_2$, jurbanite $Al(OH)(SO_4)\cdot 5H_2O$ and basaluminite $Al_4(OH)_{10}SO_4\cdot 15H_2O$.

Our data confirmed the previous result (Nordstrom, 1982) that jurbanite is stable up to the pH range of 3–5 while alunite is stable at higher pH values than jurbanite, up to 4–7, depending on the sulfate activity (Fig. 1 and 2). For pH > 7.5 gibbsite is the most stable phase. The achieved facts, based on thermodynamic predictions, elucidate the following experimental outcome: (a) the raise in sulfate concentration in soil solutions leads to a considerable diminishing of the aluminium residual concentration and (b) high concentrations of SO_4^{2-} and basic metal ions take part in an important detoxification function for aquatic organisms (Bi et al., 2001).

Figure 1. Total Gibbs energy versus pH in the system "Mineral - saturated aqueous solution". Concentrations, mol·L⁻¹: $C^0_{Org} = C^0_{PO_4} = 1 \cdot 10^{-4}; C^0_F = 5 \cdot 10^{-6}; C^0_{Al} = C^0_{SO_4} = 1 \cdot 10^{-3}$

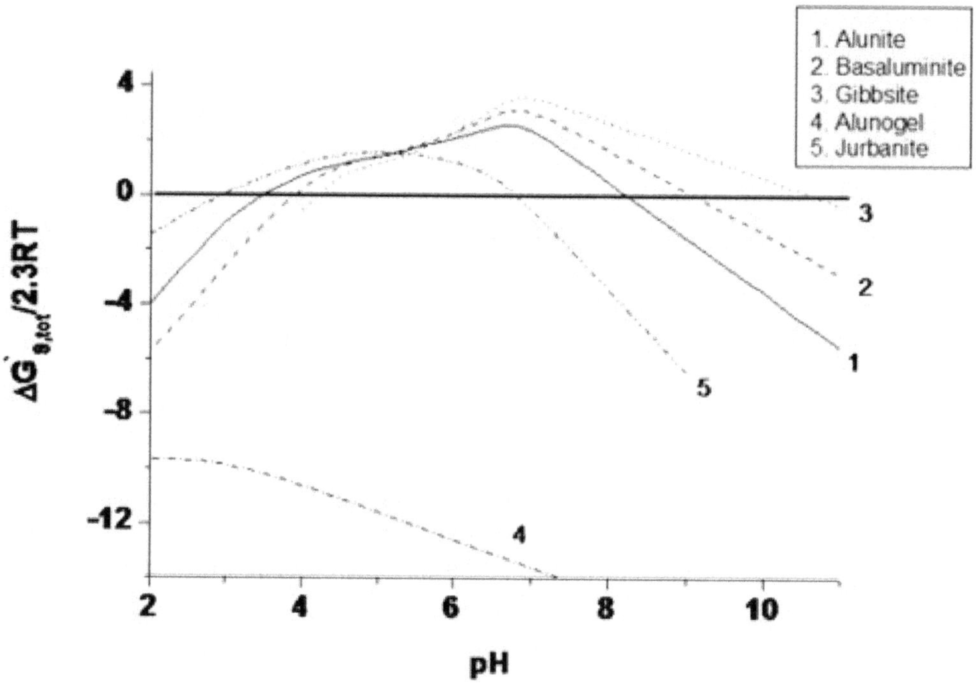

Figure 2. Total Gibbs energy versus pH in the system "Mineral - saturated aqueous solution". concentrations, mol·L⁻¹: $C^0_{Org} = C^0_{PO_4} = 1 \cdot 10^{-4}; C^0_F = 5 \cdot 10^{-6}; C^0_{Al} = C^0_{SO_4} = 1 \cdot 10^{-2}$

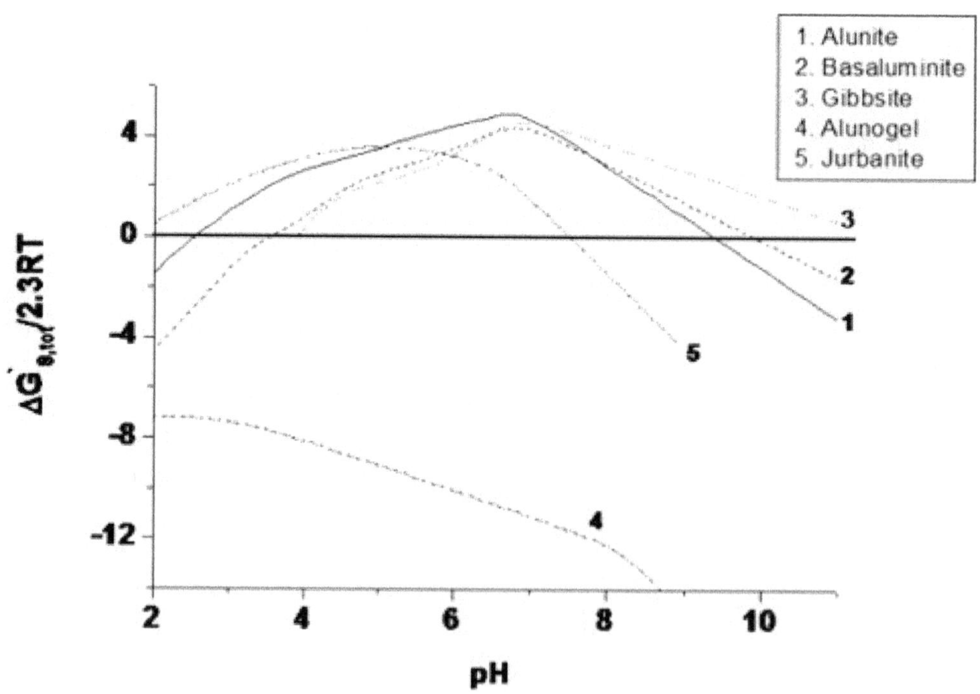

DIAGRAMS OF HETEROGENEOUS CHEMICAL EQUILIBRIA (DHCE) IN THE SYSTEMS "MINERAL – MULTICOMPONENT SATURATED AQUEOUS SOLUTIONS"

The System "Jurbanite – Aqueous Solution"

DHCE for the analyzed system are depicted in Figs. 3 and 4. The diagram reveals evidently the repartition of different aluminium chemical species both in the solid phase and saturated aqueous solutions. Partial molar fractions of soluble aluminium species are varying strongly at the pH value of dissolution – formation of the precipitate. For a certain total concentration of aluminium in acidic and neutral solutions, the equilibrium concentrations of mononuclear and polynuclear aluminium hydroxocomplexes are insignificant.

Figure 3. Diagrams of heterogeneous chemical equilibrium in the system "Jurbanite - saturated aqueous solution".

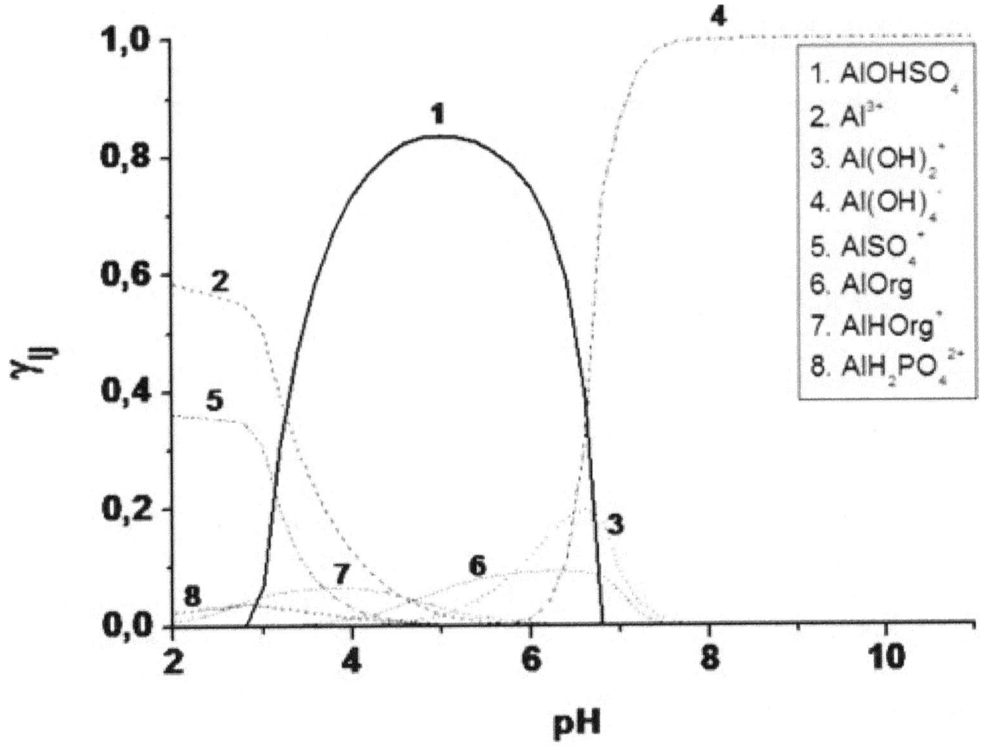

Figure 4. Diagrams of heterogeneous chemical equilibrium in the system "Jurbanite - saturated aqueous solution".

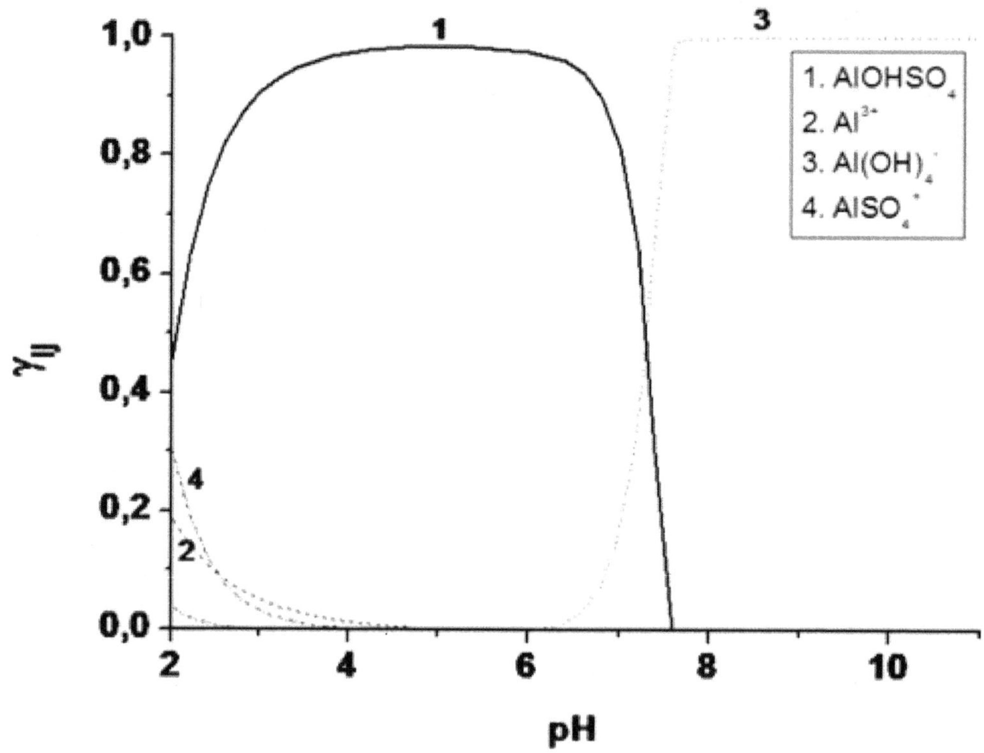

Concentrations, mol•L^{-1}: $C^0_{Org} = C^0_{PO_4} = 1 \cdot 10^{-4}; C^0_F = 5 \cdot 10^{-6}; C^0_{Al} = C^0_{SO_4} = 1 \cdot 10^{-3}$

Concentrations, mol•L^{-1}: $C^0_{Org} = C^0_{PO_4} = 1 \cdot 10^{-4}; C^0_F = 5 \cdot 10^{-6}; C^0_{Al} = C^0_{SO_4} = 1 \cdot 10^{-2}$

Al^{3+} activity is strongly interrelated with the pH soil solution and it amplifies as pH drops off. The existence of inorganic ligand, fluoride ion F^-, does not almost influence the solubility of jurbanite, while the organic ligand Org^{3-} has a modest influence. The solubility of jurbanite rises in alkaline medium. The anionic hydroxocomplex $Al(OH)_4^-$ in alkaline solutions becomes the predominant species. From Fig. 3 and 4 one can obviously distinguish that the thermodynamic region of jurbanite stability increases with the raise of the analytical concentrations of aluminium and sulfate ions in the heterogeneous mixture. In the case of $C^0_{Al} = C^0_{SO_4} = 1 \cdot 10^{-2}$ mol•L^{-1} in the strong acidic solutions, in the pH area up to 2.0 the principal species is $AlSO_4^-$.

The "Basaluminite – Aqueous Solution" System

Diagrams DHCE for diverse concentrations of sulfate are depicted in Figs. 5 - 8. The partial molar fractions of soluble aluminium species modify to a large degree at the pH value of precipitation - dissolution of the solid phase. For the considered mixture compositions in acidic and neutral solutions, the equilib-

Figure 5. Curves of distribution of the soluble species and solid phase in the system "basaluminite - saturated aqueous solution"

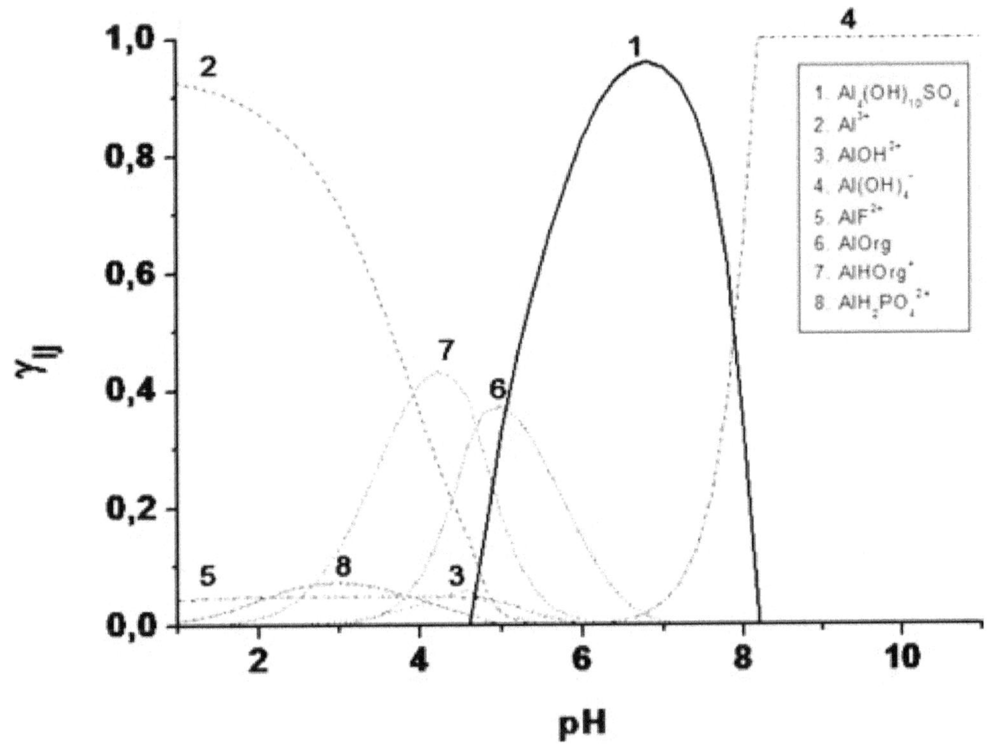

rium concentrations of mononuclear hydroxocomplexes are little, while the polynuclear complexes do not roughly form within the considered range of the variation of pH and the total sulfate concentration.

The concentration of the poisonous aluminium ion decreases considerably with rising $C^0_{SO_4}$ and for $C^0_{SO_4} = 1 \cdot 10^{-2}$ mol•L^{-1} becomes insignificant. The major species in alkaline solutions is the anionic hydroxocomplex $Al(OH)_4^-$. With the augment of the sulfate ion concentration a substantial change in the stability areas of other soluble species takes place. In the case $C^0_{SO_4} = 2.5 \cdot 10^{-5}$ mol•L^{-1} and $C^0_{SO_4} = 1 \cdot 10^{-4}$ mol•L^{-1}, within the range of pH 4.0 – 5.0 the complexes $AlOrg$ and $AlOrg^+$ predominate, as for $C^0_{SO_4} = 1 \cdot 10^{-3}$ mol•L^{-1} this range of pH narrows, whereas for $C^0_{SO_4} = 1 \cdot 10^{-2}$ mol•L^{-1} the formation of these complexes is stopped by the main complexes of aluminium with sulfate ion, $AlSO_4^+$ and $Al(SO_4)_2^-$ (Fig. 8). In Fig. 9 is depicted the variation of the partial molar fraction of the solid phase - basaluminite, in function of pH and $C^0_{SO_4}$. As one can view, the areas of the mineral formation, e.g. the areas of the heterogeneous system match with the areas of the thermodynamic stability of this mineral (Fig. 1), at the same time the values γ_S grow with the raise of $C^0_{SO_4}$.

concentrations, mol•L^{-1}: $C^0_{SO_4} = 2.5 \cdot 10^{-5}; C^0_{Al} = C^0_{Org} = C^0_{PO_4} = 1 \cdot 10^{-4}; C^0_F = 5 \cdot 10^{-6}$

Concentrations, mol•L^{-1}: $C^0_{SO_4} = 1 \cdot 10^{-4}; C^0_{Al} = C^0_{Org} = C^0_{PO_4} = 1 \cdot 10^{-4}; C^0_F = 5 \cdot 10^{-6}$

Figure 6. Curves of distribution of the soluble species and solid phase in the system "basaluminite – saturated aqueous solution".

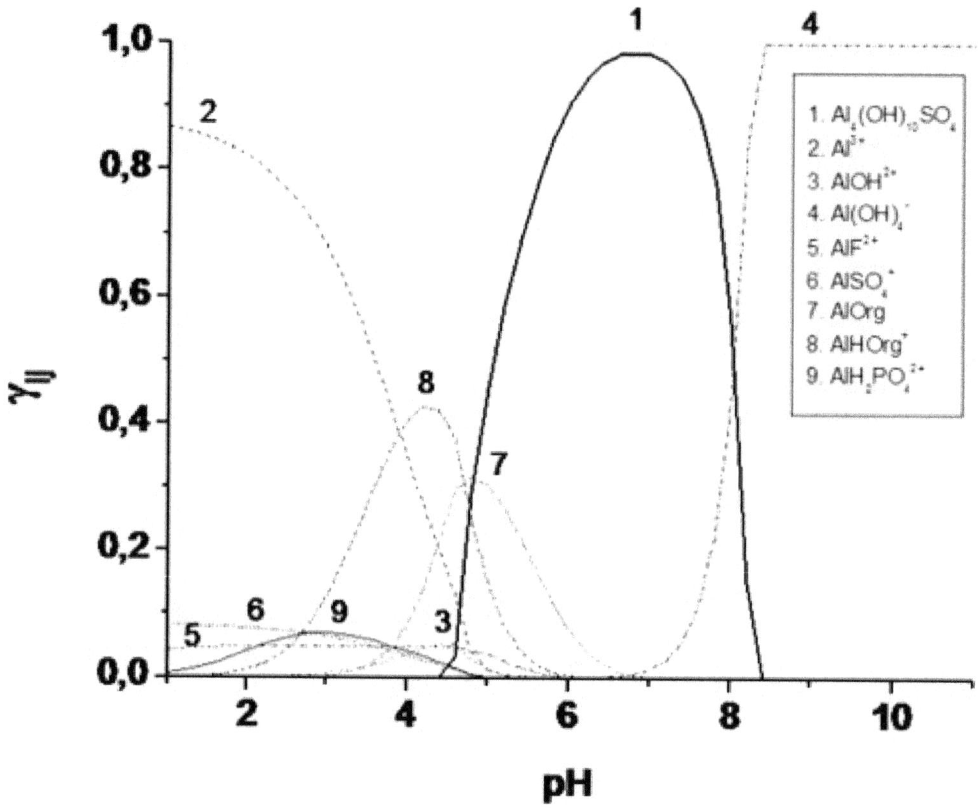

Concentrations, mol•L^{-1}: $C^0_{SO_4} = 1\cdot10^{-3}; C^0_{Al} = C^0_{Org} = C^0_{PO_4} = 1\cdot10^{-4}; C^0_F = 5\cdot10^{-6}$

Concentrations, mol•L^{-1}: $C^0_{SO_4} = 1\cdot10^{-2}; C^0_{Al} = C^0_{Org} = C^0_{PO_4} = 1\cdot10^{-4}; C^0_F = 5\cdot10^{-6}$

Concentrations, mol•L^{-1}: $C^0_{Al} = C^0_{Org} = C^0_{PO_4} = 1\cdot10^{-4}; C^0_F = 5\cdot10^{-6}$

The "Alunite – Aqueous Solution" System

DHCE for the analyzed system is illustrated in Fig. 10. The values of C^0_{Al} in acidic and neutral solutions the equilibrium concentrations of aluminium mononuclear and polynuclear hydroxocomplexes are irrelevantly small. The fraction of the free aluminium ion raises with the increase of C^0_{Al} in a narrowed area of the pH values of saturated solutions. With increasing the analytical concentration of the aluminium ion, the area of stability of the anionic hydroxocomplex $Al(OH)_4^-$ narrows, that becomes the largest species in the slightly alkaline and alkaline solutions. The analysis of the achieved results dem-

Figure 7. Curves of distribution of the soluble species and solid phase in the system "basaluminite – saturated aqueous solution".

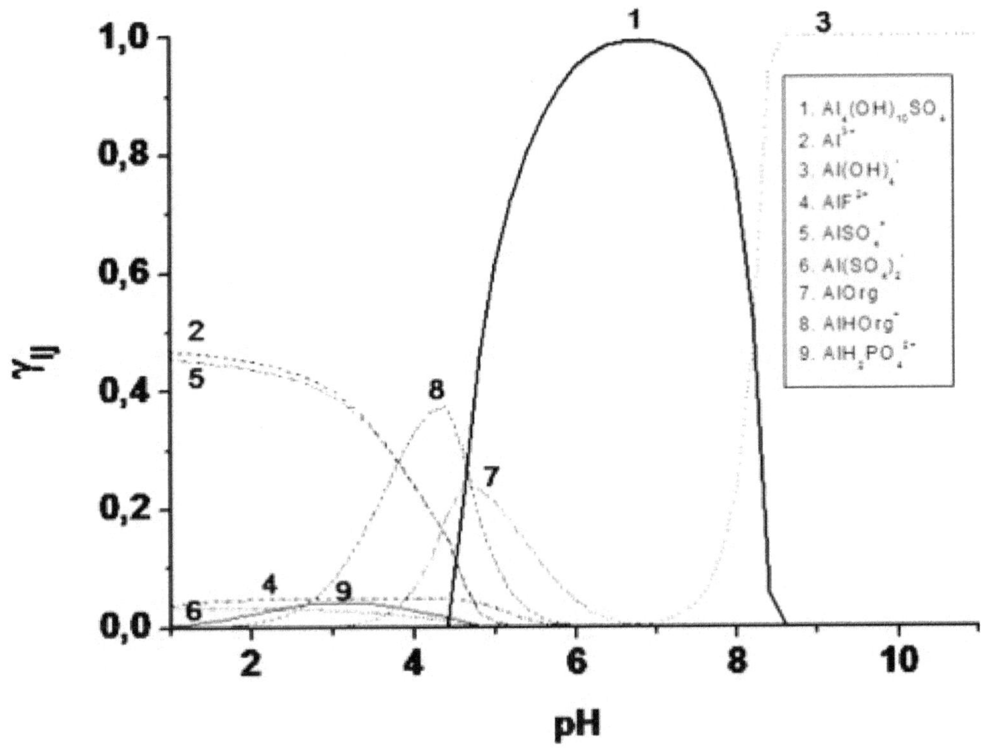

onstrate that for the concentration $C_F^0 = 5 \cdot 10^{-4}$ mol•L^{-1} the complexes AlF_2^+ and AlF_3 take a sizeable role. One can note down that the diagrams drawn here considerably diverge from diagrams (Bi et al., 2001), as the authors of cited work considered the repartition of inorganic and organic species only in homogeneous aqueous solutions. Our calculations prove that the complexes of aluminium with phosphate-ion are formed in irrelevant quantities for examined compositions. The aluminium sulfate complexes are predominant in very acidic and acidic solutions. This conclusion corresponds with that written by authors (Bi et al., 2001).

Concentrations, mol•L^{-1}: $C_{SO_4}^0 = C_K^0 = 5 \cdot 10^{-3}; C_{Al}^0 = C_{Org}^0 = C_{PO_4}^0 = 1 \cdot 10^{-4}; C_F^0 = 5 \cdot 10^{-6}$

SOLUTIONS AND RECOMMENDATIONS

The importance of chemical equilibria and chemical thermodynamics in general, results from the simple statement that many properties of compounds depend mostly on their interactions in a certain system. For example, their environmental impact can be explained by a detailed study of these interactions, whose nature and power can be evaluated by thermodynamic study of chemical equilibrium. As, the speciation modelling based on chemical equilibrium data is regularly used to investigate the mobility

Figure 8. Curves of distribution of the soluble species and solid phase in the system "basaluminite – saturated aqueous solution"

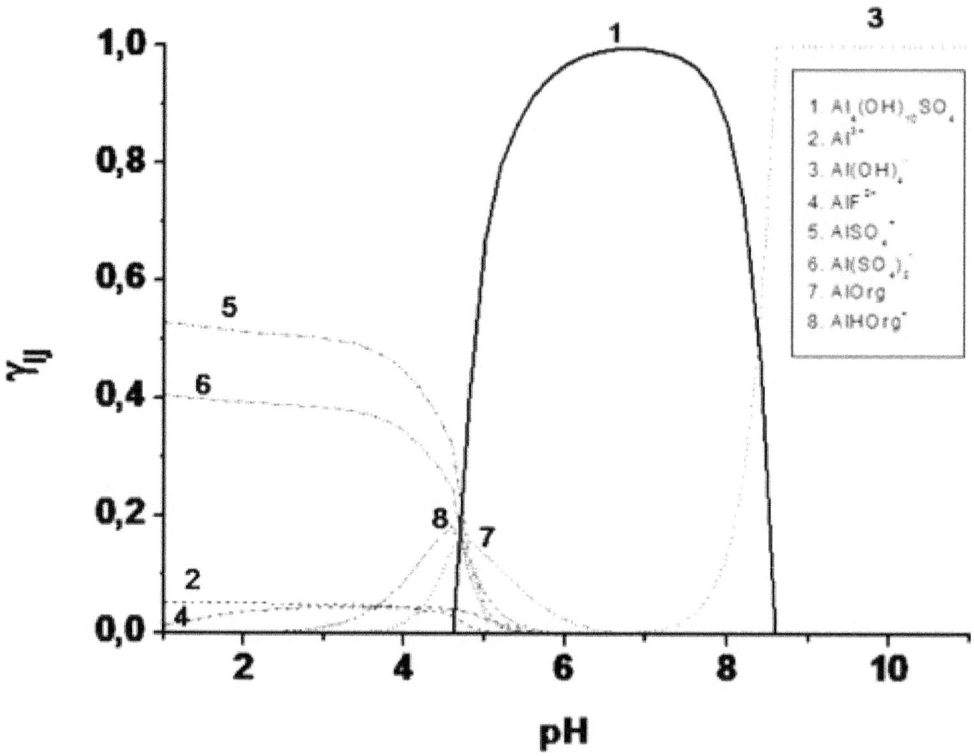

of pollutants and toxicants in the environment. Additionally, complex thermodynamic studies provide deeper insights into the mechanisms of these interactions. The developed approach in the paper requires the most reliable, accurate and comprehensive thermodynamic data as the standard formation energies of the chemical species or the equilibrium constants. As new thermodynamic data accumulates, it is possible to analyze new complex natural systems. However, a series of actual published data has not been critically analyzed and, consequently, do not ensure proper interpretations of chemical processes. Stricly speaking, it is necessary to get high-quality thermodynamic data in order to obtain high-quality outcome.

FUTURE RESEARCH DIRECTIONS

The following aspects in chemical thermodynamics, scarce considered earlier, require more profound examination, as: a) the study of competing reactions in real systems; b) an accurate analysis of literature data; c) building and/or accomplishment of appropriate general databases; and d) comparable investigation of other parameters than those of concern. Contacts and partnerships with stakeholders will amplify academia - non-academic interactions, valuable for innovative ideas and discoveries, to safeguard the intellectual property rights and to market research products.

Figure 9. Partial molar fraction of basaluminite versus pH for different $C^0_{SO_4}$ *value*

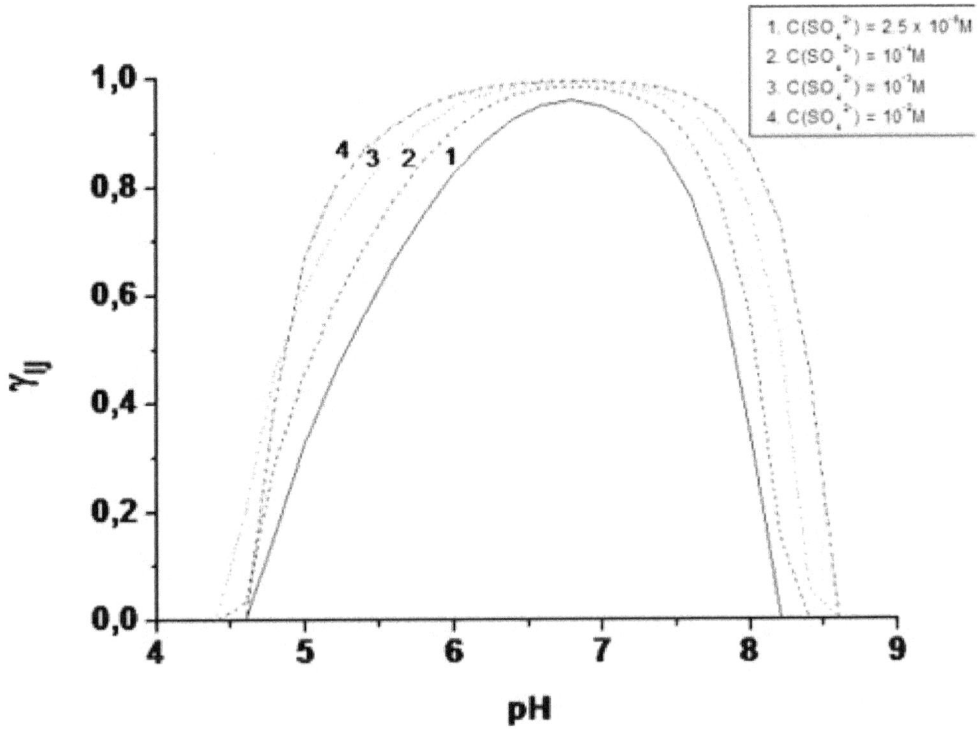

CONCLUSION

Chemical thermodynamics regulate all chemical transformations; understanding equilibrium thermodynamics of chemical processes is extremely essential for the full comprehension of the phenomena that occur in nature. In this paper, a thermodynamic approach has been developed for the study of complex chemical equilibria in heterogeneous multicomponent systems mineral – natural waters. The complex processes in the investigated systems are described by the generalized equation, which takes into account the diversity of possible chemical equilibria that occur simultaneously in large areas of variation of component concentrations and the pH of the environment. As stoichiometric coefficients in the generalized equation are the partial molar fractions of the respective species. A specific feature of these equations is that the stoichiometric coefficients of these equations are functions of temperature and chemical composition. Based on the thermodynamic analysis of the generalized equations, the expressions for the total Gibbs energy variation of the precipitation process - dissolution of the solid phases in the multicomponent systems under real conditions have been obtained for systems of different degree of complexity.

A novel type of diagrams (DHCE) based on graphical and numerical methods, which quantitatively describe the distribution of soluble and insoluble, inorganic and organic, monomeric and polymeric aluminium species in heterogeneous multicomponent systems "mineral – soil solution" has been used. This approach utilizes the thermodynamic relationships together with original mass balance restraints, where the mineral phases are clearly expressed. The factors influencing the distribution of soluble and insoluble aluminium species in aquatic systems in a large range of pH values have been investigated.

Figure 10. Diagrams of heterogeneous chemical equilibria in function of pH in the system "alunite - saturated aqueous solution".

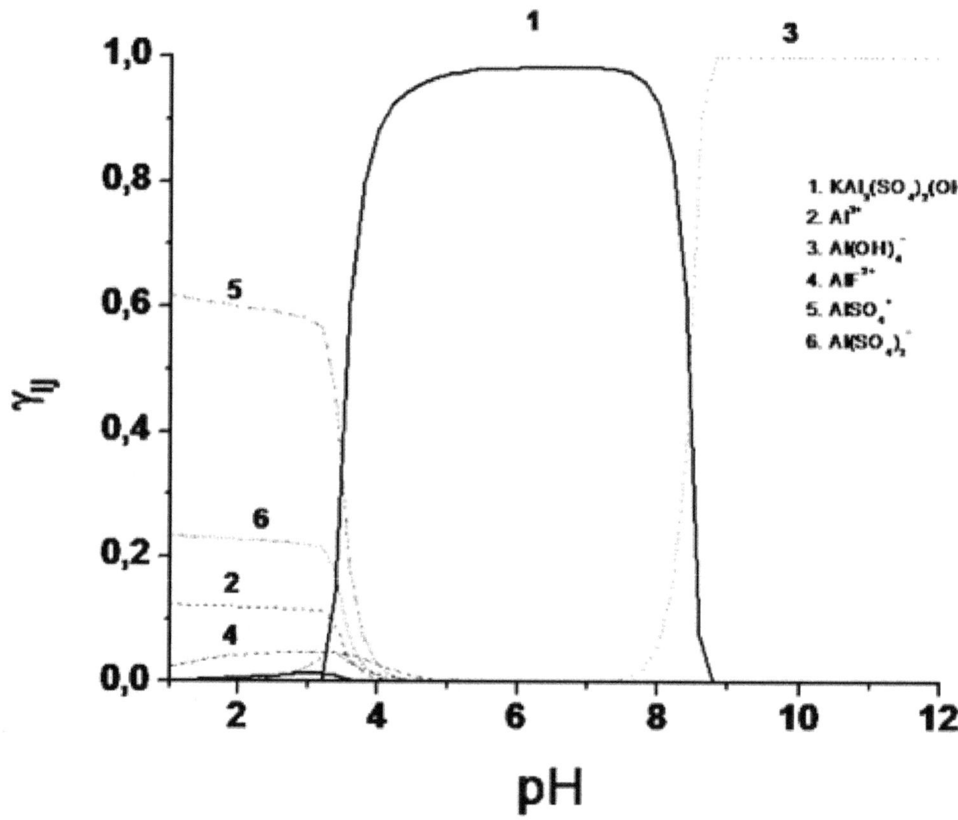

The areas of thermodynamic stability of different aluminium-containing minerals with the general composition $(K,Na)_n Al_x(OH)_y(SO_4)_z(H_2O)_m$ have been determined for a large area of pH variation and different values of aluminium and sulfate concentrations. Using data from several published sources, the developed thermodynamic approach demonstrates how the change in soil solution composition affects Al solution chemistry in equilibrium with HAS minerals. It has been proved that the SO_4^{2-} concentration has a substantial influence on the distribution of aluminium speciation in acidic soils, while the concentrations of organically complexed and fluoride complexed Al are minimal in the pH range studied. Our thermodynamic analysis for a series of different soil solution compositions shows that at decreasing the SO_4^{2-} concentration, the amounts of toxic forms of $Al^{3+}, AlOH^{2+}$ and $Al(OH)_2^+$ are greatly increased while the concentrations of $Al-F$, $Al-SO_4$ and $Al-Org$ are reduced. When the pH of a solution is raised above 4.0, Al^{3+} forms the mononuclear hydroxo-species and complexes with inorganic ligands such as sulfate and fluoride and also with organic compounds. The obtained data regarding the factors influencing Al speciation, based on the constructed DHCE diagrams, are in good agreement with the previous results obtained by other authors. The results presented here suggest that the formation of HAS minerals play a major role in sulfate retention in S-impacted soils. Based on our calculations of the soil solution composition, alunite, basaluminite and jurbanite are predicted to form in acidic soils and control the aluminium ion concentration level, for various soil environments.

It is proved that the mobility and toxicity of aluminium in soils may be enhanced by the acid deposition. As the soil pH decreases, aluminum is solubilized and the proportion of phytotoxic aluminium ions increases in the soil solution. The thermodynamic approach developed and used in this paper contributes to elucidate the mechanism of controlling aluminum concentration in dilute acid soil solutions. This approach is intended to determine the dominant processes that are responsible for the aluminium ion concentration levels in soils by HAS minerals. The developed new type of diagrams can be also used to interpret data obtained within the framework of water quality monitoring programs.

REFERENCES

Adams, F., & Rawajfih, Y. (1977). Basaluminite and Alunite: A possible cause of sulfate retention by acid soils. *Soil Science Society of America Journal, 41*(4), 686–692. doi:10.2136ssaj1977.036159950 04100040013x

Appelo, C. A. J., & Postma, D. (2005). *Geochemistry, Groundwater and Pollution* (2nd ed.). Leiden, Netherlands: CRC Press.

Bi, S., An, S. Q., Tang, W., Yang, M., Qian, H. F., & Wang, C. (2001). Modeling the distribution of aluminium speciation in acid soil solution equilibria with the mineral phase alunite. *Environmental Geology, 41*(1-2), 25–36. doi:10.1007002540100381

Bouza, P. J., Ríos, I., Idaszkin, Y. L., & Bortolus, A. (2019). Patagonian salt marsh soils and oxidizable pedogenic pyrite: Solid phases controlling aluminum and iron contents in acidic soil solutions. *Environmental Earth Sciences, 78*(1), 2. doi:10.100712665-018-7991-4

Butler, J. N. (1998). *Ionic equilibrium: solubility and pH calculations*. New York: Wiley.

Cameron, R. C., Ritchie, G. S. P., & Robson, A. D. (1986). Relative toxicities of inorganic aluminum complexes to barley. *Soil Science Society of America Journal, 50*(5), 1231–1236. doi:10.2136ssaj1986 .03615995005000050029x

Chen, F., Ai, H., Wei, M., Qin, C., Feng, Y., Ran, S., ... Chen, L. (2018). Distribution and phytotoxicity of soil labile aluminum fractions and aluminum species in soil water extracts and their effects on tall fescue. *Ecotoxicology and Environmental Safety, 163*, 180–187. doi:10.1016/j.ecoenv.2018.07.075 PMID:30053588

Delfosse, T., Elsass, F., & Delvaux, B. (2005). Direct evidence of basic aluminium sulphate minerals in an S-impacted Andosol. *European Journal of Soil Science, 56*(3), 281–286. doi:10.1111/j.1365-2389.2004.00675.x

Driscoll, C. T., & Schecher, W. D. (1990). The chemistry of Al in the environment. *Environmental Geochemistry and Health, 12*(1-2), 28–50. doi:10.1007/BF01734046 PMID:24202563

Dutova, E. M., & Nikitenko, A. N. (2010). Physico-chemical modeling of the behavior of uranium in the system "water-granitoids". [In Russian]. *Bulletin of Tomsk State University, 3*, 202–207.

Dutta, S., & Lodhari, D. R. (2018). *Extraction of nuclear and non-ferrous metals*. Singapore: Springer. doi:10.1007/978-981-10-5172-2

Fishtik, I., & Povar, I. (1987). Side reactions in calculating the thermodynamic functions for the process of formation – dissolution of salt precipitates. [Engl. Transl.]. *Russian Journal of General Chemistry*, *5*, 25–30.

Fishtik, I., Povar, I., & Vataman, I. (1987). Thermodynamics of complex chemical equilibria in the system "precipitate of salt – aqueous solution". The formation of polynuclear species. [In Russian]. *Russian Journal of General Chemistry*, *57*, 736–742.

Fishtik, I. F., Povar, I. G., & Vataman, I. I. (1986). Thermodynamics of Precipitation and Dissolution of Oxides and Hydroxides in Aqueous Solutions. *Bulletin of Academy of Sciences of MRSS* [In Russian]. *Series of Biological and Chemical Sciences*, *5*, 58–62.

Jones, A. M., Collins, R. N., & Waite, T. D. (2011). Mineral species control of aluminum solubility in sulfate-rich acidic waters. *Geochimica et Cosmochimica Acta*, *75*(4), 965–977. doi:10.1016/j.gca.2010.12.001

Karak, T., Paul, R. K., Das, D. K., & Boruah, R. K. (2016). Complexation of DTPA and EDTA with Cd 2+: Stability constants and thermodynamic parameters at the soil-water interface. *Environmental Monitoring and Assessment*, *188*(12), 670. doi:10.100710661-016-5685-5 PMID:27848112

Larssen, T., Jiling, X., Vogt, R. D., Seip, H. M., Bohan, L., & Dianwu, Z. (1998). Studies of soils, soil water and stream water at a small catchment near Guiyang, China. *Water, Air, and Soil Pollution*, *101*(1-4), 137–162. doi:10.1023/A:1004985209931

Li, X., Yang, L., Zhou, Q., Qi, T., Liu, G., & Peng, Z. (2020). A split-combination method for estimating the thermodynamic properties (Gfo and Hfo) of multicomponent minerals. *Applied Clay Science*, *185*, 105406. doi:10.1016/j.clay.2019.105406

Nordstrom, D. K. (1982). The effect of sulfate on aluminum concentrations in natural waters: Some stability relations in the system Al_2O_3-SO_3-H_2O at 298 K. *Geochimica et Cosmochimica Acta*, *46*(4), 681–692. doi:10.1016/0016-7037(82)90168-5

Povar, I. (1994). Thermodynamic calculation of the minimum solubility pH of slightly soluble oxides and hydroxides in polynuclear hydrolysis of metal ion. [In Russian]. *Ukrainian Chemistry Journal*, *60*, 371–378.

Povar, I. (1997). Method for graphic representation of heterogeneous chemical equilibria in systems sparingly soluble compound-complexing agent-aqueous solution. [Engl. Transl.]. *Russian Journal of Inorganic Chemistry*, *42*, 607–612.

Povar, I. (2000). Potentiometric determination of solubility products and equilibrium ion concentrations for sparingly soluble salts. *Russian Journal of General Chemistry*, *70*, 501–507.

Povar, I. (2001). Determination of the stability of slightly soluble complexonate from pH metric data. *Canadian Journal of Chemistry*, *79*, 1166–1172. doi:10.1139/v01-065

Povar, I., Fishtik, I., & Vataman, I. (1989). Calculation of optimal conditions of underground leaching of uranium ores. *Bulletin of Academy of Sciences of MRSS* [In Russian]. *Series of Biological and Chemical Sciences*, *5*, 57–61.

Povar, I., & Rusu, V. (2012). Aluminium heterogeneous speciation in natural waters. *Canadian Journal of Chemistry, 290*(4), 326–332. doi:10.1139/v2012-003

Sánchez-España, J., Yusta, I., & Diez-Ercilla, M. (2011). Schwertmannite and hydrobasaluminite: A re-evaluation of their solubility and control on the iron and aluminium concentration in acidic pit lakes. *Applied Geochemistry, 26*(9-10), 1752–1774. doi:10.1016/j.apgeochem.2011.06.020

Seravkin, I. B., Kosarev, A. M., & Puchkov, V. N. (2017). Geodynamic conditions of formation of massive sulfide deposits in the Magnitogorsk Megazone, Southern Urals, and prospection criteria. *Geology of Ore Deposits, 59*(3), 227–243. doi:10.1134/S1075701517030059

Singh, S., Tripathi, D. K., Singh, S., Sharma, S., Dubey, N. K., Chauhan, D. K., & Vaculík, M. (2017). Toxicity of aluminium on various levels of plant cells and organism: A review. *Environmental and Experimental Botany, 137*, 177–193. doi:10.1016/j.envexpbot.2017.01.005

Stoffregen, R. E., & Cydan, G. L. (1990). An experimental study of Na-K exchange between alunite and aqueous sulfate solution. *The American Mineralogist, 75*, 209–220.

Stumm, W., & Morgan, J. J. (1996). *Aquatic Chemistry: Chemical Equilibria and Rates in Natural Waters*. New York: John Wiley & Sons.

Vilor, N. V., & Kuzmin, L. A. (2007). Petrology, geochemistry and mineralogy, the use of physical and chemical modeling in the study of sulfoarsenid complexes in hydrothermal solutions. *Geology and Geophysics, 48*, 588. (in Russian)

Yakymchuk, C. (2017). Applying phase equilibria modelling to metamorphic and geological processes: Recent developments and future potential. *Geoscience Canada: Journal of the Geological Association of Canada, 44*(1), 27–45. doi:10.12789/geocanj.2017.44.114

ADDITIONAL READING

Astarita, G., & Ocone, R. (1989). Heterogeneous chemical equilibria in multicomponent mixtures. *Chemical Engineering Science, 44*(10), 2323–2331. doi:10.1016/0009-2509(89)85165-6

Burgot, J. L. (2012). *Ionic equilibria in analytical chemistry*. Springer Science & Business Media. doi:10.1007/978-1-4419-8382-4

Eriksson, G., & Hack, K. (1990). ChemSage - a computer program for the calculation of complex chemical equilibria. *Metallurgical Transactions. B, Process Metallurgy, 21*(6), 1013–1023. doi:10.1007/BF02670272

Ghiorso, M. S. (1994). Algorithms for the estimation of phase stability in heterogeneous thermodynamic systems. *Geochimica et Cosmochimica Acta, 58*(24), 5489–5501. doi:10.1016/0016-7037(94)90245-3

Povar, I., & Spînu, O. (2013). Thermodynamic calculations of complex chemical equilibria in soils containing aluminium minerals. *Lucrări Ştiinţifice, Universitatea de Ştiinţe Agricole Şi Medicină Veterinară" Ion Ionescu de la Brad" Iaşi. Seria Horticultură, 56*(2), 543–548.

Predel, B., Hoch, M., & Pool, M. J. (2013). *Phase diagrams and heterogeneous equilibria: a practical introduction*. Springer Science & Business Media.

Sposito, G. (2008). *The chemistry of soils*. Oxford university press.

Ung, S., & Doherty, M. F. (1995). Theory of phase equilibria in multireaction systems. *Chemical Engineering Science, 50*(20), 3201–3216. doi:10.1016/0009-2509(95)00159-3

KEY TERMS AND DEFINITIONS

Chemical Species: Atoms, molecules, molecular fragments, ions, etc., subjected to a chemical process or to a measurement.

Generalized Equation of Reaction: It describes the overall process including many simultaneous reactions. Its stoichiometric coefficients, being partial molar fractions, are variable quantities depending on the chemical composition and temperature of the multicomponent system.

Mineral: A naturally occurring chemical compound, usually of crystalline form and abiogenic in origin. A mineral has one specific chemical composition.

Molar Fraction: It is a unit of concentration, defined to be equal to the number of moles of a component divided by the total number of moles of a solution. Because it is a ratio, mole fraction is a unitless expression. The mole fraction of all components of a solution, when added together, is equal 1.

Remediation: Efforts to counteract some or all of the effects of pollution after it has been released into an environment. Also, to clean a site that has been damaged by industry that is physical, chemical, or biological in origin.

Residual Concentration: The difference between the initial concentration of species in the solution and its quantity in mol in the solid phase per one liter of saturated solution. After reaching the equilibrium state in the heterogeneous mixture, the solid phase is separated from the liquid one, and then the residual concentration is measured.

Solubility Product: The solubility product of a compound at a given temperature is equal to the product of the concentrations of its ions in the saturated solution, with each concentration term raised to the power to the number of ions produced on dissociation of one mole of the substance.

Sparingly Soluble Compound: It precipitates up to reaching equilibrium between the produced solid phase and its ions in the solution.

Stoichiometric Coefficient: A number of species appearing before the symbol for each species in the equation for a chemical reaction. By convention, it is negative for reactants and positive for products.

ENDNOTE

1 Hereinafter the charges of species for convenience are omitted.

Chapter 5
Kinetics and Mechanism of Ecochemical Processes

Sergey Olegovich Travin
Federal Research Center for Chemical Physics After N. N. Semenov of the Russian Academy of Sciences, Russia

ABSTRACT

Significant efforts of mankind and huge funds were spent to study the mechanisms of environmental processes. Recent decades have been marked by exponential growth of computer power and the accompanying decrease in the cost of computing. With regard to the mathematical modeling of physical and chemical processes that determine the quality of natural waters, atmosphere, and soil, this has led to the development of an extensive approach based on an increase in the number of components and reactions between them taken into consideration. In this chapter, the authors describe features of ecochemical systems and discuss the moments that complicates their prediction. Using the method of numerical experiment, they investigate the behavior of periodic systems with multiple stationary states. One conclusion is that the actual manifestation cannot be used to determine at what stage the impact occurred and to what stage of the food chain it relates. Another conclusion is that systems involving multiple stationary states are prone to bifurcations and chaotic jumps from one limit cycle to another.

INTRODUCTION

"The main task of a chemist, as I imagine it, is the ability to predict and control the course of reactions. However, as with any other attempt of man to master the laws of nature, two approaches can be used. One is to create general theories, from which the consequences are deduced, concerning particular properties of matter. The second, based on empirical generalizations, builds partial and approximate theories that can explain the observed phenomena or suggest an interesting direction of experimental research. Because of the nature of our science, we, chemists, are forced to go mainly on the second way."(Hammett, 1970). These words of L. Hammett are the best suited to the section of chemistry that is associated with the description of reactions in the atmosphere, natural waters and the earth's crust, i.e. to ecochemical processes.

DOI: 10.4018/978-1-7998-1241-8.ch005

Despite the rapid development of computer technology, the intensive penetration of mathematical modeling methods into practical chemical kinetics has minimally affected ecochemical systems. Indeed, the number of really theoretical works in which the deduction of the properties of the system is based on the first principles is extremely small. To such fundamental first principles, undoubtedly, we should include the law of mass action, the Arrhenius equation and the vast accumulated material on the rate and equilibrium constants of individual elementary stages. It would seem that this set of valuable information is directly intended for modeling and forecasting natural ecosystems, but a whole set of circumstances that distinguish natural processes from technological ones makes it difficult to describe them in detail.

Here are the main differences Let's enumerate the main differences between ecochemical processes and canonical ideas about how chemical reactions occur that are included in all the courses of chemical kinetics taught in universities.

EMPIRICAL REGULARITIES

1. **Endless Duration:** Almost all the cases studied in textbooks boil down to the fact that the reaction has some starting point, when the reagents come into contact and begin to interact with each other. And there is the time of the end of the reaction (theoretically infinite, and practically making up several periods of half-transformation), for which the reaction terminates more than 99%. This means that with acceptable accuracy, the one of the reagents that was in the stoichiometric deficiency was completely consumed, and the only reagent residues that was in excess and the reaction products remained in the system. Unlike such refined examples, ecochemical processes began not yesterday and will not end tomorrow. The principal feature of processes in the environment is their "eternity" in the scale of human life and the comparability of their time with the age of the Earth. In this case, the meaning of the term "half-transformation time" for most of the components involved is largely lost. It is more appropriate to talk about the characteristic response time of a natural system to an impulse load.
2. **Open Systems:** The infinity of the temporal course of natural processes is directly related to the fact that all ecochemical systems are open. For any of the components included in the natural system, in addition to endogenous reactions, i.e. occurring under the internal laws of the system, there are also exogenous processes associated with the inflow from the external reservoirs or outflow in them. If the "usual" laboratory or industrial chemical reactions proceed from the initial state to the equilibrium state (even if it is strongly shifted toward the products), in open systems the concentrations of the components relax to a steady state. And since this relaxation proceeds continuously and infinitely, it is more correct to speak of the evolution of a quasistationary state. It can be argued that on a long timeline that exceeds the period of seasonal variations of components, the system is simply stationary, and the average rate of change in the concentration of any component tends to zero:
$< W_i > = \frac{1}{T}\int_0^T w_i(t)\,dt = \frac{1}{T}\int_0^T \frac{dc_i}{dt}\,dt \to 0 ?@8 T \to \infty$. Separately should also mention the photochemical initiation. In natural aquatic environments it is, primarily, the generation of OH free radicals by photolysis of hydrogen peroxide; in the atmosphere - the generation of reactive atoms and ions. That solar radiation and flux of cosmic particles form ozone layer, trigger processes of photosynthesis and are responsible for self-purification processes of natural objects.

3. Quasistationary processes proceeding infinitely long, **must be cyclic**. Indeed, if the evolution of the concentrations were drifting, then for a long time of observation, all substances drifting toward decreasing concentrations would simply disappear, and out of all substances that tend to accumulate, eventually one would dominate. As an ideal hypothetical variant, we could consider the situation when the total (over all channels of inflow and outflow) the rate of a quasistationary process is identically zero. In such case the quasistationary process becomes truly stationary, none of the parameters change. It would seem that this was the way it was, for example, with the weather. To the undoubted joy of all the numerous forecast offices, for the multi-billion-year existence of the Earth on the whole planet, an average temperature, a continuous very shallow rain, providing an average annual amount of precipitation, and a weak wind blowing simultaneously in all directions, i.e. any. Fortunately, nature is arranged differently and this does not happen.

   All natural processes have aperiodic cyclicity. We can note the main cycles that have a periodic character. This is the diurnal variation of parameters associated with the natural change in insolation and seasonal oscillations associated with weather (temperature) changes. But they are subject to fluctuations of natural systems that do not coincide with the above-mentioned modulating factors in frequency.
4. All natural processes obey the principle of **homeostasis**. This term is understood as the maintenance of the main indicators of the environment in certain permissible limits. With the exception of relatively rare cataclysms, natural ecosystems have the property of self-regulation, i.e. being open, the system retains the ability to maintain an approximate constancy of its internal composition, despite possible, sometimes very significant, fluctuations in the rates of reactions in the channels of operating time or the death of individual components.

   In fact, it is a kind of the Le Chatelier principle, according to which impulsive action on a system in a state of equilibrium leads to a shift in the equilibrium towards the process, the course of which weakens the effect produced. It is important to emphasize that in ecological systems, speech is always not about maintaining or shifting the equilibrium, but about maintaining a dynamic quasi-stationary state through the response of coordinated reactions.

   The desire of the system to reproduce itself, restore the state lost through impulsive disturbance, overcoming the resistance of the external environment is typical for both chemical and biological components of the ecosystem. For example, the oceanic carbonate buffer maintains the pH of the sea water at approximately constant level with a multi times change in the carbon dioxide content in the atmosphere, and the biological part of the homeostasis results in the ability of the population of phyto- and zooplankton, as well as organisms that build shells from calcium carbonate, to maintain a certain number Their own individuals for a long time. At the same time, the system itself is flowing, open and far from equilibrium, so most often the response to external influences is unpredictable.

   At the same time, scientific and, especially, popular literature is replete with publications in which, when trying to model natural ecosystems, the principle of homeostasis is completely ignored, which ultimately leads to the construction of completely ridiculous scientifically not valid and non-prognostic models. Later we will return to the demonstration of the homeostasis principle by the example of generation, migration and transformation of atmospheric ozone.
5. Ecochemical reactions are always associated with **mass transfer processes**. Typically, textbooks deal with two types of reactors: ideal mixing and ideal displacement. All "test tubes" reactions, i.e. studied in the laboratory, belong to the first type. No one is interested in complicating the

study of the chemical mechanism by additional diffusion and mass transfer processes. After mixing the reactive components - the time went. And whether it is a slow reaction with sampling and chromatographic analysis of the composition of the reagents / products at selected times, or a fast reaction, which is studied by the stop-flow method, in all cases try to achieve the most uniform mixture of the reagents so that at all points of the reactor everything happens the same. In technological industrial reactors, the regime of ideal displacement is sometimes realized. The time from the beginning of the process is proportional to the distance from the entrance to the reactor. Thus, there is no fundamental difference in the construction of models and in the mathematical description of reactors of ideal mixing or displacement.

6. Even the smallest fragments of natural ecosystems are **associated with large reservoirs**. It should be noted that in the course of hundreds of millions and even billions of years, natural storage reservoirs have formed for the main elements that determine the properties of ecosystems. This is a natural consequence of the fact that reactive forms of chemical substances with high free energy inevitably transform into essentially less reactive forms, in which the main stock is contained. Consider this principle with the example of four fundamental macroelements, of which the protein form of living organisms (polyamine acids) basically consists. They are carbon, oxygen, nitrogen and hydrogen.

For example, oxygen only in a small fraction is represented in gaseous form in the atmosphere ($1,2 \cdot 10^{15}$ .ons or less than 0.01%), and the largest reservoir of oxygen on Earth is silicates and oxides in the bark and mantle (99.5%). This is the most common element available on the earth (about half by weight), the total amount of which and the earth's crust is estimated at $1,3 \cdot 10^{19}$ .ons. Oceans in the form of water have an order of magnitude less oxygen than in the crust (about $1,2 \cdot 10^{18}$ .ons), but three orders of magnitude more than in the atmosphere. We emphasize that only the atmospheric part of oxygen that is suitable for breathing takes part in the chemical circuit, whereas the water-bound form of it does not possess reactivity in most ecochemical processes, and it should be considered as a reserve stock of the element. This applies even more to inorganic oxides and silicates, in which oxygen is immobilized even more and in the exchange processes on the time scale of the duration of human life is practically not involved.

The situation with nitrogen, which makes up a large part of the earth's atmosphere (75.6% by mass or 78.1% by moles, i.e. by volume), was a little different. In the gaseous state there are about $3{,}9 \cdot 10^{15}$tons of nitrogen, whereas in the earth's crust its content is $(0,7-1,5) \cdot 10^{15}$. (in humus it is about $6 \cdot 10^{10}$ .tons), and in the Earth's mantle - $1,3 \cdot 10^{16}$ .ons. Such a mass ratio without a strong skew towards the reservoir reserve ("only" four times) is due to the extremely low reactivity of free N_2, which does not support neither burning nor breathing. The name azote which is not accidentally offered Lavoisier occurs from Greek ἄζωτος – lifeless. Apparently, the main source of nitrogen is the upper part of the mantle, from where it enters other shells of the Earth with eruptions of volcanoes.

The next of the "life-building" elements, carbon in relatively small amounts is bound in the form of organic compounds, but is abundantly represented as atmospheric carbon dioxide, oceanic carbonate buffer (solutions of carbonic acid) and limestone sedimentary rocks, mainly in the form of calcite (chalk, marble) and aragonite, of which a pearl layer of mollusk shells and an exoskeleton of corals consists of. Aragonite is an unstable phase of calcium carbonate, and "only" over a period of time of about 10 to 100 million years is modified to calcite. The ratio between the atmospheric, oceanic and lithospheric carbon dioxide content is 610:40552:14000000 (figures are given for the amounts of pure carbon in gig tons, according to Galimov (1968) with the additions by Golyshev, 2018)). Finally, hydrogen, in a free form does not occur on Earth. The mass fraction of hydrogen in the earth's crust is 1% - this is the tenth most common element. However, its role in nature is determined not by mass, but by the number of atoms whose fraction among the remaining elements is 17% (second after oxygen, whose fraction of atoms is ~ 52%). For the time of the Earth's existence, hydrogen has almost completely become oxidized. Hydrogen is part of almost all organic substances and is present in all living cells. In living cells, the hydrogen atoms fraction is almost 63%

It may be noted that the amount of carbon in the oceanic carbonate buffer, with accuracy about factor "2", is the geometric average between its amounts in the atmosphere and in solid sedimentary rocks. This is not a "mysterious magic of numbers". The result is quite natural. Since free energy is always related to the logarithm of the amount of matter, it is advisable to make a macro evaluation of global ecochemical processes exactly in order of magnitude, in logarithms of quantitative indices of content. With this representation, the dissolved in the oceans of carbonic acid (in all its dissociated forms) is approximately halfway of free energy degradation of this element (carbon). A similar approach applied to other vital elements shows that for nitrogen, with its exceptionally low reactivity, free energy did not degrade, and the amount of atmospheric N_2 is comparable to its content as solid compounds in the interior of the Earth. On the contrary, for a hydrogen that has a pronounced reducing ability, it almost completely transforms into stable compounds, primarily water and hydroxides. In any case, the element # 1 most widely distributed in the universe in the atmosphere (and other "spheres" of the Earth) is practically absent in the free elemental form. Nevertheless, in terms of atoms (that is, after correction for light atomic weight), hydrogen is 17% of the Earth's crust, although in a bound, reactive state.

7. **Presence of Biota:** The participation of living organisms in the consumption and generation of chemically significant components is a very important characteristic of ecosystems. At first glance, the biota can be considered as a quasi-reagent, the concentration of which enters the system of kinetic equations along with the concentrations of other substances. Moreover, biota usually participates in the reactions of the first, rarely of the second order. It means that the speed of biotic processes is often directly proportional to the amount of biota. Even in cases when such a dependence is more complex, it can still be described mathematically and fit the biota to a common formalism.

   Nevertheless, there are two very important, fundamental differences in biochemical reactions from "test tubes". The first is that living organisms are not required to obey the second law of thermodynamics in its classical formulation. They can conduct processes with decreasing entropy and increasing free energy due to external sources, such as, for example, solar radiation or others. Moreover, they are not only able, but even obliged to conduct processes against the reduction of free energy, while they are alive. One of the most impressive processes of this type is the construction of mollusk shells. Even in fresh water, where the product of concentrations of calcium ions and carbonate ions is much lower than the solubility limit, living organisms manage to bioconcentrate these elements, sufficient to form solid $CaCO_3$.

   The effect of living organisms on the course of chemical processes is fundamentally different from that of the catalyst. In the second case, the flow of both direct and reverse reactions is catalytically facilitated, and the equilibrium position remains unchanged. But "living matter" not only shifts the position of equilibrium or stationary state, but does it in spite of the gain of free energy.

Biological injection in general is against classical thermodynamics for closed systems. Here is a quote from the famous book by Schrödinger (1994). "What I wish to make clear in this last chapter is, in short, that from all we have learnt about the structure of living matter, we must be prepared to find it working in a manner that cannot be reduced to the ordinary laws of physics. And that not on the ground that there is any 'new force' or what not, directing the behavior of the single atoms within a living organism, but because the construction is different from a anything we have yet tested in the physical laboratory."

Similarly, knowledge, obtained in chemical laboratories is very indirectly related to the behavior of ecochemical systems. The acquired experience in the kinetic analysis of the mechanisms of chemical reactions is minimally useful for constructing prognostic models of natural systems. On the other hand, the lack of such experience is guaranteed to lead to the creation of an irrelevant model, grossly violating the basic principles of chemical physics.

The second fundamental difference between systems with biocomponents from purely chemical ones is the ability to self-organization. It is due to the consistent adjustment of the response of individual reactions to external disturbances and the system manifests the ability to self-regulation, providing homeostasis. A biota that does not even show signs of reasonable behavior ensures the stability of the environment, while maintaining optimal conditions for its existence. The very exchange of an ecosystem with powerful reservoirs of vital substances is a necessary, but insufficient, condition for homeostasis. Without self-regulation, the system would exhibit buffer properties, but would not be able to return to the previous state.

Next three paragraphs win be only nominative, due to the lack of space.

8. There is no possibility of active influence - only to contemplate.
9. Scaling is both difficult and impossible
10. Psychological extremes from man subordinated nature to we are a grain of sand in the universe
11. The most **diverse ecochemical processes take place simultaneously** (runs in parallel) in the most impure mixtures. In contrast to what we are trying to create in a laboratory test tube. Classical kinetic research implies the use of possibly pure substances. The presence of contaminants can influence and most often affect the kinetics of the process under study. E.g. ions of metals of variable valence entering into redox reactions with oxygen and/or hydrogen peroxide initiate radical chain processes of oxidation of organic substrates. In the laboratory, they are trying to get rid of them, or if they are added, in known, controlled quantities. In the natural environment, in addition to dissolved ions and iron complexes, there are certainly compounds of manganese, copper and something else. Such a batch in the laboratory will never be found. Not that the parallel flow of different processes leads to synergy, but their availability does not simplify the understanding of the system - that's for sure!
12. **The Scheme of the Ecochemical Process can Never Claim Completeness**: Even if it was possible to compile a maximally plausible model in which over a dozen elementary channels are taken into account, one can never be sure that after a while there will not be another "something" that will radically change the idea of the quality of the natural environment - water or air. A purely extensive increase in the number of reactions that are taken into account (sometimes it exceeds a hundred) is a dead end. It is easy to see that even if there are several parallel channels of transformation of an intermediate, it is difficult to imagine and, moreover, to construct a situation when at least three of them made a comparable contribution. Much more often the situation is realized when one or two channels dominate. In this sense, simplified (truncated) schemes are practically no worse than the

most exhaustive, and often even better, ones, since they require less computer memory and time. The problem, however, is somewhat deeper. The existing system of hydrochemical indicators of the quality of natural waters does not always reflect their real suitability for living and survival of hydrobionts. So in the book by Skurlatov, Duka and Miziti (1994) described in detail the concept of redox-toxication of water bodies. It turned out that in addition to the indices of the concentration of harmful substances that clearly influenced the survival of sturgeon fry, an additional integral (or complex) indicator was needed to characterize the ratio of the opposing (rival) processes: oxidant-hydrogen peroxide generation and reducing substances (most likely glutathions) appearance.

It is important that only its inhabitants can give an objective assessment of the quality of natural waters. In other words, the methods of biotesting acquire a significantly greater significance (at least for practical applications) than the whole set of instrumental methods for analyzing and detecting individual substances and compounds. The paradox is that biotesting like no other method is far from understanding the mechanism of what is happening in the natural environment.

THE SCIENTIFIC COMMUNITY AS AN ECOSYSTEM

Let us now consider the application of these principles (or rather neglect them) when creating forecast models of specific systems.

Perhaps the most expressive example of neglecting the principle of homeostasis in relation to ecochemical processes can be the evolution of the views of scientists and the public attitude to the problem of the ozone layer (Wuebbles, 2020).

In 1969, the Dutch chemist Paul Krutzen published a paper in which it was shown that nitrogen oxides, intercepting atomic oxygen, not only slow down the formation of ozone (O_3) from oxygen, but also contribute to its decomposition. Some scientists and ecologists-environmentalists in the 1970s used the results of Krutzen as an argument against the creation of a fleet of US supersonic airliners. The American program was canceled, and only a small number of French Concords and Soviet Tu-144 remained in service. Subsequently, their use was stopped as economically inexpedient, after which the effect of NOx emissions from jet aircraft on the ozone layer was recognized as insignificant. “It was found out” that the total annual NOx emission from all aircraft in the world does not exceed 2% of the current ozone content in the stratosphere, i.e. Taking into account the repeated renewal (at least 10 times a year) of the ozone “depot”, the “aircraft” contribution to the global processes of formation, transport and disappearance of ozone can be seen only in the third decimal place, which is certainly covered by the accuracy of measurements and daily fluctuations in its concentration (day/night).

In 1974, American chemists Mario Molina and F. Sherwood Rowland from the University of California showed that the man-made halogenoalkanes (CFCs, freons), under the influence of solar radiation, decompose in the stratosphere, forming atomic chlorine and chlorine monoxide (fig.1). Theoretically, these active intermediates are able to initiate the process of chain decomposition of ozone to oxygen:

More recent studies have shown that bromine and some bromine-containing compounds, such as bromine oxide (BrO), are even more effective for ozone depletion than chlorine and its reactive compounds. In subsequent laboratory measurements, the rate constants of the individual stages were established, which allowed “at the tip of the pen” to calculate the length of the chain of cyclic ozone decomposition. It turned out that it can reach hundreds and even thousands of links. Then these speculative calcula-

Figure 1. A simplified scheme for the destruction of the ozone layer in the stratosphere

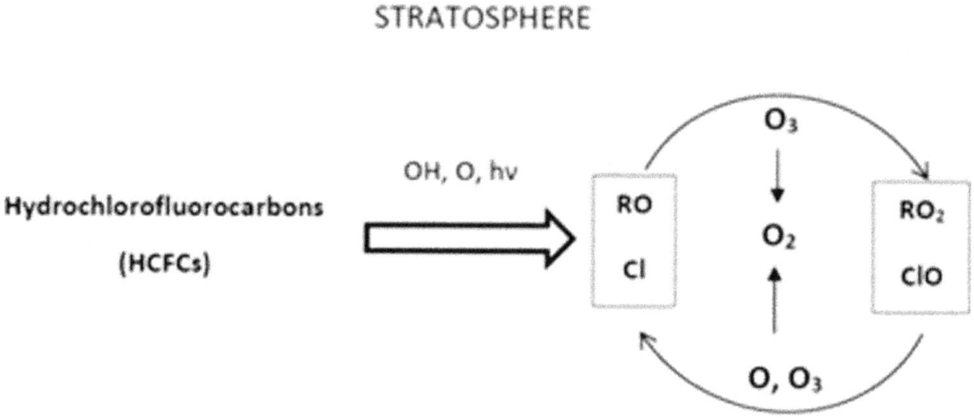

tions were mechanically transferred to chemistry of the stratosphere. The danger of destruction of the ozone layer was picked up by wide sections of the population, who had no idea of the mechanisms of odd oxygen conversion. With amazing promptness, as early as 1978, the Montreal Protocol was signed, banning the use of chlorine-containing freons.

It takes another 20 years and Krutzen, Molina and Rowland receive in 1995 the Nobel Prize in Chemistry for their efforts. The number of works devoted to the "ozone hole" is already estimated in thousands.

At the same time, two important circumstances were released from attention. The first is that the molecular mass of the most common freon R12 (CCl_2F_2) at that time is more than four times the molecular mass of air. This means that the probability of lifting such molecules to the stratospheric altitude of the ozone layer is very small. An elementary calculation using the barometric formula shows that approximately one ten-thousandth molecule (a fraction of 10^{-4}) of freon reaches the ozone layer. Therefore, the real factor in the efficiency of ozone splitting, which is usually equated with the length of the chain of the corresponding process (i.e. how many ozone molecules can destroy one molecule of halogenalkane) should be reduced in proportion to this fraction. The result is an extremely inexpressive that out of about 2000 gig tons of atmospheric ozone, hundreds of thousands of tons of fluorine-chlorine hydrocarbons produced per year can lead to the disappearance of only a thousandths of a percent.

The second circumstance follows directly from the first. It is generally accepted to estimate the lifetime of CFC in the atmosphere for about 100 years (Seinfeld and Pandis, 2006). Such a long time of life is due precisely to the fact that only a small fraction of freons that have reached stratospheric heights are subject to effective ultraviolet photolysis-below, the short-wavelength part of the solar radiation does not penetrate. But the low expenditure rate of CFC, implies a low rate of initiation of the chain process of ozone depletion. Even taking into account the long chain, the annual destruction of ozone by the "freon channel" would be tenth or hundredths of a percent, since ozone in the atmosphere is 4 orders of magnitude greater than that of freons. In other words, the estimation of the life times leads to approximately the same evaluation as the barometric formula.

After some decline in interest in the problem of the "ozone hole", Kleinman's work appears in the 1990s already, proving that the nitrogen oxides originally declared to be the main culprit of anthropogenic impact on the atmosphere (even before the investigation of freons) not only do not destroy the ozone layer, but directly contribute to the generation of this gas (Kleinman, 1994). And even proposed a linear correlation

$[O_3] = 27 + 11.4([NO_y] - [NO_x])$.Here $[NO_x]$. denotes the total concentration of nitrogen oxides with an odd number of oxygen atoms, and $[NO_y]$.the concentration of nitrogen oxides with an even number of oxygen atoms.

The main representatives of each family are NO and NO_2. respectively, and the typical ratio of their concentrations is usually 1:7. Therefore, increasing the emissions of nitroxides, as a rule, leads to an order of magnitude more significant increase in ozone concentration. This is clearly seen in the diagram in Figure 2

Figure 2. Comparison of ozone production efficiencies on high O_3 days in Phoenix (Arizona) and Houston (Texas)
(Courtesy Larry J. Kleinman, Brookhaven National Laboratory for publication in the monograph [6])

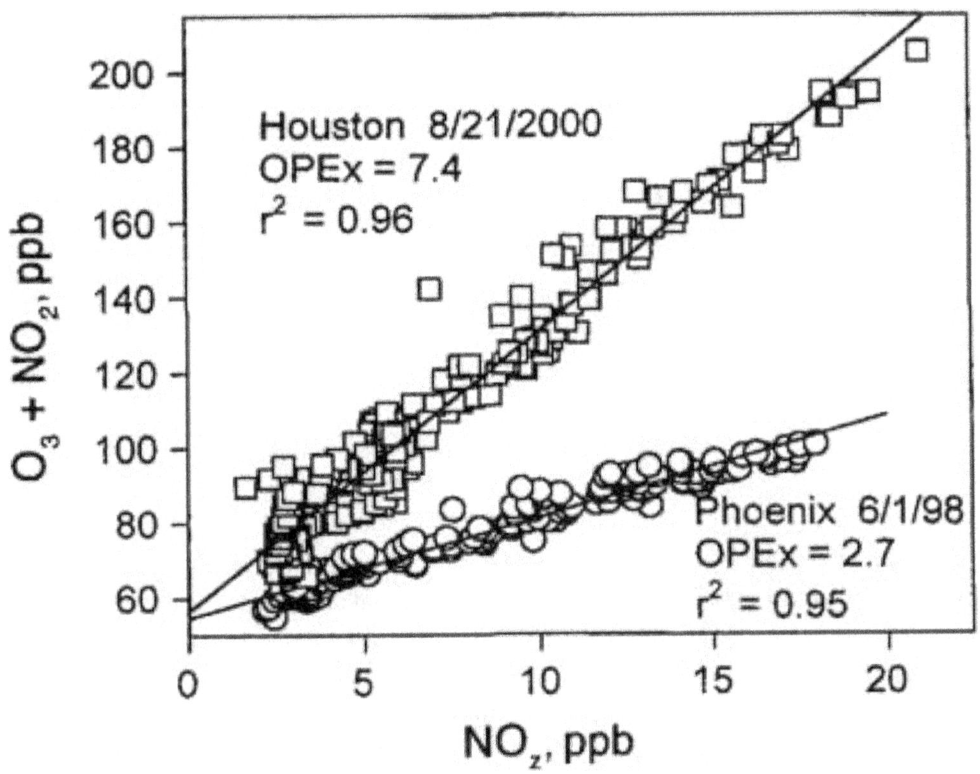

Later it was found out that pollutants not only are not "killers" of ozone but can even appear to be precursors in its generation. As such precursors, both nitrogen oxides and hydrocarbon emissions into the atmosphere can act. The limiting case of low concentrations of NOx (0.3-2 ppb or billions, such concentrations are observed in the background areas), it is customary to call the NOx-limited mode of photochemical ozone formation. The limiting case of high concentrations of NOx (> 4 ppb, such concentrations are characteristic for urbanized regions) is commonly called hydrocarbon-limited (CH-limited) or NOx-saturated mode of photochemical ozone formation. Some idea of the dependence of the concentrations of photochemically formed ozone as a function of the rates of emission of its precursors can be obtained from Figure 3.

Figure 3. Model values of the ratio of a mixture of photochemically formed ozone (ppb), depending on the emissions of nitrogen oxides and volatile organic compounds. The thick line separates the NOx-limited (upper left) and hydrocarbon-limited (right below) modes of photochemical ozone formation
The diagram is given by Zvyagintsev and Kuznetsova (2013) with reference to the original paper (Seinfeld, 1989).

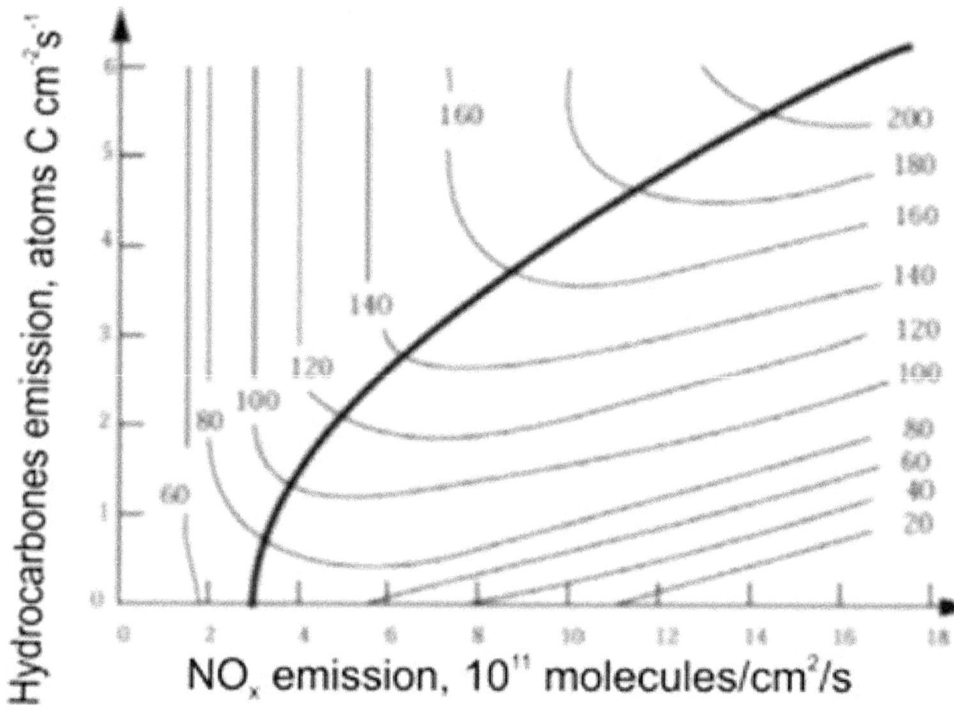

Let's return now to a principle of a homeostasis. The system of generation, depletion and redistribution of ozone has existed for millions of years, experiencing seasonal and diurnal fluctuations, the amplitude of which exceeds by orders of magnitude the possible decrease in ozone as a result of chemical interactions with anthropogenic emissions.

Moreover, since the seasonal nature of the fluctuations for the northern and southern hemispheres is in antiphase, the first role is played not by chemical processes, but by the meridional mass transfer of ozone. It is clear that in the equatorial zone the oscillations of $[O_3]$ are minimal, while the largest oscillations occur in the polar latitudes. In the middle latitudes, the maximum thickness of the ozone layer (0.43 cm) is in March, and the minimum (0.27 cm) is in October.

In general, regardless of latitude, the maximum ozone content in the atmosphere is at the end of winter and spring, and the minimum at autumn and early winter. But with the advance to the north and south, the advance of the maximum is postponed to later months. For example, in Alma-Ata, the maximum thickness of the ozone layer is observed in February, in St. Petersburg - in March, on about. Dickson - in May.

The maximum ozone concentration in the atmosphere recorded on the globe is 0.76 cm (this record value was recorded on Kerguelen island on October 20, 1967), and the minimum value (in "ozone holes") is 0.09 cm.

In Figure 4 shows the seasonal variability of the AOC (atmospheric ozone content): the maxima are recorded in the spring, and the minima are recorded in the autumn of each hemisphere. Graphs of AOC values for the Southern and Northern Hemispheres are in antiphase, they are shifted by half a year relative to each other (Kashkin, Rubleva and Hlebopros, 2015).

Figure 4. Graphs of time series of the average AOC for 1978-1993

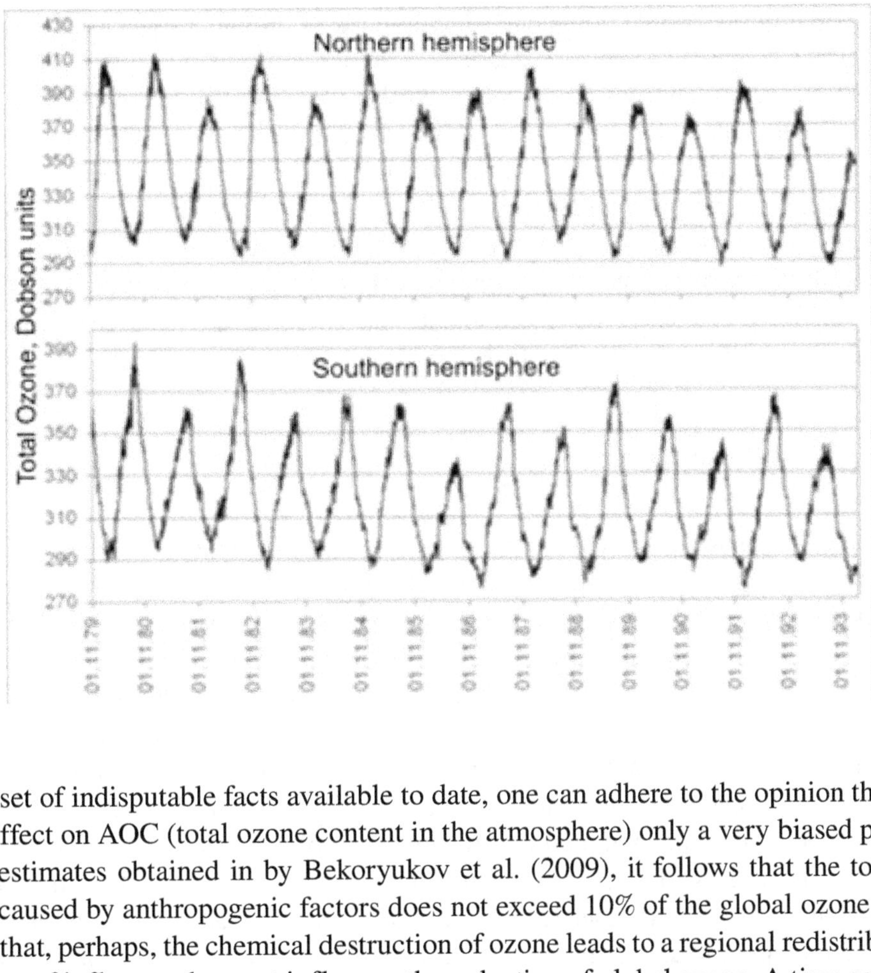

Given the set of indisputable facts available to date, one can adhere to the opinion that industry has a noticeable effect on AOC (total ozone content in the atmosphere) only a very biased person.

From the estimates obtained in by Bekoryukov et al. (2009), it follows that the total decrease in global ozone caused by anthropogenic factors does not exceed 10% of the global ozone depletion. It is hypothesized that, perhaps, the chemical destruction of ozone leads to a regional redistribution of ozone and this process of influence does not influence the reduction of global ozone. A time series of changes in ozone content over 40 years has been simulated with impressive accuracy. The basis of the model is the superposition of three dependencies: trend, impulse and oscillator. Oscillatory part, in addition to the obvious seasonal fluctuations with a period of one year, also includes frequencies corresponding to short periods of 2.37 and 11.2 years. The two-year cycle describes the effect of atmospheric circulation oscillations in the equatorial region. The eleven-year cycle describes the influence of the oscillations of the Earth's magnetic field and the activity of the Sun. The trend was estimated as linear interpolation of the long-term course of the time series, after separating the harmonic component from the empirical data. Finally, the impulse component was approximated by a sum of functions exponentially decreasing in time, the initial moment of which coincided with the most noticeable volcanic eruptions. Thus, the trifunctional model, whose empirical coefficients were adjusted by minimizing the residual for the best approximation of the experimental time series by calculation, demonstrated its adequacy. Note that no anthropogenic impacts were taken into account in this model.

Thus, convincing evidence of the large role of dynamic factors that have an impact on the spatio-temporal distribution of the global ozone field has now appeared. It should be emphasized that the study of the movement of ozone masses in certain regions has been conducted for a long time, but the real picture of zonal and meridional ozone transport on a global scale has not yet been studied in detail.

During the period 1980-2009, just during the Montreal Protocol, the area of the ozone hole has increased to almost 30 million square kilometers, which is ~ 6% of the earth's surface. For thirty years, the ozone anomaly over Antarctica has become deeper 2.5 times, and its area has increased about 30 times (NASA Ozone Watch, 2020). The reasons for such intensive variability of polar stratospheric ozone have not yet been established with certainty. The only thing that does not cause doubts now is that the ozone holes are not "man-made objects", not the result of the action of freons, whose release was reduced not less than 30 times, but a natural phenomenon.

Let's pay attention to the reader. Evolution of ideas about anthropogenic impact on the ozone layer has demonstrated all the features listed in paragraphs 1-12. Thus, we are convinced that the scientific community studying the processes in the environment is also a small ecosystem that is part of the global ecosystem and also obeys the same laws as the behavior of biota in water bodies.

MATHEMATICAL MODELING

After we formulated the main distinctive features in the description of kinetics and the recognition of the mechanisms of ecochemical processes, it is time to demonstrate what practical consequences, what surprises arise in the numerical mathematical modeling of such systems.

It has already been mentioned that an increase in the number of individual stages included in the model itself does not improve its quality. The practical application of effective algorithms for the numerical solution of a system of kinetic equations, contrary to expectations, did not bring the calculated values closer to their experimentally measured values and did not improve the predictive capabilities of the model.

Now a persistent and apparently false impression is gradually breaking down that the accumulation of primary data obtained by instrumental methods, the inclusion of chemical reactions by the processes of transformation of substances with the participation of hydrobionts and, finally, the consideration of mass transfer processes between different reservoirs of the ecosystem will lead in the foreseeable future to the creation of a comprehensive Model of the formation of a quasistationary state of a particular reservoir.

Many specialists in water use, fisheries and other areas of practical ecology have already expressed dissatisfaction with the current approach to quality control of natural waters (Shitikov, Rozenberg and Zinchenko, 2003):

1. It was not clear and still unclear until now which hydrochemical indicators should be included in the list of mandatory monitoring. A tangible shift in recent decades from purely chemical measurements to chemical-biological measurements, including biotesting, albeit improves the situation, but does not correct it radically.

2. The area of research of root causes and processes leading to the formation of a suitable or less suitable quality of the environment is even further from perfection.

At the same time, a systematic approach to the study of the whole range of ecochemical and ecological-biological processes and phenomena has already been formulated and analyzed in a very detailed manner in the works of the Brussels school of nonequilibrium thermodynamics, directed by I.R. Prigogine. However, work on the self-organization of matter was not perceived as having a direct bearing on the kinetics and mechanisms of hydroecological processes. It took at least 30 years before the beginning of the realization that Prigogine's books were written for us and about what we are trying to understand.

In 1985, in Russian was issued the book "From the existing to the emerging"(Prigogine, 1985), in which in an accessible language it was explained that dynamic systems in the general case are not integrable and not ergodic. The problem is not that there are any specific mathematical difficulties (ultimately, any system of differential equations can be integrated numerically, and modern algorithms and computational power allow solving such problems in real time). The main, fundamental complexity is caused by the fact that in classical dynamical systems the solution is invariant with respect to the reversal of the direction of time, whereas all chemical systems are fundamentally irreversible. In them, the dynamics always goes in the direction of increasing entropy, from a less likely state to a more probable state.

It would seem that this situation is completely customary for chemical kinetics, but it was missed that irreversibility in time in the general case is not compatible with Laplacian determinism. Outside the practice of hydroechemical research, such terms as bifurcation, chaos theory, the multiple stationary states remained for a long time. On the contrary, the notion that the solution of ODE systems (ordinary differential equations), based on the law of mass action, would lead to a detailed modeling of the temporal dependencies of the main ecochemical indicators, that is, will provide an opportunity to extrapolate forward and make reasonable quantitative forecasts.

Even the fact that Prigogine gave a separate paragraph "Ecology" in (Prigogine, 1985) on page 133 did not change the attitude. Without proper attention, remained approaches to the analysis of supercritical regimes and limit cycles in trophic chains of the Lotka-Volterra type, containing the "predator-prey" links. Although such processes are the main links of the trophic chain, from which in all ecosystems the processes of the motion of the phase trajectory around a certain point with the largest production of entropy, called the attractor, are built. And only in exceptional cases, which are not typical, the phase trajectory of the system reaches precisely the stationary state, and do not describe the cycles around the attractor.

The Lorenz attractor, better known as the "strange attractor", was proposed in 1963 in the "Journal of the Atmospheric Sciences" as an example of a determinate non-periodic flow. It served as a cornerstone in the foundation of the theory of chaos and radically changed the idea of modeling weather phenomena.

Initially, a nonlinear autonomous system of third-order ordinary differential equations describing the unstable behavior of a dynamical system

Originally the nonlinear autonomous system of ordinary differential equations of the third order describing the erratic behavior of the dynamic system were obtained from the system of Navier-Stokes equations describing the motion of air flows in a flat layer of constant thickness by the decomposition of velocity and temperature in a double Fourier series and then truncating to the first and second harmonics:

$\dot{x} = s(y - x)$. $\dot{y} = rx - y - xz$. $\dot{z} = xy - bz$ .Later, various real physical systems were discovered, whose behavior is described by the reduced mathematical model - from a single-mode laser to electromechanical oscillators. We are mainly interested in chemical-kinetic systems with a limiting cycle and intrinsic instability.

The set of Lorentz equations corresponds, e.g. to the scheme of reactions (1) - (8):

1 $X \rightarrow destruction$. $k_1 = 1$ .2 $Y \rightarrow X + Y$. $k_2 = s$ .3 $Y \rightarrow destruction$. $k_3 = 1$ .4 $X \rightarrow Y + X$. $k_4 = r$ .5 $X + Z \rightarrow killer + X + Z$. $k_5 = 1$. 6 $killer + Y \rightarrow destruction$. $k_6 \gg 1$. 7 $X + Y \rightarrow Z + X + Y$. $k_7 = 1$. 8 $Z \rightarrow destruction$. $k_8 = b$ .Let us pay attention to the fact that to each member of the first or second order in the system of differential equations there correspond single elementary reaction of the same order. And only a term for the disappearance of Y in a channel of zero order on Y, i.e. with a speed independent on Y, can be modeled by at least two-stage set of reactions (5) and (6). Indeed, the disappearance of Y "without its participation", without its entry into the product of the law of acting masses, requires that this elementary act occurs in an non limiting stage, i.e. it must necessarily be preceded by a speed-determining stage, depending on the products of X and Z, but not depending on Y.

We note that, unlike the disappearance reaction, the rate of buildup reaction for any reagent is most often independent of its concentration, since it occurs when absolutely other substances interact. An exception is autocatalytic reactions, where the rate of generation of the reaction product increases as it accumulates. In the above scheme, this is the fourth stage. It is the main reason why the system receives oscillatory behavior.

It may seem strange, but if the kinetic scheme of reactions (1) - (8) is unequivocal, the original system of differential equations, the attempt of numerical modeling of the Lorenz attractor by the method of numerical solution of the direct problem of chemical kinetics does not obligatory lead to the desired and well-known result. For any given accuracy and an arbitrarily small fractionation of the integration step, the system does not leave the limit cycle, but breaks down into a stable, albeit trivial solution $X(t) = Y(t) = Z(t) = 0$ for $t > t_{crit}$. (look Figures 5 and 6).

Figure 5. Time dependence of components for "kinetic variant" of the Lorenz attractor

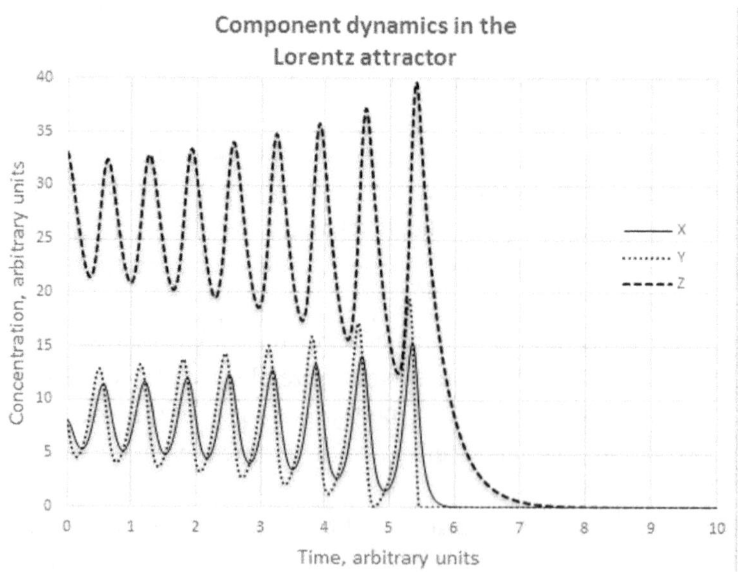

Figure 6. Phase diagram for "kinetic variant" of the Lorenz attractor

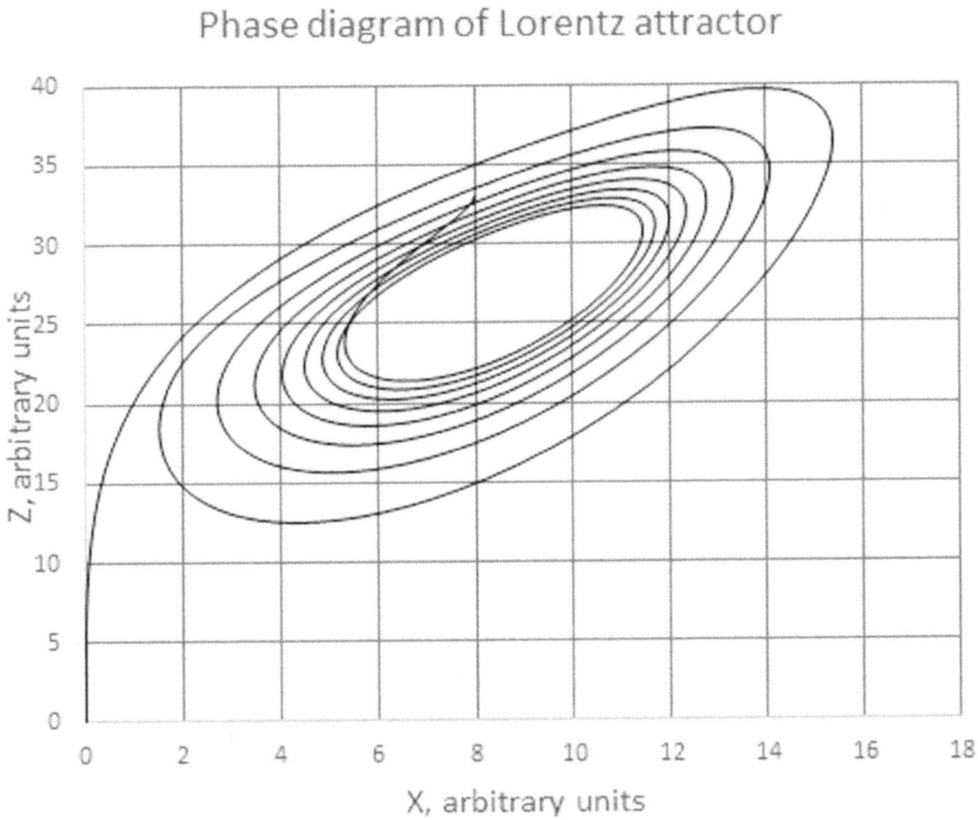

It is easy to understand the reason for the difference in the kinetic version of the reaction scheme (1) - (8) from the classical behavior of the Lorentz system. In the classical version, for the selected starting conditions, the system after eight cycles passes to another half-plane, changes the attractor, and the components X and Y change sign. In the kinetic analogue, the concentration of any component simply cannot, or at least should not (for a good solver of the ODE system) take negative values. What happens when the concentration Y ® 0 tends to zero?

No matter how great the rate constant k_6 (in our case it exceeds k_5 by 10 orders of magnitude, which certainly covers the whole conceivable range of the reactivity of the components), stage (5) ceases to be limiting, the rate of Y disappearance becomes proportional to the concentration Y and the kinetic "analog" the Lorentz scheme ceases to be so.

Nevertheless, it is possible to substitute a physically unreal value of the constant of step (6) at the level 10^{15} in the integrator (in the model calculation, we deliberately depart from the choice of units of concentrations and time and, correspondingly, the dimensionality of the constant, only that the difference of 15 orders of magnitude normally cannot be considered; the rate constants of the interconversion of the components of the ecochemical system must be within ten orders of magnitude, maximum; of course we have right to compare only rate constants of the same dimension, bimolecular with bimolecular). Another technique by which it is possible to overcome the natural counteraction of a debugged solver for the direct problem of obtaining negative concentrations is to introduce a pseudotrimolecular stage

$5'$. $X + Y + Z \rightarrow X + Z$. $k_5 = 1/Y$ .with the artificially introduced algebraic dependence of the rate constant inversely proportional to the concentration of the Y component, which will lead to its contraction in the formula of the effective mass law.

The duration of the monitoring period of the system in the Figures 7, 8 is three times higher than in Figures 5, 6.

Figure 7. Time dependence for the classical version of the Lorenz attractor

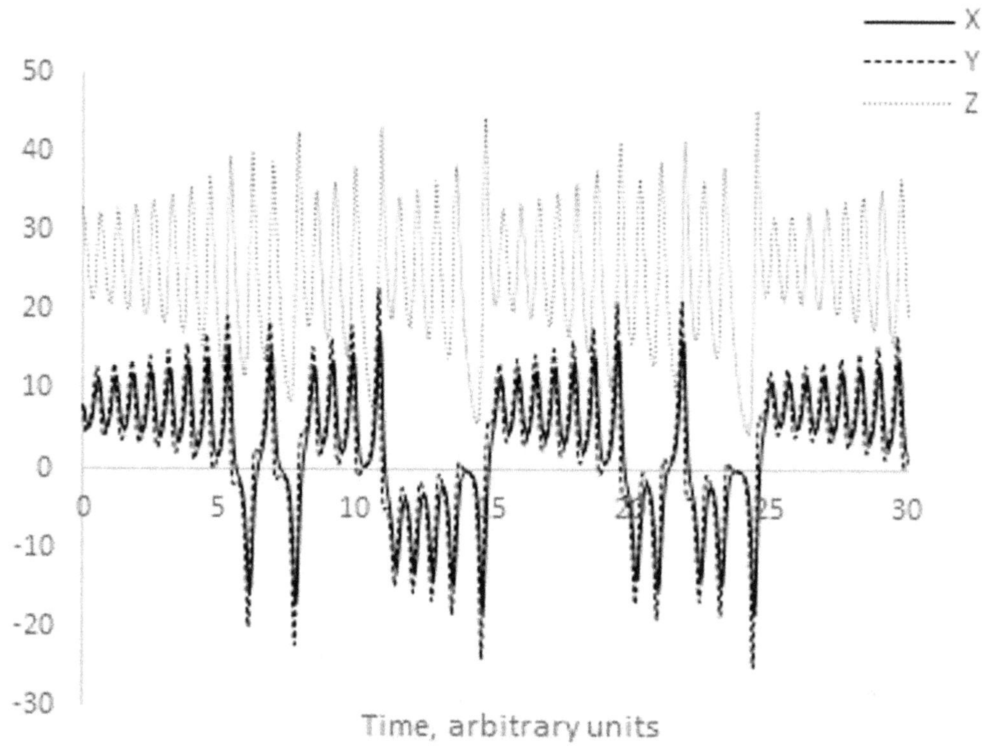

From a physical point of view, such a substitution is even more meaningless than an immeasurable increase in the rate constant of a nonlimiting stage, but for mathematical modeling such an artificial technique is quite suitable.

It is easy to see that there are two attractors in the classical version. For us, however, it is more important not that the Lorentz system changes the half-plane of accumulation of turns from time to time, but the fact that the system breaks from one trajectory to another aperiodically, at an accidental time.

It is of fundamental importance that this is an unpredictable system. An arbitrarily small difference in the initial or boundary conditions, or an insignificant change in the ratio of the rate constants of the individual stages, after just a dozen periods, leads to completely irreducible model behavior (Figure 7). And this feature can`t be attributed either to fluctuations or to the influence of hard-to-read random factors - this is an internal property of the dynamic system. It is unstable in itself.

The solid curve is given for the standard settings of the solver and is regarded as basic one.

Dashed and dotted curves correspond to a two-fold deterioration in step and a fivefold increase in accuracy. The light gray curve is obtained by shifting the initial conditions by 1%.

Figure 8. Phase diagram for the classical version of the Lorenz attractor

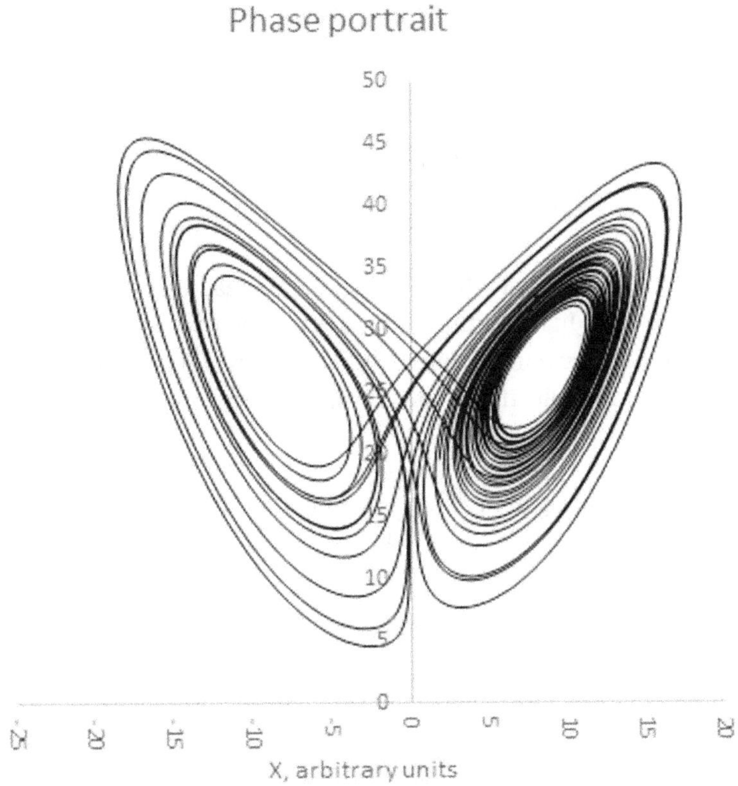

Figure 9. Model calculations of the dynamics of the Lorentz system with respect to the component X

In chemical-biological systems with transformation of components dissolved in natural water, chaotic behavior accompanies the manifestation of synergism. For example, autocatalytic stages with a high order of reaction type

$X + X \to X + X + X$. or

$X + X + Y \to X + X + X$ .have critical regimes.

In Prigogine's book [16] the example with the third order on reactions is given - otherwise the multiple stationary states will not work out. In Prigogine's scheme, an open system is meant, in which by default it is meant that the makeup of resource components A and B comes from unlimited reservoirs, so that these concentrations are kept constant. Thus, the solution is the dynamics of the only intermediate product X.

I.e. for third-order reactions for the system $A + 2X \underset{k_2}{\overset{k_1}{\rightleftarrows}} 3X$. $X \underset{k_4}{\overset{k_3}{\rightleftarrows}} B$.

There is a concentration region [*A*] and [*B*] and ratios k_3/k_2 for which the equation for the stationary concentration *X*.

$$X_{st}^3 - \frac{k_1 A}{k_2} X_{st}^2 + \frac{k_3}{k_2} X_{st} - \frac{k_4 B}{k_2} = 0$$

It has three real roots, and all three are positive (or at least not negative), i.e. have a physical sense of concentration (see Figure 10).So for the set

$A = 10$. $B = 25$. $k_1 = 1$. $k_2 = 1$. $k_3 = 30$. $k_4 = 1$.

There are three solutions: $X_{st} \approx 1{,}38; X_{st} \approx 3{,}62$ and . $X_{st} = 5{,}00$.

In this case, the first and third solutions are stable, while the second is unstable. The smallest concentration fluctuation *X* "flushes out" the solution into one of the stable states (Figure 11).

$X_{st} \approx 3{,}62$ .The system can stay unpredictably long. However, the addition of 1% of the substance of the producing *X* or absorbing *X* pushes the system into one of the stable stationary states.

A similar analysis for a system in which the maximum total reaction order for all components does not exceed two, clearly shows that the number of stationary states also decreases to two, and of them only one is stable.

$$A + X \underset{k_2}{\overset{k_1}{\rightleftarrows}} 2X \, . \, X \underset{k_4}{\overset{k_3}{\rightleftarrows}} B \, .$$

Figure 10. Dependence of the resulting production rate of component X on its concentration. The system has three stationary states, of which two are stable

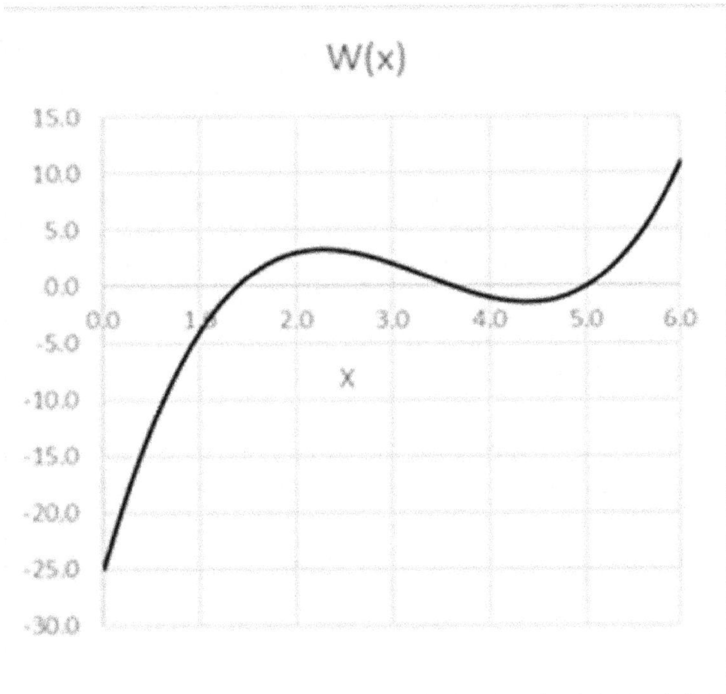

Figure 11. Computer modeling shows that in an unstable state

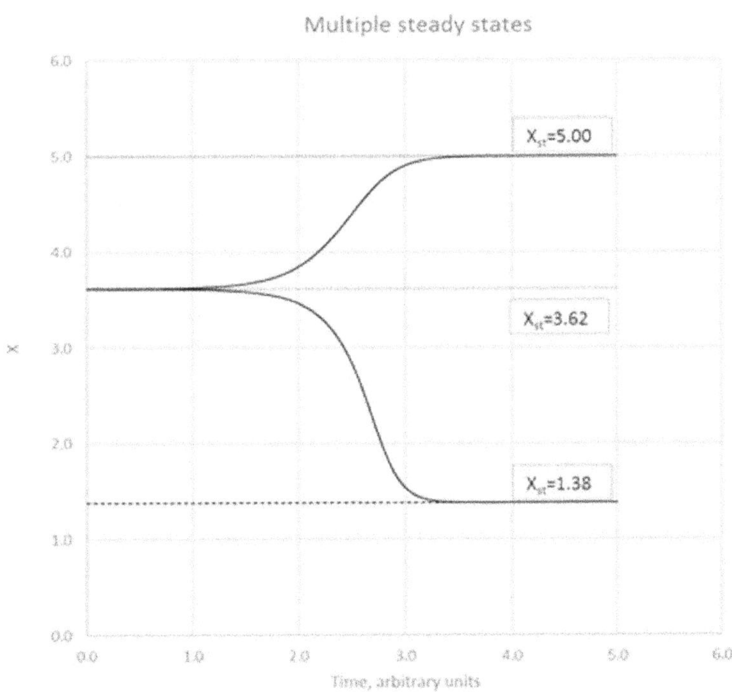

The system seems to "feel" that when $X = 1$ there should be a stationary state when passing an unstable state with a zero generation or consumption rate X. It is clear that in some neighborhood of this value the production speed of the intermediate substance is close to zero, therefore the system for a while brakes in this zone. Of course, if for the simulation to take the starting value of concentration above the unstable stationary level, the system without any delays breaks into the upper state, which is the only stable (Figure 12).

$A = 10$. $B = 1$. $k_1 = 1$. $k_2 = 2$. $k_3 = 0{,}00966$. $8; 8\,k_3 = 0{,}00977$. $k_4 = 8$ .With a slight increase in the rate constant of the death of the intermediate substance after a relatively long delay in an unstable stationary state, the system disintegrates into the trivial solution X = 0. We note that, as in the case of the Lorenz attractor, the appearance of a "nonalgebraic solution" has an understandable physical meaning - at low concentrations, the previously not limited stage becomes limiting. In other words, nature does not tolerate artificial constructions of zero-order reactions.

The possible and even the most probable explanation is that Prigogine's research in the field of supercritical behavior of chemical open systems did not cause a proper response in practicing kinetics, due to the fact that high-order reactions, especially for chemical processes, are rare. For example, the author can`t recall a single example of a liquid-phase system, where the speed would be determined by a cubic expression.

Figure 12. Computer simulation of stationary states in a system with no higher than second-order reactions.

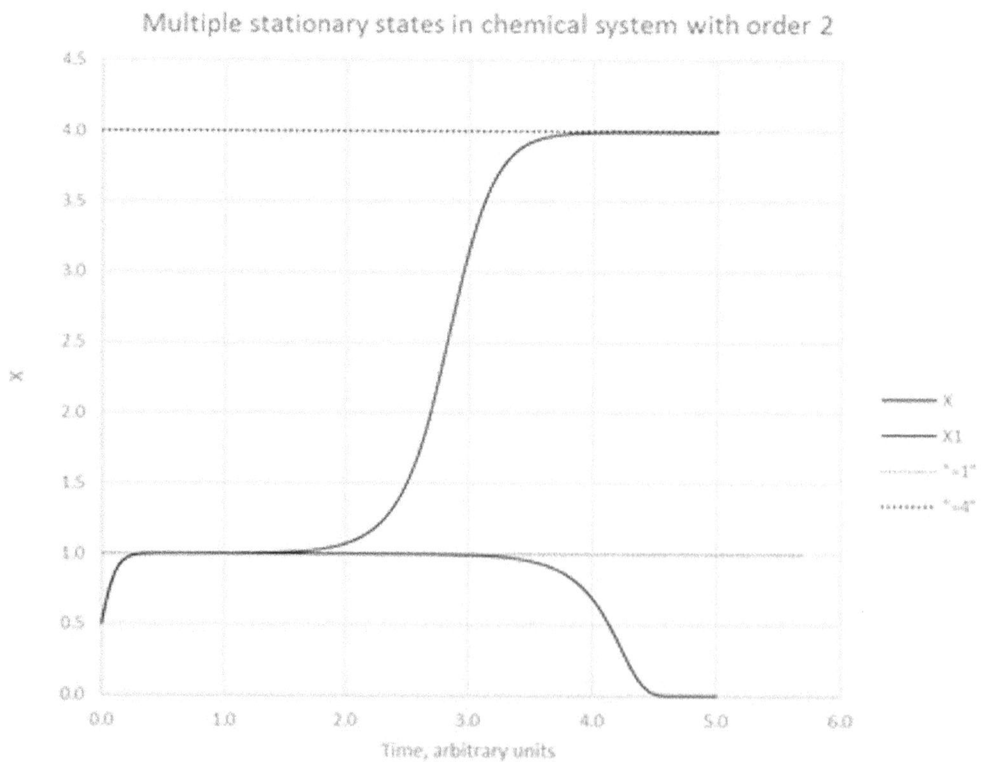

And the most "working", often occurring processes usually have an order no higher than the second, i.e. can`t have multiple stationary states, are not subject to bifurcations and, therefore, cannot (at least should not) manifest a chaotic regime.

Recent studies conducted by the author led to the conclusion that for the non-standard behavior of an open system (and all ecochemicals refer specifically to this type), not so much high order as the guaranteed cyclicity is required. In schemes like Brusselator or Lotka-Volterra, undamped oscillations are obtained precisely as a solution of kinetic equations of degree higher than the first. Despite the fact that these systems are well studied and described many times, we still need to make a short excursion into the behavior of the "predator-victim" model, since it is these processes that underlie the formation of trophic chains that are an indispensable component of ecological systems.

The simplest scheme consists of only two differential equations and can be modeled by only three chemical stages (Glansdorf and Prigogine, 1973):

1 $X + A \xrightarrow{k_1} X + X + A$. $X + Y \xrightarrow{k_2} Y + Y$. $Y \xrightarrow{k_3} destruction$ .Here A is a non-consumable substrate, a nutrient medium for the lower member of the trophic chain X, necessary for its multiplication. In turn, X serves as a food and promotes reproduction of the predator Y, stage (2). Finally, in the stage (3), the natural death of the predator occurs.

The system of differential equations has the form:

$\frac{dX}{dt} = k_1XA - k_2XY$. $\frac{dY}{dt} = k_2XY - k_3Y$ .It has only one stationary state:

$X_0 = \frac{k_3}{k_2}, Y_0 = \frac{k_1}{k_2}A$ .It is possible to represent each of the variables as a stationary part and a small addition to it: . $X(t) = X_0 + x(t); Y(t) = Y_0 + y(t)$ .In the region of small deviations from the steady state:

$\frac{x(t)}{X_0} \ll 1; \frac{y(t)}{Y_0} \ll 1$ .Substituting the values of stationary concentrations in the equations for the rates, we obtain

$\dot{x} = k_3y$. $\dot{y} = k_3Ax$ .or

$\ddot{x} = k_1k_3A \cdot x$ .The solutions of such a system are harmonic oscillations:

$x = X_1\sin(\omega t)$. $y = Y_1\cos(\omega t)$. $\omega = \sqrt{k_1k_3A}$.

In the Figure 13, 14 it is clearly seen that the oscillations are close to harmonic for small values of the deviation from the steady state level (up to 10%, circular shape), and a clearly pronounced anharmonicity appears in the system with increasing oscillation range. The circular phase diagram turns into a "triangular" one, and the dynamics of the populations themselves become more similar to the teeth characteristic of self-oscillations.

Figure 13. Lotka-Volterra system - Dynamics of predator Y and prey X populations (a)

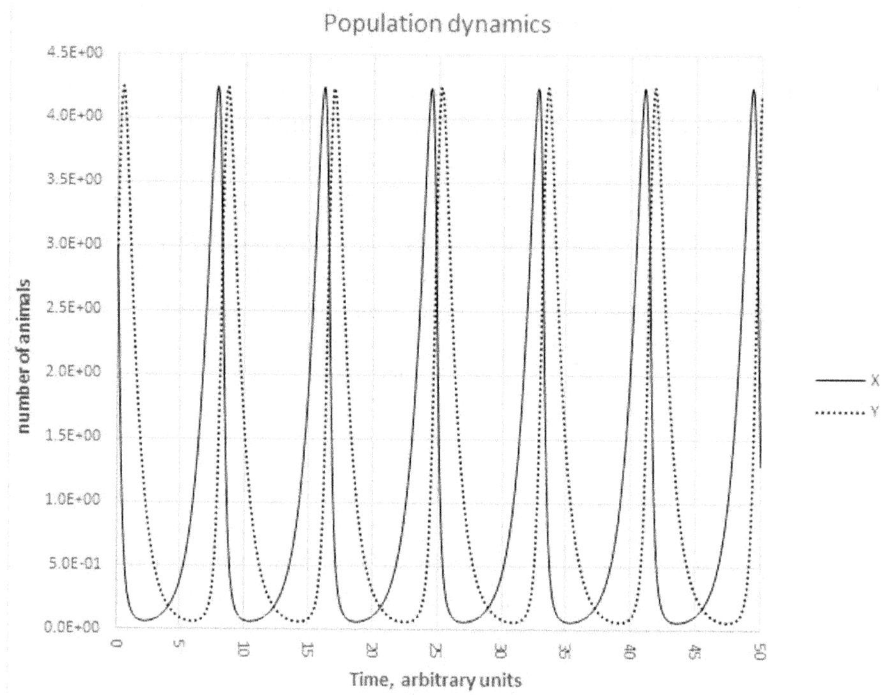

Figure 14. Lotka-Volterra system Phase portrait

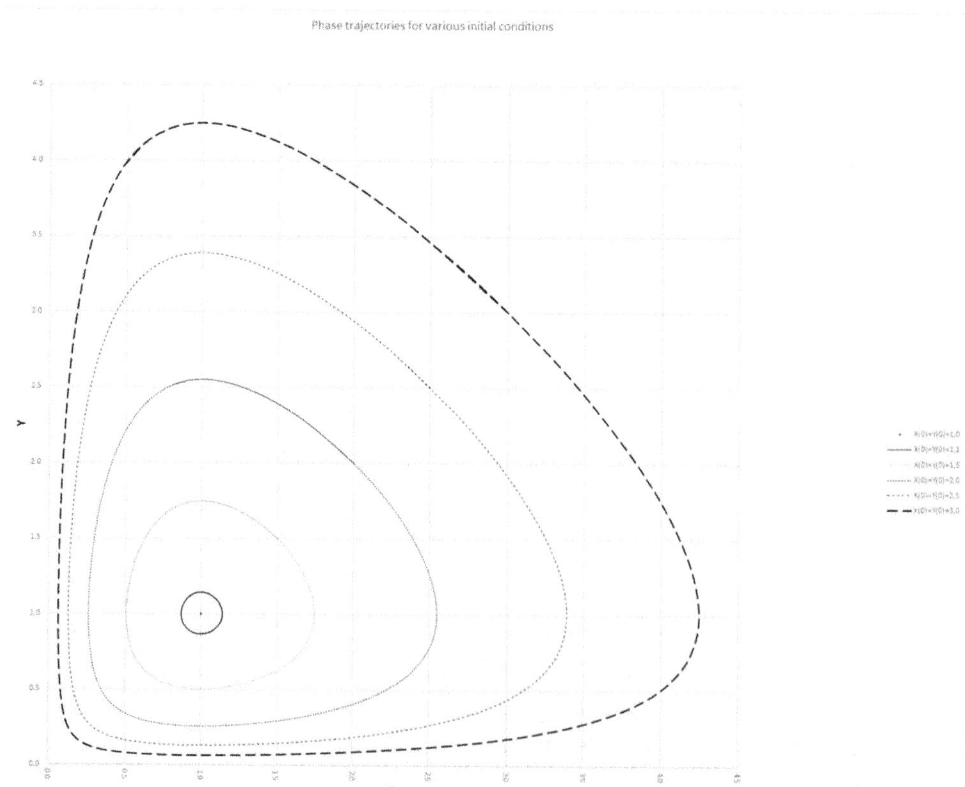

Nevertheless, the attractor of the Lotka-Volterra system is fundamentally different from the Lorenz attractor. It completely lacks an element of chaos and unpredictability, and the limit cycle is reproduced with absolute accuracy an infinite number of times. Changing the system parameters (rate constants or initial concentrations) can lead to a shift in the position of the attractor and a change in the size and shape of the limit cycle, but no more. The "entanglement" of the trajectory does not arise under any changes. Obviously, this is due to the relatively low degree of the ODE system (in the particular example, equal to 2)

Elongation of the trophic chain by only one link leads to rather unexpected consequences. In addition to the steady state (which is obvious) there are also unstable regimes that lead to a breakdown of the system down (which is not very interesting - the population was extinct). And a much more interesting option is the explosive accumulation of biomaterial.

In the second case, it is not immediately clear how it could turn out that adding a reaction of eating a predator leads (eventually) to the growth of its population, compared with the case when the Y food chain ends and this predator no one ate. In fact, the result is quite logical and understandable. Such growth occurs only in the case of accumulation of biomaterial in the next link of the chain, i.e. In the predator, more adapted to the perception of the resources of the entire range. The model of this situation is shown in Figure 15, 16.

Dynamics of populations of predators I and Y and victims X; It can be seen that with the growth of the total number, the period of oscillations decreases.

Figure 15. The appearance of the third link in the trophic chain of Lotka-Volterra system

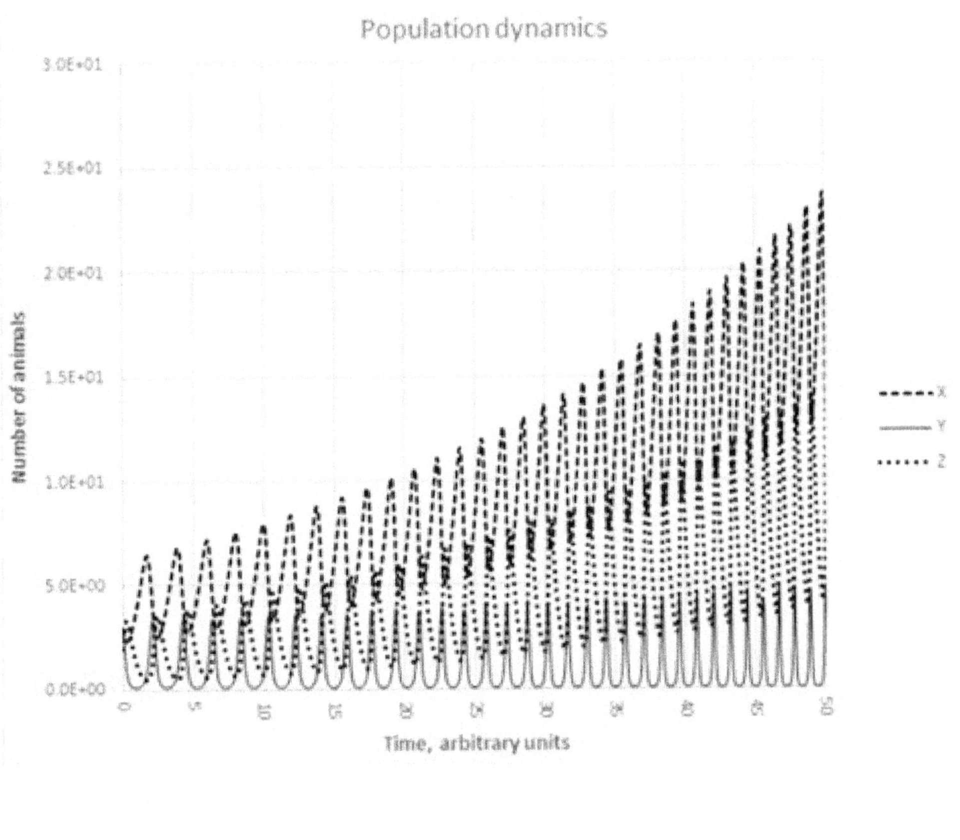

Figure 16. The appearance of the third link in the trophic chain of Lotka-Volterra system. Phase portrait with elements of chaos

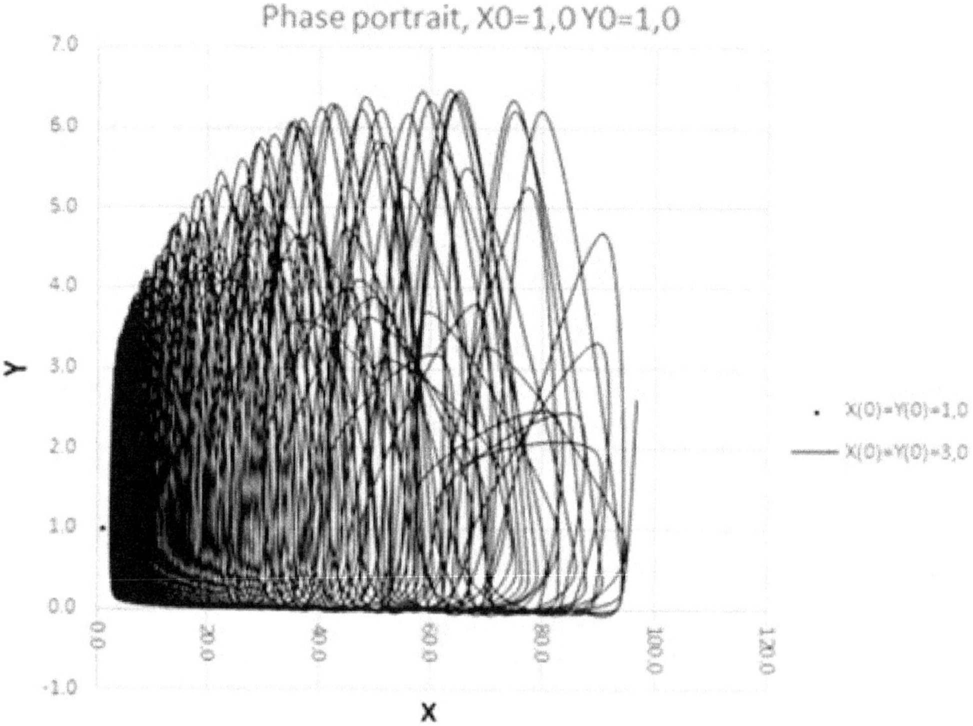

To an even greater extent, randomness (by the type of a strange attractor) is manifested in real systems, where the clock is specified by annual oscillations with a period of 12 months, and the own oscillation period, based on the rates of flow through the links of the trophic chain, is very different from it.

According to Serdyutskaya & Kameneva (2004), a system of equations is presented for the nitrogen cycle, which includes organic and inorganic nitrogen dissolved in the basin and three species of aquatic biota: phytoplankton, zooplankton and bacteria.

The authors propose the following set of differential equations:

$$\dot{P} = (k_0 - k_2)P + k_1TN_1P - \tilde{k}_4TPZ \,.\, \dot{Z} = \tilde{k}_4TPZ + k_9TN_0Z + k_{11}TBZ - (k_{10} + k_{12} - k_7)Z \,.$$

$$\dot{B} = \tilde{k}_3TN_0B - k_{11}TBZ + k_{16}TN_1B - (k_8 + k_{15} + k_{14} - k_{13})B \,.\, \dot{N}_0 = C_2 + k_2P + k_{15}B + k_7Z - \tilde{k}_3TN_0B - k_9TN_0Z - k_6N_0 \,.$$

$$\dot{N}_1 = C_1 + k_{12}Z + k_8B - k_{16}TN_1B - k_1TN_1P - k_5N_1$$

Which corresponds to a purely kinetic scheme based on the law of acting masses (Table1):

For it, it must be admitted, the phase trajectory has a complex aperiodic character without reaching the limit cycle. At the same time, although the main components have an alternation of concentration maxima and minima, these extreme points are reached at different time intervals, i.e. aperiodically.

Table 1. The reactions scheme and rate constants in format, ready-to-use in VBA-Excel module (authors design and code).

	Reagent # 1		**Reagent # 2**		**Reagent # 3**		**Product # 1**		**Product # 2**		**Product # 3**	**Rate const.**
1	T					--->	T					1.00E+00
2	Plankton					--->	Plankton	+	Plankton			2.63E-03
3	Plankton	+	N_inorg	+	T	--->	Plankton	+	Plankton	+	T	2.86E-08
4	Plankton					--->	N_org					9.72E-04
5	Plankton	+	ZooP	+	T	--->	ZooP	+	ZooP	+	T	6.01E-07
6	C_org					--->	N_org	+	C_org			1.00E+00
7	Bact	+	N_org	+	T	--->	Bact	+	Bact	+	T	7.60E-07
8	N_org					--->						8.30E-03
9	C_inorg					--->	N_inorg	+	C_inorg			1.00E+00
10	N_inorg					--->						3.24E-02
11	ZooP					--->	N_org					3.20E-01
12	ZooP					--->						1.03E-01
13	ZooP	+	N_org	+	T	--->	ZooP	+	ZooP	+	T	6.01E-07
14	Bact	+	ZooP	+	T	--->	ZooP	+	ZooP	+	T	3.01E-06
15	ZooP					--->	N_inorg					1.15E-01
16	Bact					--->	N_inorg					6.13E-05
17	Bact	+	T	+	N_inorg	--->	Bact	+	Bact	+	T	2.86E-10
18	Bact					--->	N_inorg					1.00E-01
19	Bact					--->	N_org					3.00E-01

In the Figure 17 it is clearly seen that seasonal fluctuations of organic and inorganic nitrogen (according to the model) occur annually, while there are only three outbreaks of zooplankton numbers over a ten-year period.

The ratio between organic and inorganic nitrogen also does not seem to be a harmonic cycle, but much more like a chaotically tangled trajectory (Figure 18).

The main conclusion that can be drawn from a careful analysis of the model under consideration is the lack of the possibility of a forecast. And the problem, of course, is not that the model is not complete enough or not accurate enough. Including additional components and additional reactions to the system can only complicate its already unpredictable behavior. Whereas even the most accurate set of rate constants for interconversion of substances (including "pseudo-material" - biological organisms) and initial conditions does not add a model of predictive power. Any slight weather changes, the appearance of additional biological species can change beyond recognition the phase portrait of the system, and the specific times of occurrence of these or those peaks. For example, an outbreak of zooplankton concentration may occur not on the third, but on the fourth or some other year.

Thus, we come to the conclusion that the existing approach to describing the chemical-biological processes in natural systems has become obsolete. Obtaining a pattern of ongoing reactions and measuring their rate constants is not only not a goal, but actually becomes a "distracting target", a false bait. For now, one can only talk about determining the accessibility zone for a beam of phase trajectories and

Figure 17. Modeling the dynamics of components in the Kuibyshev reservoir for 10 years

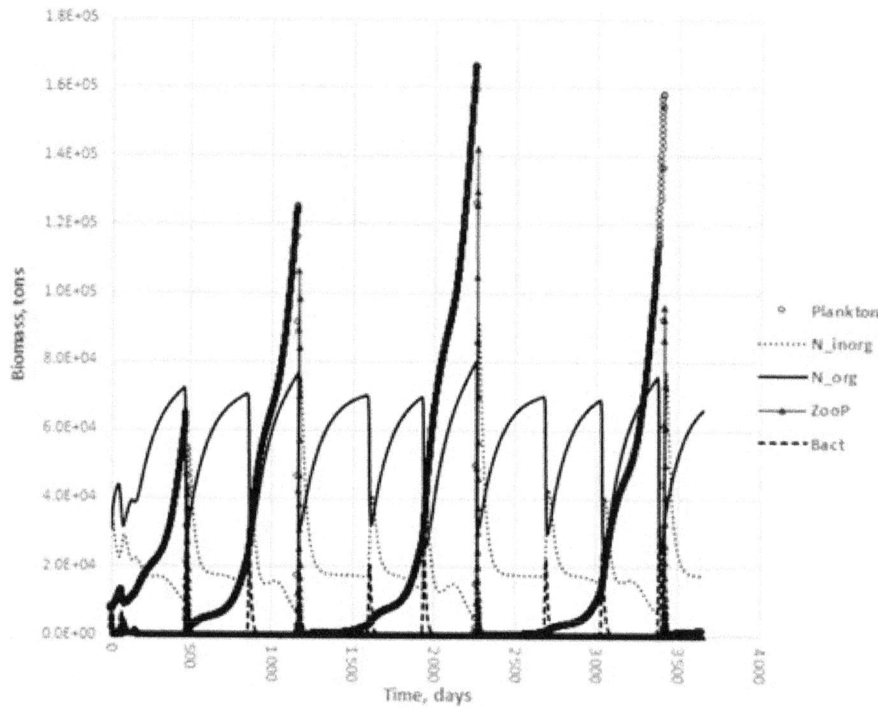

Figure 18. Phase diagram of interconversion of organic and inorganic nitrogen

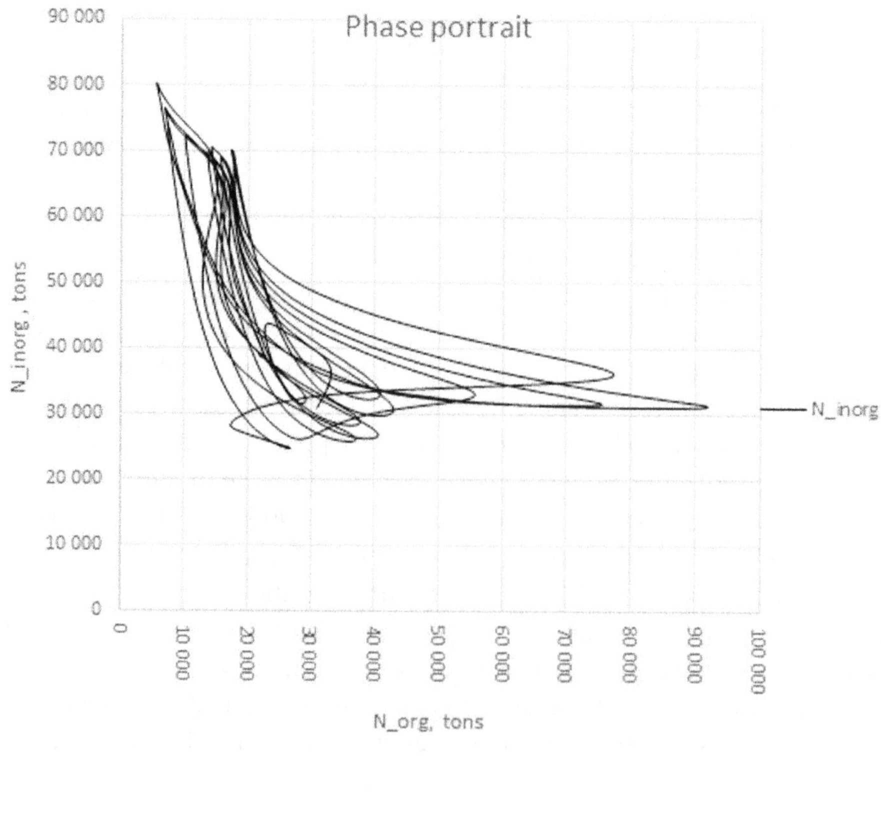

distinguish possible scenarios from impossible ones. At the same time, it is not necessary to talk about any long-term temporal prediction, just as a long-term weather forecast is impossible. We all know, without any forecast bureaus and complex mathematical programs, that the winter will inevitably come next year. But no one will say with a significant degree of certainty whether the snow will fall in November or in January. In ecochemical systems, the situation with forecasts is even tougher: we know for sure that organic life forms will cyclically change each other, but we cannot say with certainty whether this happens this year or next. And no development of the network of environmental monitoring posts the situation with forecasting will not change for the better. Unless it will allow to organize express monitoring.

Let's finalize our publication with a quotation from Prigogine 1986): "And the deterministic nature of the kinetic equations that make it possible to calculate in advance a set of possible states and determine their relative stability, and random fluctuations that" select "one of several possible states near the bifurcation point are closely interrelated. This mixture of necessity and chance is the "history" of the system.

REFERENCES

Bekoryukov, V.I., Glazkov, V.N., & Kokin, G.A. (2009). Dolgovremennyye izmeneniya globalnogo ozona okeana [Long-term changes in global ozone of the ocean]. *Izvestiya RAS. Physics of Atmosphere and Ocean, 45*(5), 607–616.

Glansdorf, P., & Prigogine, I. (1973). Termodinamicheskaya teoriya struktury, ustoichivosty I fluktuatsij [Thermodynamic theory of structure, stability and fluctuations]. Moscow: Mir.

Hammett, L. P. (1970). Physical organic chemistry. Reaction rates, equilibria and mechanisms. New York: McGraw-Hill.

Kashkin, V.B., Rubleva, T.V., & Hlebopros, R.G. (2015). *Stratosfernyi ozone: vid s kosmicheskoi orbity* [Stratospheric ozone: a view from orbit]. Krasnoyarsk: SFU.

Kleinman, L., Lee, Y.-N., Springston, S. R., Nunnermacker, L., Zhou, X., Brown, R., ... Newman, L. (1994). Ozone formation at a rural site in the southeastern United States. *Journal of Geophysical Research, 99*(D2), 3469–3482. doi:10.1029/93JD02991

NASA Ozone Watch. (2020). Retrieved from: https://ozonewatch.gsfc.nasa.gov/

Nicolis, G., & Prigogine, I. (1979). Samoorganizatsiya v neravnovesnykh sistemakh. In Dissipativnych struktur k uporyadochennosti cherez fluktuatsii [Self-organization in nonequilibrium systems: From dissipative structures to order through fluctuations]. Moscow: Mir.

Prigogine, I. (1985). *Ot suschestvuyuschego k voznikayuschemu, Vremya I slozhnost v phizicheskikh naukakh* [From Being to Becoming: Time and Complexity in the Physical Sciences]. Moscow: Nauka.

Prigogine, I., & Stengers, I. (1986). *Poryadok iz khaosa. Noviy dialog cheloveka s prirodoy* [Order out of chaos. Man's new dialogue with nature]. Moscow: Progress.

Schrödinger E. (1944). *What is life? The physical aspect of the living cell.* Based on lectures delivered under the auspices of the Dublin Institute for Advanced Studies at Trinity College, Dublin, Ireland.

Seinfeld, J. H. (1989). Urban air pollution: State of the science. *Science*, *243*(4892), 745–752. doi:10.1126cience.243.4892.745 PMID:17820421

Seinfeld, J. H., & Pandis, S. N. (2006). *Atmospheric chemistry and physics. From air pollution to climate change* (2nd ed.). John Wiley & Sons, Inc.

Serdyutskaya, L. F., & Kameneva, I. P. (2004). Modeli vodnyh ecosystem i ih phazovye portrety (na primere modeli krugovorota azota v Kuibyshevskom vodohranilische) [Models of aquatic ecosystems and their phase portraits (on the example of the nitrogen cycle in the Kuibyshev reservoir)]. *Ibid*, *1*(Appendix), 408.

Shitikov, V. K., Rozenberg, G. S., & Zinchenko, T. D. (2003). *Kolichestvennaya gidroekologiya: metody sistemnoy identifikatsii* [Quantitative Hydroecology: methods of system identification]. Tolyatti: IEVB RAS.

Skurlatov, Y.I., Duka, G.G., & Mizitty, A. (1994). *Vvedenie v ekologicheskuyu khimiyu* [Introduction to ecological chemistry]. Moscow: Vyshaya shkola.

Wuebbles, D. (2020). Ozone Depletion. *Encyclopedia Britannica*. Retrieved from: https://www.britannica.com/science/ozone-depletion

Zvyagintsev, A. M., & Kuznetsova, I. N. (2013). Formirovanie urovney ozona v prizemnom vozdukhe [The formation of ozone levels in surface air]. In Collection «Vtoroye mezhdunarodnoe soveschanie-seminar: Problemy monitoringa prizemnogo ozona i puti neytralizatsii ego vrednogo vliyaniya, Tarusa 6-7 June 2012 [Second international meeting-seminar on Problems of monitoring ground-level ozone and ways to neutralize its harmful influence." Tarusa 6-7 June 2012]. Moscow: Academic Press.

Chapter 6
Contribution of the "Great Carbon Tracker" in the Dilution of Atmospheric Radiocarbon Emissions

Sergey Olegovich Travin
Federal Research Center for Chemical Physics After N. N. Semenov of the Russian Academy of Sciences, Russia

Oleg Borisovich Gromov
High-Tech Research Institute of Inorganic Materials Named After Academician A. A. Bochvar, Russia

ABSTRACT

The general model of global carbon circulation (in particular ^{14}C) have been known for half a century and includes both natural and anthropogenic sources and sinks. The main problem was the lack of information about rates of migration from one form to another. The authors made fitting of model data concerning both ^{12}C and ^{14}C dynamics in atmosphere and in sea water. The 70-times discrepancy between generation and decay rates for radiocarbon was explained. The main channel of ^{14}C disappearance is not decay but dissolution in sea waters. By this reason, they deny the Suesse hypothesis because atmospheric carbon dioxide is diluted not by ^{14}C-free substance (after burning of fossil fuels) but from the source of oceans. Multiple numeric experiments modeling ^{14}C pulse or permanent input to the environment had demonstrated the exclusive stability of natural system. It obeys homeostatic principle and resists any attempts to shift steady state concentration. So, all speculations about extraordinary threat from technogenic ^{14}C from nuclear industry do not have any scientific base.

DOI: 10.4018/978-1-7998-1241-8.ch006

Figure 1. The long-term increasing trend for concentration of atmospheric carbon dioxide over the whole observation period in the Observatory of Mauna Loa (USA) from March 1957 to July 2015 (Dr. Pieter Tans, 1915) – pale line. Dark line – our kinetic modeling

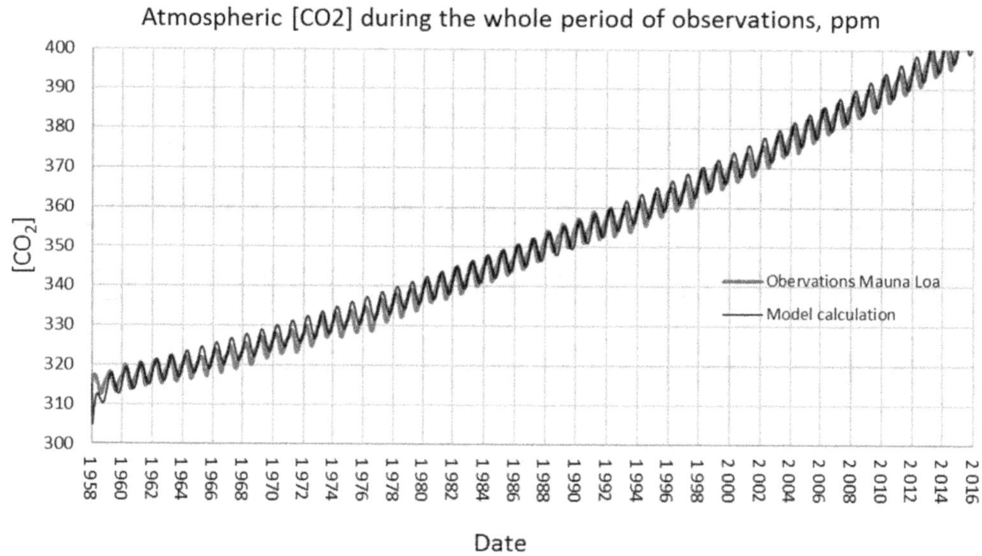

INTRODUCTION

Recent years, increasing public attention is drawn to the discussion of various aspects of the emergence, transformation and migration of radioactive isotope ^{14}C (hereinafter, radiocarbon). To some extent this is due to really existing risk of receiving high doses of radiation as a result of anthropogenic emissions of this radionuclide. Much more "popularity" of radiocarbon is connected with its fame, as a basic element used in the dating of historical artifacts by Libby method. In this paper we will be interested in primarily quantitative picture of the sensitivity of the atmosphere and biosphere of the Earth to additional emissions of ^{14}C.

The model of global circulation of carbon-14 have been known for over half a century and includes both natural and anthropogenic sources and sinks. Along with the generation of ^{14}C from atmospheric nitrogen by cosmic radiation and radiation arising from nuclear testing one should take into account also flow to the atmosphere from the emissions and discharges of the nuclear fuel cycle plants. Currently, the main source of the anthropogenic radiocarbon are not the nuclear power plants themselves (as it is sometimes presented by the media), but factories for the regeneration of spent fuel rods.

Carbon-14 exists in the atmosphere mainly in the form of $^{14}CO_2$ and partly in the form of ^{14}CO, which is then oxidized with the characteristic time $\tau \approx 5\,\text{monthes}$ (Arslanov, 1987).

It is believed that the modern nuclear industry gives wastes in the atmosphere from 0.01 to 1 TBq of ^{14}C per day (Bolin, 1972). Even for the upper limit the ratio of anthropogenic radiocarbon does not exceed 25% of its natural production and cannot play important role in the radiological impact on the Biosphere.

In addition, over the last 50 years, at the same 25% increase in the quasi-equilibrium CO_2 (stable isotope) in the atmosphere (Fig. 1). This leads to the so-called Suesse effect: the burning of fossil fuels containing no radiocarbon, lowers its relative content in the atmosphere. Really, the dilution of the "modern" carbon dioxide with "ancient" one, in which ^{14}C had existed hundreds of half-lives without

Figure 2. Annual variations in the concentration of carbon dioxide in the atmosphere

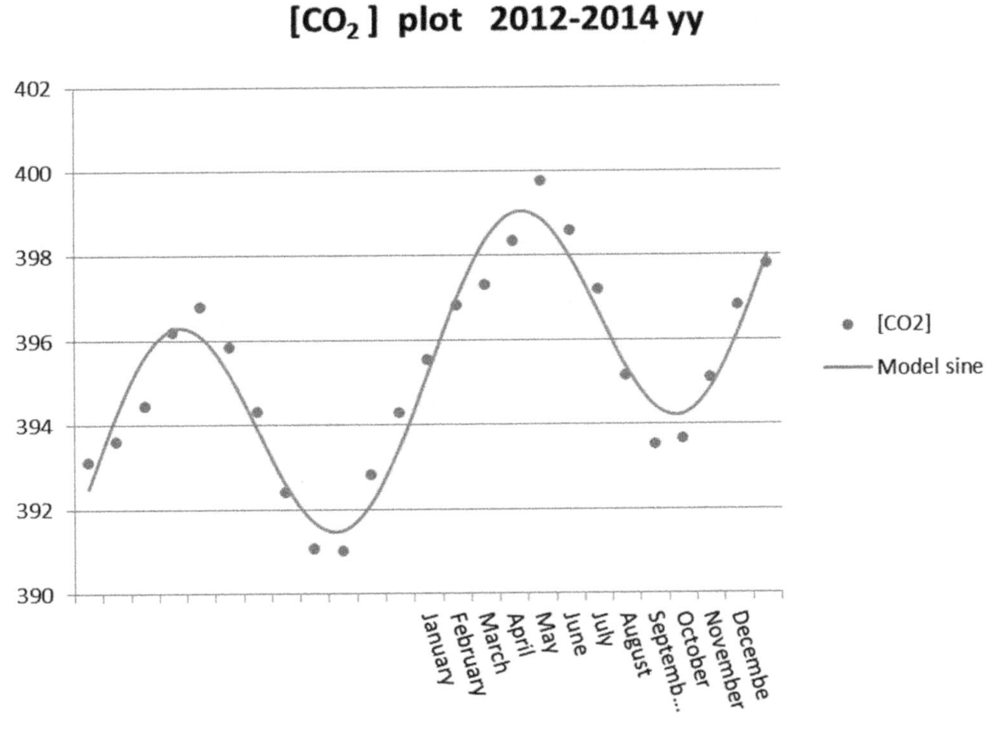

exchange with the atmosphere and completely disintegrated. The content of CO_2 in the atmosphere has increased dramatically over the past decade and now stands about 400 ppm.

To estimate the dynamic characteristics of transfer processes in the carbon cycle of the planet, one must notice the high-frequency oscillations superimposed on the growing trend.

Figure 2 shows the model curve approximation of the annual fluctuations in the concentration of atmospheric carbon dioxide. The maximum rate of recession, i.e. CO_2 consumption by plants and oceanic plankton occurs in June-July, while the maximal excess of industrial emission upon biogenic) consumption correspond to the peak of heating season (in the Northern hemisphere, where than 90% of the population is located), i.e. in December-January.

It is believed that up to 1880, anthropogenic CO_2 emissions were not significant and the was no increase in the concentration. The beginning of the twentieth century markedly increased energy production based on burning carbon-containing fossil fuels. Model calculation assumes that technogenic income of CO_2 is about 18% of total turnover and is permanently growing on 1.6 percent per year (currently).

Once more one can repeat that fossil carbon does not actually contain a radioactive isotope. In addition to the natural channels, in the 50-ies of the last century there have been significant inflow of "bomb carbon" to the atmosphere, generated during nuclear weapons tests" (Vasilenko et al., 1992). Study of the consequences of this phenomenon will be a big part of our work.

Table 1.

Geosphere	The amount of carbon billion tons
The atmosphere	**610**
Hydrosphere	**43 373**
Dissolved Organic Carbon (OC) of the ocean	2 150
Carbonate ion, bicarbonate ion ocean, marine sediments, carbonates	40 552
Carbonate-, bicarbonate-ion freshwater, OC in precipitation of lakes, bogs	671
Biosphere	**540**
Autotrophs. Marine and freshwater phytoplankton and phytobenthos	1
Woody terrestrial vegetation	537
Heterotrophs Marine and continental	2
Stratisphere	**18 665 900**
Marine sedimentary rocks	14 000 000
Freshwater sediments	110 000
OC of marine and continental deposits	3 600 000
Coal	891 000
Oil	1 900
CO_2	60 000
CH_4	3 000
Total amount of carbon in the Earth crust	**18 200 000**

BASIC VALUES FOR MODELING

Let us regard a few reference numbers.

Table 1 Distribution of carbon forms in the Earth's crust (fragments) (by E. M. Galimov (1968) with additions by S. I. Golyshev in the Chemists Handbook of the XXI century). The amount of carbon in the atmosphere corresponds to fraction 298 ppm, which is out of date. Modern value is about 400 ppm. Also we do not confirm figures for coal and oil, they have no significance for present paper.

Table 1 shows the distribution of carbon in the Earth's crust.

the surface area of the earth is 5.1_*10^{18} cm^2,
the surface area of the oceans is 3.6_*10^{18} cm^2 (over 2/3 of total),
the total rate of generation of ^{14}C of 1.22_*10^{19} atoms/s, or 0.29 mg/s,
the flow of 14C into the Earth atmosphere is 8.98 kg 14C/year,
The decay halftime $T_{1/2}$ = 5730 years corresponds to the rate constant
1.21_*10^{-4} $years^{-1}$ or 3.83_*10^{-12} s^{-1}.

These figures are fundamental for calculations of this paper.

The radionuclide ^{14}C is constantly produced in the upper atmosphere (at a height of 8-18 km) in the interaction of neutrons of cosmic origin and nuclei of nitrogen in reaction with an average speed of about 2.4 atom/s_*cm^2.

Table 2. The proportion of radiocarbon in the carbon of the atmosphere (beginning of 2000-th)

Source data	The result of the calculation
The specific activity is A=226 Bq/kg carbon $K_{-decay} = 3{,}83 \cdot 10^{-12}\ s^{-1}$ ($\tau_{1/2} = 5730$ years)	2000 y, 360 ppm CO_2) Mass of atmospheric ^{12}C is $7{,}60.10^{11}$ tons
GENERATION	
Generation rate of radiocarbon is 2,4 atom/s*cm^2 Earth surface is $5.1.10^{18}$ cm^2	Total amount of ^{14}C births is 2,4 atom/s*cm^2*$5.1.10^{18}$ cm^{2}1.22*10^{19} per sec.
DECAY	
$A = Na * \frac{1000}{12} * fraction * k_{decay}$	$fraction = \frac{12 * A}{1000 * Na * k_{decay}} = 1.18 \cdot 10^{-12}$
[^{14}C] = $1{,}18 \cdot 10^{-10}$%, The mass of ^{14}C in the atmosphere is 1040 kg	Or $4.48.10^{28}$ atoms of ^{14}C which can give $1.72.10^{17}$ decays/s ONLY!
W_gen = W_decay Must be!	
The difference between generation rate and expected decay rate is **71** times.	
It demands 74 tons of decaying 14C for steady state position or equilibrium	

According to the calculation, the total mass of carbon in the carbon dioxide gas of today atmosphere is 834 billion tons, which is almost one third higher than the value of 1958. (For comparison, the total number of explored reserves of coal (i.e. everything that the biosphere had accumulated over the last few billion years of the Earth existence) is 891000 billion tones – approximately three orders higher, then it is contained in the atmosphere. It looks like "conversion" from gas to coal was fulfilled only one thousand times since birth of life on the Earth. That means that the average duration of one cycle takes more than million years, manifold long period in comparison with exchange time for equilibrium between atmosphere and ocean. So the involvement of CO_2 in the perspiration of plants during photosynthesis, followed by its transformation into condensed forms (liquid + solid) – this is a tiny part of Great Turnover. The quasi-equilibrium between CO_2 consumption, which is the residual of rates of mainly two processes (perspiration of plants and dissolution in the oceans) and oxidation of fuels, food and organic stuff giving release of CO_2 in atmosphere plus dissolution of natural calcium carbonates, - this equilibrium is very accurate. The difference in rates of plus-process and minus-process is several orders of magnitude lower, they each item itself. Also, the oceanic DIC (dissolved inorganic carbon) is 50 to 70 times higher, then atmospheric total amount of CO_2, and mainly presented not in neutral forms of dissolved gas and hydrocarbonic acid, but in ionized form HCO_3^-. It is well known, that ionic equilibria in water are extremely fast, so the statements like "anthropogenic carbon dioxide output to atmosphere increases its concentration" are exactly the same with "The tail wagging the dog".

Thus we can avoid as well the speculation about the significant involvement of atmospheric radiocarbon into the food chain.

Figure 3. Atmospheric increase = Emissions from fossil fuels + Net emissions from changes in land use – Oceanic uptake – Terrestrial carbon sink
The diagram taken from site http://www.galacticwind.com

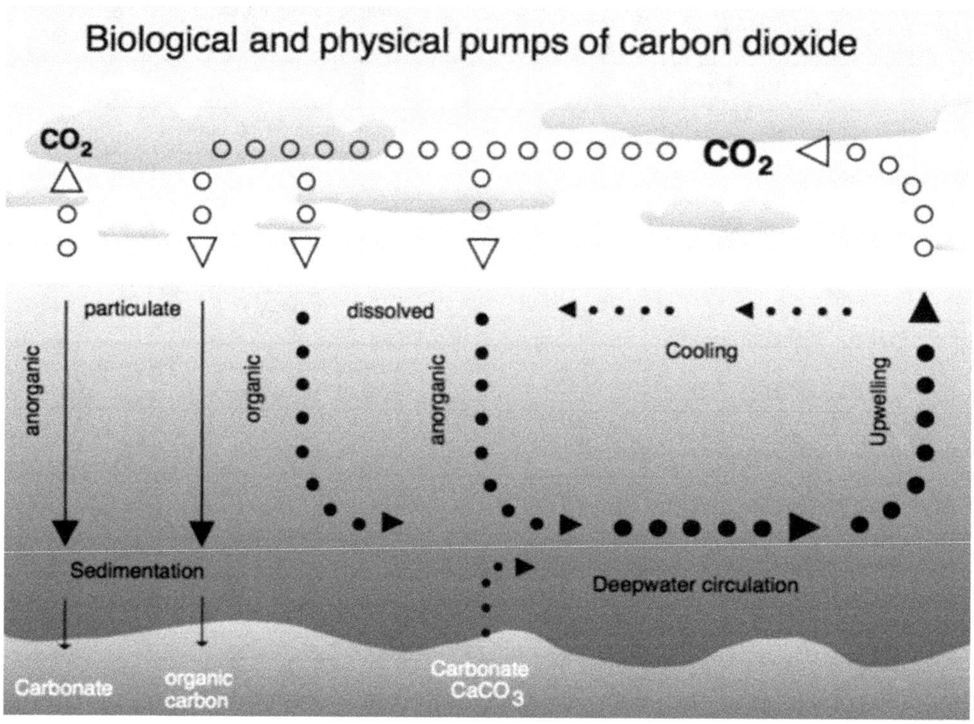

RESULTS AND DISCUSSION

The striking discrepancy between the results of two approaches of evaluation of the stationary amount of radiocarbon, reaching almost two orders of magnitude, is a very important signal. It is absolutely obvious that in fact the rates of generation and decay are equal with very high accuracy, neglecting small shifts of steady state level.

Because there is no reason to doubt neither the rate of production of ^{14}C, nor the rate constant for its spontaneous decay, we just have to admit that the actual amount of decaying radiocarbon is approximately two orders of magnitude bigger, then atmosphere contains. In other words, atmospheric source of CO_2 is attached with the oceanic capacity and there is rapid exchange (in terms of radioactive lifetime) between two reservoirs. So for namely atmospheric ^{14}C not decay is the main channel of its disappearance, but transfer to the dissolved form in the Ocean.

There is no mystery here. Looking at the distribution of carbon forms compounds in the atmosphere and the Earth's crust is easy to guess what kind of a channel it is. Because over the history of the Earth the main amount of carbon was stored in the form of marine sedimentary rocks (carbonates of calcium and in minor part magnesium, i.e. mainly chalk, marble, and other varieties of limestone) that is binding it in inorganic compounds is the main route of withdrawal of radiocarbon from circulation.

This diagram of the carbon cycle shows the movement of carbon between land, atmosphere, and oceans. Yellow numbers are natural fluxes, and red are human contributions in gig tons of carbon per year. White numbers indicate stored carbon. (Diagram adapted from U.S. DOE, Biological and Environmental Research Information System by NASA Earth Observatory, (https://genomicscience.energy.gov/).

Only involvement in the isotope exchange strongly aged million-year-old substances, such as rocks or coal, almost not containing the radioactive isotope, can lead to an effective decrease in the observed steady-state concentration of radiocarbon.

Quite often in the published papers one can meet something like the following text: "Under the influence of wind the atmosphere is constantly mixed, and finally the radioactive carbon dioxide is evenly distributed in usual atmospheric carbon dioxide. Thanks to the constant flow of cosmic rays bombarding the Earth's atmosphere, the rate of formation of ^{14}C is constant. Emitting an electron and an antineutrino, ^{14}C is transformed into stable nitrogen. ... Joint effect of radioactive losses and new formations in the stratosphere leads to a constant, albeit negligible, **equilibrium** (actually steady-state) **concentration** of ^{14}C in the biosphere"(Kulkova, 2011, bold font added by chapter authors).

If the steady-state level of ^{14}C was achieved by Oceanic transfer, not due to radioactive decay, the input of the radiocarbon into the food chains is actually two orders of magnitude lower than it is usually regarded, because only one percent of radiocarbon goes to trophic chains. In other words, in any case one must not forget about the non-radiative channel which is a hundredfold more powerful then radioactive decay in the attenuation of the radiocarbon level. This fact is always overlooked in all publications without any exclusion.

The correct statement is as follows:

Every year cosmic rays in the atmosphere produce about 9 kg of ^{14}C. The nuclide ^{14}C is radioactive and decays continuously. The whole Earth crust contains about 60 tons of radiocarbon, which decays annually on 9 kg — the same as it is formed (radioactive equilibrium or steady-state).

In the process of photosynthesis ^{14}C assimilated by plants, through which it goes later to the organism of animals and humans. It is necessary to stress that this is only residual ^{14}C compared to its total production, because over 98% of generated radiocarbon is absorbed by the oceans and is carried away. Mainly in the form of inorganic sedimentary deposits (94%) and only about 6% involved in organic turnover.

The emission of ^{14}C from nuclear power plants are additional, but highly insignificant factor in the input of this radionuclide in the atmosphere. Larger quantities of ^{14}C, which are comparable with its production in the atmospheric air are emitted during the processing of fuel rods in which it accumulates as a result of neutron activation of impurities of fuel and coolant, as well as due to the "bomb" radiocarbon.

Figure 4 confirms the stated point of view.

Solid lines show the exponential approximation for the Northern hemisphere (upper) and for the South (lower).

After the atmospheric thermonuclear tests of the mid-sixties the content of radiocarbon in the atmosphere of the Northern hemisphere (in the South we just were able to find only approximate data) has increased eightfold. This figure surpassed almost all the periodicals of the world in which there were some calls for the struggle for environmental cleanliness of the planet. However, again all authors had overlook the time constant of "decay" of a radiocarbon. In the figure the blue curve corresponds to the disappearance of ^{14}C with rate constant 0,066 year^{-1}, which corresponds to a characteristic time $\tau \approx 15\,\text{years}$ ($\tau_{1/2} = ln2 \cdot \tau \approx 10\,\text{years}$).

Figure 4. The decline of radiocarbon background after the nuclear tests in the atmosphere (dot points)

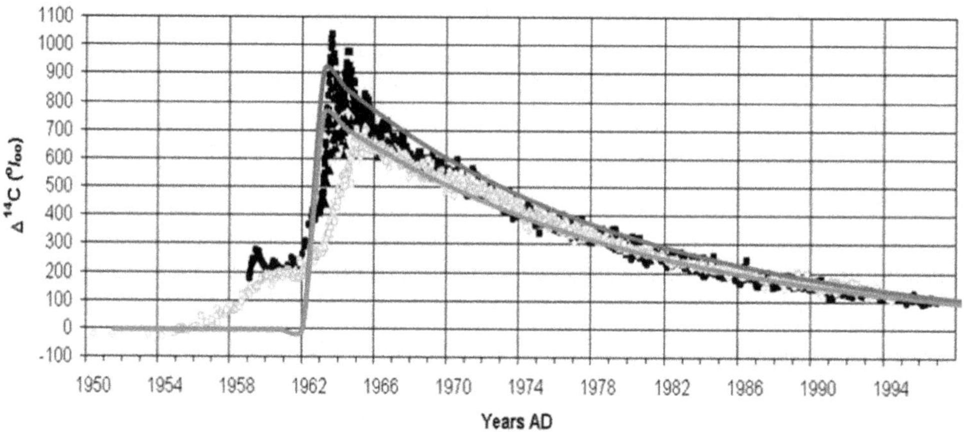

Such dynamics of "resolution" of volley ejection is absolutely unequivocal proof of the withdrawal of radiocarbon from the cycle of atmospheric circulation in the non-radiative channel, which is more than 600 times rapid then traditional radiation. In fact, it can only mean that in this channel there is an intensive exchange between the mixture with natural content of isotope, and huge source with radioactive-to-lean carbon. The latter is possible if and only if carbon from "exchange fund" did not participate in atmospheric circulation tens and hundreds of half-lives of ^{14}C, i.e. millions of years. It is obvious that again we deal with inorganic sedimentary rocks, which are not involved in trophic chains.

It should be noted that a very similar approach recently published work of A. V. Byalko, 2013. Here is an excerpt from his article:

"There are several carbon pools in the ocean and on land with different mass M, with different concentrations of 14C and varying intensity of gas exchange with the atmosphere. On land carbon is fossil fuels and biomass in the ocean is carbon in the form of dissolved CO_2, organic carbon, calcite and sediment methane hydrates. The biomass of ocean and precipitation differ in isotopic composition of carbon. A substantial portion of dissolved carbon dioxide merges with the "Great Conveyor" (GC) in the depths of the ocean and transferred them to great distances, where it is released into the atmosphere with a markedly reduced concentration of radiocarbon. Unfortunately, justified estimates for the flow of water masses GC does not yet exist, the more unable to estimate the amount of the transferred they are the mass of carbon".

All data obtained can be quantitatively modeled according to scheme from the fig.5.

Somewhat earlier, in 1958, the academician A. D. Sakharov pointed out negative biological consequences of the increased concentration of radiocarbon (Sakharov, 1958).

His article played a major role in the conclusion of an International Treaty banning nuclear tests in the atmosphere; it entered into force in October 1965.

"By this time, natural background levels of radiocarbon were exceeded twice. Then followed a rapid drop in the atmospheric concentration of 14C". More important conclusion which comes A. V. Byalko and with which we fully agree:

"The conclusion of Sakharov about the long-term impact of radiocarbon on the biota did not take into account the role of gas exchange of the atmosphere with the ocean, the parameters of which 50 years ago were unknown"

Figure 5. Rate constants and masses in mass action law for carbon turnover modeling

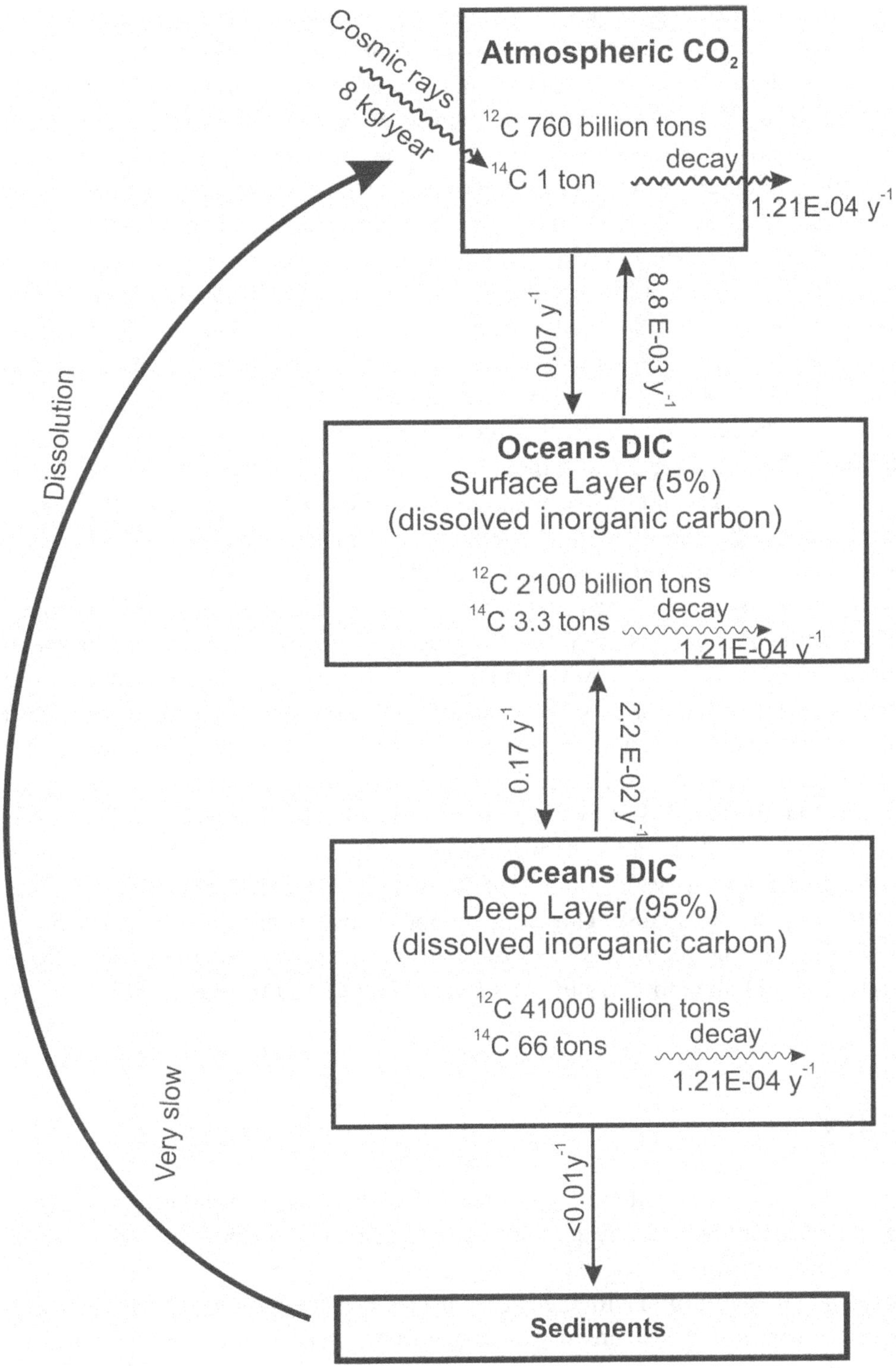

Now we had calculated them from the model.

Evaluating the time constant (or, equivalently, opposite to it the sum of the all pseudo-first order rate constants of all reactions) to substitute the radioactive carbon by stable isotope, it can be noticed that this factor attenuates the biological effects of ^{14}C at least two orders of magnitude (up to 600 times).

The damaging effects of ^{14}C included in the composition of protein molecules, and particularly high in DNA and RNA of a living organism caused by radiation exposure of β-particles and recoil nuclei of nitrogen and changes in the chemical composition of molecules in the result of the conversion of carbon into nitrogen. A significant portion of DNA damage in the decay of ^{14}C predominantly leads to gene mutations; it is argued that such changes are almost not recovered by the repair system and are irreversible.

The mention that "the apparent lack of genetic load as a result of irradiation of natural radiocarbon, apparently, is due to the generation during the evolution of protective mechanisms that remove mutational damage in different stages of development of organisms" (Vasilenko et al., 1992), to a certain extent confirm our point of view on the global elimination of radiocarbon from the atmosphere.

In addition to the already said, it is impossible to agree with the reasoning (Vasilenko et al., 1992) on the kinetics of carbon exchange. The authors of this work believe that the "oxidized in the environment to $^{14}CO_2$ carbon-14 through photosynthesis, accumulates in plants, then via the food chains goes to animals and humans. The transition rate in the chain "of atmospheric carbon — the carbon of plants" is equal to few percent of Great Turnover.

According to our calculations, this factor is two, if not three orders of magnitude below one. Moreover, the balance of direct and inverse processes is always for one and the same time equal to the reciprocal of the sum of the rate constants for forward and backward processes is known from a basic course of chemical kinetics. In fact, for the stage of "cleansing food" a stationary state is reached for the same a couple of months, and their pollution, and a slower coming years, the process is modulated by a decrease in radioactivity (and not separate products) related to oceanic mass and a Great Turnover.

One can assess the doses received by the human body from the presence in his cells (and the cells they are consumed food) background level of radiocarbon. This can be done by following the clear and visual technique (Broeker W. S. & Walton A., 1959.). In fact, we received result of 5 μSv (microsievert) instead of usually calculated 200 μSv. This is the obvious consequence from established fact that from formed 9 kg of ^{14}C, all plants, animals and humans can consume only 120 grams, and the rest is falling apart and carried away by the processes of mass transfer in GC and further disintegrates in the depths of the oceans.

CONCLUSION

Thus, we can say that the participation of atmospheric CO_2 in the global carbon pipeline is much more significant than its absorption in the respiratory activity of plants (absorption) and animals (exhaust), as well as its receipt from energy activities. The greatest contribution to mass transfer is made by an unusually large area of equilibrium with the ocean surface and the carbonate buffer located in it.

The observed and unmistakable increase in atmospheric carbon dioxide concentration is not due to an increase in the amount of fossil fuel burned annually, an increase in the number of cattle farms, or other anthropogenic causes. In fact, the whole picture completely fits into the shift of equilibrium in the ocean carbonate buffer, which in turn is associated with a barely noticeable (but measurable) shift in the pH of ocean water to the acidic side with a simultaneous increase (within 1-2 degrees) in the temperature of the surface layers. The main fate of the additional portions of CO2 that appeared in the atmosphere is to be absorbed not by plants, but by the ocean, with the final fate of finding its place among the Cretaceous deposits, which contain more than 99.8% of the planet's carbon. This applies equally, if not more, to the radioactive isotope ^{14}C.

Also we deny the Suesse hypothesis, because atmospheric carbon dioxide is diluted not by ^{14}C-free substance (after burning of fossil fuels), but from the source of oceans, which have exactly the same composition and ratio ^{14}C / ^{12}C with atmospheric carbon compounds.

And the very last. Multiple numeric experiments with imitation modeling of ^{14}C pulse or permanent input to the environment, had demonstrated the exclusive stability of natural system. It obeys homeostatic principle and resists to any attempts to shift steady state concentration. For instance, even in case of twofold increase of cosmic irradiation, the amount of ^{14}C in atmosphere will rise only few percent during nearest 100 years. So all speculations about extraordinary threat from technogenic ^{14}C from nuclear industry do not have any scientific base.

REFERENCES

Arslanov, K. A. (1987). Radiocarbon: Geochemistry and geochronology. Publishing house Leningrad State University.

Bolin, B. (1972). *The carbon Cycle. Biosphere.* Izd-vo "Mir".

Broeker, W. S., & Walton, A. (1959). Radiocarbon from Nuclear Tests. *Science, 130*(3371), 309–314. doi:10.1126cience.130.3371.309 PMID:13668560

Byalko, A. V. (2013). Variations in the concentration and gas exchange atmosphere-ocean. Proc. Russian Acad. of Sciences, 451(1), 28-32.

Chemist's Handbook of the XXI century. (n.d.). Retrieved from https://chem21.info/page/1511721551521860500482321621512370381340480221 08/

Kulkova, M. A. (2011). Radiocarbon (14C) in the environment and the radiocarbon Dating method. In *Training Handbook.* Izdatel'stvo RGPU im.

NOAA/ESRL. (n.d.). Retrieved from https://genomicscience.energy.gov/

Sakharov, A. D. (1958). Radioactive carbon in nuclear explosions and biological non threshold effects. Atomic Energy, 4(6).

Vasilenko, I. Y., Osipov, V. A., & Rublevsky, V. P. (1992). Radioactive carbon. *Nature, 12*, 59-65.

Section 2

Water, Aqueous Systems, Soil

Chapter 7
Role of Oxidation–Reduction Processes in Formation of Toxic Properties of Natural Aqueous Environment

Yuri Ivanovich Skurlatov
Federal Research Center for Chemical Physics After N. N. Semenov of the Russian Academy of Sciences, Russia

Elena Valentinovna Shtamm
Federal Research Center for Chemical Physics After N. N. Semenov of the Russian Academy of Sciences, Russia

Sergey Olegovich Travin
Federal Research Center for Chemical Physics After N. N. Semenov of the Russian Academy of Sciences, Russia

Vyacheslav Olegovich Shvydkiy
N. M. Emanuel Institute of Biochemical Physics of the Russian Academy of Sciences, Russia

Lyudmila Vasilevna Semenyak
All-Russian Research Institute of Fisheries and Oceanography (VNIRO), Russia

ABSTRACT

Redox processes involving hydrogen peroxide, as oxidizing agent, and compounds of sulfur as carrier of reducing equivalents form the quality of natural waters. The inflow of reductants interacting with H_2O_2 can lead to a toxic quasi-reductive state. Dynamic change of redox in a natural aqueous medium is pernicious for organisms with intensive water-exchange, such as larvae of fishes, despite the concentration of dissolved oxygen being normal. Favorable conditions for "flowering" of emerging toxins blue-green alga are formed. In reductive state, copper ions become biologically unavailable. Sewage after biological cleaning are the main anthropogenic source of the reductants, mainly hydrosulphide. The natural sources of reductants are blue-green alga and the bottom sediments. The ions of Cu(I) and Fe(II) form high-strength 1:1 complexes with reduced sulfur compounds that are stable to O_2 but efficiently react with H_2O_2. The increased content of manganese can form mixed-valence manganese species Mn(III,IV) giving super-oxidizing state of the aquatic environment, which is also toxic.

DOI: 10.4018/978-1-7998-1241-8.ch007

INTRODUCTION

Most of natural redox processes in the environment use dioxygen O_2, which transforms to water via intermediate formation of hydrogen peroxide H_2O_2. In the aerobic water environment and in living cell as the natural carrier of oxidizing equivalents the main role is played by the oxygen dissolved in water which content "normal" makes »2.10^{-4} M, varying in rather narrow limits depending on temperature and the content in water of the easily oxidized substances - carriers of reducing equivalents, DH_2.

Typical reactions with participation of the molecular oxygen dissolved in water in the ground state (triplet) and in electron-excited (singlet) states are shown in Schemes 1, 2.

In the aqueous environment molecular oxygen is chemically inert, effectively it interacts mainly with organic free radicals •R with formation of peroxide-radicals ROO •, and also with the electronically excited molecules S* with formation or the superoxide radical $\bullet O_2^-$ – at transfer of an electron, or singlet oxygen - when electronic excitation is transferred (Duca Gh, Skurlatov Iu., Miziti A.,1994):

$$\mathbf{O_2} + \{2H\}\ (DH_2) \overset{M^+}{\rightleftarrows} \mathbf{H_2O_2}\ (+ D)$$

$$\mathbf{O_2} + 2\{2H\} \overset{M^+}{\rightleftarrows} 2\ H_2O$$

$$\mathbf{O_2} + RH \overset{M^+}{\rightleftarrows} OH + H_2O$$

$$\mathbf{O_2} + {}^{\bullet}R \rightleftarrows ROO^{\bullet}$$

$$\mathbf{O_2} + e^-\ (S^*) \rightleftarrows \mathbf{{}^{\bullet}O_2^-}\ (+ {}^{\bullet}S^+)$$

$$\mathbf{O_2} + h\nu\ ({}^3S^*) \rightleftarrows \mathbf{{}^1O_2}\ (+ S)$$

Scheme 1

Oxidation of H-donors by molecular oxygen is possible only at its activation in the coordination sphere of an ion of metals of a variable valence which is in the reduced form (M^+) (Sychev et al., 1983).

As a rule, in the oxidase processes oxidation of H-donors is followed by the reducing of O_2 to H_2O_2. In the processes of cellular respiration there is mainly four- electronic reduction of O_2 to the water molecule, and in the oxygenase processes - introduction of the O atom of the O_2 molecule into the C-H bond

Singlet oxygen in water either relaxes quickly to the ground state, or interacts either with donors of H with formation of H_2O_2, or with donors of an electron with formation of the superoxide radicals ${}^{\bullet}O_2^-$:

$$\mathbf{{}^1O_2} + H_2O \rightleftarrows O_2 + H_2O$$

$$\mathbf{{}^1O_2} + \{2H\}\ (DH_2) \rightleftarrows \mathbf{H_2O_2}\ (+ D)$$

$$\mathbf{{}^1O_2} + e^-\ ({}^{\bullet}D^-) \rightleftarrows \mathbf{{}^{\bullet}O_2^-}\ (+D)$$

Scheme 2

In turn superoxide radicals in the neutral aqueous environment are partially protonated and easily disproportionate with formation of hydrogen peroxide: pK_a=4,88

$${}^{\bullet}O_2^- + H^+ \rightleftarrows {}^{\bullet}HO_2$$

$${}^{\bullet}O_2^- + {}^{\bullet}HO_2 + H^+ \rightleftarrows \mathbf{H_2O_2} + O_2$$

This reaction in natural waters is competed by interaction of the superoxide radical with the ions of bivalent copper dissolved in water:

H^+

$•O_2^- + M^{2+} \rightleftarrows M^+ + \mathbf{O_2}$

Anyway, oxidation processes involving O_2 are followed by the formation of hydrogen peroxide in an aqueous medium.

In the "norm" when the intra-reservoir redox-processes are balanced, the natural content of H_2O_2 in aquatic environment is about 2 orders less, than the oxygen dissolved in water, and varies within the limits of the order of magnitude depending on a season, weather factors and time of day (Shtamm et al., 1991).

In the absence, of catalysts hydrogen peroxide, like O_2, is kinetically inert. When it interacts with electron donors or under the influence of UV-light the O-O bond breaks to form OH radicals. Oxidation of H-donors by hydrogen peroxide, as well as in case of O_2, is possible at activation of H_2O_2 at entry into the coordination sphere of an ion of metal of variable valence in a reduced form, however unlike O_2, hydrogen peroxide interacts with the ions of metals of variable valence which are not only in a reduced form (M $^+$), but also in the oxidized form - M^2+ (Sychev et al., 1983), scheme 3:

M^+ (Cu^+)

$\mathbf{H_2O_2}$ + {2H} (DH_2) $\rightleftarrows$ $2H_2O$ (+ D)

$\mathbf{H_2O_2}$ + e^- (M^+) $\rightleftarrows$ $\mathbf{•OH}$ + OH^- (MOH^+)

$\mathbf{H_2O_2}$ + { }

$\mathbf{H_2O_2}$ + hν $\rightleftarrows$ $\mathbf{2•OH}$

$\mathbf{H_2O_2}$ + $M^{2+} \rightleftarrows M^{2+}(HO_2^-) \rightleftarrows M^+ + \mathbf{•O_2^-}$

Scheme 3

Generally, metal ions of variable valence can be in water in different states of oxidation: oxidized (M^2+), reduced (M^+) or super- oxidized (M^{3+}). The main role in the intra-reservoir and intracellular oxygen – dependent redox-processes play ions of copper, iron and manganese. According to the general contents in the aquatic environment iron and manganese prevail. Their concentration can reach 0,3 mg/dm^3 and higher.

At the same time, from all ions of metals of variable valence in the natural water environment only copper ions can be dissolved in the oxidized and reduced aqua-forms (Skurlatov & Purmal, 1984), that explains their broad participation in the realization of catalytic redox- processes both in natural waters, and in intracellular processes, despite the relatively low contents in the aquatic environment - at the level of »10^{-7} M and lower.

Unlike copper ions, iron ions in a neutral aqueous environment in the oxidized form (Fe^{3+}) can be in the dissolved form, mainly as a part of complexes with the ligands (L) which usually form chelate complexes. In the reduced form, iron ions (Fe^{2+}) are hydrolyzed and easily are oxidized by oxygen dissolved in water. In contrast, manganese ions in the reduced form (Mn^{2+}) are resistant against oxidation, in the oxidized state (Mn^{3+}) its oxidize water, and in the super oxidized state (Mn^{4+}) form water-insoluble manganese dioxide MnO_2.

The main redox- processes with participation of O_2, H_2O_2 and metal ions of variable valence are presented in the scheme 4:

e- e-

$M^{3+} \rightleftarrows M^{2+} \rightleftarrows M^+$

$L + (M^{2+/+})_{aq} \rightleftarrows (LM)^{2+/+}$

$(M^+ O_2 \rightleftarrows M^{2+}O_2^-)$

ì
$LM^{+} + O_2 \rightleftarrows \mathbf{LMO_2^{+}}$
î
$(M^{+} O_2 \rightleftarrows M^{3+}O_2^{=} \rightleftarrows M^{3+}O^{=} ..\mathbf{O})$
$(M^{+} H_2O_2) \rightleftarrows M^{2+}OH^{-}\ \mathbf{{}^{\bullet}OH}$
ì
$LM^{+} + H_2O_2 \rightleftarrows \mathbf{LMH_2O_2^{+}}$
î
$(M^{+} H_2O_2 \rightleftarrows \mathbf{M^{3+}}(OH^{-})_2 \rightleftarrows M^{+}\mathbf{O}\ ..H_2O)$

Scheme 4

Being in the reduced form, ions of metals, in particular copper ions, can interact both with O_2, and with H_2O_2 with formation of reactive intermediate complexes with partial one - or two- (and more) electronic intrasphere charge transfer (Sychev et al., 1983).

Intracellular and intra-reservoir redox- processes involving H_2O_2 can be conditionally divided into 2 types – catalase and peroxidase.

In peroxidase process hydrogen peroxide participates in the oxidation of donors of H, in catalase process the second molecule of H_2O_2 acts as the donor of H. The principal difference between these two processes is that the rate of catalase process is proportional to the concentration of hydrogen peroxide in any range of concentrations. The rate of the peroxidase process may not depend on concentration of H_2O_2 (Scheme 5):

Catalase process: $H_2O_2 + H_2O_2 \rightleftarrows O_2 + 2H_2O$
Peroxidase process: $H_2O_2 + DH_2 \rightleftarrows D + 2H_2O$
catalase process: $W_{Cat} = k_d[H_2O_2]$
ì
$\mathbf{W_f} \rightleftarrows H_2O_2$
î
$\mathbf{W_{Red}} \rightleftarrows DH_2 \rightleftarrows$ peroxidase process: $W_{Per} = W_{Red}$
$[H_2O_2] = (\mathbf{W_f} - \mathbf{W_{Red}}) / k_d$

Scheme 5

Zafiriou, Zepp & Zika (1984) had demonstrated the intriguing evidence for important couplings between photoprocesses and the biosphere. They had discussed an algal-induced photolysis. Microorganisms also can excrete free radicals: NO is emitted in the oceans by nitrifying and denitrifying bacteria and some specialized cells excrete superoxide anion-radical O_2^- and HOOH.

In an open type system, which are both a living cell, and natural water objects, the rate of peroxidase process can be determined by the rate of entering into medium the substances possessed reducing activity which effectively interact with H_2O_2 - W_{Red}. If at the same time the rate of formation of H_2O_2 (W_f) will be less than the rate of involve of reducing substances, W_{Red}, in water the *quasi*- reductive state is formed (Skurlatov et al., 1984). The prefix *quasi*- means the presence of the dissolved O_2 in water, so in thermodynamic sense the aqueous medium remains in an oxidizing state.

As a result of formation of a *quasi*-reductive state, water becomes toxic concerning aquatic aerobic organisms with intensive water circulation. In particular, with respect to larvae of fish in the early stages of their development (Shtamm, 1988). There are favorable conditions for mass reproduction of blue-green algae and pathogenic microorganisms.

In this regard it should be noted that the water protection requirements existing in Russia are focused mainly on control of the pollutants in water in comparison with standards of maximum allowable or maximum-permissible concentration (MPC).

It is believed that if the MPC is respected for all controlled pollutants, natural water is harmless in chemical composition. However:

1. The integral factor of toxic effect on the aquatic ecosystem is actually ignored: direct to aquatic ecosystems or indirect - through the influence of biogenic substances on the intra-reservoir cycle.

At the same time the integrated factor of toxic impact on aquatic ecosystem is actually ignored: both direct interaction on hydrobionts and indirect – through influence on the intra-reservoir circulation of biogenic substances.

2. It is practically impossible to control the entire list of the standardized normalized chemicals in water especially because for many of them the accepted value of maximum permissible concentration is lower than a limit of sensitivity of analytical control methods. Besides, both in waste water, and at natural water there are possible substances for which maximal allowable concentrations aren't established.

In practice the control of the chemical composition natural water and sewage is limited to the list of priority pollutants, mainly, of chemically resistant organic compounds and heavy metals, and also the main ions, the suspended substances, oil products and the biogenic elements.

3. Results of the toxicological researches conducted as us, and many other researchers on a large number of water objects with application of various methods of biotesting (Skurlatov et al., 2011) testify to inadequacy of this approach.

On biotesting indicators correlation between the level of toxicity of tests of water samples and contents of controlled pollutants in them - water of practically all surveyed water objects showed toxicity at rather low level of pollution of water on controlled chemical indicators. At the same time the increased level of toxicity is observed in areas of dumping of city sewage after their biological cleaning.

It testifies, at least, to two basic shortcomings of the operating control and management system for quality of waters.

Firstly, the control is carried out with not of those substances that play the defining role in the formation of the biological fullness of the aquatic environment. Respectively, the accepted indicators don't reflect quality of water as habitats of aquatic organisms.

Secondly, the technology of biological wastewater treatment existing in Russia in the aerotanks not only doesn't provide protection of water ecosystems against toxic influence, and, on the contrary, strengthens it.

As it has turned out (Shtamm et al, 2015) one of the main ways of anthropogenic impact on a redox-state of aquatic ecosystems is the discharge into surface water objects of sewage after their biological treatment.

Still it has been in the mid-eighties established (Duka, 1988) that the sewage entering to the aeration tanks of city treatment facilities from primary settlers contains substances of the reducing nature (**Red**) which aren't oxidized by air oxygen, and at the same time effectively interact with H_2O_2.

These reductants are not volatile substances, aren't sorbed by suspended particles, don't accumulate by the biota, aren't retain in the aeration tank and together with sewage come to water objects where interact with H_2O_2 of a natural origin.

Table 1. Waste water treatment sites as a negative influence factor on the internal oxidation-reduction processes in aquatic environment

Waste waters	**Primary tank**	**Aerotank**	**Secondary tank**	**Wastewater discharge**
Formation of **Red** in anaerobic environment		Biological oxidation of organic compounds, bio-sorption of low-soluble compounds	Separation of suspended particles of activated sludge	Influx of **Red** into reservoir with waste waters

Definition method of these reductive substances is based on titration of water sample by small additions of hydrogen peroxide in the presence of peroxidase and a leuco-base of trimetylphenolic dye. The sensitivity of the method is 10^{-7}M (Zepp et al., 1988).

There is an extensive literature on photoprocesses in natural waters and application of photo processes to the treatment of waters and wastes. Since nineteen eighties numerous studies have yielded considerable insight into the range of light-driven processes that occur in the natural aquatic environment or that result in degradation of toxic or troublesome substances in waters and wastes. In addition, detailed laboratory studies in simple, well defined systems have provided information on the likely kinetics and mechanism of photoprocesses in complex natural waters and wastes (Waite, 1990, Pelizzetti et al., 2002).

Similar substances of the reducing nature, titrated by small H_2O_2 additives, have been found by us in sewage of the pulp-and-paper mills (PPM) (Maksimenko et al., 2000).

In technological process these substances are formed when cooking cellulose and are a part of black liquor part of which (about 1%) falls into the sewage (Shtamm et al, 2015).

Table 2. Content of the substances titrated by hydrogen peroxide (Red) in sewage of pulp-and-paper production

Sampling point	**10^4[Red], M**		
	Kotlass PPM	**Sokol PPM**	**Syktyvkar PPM**
Balancing tank (after primary setting, before biological treatment)	1.7	1.5	0.9
Gravity canal (after secondary tank)	1.3	2.3.	0.75±0.05
Tail drain (after aerating pond)	3.2	1.8	0.79±0.01

Apparently from the data provided in the table, for the different PPM when the sewage passes through the biological treatment installations, the content of reducing substances in them does not decrease and makes about 10^{-4} M in magnitude which is about 2 orders more than the natural content of H_2O_2 in natural water objects. Respectively, even at 100-fold dilution of sewage in a reservoir receiver natural hydrogen peroxide will be destroyed.

In many cases the rate of natural photodegradation of pollutants can be more important than even industrial treatment. The rates of the photochemical processes near the surface are dependent on several environmental factors such as season, latitude, time-of-day, depth of the water body (Kouassi et al, 1990).

The decrease of the photochemical rates with depth is largely dependent on water type. The rates decrease sharply with depth in high organic-content coastal water. The photochemical processes extend deeper in the open ocean (where mixing processes are intense and the mean depth of the photic zone extends deeper).

Natural water fluorescence is mainly influenced by the intense solar radiation in the surface layers and relatively less in the deeper layers

Based on indirect data, the reductive substances interacting effectively with H_2O_2 and inert with respect to O_2 are *water-soluble reduced sulfur compounds* (Frog et al, 2012).

Feature of these compounds is their lack of pronounced analytical properties.

The very peculiar study of photolysis of herbicides in natural and mineral waters was reported by Benitez, 2004. In this work the photooxidation of the selected herbicides in some natural and mineral waters was carried out to analyze the influence of water quality parameters and to propose a general kinetic model for the elimination of herbicides in natural waters, which uses the kinetic parameters and quantum yields.

Photogeneration of free radicals (sometimes via singlet oxygen) can occur with participation of dissolved organic matter (Sandvik et al., 2000) as well as on heterogeneous suspensions of photo sensitive nano-particles, such as TiO_2 (Salvador, 2007).

The controversy on whether photogenerated free holes in valence band are trapped either at adsorbed water species to produce adsorbed OH radicals or at terminal oxygen ions is still on the level of hypothesis. Nevertheless, the author of this chapter agrees with Salvador`s point of view:

OH radicals in the water layers close to the TiO_2 surface can`t be photogenerated, moreover, adsorbed OH radicals may not be desorbed and diffuse into the aqueous solution; such mechanism must be ruled out because we see only pollutants degradation, but not photoinduced oxidation of water.

Sources and sinks of hydroxyl radicals upon irradiation of natural water solutions are in details discussed in the paper by Vion et al, 2006.

Also the author must emphasize, that not in all published papers are used proper conditions for water treatment with hydrogen peroxide and/or photoinitiated catalytic processes. For instance, in paper by Horvat and Huszank (2003) the concentration conditions and pH are four – five orders of magnitude different from what one can expect in natural waters, so it is difficult to extrapolate their results to environment.

By methods of biotesting we found that the treatment of municipal waste water or PPM sewage, that show toxic influence on various test organisms, by the small additives of H_2O_2 comparable to the content of reductive substances in water, leads to their detoxification (Shtamm et al, 2015).

The similar effect of a detoxication with small additives of hydrogen peroxide is also observed with respect to the effect of MFO- induction – a multiple increase in monooxygenase activity of a liver of the fishes living in the area of wastewater discharges of PPM (Shtamm et al, 2015), see Figure 1.

Test organisms were the young of rainbow trout. Dotted line is normal level of the monooxygenase activity in control (under normal conditions).

Figure 1 . Effects of hydrogen peroxide on chronic toxicity (MFO- induction of black liquor (1: 5000 dilution).

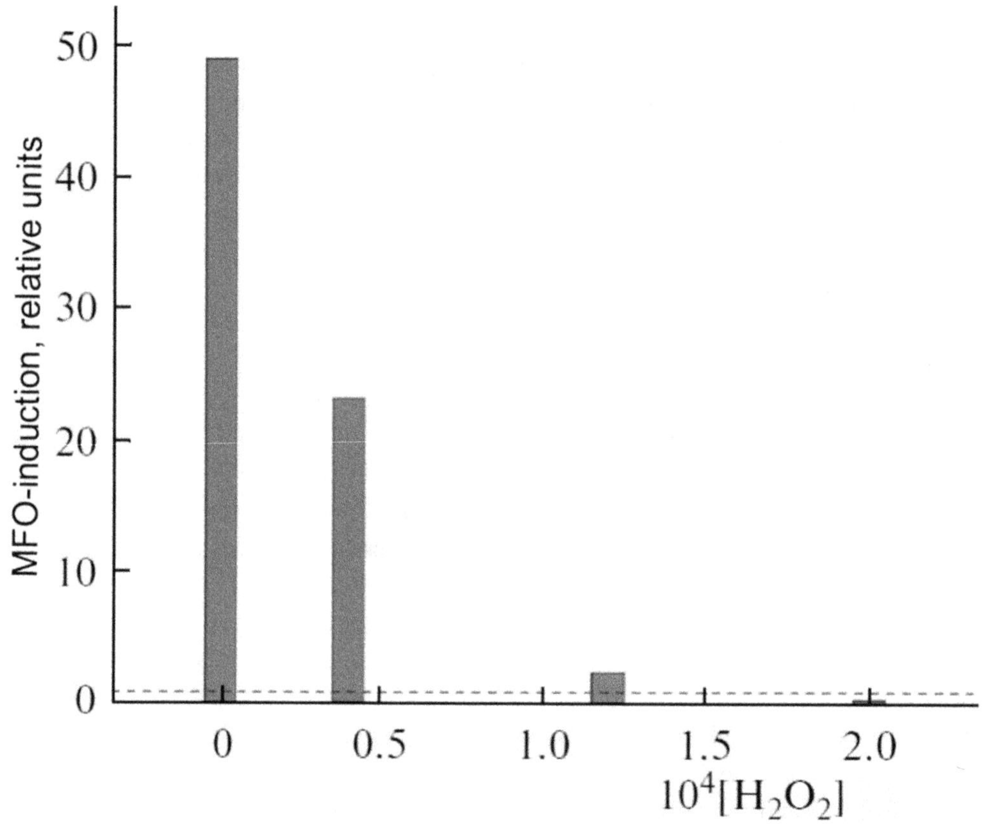

By no other methods, except a nano-filtration (Hewitt et al, 1996) the effect of MFO- induction isn't removed. Toxicity of sewage is approximately the same as for the enterprises with a cellulose bleaching by chlorine, and the enterprises using the newest technologies of a bleaching of cellulose and sewage treatment. At the same time no correlation between the level of MFO-induction and content in fish tissues of such "priority" toxicants as the polychlorinated dibenzodioxins, dibenzofurans or phenols were revealed.

This circumstance demonstrates that toxicity of both municipal, and PPM waste waters is caused, mainly, by *uncontrollable water-soluble substances* which are effectively interacting with hydrogen peroxide.

Sources of soluble reducing substances inert towards O_2, but effectively interacting with H_2O_2 are:

- Urban waste waters during transportation through the sewage network and settling tanks;
- Industrial waste waters from pulp and paper mills (PPMs), as well as other industries associated with processing of raw materials containing sulfur compounds;

- Blue-green algae during blooming period;
- Sediments rich in organic matter in the periods preceding blooming of blue-green algae.

Similar reductive substances are formed also in the natural aquatic environment, for example, at mass development of blue-green algae, and also can enter to water column from the bottom in which anaerobic conditions were formed.

Presumably, water-soluble reduced sulfur compounds are derivatives of hydrosulfide and thiolic compounds as a complexes with ions of metals of variable valence in a reduced state, mainly, with ions of iron and copper.

Interaction of HS^- with Fe^{2+} ions in the ratio 1:1 is followed by the formation of high-strength *nano*-dimensional clusters of iron sulfides $(FeS)_n$ that interact effectively with H_2O_2 and at the same time are rather steady against oxidation by O_2 due to quadratic character of dependence of the reaction rate on the concentration of clusters. The mechanism of formation and interaction $(FeS)_n$, with O_2 and H_2O_2 is presented on the scheme 6.

$pK_a = 6.98$

$H_2S^- \rightleftarrows HS- + H+$

$Fe^{2+}(OH^-) + HS^- \rightleftarrows 1/n\ (FeS)_n + H_2O$

$+4H^+$

$2(FeS)_n + O_2 \rightleftarrows 2(FeS)_{n-1}Fe^{2+} + 2S^0 + 2H_2O$

$(FeS)_n + H_2O_2 \rightleftarrows (FeS)_{n-1}Fe^{2+}(OH^-)_2 + S^0$

Scheme 6

Besides iron ions, the effective catalysts of oxidation of compounds of the reduced sulfur compounds are copper ions Cu^+ which form high-strength complexes of structure 1:1 with thiolic compounds (RSH) (Bagiyan et al, 2005 and Dyatchina et al., 1992).

Humic acids are well known sensitizers of photocatalytic processes, able to promote the degradation of contaminants in natural waters. On this basis, the use of humic acids as additives in the photochemical treatment of waters has been considered (Carlos et al. 2012). Surprisingly it was found, that binding of the humic acids to magnetite nanoparticles Fe_3O_4 resulted in lower yields of reactive oxygen species (ROS) generation by the two different humic acids tested. The data also indicate that the production of ROS both by the free and bonded humic acids depends on the particular humic acid involved.

Mono-complexes CuSR are non-reactive with respect to O_2 and are rapidly oxidized by hydrogen peroxide. Only *bis*-complexes Cu^+ - with two thiolic ligands interact with O_2, and as in the case of iron sulphide, the reduction of O_2 proceeds via a 4-electron mechanism - with the cooperative action of two thiolic bis-complexes.

In Figure 2 the plot of titration of a mono-complex of monovalent copper with mercaptoacetic acid is submitted.

Under conditions of low natural concentrations, the rate of oxidation of *bis*-complexes by dissolved molecular oxygen is negligible, while thiol *mono*-complexes of monovalent copper effectively interact with hydrogen peroxide.

1(●) – distilled water, 2(▲) - mercaptoacetate, 3(■) – complex CuSR.

$[Cu^+]_0 = [HSCH_2COOH]_0 = 10^{-5}M$, D_{592} – absorbance of the dye crystal violet («indicator» of H_2O_2), $[CuRS]: [H_2O_2] = 2:1$.

Figure 2. Titration of mono-thiolate complexes of copper (I) by hydrogen peroxide

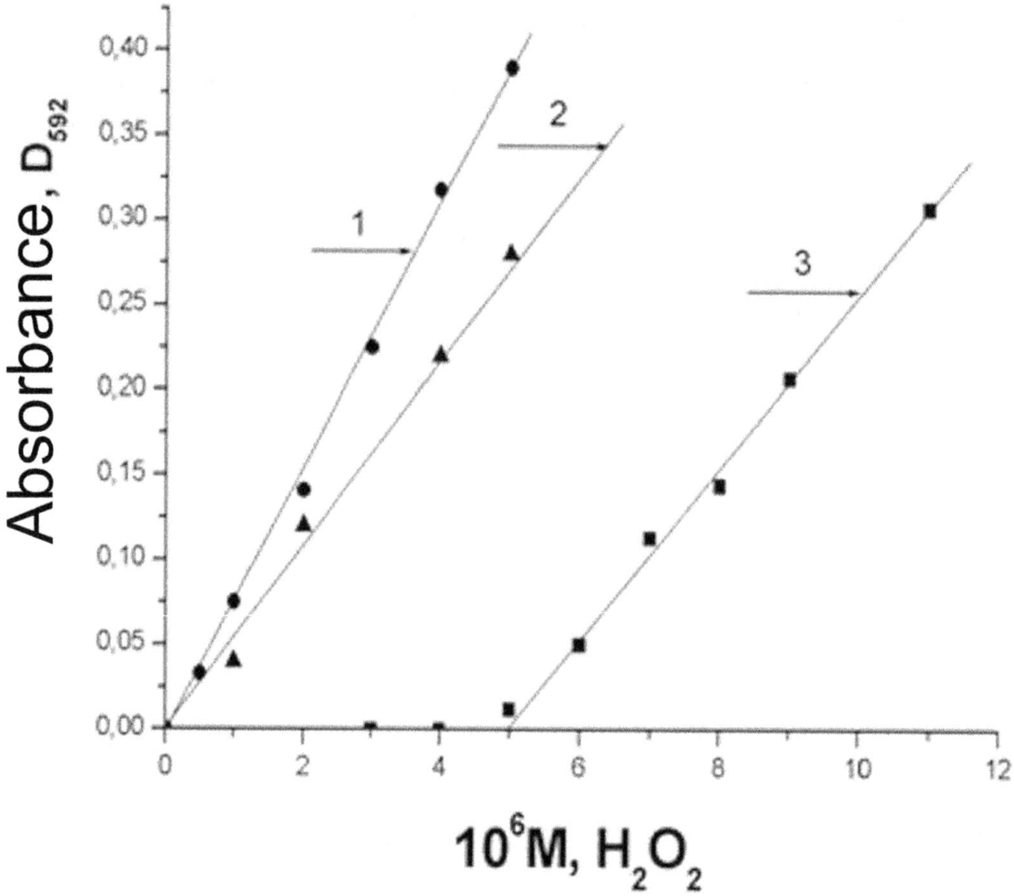

From the value of the segment cut off on the abscissa axis, it follows that oxidation of thiols by hydrogen peroxide proceeds in stoichiometric relationship 2:1, that indicates the intermediate formation of thiolic radicals and their subsequent recombination.

The mechanism of formation and interaction $(FeS)_n$, with O_2 and H_2O_2 is presented on the Scheme 7.

$RSH \rightleftarrows RS^- + H^+$

$Cu^+ + RS^- \rightleftarrows CuSR$

$+H^+$

$CuSR + H_2O_2 \rightleftarrows Cu^+ + RS^\bullet + OH + H_2O$

$+H^+$

$OH + RS^- \rightleftarrows RS^\bullet + H_2O$

$2RS^\bullet \rightleftarrows RSSR$

Cu^+

$H_2O_2 + 2RSH \rightleftarrows RSSR + 2H_2O$

$CuSR + RS^- \rightleftarrows Cu(RS^-)_2$

$+4H^+$

$2Cu^+(RS^-)_2 + O_2 \rightleftarrows 2Cu^+ + 2RSSR + 2H_2O$

Scheme 7

When monovalent copper ions are bound to high-strength *mono*-thiolic complexes, copper as a trace biogenic element becomes biologically inaccessible, that leads to intoxication of the natural aquatic environment, in particular, with respect to larvae of fish at the stage of formation of respiratory enzymes.

Oxidation-reduction processes in the natural water environment are closely intertwined with intra-reservoir free radical processes with participation of superoxide and hydroxyl free radicals.

In particular, catalytic processes involving copper ions and H_2O_2 lead to the formation of OH radicals (Sychev et al., 1983):

$Cu^+ + H_2O_2 \rightleftarrows Cu(OH)^+ + \mathbf{OH}$

Cu^{2+}/Cu^+

$\mathbf{{}^{\bullet}O_2^- + H_2O_2} \rightleftarrows O_2 + \mathbf{OH} + OH^-$

$Cu^{2+} + DH^-$ Ç $CuDH^+ (\rightleftarrows Cu^+ + \bullet DH)$

$CuDH^+ + H_2O_2 \rightleftarrows Cu(OH)^+ + \bullet DH + \mathbf{\bullet OH}$

Cu^{2+}

$DH^- + H_2O_2 \rightleftarrows {}^{\bullet}D^- + \mathbf{OH} + H_2O$

Scheme 8

This is the process of the Fenton type, the catalytic transfer of an electron from the superoxide radical to H_2O_2, processes involving partial charge transfer complexes of copper ions with H-donors ligands $CuDH^+$ donors. In the latter case, the copper ion does not formally change its valence, remaining in the oxidized form.

Under normal conditions, $[OH]_{st}$ varies from **10^{-16} to 10^{-15}M** (Ernestova & Skurlatov, 1995).

Marchisio et al. (2015) had studied photogeneration of reactive transient species upon irradiation of natural water samples. It was found, that chromophoric dissolved organic matter (CDOM) in surface waters is a photochemical source of several transient species such as CDOM triplet excited states ($^3CDOM^*$), singlet oxygen($^1O_2$) and the hydroxyl radical OH·. By irradiation of lake water samples, it was shown that the quantum yields for the formation of active transients by CDOM vary depending on the irradiation wavelength range, in the order UVB > UVA > blue.

In case of excessive intensification of free-radical processes and an increased content of manganese ions in water, a super-oxidative state of the natural aquatic environment that is toxic for aquatic biota can be formed. Carriers of super-oxidative equivalents are microcolloidal manganese particles of manganese in the mixed-valence state of Mn (III, IV) formed by the interaction of OH radicals with Mn^{2+} ions (Baral et al., 1986). In the composition of these nanodimensional, metastable particles, the ion Mn(III) is surrounded by Mn(IV) ions. This is a kind of "repository" of OH-radicals in the form of Mn(III), surrounded by Mn(IV) in the composition of micro-colloidal particles Mn(III, IV).

Feature of these microcolloid particles is the avalanche-like nature of their interaction with potential donors of hydrogen and an electron: the ion of Mn (III) interacts with the donor of an electron with formation the Mn (II) ion surrounded by Mn(IV) ions. As a result of interatomic transfer of a charge in the composition of a microcolloidal particle instead of single ion of Mn (III) 2 such ions and so on accruing in a geometrical progression appear until all manganese is restored to a bivalent state.

The mechanism of formation and transformation of microcolloidal particles Mn (III, IV) is presented in Scheme 9.

$nMnOH^{2+}$

$\mathbf{OH} + Mn^{2+} \rightleftarrows MnOH^{2+} \rightleftarrows \{\mathbf{Mn(III)}, Mn(IV)n\}$

$\{\mathbf{Mn(III)}, Mn(IV)n\} + e^{-} \rightleftarrows \{Mn(II), Mn(IV)_{n}\} \rightleftarrows$

$\rightleftarrows \{\mathbf{2Mn(III)}, Mn(IV)_{n-1}\} + 2e^{-} \rightleftarrows \{2Mn(II), Mn(IV)_{n-1}\} \rightleftarrows$

$\rightleftarrows \{\mathbf{4Mn(III)}, Mn(IV)_{n-3}\} \rightleftarrows +\ldots (n-2)e^{-} \rightleftarrows (n+1)Mn^{2+}$

Scheme 9

The super-oxidative state has been for the first time noted by us in the mid-eighties on the Kuchurgansky reservoir where in fish-breeding ponds of the local fish farm located near the Kuchurgansky state district power plant, mass death of adult herbivorous fishes was observed. Additives of technical ascorbic acid have led to a detoxication of the fish-breeding pond.

The most accurately, the super-oxidative state of the aquatic environment was manifested in the Kiev reservoir after the Chernobyl accident (Skurlatov., Sirenko & Shtamm, 1992).

Of course OH radicals are not unique active species, which can be readily photogenerated. Wu et al. (2015) studied the activation of persulfate induced by Fe(III)-ethylenediamine-N,N'-disuccinic acid (EDDS) under dark and irradiation conditions and the reactivity of the generated sulfate radical under a wide range of experimental conditions were investigated by means of experimental kinetic analyses and modelling.

From our experience of researches we have come to the conclusion that formation of a *super*-oxidative state can serve as the evidence of radiation contamination of the water objects. The example of the Kuchurgansky reservoir indicates a possibility of the hidden radiation contamination of the environment at large volumes of the coal burned on state district power plant.

CONCLUSION

1. The main anthropogenic factor leading to disbalance of internal oxidation-reduction processes is the discharge of waste waters containing reducing substances which effectively interact with H_2O_2 and are inert towards the oxygen dissolved in water.
2. The key role in formation of toxic properties in natural and waste waters is played by soluble reduced sulfur compounds, falling beyond control of conventional analytical methods.
3. The major technogenic factor forming a *super*-oxidizing condition of a natural aqueous medium is the excessive intensification of free-radical processes caused by radiation pollution of an aqueous medium at the increased maintenance of manganese ions in it.
4. The current control system of water quality does not ensure ecological safety of the natural aqueous environment. Inclusion in the control system of such indexes as toxicity and content in water of hydrogen peroxide and the substances reducers which are efficiently interacting with H_2O_2, or, opposite – substances with the expressed oxidizing ability is necessary.

REFERENCES

Bagiyan, G. A., Koroleva, I. K., Ufimtsev, A. V., & ... . (2005). Ion-molekulyarnye mekhanizmy kataliticheskogo okisleniya tiol'nyh soedinenij v prisutstvii ionov medi [Ion-molecular mechanisms for the catalytic oxidation of thiol compounds in the presence of copper ions] Himicheskaya fizika. *Chemical Physics*, *24*(6), 51–62.

Bajkova, I. S., Shtamm, E. V., Vichutinskaya, E. V., & Skurlatov, Yu. I. (2009). Mekhanizm okisleniya nanochastic FeS molekulyarnym kislorodom i peroksidom vodoroda v razbavlennyh vodnyh rastvorah [Mechanism of oxidation of FeS nanoparticles with molecular oxygen and hydrogen peroxide in dilute aqueous solutions] Himicheskaya fizika. *Chemical Physics*, (4): 86–91.

Baral, S., Lume-Pereeira, C., Janata, E., & Henglein, A. (1986). Chemistry of colloidal manganese dioxide. III. Formation in the reaction of hydroxyl radicals with Mn2+ ions. *Journal of Physical Chemistry*, *90*(22), 6025–6028. doi:10.1021/j100280a113

Benitez, F.J., Acero, J.L., Real, F.J., & Maya, C. (2004). Modeling of photooxidation of acetamide herbicides in natural waters by UV radiation and the combinations UV/H2O2 and UV/O3. *Journal of Chemical Technology and Biotechnology, 79*, 987–997. DOI: doi:10.1002/jctb.1073

Carlos, L., Cipollone, M., Soria, D., Moreno, M. S., Ogilby, P. R., & Einschlag, G. F. (2012). S a, Mártire D.O. (2012) The effect of humic acid binding to magnetite nanoparticles on the photogeneration of reactive oxygen species. *Separation and Purification Technology*, *91*, 23–29. doi:10.1016/j.seppur.2011.08.028

Duca, G., Skurlatov, I., & Miziti, A. (1994). *Introduction in Ecological Chemistry*. Moscow: Vysshaia shcola. (in Russia)

Duka, G. G. (1988). Ekologicheskaya himiya stochnyh vod [Ecological chemistry of wastewater]. In *Materialy II Vsesoyuz. shkoly po ekologicheskoj khimii vodnoj sredy* [Materials II All-Union. School on Environmental Chemistry of the Aquatic Environment]. Moscow: Institute of Chemical Physics of the Academy of Sciences of the USSR.

Dyatchina, O. V., Semenyak, L. V., & Skurlatov, Yu. I. (1992). Okislenie glutationa kislorodom i peroksidom vodoroda v prisutstvii ionov medi [Oxidation of glutathione with oxygen and hydrogen peroxide in the presence of copper ions] Himicheskaya fizika. *Chemical Physics*, *11*(9), 1255–1259.

Ernestova, L. S., & Skurlatov, Yu. I. (1995). Obrazovanie i prevrashchenie svobodnyh radikalov OH i O2- v prirodnyh vodah [Formation and transformation of free radicals OH and O2- in natural waters] Zh. Fiz. himii. *Journal of Physical Chemistry*, *69*(7), 1157–1164.

Frog, B. N., Skurlatov, Yu. I., Shtamm, E. V., & Vichutinskaya, E. V. (2012) Vliyanie vodorastvorimyh soedinenij sery na toksicheskie svojstva prirodnyh i stochnyh vod [The influence of water-soluble sulfur compounds on the toxic properties of natural and waste waters]. Vestnik MGSU, 6, 105-113.

Hewitt, L. M., Carey, J. H., Dixon, D. G., & Munkittrick, K. R. (1996). Examination of bleached kraft mill effluent fraction for potential inducers of mixed function oxygenase activity in rainbow trout. In M. R. Servos, K. R. Munkittrick, J. H. Carey, & G. J. van der Kraak (Eds.), *Environmental fate and effects of pulp and paper mill effluents* (pp. 79–94). Delray Beach, FL: St. Lucie Press.

Horváth, O., & Huszánk, R. (2003). Degradation of surfactants by hydroxyl radicals photogenerated from hydroxoiron(III) complexes. *Photochem. Photobiol. Sci., 2003*(2), 960–966. DOI: doi:10.1039/b303697a

Kouassi, A. M., Zika, R. G., & Plane, J. M. C. (1990). Light-induced alteration of the photophysical properties of dissolved organic matter in seawater Part II. Estimates of the environmental rates of the natural water fluorescence. *Netherlands Journal of Sea Research, 27*(1), 33–41. doi:10.1016/0077-7579(90)90032-C

Maksimenko, P. Y., Skurlatov, Y. I., & Shtamm, E. V. (2000) Veroyatnaya rol' serosoderzhashchih soedinenij v formirovanii toksicheskih svojstv stochnyh vod cellyulozno-bumazhnogo proizvodstva [The probable role of sulfur-containing compounds in the formation of the toxic properties of wastewater from pulp and paper production]. Vestnik RUDN, Ser. Ekologiya i bezopasnost' zhiznedeyatel'nosti, 4, 63-70.

Marchisio, A., Minella, M., Maurino, V., Minero, C., & Vione, D. (2015). Photogeneration of reactive transient species upon irradiation of natural water samples: Formation quantum yields in different spectral intervals, and implications for the photochemistry of surface waters. *Water Research, 73*, 145–156. doi:10.1016/j.watres.2015.01.016 PMID:25655321

Pelizzetti, E., & Calza, P. (2002). Photochemical processes in the euphotic zone of sea water: progress and problems. In *Chemistry of Marine Water and Sediments. Springer-Verlag*. doi:10.1007/978-3-662-04935-8_3

Salvador, P. (2007). On the nature of photogenerated radical species active in the oxidative degradation of dissolved pollutants with TiO2 aqueous suspensions: A revision in the light of the electronic structure of adsorbed water. *The Journal of Physical Chemistry C, 2007*(111), 17038–17043. doi:10.1021/jp074451i

Sandvik, S. L. H., Bilski, P., Pakulski, J. D., Chignell, C. F., & Coffin, R. B. (2000). Photogeneration of singlet oxygen and free radicals in dissolved organic matter isolated from the Mississippi and Atchafalaya River plumes. *Marine Chemistry, 69*(1-2), 139–152. doi:10.1016/S0304-4203(99)00101-2

Shtamm, E. V. (1988). Redoks-sostoyanie vodnoj sredy i voprosy vosproizvodstva rybnyh resursov [Redox state of the aquatic environment and issues of reproduction of fish resources]. In Digest Ekologicheskaya himiya vodnoj sredy. Materialy 1 Vsesoyuznoy shkoly [Ecological chemistry of water environment]. Materials of the I All-Union School.

Shtamm, E. V., Purmal, A. P., & Skurlatov, Y. I. (1991). Rol' peroksida vodoroda v prirodnoj vodnoj srede [The role of hydrogen peroxide in a natural water environment]. *Uspekhi Khimii, 60*(11), 2373–2411.

Shtamm, E. V., Skurlatov, Yu. I., & Shvydkij, V. O. (2015). Priroda toksicheskogo vozdejstviya stochnyh vod predpriyatij cellyulozno-bumazhnogo proizvodstva na vodnye ekosistemy [The nature of the toxic effects of wastewater from pulp and paper production enterprises on aquatic ecosystems] Himicheskaya fizika. *Chemical Physics, 34*(6), 22–29.

Skurlatov, Yu. I., Ernestova, L. S., & Shtamm, E. V. (1984). Redoks-sostoyanie i sezonnaya toksichnost' prirodnoj vody [Redox State and Seasonal Toxicity of Natural Water] [Proceedings of Academy of Sciences of the USSR]. *Dokl. AN SSSR, 276*(4), 1014–1016.

Skurlatov, Y. I., & Purmal, A. P. (1984). Ekologicheskaya khimiya vodnoj sredy [Ecological chemistry of the aquatic environment]. Priroda, 10, 94-103.

Skurlatov, Y. I., Shtamm, E. V., Kozlova, N. B., Zajceva, N. I., & Aleksandrova, E. V. (2011). Biotestirovanie v ocenke effektivnosti tekhnologij ochistki stochnyh vod [Biotesting in assessing the effectiveness of wastewater treatment technologies]. *Водные ресурсы*, *38*(2), 232–238.

Skurlatov, Yu. I., Sirenko, L. A., & Shtamm, E. V. (1992) Svobodno-radikal'nye processy v vode Dneprovskih vodohranilishch posle avarii na ChAES [Free radical processes in the water of the Dnieper reservoirs after the Chernobyl accident]. In Radioaktivnoe i himicheskoe zagryaznenie Dnepra i ego vodohranilishch posle avarii na Chernobyl'skoj AES [Radioactive and chemical pollution of the Dnieper and its reservoirs after the accident at the Chernobyl nuclear power plant, Kiev]. Kiev: Naukova Dumka.

Sychev, A. Y., Travin, S. O., Duka, G. G., & Skurlatov, Y. I. (1983). *Kataliticheskie reakcii i ohrana okruzhayushchej sredy* [Catalytic reactions and environmental protection]. Chisinau: Shtiintsa.

Vione, D., Falletti, G., Maurino, V., Minero, C., Pelizzetti, E., Landrino, M., ... Lariu, R.-G. (2006). And Arsene C. (2006) Sources and Sinks of Hydroxyl Radicals upon Irradiation of Natural Water Samples Environ. *Sci. Technol.*, *40*(12), 3775–3781. doi:10.1021/es052206b PMID:16830541

Waite, T. D. (1990). Photochemistry of colloids and surfaces in natural waters and water treatment. In R. Beckett (Ed.), Surface and Colloid Chemistry in Natural Waters and Water Treatment. Springer Science+Business Media.

Wu, Y., Bianco, A., Brigante, M., Dong, W., Sainte-Claire, D. P., Hanna, K., & Mailhot, G. (2015). Sulfate Radical Photogeneration using Fe-EDDS: Influence of Critical Parameters and Naturally Occurring Scavengers. *Environ. Sci. Technol.* DOI: doi:10.1021/acs.est.5b03316

Zafiriou, O. C., Joussot-Dubien, J., Zepp, R. G., & Zika, R. G. (1984). Photochemistry in natural waters. *Environmental Science & Technology*, *18*(12), 1984. doi:10.1021/es00130a711

Zepp, R. G., Skurlatov, Yu. I., & Rutmiller, L. F. (1988). Effects of aquatic humic substances on analysis for hydrogen peroxide using peroxidase catalyzed oxidation of triarylmethanes or p-hydroxyphenylacetic acid. *Environmental Technology Letters*, *9*(4), 287–298. doi:10.1080/09593338809384569

Chapter 8
Buffer Properties of Heterogeneous Mixtures "Mineral – Natural Waters"

Igor Povar
Institute of Chemistry, Moldova

Oxana Spinu
https://orcid.org/0000-0002-3166-1931
Institute of Chemistry, Moldova

ABSTRACT

The provision of plants is generally assessed by the potential buffering capacity of soils in relation to the elements of plant nutrition. The main focus in this chapter is to review the developed earlier buffer theory and show how to use the for assessing and predicting the long-term phenomena of attenuation and natural remediation of ionic pollutants in contaminated aquatic ecosystems. The buffer theory is based on the rigorous thermodynamic analysis of complex chemical equilibria under environmental conditions in aquatic ecosystems, developed in the previous chapter. It has been proved that both homogeneous and heterogeneous systems manifest a buffer action towards all their components. It has been established that the buffer capacities of components are mutually proportional, whereas for heterogeneous systems these relationships depend on the stoichiometric composition of solid phases. The use of the developed buffer approach yields deeper understanding of the processes that control the concentrations of various components.

INTRODUCTION

In view of the increasing anthropogenic impact on aquatic ecosystems, our concern increases about the occurrence of some irreversible and undesirable changes. The tasks regarding the prediction and assessment of the aquatic ecosystems quality along with the determination of its resistance to anthropogenic impacts come out, as the ability to restore itself is not limitless. Several environmental factors, such as

DOI: 10.4018/978-1-7998-1241-8.ch008

speciation of heavy metals (HMs) and solubility, and the solution pH can all contribute to the release of HMs in contaminated soil (Lu et al., 2017). For aquatic ecosystems - buffer open dynamical systems, associated with environmental flows of matter and energy, the ecosystem stability can be defined as its ability to come back to initial states after external disturbances, maintaining the productive function in the socio-economic system.

For the modern ecological sciences, the mathematization of research is a general trend. If the application of mathematics in ecological science was earlier limited to the use of preponderantly statistical methods for the analysis and treatment of experimental data, now it is more emphasized on mathematical modeling using the computer technology. These days, the specialists in the field of natural sciences need to learn how to answer the questions: "*How to make a quantitative assessment of the effects of external (including anthropogenic) impacts on ecosystems or how to get the scenarios for the development of natural systems*?" For assessments and forecasts of the natural systems development in time, a mathematical model of the process is essential.

BACKGROUND

Buffer capacity is an important concept in many areas of science: analytical and electroanalytical chemistry (Roy et al., 2011a), geochemistry (Bethke, 2007), biochemistry (Roy, 2011b), medicine and pharmaceutical industry (Kay et al., 2008), water treatment (Totsche et al., 2006), agriculture (Doughertyet al, 2006), environment (Upadhyay et al., 2011), etc. The capacity of buffer systems to oppose (resist) the variation of their composition (usually pH changes) caused by the influence of external fluxes of chemical compounds of natural or anthropogenic character, that shift chemical equilibria, is called *buffer property*, and its efficiency – *buffer action*. The concept of *buffer action* helps to find out which reactions control the natural waters composition, including soil solutions (Povar & Spinu, 2015; Stumm & Morgan, 2012; Pertusatti & Prado, 2007). The parameters of buffer action are integrated functions of all the soil chemical components by virtue of their capacity, by means of chemical reactions and sorption-desorption processes, to extinguish or strengthen the effect of entered pollutants (Stutter & Richards, 2012).

Despite the abundant information on buffer systems, the quantitative theory of buffer action has only been developed for mono-phase systems, predominately for pH buffers (Komari, 1984; Perrin & Dempsey, 2012).Unlike the classical mono-phase buffer systems in which the buffer components are dissolved in a unique phase, in two-phase (heterogeneous) buffers they are distributed between two phases: in aqueous phase and solid (gaseous or liquid), insoluble phase. The aqueous (buffer) phase contains all the charged particles and a restricted quantity of electro-neutral species. The solid phase contains in significant quantities only electro-neutral particles and serves as their reservoir by means of which the equilibrium is adjusted and one of parameters of buffer system is maintained constantly (Charykov & Osipov, 1991; Fishtik & Povar, 2006; Komari, 1984; Perrin & Dempsey, 2012; Povar, 2000a; Povar & Luca, 2003). The buffer action of two-phase systems is based on the shift of complex equilibria, both homogeneous and heterogeneous in the aqueous phase and between phases, respectively. For pH buffers, by increasing the acidity of solution the role of simultaneously occurring protolytic reactions with participation of salt anions is strengthened; with an increase of alkalinity the contribution of complex formation reactions with participation of salt cations amplifies. Authors (Fishtik & Povar, 2006) proved that the buffer capacity is a special case of the sensitivity analysis, as a more general theoretical approach, which studies the response of a system to various external perturbations (as for example, the variation of the certain component concentration, the change of chemical composition). The authors believe that it is more correct to consider, as a potential reservoir of heterogeneous buffer systems, the complex chemical heterogeneous equilibria with participation of solid phases (Povar, 2000a). Unfortunately, so far there are a limited number of studies dedicated to the systematic investigations and the development of theoretical aspects of the buffer action for two-phase systems (Charykov & Osipov, 1991; Fishtik & Povar, 2006; Komari, 1984; Perrin & Dempsey, 2012; Pfendt, 1995; Povar, 2000a; Povar & Luca, 2003, Povar & Rusu, 2012). Besides, in the majority of these studies only the *pH* – buffer properties of heterogeneous systems have been investigated.

MAIN FOCUS OF THE CHAPTER

Issues, Controversies, Problems

Determination of the buffer capacity of natural waters and soil allows calculating the pollutant amounts which delivery does not essentially break the natural character of their functioning. This question is the major one in the soil sciences, ecology and wildlife management. The complexity of the soil solution composition containing mineral phases and a large set of the involved chemical compounds determine the possibility of simultaneous chemical reactions occurrence along with the capacity of solid phases of minerals to maintain relatively constant composition of aqueous solution. Under the real conditions, the buffer action of natural heterogeneous aqueous systems is expressed as the consumption of an element from the solution, which causes the partial dissolution of solid phases. As a result the chemical composition of the solution is restored.

Currently, the extensive information has been gathered on negative (harmful) transformations in natural waters and soils, due to the progressing acid and alkaline loads (in the form of mineral fertilizers, plant protection chemicals and the industrial emissions dropping out with atmospheric precipitations). Thus, the quantitative assessment of the acid-base buffer action, revealing the degree of influence of the systematic use of fertilizers and long-term technogenic pollution by substances of acid and alkaline nature, is an actual challenge of the agrology. Besides, the buffer action of soils contains an important information on the soil formation processes (their direction and intensity), which is used for the soil diagnostics and classification.

The aim of the present paper was to review the existent theory and develop the quantitative aspects of the buffer action of various components of homogeneous and heterogeneous natural systems as well as to establish their interrelation.

The Buffering Properties of Sparingly Soluble Hydroxides and Acids in Saturated Solutions

The concept of buffering is directly associated to routing the solution composition. Several environmental factors, such as speciation of heavy metals (HMs) and solubility, and the solution pH can all contribute to the release of HMs in contaminated soil (Lu et al., 2017).

Although there is an abundance of information on buffer systems relative to pH, the theory of buffering has been developed only for homogeneous solutions, especially for acid and conjugate base mixtures. (Here, as usual, the heterogeneous system means the "*liquid phase - solid phase*" system).

Systematic investigations of heterogeneous buffering properties are practically absent, although these problems have a theoretical and practical importance. In connection with this it is necessary to highlight the works of (Komari, 1984), in which the calculation of equilibria in simple heterogeneous systems gives a qualitative estimation of their buffering properties. As such a system, the saturated solution of calcium hydroxide is examined (Komari, 1984). In the paper (Charykov & Osipov, 1991), based on the approximate calculations, authors shown that compared to heterogeneous buffer systems, at one and the same concentration of base C_b or acid C_b, the classical homogeneous buffer systems have a two-fold lower capacity. In this paper, heterogeneous buffer systems are empirically created based on sparingly soluble fatty acids.

In the paper (Povar, 1996), the authors derived the quantitative characteristics of the buffers of heterogeneous systems containing precipitates of the sparingly soluble hydroxides and acids. For their full description account, the acid-base equilibrium in solution with the hydroxide metal ion and acid anion participation was taken into account. As a quantitative measure of the buffer action, the classical notion of buffering capacity was used:

$$\beta = \frac{dC_{OH}^{0}}{dpH} = -\frac{dC_{H}^{0}}{dpH} = 2.3\frac{dC_{H}^{0}}{d\ln[H]} \equiv 2.3\beta_{H} \tag{1}$$

In the deduction of formulas for calculating the basic principles, the method of residual concentrations (CR) was used (Povar, 1996; Beresnev, 1992). Within the CR method, the basic mass balance (BM) equation is written as follows:

$$C_i^0 = C_i^r + \Delta C_i$$

Here by C_i^0 and C_i^r (mol•L^{-1}) the initial and residual concentration of ion *i* in the solution are respectively noted, while ΔC_i symbolizes its quantity in mol in the solid phase in one liter of saturated solution (Beresnev, 1992). Firstly, the system consisting of the precipitation of a sparingly soluble hydroxide and aqueous solution will be examined, in which the following equilibria are established:

$$M(OH)_{m(S)} + mH^+ = M^{m+} + mH_2O,\ K_S = [M][H]^{-m} \tag{2}$$

$$H_2O = H^+ + OH^-,\ K_w = [H][OH] \tag{3}$$

It is also necessary to take into account the process of hydrolysis of the metal ion over a wide range of pH:

$$M^{m+} + iH_2O = M(OH)_i^{(m-i)+} + iH^+, \qquad K_i = [M(OH)_i][H]^i / [M] \tag{4}$$

Near the equations of reactions, the respective equilibrium constants are listed. For convenience, ion charges are omitted. With square brackets, the equilibrium concentrations of the species are denoted.

The *MB* equations for reactions (2) - (4) take the following form:

$$C_M^0 = C_M^r + \Delta C_M \tag{5}$$

$$C_{OH}^0 = C_{OH}^r + \Delta C_{OH} = -C_H^0 \tag{6}$$

Additionally, within the RC method, the following relations are valid:

$$m = \Delta C_{OH} / \Delta C_M, \qquad C_{OH}^0 = -C_H^0, \qquad C_{OH}^0 = -C_H^r$$

Given past relationships, after some transformations, the equation (6) gets the form:

$$C_H^0 = -m\Delta C_M + C_H^r \tag{7}$$

The residual concentrations of ions C_M^r and C_H^r constitute the total solution concentrations of all species containing ions of the given type:

$$C_M^r = [M] + \sum_{i=1}[M(OH)_i] = [M]\left(1 + \sum_{i=1} K_i[H]^{-i}\right) = [M]\alpha_M \tag{8}$$

$$C_H^r = [H] - [OH] - \sum_{i=1} iM(OH)_i] = [H] - [OH] - \sum_{i=1} iK_i[M][H]^{-i} \tag{9}$$

When writing the last relationships, the equations of law of mass action were used (Beresnev, 1992). By α_M the degree of complex formation is noted (Burgot, 2012):

$$\alpha_M = 1 + \sum_{i=1} K_i[H]^{-i} \tag{10}$$

The equilibrium concentration of the metal ion [*M*] is determined from the equilibrium constant (2) and is equal:

$$[M] = K_S[H]^m \tag{11}$$

Based on the equilibria (2) - (4) and the mass balance equations (6) - (9), the equation for computing the buffering capacity can be deduced. At the same time, obviously, the quantity C_M^0 does not depend on the solution pH. By differentiating in respect to [*H*] the equations (10) and (11), the following derivatives are obtained:

$$\frac{d\ln[M]}{d\ln[H]} = m, \qquad \frac{d\ln\alpha_M}{d\ln[H]} = \frac{\sum_{i=1} i[M(OH)_i]}{C_M^r}$$

Here it is taken into account that:

$$\frac{d[H]}{d\ln[H]} = [H], \qquad \frac{d[OH]}{d\ln[H]} = -[OH]$$

By differentiating (7) in respect to ln[*H*], taking into account the equations of mass action law and *MB* conditions, and derived derivatives, after a series of transformations, the calculation formula for the buffer capacity is derived:

$$\beta_H = [H] + [OH] + mC_M^r(m - \overline{m}) + \sum_{i=1} i^2[M(OH)_i] - m\sum_{i=1} i[M(OH)_i] \tag{12}$$

Here, through $\overline{m}$ the Bjerrum function is noted:

$$\overline{m} = \frac{\sum_{i=1} i[M(OH)_i]}{C_M^r}. \tag{13}$$

The equation (12) can be formally represented as a sum of three terms:

$$\beta_H = \beta_w + \beta_S + \beta_h \tag{14}$$

where:

$$\beta_w = [H] + [OH]$$

$$\beta_S = mC_M^r(m - \overline{m})$$

$$\beta_h = \sum_{i=1} i^2[M(OH)_i] - m\sum_{i=1} iM(OH)_i]$$

The first term β_w describes the buffering action of water (equation 3), the second one β_S denotes the buffering action of the solid phase (reaction 2), and the third term β_h is the contribution of hydrolysis reactions 4). Our analysis shows that for most systems β_h can be neglected.

If the complete set of equilibrium constants (2) - (4) is known, the pH of the heterogeneous mixture can be calculated for certain initial concentrations of components that form precipitate, based on the *MB* equations.

It is worthy to mention that the equation (12) is applicable only under conditions of the solid phase formation. This condition can be calculated based on a thermodynamic analysis of buffer system (***BS***) balancing. As a criterion of sparingly soluble metal hydroxide formation, the variation of the Gibbs energy of the heterogeneous system is used (reactions (2) - (4)) (Povar & Spinu, 2014b):

$$\Delta G_S = RT \ln K_S \alpha_M + RT \ln C_M^0 [H]^{-m}$$

The precipitate deposits under the condition $\Delta G_s > 0$, that is, when $C_M^0 > K_S \alpha_M [H]^m = C_M^r$. Using the relations (8) and (13), the modified form of the equation (14) is derived:

$$\beta_H = m^2[M] + \sum_{i=1} (m-i)^2 [M(OH)_i] + \beta_w \quad (15)$$

The last term in equation (15) within the range of pH 3-11 can be neglected. Here three situations are possible:

1. Hydrolysis of the metal ion (4) does not occur, $\alpha_M = 1$, $C_M^r = [M]$. Then $\beta_H = m^2[M]$; therefore, the value of the buffer capacity depends only on the equilibrium concentration of metal ion.
2. 2. Since $[M] = K_S[H]^m$, then β_H is a function of the composition and stability of the poorly soluble hydroxide. The quantity $[M]$ is negligibly small. Then

$$\beta_H = \sum_{i=1} (m-i)^2 [M(OH)_i] = K_S \sum_{i=1} (m-i)^2 K_i [H]^{m-i} \quad (16)$$

Under the conditions of predominance of one hydroxocomplex $M(OH)_i$, the equation (16) is simplified:

$$\beta_H = (m-i)^2 K_S K_i [H]^{m-i} = (m-i)^2 C_M^r$$

Then, using the condition of the hydroxide precipitation $C_M^0 > K_S K_i [H]^{m-i}$, the relationship between $\beta \equiv 2.3\beta_H$ and C_M^0 can be determined:

$$\beta < 2.3(m-i)^2 C_M^0$$

The last inequality allows the estimation of the maximum value of the buffer capacity from a certain value of the initial concentration of the metal ion under the conditions of the formulated restrictions.

3. The same conditions as in b) and an additional condition $m=i$. In this case $\beta_H = 0$. Relationships (12) and (14) constitute the basis of the quantitative analysis of the buffer properties of the studied systems.

It should be mentioned that the equations for calculating the buffer capacity β in heterogeneous systems differ radically from the equations for single-phase buffer systems (Burgot, 2012). In the case $i = 0$ (Komari, 1984), the calculation formula for the "*calcium hydroxide - saturated aqueous solution*" system was obtained in form similar to the equation (12).

Analogous equations have been obtained for heterogeneous ***BS***s, which contain sparingly soluble acid $H_nA_{(S)}$. It is well-known that humus or generally classified as "humic acids", being practically insoluble in water, exerts various other important effects upon the plant growth. The acid–base properties of humic substances are of the intrinsic interest because these substances contribute to the acid–base balance of natural waters (Ritchie et al., 2003). Certain constituents of humus may also be dissolved in acid solutions and be precipitated at the isoelectric point, which is at a pH of about 4.8.

For these systems, apart from water dissociation (3), the following set of equilibria was considered:

$$\begin{aligned} &H_nA_{(S)} = nH + A, \qquad K_{Sa} \\ &A + jH = H_jA, \qquad K_j \end{aligned} \tag{17}$$

The equation for calculating the buffer capacity of the system can be represented as follows:

$$\beta_H = \beta_w + \beta_S + \beta_a, \tag{18}$$

where

$$\beta_S = n^2 C_A^r, \qquad \beta_a = \sum_{j=1} j^2[H_jA] - 2n\sum_{j=1} j[H_jA]$$

Equation (18) can be transformed into the form:

$$\beta_H = n^2[A] + \sum_{j=1}(n-j)^2[H_jA] + \beta w\,.$$

From the the Gibbs relationship of energy variation of the heterogeneous system, deduced by one of the authors (Povar, 1996; Povar, 2000a; Povar, 2000b):

$$\Delta G_S - RT\ln K_S\alpha_M + RT\ln C_M^0[H]^{-m},$$

Where

$$\alpha_A = 1 + \sum_{j=1} K_j[H]^j$$

The conditions for formation of the sparingly soluble acid precipitate $\Delta G_S > 0$, corresponding to the inequality $C_A^0[H]^n > K_{Sa}\alpha_A$ are calculated and thus the applicability range of deduced equations is determined. When a single protonated species H_jA predominates, the following equation is valid:

$$\beta < 2.3(n-j)^2 C_A^0 \tag{19}$$

If the acid anion is not protonated, therefore $C_A^r = [A]$, the following inequality is realized:

$$\beta < 2.3n^2 C_A^0 \tag{20}$$

Further, taking as an example the slightly soluble hydroxide, the mechanism of the buffering action of heterogeneous systems will be considered. When adding a strong acid, depending on the relationship between m and i, the following processes take place:

$i = 0,\quad M(OH)_{m(S)} + mH = M + mH_2O;$

$m > i,\quad M(OH)_{m(S)} + (m-i)H = M(OH)_i + (m-i)H_2O;$

$m < i,\quad M(OH)_{m(S)} + (i-m)H_2O = M(OH)_i + (i-m)H.$

According to the principle of equilibrium shift, in the first two cases the acid action is "attenuated" by the dissolving a certain amount of hydroxide precipitate. In the latter case, the addition of small portions of strong acid reduces the residual concentration of metal ion.

If a strong base is added to the heterogeneous system, the following processes are possible:

$i = 0\ M(OH)_{m(S)} = M + mOH$

$m > i\ M(OH)_{m(S)} = M(OH)_i + (m - i)OH$

$m < i\ M(OH)_{m(S)} + (i - m)OH = M(OH)_i$

According to the Le Châtelier principle, in the first two cases, when adding a small amount of base, the equilibrium shifts in the direction of the precipitate formation, in the third case its partial dissolution occurs.

Any buffer system possesses a certain potential reserve, by which the equilibrium is controlled and one of the parameters is maintained constant. The author (Komari, 1984) considers the solid phase serves as such reserve in the heterogeneous buffer systems. Considering the mechanism discussed above, in the opinion of the author (Povar, 1996; Povar, 2000a; Povar, 2000b), it would be more correct to talk about complex chemical equilibria with the participation of solid phase as a potential reserve of these buffer systems.

Table 1. Equilibrium constant values in the studied systems

The equation of reaction	*logK*
$Zn^{2+} + H_2O = ZnOH^{+} + H^{+}$	-8.96
$Zn^{2+} + 2H_2O = Zn(OH)_2 + 2H^{+}$	-16.90
$Zn^{2+} + 3H_2O = Zn(OH)_3^{-} + 3H^{+}$	-28.40
$Zn^{2+} + 4H_2O = Zn(OH)_4^{2-} + 4H^{+}$	-41.20
$Zn(OH)_{2(S)} + 2H^{+} = Zn^{2+} + 2H_2O$	12.48
$A^{-} + H^{+} = HA^{*}$	5.28
$A^{-} + 2H^{+} = H_2A^{+}$	9.69
$A^{-} + H^{+} = HA^{*}$	-10.53

* *HA*- adipic acid.

Next, two examples of numerical calculations using the deduced equations the following systems: "*zinc hydroxide - aqueous solution*" and "*adipic acid - aqueous solution*" have been analyzed (Table 1). Zn solubility in calcareous or alkaline soils was often controlled by $ZnCO_3$ or $Zn(OH)_2$ (McBride & Blasiak, 1979). Nucleation of Zn hydroxide on clay mineral surfaces may have produced the strongly pH dependent retention in the soil. On the other hand, adipic acid, representing a type of structures that according to existing knowledge might be present in humic and fulvic acids (Schnitzer et al., 1965), was detected in soil samples (215 - 568 and 2,050 μg/kg). Figures 1 and 2 show the curves of dependency $\beta = f(pH)$ for both systems.

Also there, for comparison, the pH dependence of the precipitation degree $\gamma = \Delta C_i / C_i^0$ (Povar, 1996) is displayed. Figures 1 and 2 show that the buffer capacity increases with the growth in the precipitate solubility, i.e. with the increasing residual concentration in the solution, and it is quite high. $Zn(OH)_2$ can be combined with OH^- to form hydroxocomplex $Zn(OH)_3^-$, thus increasing its solubility.

Table 2 presents the basic characteristics of the "*Adipic acid-aqueous solution*" heterogeneous buffer system for different initial concentrations of C_A^0 in the vicinity of the pH of the complete dissolution of acid precipitate. With the C_A^0 decrease (dilution), the precipitation deposition range is narrowed and the maximum values of buffer capacity β_{het} decrease too (see also Fig. 3). Also, the values of buffer capacity β_{hom} for the homogeneous buffer system "*adipic acid - aqueous solution*" are displayed here, which even for $C_A^0 = 0.001$ mol•L^{-1} are two orders less than β_{het}. From the data of Table 2, one can easy observe that for all the concentrations the inequality (20) is fulfilled, since $j = 0$, $n = 1$ and $2.3C_A^r > \beta_{het}$. As it is seen from Fig. 3, the Gibbs energy varies continuously when passing from the heterogeneous system, containing the solid and liquid phases, to the homogeneous aqueous solution, while at the mutual transition point (the point of intersection of the curve ΔG_S *(pH)*with the pH axis) $\Delta G_S = 0$. In this

Figure 1. The dependence of buffer capacity β and degree of precipitation γ on pH for the system „$Zn(OH)_{2(S)}$ - aqueous solution", $C^0_{Zn} = 0.01\,mol{\cdot}L^{-1}$

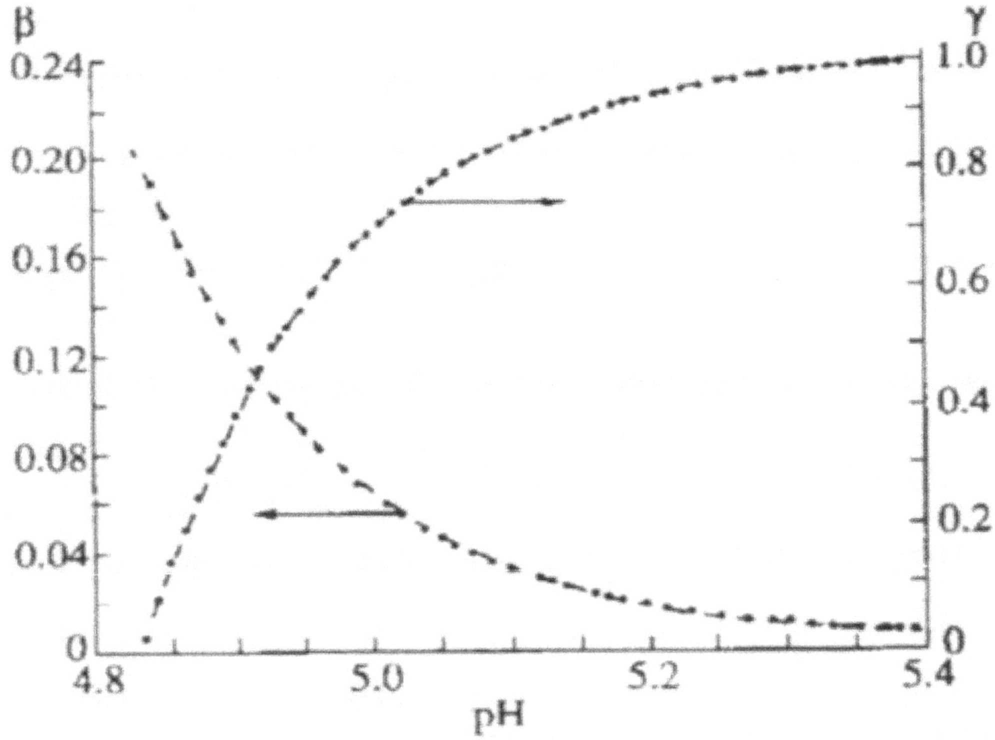

case, the initial ion concentrations are directly dependent on the function ΔG_S (Povar & Spinu, 2014b). At the same time, the analysis and calculations for a large number of systems show that the buffer capacity, which is in essence a partial derivative of the initial concentration of C^0_H (for $C^0_i = const$), and therefore as well as of ΔG_S, at the point of dissolution of the solid phase decreases by jumps. The totality of these peculiarities is characteristic for the abrupt transformations of the first order.

Hence, based on the residual concentration method, the basic principles of the buffering action theory of "*precipitated hydroxide - solution*" and "*sparingly soluble acid - solution*" systems are exposed. A simple reciprocity relationship has been deduced between the buffer capacity and the initial concentration of the metal ion (anion) in the heterogeneous mixture, which allows simply determining the maximum buffer capacity. It has been established that the buffer capacity increases with the growth in the solubility of the precipitate, thus increasing the residual concentration of the precipitate component. It has been proved that the totality of the properties upon the complete dissolution of the precipitate, including the Gibbs energy variation ΔGS, as well as the decrease by jumps of the buffer capacity, is characterized by the phase transition of the first order.

Figure 2. The curves of dependencies of the buffer capacity β and degree of precipitation γ on pH for the system "Adipic acid - aqueous solution", $C_A^0 = 0.01\,mol{\cdot}L^{-1}$

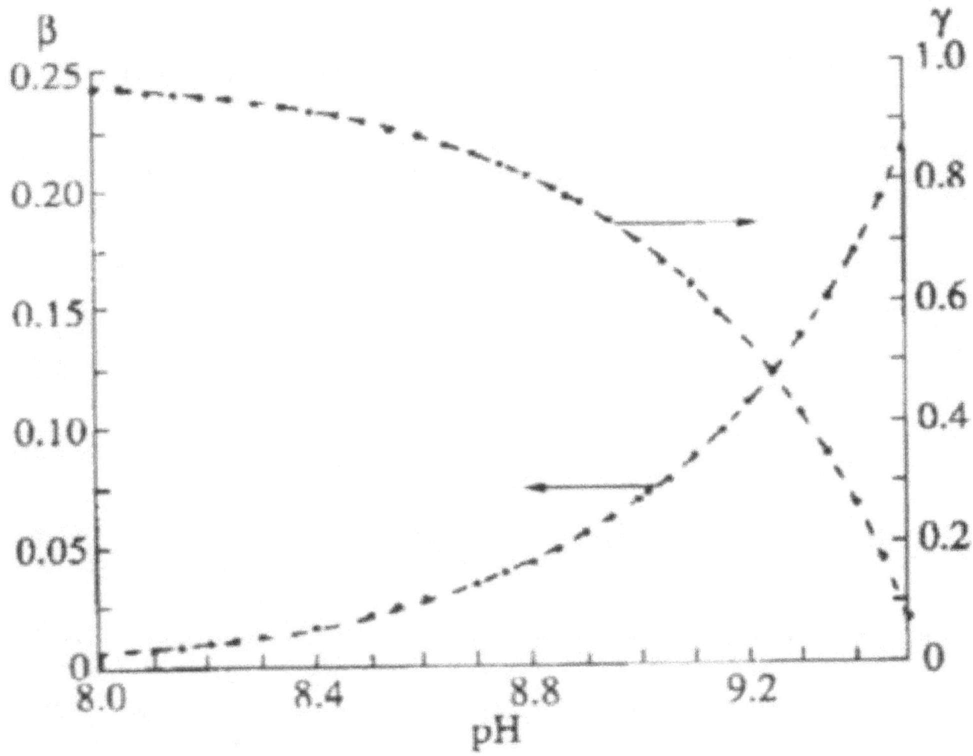

Table 2. The basic characteristics of the "Adipic acid-aqueous solution" system in the vicinity of the pH of complete dissolution of the acid precipitate

C_A^0 mol•L-1	C_A^{r} mol•L-1	pH	β_{het}	β_{hom}	$2.3n^2C_A^0$
0.5	0.468	10.2	1.077	$1.30 \cdot 10^{-5}$	1.151
0.1	$9.33 \cdot 10^{-2}$	9.5	0.215	$1.30 \cdot 10^{-5}$	0.230
0.01	$9.39 \cdot 10^{-3}$	8.5	$2.15 \cdot 10^{-2}$	$1.30 \cdot 10^{-5}$	0.023
0.001	$9.39 \cdot 10^{-4}$	7.5	$2.15 \cdot 10^{-3}$	$1.30 \cdot 10^{-5}$	$2.3 \cdot 10^{-3}$

The Buffering Properties of the Sparingly Soluble Salt in Saturated Solutions

As for previous systems, as a criterion for quantitative assessment of the intensity of buffer action of the studied multicomponent heterogeneous systems is used the value of the buffer capacity β_i^S (the superscript index "*S*" specifies the presence of solid phases), which can be defined as a partial derivative:

*Figure 3. Dependence of the Gibbs energy variation ΔG_S on pH for the heterogeneous system „Adipic acid – aqueous solution", C_A^0 mol·L^{-1}: **1**– 0.5; **2**– 0.1; **3** - 0.01; **4** – 0.001. ΔG_S=0*

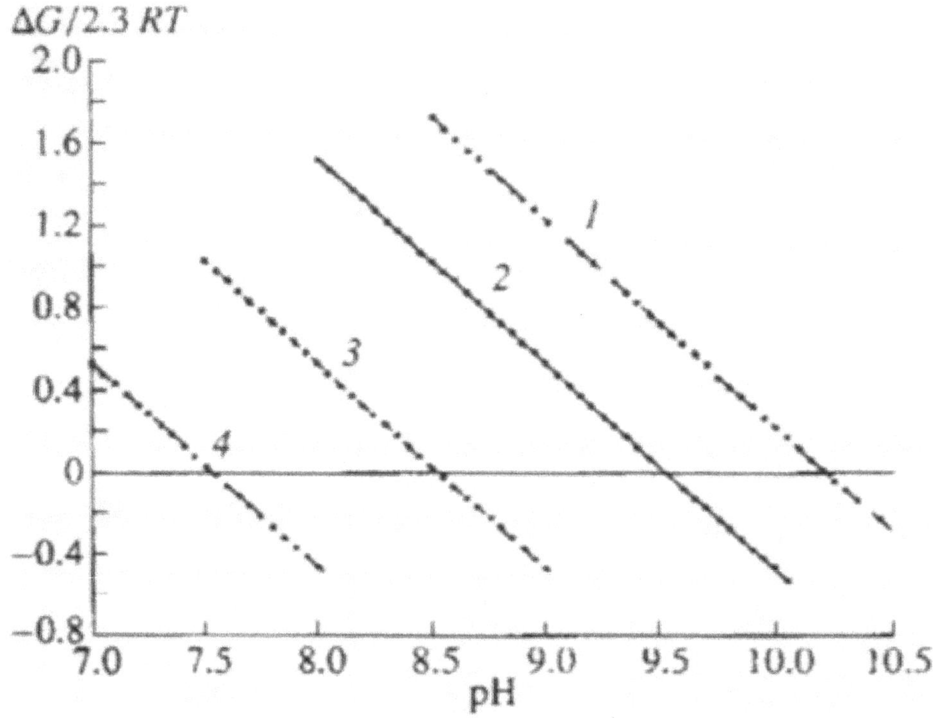

$$\beta_i^S = \left(\frac{\partial C_i^0}{\partial \ln[i]}\right)_{C_j^0(j\neq i)},$$

where C_i^0 and $[i]$ denote the initial (analytical) concentration in mixture and equilibrium concentration of the component "*i*" of solid phase, correspondingly, the subscript index shows that the initial concentrations of other components of the mixture are maintained constant.

Further, the process of the sparingly soluble salt formation of arbitrary stoichiometric composition $M_mA_{n(S)}$ (*M* - metal ion, *A* - anion of salt) is investigated:

$$M_mA_{n(S)} = mM + nA, \qquad K_S = [M]^m[A]^n. \tag{21}$$

The following set of possible simultaneous reactions in the saturated solution is taken into account:

$$M + iH_2O = M(OH)_i + iH, \qquad K_i = [M(OH)_i][H]^i/[M] \tag{22}$$

$$A + jH = H_jA, \qquad K_j = [H_jA]/\left([A][H]^j\right) \tag{23}$$

$$M + qA + rH = MH_rA_q, \qquad K_q = [MH_rA_q]/\left([M][H]^r[A]^q\right) \tag{24}$$

$$H_2O = H + OH, \qquad K_w = [H][OH]. \tag{25}$$

For the simplicity, the charges of species are omitted. Near to the reaction equations the corresponding equilibrium constants are specified. The mass balance (*MB*) conditions in this system are formulated by the following equations:

$$C_M^0 = \Delta C_M + C_M^r = \Delta C_M + \sum_{i=1}\sum_{j=0} i[M_i(OH)_j] + \sum_{q=1}\sum_{r=0}[MH_rA_q] \tag{26}$$

$$C_A^0 = \Delta C_A + C_A^r = \Delta C_A + \sum_{l=0}[H_lA] + \sum_{q=1}\sum_{r=0} q[MH_rA_q] \tag{27}$$

$$C_H^0 = [H] - [OH] + \sum_{l=1} l[H_lA] - \sum_{i=1}\sum_{j=1} j[M_i(OH)_j] + \sum_{q=1}\sum_{r=1} r[MH_rA_q] \tag{28}$$

The quantity C_i^r represents the residual concentration in solution of the ion "i", e.g. the total concentration of all the species, containing a given ion, while ΔC_i is its molar quantity in the solid phase in 1 L of solution (Beresnev, 1992). In the equations (28) C_H^0 denotes the excess of H^+ ions in relation to hydroxyl ions in two-phase mixtures $(C_H^0 = -C_{OH}^0)$. The square brackets designate the equilibrium concentrations of species in solution.

Proceeding from the stoichiometric composition of the solid phase, the following ratio is valid:

$$\frac{\Delta C_M}{m} = \frac{\Delta C_A}{n} \text{ or } \Delta C_A = \frac{n}{m}\Delta C_M \tag{29}$$

On the basis of the equations written, it is possible to deduce the formulas for calculating the buffer capacity expression in relation to any component of the mixture. Firstly, the analysis of the deduction of the expression for the buffer capacity towards the metal ion (or the metal ion buffer capacity) was performed. Taking the logarithm and differentiating the expression for the solubility product on $\ln[M]$, one can find:

$$\left(\frac{\partial \ln K_S}{\partial \ln[M]}\right)_{C_H^0, C_A^0} = 0 = m + n\frac{\partial \ln[A]}{\partial \ln[M]} \text{ and } \frac{\partial \ln[A]}{\partial \ln[M]} = -\frac{m}{n} \tag{30}$$

The subscript indices symbolize the conditions $C_H^0 = const, C_A^0 = const$, which are further omitted. Differentiating the *MB* equation (26) on ln[*M*], one gets the partial derivative:

$$\left(\frac{\partial C_M^0}{\partial \ln[M]}\right)_{C_H^0,C_A^0} \equiv \beta_M^S = \frac{\partial \Delta C_M}{\partial \ln[M]} + \sum_{i=1}\sum_{j=0} i^2[M_i(OH)_j] + \sum_{q=1}\sum_{r=0}[MH_rA_q] +$$
$$+\frac{\partial \ln[H]}{\partial \ln[M]}\left(-\sum_{i=1}\sum_{j=1} ij[M_i(OH)_j] + \sum_{q=1}\sum_{r=0} r[MH_rA_q]\right) - \frac{m}{n}\sum_{q=1}\sum_{r=0} q[MH_rA_q] \quad (31)$$

The deduced equation (31) contains two unknown terms:

$$\beta_M^S = f\left(\frac{\partial \ln[H]}{\partial \ln[M]}, \frac{\partial \Delta C_M}{\partial \ln[M]}\right)$$

For their determination, the MB equations (27) and (28) are differentiated on ln[*M*]:

$$\left(\frac{\partial C_A^0}{\partial \ln[M]}\right)_{C_M^0,C_H^0} = 0 = \frac{n}{m}\frac{\partial \Delta C_M}{\partial \ln[M]} + \frac{\partial \ln[H]}{\partial \ln[M]}\left(\sum_{l=1} l[H_lA] + \sum_{q=1}\sum_{r=0} rq[MH_rA_q]\right) -$$
$$-\frac{m}{n}\sum_{l=0}[H_lA] - \frac{m}{n}\sum_{q=1}\sum_{r=0} q^2[MH_rA_q] + \sum_{q=1}\sum_{r=0} q[MH_rA_q],$$

whence

$$\frac{\partial \Delta C_M}{\partial \ln[M]} = -\frac{m}{n}\frac{\partial \ln[H]}{\partial \ln[M]}\left(\sum_{l=1} l[H_lA] + \sum_{q=1}\sum_{r=0} rq[MH_rA_q]\right) + \frac{m^2}{n^2}\sum_{l=0}[H_lA] + \frac{m^2}{n^2}\sum_{q=1}\sum_{r=0} q^2[MH_rA_q] -$$
$$-\frac{m}{n}\sum_{q=1}\sum_{r=0} q[MH_rA_q]. \quad (32)$$

After that,

$$\left(\frac{\partial C_H^0}{\partial \ln[M]}\right)_{C_M^0,C_A^0} = 0 = \frac{\partial \ln[H]}{\partial \ln[M]}([H]+[OH]) +$$
$$+\frac{\partial \ln[H]}{\partial \ln[M]}\left(\sum_{l=1} l^2[H_lA] + \sum_{i=1}\sum_{j=0} j^2[M_i(OH)_j + \sum_{q=1}\sum_{r=1} r^2[MH_rA_q]\right) - \quad (33)$$
$$-\sum_{i=1}\sum_{j=0} ij[M_i(OH)_j] - \frac{m}{n}\sum_{l=0} l[H_lA] + \sum_{q=1}\sum_{r=1} r[MH_rA_q] - \frac{m}{n}\sum_{q=1}\sum_{r=1} rq[MH_rA_q]$$

From the equation (33), it follows:

$$\frac{\partial \ln[H]}{\partial \ln[M]}=\frac{\sum_{i=1}\sum_{j=0} ij[M_i(OH)_j]+\frac{m}{n}\sum_{l=1} l[H_lA]-\sum_{q=1}\sum_{r=1} r[MH_rA_q]+\frac{m}{n}\sum_{q=1}\sum_{r=1} rq[MH_rA_q]}{[H]+[OH]+\sum_{l=1} l^2[H_lA]+\sum_{i=1}\sum_{j=0} j^2[M_i(OH)_j]+\sum_{q=1}\sum_{r=1} r^2[MH_rA_q]} \tag{34}$$

Substituting the expression (34) in the equation (33), after some rearrangements, one can obtain:

$$\frac{\partial \Delta C_M}{\partial \ln[M]}=-\frac{m}{n}\left(\frac{\sum_{i=1}\sum_{j=0} ij[M_i(OH)_j]+\frac{m}{n}\sum_{l=1} l[H_lA]-\sum_{q=1}\sum_{r=1} r[MH_rA_q]+\frac{m}{n}\sum_{q=1}\sum_{r=1} rq[MH_rA_q]}{[H]+[OH]+\sum_{l=1} l^2[H_lA]+\sum_{i=1}\sum_{j=0} j^2[M_i(OH)_j]+\sum_{q=1}\sum_{r=1} r^2[MH_rA_q]}\right)\times$$

$$\times\left(\sum_{l=1} l[H_lA]+\sum_{q=1}\sum_{r=0} rq[MH_rA_q]\right)+\frac{m^2}{n^2}\sum_{l=0}[H_lA]+\sum_{q=1}\left(\frac{m^2q^2}{n^2}-\frac{mq}{n}\right)\sum_{q=1}\sum_{r=0}[MH_rA_q] \tag{35}$$

Further, substituting the obtained expressions for $\partial \ln[H]/\partial \ln[M]$ (34) and $\partial \Delta C_M/\partial \ln[M]$ (35) into the equation (31), after a series of transformation, one can finally get:

$$\beta_M^S=-\frac{\left(\sum_{i=1}\sum_{j=0} ij[M_i(OH)_j]+\frac{m}{n}\sum_{l=1} l[H_lA]-\sum_{q=1}\sum_{r=1} r[MH_rA_q]+\frac{m}{n}\sum_{q=1}\sum_{r=1} rq[MH_rA_q]\right)^2}{[H]+[OH]+\sum_{l=1} l^2[H_lA]+\sum_{i=1}\sum_{j=0} j^2[M_i(OH)_j]+\sum_{q=1}\sum_{r=1} r^2[MH_rA_q]}+$$

$$+\frac{m^2}{n^2}\sum_{l=0}[H_lA]+\sum_{q=1}\left(\frac{m^2q^2}{n^2}-2\frac{mq}{n}+1\right)\sum_{q=1}\sum_{r=0}[MH_rA_q]+\sum_{i=1}\sum_{j=0} i^2[M_i(OH)_j] \tag{36}$$

$$\text{or } \beta_M^S=\phi_3-\frac{\phi_1^2}{\phi_2}, \tag{37}$$

where ϕ_1, ϕ_2 and ϕ_3 denote the following concentration functions:

$$\phi_1=\sum_{i=1}\sum_{j=0} ij[M_i(OH)_j]+\frac{m}{n}\sum_{l=1} l[H_lA]-\sum_{q=1}\sum_{r=1} r[MH_rA_q]+\frac{m}{n}\sum_{q=1}\sum_{r=1} rq[MH_rA_q]$$

$$\phi_2=[H]+[OH]+\sum_{l=1} l^2[H_lA]+\sum_{i=1}\sum_{j=0} j^2[M_i(OH)_j]+\sum_{q=1}\sum_{r=1} r^2[MH_rA_q] \tag{38}$$

$$\phi_3 = \frac{m^2}{n^2}\sum_{l=0}[H_l A] + \sum_{q=1}\left(\frac{m^2 q^2}{n^2} - 2\frac{mq}{n} + 1\right)\sum_{q=1}\sum_{r=0}[MH_r A_q] + \sum_{i=1}\sum_{j=0} i^2 [M_i (OH)_j]$$

Similarly, it is possible to prove that, for the buffer capacity towards proton, the following expression is valid:

$$\left(\frac{\partial C_H^0}{\partial \ln[H]}\right)_{C_M^0, C_A^0} \equiv \beta_H^S = \phi_2 - \frac{\phi_1^2}{\phi_3}. \quad (39)$$

For the buffer capacity with regard to the anion of the solid phase one can deduce:

$$\beta_A^S = \frac{n^2}{m^2}\phi_3 - \frac{n^2}{m^2}\frac{\phi_1^2}{\phi_2} = \frac{n^2}{m^2}\left(\phi_3 - \frac{\phi_1^2}{\phi_2}\right) = \frac{n^2}{m^2}\beta_M^S \quad (40)$$

On the basis of equations obtained (37), (39) and (40), the following remarkable conclusion follows: the buffer capacities towards different components are reciprocally proportional, while the buffer capacities in relation to the ions of the solid phase are interconnected through its stoichiometric coefficients:

$$\frac{\beta_A^S}{n^2} = \frac{\beta_M^S}{m^2} \quad (41)$$

It is worthy to mention that the obtained relations are only valid in the presence of the mineral (solid phase) $M_m A_{n(S)}$. The thermodynamic stability area of the latter is determined by the value of the Gibbs energy of the overall process (21)-(25) (Povar & Spinu, 2014a):

$$\Delta G_{S,tot} = -mRT\ln\frac{C_M^r}{C_M^0} - nRT\ln\frac{C_A^r}{C_A^0} \quad (42)$$

The solid-phase is stable if $\Delta G_{S,tot} > 0$. The condition $\Delta G_{S,tot} = 0$ corresponds to the beginning of its dissolution and (or) sedimentation.

The analysis of the derived equations shows that, as in previous cases, the buffer capacities β_i^S grow with the increase of the precipitate solubility, e.g. by rising the residual concentration of the component of minerals. The buffer action of heterogeneous system "*Mineral - saturated solutions*" depends on the chemical composition of the water solutions, as well as on the composition and properties of the mineral phases. The system "*Iron (III) minerals - saturated solution*" was examined (Spinu & Povar, 2015), although the approach developed can be applied for other cases. The oxides of iron, particularly the poorly crystallized or microcrystalline forms, are certainly ones the most reactive components of acidic soils. The elements are initially liberated into solution as metal ions by acidic weathering of soil minerals, and then re-precipitated, as hydrous oxide species of high specific surface area and reactivity (Paterson, 1991). As iron occurs in many minerals and materials, it is mostly present in all superficial waters. The average concentration of iron in river waters is 0.7 mg/L (WHO, 2003). With an increase of the acidity of waters (up to a critical threshold of water biota survival; for example, for mollusks this threshold is *pH* 6.0 and for perches it is *pH* 4.5), the contents in iron (III) increases rapidly because of the interaction of iron (III) hydroxide of natural materials with acid:

$$\tfrac{1}{2}Fe_2O_{3(S)} + 3H^+ \Leftrightarrow Fe^{3+} + \tfrac{3}{2}H_2O, K_{S1} = [Fe^{3+}][H^+]^{-3}$$

$$FeOOH_{3(S)} + 3H^+ \Leftrightarrow Fe^{3+} + 2H_2O, K_{S2} = [Fe^{3+}][H^+]^{-3}$$

$$Fe(OH)_{3(S)} + 3H^+ \Leftrightarrow Fe^{3+} + 3H_2O, K_{S3} = [Fe^{3+}][H^+]^{-3}$$

The following important relation of proportionality between the buffer capacities of heterogeneous systems, "$Fe(OH)_{3(S)}$ *- saturated aqueous solution* " is deducted:

$$\beta_H^S = 3^2 \beta_{Fe}^S \tag{33}$$

In case of formation of poorly soluble oxy-hydroxides of the stoichiometric composition $M(OH)_{n(S)}$, $MOOH_{(S)}$ or $1/2M_2O_{n(S)}$, the relation (40) can be generalized (Povar & Rusu, 2012):

$$\frac{\beta_H^S}{n^2} = \frac{\beta_M^S}{1^2} \tag{34}$$

Besides the process of dissolution of the minerals containing iron, a set of possible equilibria in the system "*Mineral phase – soil solution*" (see Table 3) is considered (MINTEQA2/PRODEFA2, 1998). For the calculations the following composition heterogeneous mixture is used (mol•L^{-1}).

$$C_{Fe}^0 = 1\cdot10^{-6} \div 1\cdot10^{-4},\ C_F^0 = 5\cdot10^{-6},\ C_{Org}^0 = C_{PO_4}^0 = C_{SO_4}^0 = 1\cdot10^{-4},\ C_{CO_3}^0 = 1\cdot10^{-3}.$$

Figure 4 shows the calculation results of the buffer capacity β_{Fe}^S as a function of *pH* for the various iron (III) minerals. The analysis of the derived equations for the heterogeneous system shows that the increase in the total concentration C_{Fe}^0 for the *pH* values above 4.5, as well as the nature of the mineral have an insignificant effect on the area of the buffer action of studied system due to the very low solubility of iron oxy - hydroxides minerals. The appearance of maxima on the curves $\beta_H^S(pH)$ and $\beta_{Fe}^S(pH)$ (Figures 4-6) is due to the dissociation process of carbonic acid with formation of HCO_3^- ions (pH = 6.36) and CO_3^{2-}(pH = 10.34).

$$C_{Fe}^0 = 1\cdot10^{-4}, C_F^0 = 5\cdot10^{-6}, C_{Org}^0 = C_{PO_4}^0 = C_{SO_4}^0 = 1\cdot10^{-4}, C_{CO_3}^0 = 1\cdot10^{-3}$$

Table 3. The equilibrium constants of the analyzed reactions

The equations of reactions	logK
$Fe^{3+} + H_2O = FeOH^{2+} + H^+$	-2.187
$Fe^{3+} + 2H_2O = Fe(OH)_2^+ + 2H^+$	-4.59
$Fe^{3+} + 3H_2O = Fe(OH)_3^0 + 2H^+$	-12.56
$Fe^{3+} + 4H_2O = Fe(OH)_4^- + 4H^+$	-21.59
$1/2\ \alpha - Fe_2O_{3(S)}hematite + 3H^+ = Fe^{3+} + 3/2H_2O$	-0.85
$1/2\ \gamma - Fe_2O_{3(S)}maghemite + 3H^+ = Fe^{3+} + 3/2H_2O$	+1.75
$1/2\ \varepsilon - Fe_2O_{3(S)} + 3H^+ = Fe^{3+} + 3/2H_2O$	+2.35
$\alpha - FeOOH_{(S)goethite} + 3H^+ = Fe^{3+} + 3H_2O$	+0.40
$\gamma - FeOOH_{(S)lepidocrocite} + 3H^+ = Fe^{3+} + 3H_2O$	+3.70
$2 - line - Fe(OH)_{3(S)} + 3H^+ = Fe^{3+} + 3H_2O$	+3.00
$6 - line - Fe(OH)_{3(S)} + 3H^+ = Fe^{3+} + 3H_2O$	+3.40
$Fe^{3+} + F^- = FeF^{2+}$	6.04
$Fe^{3+} + 2F^- = FeF_2^+$	10.47
$Fe^{3+} + SO_4^{2-} = Fe(SO_4)^+$	4.05
$Fe^{3+} + 2SO_4^{2-} = Fe(SO_4)_2^-$	5.38
$Fe^{3+} + H^+ + 2SO_4^{2-} = FeH(SO_4)_2$	8.10
$Fe^{3+} + Org^{3-} = FeOrg$	8.00
$Org^{3-} + H^+ = HOrg^{2-}$	4.30
$2Fe^{3+} + 2H_2O = Fe_2(OH)_2^{4+} + 2H^+$	-2.85

continued on following page

Table 3. Continued

The equations of reactions	*logK*
$3Fe^{3+} + 4H_2O = Fe_3(OH)_4^{5+} + 4H^+$	-6.29
$Fe^{3+} + H^+ + PO_4^{3-} = FeHPO_4^+$	19.87
$Fe^{3+} + 2H^+ + PO_4^{3-} = FeH_2PO_4^{2+}$	21.70
$Fe^{3+} + 3H^+ + PO_4^{3-} = FeH_3PO_4^{3+}$	26.61
$CO_3^{2-} + H^+ = HCO_3^-$	10.329
$CO_3^{2-} + 2H^+ = H_2CO_3$	16.681
$H^+ + PO_4^{3-} = HPO_4^{2+}$	12.38
$2H^+ + PO_4^{3-} = H_2PO_4^+$	19.57
$3H^+ + PO_4^{3-} = H_3PO_4$	21.72
$H^+ + F^- = HF$	3.17

Figure 4. The curves of the buffer capacity β_H dependence on pH for the system "Iron (III) mineral - saturated aqueous solution". The used concentrations (mol·L^{-1}):

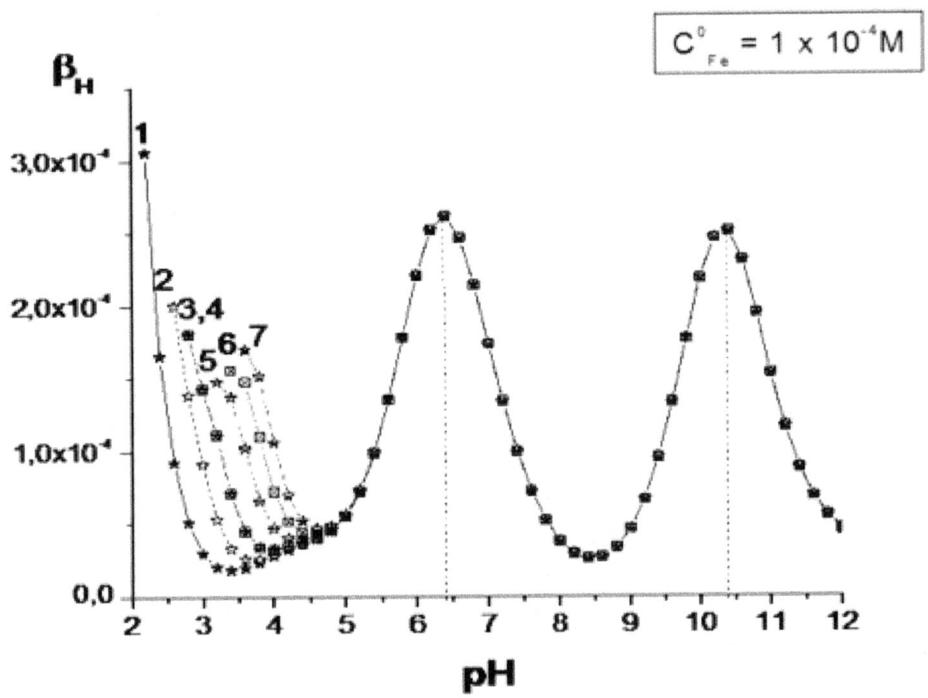

Figure 5. The curves of the buffer capacity β^S_{Fe} dependence on pH for the system "Iron (III) mineral - saturated aqueous solution"

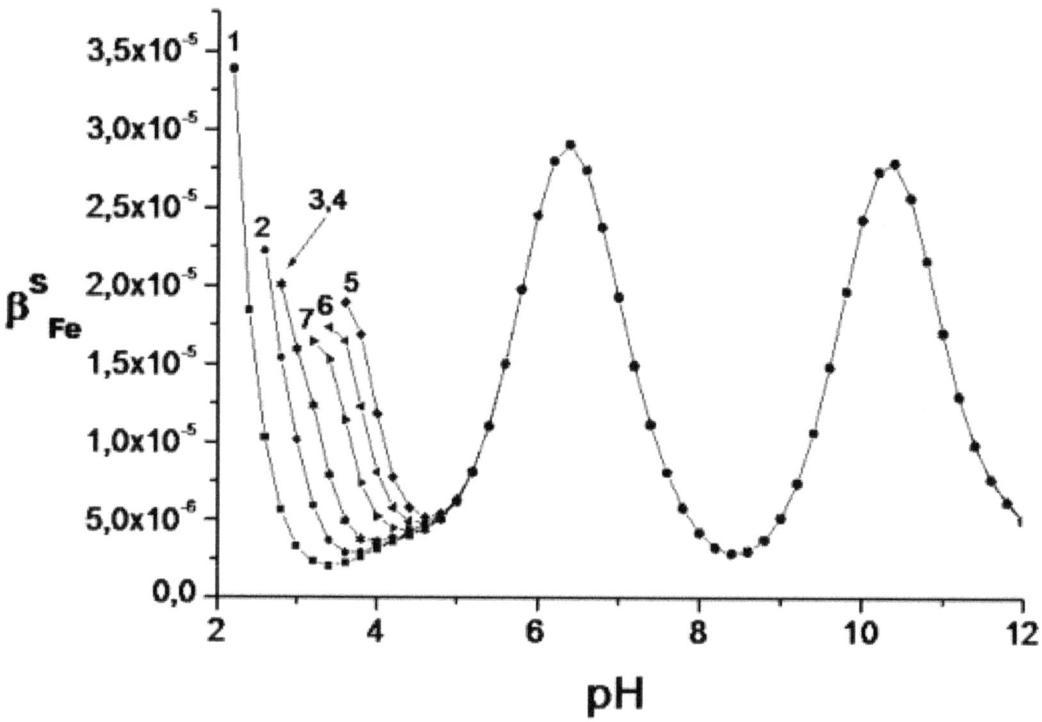

1 - α-Fe_2O_3 (hematite), **2** - γ-Fe_2O_3 (maghemite), **3** – ε-Fe_2O_3, **4** - α-FeOOH (goethite), **5** -γ-FeOOH (lepidocrocite), **6** – $Fe(OH)_3$ (2-line ferrihydrite), **7** – $Fe(OH)_3$ (6-line ferrihydrite)

The used concentrations (mol•L^{-1}):

$$C^0_{Fe} = 1 \cdot 10^{-4}, C^0_F = 5 \cdot 10^{-6}, C^0_{Org} = C^0_{PO_4} = C^0_{SO_4} = 1 \cdot 10^{-4}, C^0_{CO_3} = 1 \cdot 10^{-3}$$

1 - α-Fe_2O_3 (hematite), **2** - γ-Fe_2O_3 (maghemite), **3** – ε-Fe_2O_3, **4** - α-FeOOH (goethite), **5** - γ-FeOOH (lepidocrocite), **6** – $Fe(OH)_3$ (2-line ferrihydrite), **7** – $Fe(OH)_3$ (6-line ferrihydrite)

The used concentrations (mol•L^{-1}):

$$C^0_{Fe} = 1 \cdot 10^{-4}, C^0_F = 5 \cdot 10^{-6}, C^0_{Org} = C^0_{PO_4} = C^0_{SO_4} = 1 \cdot 10^{-4}, C^0_{CO_3} = 1 \cdot 10^{-3}$$

The iron (III) ions and its complexes (Table 3) have a significant contribution only in very acidic solutions up to *pH* 4.5. The analysis of the data presented in Figures 4 and 5 allows to conclude that the studied heterogeneous system has a buffer capacity towards ions of hydrogen and iron (III). These results were found experimentally by the authors (Cabrera, 2016). It is worthy to note down again that these relations are valid only in the presence of solid phases.

Another example of real natural system is "*Gibbsite – saturated aqueous solution*" was considered (Adekola et al., 2011; Povar & Rusu, 2012). By increasing the acidity in aqueous solutions the content of aluminum grows quickly due to the interaction of the gibbsite with an acid:

*Figure 6. Buffer capacities β_H^S (**1**) and β_{Fe}^S (**2**) versus pH for the system "Hematite - saturated aqueous solution"*

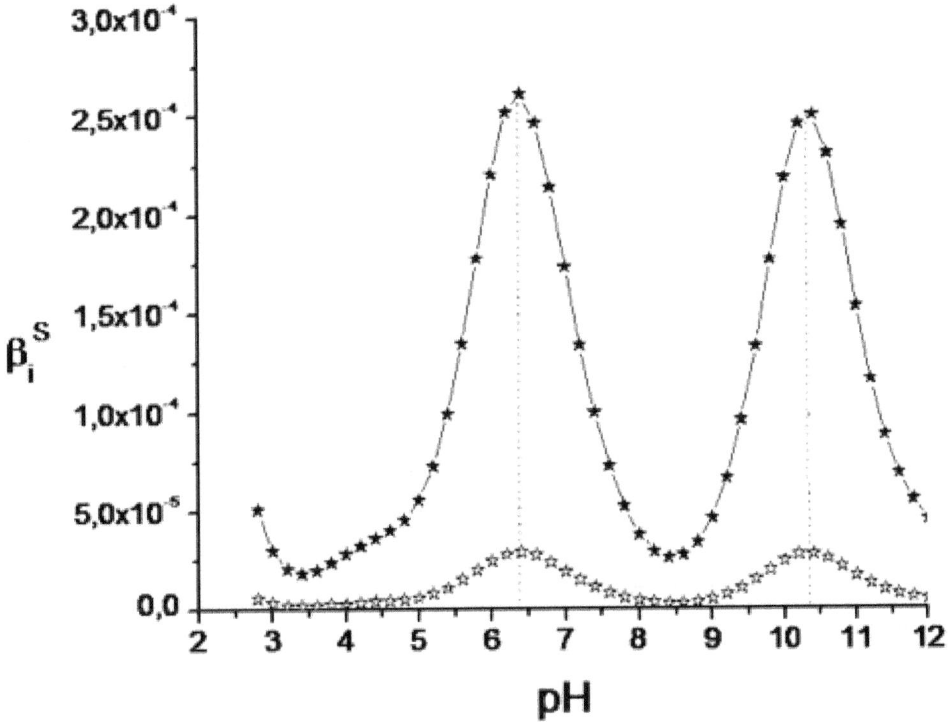

$$Al(OH)_{3(S)} + 3H^+ = Al^{3+} + 3H_2O, K_S = [Al^{3+}][H^+]^{-3}$$

An important interrelation between the buffer capacities of the heterogeneous system "*Gibbsite $Al(OH)_{3(s)}$ - saturated aqueous solution*" system has been identified:

$$\beta_H^S = 9\beta_{Al}^S \quad (45)$$

Besides the process of gibbsite dissolution, a set of possible equilibria in the system "*Mineral phase – natural water*", listed in Table 4 has been taken into account (MINTEQA2/PRODEFA2, 1998). The following composition of the heterogeneous system was used for the calculations,mol•L^{-1}: $C_{Al}^0 = 1{\bullet}10^{-4}$, $C_F^0 = 5{\bullet}10^{-6}$, $C_{Org}^0 = 1{\bullet}10^{-4}$, $C_{PO_4}^0 = 1{\bullet}10^{-4}$, $C_{SO_4}^0 = 1{\bullet}10^{-4}$ and $C_{CO_3}^0 = 1{\bullet}10^{-3}$. One can state that the influence of H^+ in the studied *pH* interval may be neglected, and the OH^- ions exercise influence on β_H^S only at *pH*> 8.5. The aluminum ions and their hydroxocomplexes make a significant contribution in acid solutions up to *pH* 5.5 and at pH> 8 because of the predominance of the stable anionic hydroxo-complex $Al(OH)_4$ The results of the calculations of the buffer capacities β_H^S as a function of *pH* for

Table 4. Equilibrium constants and values of enthalpies (ΔH)

The equations of reactions	log K	ΔH (cal mol^{-1})
$Al^{3+} + H_2O = AlOH^{2+} + H^+$	-4.99	11900
$Al^{3+} + 2H_2O = Al(OH)_2^+ + 2H^+$	-10.00	22000
$Al^{3+} + 4H_2O = Al(OH)_4^- + 4H^+$	-23.00	44060
$Al(OH)_{3(S)} + 3H^+ = Al^{3+} + 3H_2O$	9.35	–22800
$Al^{3+} + F^- = AlF^{2+}$	7.02	1100
$Al^{3+} + 2F^- = AlF_2^+$	12.76	2000
$Al^{3+} + 3F^- = AlF_3$	17.03	2500
$Al^{3+} + 4F^- = AlF_4^-$	19.73	2200
$Al^{3+} + 5F^- = AlF_5^{2-}$	20.92	1800
$Al^{3+} + SO_4^{2-} = AlSO_4^+$	3.01	2150
$Al^{3+} + 2SO_4^{2-} = Al(SO_4)_2^-$	4.90	2840
$Al^{3+} + Org^{3-} = AlOrg$	8.39	—
$Al^{3+} + H^+ + Org^{3-} = AlHOrg^+$	13.09	—
$Org^{3-} + H^+ = HOrg^{2-}$	6.83	—
$Org^{3-} + 2H^+ = H_2Org^-$	12.73	—
$Org^{3-} + 3H^+ = H_3Org$	14.49	—
$H^+ + F^- = HF$	3.17	3460
$H_2O = H^+ + OH^-$	-14.00	13340
$2Al^{3+} + 2H_2O = Al_2(OH)_2^{4+} + 2H^+$	-6.3	—
$3Al^{3+} + 4H_2O = Al_3(OH)_4^{5+} + 4H^+$	-12.1	—

continued on following page

Table 4. Continued

The equations of reactions	*log K*	ΔH *(cal mol⁻¹)*
$Al^{3+} + H_2PO_4^- = AlH_2PO_4^{2+}$	3.1	—
$PO_4^{3-} + H^+ = HPO_4^{2-}$	12.0	—
$PO_4^{3-} + 2H^+ = H_2PO_4^-$	19.21	—
$PO_4^{3-} + 3H^+ = H_3PO_4$	21.36	—

Note: 1 cal = 4.184 J, Org - organic ligand, the „—" specifies the absence of experimental data.

Figure 7. Buffer capacity β_H^S versus pH for the "Gibbsite - saturated aqueous solution" system

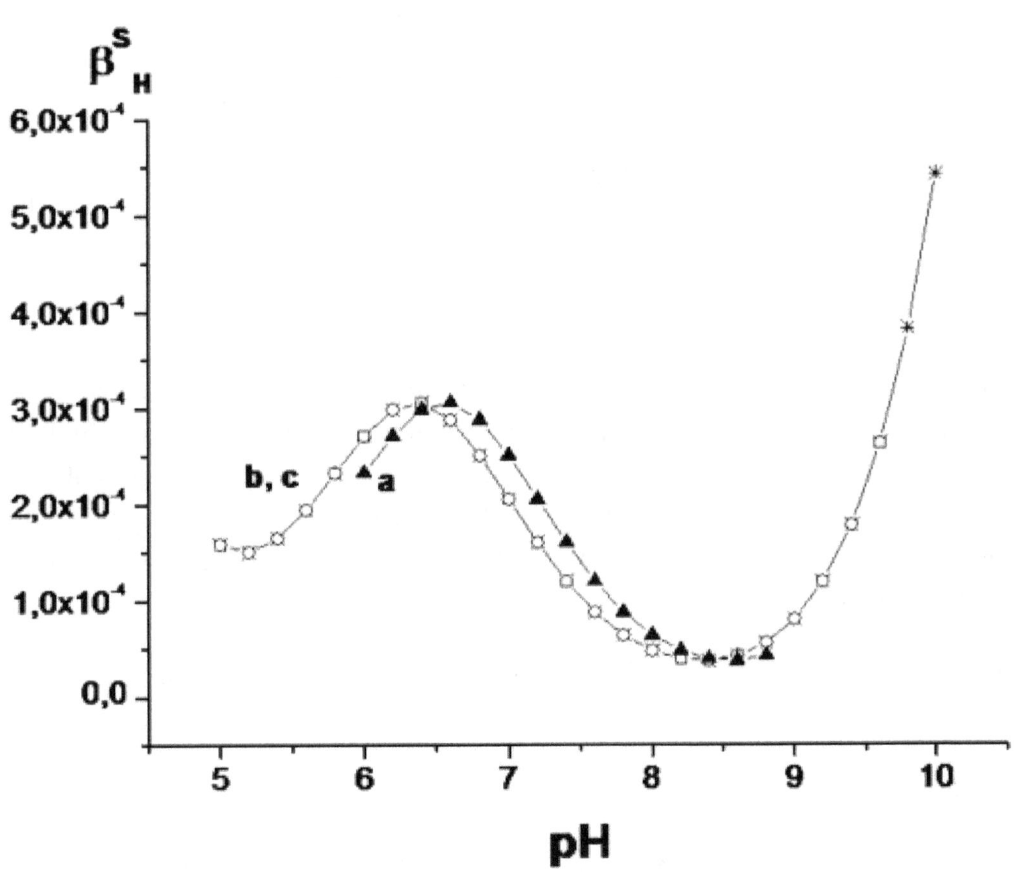

Figure 8. Buffer capacity β_H^S versus pH for the "Gibbsite - saturated aqueous solution" system

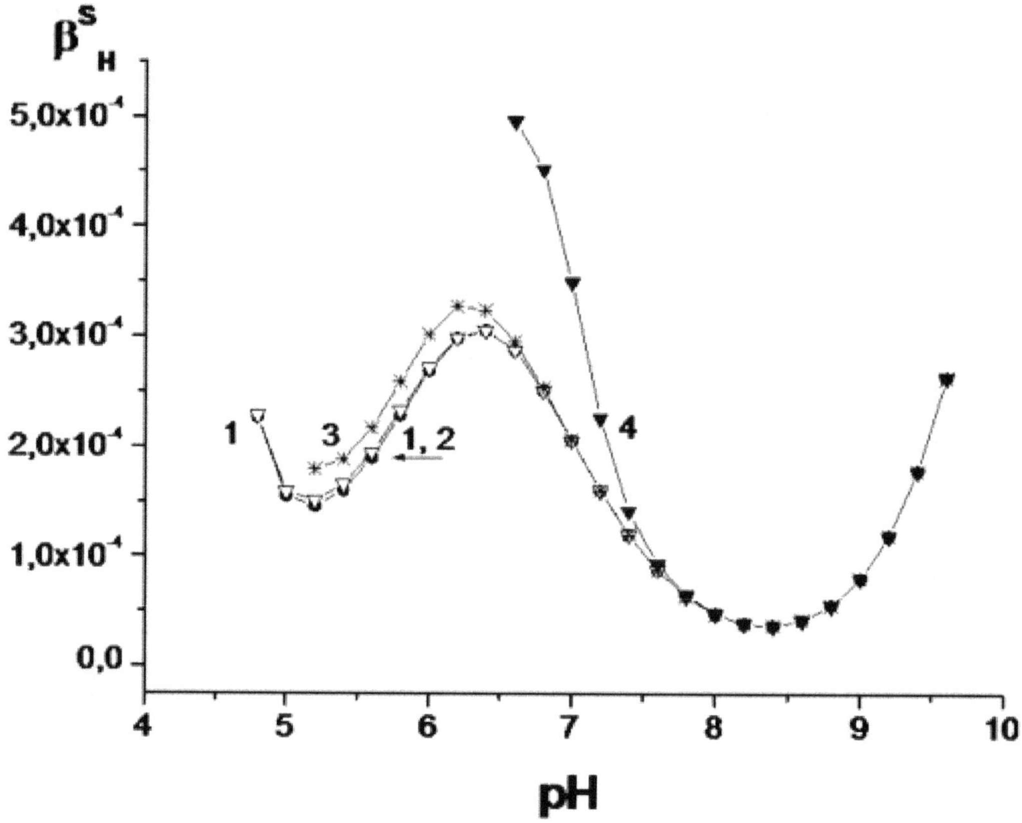

different compositions of the heterogeneous mixture are shown in Figures 7–8. Obviously, an increase of the total concentration C_{Al}^0 (Figure 7) expands the area of the buffer action because of the extension of the *pH* interval of the thermodynamic stability of the solid phase. At concentrations of fluoride C_F^0 = 5•10-4 mol•L-1, the β_H^S value increases sharply in the range of *pH* values of the gibbsite dissolution – formation with a simultaneous narrowing of the total *pH* range of the buffer action (Figure 8).

Concentrations (mol•L-1):

$$C_{Al}^0 : a-1\cdot10^{-3},\ b-1\cdot10^{-4},\ c\text{-}1\cdot10^{-5}, C_F^0 = 5\cdot10^{-6}, C_{Org}^0 = C_{PO_4}^0 = C_{SO_4}^0 = 1\cdot10^{-4}, C_{CO_3}^0 = 1\cdot10^{-3}$$

Concentrations (mol•L-1):

$$C_F^0 : 1-5\cdot10^{-7}, 2-5\cdot10^{-6}, 3\text{-}5\cdot10^{-5}, 4\text{ - }5\cdot10^{-4}, C_{Al}^0 = C_{Org}^0 = C_{PO_4}^0 = C_{SO_4}^0 = 1\cdot10^{-4}, C_{CO_3}^0 = 1\cdot10^{-3}$$

The dependences of β_H^S (pH) for different concentrations of carbonate ion are presented in Figure 9. By taking into account the contribution of equilibria including carbonate ion in the total buffer capacity β_H^S, the $\beta_{CO_3}^H$ value was calculated separately. A comparison of the calculated curves shows that equilibria with CO_3^{2-} participation have a substantial contribution to β_H^S at a $C_{CO_3}^0 > 1 \cdot 10^{-4}$ mol•L^{-1} (Figures 10a and 10b). The analysis of the obtained data presented in Figures 7–10, taking into account Eq.(45) allows to conclude that the investigated heterogeneous system has a considerable buffer capacity towards to aluminum as well.

Figure 9. Buffer capacity β_H^S versus pH for the "Gibbsite - saturated aqueous solution" system

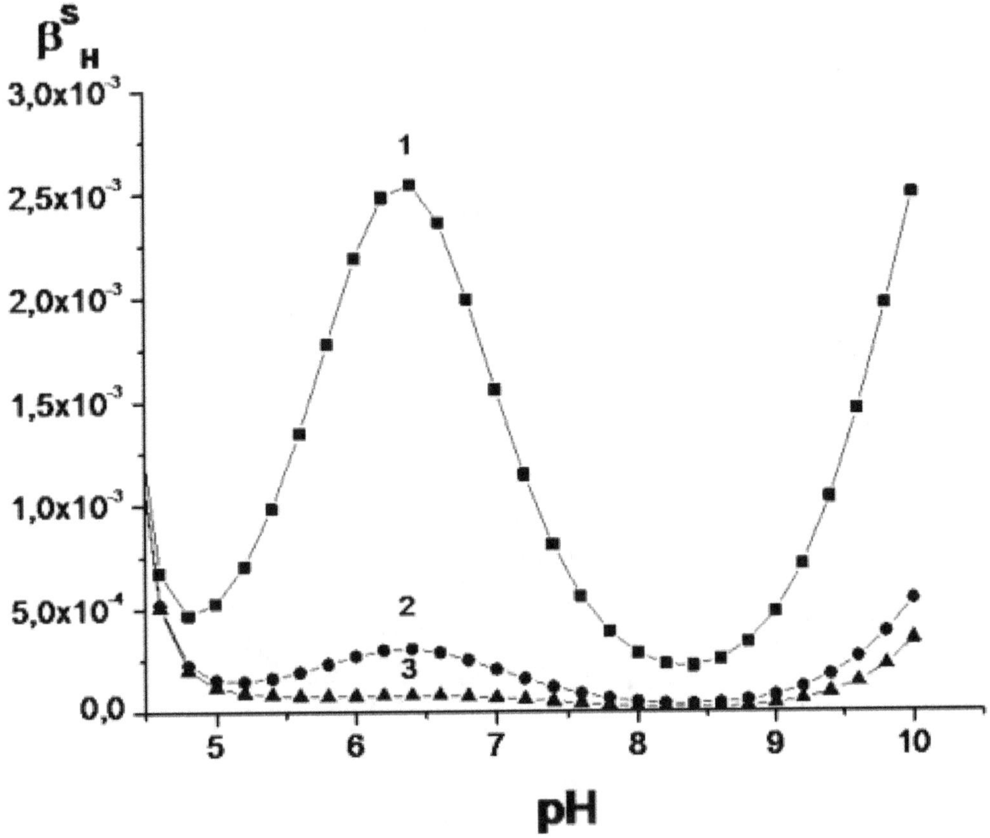

Concentrations (mol • L^{-1}): $C_{CO_3}^0 : 1 - 1 \cdot 10^{-2}, 2 - 1 \cdot 10^{-3}, 3 - 1 \cdot 10^{-4}, C_{Al}^0 = C_{PO_4}^0 = C_{SO_4}^0 = 1 \cdot 10^{-4}, C_F^0 = 5 \cdot 10^{-6}$

pH and their difference $\beta_H^S - \beta_{CO_3}^H$ for the "*Gibbsite - saturated aqueous solution*" system

Concentrations (mol•L^{-1}): $C_{CO_3}^0 : a - 1 \cdot 10^{-3}$, b-$1 \cdot 10^{-4}, C_{Al}^0 = C_{PO_4}^0 = C_{SO_4}^0 = 1 \cdot 10^{-4}, C_F^0 = 5 \cdot 10^{-6}$

Figure 10. Total buffer capacity β_H^S versus pH, carbonate buffer capacity $\beta_{CO_3}^H$ versus

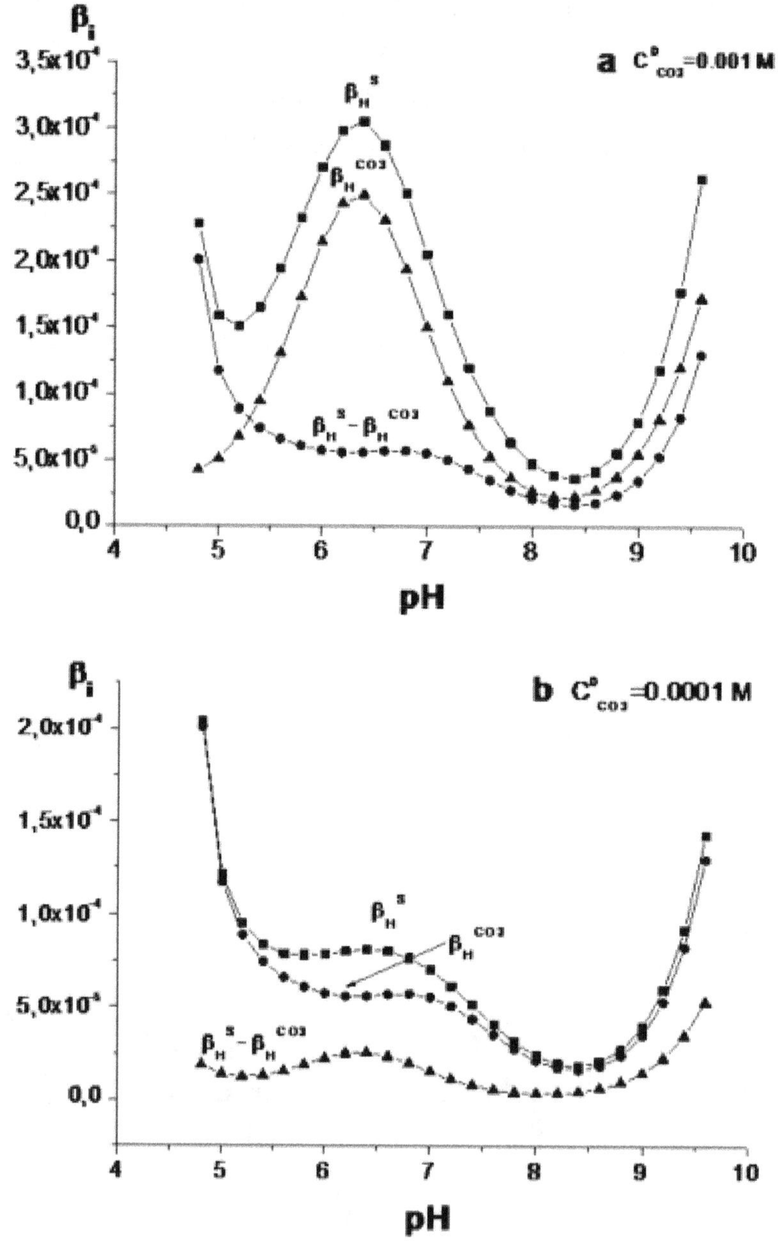

The influence of temperature on the value of the buffer capacity was also studied. The results of the calculations are presented in Figure 11. The equilibrium constants for different temperatures have been estimated by the van't Hoff equation.

$logK_2 = logK_1 + (1/T_1 - 1/T_2)\ \Delta H/2.303R$

Concentrations (mol•L^{-1}): $C^0_{CO_3} = 1\cdot10^{-3}, C^0_{Al} = C^0_{PO_4} = C^0_{Org} = C^0_{SO_4} = 1\cdot10^{-4}, C^0_F = 5\cdot10^{-6}$

*Figure 11. Buffer capacity β_H^S versus pH for the "Gibbsite - saturated aqueous solution" system at different temperatures (t, °C): (**1**) –5, (**2**) +10, (**3**) +25 and (**4**) +40*

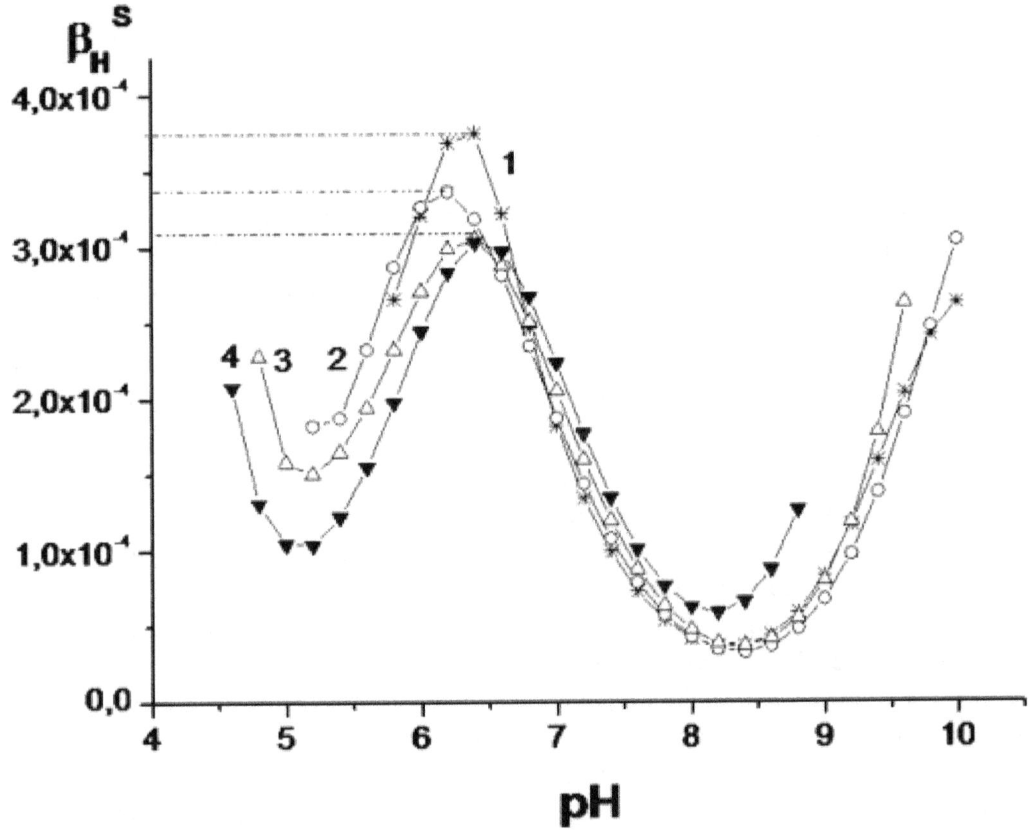

The necessary values of enthalpies (ΔH) are listed in Table 4. T_1 was set to 298 K = 25 °C. It was assumed that the temperature insignificantly influences the ΔH values inside the investigated temperature range. An analysis of the curves in Figure 11 shows that the buffer capacity increases with a temperature decrease, whereas the *pH* interval of 5.5 – 7.0 of the maximum values of the buffer capacity displaces slightly. The theoretical results obtained here concerning the buffer capacity of the system gibbsite – soil solution describe very well existing experimental evidences (Bleam, 2016; Goulding, 2016; Hong et al., 2018; Sartz, 2016; Uhlmann et al., 2006).

FUTURE RESEARCH DIRECTIONS

The use of the buffer approach developed may yield the extended knowledge and a deeper understanding of the processes that control the concentrations of components. The obtained results are indented to provide researchers with a tool needed to help them to set reliable limits of ion (metal) levels in the environment. It is of great interest to analyze the buffer properties of heterogeneous systems, which contain two or more solid phases with the common metal ions or anions and compare with the buffer systems analyzed in this paper. It is expected that the proportionality between buffer capacities of different components in multicomponent systems occurs for the two solid buffer systems too.

CONCLUSION

The thermodynamic study within the described buffer approach allows and explains the following: (a) To extend our mechanistic comprehending of natural processes and of accelerate natural processes that may attenuate certain contaminants in groundwater via complex formation (chelating) and precipitation reactions; (b) To predict and identify the exact conditions and to what extent these processes are reversible; (c) Improve basic understanding of ecological mechanisms and processes; (d) Presence of ligands that form stable complexes with ions of the solid phase promotes the amplification of the buffer actions of the heterogeneous systems; (e) Use of results of this research allows predicting the variations of the composition and response of the environment at the technogenic loads increasing; (f) On the basis of the residual concentrations method, the original mathematical expressions for the quantitative calculation of buffer capacities towards all the components of heterogeneous mixtures, including those of the mineral phase have been deduced. The influence of different factors (temperature, pH, and concentrations of the main natural water components) on the values of buffer capacities of the investigated heterogeneous systems has been revealed; (g) Quantitative basis of the theory of buffer action for heterogeneous systems has been stated. It is established that buffer properties in relation to components of the solid phase amplify with the increase of the solubility. The buffer capacity of any component is a complex function of the equilibrium composition of multicomponent heterogeneous systems; (h) It is proved that the buffer capacities of components are mutually proportional and for heterogeneous systems these relations depend on the stoichiometric composition of the solid phases (minerals). It is established that natural waters in equilibrium with gibbsite have a certain potential reserve, by means of which the equilibrium is adjusted and one of its parameters is maintained constant; (i) Use of the results of this research allows predicting the variations of the composition and response of the soil environment at the technogenic loads increasing;

REFERENCES

Adekola, F., Fédoroff, M., Geckeis, H., Kupcik, T., Lefèvre, G., Lützenkirchen, J., ... Schild, D. (2011). Characterization of acid–base properties of two gibbsite samples in the context of literature results. *Journal of Colloid and Interface Science*, *354*(1), 306–317. doi:10.1016/j.jcis.2010.10.014 PMID:21044786

Beresnev, E. N. (1992). *Method of residual concentrations*. Moscow: Nauka.

Bethke, C. M. (2007). *Geochemical and biogeochemical reaction modeling*. Cambridge: University Press. doi:10.1017/CBO9780511619670

Bleam, W. (2016). *Soil and Environmental Chemistry* (2nd ed.). New York: Academic Press.

Burgot, J. L. (2012). *Ionic equilibria in analytical chemistry*. Springer Science & Business Media. doi:10.1007/978-1-4419-8382-4

Cabrera, J. M., Diaz, M. M., Schultz, S., Temporetti, P., & Pedrozo, F. (2016). Iron buffer system in the water column and partitioning in the sediments of the naturally acidic Lake Caviahue, Neuquén, Argentina. *Journal of Volcanology and Geothermal Research, 318*, 19–26. doi:10.1016/j.jvolgeores.2016.03.001

Charykov, A. K., & Osipov, N. N. (1991). *Carbonic acids and carboxylate complexes in chemical analysis*. Leningrad: Chemistry. (in Russian)

Dougherty, D. P., Neta, C. E., McFeeters, R. F., Lubkin, S. R., & Breidt, F. J. (2006). Semi-mechanistic partial buffer approach to modelling pH, the buffer properties, and the distribution of ionic species in complex solutions. *Journal of Agricultural and Food Chemistry, 54*(16), 6021–6029. doi:10.1021/jf0531508 PMID:16881711

Fishtik, I., & Povar, I. (2006). Buffer capacity in multiple chemical reaction systems involving solid phases. *Canadian Journal of Chemistry, 84*(8), 1036–1044. doi:10.1139/v06-127

Goulding, K. W. T. (2016). Soil acidification and the importance of liming agricultural soils with particular reference to the United Kingdom. *Soil Use and Management, 32*(3), 390–399. doi:10.1111um.12270 PMID:27708478

Hong, S., Piao, S., Chen, A., Liu, Y., Liu, L., Peng, S., ... Zeng, H. (2018). Afforestation neutralizes soil pH. *Nature Communications, 9*(1), 520. doi:10.103841467-018-02970-1 PMID:29410472

Kay, J. W., Steven, R. J., McGuigan, J. A. S., & Elder, H. Y. (2008). Automatic determination of ligand purity and apparent dissociation constant (Kapp) in Ca^{2+}/Mg^{2+} buffer solutions and the Kapp for Ca^{2+}/Mg^{2+} anion binding in physiological solutions from Ca^{2+}/Mg^{2+} - macroelectrode measurements. *Computers in Biology and Medicine, 38*, 101–110. doi:10.1016/j.compbiomed.2007.07.008 PMID:17854794

Komari, N. P. (1984). *Chemical metrology. Ionic heterogeneous equilibria*. Kharkov: Vishcha Shkola. (in Russian)

Lu, K., Yang, X., Gielen, G., Bolan, N., Ok, Y. S., Niazi, N. K., ... Wang, H. (2017). Effect of bamboo and rice straw biochars on the mobility and redistribution of heavy metals (Cd, Cu, Pb and Zn) in contaminated soil. *Journal of Environmental Management, 186*(part 2), 285–292. doi:10.1016/j.jenvman.2016.05.068 PMID:27264699

McBride, M. B., & Blasiak, J. J. (1979). Zinc and Copper Solubility as a Function of pH in an Acid Soil. *Soil Science Society of America Journal, 43*(5), 866–870. doi:10.2136ssaj1979.03615995004300050009x

MINTEQA2/PRODEFA2, A Geochemical Assessment Model for Environmental Systems: User Manual Supplement for Version 4.0. (1998). = Allison Geoscience Consultants. Inc. Flowery Branch.

Paterson, E., Goodman, B. A., & Farmer, V. C. (1991). The Chemistry of Aluminium, Iron and Manganese Oxides in Acid Soils. In B. Ulrich & M. E. Sumner (Eds.), *Soil Acidity*. Berlin: Springer. doi:10.1007/978-3-642-74442-6_5

Perrin, D. D. (2012). *Buffers for pH and metal ion control*. Springer Science & Business Media.

Pertusatti, J., & Prado, A. G. (2007). Buffer capacity of humic acid: Thermodynamic approach. *Journal of Colloid and Interface Science, 314*(2), 484–489. doi:10.1016/j.jcis.2007.06.006 PMID:17603069

Pfendt, L. B. (1995). Two-phase liquid and solid-liquid pH buffers based on solubility equilibria: Theoretical considerations. *Analyst (London)*, *120*(8), 2129–2144. doi:10.1039/AN9952002129

Povar, I. (1996). Buffer properties of heterogeneous systems containing sparingly soluble acids and metal hydroxides. *Russian Journal of Inorganic Chemistry*, *41*, 1167–1171.

Povar, I. (2000a). Buffer properties of heterogeneous water-salt systems in relation to the components of a low-solubility precipitate. *Russian Journal of Inorganic Chemistry*, *45*, 1632–1636.

Povar, I. (2000 b). On Homogeneous Ion Buffer Systems. *Russian Journal of Inorganic Chemistry*, *45*, 1628–1631.

Povar, I., & Luca, C. (2003). Considerations regarding ionic-molecular buffer solutions. *Revista de Chimie*, *54*, 312–316.

Povar, I., & Rusu, V. (2012). Buffer capacity of heterogeneous chemical equilibria in natural waters. *Canadian Journal of Chemistry*, *90*(4), 395–402. doi:10.1139/v2012-008

Povar, I., & Spinu, O. (2014a). The role of hydroxy aluminium sulfate minerals in controlling Al^{3+} concentration and speciation in acidic soils. *Open Chemistry*, *12*(8), 877–885. doi:10.247811532-014-0540-4

Povar, I., & Spinu, O. (2014b). *Thermodynamics of Complex Chemical Equilibria in Heterogeneous Multicomponent Systems*. Chisinau: Printing House of the Academy of Sciences of Moldova. (in Romanian)

Povar, I., & Spinu, O. (2015). Application of the buffer theory for evaluating attenuation and natural remediation of ionic pollutants in aquatic ecosystems. *Ecological Processes*, *4*(1), 17. doi:10.118613717-015-0043-z

Ritchie, J. D., & Perdue, E. M. (2003). Proton-binding study of standard and reference fulvic acids, humic acids, and natural organic matter. *Geochimica et Cosmochimica Acta*, *67*(1), 85–96. doi:10.1016/S0016-7037(02)01044-X

Roy, L. N., Roy, R. N., Wollen, J. T., Harmon, M. A., Stegner, J. M., Shah, A. A., & Henson, I. B. (2011b). Buffer standards for the biological pH of the amino acid n-[2-hydroxyethyl] piperazine-n¢-[3-propanesulfonic acid], Hepps, from (278.15 to 328.15) K. *Journal of Chemical & Engineering Data*, *56*(11), 4126–4132. doi:10.1021/je200629y PMID:22096257

Roy, R. N., Roy, L. N., Stegner, J. M., Sechler, S. A., Jenkins, A. L., Krishchenko, R., & Henson, I. B. (2011a). Buffer standards for the physiological pH of N, N - bis (2- hydroxyethyl) - 2-aminoethanesulfonic acid (BES) from (278.15 to 328.15) K. *Electroanalytical Chemistry*, *663*(1), 8–13. doi:10.1016/j.jelechem.2011.09.019 PMID:22247703

Sartz, L., Bäckström, M., Karlsson, S., & Allard, B. (2016). Mixing of acid rock drainage with alkaline ash leachates: Formation of solid precipitates and pH-buffering. *Mine Water and the Environment*, *35*(1), 64–76. doi:10.100710230-015-0347-3

Schnitzer, M., & Gupta, U. C. (1965). Determination of Acidity in Soil Organic Matter. *Soil Science Society of America Journal*, *29*(3), 274–277. doi:10.2136ssaj1965.03615995002900030016x

Spinu, O., & Povar, I. (2015). Buffer capacity in heterogeneous multicomponent systems. [Review]. *Chemistry Journal of Moldova*, *10*(2), 8–25. doi:10.19261/cjm.2015.10(2).01

Stumm, W., & Morgan, J. J. (2012). *Aquatic Chemistry: Chemical Equilibria and Rates in Natural Waters*. New York: Wiley-Interscience.

Stutter, M. I., & Richards, S. (2012). Relationships between soil physicochemical, microbiological properties, and nutrient release in buffer soils compared to field soils. *Journal of Environmental Quality*, *41*(2), 400–409. doi:10.2134/jeq2010.0456 PMID:22370402

Totsche, O., Fyson, A., Kalin, M., & Steinberg, C. E. W. (2006). Titration Curves: A useful instrument for assessing the buffer systems of acidic mining waters. *Environmental Science and Pollution Research International*, *13*(4), 215–224. doi:10.1065/espr2005.09.284 PMID:16910118

Uhlmann, W., Büttcher, H., Totsche, O., & Steinberg, C. E. W. (2004). Buffering of acidic mine lakes: The relevance of surface exchange and solid-bound sulphate. *Mine Water and the Environment*, *23*(1), 20–27. doi:10.100710230-004-0032-4

Upadhyay, N., Majestic, B. J., & Herckes, P. (2011). Solubility and speciation of atmospheric iron in buffer systems simulating cloud conditions. *Atmospheric Environment*, *45*(10), 1858–1866. doi:10.1016/j.atmosenv.2011.01.010

World Health Orgranization. (2003). *Guidelines for Drinking Water Quality. Iron in Drinking Water.* Retrieved from https://www.who.int/water_sanitation_health/dwq/chemicals/iron.pdf

ADDITIONAL READING

Butler, J. (1998). *Ionic Equilibrium: Solubility and pH Calculations*. New York: John Wiley & Sons.

Jalali, M., &Moradi, A. (2019). Measuring and simulating pH buffer capacity of calcareous soils using empirical and mechanistic models. *Archives of Agronomy and Soil Science*, (just-accepted). doi:10.1080/03650340.2019.1628344

Povar, I., & Rusu, V. (2012). Aluminium heterogeneous speciation in natural waters. *Canadian Journal of Chemistry*, *90*(4), 326–332. doi:10.1139/v2012-003

Povar, I., & Spinu, O. (2013). Buffer properties of soil minerals. Part.1 Theoretical Aspects. *Chemistry Journal of Moldova*, *8*(2), 67–72. doi:10.19261/cjm.2013.08(2).07

Povar, I., & Spinu, O. (2014). Application of the theory to assess buffering natural attenuation of metal ions and rehabilitation of aquatic ecosystems. *Water: Chemistry and Ecology*, *4*, 89-93.

Povar, I., & Spinu, O. (2015). Acid-Base Buffer Properties of Heterogeneous Multicomponent Extraction Systems. *Solvent Extraction and Ion Exchange*, *33*(2), 196–209. doi:10.1080/07366299.2014.951284

Povar, I., & Spinu, O. (2016). Correlation between global thermodynamic functions and experimental data in multicomponent heterogeneous systems. *Canadian Journal of Chemistry*, *94*(2), 113–119. doi:10.1139/cjc-2015-0411

Wang, Y., Cheng, P., Li, F., Liu, T., Cheng, K., Yang, J., & Lu, Y. (2018). Variable charges of a red soil from different depths: Acid-base buffer capacity and surface complexation model. *Applied Clay Science*, *159*, 107–115. doi:10.1016/j.clay.2017.08.003

KEY TERMS AND DEFINITIONS

Buffer Capacity: The capacity of buffer system to oppose (resist) to the variation of its chemical composition (i.e., to changes of concentration of any component) by influence of external fluxes of chemical compounds of natural or anthropogenic characters.

Chemical Equilibrium: The state of chemical reaction in which both reactants and products are present in concentrations which have no further tendency to change with time, so that there is no observable change in the properties of the system.

Chemical Thermodynamics: The branch of physical chemistry concerned with thermodynamic phenomena in chemistry, as well as with the dependence of the thermodynamic properties of substances on the substances' composition and aggregation state.

Gibbs Energy: One of the main thermodynamic characteristic functions of a system, denoted by G, widely used in examining various thermodynamic processes occurring at constant temperature and pressure.

Heterogeneous System: A non-uniform physicochemical system comprising two or more phases, separated from an adjacent phase by an interface where one or more physical properties or chemical composition of the system change abruptly.

Homogeneous System: A thermodynamic system in which the components do not differ from one another in chemical composition or physical properties and are not separated from one another by surfaces of separation.

Mass Balance: Also called a material balance, is an application of conservation of mass to the analysis of thermodynamic systems. Specifically, it is a sum on equilibrium concentrations of all chemical species containing a certain element or ion.

Soil Solution: The liquid phase of soil, which contains dissolved gases, minerals, and organic matter.

Chapter 9
Ecological Chemistry of the Natural Waters

Viorica Gladchi
Moldova State University, Moldova

Elena Bunduchi
Moldova State University, Moldova

Lidia Romanciuc
https://orcid.org/0000-0002-4555-1194
Institute of Chemistry, Moldova

ABSTRACT

Hydrosphere collects and accumulates the most anthropogenic and natural pollutants released to atmosphere or lithosphere. Polluted water affects life activity of both individual organisms and the water ecosystems. For adequate estimating of natural waters state and quality, the different methods of research, including the study of physical, chemical, and biological processes that occur in the aquatic ecosystems were used. The kinetic approach was applied to study the complexity of redox processes in the aquatic medium, redox state of water, and its ability to self-purify both in the surface water objects of the Republic of Moldova and modeling systems. For this purpose, the redox active components in natural waters and their seasonal dynamics were studied by detecting the hydrogen peroxide, OH radicals, and inhibition capacity of natural waters.

INTRODUCTION

The natural waters state and quality monitoring in the Republic of Moldova started in the 80s of the last century with the scope to determine the water chemical composition and various types of biogenic elements, taking into account that natural waters constitute a living environment for hydrobionts (Zepp, 1980, Sychev et al., 1983). Later more important issues arose related to the necessity to study the anthropogenic pollution of environment and, primarily, of natural waters (Zepp and Cline, 1977, 1980;

DOI: 10.4018/978-1-7998-1241-8.ch009

Turro et al., 1978; Scurlatov and Duca, 1985; Baud et al., 1986; Scurlatov et al., 1987; Litchfield, 1989; Murti and Nag, 1991). As a result of studies appeared necessity to conduct the extensive and thorough scientific research of the redox processes in natural systems, and in particular, to reveal their role in the estimation of quality of the ambient environment (Zanoaga, 1995; Duca et al., 1999, 2001, 2008; Duca, 2012; Skurlatov and Shtamm, 2012).

BACKGROUND

By the chemical point of view, natural water represent very dilute solutions, which include the following components: dissolved gases; macro-elements that determine mineralization; different mineral forms of biogenic elements; micro-elements and various types of organic substances and pollutants (Miller and Zepp, 1979; Travin et al., 1986). All these components were studied usually by hydro-chemical methods and represent the well known traditional approach.

By the physicochemical point of view, natural waters represent an open photochemical and catalytic system with several components and phases. In natural waters the redox processes take place, in the presence of metal ions with variable valence (Duca et al., 1989; Andrews et al., 1998). The composition of natural waters is not constant, being determined by several factors such as climate conditions, variety of microflora, duration and intensity of solar radiation, anthropic factor, etc. As a result of the redox processes, a large number of active substances are formed: hydrogen peroxide, singlet oxygen, ozone, free radicals, etc. (Skurlatov et al., 1989; Alvarez et al., 1998; Thannickal and Fanburg, 2000; Duca, 2012; Gonciaruc, 2012).

Therefore, according to the theory proposed by Duca et al (1999), the complex estimating of the aquatic objects status can not be achieved through traditional hydro analysis parameters only, but also by applying of advanced methods, which describe more completely the chemical processes that occur in natural waters and characterize most appropriately water as living environment for hydrobionts. The proposed theory is based on the concept of the waters redox state and is resulted from the analysis of the redox equivalents balance in aquatic environment.

The proposed concept denotes the necessity of introducing and implementation of new methods for the characterization of natural water status and estimating of their self-purification capacity.

The most important oxidative equivalents in water systems are dissolved oxygen and hydrogen peroxide (Duca et al, 2002). The specifics of these oxidants is based on a quite low reactivity, which could be increased by their activation by some metal ions, like copper or iron.

Over the years, the research has been focused on study the mechanisms of organic substances oxidation in aquatic systems similar to natural waters. The scientific results were included in numerous monographs and articles at the national and international level (Skurlatov et al., 1994; Korte et al.,1997; Duca and Skurlatov, 2002; Bahadir and Duca, 2008; Duca, 2017).

MAIN FOCUS OF THE CHAPTER

The ecological chemistry of natural waters is studied through the aspects of physical, chemical and biological processes, which take place in the aquatic ecosystems, for adequate estimating of the aquatic medium redox state with application of different research methods. The basic principles proposed for discussion in this Chapter are based on application of the specific methods and kinetic approach for complex estimating of the water bodies state and their ability to self-purification. The research was performed on the model systems and natural surface waters of the Republic of Moldova.

ISSUES, CONTROVERSIES, PROBLEMS

The important oxidant in the aquatic environment is considered the dissolved oxygen, which under the normal circumstances is a fairly inert substance. For normal activity of hydrobiontes, the cycle of oxygen in water is a fundamental factor and is closely linked with other biogeochemical cycles. In case of dissolved oxygen participation in the water chemical self-purification processes, due to its reducing, several intermediates are formed with the advanced reactivity, contributing to the more effective and wider development of these processes (Duca, 1997; Korte et al., 1997; Duca et al.,2002; Skurlatov et al., 2017).

Presence of the hydrogen peroxide in natural waters is of great importance for self-purification. It practically represents a carrier of oxidative equivalents in the aquatic environment, being an intermediate in the reducing of molecular oxygen to water:

$$O_2 \xrightarrow[+2e^-]{+2H^+} H_2O_2 \xrightarrow[+2e^-]{+2H^+} 2H_2O$$

Activation of O_2 and H_2O_2 in the real water conditions can be carried out in presence of the reduced forms of transition metal ions - Fe (II) and Cu (I), which have catalytic properties. Iron and copper ions and their complexes interacting with H_2O_2 form the hydroxyl radicals, possessing the considerable chemical and biological activity in mono-electronic activation of oxygen (Duca, 2012).

In principle, two redox subsystems are well known in the water medium. The first one is the "slow" subsystem, where the molecular oxygen possesing the low reactivity serves as oxidative equivalent. In this case, the redox equilibrium is established slowly. The second is the "fast" subsystem, where hydrogen peroxide with its advanced reactivity participates as an oxidant. In this case the redox equilibrium is established quickly.

The processes which occur in the "slow" subsystem are characterized by redox potential- Eh, which represents the thermodynamic parameter reflecting the total ratio of the oxidative and reducing equivalents. The thermodynamic character of Eh is a negative factor in the hydro-chemical determinations because it requires much time and does not indicate concretely the dominance of certain classes of redox equivalents. Another shortcoming of the Eh parameter consists in dependence on several factors such as pH, temperature, etc.

To enhance the effectiveness of redox potential was introduced the concept of rH parameter needed for Eh correction. It consists in using the pH as a factor of influence on the Eh values (Zanoaga, 1995; Rabotnova, 1996; Duca & Skurlatov, 2002).

Unlike Eh, which can be determined experimentally, the value of rH is calculated based on the measurements of Eh and pH:

$$H = \frac{Eh + 0.058\,pH}{0.029}$$

The most frequently, parameter rH is used in the biochemical research to estimate the abiotic environment in relation to living organisms.

Unlike the "slow" subsystem, in the "fast" subsystem the hydrogen peroxide plays the main role as an oxidizing agent, which requires the smaller activation energy compared to oxygen, and serves as a much more active oxidative equivalent. Therefore, the presence or absence of hydrogen peroxide in the environment leads to formation of essentially different redox states of water medium.

The essential particularity of the redox state represents a rapidly established kinetic parameter that easily varies due to the increased catalytic activity of hydrogen peroxide. The redox state depends directly on the presence of peroxidase-type substances in the environment, which can be easily oxidized with hydrogen peroxide. Apart from these advantages, there is a practical priority - hydrogen peroxide can be easily and quickly determined in the studied samples (Duca et al., 1996; Gladchi et al., 2008).

If the rate of hydrogen peroxide penetration in natural waters is higher than that of the reductants, hydrogen peroxide concentration is higher than the stationary concentration of reductant. In this case, natural waters are in the oxidation state. Oxidation state of the environment is considered as favorable for the existence and development of hydrobiontes and for development of its chemical self-purification process (Duca et al, 2008).

But if the stationary concentration of the reductants is higher than hydrogen peroxide, or it is missing, than the reducing state of the environment is forming, which leads to the diminishing of the aquatic environment self-purification capacity. Thus, as a result of the redox equivalents determination by the titration of water samples, the reductants will be detected, instead of the oxidants. The reducing state is temporary, because in the presence of dissolved oxygen in the natural waters, the oxidation of all reducing equivalents will occur (Bunduchi et al., 2006; Gladchi et al., 2008).

The third case could happen when, apart from peroxide hydrogen in the environment, the other oxidative equivalents are detected in large amounts. It leads to the formation of superoxidation state, which is dangerous for ecosystems because of the enormous intensification of chemical and biochemical processes and negative consequences. This state is conditioned by the presence of increased amounts of free radicals.

The research initiated by academician Gheorghe Duca, in collaboration with partners from Russia, professor Iuriy Skurlatov, Elena Stamm, Ludmila Ernestova and others, and continued by his disciples, has revealed that in the natural waters, concomitant with peroxidase oxidation processes, occur the oxidation processes with participation of $^{\bullet}OH$ radicals (Skurlatov et al., 1983; Duca, 1997, 2002, 2012, 2017; Katunina, 2015). At the interaction of these radicals with substances with reducing properties two types of secondary radicals can be formed:

- High reactivity radicals, which participate in the regeneration of the $HO^{\bullet}$ particle;
- Lower reactivity radicals, which are usually recombined with each other, breaking the chain of $^{\bullet}OH$ radicals generation.

The presence of radicals from the second group decreases the intensity of water self-purification water. These substances determine the inhibition capacity of natural waters ($\Sigma k_i[S_i]$), which is defined as the effective constant of the one-pseudo order of the oxidation chain interruption with radicals in natural waters. The direct measurement of $^{\bullet}OH$ radicals quantities is difficult in the dynamic conditions of natural waters. In this case, the universal kinetic method is easy to implement, based on using of the acceptors competitors as a "radical traps" such as 4-nitroso-N, N-dimethylaniline (PNDMA). The study of self-purification processes with participation of $^{\bullet}OH$ radicals assumes UV irradiation of the systems containing hydrogen peroxide (the source of $^{\bullet}OH$ radicals) and PNDMA (radicals acceptor).

The existence and importance of free radicals in natural waters is discussed and studied for a long time. It is known (Skurlatov et al., 2017) that the stationary concentration of ˙OH radicals, the strongest oxidant in the natural waters is very low (»10^{-15}–10^{-17}M), but thanks to their reactivity, they play an important role in the regulation of the composition and self-purification of aquatic environment. The determination method of the ˙OH radicals stationary concentration implies the detecting of ratio between two kinetic parameters - rate of initiation (W_{iOH}) and effective constant of disappearance rate of ˙OH radicals ($k_{ef\,OH}$) following their interaction with components of the aquatic environment [S_i].

Therefore, the application of the kinetic parameters for determination of the water redox state through the content of hydrogen peroxide, its inhibition capacity and stationary concentrations of ˙OH radicals, estimation the dynamic balance between oxidative equivalents and reductants in the natural waters can serve as an integral parameter of the redox-catalytic processes balance within the water body and for intensification of self-purification processes with the participation of ˙OH radicals to characterize the ecological state of this environment (Duca and Skurlatov, 2002).

The matters set out above confirm the need of introduction and implementation of the new methods for characterization of the natural waters state and estimation of their self-purification capacity. Therefore, during the last years, particular attention was given to the kinetic regularities of chemical and photo-chemical self-purification of water systems in the case of increased amounts of pollutants presence in them. The research performed has demonstrated that such approach plays an essential role in the natural waters self-purification understanding.

SOLUTIONS AND RECOMMENDATIONS

To correlate the results obtained on the model system in report to the natural systems, an extensive research was performed of various water bodies in Moldova, in order to implement the complex eco-logical monitoring of the natural waters. For this purpose the methods of determination the redox active components in natural waters and their seasonal dynamics by detecting the hydrogen peroxide, ˙OH radicals and inhibition capacity of the natural waters in the processes with participation of ˙OH radicals were developed, perfected and applied. For the first time these methods were applied to estimate the state of the Dniester, Prut and its tributaries within the country (Bunduchi et al., 2006; Gladchi et al., 2008; Sandu et al, 2013).

The method of the kinetic parameters applying along with the traditional hydro-chemical parameters was approved in Moldova and currently is used for the estimation of the chemical composition and processes of self-purification of waters within the Dniester and Prut rivers basins (Gladchi et al., 2008; Bahadir and Duca, 2008; Lis et al., 2010).

To estimate the redox state of waters, to determine the stationary concentration of ˙OH radicals and the inhibition capacity in natural waters ($\Sigma k_i[S_i]$), the analyzes were made in the period May-August 2000 on the segment of Prut river, between Ungheni city (387 km from the mouth of the Danube river) and Leova city (200 km from the mouth of the Prut river). The results obtained in the expeditions have demonstrated that changing of the redox state from oxidant (in May $[H_2O_2] = 4{,}8 \cdot 10^{-6}$ M) to reducing (in July - August [Red] = $2{,}6 \cdot 10^{-6}$ M) determines the quality of waters of the Prut river. The dynamics of the redox state is accompanied by increased of the concentration of thiol substances in water (from $1{,}5 \cdot 10^{-6}$ M to $4{,}2 \cdot 10^{-6}$ M), of the inhibition capacity (from $2{,}7 \cdot 105$ s^{-1} to $1{,}26 \cdot 10^{6}$ s^{-1}), as well as of the diminishing of the content of ˙OH free radicals (from $5{,}3 \cdot 10^{-16}$ M to $0{,}6 \cdot 10^{-16}$ M).

The complex research was performed on the main aquatic artery from Moldova - Dniester river as well. The data obtained (Tab.3, 4) show the importance of the kinetic parameters applying for adequate and complex characterization of the state of the Dniester waters.

Table 1. The average content of hydrogen peroxide, inhibition capacity of waters ($Sk_i[S_i]$, s^{-1}) and stationary concentration of $^{\bullet}OH$ radicals, M in the river Dniester, during 2005-2010

	H_2O_2, M	[$^{\bullet}$OH], M	$Sk_i[S_i]$, s^{-1}
Winter	$(3,50\pm0,10)\cdot10^{-7}$	$(2,87\pm0,46)\cdot10^{-17}$	$(4,17\pm0,30)\cdot10^{5}$
Spring	$(4,19\pm0,58)\cdot10^{-7}$	$(4,12\pm0,38)\cdot10^{-17}$	$(2,83\pm0,74)\cdot10^{5}$
Summer	$(1,35\pm0,43)\cdot10^{-6}$	$(9,66\pm2,69)\cdot10^{-17}$	$(3,19\pm0,29)\cdot10^{5}$
Autumn	$(6,82\pm1,79)\cdot10^{-7}$	$(3,46\pm0,31)\cdot10^{-17}$	$(6,01\pm0,20)\cdot10^{5}$

Table 2. Correlations between the redox parameters and coefficients of correlation r for the Dniester river in the period 2005-2010

Correlation parameters	The correlation relationship	Coefficients of correlation, *r*
	Naslavcea	
$[H_2O_2] = f(rH)$	$[H_2O_2] = 1,45\cdot10^{-6}\cdot rH - 3,94\cdot10^{-5}$	r = 0,7
$[H_2O_2] = f(BOD)$	$[H_2O_2] = 6,68\cdot10^{-7}\cdot BOD - 6,24\cdot10^{-7}$	r = 0,7
	Meresheuca	
$BOD = f([O_2])$	$BOD = 0,49\bullet[O_2] - 2,345$	r = 0,7
	Boshernitsa	
$rH = f(pH)$	$rH = 1,76\bullet pH + 14,17$	r = 0,9
$[H_2O_2] = f(\%O_2)$	$[H_2O_2] = 1,73\cdot10^{-8}(\%O_2) - 1,54\cdot10^{-6}$	r = 0,8

The analysis of the obtained results have demonstrated that for the Dniester waters the seasonal change of the redox state was detected.

The general tendency for the autumn-spring is based on decreasing of hydrogen peroxide concentration and creation of the unstable or even quasi-reducing state of water with regard to the redox equivalents balance, including hydrogen peroxide and the peroxidase type reductants.

During the summer period, hydrogen peroxide content in water increases to order 10^{-6}M. But this increasing is registered for the Naslavcea-Cosăutsi section only, whereas for the Dubasari - Boshernitsa section, downstream the dam, hydrogen peroxide was missing.

Diminishing of H_2O_2 concentration from September to November and its insufficiency in March can be explained by, on one hand, decreasing of photosynthetic activity, which plays the dominant role in the formation of H_2O_2 in the natural waters being held by sunlight radiation. Under this action the superoxide anion radical $O_2^{\cdot-}$ is forming, which is the precursor of hydrogen peroxide. On the other hand, in the autumn months the water temperature decreases, being one of the physico-chemical parameters that influence the rate of the redox-catalytic processes.

The registered values of H_2O_2 concentrations are predominantly of the 10^{-7}M order that is insufficient for the effective realization of waters chemical self-purification.

According to the results obtained, it could be considered that chemical self-purification with the participation of $^{\cdot}OH$ radicals in the Dniester waters are more effective during the summer period.

The research have highlighted that the Dniester waters continue to be polluted with reducing substances that decrease hydrogen peroxide content in the water basin, and create the conditions of the reducing state of aquatic ecosystems formation, being unfavorable for hydrobionts.

The State Program entitled "Scientific research and management of water quality", started in 2009, was aimed to create the legislative base for the adjustment of scientific-methodological, cadastral and technological methods and activities for sustainable exploitation and management of water resources, drinking water supplies, implementation of the modern technologies for water and wastewater treatment, reuse of the wastewaters and irrigation (Duca, 2014). The results obtained constitute an essential contribution to the elucidation of some aspects with fundamental and applied character aiming environmental protection and rational use of water resources. The program contributed to the training of technical and scientific staff in the field and forestablishment of cooperation with partners from Ukraine and Romania, given the transboundary status of the Dniester, Prut and aquifers layers (Sandu et al., 2010; Bunduchi et al., 2012; Sandu et al., 2013; Gladchi et al., 2014).

In the State Program were included the scientific projects dealing with the study of ecological state of Dniester and its tributaries and estimation of their impact on the river. The tributaries of a river may present a risk factor in cases when their chemical composition and the degree of pollution is essentially different from the chemical composition of river waters. The tributaries can contribute to temporarily or permanently changing of the chemical composition of the main water course increasing the pollutants content, to the formation of the toxic factors etc (Goreaceva et al., 2012; Gladchi et al., 2013).

For many years a high degree of pollution of the Raut, Ikel, Bic and Botna rivers was recorded, whose waters, reaching the Dniester, considerably change its water quality. That is why the continuous monitoring of tributaries' impact on water quality in Dniester represents a major scientific and practical interest. The chemical composition of tributaries' waters is determined by the climate and physico-geographical particularities, existence of karst areas, activities of adjusting the flow and riverbed, technogenic influence, etc. Therefore, to estimate the influence of these small rivers on chemical composition of Dniester waters, the analyses were performed of tributaries' water chemical composition at the mouth of inflowing directly into the river.

The field research was carried out in several sampling points: Raut river - Ustia village; Ikel river - Coshernitsa village; Bic river - Gura Bicului village; Botna river - mouth of shedding near the Kitscani village. Concomitantly, the water samples from Dniester upstream and downstream of the mouth of tributaries inflow was taken. Physico-chemical parameters were analysed, such as: t (°C), pH, concentration of the main ions and waters mineralization, concentrations of biogenic elements, dissolved oxygen, organic substances (by BOD_5, COD_{Cr}, COD_{Mn} indicators), buffering capacity of the water, redox state according to the content of H_2O_2, the reductants of peroxidase nature and ability to proceedthe chemical self-purification processes involving $^{\cdot}OH$ radicals.

The obtained results demonstrated that the content of metalions, mineralization and total hardness of water in the Dniester tributaries differ essentially from the main river waters, being much higher (Gladchi et al., 2008, 2012, 2014).

Table 3. Physico-chemical parameters, content of the main ions and mineralization of Dniester waters tributaries (annual averages)

The place of sampling	Total hardness,mmol/L	The content of the main ions, mg/L						I S, mg/L
		Ca^{2+}	Mg^{2+}	$Na^{+}+K^{+}$	HCO_3^-	SO_4^{2-}	Cl^-	
Dniester upstream of the mouth of the Raut	4,47	63,8	15,5	33,1	171,8	86,8	41,8	413,0
Raut river	11,35	71,0	94,8	76,5	493,6	176,4	96,3	1008,7
Dniester downstream of the mouth of the Raut	5,65	55,5	35,0	65,0	239,4	134,0	54,9	583,8
Dniester upstream of the mouth of the Ikel	4,52	54,0	22,2	34,5	183,0	79,2	44,9	417,8
Ikel river	11,8	75,7	97,1	40,9	451,9	174,4	86,4	926,2
Dniester downstream of the mouth of the Ikel	4,48	53,3	22,1	37,6	185,5	81,6	44,9	425,0
Dniester upstream of the mouth of the Bic	4,41	61,2	16,4	32,2	189,8	63,0	45,2	407,9
Bic river	8,41	87,6	49,0	115,0	433,7	140,6	106,8	937,3
Dniester downstream of the mouth of the Bic	4,26	52,7	19,9	33,3	177,9	68,0	44,9	396,7
Dniester upstream of the mouth of the Botna	4,75	68,0	16,4	39,9	187,6	97,2	44,9	454,0
Botna river	12,05	71,2	103,2	102,3	382,3	152,4	242,3	1053,7
Dniester downstream of the mouth of the Botna	4,72	66,0	17,2	35,4	190,6	78,8	48,4	435,0

At the mouth of inflow, the Raut waters are typical for hydrogen carbonate class, Mg group and type II after the concentration of anions and cations ($[HCO_3^-]<[Ca^{2+}]+[Mg^{2+}]<[HCO_3^-]+[SO_2^{2-}]$) – C^{Mg}_{II}. Mineralization of these waters constituted in average 1008.7 mg/L and total hardness 11.35 mmol/L (tab. 5). After their penetration in the Dniester, the increasing of all the main ions content (except Ca^{2+} ions) with 41% and total hardness with 26% was observed. Besides, chemical composition has shown showing a slight dominance of Mg^{2+} ions.

In the Ikel river waters at the mouth of inflowing, the mineralization made 926.2 g/L and total hardness 11.8 mmol/L. The river waters have characteristics available for hydrogen carbonate class, Mg^{2+} group, II–III type (C^{Mg}_{II-III}).

Mineralization of the Bic and Botna river waters had an average 937.3 mg/L, respectively 1053.7 mg/L; the total hardness 8.4 mmol/L, respectively 12.05 mmol/L. The Bic river waters are part of hydrogen carbonate class, sodium or magnesium group, II or III type ($C^{Na},{}^{Mg}_{II-III}$), and the waters at the mouth of shedding of the Botna river of chlorine class, sodium or magnesium group, III type, (Cl^{Na}_{III} sau Cl^{Mg}_{III}).

It was found that the content of mineral forms of biogenic elements and their concentrations in the Dniester tributaries are much higher in comparison to the main river (tab. 4).

The most advanced level of water pollution with various forms of nitrogen and PO_4^{3-} ions was detected in Bic river. The waters of this river contain on average 12,54 mg N-NH_4^+/L; 0,064 mg N-NO_2^- mg/L; 9,15 mg PO_4^{3-}/L.

Table 4. The average values of the content of biogenic elements in the rivers from Dniester basin, mg/L

The place of sampling	$N\text{-}NH_4^+$	NH_3	$N\text{-}NO_2^-$	$N\text{-}NO_3^-$	PO_4^{3-}
Dniester upstream of the mouth of the Raut	0,08	0,10	0,0085	1,30	1,41
Raut river	0,55	0,69	0,021	2,18	1,57
Dniester downstream of the mouth of the Raut	0,12	0,147	0,01	1,37	1,67
Dniester upstream of the mouth of the Ikel	0,11	0,138	0,005	1,45	1,33
Ikel river	1,33	1,61	0,0995	2,13	3,27
Dniester downstream of the mouth of the Ikel	0,09	0,113	0,007	5,02	1,76
Dniester upstream of the mouth of the Bic	0,043	0,53	0,004	1,12	1,86
Bic river	12,54	15,24	0,064	2,26	9,15
Dniester downstream of the mouth of the Bic	0,92	1,12	0,009	0,93	3,90
Dniester upstream of the mouth of the Botna	0,18	0,22	0,012	1,40	1,69
Botna river	0,3	0,36	0,07	1,32	1,02
Dniester downstream of the mouth of the Botna	0,44	0,54	0,014	1,52	1,67

The study has confirmed the growth trend of biogenic elements content in the Dniester waters downstream of the mouth of the tributaries inflowing

The Dniester tributaries waters are characterized by the increased content of oxidizable organic substances producted by chemical and biochemically way. The average of their content in the tributaries, according to BOD_5 parameter, enroll in the ranges 3.8-16.0 mgO_2/L. Each time, the highest values were recorded in the Bic river waters.

The average values of bichromatic oxidizability of the Dniester tributaries are 24.2- 29.0 mgO_2/L in the Raut and Botna rivers and respectively 42.2-43.5 mgO_2/L in Bic and Ikel rivers, the latter constituting a highly advanced pollution degree. In the Dniester waters, downstream from the mouth of tributaries inflow, a rising trend of these parameters was registered, compared with the data obtained upstream.

The saturation degree of the tributaries' water with the dissolved oxygen ranged on average between 101% and 119%. The most deplorable situation was recorded every time in the Bic river waters, in which, at the mouth of inflowing in the Dniester, the content of dissolved oxygen ranged between 90% of saturation level (a unique case) and total absence of the oxygen, the average of this parameter being of 64% of saturation level.

Determination of the redox state according to the kinetic parameters indicated the permanent absence of hydrogen peroxide in the tributaries' waters. At the same time, the substrates of the peroxidase nature, which are actively oxidized by hydrogen peroxide, were recorded permanently in the Ikel, Bic and Botna, rivers waters, contributing to the balance of the redox processes moving and creating the quasi-reducing conditions of waters. In the Raut river waters the presence of the reducing substances was detected periodically, contributing to the state ranging from the unstable to reducing. The amount of the reducing substances in Ikel, Bic and Botna rivers made an average 0.80–1.60 μg-ech/L (tab. 7).

Table 5. The content of the dissolved oxygen and average kinetic parameters of the Dniester river tributaries

Tributary	Degree of water saturation with oxygen,%	H_2O_2, µg/L	DH_2, µg-ech/L	[$^{\cdot}$OH], 10^{17}, M	[$^{\cdot}$OH]hν, 10^{14}M	$Sk_i[S_i]$, $\cdot 10^{-5}$,s^{-1}
Raut	101	0,0	0,08	2,36	1,87	4,94
Ikel	119	0,0	1,60	5,27	1,21	4,75
Bic	64	0,0	1,30	6,26	1,40	3,38
Botna	106	0,0	0,80	6,2	1,72	3,11

FUTURE RESEARCH DIRECTIONS

Generalizing the aforesaid, it is evident that self-purification of natural waters under the favorable conditions is caused by a combination of physical, biological and chemical processes. To assess and forecast the physical, chemical and biological processes of natural water self-purification capacity, the hydrodynamic, physico-mathematical, biological and biochemical methods should be applied,

CONCLUSION

The long term data related to the chemical composition formation of the downstream and upstream of Dniester river under the influence of the tributaries (Raut, Ikel, Bic, Botna) in the sections of their inflow into the river were accumulated and analyzed. The data obtained have demonstrated the considerable influence of the Raut river on the polluting substances content in the Dniester waters.

The Dniester tributaries are polluted with mineral forms of biogenic elements - ammonium, nitrite, nitrate and phosphate ions. Maximum concentrations were found at the mouth of the Bic river inflow. The growing trend was established of the various forms of biogenic elements concentration appearance in Dniester waters downstream of the Bic inflowing.

The redox state and intensity of chemical self-purification process of the Dniester and its tributaries were estimated in the places of their inflowing. The oxidation processes mediated by $^{\cdot}$OH radicals arising intensely in the Dniester waters and undergoing slowly in the tributaries were observed. The dynamics of the kinetic indicators demonstrates that self-purification capacity of the tributaries' waters is lower than that of of the river waters.

Follow-on research was proposed to determine what measures might be proposed to improve the situation in this field.

To evaluate and forcast the physical and biological processes of water self-purification capacity, the determining role is played by the hydrodynamic, physico-mathematical, biological and biochemical methods, elaborated by scientists and practitioners. The chemical processes of self-purification are relatively lesser studied in the world, but in Moldova they were studied intensively during the recent decades based on a complex approach in assessment and exploiting the traditional methods of measurement of the pollutants chemical transformations in the aquatic environment (Duca et al., 2002, 2008, 2012, 2014, 2017).

As a result, more convincing arguments were accumulated in favor of the significant role of chemical processes for the reestablishment of environmental quality, including the hydrosphere. This conclusion is confirmed by the results obtained: the research of the kinetic legalities of the redox- catalytic transformations of the pollutants in aquatic systems in the presence of ions and complex compounds of copper and iron; examination of chemical self-purification processes of aquatic systems; elaboration of the new parameters for estimation the quality of the natural waters;creationa database on qualitative characteristics of chemical transformations of pollutants; study the redox state of surface waters and proposal of the complex estimation methods for the water bodies state.

The main priority in scientific research in the field of ecological chemistry of natural waters is given to evaluation of chemical self-purification processes of aquatic systems. For this reason the research could be carried out on the base of the model systems to study kinetics and mechanisms of redox-catalytic and photochemical processes with participation of free radicals that lead to the transformation of a wide variety of pollutants. Besides, the self-purification capacity of surface waters in Moldova serve as the methods of the redox-state and kinetic parameters evaluation, concomitant with the traditional methods of water analysis. Important role play the new indicators, which significantly contribute to the estimation of the natural waters quality and theirs self-purification capacity. Elaboration of recommendations for their application and creating the database on qualitative characteristics of the chemical transformations of the pollutants is important stage on practical implementation of this approach.

The novelty of this research lies in the application of the kinetic methods to study the processes and to estimate the level of the aquatic environment self-purification, as well as to determine the role of the investigated substances in the formation of the redox state of the aquatic environment; the action of the substances participated in the water self-purification processes with forming of free radicals; the study of the seasonal dynamics of the redox state of surface waters and correlation of this parameter with the dynamic of the waters inhibition capacity, the concentration of $^{\bullet}OH$ radicals and the presence of substances that have been investigated in the laboratory conditions. Conducting this research makes possible the detailed description of the processes in chemical and biochemical systems and elucidation of a number of phenomena in the living nature.

The research results could be applied to reveal the phenomena of redox-catalytic transformation of different classes of substances in natural waters. The constants values of chemical transformation reactions of studied substances in the laboratory conditions are used to estimate the effectiveness of self-purification processes of natural waters. The broad spectrum of physico-chemical parameters (such as organic matter content, $^{\bullet}OH$ radicals, hydrogen peroxide and the inhibition capacity of the natural waters) is important for a more complex assessment of ecological state of the natural waters and forprevention of adverse effects.

The scientific value of the results achieved under the leadership of academician Gheorghe Duca have brought a contribution to the development of the theory of homogeneous catalysis through new ideas on redox oxidation of pollutants in aqueous solutions. The introduction of the theory of homogeneous redox catalysis in the training process helps to deepen the knowledge on the processes in the environment and will make possible forecasting of ecological consequences according to the anthropogenic impact on the natural water.

Therefore, the research discipline on ecological chemistry of natural waters demonstrates in a reasoned and sure mode, that application of kinetic parameters for estimation of water bodies' state is a necessary complement to the traditional hydro-chemical parameters. In the cases of critical and emergent situations of various water bodies (the mass death of fish, discharges increased amounts of pollutants, etc.), application of methods for determining the hydrogen peroxide in the waters, redox state, inhibition capacity and stationary concentration of free radicals may lay the necessary ground to implement the urgent actions to improve the situation.

REFERENCES

Alvarez, M. E., Pennell, R. I., Meijer, P.-J., Ishikawa, A., Dixon, R. A., & Lamb, C.ALVAREZ M.E. (1998). Reactive oxygen intermediates mediate a systemic signal network in the establishment of plant immunity. *Cell*, *92*(6), 773–784. doi:10.1016/S0092-8674(00)81405-1 PMID:9529253

Andrews, J. E., Brimblecombe, P., Jickells, T. D., & Liss, P. S. (1998). *An introduction to environmental chemistry*. Oxford: Blackwell Science Limited, UK.

Bahadir, A. M., & Duca, Gh. (2008). The role of ecological chemistry în pollution research and sustainable development. In *Proc. of the NATO Advanced Research Workshop*. Chisinau, Moldova: Springer.

Baud, L., & Ardaillou, R. (1986). Reactive oxygen species: Production and role in the kidney. *American Journal of Physiology. Renal, Fluid and Electrolyte Physiology*, *251*(5), F765–F776. doi:10.1152/ajprenal.1986.251.5.F765 PMID:3022602

Bunduchi, E., Duca, Gh., Gladchi, V., Goreaceva, N., & Mardari, I. (2006). Assessment of kinetic parameters în the waters of the Nistru course în the section Naslavcea – Dubasari. *Chemistry Journal of Moldova. General. Industrial and Ecological Chemistry*, *1*(1), 68–73.

Bunduchi, E., Gladchi, V., Duca, Gh., & Goreacheva, N. (2012). Kinetic indicators based evaluation of water quality of some tributaries of Dniester river. In *Abstract Book of the V Intern. conf.-simpos. Ecological Chemistry*. Chişinău, Moldova: CEP USM.

Duca, G. (1997). Free radicals in natural water. Free Radicals în Biology and Environment Journal, 27, 475–489.

Duca, G. (2012). Homogeneous catalysis with metal complexes: Fundamentals and applications. Berlin: Springer Series in Chemical Physics. 102, 478 pp. doi:10.1007/978-3-642-24629-6

Duca, G., Romanciuc, L., & Porubin, D. (2012). Groundwater resources of Moldova and transboundary impact. In: Transboundary aquifers in the eastern borders of the European Union. In Regional cooperation for effective management of water resources. Dordrecht: Springer Series in Environmental Security.

Duca, G. (2014). Management of water quality in Moldova. Dordrecht: Springer. doi:10.1007/978-3-319-02708-1

Duca, G. (2017). Ecological and environmental chemistry. *Chemistry Journal of Moldova. General. Industrial and Ecological Chemistry.*, *12*(1), 9–19.

Duca,, G., & Bunduchi,, E., Romanciuc L., & Goreaceva, N. (2008). Estimation of the natural water self-purification capacity from the kinetic standpoint. *Chemistry Journal of Moldova, Series General. Industrial and Ecological Chemistry*, *3*(1), 10–21.

Duca, G., Ciub, L., Skurlatov, Yu., & Sychev, A. (1989). Redox transformation of iron in the natural water. *Journal of Physical Chemistry (Rus)*, *63*(9), 2366–2371.

Duca, G., Goreaceva, N., & Romanciuc, L. (1996). Investigation of the natural water self-purification capacity under simulated conditions [Rus]. *Water Resources Journal*, *23*(5-6), 523–536.

Duca, G., & Skurlatov, Y. (2002). *Ecological chemistry*. Chisinau, Moldova: CE USM.

Duca, G., Skurlatov, Y., Misiti, A., Macoveanu, M., & Surpateanu, M. (1999). *Ecological chemistry*. Bucharest, Romania: Matrix Rom.

Duca, G., Skurlatov, Y., & Sychev, A. (2002). *Redox catalysis and ecological chemistry*. Chisinau, Moldova: CE USM.

Duca, G., Zanoagă, C., Duca, M., & Gladchi, V. (2001). Redox processes in the environment. Chisinau, Moldova: Moldovan State University.

Gladchi, V., Duca, G., Goreaceva, N., Bunduchi, E., & Lis, A. (2013). Chemical composition from the Dniester river tributaries. *Chemistry Journal of Moldova, 8*(1), 23-32.

Gladchi, V., Goreaceva, N., Duca, Gh., Bunduchi, E., Borodaev, R., Surighina, O., & Lis, A. (2014). Chemical compozition of right bank tributaries of Nistru river and their general impact. In Management of water quality in Moldova. Springer International Publishing.

Gladchi, V., Goreaceva, N., Duca, G., Bunduchi, E., Romanciuc, L., Mardari, I., & Borodaev, R. (2008). The study of redox conditions în the Dniester river. *Chemistry Journal of Moldova. General. Industrial and Ecological Chemistry*, *3*(1), 70–76.

Goncharuk, V. (2012). Chemistry, physics and biology of water. In *Abstract Book of the V Intern. conf.-simpos. Ecological Chemistry, Chisinau, March 2-3, 2012*. Chişinău, Moldova: CEP USM.

Goreacheva, N., Gladchi, V., Bunduchi, E., Borodaev, R., & Romanciuc, L. (2012). On the ecological state of Dniester. In *Abstract Book of the V Intern. conf.-simpos. Ecological Chemistry, Chisinau, March 2-3, 2012*. Chişinău, Moldova: CEP USM.

Katunina, E. (2015). Kinetic parameters of sea water quality in assessing the ecological state of the Lakspi Bay water area in the Black Sea. In Abstract Book on Modern problems of hydrochemistry and monitoring of surface water quality. Rostov on Don, Russia: Academic Press.

Korte, F., Bahadir, M., & Klein, W. (1997). Ecological Chemistry. In Basics and Concepts. Mir.

Lis, A., Duca, Gh., Bunduchi, E., Gladchi, V., & Goreaceva, N. (2010). The study of the buffering capacity of several water objects in the Republic of Moldova. *Chemistry Journal of Moldova. Series General. Industrial and Ecological Chemistry.*, *5*(2), 30–36.

Litchfield, M. H. (1989). *Toxicology evaluation of pesticides in drinking water*. Brussels: GIFAP.

Miller, G. C., & Zepp, R. G. (1979). Effects of dissolved pollutants. *Water (Basel)*, 453.

Murti, K. C. R., & Nag, D. (1991). *Human health impact of pesticide in the environment*. Cambridge.

Rabotnova, I. L. (1996). *Die Bedentung physikalish-chemischer Factoren (pH und rH$_2$) fur die Leben-Statigkeit der Microorganismen*. Jena: Fischer-Verlag.

Sandu, M., Moşanu, E., Gladchi, V., Tarita, A., Duca, Gh., Spataru, P., ... Turcan, S. (2010). Study of springs water quality as sources of potable water and for irrigation în Rezina district. *Chemistry Journal of Moldova: General. Industrial and Ecological Chemistry*, *5*(1), 84–89.

Sandu, M., Tarita, A., Lozan, R., Gladchi, V., Duca, G. H., Turcan, S., ... Prepelita, A. (2013). Water of springs – sources for water supply and irrigation in the Nistru river basin. *Chemistry Journal of Moldova. Series General. Industrial and Ecological Chemistry.*, *8*(2), 42–50.

Skurlatov, Y., Duca, G., Batir, D., & Travin, S. (1989). Coordination compounds of the partial charge transfer complexes in the redox processes. *Coordination Chemistry Journal, 15*(3), 291–307.

Skurlatov, Y., Duca, G., & Ernestova, L. (1983). The processes of toxication and mechanisms of natural water self-purification in the condition of anthropogenic impact. *News of the Academy of Sciences of Moldova Series of Biological and Chemical Sciences*, *5*, 3–20.

Skurlatov, Y., Duca, G., & Misiti, A. (1994). *Introduction in ecological chemistry: Manual for chem. and chem.-tech.* VyshaiaShkola.

Skurlatov, Y., & Shtamm, E. (2012). Redox and free-radical processes balance in natural aquatic environment as a factor for the stability of the aerobic aquatic ecosystems under anthropogenic impacts. In *Abstract Book of the V Intern. conf.-simpos. Ecological Chemistry, Chisinau, March 2-3, 2012*. Chişinău, Moldova: CEP USM.

Skurlatov, Y., Stamm, E., & Shvydkii, V. (2017). Role of oxidation-reduction processes in formation of toxic properties of a natural aqueous environment. In Abstract Book on Ecological and Environmental Chemistry. Chisinau, Republic of Moldova: Academy of Sciences of Moldova.

Skurlatov, Y., Ernestova, L.S., & Travin, S.O. (1987). Oxidation processes in natural waters. Water Resources, 5, 66-72.

Sychev, A., Travin, S., Duca, G., & Skurlatov, Y. (1983). Catalitical reactions and environmental protection. Chisinau, Moldova: Stiinta.

Thannickal, V. J., & Fanburg, B. F. (2000). Reactive oxygen species in cell signaling. *American Journal of Physiology. Lung Cellular and Molecular Physiology*, *279*(6), L1005–L1028. doi:10.1152/ajplung.2000.279.6.L1005 PMID:11076791

Travin, S., Duca, G., Skurlatov, Y., & Sychev, A. (1986). Kinetics and mechanisms of intermediary complexes formation in the redox system. *Journal of Physical Chemistry (Rus)*, *60*(6), 1394–1398.

Turro, N. J., Ramamurthy, V., Cherry, W., & Farneth, W. (1978). The effect of wavelength on organic photoreactions in solution reaction from upper excited states. Chem. Rew., 78, 125. doi:10.1021/cr60312a003

Yu, S., & Gh, D. (1985). *Catalitical reactions in the natural waters/* Chisinau, Moldova: News of the Academy of Sciences of Moldova. *Series of Biological and Chemical Sciences (Rus)*, *3*, 74–76.

Zănoagă, C. (1995). Role of the redox state in informing the biological quality of natural waters. Self-Purification Processes in Natural waters, 218-245.

Zepp, R. (1980). Assessing the photochemistry of organic pollutants in aquatic environments in dynamics, eposure and hazard assessment of toxic chemical. Arbor Science Publishers.

Zepp, R. G., & Cline, D. M. (1977). Rates of direct photolysis in aquatic environments. *Environmental Science & Technology*, *11*(4), 359–366. doi:10.1021/es60127a013

Chapter 10
Assessment of Surface Water Pollution With Heavy Metals

Duca Gheorghe
Institute of Chemistry, Moldova

Mereuta Aliona
State University "Dimitrie Cantemir", Moldova

Claudiu Tănăselia
https://orcid.org/0000-0001-8931-787X
INCDO INOE 2000 Research Institute for Analytical Instrumentation (ICIA), Romania

Erika Levei
INCDO INOE 2000 Research Institute for Analytical Instrumentation (ICIA), Romania

Velisco Natalia
State University "Dimitrie Cantemir", Moldova

ABSTRACT

Water pollution represents the direct or indirect change of its normal composition as a result of human activity in such extent that is affecting all other possibilities to use the water in its natural state. This study was carried out to investigate the seasonal variability and distribution of heavy metals in the waters of Somesul Mic (Romania), Bic (Republic of Moldova), and Prut (Romanian-Moldovan natural border) rivers. Water samples were collected from nine sampling sites in Autumn 2016, Spring and Summer 2017. Dissolved metals concentrations were measured using inductively coupled plasma mass spectrometry (ICP-MS) and flame atomic absorption spectrometry (AAS). The metal index (MI) was calculated to assess the contaminations of the Somesul Mic, Bic, and Prut rivers. MI values revealed that Somesul Mic and Prut rivers are pure to moderately affected by metal pollution, while Bic river is slightly to strongly affected by metal pollution.

DOI: 10.4018/978-1-7998-1241-8.ch010

INTRODUCTION

The small rivers are the focus of scientists around the world, as they have become the main receptors of the most diverse pollutants. According to the numerous bibliographic data, the pollution induces considerable changes in both the hydrochemical composition of the rivers and the future water use options (Xie et al., 2013; Cao, 2016; Khatri, 2015; Begu, 2006). Permanent sources of pollution, which primarily modify the composition of biocenoses, include the household and agricultural waste waters, industrial effluents, meteoric water, as well as the unauthorized landfills near the river meadows (Mustea, 2002; Duca, 2014).

In the Republic of Moldova domestic waste waters represent the main and most serious pollution source of small rivers, as most of these waters are discharged untreated or treated inappropriately. The discharge of untreated waters with high loads of pollutants and nutrients directly into the natural basins cause their eutrophication, a phenomenon detected in the Bic and other small rivers, as described in "The state of the environment in the Republic of Moldova" for 2004 and 2006 years.

River water pollution in Romania is mainly due to the agricultural activities, urban agglomerations and industrial activities. In the Somes river basin extraction and processing of non-ferrous ores causes the metal pollution in the Rodna and Baia Mare mining areas, along with the agro-zoo-technical activities (fertilizers, animal manure and untreated wastewater) in the rural areas, and insufficiently treated wastewater discharges in the urban areas. Levei (2008a, 2011b), Ani (2011) and Smical (2015) showed that in the Somesul Mic river the main source of pollution is the discharge of insufficiently or partially treated wastewaters.

Worldwide, the problem of water pollution is of a big concern. A large number of researches demonstrate the worsening of the river water quality and ecological status around the world (Nienhuis, 2002; Barinova, 2004; Scierbak, 2006; Ivanova, 2007; Goher, 2014; Zhang, 2014; Hasan, 2016; Bilgin, 2016).

Some of the most toxic inorganic pollutants are heavy metals. Generally, trace amounts of metals are always present in freshwaters from the weathering of rocks and soils. In addition, particularly in developed countries, industrial wastewater discharges and mining are major sources of metals in freshwaters. Significant amounts of metal pollutants penetrate the surface waters in sewage, as well as with atmospheric deposition (e.g. lead). Lead is still widely used as an additive in petroleum for automobiles and is emitted to the atmosphere in their exhaust gases, thereby entering the hydrological cycle (Chapman, 1996).

Thus, this chapter presents a study of heavy metals concentrations in the water samples collected from the Somesul Mic, Bic and Prut rivers. For the first time the water quality of these rivers was evaluated and compared using metal index (MI). The MI values of water samples were calculated according to Tamasi and Cini (2004), and for assessment of water quality was used classification applied by Yusuf et al. (2018). For calculations were used the maximum allowable concentration for surface water regulated by the Moldavian standards ("The Regulation on Environmental Requirements for Surface Waters. GD Nr.890/2013", 2013). The concentrations of heavy metals were monitored by applying inductively coupled plasma mass spectrometry method (ICP-MS), a modern and powerful characterization tool used in elemental analysis, along with the atomic absorption spectrometry (AAS).

DETERMINATION OF THE HEAVY METAL CONCENTRATION IN SOMESUL MIC, BIC, AND PRUT RIVERS

Prut river is a transboundary basin shared by three countries: Republic of Moldova, Romania and Ukraine. Of the total basin area, 28% of the Prut River Basin is located in the territory of Republic of Moldova, 33% in the territory of Ukraine, and 39% in the territory of Romania. Prut River is one of the most important rivers that cross the territory of Republic of Moldova and constitutes an important source of water supply for the country; therefore its water quality must be continuously monitored (Duca, 2019).

Different teams of researchers are actively involved in the assessment of the quality of water of Prut river based on various physicochemical parameters. Thus, Matache et al. (2018) studied the concentration levels of Cu, Zn, Pb, and Cd by inductively coupled plasma – optical emission spectrometry in sediments and water samples from six locations situated along the inferior reach of the Prut River on the Romanian shore. Their investigation demonstrated that these trace elements were present in water samples at concentration levels below the maximum allowable concentration (MAC) imposed by the Romania regulations for freshwater bodies. Another study performed by Ene et al. (2010) on the detection of heavy metals concentration in water samples collected from different rivers, including Prut (Giurgulesti village, Romanian shore of the river) using atomic adsorption spectrometry also showed that the concentration was within the MAC limits imposed by the Romania regulations for freshwater bodies.

The Bic River is the third largest inland river of the Republic of Moldova, after Raut and Kogilnik rivers. The river has a length of 155 km, covering an area of about 2040 km^2. In this territory lives a quarter of the population of Moldova. The population density here is four times higher (460 inhabitants per 1 km^2), than the average per country (130 inhabitants per 1 km^2). The river and its tributaries pass through 5 cities, including the capital of the Republic of Moldova – Chisinau (Lavric & Ceban, 2011).

The water quality of Bic river is monitored monthly by the State Hydrometeorological Service. According to the annual report (2016), the operational monitoring program established for 4 monitoring sections on the Bic region, in 2015 included, as well, the quarterly monitoring of the concentrations of heavy metals, organochlorine pesticides and polyaromatic hydrocarbons. Thus, the maximum concentration for copper, in the dissolved form, constituted 68.58 µg/L, detected in summer at the station downstream of Gura Bicului village. The maximum values for dissolved zinc oscillated within the limits 27.16-120.92 µg/L, the maximum being detected in the summer downstream of Chisinau municipality. The average of dissolved nickel concentration was between 2.91-4.15 µg/l. The concentration of dissolved lead and cadmium in the water samples was below the detection limit, except for one sample collected in October at the downstream section of the village Gura Bicului where the concentration of dissolved cadmium constituted 1.87 µg/L. The quality of the Bic river water has been characterized as highly polluted at all monitoring points and attributed to the class V of water quality. The results of the researches obtained by Bulat et al. in 2018 showed a radical deterioration of the ecological status of the Bic district.

The Somesul Mic river is part of the Somes-Tisa Cathment from the NW Romania. The Somesul Mic river is formed by the confluence of Somesul Cald river and Somesul Rece river near Gilau village. The Somesul Mic flow through Cluj-Napoca, Apahida and Gherla towns before the confluence with the Somesul Mare river in Manastirea, near Dej town and form the Somes River. The length of the watercourse is 178 km and have an average debt of 21.2 m^3/s and drain about 3733 km^2 (Badaluta-Minda & Dunca, 2016; Voicu & Banaduc, 2016; Barhoumi, 2019). Several hydrotechnical facilities consisting in water dams and reservoirs exist in the upper basin of Somesul Mic, assuring the water supply for the downstream localities, the regularization of water discharge, production of electricity, tourism and recreational activities. Agriculture is the main anthropic activity upstream Cluj-Napoca, while downstream, agricultural, industrial and urban activities represent a potential pressure for the water body (Barhoumi, 2019; Moldovan, 2009).

Despite the fact that several studies (Smuleac, 2017; Moldovan, 2009) present the level of various organics in the Somesul Mic river water, especially downstream Cluj-Napoca town, the studies on metal pollution of Somesul Mic river water are scarce. Voica et al. (2012) showed that Somesul Mic have low levels of metals. The low metal contents in water could be explained by the absence of highly polluting industry that could discharge metal polluted waters into the river. Barhoumi et al. (2019) investigated sources and ecological risk of trace metals in surface sediments from Somesul Mic river using geoaccumulation index and enrichment factor, which indicate widespread pollution by Pb, Cd, Zn, Ni and Cu.

The concentration of different metals in waters varies over a wide range (0.1-0.001 µg/L) at background sites and can rise to concentrations which are dangerous for human health in some water bodies influenced by human activities. Dissolved metal concentrations are particularly difficult to measure due to possible contamination during sampling, pre-treatment and storage. The low concentrations of metals in natural waters require determination by instrumental methods. Photometric methods, sometimes in combination with extraction, are the oldest and most inexpensive methods. However, as these have high detection limits, they can only be used for analysis of comparatively polluted waters (Chapman, 1996; Ene, 2010).

Atomic spectrometry is the most commonly applied method for determining the major (Ozbek & Akman, 2016) and minor constituents, traces (Manousakas, 2014) or ultra-traces (Li, 2016). At present, among the spectrometric methods, the best analytical performances possesses inductively coupled plasma mass spectrometry (ICP-MS) and atomic absorption spectroscopy (AAS), both used to determine the concentration of heavy metals in the water. ICP-MS is suitable for Cr, Co, Ni, Cd, As, Hg, Pb, Mn, Cu, Zn, etc. determination in the range of µg/L to mg/L, or even below. In addition to the low detection limits, another advantage of ICP-MS is the possibility of analyzing a high number of elements in a very short time.

Atomic absorption spectrometry (AAS) is the spectroanalytical method applied in the quantitative determination of elements (about 70 metals and semi-metals, less recommended for non-metals). The AAS disadvantages are the limitation of the number of elements that can be determined by the instrument configuration and the possibility of performing monoelement analyzes, which lead to higher consumption of resources (reagents and time).

Application of modern methods for determination of heavy metals concentration in waters, which is a worldwide problem, confers actuality and importance to the study. Water pollution by heavy metals as a result of human activities is causing serious ecological problems in many parts of the world. This situation is aggravated by the lack of natural elimination processes for metals. As a result, metals shift from one compartment within the aquatic environment to another, including the biota, often with detrimental effects (Chapman, 1996). Thus, the assessment of heavy metal in water samples collected from the Somesul Mic, Bic and Prut rivers is very important for determination of the water quality class and for elaboration of water use recommendations.

Figure 1. Water sampling points in the Somesul Mic River

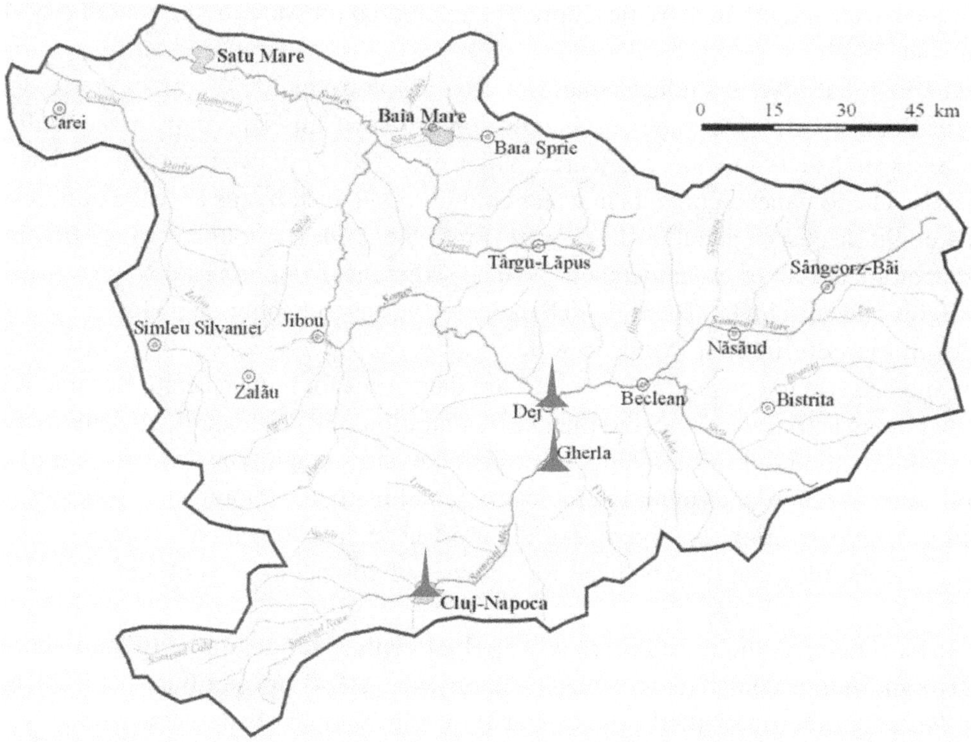

Study Area and Laboratory Methods

Water samples were collected from the Somesul Mic, Bic and Prut rivers in the autumn of 2016, spring and summer 2017. A number of 9 sampling sites have been established as follows: 3 sites on the Somesul Mic river (figure 1), 3 sites on the Prut (figure 2) and Bic Rivers (figure 3), respectively. For the Somesul Mic river the sampling sites were upstream the Cluj-Napoca (S_1), upstream Gherla (S_2) and downstream the Dej town (S_3). The sampling sites for the Bic river were upstream Straseni (B_1), in Chisinau city (B_2) and downstream Bic village (B_3), while for the Prut river they were upstream Criva (P_1), upstream Sculeni (P_2) and upstream Giurgiulesti (P_3) villages (at the Moldavian side of the river).

The water samples were taken with the 2 L scoop angular polypropylene beaker with telescopic rod. Immediately after sampling, water was filtered using a 0.45 µm pore size cellulose acetate membrane filter and acidified with 65% ultra-pure HNO_3. The samples were transported to the laboratory in 20% HNO_3 treated flasks rinsed with ultrapure water. Two liters of water sample was taken from each sampling point, and the sample was divided into two parts for parallel analysis by ICP-MS and AAS.

For ICP-MS quantitative analyzes, calibration was done using the 4 point calibration curve (10, 20, 50, 100 µg/L) prepared using a Perkin Elmer Elan Standard III multielement standard solution by appropriate dilutions, using a Perkin Elmer Elan DRC II ICP-MS instrument. The detection limit for ICP-MS method is listed in Table 1.

Figure 2. Water sampling points in the Prut River

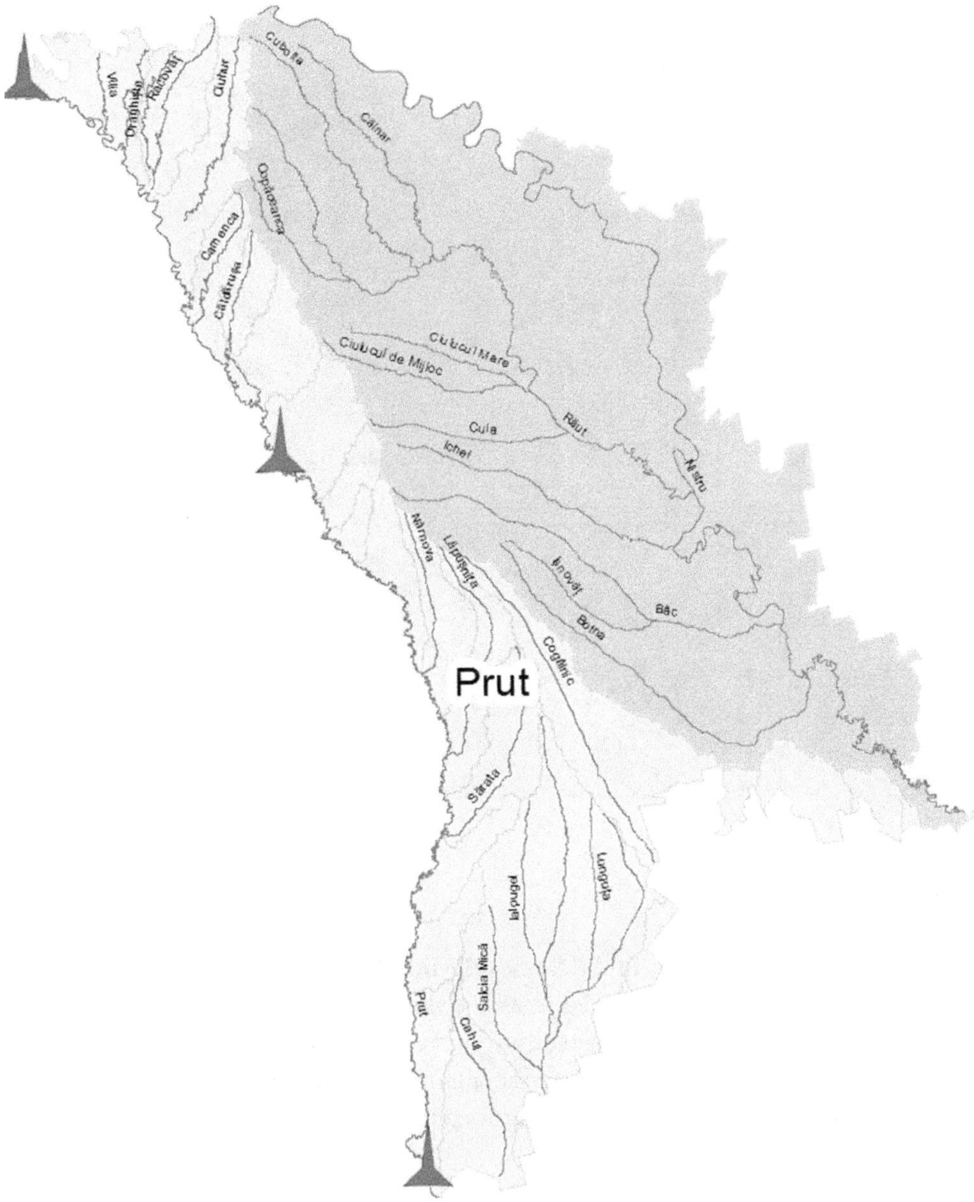

Quantitative AAS determinations were performed using AAS-1N and AAS-1 atomic absorption spectrometers (Germany). The conditions for heavy metals determination by the AAS method are presented in Table 2.

Figure 3. Water sampling points in the Bic River

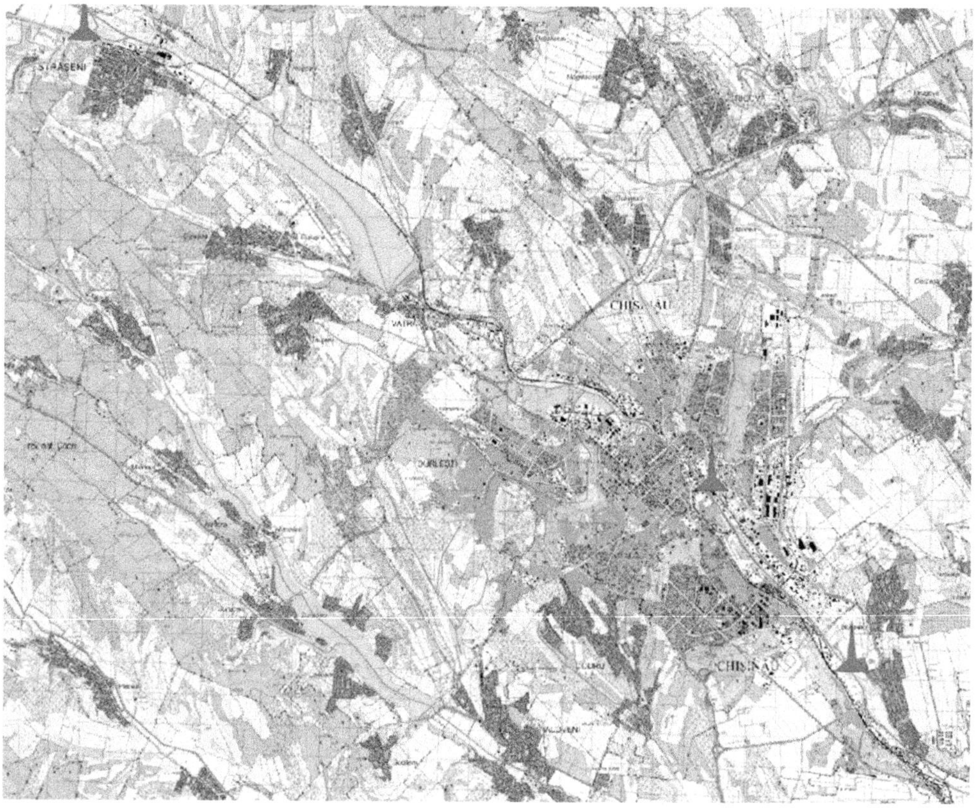

The Metal Index

The metal index (MI) of water samples in the Somesul Mic, Bic and Prut rivers was calculated to determine the level of metal contamination in order to assess the suitability of the water quality for drinking, irrigation, transport, etc. The higher the concentration of a metal compared to its respective maximum allowable concentration (MAC) value, the worse the quality of the water is. According to Tamasi & Cini (2004), the MI is calculated by using the following formula:

Table 1. Conditions of metal determination by ICP-MS method

Element	Isotope	Detection limit, µg/L
Cr	^{53}Cr	0.9
Mn	^{55}Mn	1.1
Cu	^{63}Cu	1.2
Zn	^{64}Zn	2.0
Ni	^{58}Ni	0.8
Pb	^{208}Pb	1.2
Cd	^{114}Cd	0.8

Table 2. Conditions of metal determination by AAS method

Element	Analytical technique	Atomization	Wavelength, nm	Detection limit, µg/L
Cr	AAS-1N	Acetylene-air	357.9	10
Mn	AAS-1N	Acetylene-air	279.5	5
Cu	AAS-1	Propane-butane-air	324.7	7
Zn	AAS-1	Propane-butane-air	213.8	8
Ni	AAS-1N	Acetylene-air	232.0	10
Pb	AAS-1N	Acetylene-air	283.3	5
Cd	AAS-1N	Acetylene-air	228.8	0,5

Table 3. Categories of Water in function of Metal Pollution Index

Class	MI	Categories
1	<0.3	Very pure
2	0.3-1.0	Pure
3	1.0-2.0	Slightly affected
4	2.0-4.0	Moderately affected
5	4.0-6.0	Strongly affected
6	>6.0	Seriously affected

Table 4. Limit values of heavy metals for each class of surface water quality

Element	Unit	Quality class				
		I	II	III	IV	V
Cd	µg/l	< 0.2 or NF*	0.2	0.5	1.0	>1.0
Pb	µg/l	< 2.5 or NF	2.5	5.0	7.5	>7.5
Ni	µg/l	8.0 or NF	20.0	30.0	40.0	>40.0
Cu	µg/l	< 5.0 or NF	10.0	15.0	30.0	>30.0
Zn	µg/l	< 20.0 or NF	30.0	50.0	120	>120
Cr	µg/l	< 0.5 or NF	2.0	4.0	10.0	>16.0
Mn	µg/l	< 100 or NF	100	1000	2000	>2000

*NF – the natural background

Table 5. Water use types based on surface water quality classes

Utilization	Differentiating utility	Water quality class*				
		I	II	III	IV	V
Operation of ecosystems		+	+	-	-	-
Fisheries / protection of ichthyofauna	Salmonids	+	+	-	-	-
	Cyprinids	+	+	+	-	-
Drinking water supply, supplying water to industries requiring equivalent water quality	Simple treatment	+	+	-	-	-
	Normal treatment			+	-	-
	Advanced treatment				+	-
Recreational activities		+	+	+	-	-
Irrigation		+	+	+	-	-
Use of water for industrial purposes (in the process, for cooling)		+	+	+	+	-
Hydroelectric power generation		+	+	+	+	+
Transport		+	+	+	+	+

* + Water quality ensures use, - Water quality does not ensure use, water use is not allowed or limited.

$$MI = \sum_{i=1}^{n} \frac{C_i}{(MAC)_i}$$

C_i is the concentration of each element in solution and i is the ith sample; MAC is the maximum allowable concentration.

Bakan et al. (2010) mentioned in their work that the higher the concentration of a metal compared to its respective MAC value, the worse the quality of the water, and MI value > 1 is a threshold of warning. Later, Adullah (2013) and Yusuf et al. (2018) used the values of MI for water classification into six classes (Table 3).

According to the dissolved metals concentration surface waters are classified into five classes ("The Regulation on Environmental Requirements for Surface Waters. GD Nr.890/2013", 2013, "The Norms on Classification of Surface Water Quality in order to determine the ecological status of water bodies", 2006; Sandu, 2017). The metals concentration for each class of surface water quality according Moldavian regulation is presented in Table 4.

Surface water quality classes, as stated in "The Regulation on Environmental Requirements for Surface Waters. GD Nr.890/2013" (2013), determine the types of water use and their correlation (Table 5).

Assessment of Water Quality

Water samples were subjected to quantitative analysis for 7 heavy metals (Cr, Zn, Mn, Pb, Ni, Cu, Cd) using ICP-MS and AAS. The results obtained for each of the 9 sampling sites are shown in Table 6.

The results have revealed that in most of the sampling sites, water was slightly alkaline except in Bic river where water was found to be slightly acidic. The high pH of the River water may result in the reduction of heavy metal toxicity.

Table 6. Comparison of dissolved metal concentration in Somesul Mic, Bic and Prut Rivers (μg/L)

Sampling site	Metal concentration, μg/L						
	Cr	*Zn*	*Mn*	*Pb*	*Ni*	*Cu*	*Cd*
Autumn 2016							
S_1	<LOD	<LOD	1.1	<LOD	<LOD	<LOD	<LOD
S_2	3.1	<LOD	27.0	<LOD	<LOD	<LOD	<LOD
S_3	1.5	<LOD	30.7	<LOD	<LOD	<LOD	<LOD
P_1	9.1	<LOD	3.1	<LOD	2.1	1.7	<LOD
P_2	6.4	<LOD	<LOD	<LOD	2.1	5.3	2.2
P_3	9.5	<LOD	<LOD	<LOD	3.4	15.9	<LOD
B_1	19.0	<LOD	119	<LOD	4.8	17.1	<LOD
B_2	21.7	1.0	114	<LOD	4.6	1.6	<LOD
B_3	32.1	1.5	716	<LOD	7.8	5.2	<LOD
Spring 2017							
S_1	2.5	<LOD	38.9	1.1	1.8	2.3	1.8
S_2	3.4	<LOD	1.1	1.1	3.1	3.0	1.8
S_3	3.1	<LOD	1.4	1.2	2.7	3.5	1.9
P_1	3.1	9.8	9.3	<LOD	1.8	6.1	1.0
P_2	3.1	<LOD	4.0	<LOD	2.2	3.6	0.8
P_3	3.0	<LOD	<LOD	<LOD	2.6	3.2	0.8
B_1	11.4	<LOD	7.9	<LOD	4.5	7.4	0.8
B_2	10.7	3.1	50.1	<LOD	5.5	9.3	0.8
B_3	2.1	6.6	601	<LOD	8.3	10.1	0.8
Summer 2017							
S_1	<LOD	2.5	15.4	<LOD	2.1	3.3	<LOD
S_2	1.5	1.9	<LOD	<LOD	4.9	3.0	<LOD
S_3	2.2	<LOD	31.5	<LOD	5.5	6.5	<LOD
P_1	2.3	2.3	1.2	<LOD	5.9	5.5	<LOD
P_2	<LOD	1.5	2.1	<LOD	5.2	7.1	<LOD
P_3	<LOD	7.6	2.6	<LOD	5.9	10.3	<LOD
B_1	<LOD	2.2	683	<LOD	9.2	2.3	<LOD
B_2	<LOD	6.5	102	<LOD	10.1	6.8	<LOD
B_3	<LOD	5.7	943	<LOD	20.2	4.8	<LOD

Figure 4. Seasonal variation in heavy metal concentration in Somesul Mic, Bic and Prut Rivers

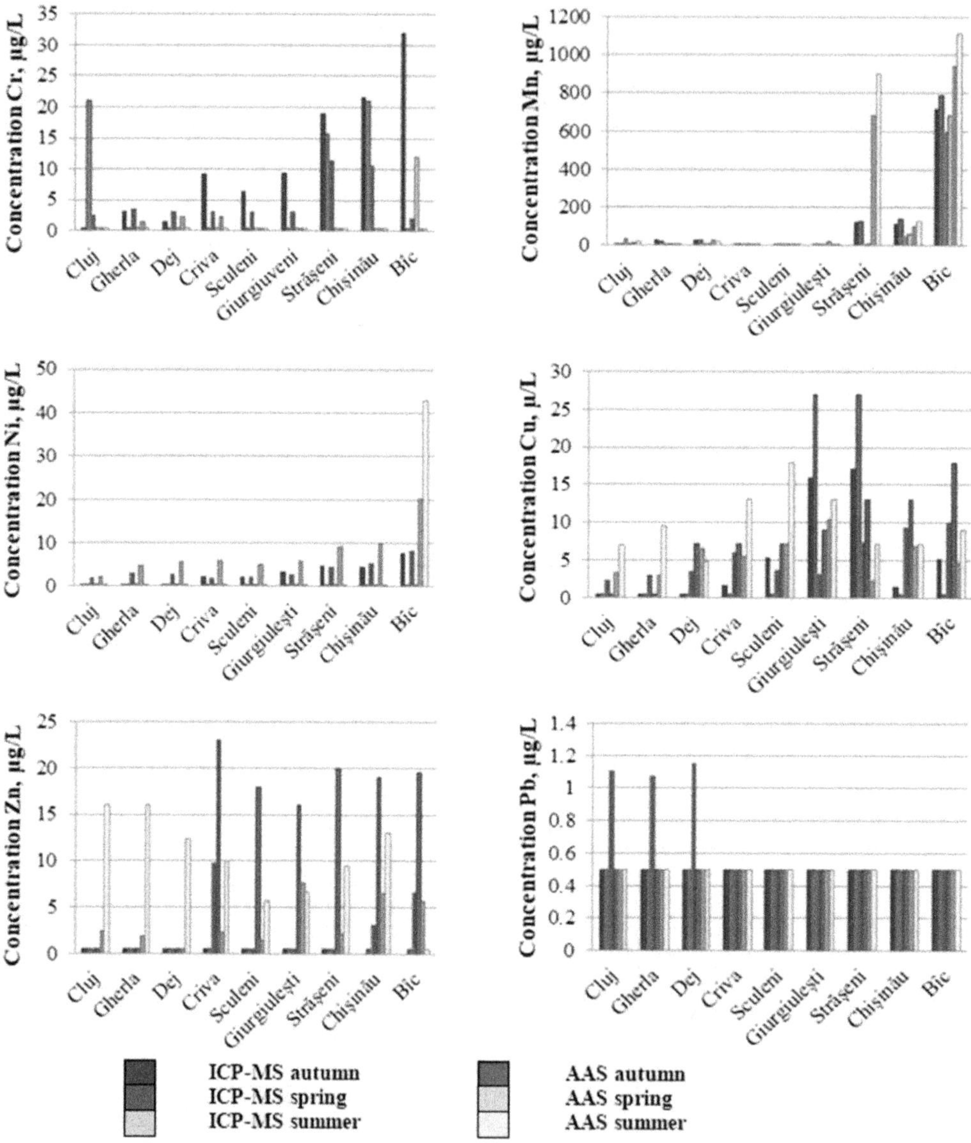

The highest Pb concentration observed in Somesul Mic River was 1.2 µg/L during spring, while in the Bic and Prut rivers it was below 1.0 µg/L in all three seasons. The highest Cd concentration in Somesul Mic river was 1.9 µg/L in spring, while in the Prut river it was 2.2 µg/L in autumn, probably due to coal combustion for industrial and domestic purposes. The low values of Pb and Cd indicate that there is no significant pollution source with these metals. High concentrations of Zn were found in Prut river during spring (9.8 µg/L) and summer (7.6 µg/L), in Somesul Mic, only in summer (2.5 µg/L), while in the Bic ranged from 1.0 µg/L in autumn to 6.6 µg/L in spring. It is attributed to the presence of unused remains of zinc sulphate in fertilizers. The maximum concentration of Cu was found during summer in Somesul Mic (6.5 µg/L) and during autumn in Bic (17.1 µg/L) and Prut rivers (15.9 µg/L). It may be attributed to domestic sewage and run-off from extensive farmed areas. High concentration of Ni was found in the Bic river during autumn (7.8 μg/L), spring (8.3 μg/L) and summer (20.2 μg/L). In the Somesul Mic and Prut rivers the maximum concentration of Ni varied between 5.5 μg/L and 5.9 μg/L, respectively during summer. The highest concentrations of Cr in Bic river were 32.1 μg/L in autumn and 11.4 μg/L in spring. In Prut chrome was found in autumn at a maximum concentration of 9.5 μg/L. In Somes river the highest concentrations were observed in spring (3.1-3.4 μg/L). Of the 7 heavy metals analyzed, Mn was found predominantly in all rivers monitored in all three seasons. In the Prut rivers the maximum Mn concentration was 9.3 μg/L in spring. The concentration of Mn in the Somesul Mic river was high in all three seasons, ranging from 30.7 μg/L in autumn to 38.9 μg/L in spring. High concentration of Mn was found in the Bic river (716.0 μg/L in autumn, 601.0 μg/L in spring and 943.2 μg/L in summer). The contents in Mn may be attributed to the use of manganese fertilizers.

Comparison of the dissolved heavy metals concentration in Somesul Mic, Bic and Prut rivers (μg/L) in autumn, spring and summer is shown in figure 4.

The graphical representation of heavy metals pollution of the rivers reveals the maximum pollution of the Bic river. There is no particular seasonal peak, maximal concentrations of heavy metals being detected over the entire period.

Based on the water quality classes presented, we can conclude: 1. the waters of the Somesul Mic river can be used in fisheries (Cyprinids); drinking water supply, supplying water to industries requiring equivalent water quality after normal treatment; recreational activities; irrigation; use of water for industrial purposes (in the process, for cooling); hydroelectric power generation and transport; 2. the waters of Prut river can be used for industrial purposes (in the process, for cooling); hydroelectric power generation and transport; 3. the waters of Bic river can be used for hydroelectric power generation and transport.

The metal index (MI) of water samples (Table 7) showed that the rivers were most frequently polluted with Cr and Cd. According to MI the highest state of pollution is class 4, moderately affected. To class 4 were corresponded the water samples collected in autumn from the Prut river near Sculeni village, from the Bic river near Bic village, and in spring from Somes river along all length. In the case of Somes river an ecological accident is evident. In autumn and summer the water of Somes river was very pure and pure, respectively. The quality of Prut river water was mostly slightly affected (class 3), with the exception of summer when water was pure (class 2). The Bic river water corresponded to class 3 in all the seasons.

CONCLUSION

The present study have revealed that most of the water samples of Somesul Mic, Bic and Prut rivers, collected from 9 points in autumn 2016, spring and summer seasons in 2017, were unpolluted or slightly polluted with heavy metals, while the pollution profile shows a seasonal variation. According to the concentrations of dissolved metals, measured using ICP-MS and AAS, water pollution with metals followed a descending order: Cr>Cd>Cu>Mn> Ni>Zn>Pb.

The metal index (MI) was calculated to assess the pollution of the Somesul Mic, Bic and Prut rivers. The water quality corresponded to class 2 for Somes river and to class 3 for the Prut and Bic. The most affected by anthropogenic activities was Bic river.

Table 7. Metal index (MI) of surface water in Somesul Mic, Bic and Prut Rivers

Sampling site	MI	*Cr*	*Zn*	*Mn*	*Pb*	*Ni*	*Cu*	*Cd*
MAC, µg/L		16	120	2000	7.5	40	30	1
				Autumn 2016				
S_1	0.0006	0	0	0.0006	0	0	0	0
S_2	0.21	0.19	0	0.01	0	0	0	0
S_3	0.11	0.09	0	0.02	0	0	0	0
P_1	0.68	0.57	0	0.0016	0	0.05	0.06	0
P_2	2.83	0.40	0	0	0	0.05	0.18	2.20
P_3	1.21	0.59	0	0	0	0.09	0.53	0
B_1	1.94	1.19	0	0.06	0	0.12	0.57	0
B_2	1.59	1.36	0.01	0.06	0	0.12	0.05	0
B_3	2.75	2.01	0.01	0.36	0	0.20	0.17	0
				Spring 2017				
S_1	2.24	0.16	0	0.02	0.15	0.05	0.08	1.80
S_2	2.34	0.21	0	0.0006	0.15	0.08	0.10	1.80
S_3	2.44	0.19	0	0.0007	0.16	0.07	0.12	1.90
P_1	1.53	0.19	0.08	0.0047	0	0.05	0.20	1.00
P_2	1.17	0.19	0	0.0020	0	0.06	0.12	0.80
P_3	1.16	0.19	0	0	0	0.07	0.11	0.80
B_1	1.88	0.71	0	0.0040	0	0.11	0.25	0.80
B_2	1.97	0.67	0.03	0.03	0	0.14	0.31	0.80
B_3	1.83	0.13	0.06	0.30	0	0.21	0.34	0.80
				Summer 2017				
S_1	0.19	0	0.02	0.01	0	0.05	0.11	0
S_2	0.33	0.09	0.02	0.00	0	0.12	0.10	0
S_3	0.51	0.14	0	0.02	0	0.14	0.22	0
P_1	0.49	0.14	0.02	0.0006	0	0.15	0.18	0
P_2	0.38	0	0.01	0.0011	0	0.13	0.24	0
P_3	0.67	0.11	0.06	0.0013	0	0.15	0.34	0
B_1	0.67	0	0.02	0.34	0	0.23	0.08	0
B_2	0.58	0	0.05	0.05	0	0.25	0.23	0
B_3	1.18	0	0.05	0.47	0	0.51	0.16	0

ACKNOWLEDGMENT

The research carried out was financially supported by the Executive Agency for Higher Education, Research Development and Innovation Funding, Academy of Sciences of Moldova, under the Moldovan-Romanian bilateral project "Improving surface water quality assessment by semiquantitative and multielemental spectrometric methods" (20BM/2016 and 02/ro/2016).

REFERENCES

Begu, A., Lungu, A., & Obuh, P. (2006). Issues of Taxonomy, Ecology, and the Ecobio Indicator Role of Algal Vegetation in the Left Prut of the Middle Prut. *Bulletin ASM*, *2*(299), 172–178.

Cao, Y., Tang, C., Song, X., Liu, C., & Zhang, Y. (2016). Identifying the hydrochemical characteristics of Rivers and groundwater by multivariate statistical analysis in the Sanjiang Plain, China. *Applied Water Science*, *6*(2), 169–178. doi:10.100713201-014-0215-5

Duca, Gh. (2014). Management of Water Quality in Moldova. In *Water Science and Technology Library*, *69*. Springer International Publishing. doi:10.1007/978-3-319-02708-1

Khatri, N., & Tyagi, S. (2015). Influences of natural and anthropogenic factors on surface and groundwater quality in rural and urban areas. *Frontiers in Life Science*, *8*(1), 23–39. doi:10.1080/21553769.2014.933716

Mustea, M., & Holban, V. (2002). The main causes of pollution of water resources. *Management Resources Water*, 57-59.

The state of the environment in the Republic of Moldova 2004. (National report). (2005). Chisinau, Republic of Moldova.

Xie, Y., Li, X., Wang, H., & Li, W. (2013). Spatio-temporal variation analysis of hydrochemical characteristics in the Luanhe River Basin, China. *Water Science and Technology*, *67*(6), 1332–1338. doi:10.2166/wst.2013.007 PMID:23508159

The state of the environment in the Republic of Moldova 2006. (National report). (2007). Chisinau, Republic of Moldova.

Levei, E., Senila, M., Miclean, M., Roman, C., Abraham, B., & Cordos, E. (2008, May). Surface water pollution with heavy metals in Baia Mare mining basin. *Proceedings of the 15 th International Soils Conservation Organization (ISCO) Congress*.

Ani, E., Avramenko, Y., Ktaslawski, A., & Agachi, P. (2011). Identification of pollution sources in the Romanian Somes River using graphical analysis of concentration profiles. *Asia-Pacific Journal of Chemical Engineering*, *6*(5), 801–812. doi:10.1002/apj.522

Barinova, S., Anissimova, O., Nevo, E., Jarygin, M., & Wasser, S. (2004). Diversity and ecology of algae from the Nahal Qishon River, northern Israel. *Plant Biosystems*, *138*(3), 245–259. doi:10.1080/11263500400006985

Goher, M., Hassan, A., Abdel-Moniem, I., Fahmy, A., & El-Sayed, S. (2014). Evaluation of surface water quality and heavy metal indices of Ismailia Canal, Nile River, Egypt. *Egyptian Journal of Aquatic Research*, *40*(3), 225–233. doi:10.1016/j.ejar.2014.09.001

Ivanova, E., Kravciuk, E., & Kommakova, O. (2007). Ecologist floristic characteristics of phytoplankton of small water bodies in Krasnoyarsk (Russian). *Algol.*, *17*(1), 3–14.

Levei, E., Senila, M., Miclean, M., Abraham, B., Roman, C., Stefanescu, L., & Moldovan, O. T. (2011). Influence of Rosia Poieni and Rosia Montana mining areas on the water quality of the Aries River. *Environmental Engineering and Management Journal*, *10*(1), 23–29. doi:10.30638/eemj.2011.004

Nienhuis, P., Buijse, A., Leuven, R., Smits, A. J. M., de Nooij, R. J. W., & Samborska, E. M. (2002). Ecological rehabilitation of the lowland basin of the River Rhine (NW Europe). *Hydrobiologia, 478*(1/3), 53–72. doi:10.1023/A:1021070428566

Scierbak, V., & Kuzminciuk, V. (2006). The occurrence of species and the diversity of plankton algae of the Teterev River and its reservoirs (Ucraine). *Algol., 6*(1), 336–344.

Smical, I., Muntean, A., & Nour, E. (2015). Research on the Surface Water Quality in Mining Influenced Area in North-Western part of Romania. *Geographica Pannonica, 19*(1), 20–30. doi:10.5937/GeoPan1501020S

Zhang, X. (2014). A study on the water environmental quality assessment of fenjiang River in Yaan city of Sichuan Province in China. *ScienceDirect IERI Procedia, 9*, 102–109. doi:10.1016/j.ieri.2014.09.048

Hasan, M., Khan, M. Z. H., Khan, M., Aktar, S., Rahman, M., & Hossain, F. (2016). Heavy metals distribution and contamination in surface water of the Bay of Bengal coast. Cogent Environmetal Science, *2*(1).

Bilgin, A., & Konanc, M. (2016). Evaluation of surface water quality and heavy metal pollution of Coruh River Basin (Turkey) by multivariate statistical methods. *Environmental Earth Sciences, 75*(12), 1029. doi:10.1007/s12665-016-5821-0

Abdullah, E. J. (2013). Quality assessment for Shatt Al-Arab River using heavy metal pollution index and metal index. *Journal of Environment and Earth Science, 3*(5), 114–120.

Annual report. The surface water quality status according to hydrochemical indices on the territory of the Republic of Moldova in 2015 year. (2016). Chisinau, Republic of Moldova.

Badaluta-Minda, C., & Dunca, A.M. (2016). The analyse of the maximum runoff in Someşul Mic catchment area. *Scientific Bulletin of Politehnica University of Timişoara, Transactions on Hydrotechnics, 61*(75), 39-42.

Bakan, G., Boke-Ozkoc, H., Tulek, S., & Cuce, H. (2010). Integrated environmental quality assessment of Kizilirmak River and coastal environment. *Turkish Journal of Fisheries and Aquatic Sciences, 10*(4), 453–462. doi:10.4194/trjfas.2010.0403

Barhoumi, B., Beldean-Galea, M. S., Al-Rawabdeh, A. M., Roba, C., Martonos, I. M., Kahlaoui, M., ... Baciu, C. (2019). Occurrence, distribution and ecological risk of trace metals and organic pollutants in surface sediments from a Southeastern European river (Someşu Mic River, Romania). *The Science of the Total Environment, 660*(10), 660–676. doi:10.1016/j.scitotenv.2018.12.428 PMID:30641395

Bulat, D., Ciornea, V., & Bulat, D. (2018). The state of ichthyofauna of the Bic river in 2018 year. *Studia Universitatis Moldaviae, 6*(116), 14–19.

Chapman, D. (1996). *Water Quality Assessments - A Guide to Use of Biota, Sediments and Water in Environmental Monitoring* (2nd ed.). University Press, Cambridge.

Duca, G., Mereuta, A., Velisco, N., Tanaselia, C., & Mitina, T. (2019). Determination of quality indicators of Prut River water. *Chemistry Journal of Moldova, 14*(1), 61–67. doi:10.19261/cjm.2018.541

Ene, A., Popescu, I. V., Stihi, C., Gheboianu, A., Radulescu, C., Tigau, N., & Gosav, S. (2010). Assessment of river water quality in central and eastern parts of Romania using atomic and optical methods. *Journal of Science and Arts. Section C-Physics*, *1*(12), 113–118.

Lavric, M., & Ceban, E. (2011). Ecological problems of Bic river. *Mediul Ambiant*, *2*(56), 21–23.

Li, Y., Guo, W., Wu, Z., Jin, L., Ke, Y., Guo, Q., & Hu, S. (2016). Determination of ultra-trace rare earth elements in high-salt groundwater using aerosol dilution inductively coupled plasma-mass spectrometry (ICP-MS) after iron hydroxide co-precipitation. *Microchemical Journal*, *126*, 194–199. doi:10.1016/j.microc.2015.12.006

Manousakas, M., Papaefthymiou, H., Eleftheriadis, K., & Katsanou, K. (2014). Determination of water-soluble and insoluble elements in PM2.5 by ICP-MS. *The Science of the Total Environment*, *493*, 694–700. doi:10.1016/j.scitotenv.2014.06.043 PMID:24992462

Matache, M. L., David, I. G., Dinu, C., & Radu, L. G. (2018). Trace metals in water and sediments of the Prut River, Romania. *Environmental Engineering and Management Journal*, *17*(6), 1363–1371. doi:10.30638/eemj.2018.135

Moldovan, Z., Chira, R., & Alder, A. C. (2009). Environmental exposure of pharmaceuticals and musk fragrances in the Somes River before and after upgrading the municipal wastewater treatment plant Cluj-Napoca, Romania. *Environmental Science and Pollution Research International*, *16*(46), 46–54. doi:10.100711356-008-0047-7 PMID:18972147

Ozbek, N., & Akman, S. (2016). Method development for the determination of calcium, copper, magnesium, manganese, iron, potassium, phosphorus and zinc in different types of breads by microwave induced plasma-atomic emission spectrometry. *Food Chemistry*, *200*, 245–248. doi:10.1016/j.foodchem.2016.01.043 PMID:26830585

Sandu, M., Tarita, A., Mosanu, E., & Turcan, S. (2017). *Surface water pollution index. Case study - water from the forest bypass Hârjauca. Scientific-practical guide.* Chisinau, Republic of Moldova: Primex - Com.

Smuleac, L., Imbrea, F., Corodan, I., Ienciu, A., Smuleac, A., & Manea, D. (2017). The influence of anthropic activity on Somes river water quality. *AgroLife Scientific Journal*, *6*(2), 178–182.

Tamasi, G., & Cini, R. (2004). Heavy metals in drinking waters from Mount Amiata (Tuscany, Italy). Possible risk from arsenic for public health in the Province of Siena. *The Science of the Total Environment*, *327*(1-3), 41–51. doi:10.1016/j.scitotenv.2003.10.011 PMID:15172570

The Norms on Classification of Surface Water Quality in order to determine the ecological status of water bodies. Order of the Romanian Ministry nr. 161/2006.

The Regulation on Environmental Requirements for Surface Waters. GD Nr.890/2013. MO 262-267 (22.11.2013), art. Nr.1006.

Voica, C., Kovacs, M. H., Dehelean, A., Ristoiu, D., & Iordache, A. (2012). ICP-MS determinations of heavy metals in surface waters from Transylvania. *Romanian Journal of Physics*, *57*(7-8), 1184–1193.

Voicu, R., & Banaduc, D. (2016). Suggestions of scenarios for restoring longitudinal connectivity to sustain fish fauna migration upstream and downstream of Apahida bottom sill. *Lakes, Reservoirs and Ponds*, *10*(2), 126–138.

Yusuf, A., Olasehinde, A., Mboringong, M. N., Tabale, R. P., & Daniel, E. P. (2018). Evaluation of heavy metals concentration in groundwater around Kashere and its environs, upper Benue trough, northeastern Nigeria. *Global Journal of Geological Sciences*, *16*(1), 25–36. doi:10.4314/gjgs.v16i1.4

Chapter 11
Nano-Catalysis Process for Treatment of Industrial Wastewater

Manoj Kumar Karnena
https://orcid.org/0000-0001-9241-2990
Department of Environmental Science, GITAM Institute of Science, GITAM University (Deemed), India

Madhavi Konni
https://orcid.org/0000-0002-2022-5528
Department of Basic Science and Humanities, Vignan's Institute of Engineering for Women, India

Vara Saritha
https://orcid.org/0000-0003-0473-993X
Department of Environmental Science, GITAM Institute of Science, GITAM University (Deemed), India

ABSTRACT

The rapid increase in population and urbanization leads to the scarcity of water resources in the present era. Therefore, effective wastewater treatment is a prerequisite for a growing economy. Development and implementing the advanced treatment technologies of wastewater with high efficiency and low capital is difficult. In the recent advancements among various treatment processes, nanomaterial science has been attracting the attention of researchers. However, limited collective knowledge is available in this context. The chapter reviews the potential of nano catalysis's process, mechanism, and current drawbacks in treatment technologies. It explains the different nano catalysts that are widely utilized for the treatment and removal of organic and inorganic pollutants in water and wastewater and discusses the nano-based photocatalytic, nano-based electrocatalysis, nano-based Fenton catalysis and their efficiency in various removal of pollutants from wastewater.

DOI: 10.4018/978-1-7998-1241-8.ch011

INTRODUCTION

Water is one of the most abundant natural resources available on the earth, but for human consumption, only about 1% of that resource is available (Grey, 2013; Adeleye et al., 2016). Due to the rising cost, growing populations and a variety of climatic conditions, it is estimated that over 1.1 billion people (WHO, 2015), lack the supply of adequate drinking water (Adeleye et al., 2016). The water supply chain is one of the significant challenges due to the continued contamination of freshwater resources by a variety of inorganic and organic pollutants (Schwarzenbach et al., 2006). The treatment of wastewater and drinking water can reduce (Xia et al., 2011), these concerns; however, the outdated methods of treatment are not competent enough to completely remove the emergent pollutants and meet the water quality standards.

Moreover, the current technologies of wastewater treatments have several downsides such as high-energy requirement, incomplete pollutant removal and generation of toxic sludge (Qu et al., 2012). The biological wastewater treatment is effective and widely applied, but these are usually sluggish, inadequate due to the presence of the non-biodegradable contaminants and yet cause toxicity to microorganisms due to the toxic pollutants (Reddy et al., 2017; Zelmanov et al., 2008). Physical processes like filtration can remove the contaminants by transmuting one phase to another but, produces highly concentrated sludge which is toxic and difficult to dispose of (Çatalkaya & Bali, 2003).

In the above context, there is a real necessity for more efficient and robust technologies for the treatment of municipal and industrial wastewaters which can be achieved either by the development of entirely new methods or by improving the existing processes through some interventions. Among the various developing technologies, the advancement in nanotechnology has proved a farfetched potential for the treatment of wastewater and various other environmental concerns (Burkhard et al., 2000; Parsons et al., 2006; Crini et al., 2007; Ferroudj et al., 2013).

Nanotechnology is the field of nanoscience, and the phenomena applied on a nanometre scale level (Chaturvedi et al., 2012). Nanomaterials are the smallest structures that humans have developed, having the size of a few nanometres. More precisely, the nanoparticles are those that have structured components with one dimension at least less than 100 nm (Amin et al., 2014). Nano-materials have been developed in a variety of forms such as nanowires, nanotubes, films, particles, quantum dots and colloids in wastewater treatment application, a variety of efficient, eco-friendly and cost-effective nano-materials have been developed having unique functionalities for potential decontamination of industrial effluents, surface water, groundwater, and drinking water (Edelstein, 1998; Lubick et al., 2008). Nanotechnology has been cited, in different kinds of literature, as one of the most advanced processes for wastewater treatment. Nano catalyst is the nanomaterial such as metal oxides and semiconductors have gained considerable attention of the scientists in developing wastewater treatment technologies. Different types of nano-catalyst employed for degradation of pollutants in wastewater, for instance, electrocatalysts, fenton based catalysts for improving chemical oxidation of organic pollutants and catalysts having antimicrobial properties (Zhang et al., 2014; Zhang et al., 2014; Tang et al., 2014; Shamsizadeh et al., 2014).

Present chapter focusses on mechanism of nano catalysis process, their contemporary drawbacks in treatment technologies. Further, different nano catalysts that are widely utilized for treatment and removal of organic and inorganic pollutants in water and wastewater using nano-based photocatalytic, nano-based electrocatalysis, nano-based fenton catalysis are discussed in detail.

CATALYSIS: BACKGROUND

The field of nano catalysis isn't as new as could be expected from the existing nano hype (Kyzas et al., 2015). Indeed, its principle is known considering that the 1950s when the term nanotechnology was not also known (Pacchioni et al., 2009; Gellman et al., 2009; Zhou et al., 2005). Nano catalysis combines the advantageous characteristics of both homogenous and heterogeneous catalyses while reducing their respective drawbacks (Polshettiwar et al., 2010). In homogeneous catalysis, the starting materials, and the catalytic substance are brought together in the same phase, which ensures high catalytic activity and selectivity (Behr et al., 2012). The former can be expressed with the turnover frequency (TOF), defined as the number of substrate molecules that catalytically converted into the product during a specific period. The latter is indicated by the yield of the desired outcome (Van Leeuwen et al., 2006).

However, the practical application of homogeneous catalysis is limited by the difficulties to separate the catalyst from the product after completion of the reaction (Cole-Hamilton et al., 2003). In heterogeneous catalysis, the starting materials and the catalytic substance reside in different phases (Jacoby et al., 2009), thereby alleviating the separation of products and catalyst. The main drawback of traditional heterogeneous catalyst systems compared to their homogeneous counterparts is the reduced surface area that is accessible to reactant molecules, thereby limiting their catalytic activities (Zäch et al., 2006) and leading to unnecessarily high consumption of expensive catalyst materials (Zäch et al., 2006). One possible way to solve this problem is to increase the surface to volume ratio (S/V) by decreasing the size of the catalytically active material (Campelo, 2009; Teunissen et al., 1999). A high S/V can be achieved by synthesizing specifically engineered catalysts on the nanoscale. Nanosized materials show additional unique properties compared to the macroscale (Jiao et al., 2014). A prominent example is the unexpected catalytic activity of gold nanoparticles, which is not found with bulk gold (Campbell et al., 2004; Haruta et al., 2003; Ma et al., 2011; Carabineiro et al., 2011).

Nano catalysis, finally, combines the positive aspects of the two conventional catalytic methods described above. The principal targets that are pursued with nano catalysis are close to 100% selective reactions, extremely high activity, and excellent yield all traditionally related to homogeneous catalysis—as well as enhanced products separation and catalyst recovery typically associated with heterogeneous catalysts. Some authors consider nanocatalysis as "semi heterogeneous" (Astruc et al., 2005) or "soluble heterogeneous" (Widegren et al., 2003) catalysis. In contrast to them, we want to define nanocatalysis as a distinct category, because it closes the gap between hetero- and homogeneous catalysis (Figure 1), while concomitantly requiring new synthetic approaches (Jacoby et al., 2009) and displaying unique characteristics (Domínguez-Quintero et al., 2003). As nano catalysts are further expected to contribute to lower process energy consumption, longer lifetime of the catalyst systems, and enhanced possibilities to isolate and reuse the active nanomaterials.

SYNTHESIS OF NANO CATALYSIS

Progress in Preparation of Nano Catalysts

In the past, particularly the characteristics activity and selectivity of the catalysts were optimized by a simple trial and error approach (Zhou et al., 2005) without a clear understanding of the underlying influencing factors. The traditional synthesis methods did not allow for precise control of nano catalyst

Figure 1. Comparative efficiency of homogeneous, heterogeneous and nano catalysis (Redrawn with permission from Domínguez-Quintero et al., 2004)

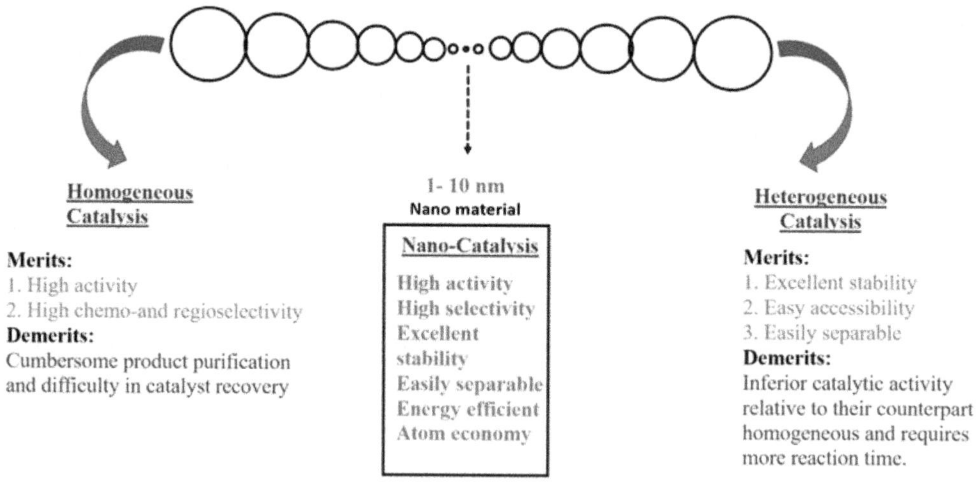

size and shape (Liu et al., 2005), and therefore the field of catalysis science is often criticized as being ad hoc and empirical" (Gellman et al., 2009). Today, activity, selectivity, durability, and recoverability can be influenced owing to the advances in nanoscience (Chang et al., 2004). The size, shape, and surface compositions of nano catalysts are designed and synthesized in a more precise way by adjusting the reaction conditions, such as reaction time, temperature and reactant concentrations (Glaspell, 2008; Dahl et al., 2007). In general, nano catalysts can be synthesized by either a top-down or bottom-up approach (Islam et al., 2007). As the name implies, the idea behind the top-down approach is to break bulk material down mechanically (Schmidt, 2001; Corain et al., 2011), thermally (Zhang et al., 2003), or chemically (Garrigue et al., 2004), into smaller and smaller particles. The top-down approach is criticized because of its inability to yield particles with uniform characteristics (Hussain et al., 2013). However, there are already improved procedures to control the size and surface composition more precisely.

The bottom-up approach involves the formation of nano catalysts by reaction or agglomeration of suitable starting molecules with or without structure-directing agents (Yilmaz et al., 2009). This principle is used more commonly than the former approach, although it is rather disadvantageous both from an economical and environmental point of view, due to harsh reaction conditions employed and the use of expensive precursors and structure-directing agents (Nadagouda et al., 2006). Nevertheless, it allows the synthesis of well-defined catalysts on the nanoscale, regarding size, shape and surface composition (El-Sayed et al., 2004).

In the following part, we distinguish between nanoparticle catalysts and porous nano catalysts, like other authors (Faraday et al., 1857), and discuss their respective production methods in more detail.

SYNTHESIS OF NANOPARTICLE CATALYSTS

Already in the 1850s, Faraday produced metal nanoparticles by the chemical reduction of the respective metal salts, which is still a common way in nanoparticle synthesis (Faraday et al., 1857). Later, in the 1920s, metal nanoparticles were introduced to catalyse chemical reactions (Somorjai et al., 2001). In a desire to reduce costs of large-scale catalytic systems, for example, in the refinery industry, researchers in the 1950s began to produce smaller catalytic particles to profit from the resulting advantageous characteristics described above. Thereby, the size was lowered from 100 nm in the beginning to less than 1 nm today (Mahajan et al., 2003). Top-down and bottom-up approaches (Table 1)are conceivable for the synthesis of nanoparticle catalysts. Numerous methods were developed, including rather conventional techniques like mechanical grinding or chemical breakdown of bulk material or electrochemical or solvothermal processing of precursor solutions (Bönnemann, 2003; Toshima et al., 1998). Further selected examples and innovative alternatives, such as microwave irradiation processing, resulting in catalysts of more precisely defined size and shape are listed (Patzke et al., 2011; Cushing et al., 2004; Chen et al., 2002).

Table 1. Approaches for preparation of Nano-Materials.

Top-Down Technologies	Bottom-Up technologies
1. Mechanical Grinding	1. Sol-gel
2. Metal Vapour	2. Chemical Vapour Deposition
3. Thermal Breakdown	3. Electro-Chemistry
4. Chemical Breakdown	4. Solvothermal process
5. Spontaneous Chemisorption	5. Template-Directed
	6. Precipitation
	7. Microemulsion
	8. Microwave radiation
	9. Sonochemistry

The two major problems in the synthesis and application of nanoparticle catalysts are their tendency to agglomerate associated with deterioration in their unique characteristics and the difficulties encountered in the catalyst recovery from the reaction mixture (Roucoux et al., 2002). To overcome, aggregation problem, the nanoparticle catalysts can be deposited on different kinds of supports, such as carbon, graphite, and hydrogels. Li et al., 2016 used carbon nanotubes to support palladium nanoparticles in the catalytic reduction of 4-nitrophenol. Alternatively, support-free nano catalysts can be stabilized by using one of the three following methods, generally referred to as electrostatic, steric, and electrostatic stabilization.

Electrostatic stabilization is achieved by creating an electrical double layer around the nanoparticle catalyst with salts dissolved in the reaction mixture or by ionic liquids (Roucoux et al., 2002). These double-layers hinder the particles from coming too close to each other and therefore prevent aggregation. For an instance, rhodium nanoparticles used in the catalytic hydrogenation of functionalized aromatic compounds were stabilized by ionic liquids (Denicourt-Nowicki et al., 2011). Such a protective layer can also be formed by adsorbing macromolecules or other ligands, such as phosphines or thiols to the particle surface, leading to steric repulsion. Léger et al., for example, stabilized rhodium nanoparticles with various bipyridine ligands, for their use as hydrogenation catalysts for arenes (Léger et al., 2008).

The third method, electrostatic stabilization, results from combining the effects of the two previous ones. Thereby, the surfaces are covered with surfactants containing polar head groups (for the electric double layer) and lipophilic side chains (for the steric repulsion) (Yan et al., 2011) or polyoxoanions.

Mévellec et al., for example, stabilized iridium (0) nano catalyst for the hydrogenation of arenes in a biphasic medium by using N, N-dimethyl-N-cetyl-N-(2-hydroxyethyl) ammonium chloride salt as a surfactant (Mevellec et al., 2004). Hornstein and Finke, on the other hand, prepared a tetrabutylammonium- and polyoxoanion ($P_2W_{15}Nb_3$)-stabilized Ir (0) nano catalyst (Hornstein et al., 2003). Despite all the advantages of the above-described stabilization methods, it has to be borne in mind that stability and activity of the catalytic nanoparticle are reciprocally related; that is, a nanoparticle withstanding even the harshest conditions is likely to be catalytically inactive (Stowell et al., 2005). Within this trade-off, it is nevertheless possible to obtain a stable and active nano catalyst, by choosing stabilizers that are weakly bound to the active site but still offer proper protection, for example, by electrostatic repulsion. Yan et al.2011, proved that a suitable stabilizer can optimize catalyst stability while retaining high activity.

As mentioned above, besides nanoparticle aggregation, the second challenge in the successful application of nano catalyst particles is their efficient recovery from the reaction mixture. The recovery of used catalysts would be beneficial in two respects. On the one hand, catalytic nanomaterials, precious metals are very expensive. On the other hand, unrecovered catalysts are a significant source of unwanted impurities, for example, in pharmaceutical products. The efficient separation of the catalyst from the reaction mixture could help to meet legal standards regarding impurities more efficiently (Olveira et al., 2014). For solving the recovery problem, the use of magnetic support for the catalytically active nanoparticles was proven to be effective (Yoon et al., 2003; Tang et al., 2008; Ge at al., 2008), for an example, Schätz et al., 2010 synthesized a novel magnetic hybrid material as support for a palladium catalyst that can be quickly recovered after use.

The nano-catalysts, especially those of inorganic materials such as semiconductors and metal oxides, are gaining considerable attention of the researchers in the application of wastewater treatment. Various kinds of nano-catalysts are employed for wastewater treatment such as photocatalysts (Dutta et al., 2014) electrocatalysts (Dutta et al., 2014) and Fenton based catalysts (Kurian et al., 2015) for improving chemical oxidation of organic pollutants (Ma et al., 2015) and antimicrobial actions (Pranjali et al., 2013).

NANOMATERIALS AS PHOTOCATALYSTS

The simple mechanism(Figure 2) of the working of photocatalysis is based on the photoexcitation of the electron in the catalyst. The irradiation with light (UV in case of TiO_2) generates holes (H+) and excited electrons (e-) in the conduction band. In an aqueous media, the holes (H+) are trapped by water molecules (H_2O) and generate hydroxyl radicals (-OH) (Akhavan et al., 2009). The radicals are indiscriminate and powerful oxidization agent. These hydroxyl radicals on reaction oxidize the organic pollutants into water and gaseous degradation products (Samsudin et al., 2015).

Nanoparticle photocatalytic reactions are based on the interaction of light energy with metallic nanoparticles and are of great interest due to their broad and high photocatalytic activities for various pollutants (Akhavan et al., 2009). Usually, these photocatalysts are comprised of semiconductor metals that can degrade a variety of persistent organic pollutants in wastewater such as dyes, detergents, pesticides, and volatile organic compounds (Lin et al., 2014). Furthermore, semiconductor nano-catalysts are also highly useful for the degradation of halogenated and non-halogenated organic compounds, PCPPs and also heavy metals in a specific situation (Adeleye et al., 2016). Semiconductor nanomaterials require mild operation conditions and are very useful even at a small concentration. Among various nano photocatalysts developed till now, TiO_2 is one of the most widely applied in photocatalysis due to its high

Figure 2. Mechanism of photocatalytic dye degradation on TiO_2 based photocatalysts

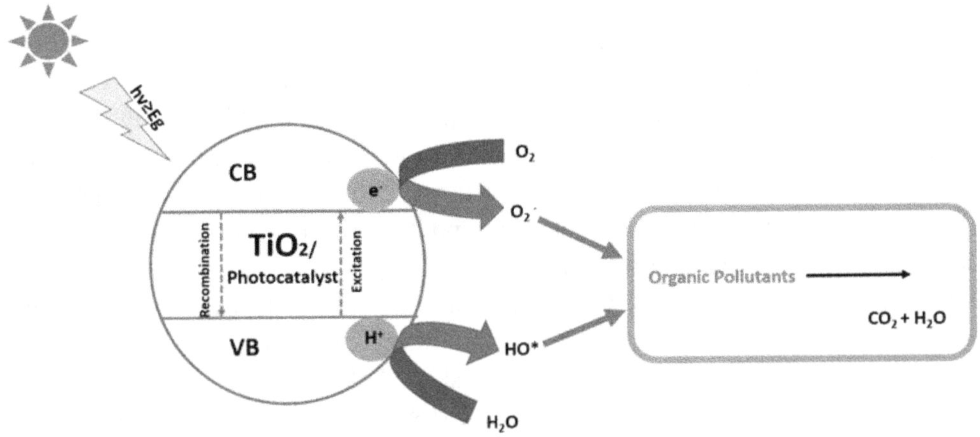

reactivity under ultraviolet light (k< 390 nm) and chemical stability (Akhavan et al., 2009). Numerous studies have shown the photocatalytic activity of various synthesized catalysts. Their efficiency depends on different factors such as bandgap energy, particle size, dose, pollutant concentration, and pH.

For instance, Hayat et al. (2011) found that the photocatalytic degradation efficiency of ZnO decreased with high calcination temperature that increases the particle size due to agglomeration. CdS is also a well-known semiconductor having a bandgap of 2.42 eV and can be operated at wavelength <495 nm (Khallaf et al., 2008). CdS nanoparticles have been attracting intense interest as photocatalyst for treatment of industrial dyes in wastewater (Tristao, 2006; Zhu et al., 2009). However, the catalysts mentioned above are well known for their photocatalytic activity, but they are active only under ultraviolet radiations (1 <387 nm) due to the comprehensive bandgap energy, i.e. 3.2 eV as in case of TiO_2. Thus, further modifications for the catalysts have been studied to increase their activities under visible light source (sunlight) for degradation of organic pollutants (Dutta et al., 2014).

Treatment of wastewater with the help of visible light photocatalytic action is contemporary in research interest. For reaching the goal, some modifications are required to be made to nanomaterials/ Semiconductor particles which will help to decrease the band energy from UV to Visible Region (Anjum et al., 2017).

Numerous studies are conducted by various researchers to evaluate the photocatalytic activity by modified nano-catalyst under visible light. Dye sensitization, doping metal impurities, Hybrid nanoparticles, or anions are the general methods used for modification of nanocatalyst (Ni et al., 2007). The narrow bandgap was created by new metals and anions also called as impurity energy levels which upon exposure to visible light conducts electrons into semiconductor enabling initiation of catalytic reaction (Qu et al., 2013). The wide bandgap of 3.2 eV has been extensively identified in ZnO, and TiO_2 Nanomaterials have been extensively investigated for the activity of Photocatalysis.

A small portion of the UV region can be absorbed by the Catalysts in Solar Spectrum, which decreases the efficiency (Chen et al., 2004). However, a slight modification in catalysts by loading metals on the surface can solve this problem. Under the illumination of solar radiation, modified composite materials decrease the bandgap energy and subsequently transfers the excited electron to the semiconductor. Furthermore, not all conductive metals are effective for doping to improve photocatalytic activity, e.g. Pt and

Ru are ineffective for doping, while other metals such as Au, Ag and Pd showed excellent photocatalytic activities (Sathishkumar et al., 2011; Barakat et al., 2013; Barakat et al., 2013).

In the recent years, various doped nano-catalysts have been developed such as ZnO: Co, Ni, ZnS: Mn, ZnS: Cu, CdS Eu, CdS: Mn, ZnSe: Mn, ZnS: Pb, and Cu (Chandrakar et al., 2015). For enhancing the surface area of metal oxide nanoparticles, many dopants such as Cr, Si, Co, Mg, Mn, Fe, Fe, Al, In and Ga are used (Jamal et al., 2012). Anions such as nitrogen are most feasible and cost-effective for industrial application amongst various dopants (Fujishima, 2008; Qu et al., 2013). The surface area of the catalyst can be elevated by doping nanomaterials, and it is useful to protect the nanocomposite from size reduction, change in shape and morphology.

Sathish kumar et al., 2011 synthesized CuO/ ZnO photocatalyst for the degradation of a textile wastewater dye (Acid Red 88) under visible light. They achieved two times higher photocatalytic degradation with CuO/ZnO compared to that of unmodified ZnO. Eskizeybek et al., 2012 synthesized modified ZnO nanocomposite using organic homopolymer polyaniline (PANI). The modified PANI/ZnO nano-catalyst showed 99% of removal of organic pollutants in wastewater such as methylene blue and malachite green dyes during photocatalysis, even with a little dose of catalyst, i.e. 0.4 g/L of wastewater. Similarly, in another study, the modified CdS cross-linked (CS/n-CdS) chitosan found to be highly effective for the degradation of model organic pollutant CR Congo Red where 85.9% of degradation was observed in just three hours of photocatalytic process under visible light (Zhu et al., 2009). Dutta et al., 2014, synthesized c-Fe_2O_3 nanoparticles by thermal decomposition method and observed a high photocatalytic activity toward degradation of rose Bengal and methylene blue dyes under irradiation of visible light.

The use of graphene for modification of catalyst has also got the attention of the researcher due to its unique characteristics. The composites of graphene with other semiconductor materials could significantly enhance the electron mobility through the interfacial electron transfer process during photocatalysis. Moreover, graphene can promote the separation efficiency of photo-induced electrons and holes in photocatalyst (Li et al., 2016). Anitha et al., 2017 synthesized sulphur-doped TiO_2 (S-TiO_2) nanomaterial for degradation of malachite green dye under visible light. Mushtaq et al., 2018 synthesized $BiFO_3$ NS and NW Nanoparticles for degradation of dye (Rhodamine B) under visible light and achieved a removal to 66%. Xia et al., 2019, synthesized TiO_2@ g-C3N4 nano catalyst for the degradation of Rhodamine B and achieved more removal efficiency than Mushtaq et al., 2018.

Advantages

1. Low-Cost treatment process.
2. Reusable.
3. Complete Degradation.
4. Eco-friendly and will not release more toxicants into the environment.

Disadvantages

1. Spectral sensitization is essential for the utilization of a broad range of photonic energy.

Table 2. A comprehensive view on the photocatalytic efficiency of various modified nano-catalysts for removal of organic pollutants from wastewater is summarized.

Catalyst	Light region	Pollutant	Results	Reference
TiO_2@ g-C3N4	Visible Light	Rhodamine B	Up to 70% of degradation was observed	Xia et al., 2019
$BiFO_3$ NS and NW	Visible Light	Rhodamine B	Up to 66% of degradation was observed	Mushtaq et al., 2018
Sulfur-doped TiO_2 (S-TiO_2)	Visible Light	malachite green	Up to 98% of degradation was observed	Anitha et al., 2017
TiO_2CO coatings	420 nm (Vis)	Methylene blue	Removal of Methylene blue form 10 μmol/L to around 8.5 μmol/L was observed	Guan et al., 2016
Carbon nanorods-TiO_2	³420 nm (Vis)	2,4 dichlorophenol	Up to 96% of photocatalytic degradation of 2,4-dichlorophenol was achieved within 8 h	Ortega-Liebana et al., 2016
Zero-valent nano copper	Visible region	Methyl orange	Up to 35% degradation of methyl orange was achieved with in 80 min.	Liu et al., 2016
Immobilized TiO_2	300-800 nm	Pharmaceutically active compounds (propranolol, diclofenac, and carbamazepine)	The photocatalytic efficiency of 100% for propranolol, 100% for diclofenac and 76% for carbamazepine was achieved	He et al., 2016
Y_2O_3: Eu^{3+}	350-400 nm (UV)	Methylene blue	Up to 90% degradation efficiency of methylene blue was observed	Kumar et al., 2015
TiO_2/trititanate	UV-Vis region	Rhodamine B	The degradation efficiency of Rhodamine B of >91% was achieved	Chen et al., 2015
AgBr/ZnO	410 nm (Vis)	Methylene blue	Up to 87% of methylene blue was decomposed by AgBr/ZnO after 240 min Up to 87% of methylene blue was decomposed by AgBr/ZnO after 240 min	Dai et al., 2014
3D SnO	365 nm (UV)	Methyl orange	Photocatalytic degradation of methyl orange was 83% after 150 min	Cui et al., 2015
ZnO nanorods	365 nm (UV)	Rhodamine B	The quenched catalysed greatly improved the photocatalytic degradation of Rhodamine B	Fang et al., 2015
Graphene-CoxZn1-xFe2O4	Visible region	Methylene blue	The Graphene-CoxZn1-xFe_2O_4 heterostructures showed more photocatalytic efficiency compared to CoxZn1-xFe_2O_4 nanoparticles (without graphene)	Nazim et al., 2016
$CuFe_2O_4$@C3N4	>420 nm (Vis)	Orange II	Nearly 98% of Orange II is degraded within 210 min photocatalysis	Yao et al., 2015
Ag_3PO_4/$BiPO_4$/Cu (top). grapheme	³420 nm (Vis)	Atrazine herbicide	The nanocomposite showed more 80% photocatalytic degradation of Atrazine herbicide	Mohaghegh et al., 2015
Fe-TiO_2 coated with side-glowing optical fibres	Visible region	Rhodamine B	Up to 90% of Rhodamine b removal can be achieved within 8 h of photocatalysis	Lin et al., 2015
Polyaniline/ZnO	Visible region	Methylene blue and malachite green dyes	Dyes removal with 99% efficiency was achieved under natural sunlight	Eskizeybek et al., 2012
ZnO/Zn	365 nm (UV)	Methylene blue	Complete degradation of methylene blue was achieved (98%)	Lin et al., 2014
ZnO-FeOclinoptilolite	Solar light	Real Fishpond wastewater	More than 80% degradation efficiency was achieved	Bahrami and Nezamzadeh-Ejhieh et al.,2014
CuO/ZnO	Visible region	Acid Red 88 dye	In the presence of external electron acceptors such as peroxomonosulfate, peroxodisulfate, and H2O2, CuO-ZnO shows approximately 2 times higher photocatalytic degradation of Acid Red 88 compared to that ZnO.	Sathishkumar et al., 2011
Ag-TiO_2/Ag/a-TiO_2	425 nm (Vis)	Escherichia coli bacteria	The relative rate of reduction in the bacteria was improved to 4.7 _ 10_2/min corresponding to 110 min in the visible	Akhavan et al.,2009

Nano-materials as Electro-Catalysts

The process of electrocatalysis in a microbial fuel cell is an emerging topic of discussion for wastewater treatment and direct electricity generation. In a microbial fuel cell, electrocatalyst plays a key role in working of a fuel cell (Chen et al., 2015). The use of nanomaterial as electrocatalyst can improve the performance of fuel cell by achieving a larger surface area and uniform distribution of catalyst in the reaction media (Liu et al., 2016). Tremendous research has been conducted on the development of carbon-supported nano electrocatalysts for application in fuel cells (Tang, 2014; Pranjali et al., 2013). It is reported the Pt nano catalyst supported by carbon black XC72 showed potential up to 6.2 mA cm^2 of current density in glucose oxidation electro-catalysis reaction (Chen et al., 2015). Moreover, Pt electrocatalysts also showed high potential for ethanol oxidation reaction in a fuel cell in several studies (Kamarudin, 2013; Habibi et al., 2015).

Although Pt can be used as electro-catalyst, yet, it has various disadvantages which limit its application. For instance, Pt is a precious metal, and it has limited availability its high cost narrowed the interest to use it as a catalyst. Also, during electro-catalysis Pt may restrict the reaction due to the poisoning of intermediate compounds (Zhou et al., 2003). However, these problems can be overcome by replacing Pt with Pd nanoparticles. For instance, in ethanol fuel cell Pd nano-catalyst can reduce the cost of anodes due to its abundance in the earth, i.e. fifty times more compared to Pt (Savadogo et al., 2004). Recently, Habibi and Mohammadyari et al., 2015, have prepared the Pd nanoparticles and functionalized carbon nanomaterials by a simple, spontaneous redox process. The PdNPs with functionalized single-wall carbon nanotubes showed the high electrocatalytic activity and thus could be used as supporting nanocarbon material for the synthesis of the nanocomposite.

Similarly, Pd can also form composites on six-carbon nano-materials having different functionalities. These nano-materials include pristine multi-walled carbon nanotubes (SWCNT), amine-modified MWCNT, carboxylated MWCNT, hydroxyl-modified pristine multi-walled carbon nanotubes carboxylated graphene and XC72 carbon black (Chen et al., 2015). Cai et al., 2013, demonstrated that MWCNT is ideal support material for Pd in glucose oxidation reaction that showed that MWCNT-supported Pd catalyst could effectively catalyse GOR in alkaline medium and resulted in a catalytic current density of 2.7 mA cm^2. The newly developed electro-catalyst Au supported on the imidazolium-type ionic liquids/Polypyrrole nanotubes surface was also evaluated for its activity in a fuel cell for wastewater treatment (Figure 3). It was found that this hybrid electro-catalyst has smaller particle size and better dispersion, and thus showed a high catalytic reduction in model pollutant 4-nitrophenol due to efficient electrocatalysis of dioxygen reduction and 4- nitrophenol (Qiu et al., 2012).

Advantages

1. More efficient than photocatalysis.
2. Bromated compounds are not formed while treating wastewater.
3. Less heat transfer relative to Photocatalysis.

Figure 3. Mechanism of electrocatalysis wastewater degradation on Pt-based nano electrocatalyst (Re-drawn with permission from Feng et al., 2019)

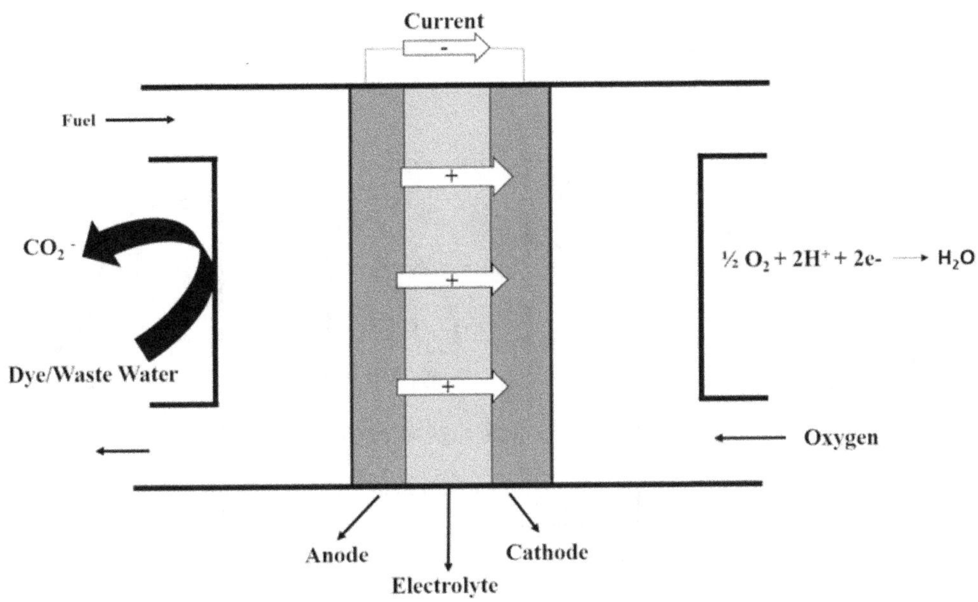

Disadvantages

1. No full-scale application exists.
2. Low rate of degradation.
3. Energy and Cost intensive process.

Nanomaterial-Based Fenton Catalyst

Oxidation of organic pollutants by using Fenton's reaction has been widely applied in wastewater treatment (Neyens et al., 2003). The main drawback of Fenton's reaction is the continuous loss of catalyst material with effluent and requirement of acidic conditions (pH = 3) for optimum function (Kurian, 2015; Ferroudj et al., 2013). The use of nano-material based Fenton's reagent has been employed to overcome these problems. The nano-ferrites with controlled crystalline size, distribution, and chemical structure can be obtained through the sol-gel and auto combustion method (Kurian, 2016; Kurian et al., 2014). The spinel ferrites containing Ni, Zn, Co, and Cu have significant importance as catalysts due to their select magnetic and electronic properties. The presence of these metal lattice modifies the stability and redox properties of ferrites, which further increases the catalysis efficiency. It is reported that the heterogeneous MFe_2-O_4 is extensively used as a catalyst due to its high chemical and thermal stability (Kurian et al., 2014).

Magnetically separable nano-particles of iron oxide can be used as Fenton catalysts for removal of several types of pollutants (Weng, 2013 and Chen et al., 2019). Divided magnetic materials containing iron oxide phases such as carbonaceous materials, ferrite with Ba, Co and Mn, and maghemite have high

Figure 4. Mechanism of fenton catalysts on wastewater degradation (Redrawn with permission from Illés et al., 2019)

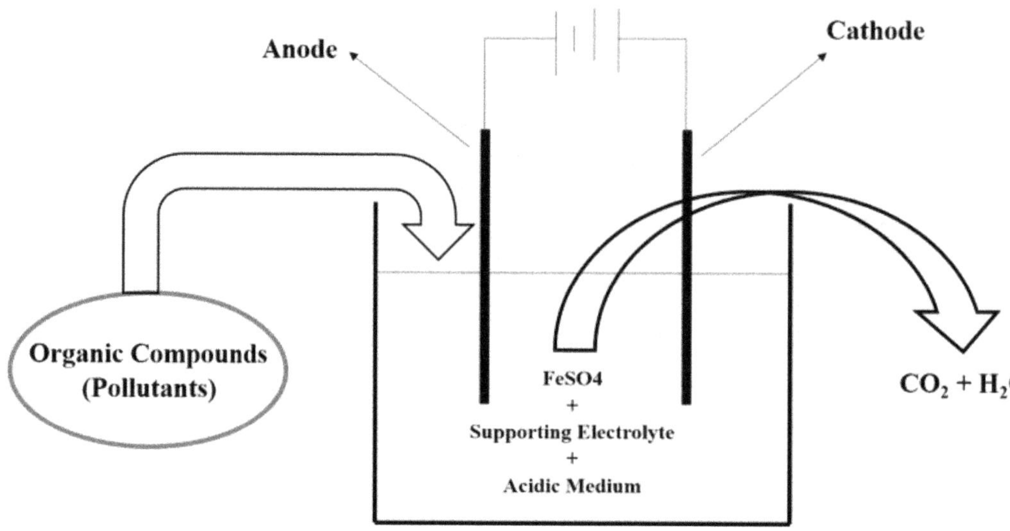

capability to recover from water through the magnetic field. Magnetic separation is quicker and more efficient compared to the decantation and filtration methods (Bhateria, 2019; Boussouf et al., 2019) found that maghemite nanoparticles (Fe_2O_3) and maghemite/silica nano-composite (Fe_2O_3/SiO_2) have controlled mineralization rates during Fenton reaction(Figure 4).

This indicates that these nano-catalysts have stability toward uncontrolled oxidation of pollutant and organic intermediate products. Similarly, in another study nano nickel-zinc, ferrite catalysts were evaluated for the degradation of 4-chlorophenol. Experimental results revealed the complete degradation of the target pollutant with a considerable reduction of COD in just 75 min of reaction (Reddy et al., 2017; Illés et al., 2019; Hu et al., 2019; Hu et al., 2019).

Advantages

1. No energy input is necessary to activate hydrogen peroxide.
2. Relatively inexpensive and the process is easy to operate and maintain.
3. Shortest reaction time amongst all advanced methods available in the current treatment technologies.

Disadvantages

1. Limited by narrow pH range (pH 2-3).
2. Forms intermediate products in the reaction and causes pollution to the environment.

CONCLUSION

Advanced water and wastewater treatment technologies is the need of hour towards achieving superior quality of treated waters through removing pollutants of chemical and biological nature. Nanotechnology has evolved as a potential and ideal option in achieving such treatment. Researchers around the world have studied, developed and applied various types of nanomaterials successfully for various wastewaters. Nano-particle photocatalysts have been applied for treatment of wide variety of wastewaters consisting of toxic pollutants as and heavy metals. Modifications in these catalysts enable their capabilities of using visible region of solar radiation instead of costly artificial ultraviolet radiation.

Electrocatalytic process of wastewater treatment can be enhanced by utilizing nanoparticles in order to achieve large surface area along with uniform distribution of catalyst in the reaction media. Further, nanomaterials are used in fenton reactions for creating stability during the oxidation of pollutants leading to enhanced removal of contaminants at faster rate. Although nanoparticles have proved their potentiality as promising solutions for water and wastewater treatment, these also come with some disadvantages which are to be addressed owing to the potential of nanoparticles to enter environment, causing accumulation in the process of preparation and treatment.

In order to minimize the health effects, there is a need for future research to prepare a catalyst having less toxicity to the environment. Further research is required to re-evaluate the ecotoxicity potential for modification of catalyst. Life cycle assessment is significantly needed while addressing the risks and benefits of catalyst. Nanotechnology has not been yet adopted at large scale owing to cost involved in comparison with standard materials like activated charcoal, and therefore the future application can specialize in cost-effectively processed nanomaterials wherever little quantities are needed. Moreover, much work is needed on developing a compelling value strategy of synthesizing nanomaterials and testing the potency at massive scale for triple-crown application.

REFERENCES

Adeleye, A. S., Conway, J. R., Garner, K., Huang, Y., Su, Y., & Keller, A. A. (2016). Engineered nanomaterials for water treatment and remediation: Costs, benefits, and applicability. *Chemical Engineering Journal*, *286*, 640–662. doi:10.1016/j.cej.2015.10.105

Akhavan, O. (2009). Lasting antibacterial activities of Ag–TiO2/Ag/a-TiO2 nanocomposite thin film photocatalysts under solar light irradiation. *Journal of Colloid and Interface Science*, *336*(1), 117–124. doi:10.1016/j.jcis.2009.03.018 PMID:19394952

Amin, M. T., Alazba, A. A., & Manzoor, U. (2014). A review of removal of pollutants from water/wastewater using different types of nanomaterials. *Advances in Materials Science and Engineering*, *2014*, 2014. doi:10.1155/2014/825910

Anitha, B., Ravidhas, C., Venkatesh, R., Raj, A. M. E., Ravichandran, K., Subramanian, B., & Sanjeeviraja, C. (2017). Self-assembled sulfur induced interconnected nanostructure TiO2 electrode for visible light photoresponse and photocatalytic application. *Physica E, Low-Dimensional Systems and Nanostructures*, *91*, 148–160. doi:10.1016/j.physe.2017.04.017

Anjum, M., Al-Makishah, N. H., & Barakat, M. A. (2016). Wastewater sludge stabilization using pre-treatment methods. *Process Safety and Environmental Protection, 102*, 615–632. doi:10.1016/j.psep.2016.05.022

Anjum, M., Oves, M., Kumar, R., & Barakat, M. A. (2017). Fabrication of ZnO-ZnS@ polyaniline nanohybrid for enhanced photocatalytic degradation of 2-chlorophenol and microbial contaminants in wastewater. *International Biodeterioration & Biodegradation, 119*, 66–77. doi:10.1016/j.ibiod.2016.10.018

Astruc, D., Lu, F., & Aranzaes, J. R. (2005). Nanoparticles as recyclable catalysts: The frontier between homogeneous and heterogeneous catalysis. *Angewandte Chemie International Edition, 44*(48), 7852–7872. doi:10.1002/anie.200500766 PMID:16304662

Bahrami, M., & Nezamzadeh-Ejhieh, A. (2014). Effect of supporting and hybridizing of FeO and ZnO semiconductors onto an Iranian clinoptilolite nano-particles and the effect of ZnO/FeO ratio in the solar photodegradation of fish ponds waste water. *Materials Science in Semiconductor Processing, 27*, 833–840. doi:10.1016/j.mssp.2014.08.030

Bali, U., Çatalkaya, E. Ç., & Şengül, F. (2003). Photochemical degradation and mineralization of phenol: A comparative study. *Journal of Environmental Science and Health. Part A, Toxic/Hazardous Substances & Environmental Engineering, 38*(10), 2259–2275. doi:10.1081/ESE-120023373 PMID:14524680

Barakat, M. A., Al-Hutailah, R. I., Hashim, M. H., Qayyum, E., & Kuhn, J. N. (2013). Titania-supported silver-based bimetallic nanoparticles as photocatalysts. *Environmental Science and Pollution Research International, 20*(6), 3751–3759. doi:10.100711356-012-1306-1 PMID:23161501

Barakat, M. A., Ramadan, M. H., Alghamdi, M. A., Algarny, S. S., Woodcock, H. L., & Kuhn, J. N. (2013). Remediation of Cu (II), Ni (II), and Cr (III) ions from simulated wastewater by dendrimer/titania composites. *Journal of Environmental Management, 117*, 50–57. doi:10.1016/j.jenvman.2012.12.025 PMID:23353877

Behr, A., & Neubert, P. (2012). *Applied homogeneous catalysis*. John Wiley & Sons.

Bhateria, R., & Singh, R. (2019). A review on nanotechnological application of magnetic iron oxides for heavy metal removal. *Journal of Water Process Engineering, 31*, 100845. doi:10.1016/j.jwpe.2019.100845

Bönnemann, H., & Richards, R. M. (2001). Nanoscopic metal particles– synthetic methods and potential applications. *European Journal of Inorganic Chemistry, 2001*(10), 2455–2480. doi:10.1002/1099-0682(200109)2001:10<2455::AID-EJIC2455>3.0.CO;2-Z

Boussouf, I., Medjram, M. S., Ferroudj, N., & Abramson, S. (2019). Unexpected activity of magnetically separable Fenton catalyst in clay slurries. *Environmental Technology*. (accepted)

Burkhard, R., Deletic, A., & Craig, A. (2000). Techniques for water and wastewater management: A review of techniques and their integration in planning. *Urban Water, 2*(3), 197–221. doi:10.1016/S1462-0758(00)00056-X

Cai, Z. X., Liu, C. C., Wu, G. H., Chen, X. M., & Chen, X. (2013). Palladium nanoparticles deposit on multi-walled carbon nanotubes and their catalytic applications for electrooxidation of ethanol and glucose. *Electrochimica Acta, 112*, 756–762. doi:10.1016/j.electacta.2013.09.040

Campbell, C. T. (2004). The active site in nanoparticle gold catalysis. *Science*, *306*(5694), 234–235. doi:10.1126cience.1104246 PMID:15472065

Campelo, J. M., Luna, D., Luque, R., Marinas, J. M., & Romero, A. A. (2009). Sustainable preparation of supported metal nanoparticles and their applications in catalysis. *ChemSusChem: Chemistry & Sustainability Energy & Materials*, *2*(1), 18–45. doi:10.1002/cssc.200800227 PMID:19142903

Carabineiro, S. A., Bogdanchikova, N., Avalos-Borja, M., Pestryakov, A., Tavares, P. B., & Figueiredo, J. L. (2011). Gold supported on metal oxides for carbon monoxide oxidation. *Nano Research*, *4*(2), 180–193. doi:10.100712274-010-0068-7

Çatalkaya, E. Ç., Bali, U., & Şengül, F. (2003). Photochemical degradation and mineralization of 4-chlorophenol. *Environmental Science and Pollution Research International*, *10*(2), 113–120. doi:10.1065/espr2002.10.135 PMID:12729044

Chandrakar, R. K., Baghel, R. N., Chandra, V. K., & Chandra, B. P. (2015). Synthesis, characterization and photoluminescence studies of Mn doped ZnS nanoparticles. *Superlattices and Microstructures*, *86*, 256–269. doi:10.1016/j.spmi.2015.07.043

Chang, J. S., Hwang, J. S., Jhung, S. H., Park, S. E., Férey, G., & Cheetham, A. K. (2004). Nanoporous Metal-Containing Nickel Phosphates: A Class of Shape-Selective Catalyst. *Angewandte Chemie International Edition*, *43*(21), 2819–2822. doi:10.1002/anie.200353502 PMID:15150758

Chaturvedi, S., Dave, P. N., & Shah, N. K. (2012). Applications of nano-catalyst in new era. *Journal of Saudi Chemical Society*, *16*(3), 307–325. doi:10.1016/j.jscs.2011.01.015

Chen, C. C., Lin, C. L., & Chen, L. C. (2015). Functionalized carbon nanomaterial supported palladium nano-catalysts for electrocatalytic glucose oxidation reaction. *Electrochimica Acta*, *152*, 408–416. doi:10.1016/j.electacta.2014.11.116

Chen, C. Z., & Zhou, Z. W. (2004). The preparation of nano-ZnO and its middle infrared-ultraviolet-visible light absorption properties. *Journal of Functional Materials*, *35*(1), 97–98.

Chen, Q., Rao, P., Cheng, Z., Yan, L., Qian, S., Song, R., & Shen, G. (2019). Novel soil remediation technology for simultaneous organic pollutant catalytic degradation and nitrogen supplementation. *Chemical Engineering Journal*, *370*, 27–36. doi:10.1016/j.cej.2019.03.179

Chen, W. X., Lee, J. Y., & Liu, Z. (2002). Microwave-assisted synthesis of carbon supported Pt nanoparticles for fuel cell applications. *Chemical Communications*, (21): 2588–2589. doi:10.1039/b208600j

Cole-Hamilton, D. J. (2003). Homogeneous catalysis--new approaches to catalyst separation, recovery, and recycling. *Science*, *299*(5613), 1702–1706. doi:10.1126cience.1081881 PMID:12637737

Corain, B., Schmid, G., & Toshima, N. (Eds.). (2011). *Metal nanoclusters in catalysis and materials science: the issue of size control*. Elsevier.

Crini, G., & Badot, P. M. (2007). *Traitement et épuration des eaux industrielles polluées: Procédés membranaires, bioadsorption et oxydation chimique*. Presses Univ. Franche-Comté.

Cui, Y., Wang, F., Iqbal, M. Z., Wang, Z., Li, Y., & Tu, J. (2015). Synthesis of novel 3D SnO flower-like hierarchical architectures self-assembled by nano-leaves and its photocatalysis. *Materials Research Bulletin, 70*, 784–788. doi:10.1016/j.materresbull.2015.06.021

Cushing, B. L., Kolesnichenko, V. L., & O'Connor, C. J. (2004). Recent advances in the liquid-phase syntheses of inorganic nanoparticles. *Chemical Reviews, 104*(9), 3893–3946. doi:10.1021/cr030027b PMID:15352782

Dahl, J. A., Maddux, B. L., & Hutchison, J. E. (2007). Toward greener nanosynthesis. *Chemical Reviews, 107*(6), 2228–2269. doi:10.1021/cr050943k PMID:17564480

Dai, K., Lv, J., Lu, L., Liu, Q., Zhu, G., & Li, D. (2014). Synthesis of micro-nano heterostructure AgBr/ZnO composite for advanced visible light photocatalysis. *Materials Letters, 130*, 5–8. doi:10.1016/j.matlet.2014.04.190

Denicourt-Nowicki, A., Léger, B., & Roucoux, A. (2011). N-Donor ligands based on bipyridine and ionic liquids: An efficient partnership to stabilize rhodium colloids. Focus on oxygen-containing compounds hydrogenation. *Physical Chemistry Chemical Physics, 13*(30), 13510–13517. doi:10.1039/c1cp20602h PMID:21660329

Domínguez-Quintero, O., Martínez, S., Henríquez, Y., D'Ornelas, L., Krentzien, H., & Osuna, J. (2003). Silica-supported palladium nanoparticles show remarkable hydrogenation catalytic activity. *Journal of Molecular Catalysis A Chemical, 197*(1-2), 185–191. doi:10.1016/S1381-1169(02)00583-6

Dutta, A. K., Maji, S. K., & Adhikary, B. (2014). γ-Fe2O3 nanoparticles: An easily recoverable effective photo-catalyst for the degradation of rose bengal and methylene blue dyes in the waste-water treatment plant. *Materials Research Bulletin, 49*, 28–34. doi:10.1016/j.materresbull.2013.08.024

Edelstein, A. S., & Cammaratra, R. C. (1998). *Nanomaterials: synthesis, properties and applications*. CRC Press. doi:10.1201/9781482268591

El-Sayed, M. A. (2004). Small is different: Shape-, size-, and composition-dependent properties of some colloidal semiconductor nanocrystals. *Accounts of Chemical Research, 37*(5), 326–333. doi:10.1021/ar020204f PMID:15147173

Eskizeybek, V., Sarı, F., Gülce, H., Gülce, A., & Avcı, A. (2012). Preparation of the new polyaniline/ZnO nanocomposite and its photocatalytic activity for degradation of methylene blue and malachite green dyes under UV and natural sun lights irradiations. *Applied Catalysis B: Environmental, 119*, 197–206. doi:10.1016/j.apcatb.2012.02.034

Fang, J., Fan, H., Ma, Y., Wang, Z., & Chang, Q. (2015). Surface defects control for ZnO nanorods synthesized by quenching and their anti-recombination in photocatalysis. *Applied Surface Science, 332*, 47–54. doi:10.1016/j.apsusc.2015.01.139

Faraday, M. (1857). X. The Bakerian Lecture.—Experimental relations of gold (and other metals) to light. *Philosophical Transactions of the Royal Society of London*, (147): 145–181.

Feng, J., Sun, J., Liu, X., Zhu, J., Xiong, Y., & Tian, S. (2019). Enhancement and mechanism of nano-BaTiO 3 piezo-catalytic degradation of tricyclazole by co-loading Pt and RuO 2. *Environmental Science: Nano*.

Ferroudj, N., Nzimoto, J., Davidson, A., Talbot, D., Briot, E., Dupuis, V., ... Abramson, S. (2013). Maghemite nanoparticles and maghemite/silica nanocomposite microspheres as magnetic Fenton catalysts for the removal of water pollutants. *Applied Catalysis B: Environmental*, *136*, 9–18. doi:10.1016/j.apcatb.2013.01.046

Fujishima, A., Zhang, X., & Tryk, D. A. (2008). TiO2 photocatalysis and related surface phenomena. *Surface Science Reports*, *63*(12), 515–582. doi:10.1016/j.surfrep.2008.10.001

Garrigue, P., Delville, M. H., Labrugère, C., Cloutet, E., Kulesza, P. J., Morand, J. P., & Kuhn, A. (2004). Top– down approach for the preparation of colloidal carbon nanoparticles. *Chemistry of Materials*, *16*(16), 2984–2986. doi:10.1021/cm049685i

Ge, J., Zhang, Q., Zhang, T., & Yin, Y. (2008). Core–satellite nanocomposite catalysts protected by a porous silica shell: Controllable reactivity, high stability, and magnetic recyclability. *Angewandte Chemie International Edition*, *47*(46), 8924–8928. doi:10.1002/anie.200803968 PMID:18925600

Gellman, A. J., & Shukla, N. (2009). Nanocatalysis: More than speed. *Nature Materials*, *8*(2), 87–88. doi:10.1038/nmat2363 PMID:19165207

Glaspell, G., Hassan, H. M., Elzatahry, A., Abdalsayed, V., & El-Shall, M. S. (2008). Nanocatalysis on supported oxides for CO oxidation. *Topics in Catalysis*, *47*(1-2), 22–31. doi:10.100711244-007-9036-1

Grey, D., Garrick, D., Blackmore, D., Kelman, J., Muller, M., & Sadoff, C. (2013). Water security in one blue planet: twenty-first-century policy challenges for science. *Philosophical Transactions of the Royal Society A: Mathematical, Physical and Engineering Sciences, 371*(2002), 20120406.

Guan, S., Hao, L., Lu, Y., Yoshida, H., Pan, F., & Asanuma, H. (2016). Fabrication of oxygen-deficient TiO2 coatings with nano-fiber morphology for visible-light photocatalysis. *Materials Science in Semiconductor Processing*, *41*, 358–363. doi:10.1016/j.mssp.2015.09.018

Habibi, B., & Mohammadyari, S. (2015). Facile synthesis of Pd nanoparticles on nano carbon supports and their application as an electrocatalyst for oxidation of ethanol in alkaline media: The effect of support. *International Journal of Hydrogen Energy*, *40*(34), 10833–10846. doi:10.1016/j.ijhydene.2015.07.021

Haruta, M. (2003). When gold is not noble: Catalysis by nanoparticles. *Chemical Record (New York, N.Y.)*, *3*(2), 75–87. doi:10.1002/tcr.10053 PMID:12731078

Hayat, K., Gondal, M. A., Khaled, M. M., Ahmed, S., & Shemsi, A. M. (2011). Nano ZnO synthesis by modified sol gel method and its application in heterogeneous photocatalytic removal of phenol from water. *Applied Catalysis A, General*, *393*(1-2), 122–129. doi:10.1016/j.apcata.2010.11.032

He, Y., Sutton, N. B., Rijnaarts, H. H., & Langenhoff, A. A. (2016). Degradation of pharmaceuticals in wastewater using immobilized TiO2 photocatalysis under simulated solar irradiation. *Applied Catalysis B: Environmental*, *182*, 132–141. doi:10.1016/j.apcatb.2015.09.015

Hornstein, B. J., & Finke, R. G. (2003). Transition-metal nanocluster catalysts: Scaled-up synthesis, characterization, storage conditions, stability, and catalytic activity before and after storage of polyoxoanion-and tetrabutylammonium-stabilized Ir (0) nanoclusters. *Chemistry of Materials*, *15*(4), 899–909. doi:10.1021/cm020856y

Hu, J., Zhang, P., An, W., Liu, L., Liang, Y., & Cui, W. (2019). In-situ Fe-doped g-C3N4 heterogeneous catalyst via photocatalysis-Fenton reaction with enriched photocatalytic performance for removal of complex wastewater. *Applied Catalysis B: Environmental*, *245*, 130–142. doi:10.1016/j.apcatb.2018.12.029

Hussain, S. T., & Khan, M. B. (2013). *U.S. Patent No. 8,487,147*. Washington, DC: U.S. Patent and Trademark Office.

Illés, E., Mizrahi, A., Marks, V., & Meyerstein, D. (2019). Carbonate-radical-anions, and not hydroxyl radicals, are the products of the Fenton reaction in neutral solutions containing bicarbonate. *Free Radical Biology & Medicine*, *131*, 1–6. doi:10.1016/j.freeradbiomed.2018.11.015 PMID:30458276

Islam, N., & Miyazaki, K. (2007, August). Nanotechnology systems of innovation: investigation of scientific disciplines' fusion trend into nanotech. In *PICMET'07-2007 Portland International Conference on Management of Engineering & Technology* (pp. 2922-2931). IEEE.

Jacoby, M. (2009). *Perfecting solid-catalyst synthesis*. Science & Technology. doi:10.1021/cen-v087n017.p037

Jamal, A., Rahman, M. M., Khan, S. B., Faisal, M., Akhtar, K., Rub, M. A., ... Al-Youbi, A. O. (2012). Cobalt doped antimony oxide nano-particles based chemical sensor and photo-catalyst for environmental pollutants. *Applied Surface Science*, *261*, 52–58. doi:10.1016/j.apsusc.2012.07.066

Jiao, Y., Zhu, M., Feng, C., & Zhang, J. (2013). Hydrothermal preparation of lanthanum-doped titanium dioxide nanocrystals with high photocatalytic activity (English). *Chinese Journal of Catalysis*, (3): 585–592. doi:10.1016/S1872-2067(11)60483-X

Kamarudin, M. Z. F., Kamarudin, S. K., Masdar, M. S., & Daud, W. R. W. (2013). Direct ethanol fuel cells. *International Journal of Hydrogen Energy*, *38*(22), 9438–9453. doi:10.1016/j.ijhydene.2012.07.059

Khallaf, H., Oladeji, I. O., & Chow, L. (2008). Optimization of chemical bath deposited CdS thin films using nitrilotriacetic acid as a complexing agent. *Thin Solid Films*, *516*(18), 5967–5973. doi:10.1016/j.tsf.2007.10.079

Kurian, M., & Nair, D. S. (2015). Heterogeneous Fenton behavior of nano nickel zinc ferrite catalysts in the degradation of 4-chlorophenol from water under neutral conditions. *Journal of Water Process Engineering*, *8*, e37–e49. doi:10.1016/j.jwpe.2014.10.011

Kurian, M., & Nair, D. S. (2016). Effect of preparation conditions on nickel zinc ferrite nanoparticles: A comparison between sol–gel auto combustion and co-precipitation methods. *Journal of Saudi Chemical Society*, *20*, S517–S522. doi:10.1016/j.jscs.2013.03.003

Kurian, M., Nair, D. S., & Rahnamol, A. M. (2014). Influence of the synthesis conditions on the catalytic efficiency of NiFe 2 O 4 and ZnFe 2 O 4 nanoparticles towards the wet peroxide oxidation of 4-chlorophenol. *Reaction Kinetics, Mechanisms and Catalysis*, *111*(2), 591–604. doi:10.100711144-013-0667-x

Kyzas, G. Z., & Matis, K. A. (2015). Nanoadsorbents for pollutants removal: A review. *Journal of Molecular Liquids*, *203*, 159–168. doi:10.1016/j.molliq.2015.01.004

Léger, B., Denicourt-Nowicki, A., Olivier-Bourbigou, H., & Roucoux, A. (2008). Rhodium nanocatalysts stabilized by various bipyridine ligands in nonaqueous ionic liquids: Influence of the bipyridine coordination modes in arene catalytic hydrogenation. *Inorganic Chemistry*, *47*(19), 9090–9096. doi:10.1021/ic8010713 PMID:18781733

Li, W., Zhang, Y., Tian, G., Xie, S., Xu, Q., Wang, L., ... Bu, Y. (2016). Fabrication of graphene-modified nano-sized red phosphorus for enhanced photocatalytic performance. *Journal of Molecular Catalysis A Chemical*, *423*, 356–364. doi:10.1016/j.molcata.2016.07.039

Lin, L., Wang, H., Luo, H., & Xu, P. (2015). Enhanced photocatalysis using side-glowing optical fibers coated with Fe-doped TiO2 nanocomposite thin films. *Journal of Photochemistry and Photobiology A Chemistry*, *307*, 88–98. doi:10.1016/j.jphotochem.2015.04.010

Lin, S. T., Thirumavalavan, M., Jiang, T. Y., & Lee, J. F. (2014). Synthesis of ZnO/Zn nano photocatalyst using modified polysaccharides for photodegradation of dyes. *Carbohydrate Polymers*, *105*, 1–9. doi:10.1016/j.carbpol.2014.01.017 PMID:24708945

Liu, H., Gong, C., Wang, J., Liu, X., Liu, H., Cheng, F., ... Wen, S. (2016). Chitosan/silica coated carbon nanotubes composite proton exchange membranes for fuel cell applications. *Carbohydrate Polymers*, *136*, 1379–1385. doi:10.1016/j.carbpol.2015.09.085 PMID:26572483

Liu, Z., Gan, L. M., Hong, L., Chen, W., & Lee, J. Y. (2005). Carbon-supported Pt nanoparticles as catalysts for proton exchange membrane fuel cells. *Journal of Power Sources*, *139*(1-2), 73–78. doi:10.1016/j.jpowsour.2004.07.012

Lubick, N., & Betts, K. (2008). *Silver socks have cloudy lining| Court bans widely used flame retardant*. Academic Press.

Ma, H., Wang, H., & Na, C. (2015). Microwave-assisted optimization of platinum-nickel nanoalloys for catalytic water treatment. *Applied Catalysis B: Environmental*, *163*, 198–204. doi:10.1016/j.apcatb.2014.07.062

Ma, Z., & Dai, S. (2011). Development of novel supported gold catalysts: A materials perspective. *Nano Research*, *4*(1), 3–32. doi:10.100712274-010-0025-5

Mahajan, D., Gütlich, P., Ensling, J., Pandya, K., Stumm, U., & Vijayaraghavan, P. (2003). Evaluation of nanosized iron in slurry-phase Fischer– Tropsch synthesis. *Energy & Fuels*, *17*(5), 1210–1221. doi:10.1021/ef0300343

Mévellec, V., Roucoux, A., Ramirez, E., Philippot, K., & Chaudret, B. (2004). Surfactant-stabilized aqueous iridium (0) colloidal suspension: An efficient reusable catalyst for hydrogenation of arenes in biphasic media. *Advanced Synthesis & Catalysis*, *346*(1), 72–76. doi:10.1002/adsc.200303157

Mohaghegh, N., Tasviri, M., Rahimi, E., & Gholami, M. R. (2015). Comparative studies on Ag3PO4/BiPO4–metal-organic framework–graphene-based nanocomposites for photocatalysis application. *Applied Surface Science*, *351*, 216–224. doi:10.1016/j.apsusc.2015.05.135

Mushtaq, F., Chen, X., Hoop, M., Torlakcik, H., Pellicer, E., Sort, J., ... Pané, S. (2018). Piezoelectrically Enhanced Photocatalysis with BiFeO3 Nanostructures for Efficient Water Remediation. *iScience, 4*, 236-246.

Nadagouda, M. N., & Varma, R. S. (2006). Green and controlled synthesis of gold and platinum nanomaterials using vitamin B2: Density-assisted self-assembly of nanospheres, wires and rods. *Green Chemistry*, *8*(6), 516–518. doi:10.1039/b601271j

Nazim, S., Kousar, T., Shahid, M., Khan, M. A., Nasar, G., Sher, M., & Warsi, M. F. (2016). New graphene-CoxZn1− xFe2O4 nano-heterostructures: Magnetically separable visible light photocatalytic materials. *Ceramics International*, *42*(6), 7647–7654. doi:10.1016/j.ceramint.2016.01.177

Neyens, E., & Baeyens, J. (2003). A review of classic Fenton's peroxidation as an advanced oxidation technique. *Journal of Hazardous Materials*, *98*(1-3), 33–50. doi:10.1016/S0304-3894(02)00282-0 PMID:12628776

Ni, M., Leung, M. K., Leung, D. Y., & Sumathy, K. (2007). A review and recent developments in photocatalytic water-splitting using TiO2 for hydrogen production. *Renewable & Sustainable Energy Reviews*, *11*(3), 401–425. doi:10.1016/j.rser.2005.01.009

Olveira, S., Forster, S. P., & Seeger, S. (2014). Nanocatalysis: Academic discipline and industrial realities. *Journal of Nanotechnology*.

Ortega-Liebana, M. C., Hueso, J. L., Ferdousi, S., Yeung, K. L., & Santamaria, J. (2016). Nitrogen-doped luminescent carbon nanodots for optimal photo-generation of hydroxyl radicals and visible-light expanded photo-catalysis. *Diamond and Related Materials*, *65*, 176–182. doi:10.1016/j.diamond.2016.03.021

Pacchioni, G. (2009). Nanocatalysis: Staying put. *Nature Materials*, *8*(3), 167–168. doi:10.1038/nmat2394 PMID:19229260

Parsons, S. A., & Jefferson, B. (2006). *Introduction to potable water treatment processes*. Blackwell Publishing. doi:10.1002/9781444305470

Patzke, G. R., Zhou, Y., Kontic, R., & Conrad, F. (2011). Oxide nanomaterials: Synthetic developments, mechanistic studies, and technological innovations. *Angewandte Chemie International Edition*, *50*(4), 826–859. doi:10.1002/anie.201000235 PMID:21246678

Polshettiwar, V., & Varma, R. S. (2010). Green chemistry by nano-catalysis. *Green Chemistry*, *12*(5), 743–754. doi:10.1039/b921171c

Pranjali, G., Deepa, M., & Nair, A. B. (2013). Nanotechnology in waste water treatment: A review. *International Journal of Chemtech Research*, *5*(5), 2303–2308.

Qiu, L., Peng, Y., Liu, B., Lin, B., Peng, Y., Malik, M. J., & Yan, F. (2012). Polypyrrole nanotube-supported gold nanoparticles: An efficient electrocatalyst for oxygen reduction and catalytic reduction of 4-nitrophenol. *Applied Catalysis A, General*, *413*, 230–237. doi:10.1016/j.apcata.2011.11.013

Qu, X., Alvarez, P. J., & Li, Q. (2013). Applications of nanotechnology in water and wastewater treatment. *Water Research*, *47*(12), 3931–3946. doi:10.1016/j.watres.2012.09.058 PMID:23571110

Qu, X., Brame, J., Li, Q., & Alvarez, P. J. (2012). Nanotechnology for a safe and sustainable water supply: Enabling integrated water treatment and reuse. *Accounts of Chemical Research*, *46*(3), 834–843. doi:10.1021/ar300029v PMID:22738389

Ramgopal, G., Vidya, Y. S., Anantharaju, K. S., Prasad, B. D., Sharma, S. C., Prashantha, S. C., ... Nagabhushana, H. (2015). Bio-inspired synthesis of Y2O3: Eu3+ red nanophosphor for eco-friendly photocatalysis. *Spectrochimica Acta. Part A: Molecular and Biomolecular Spectroscopy*, *141*, 149–160. doi:10.1016/j.saa.2015.01.055

Reddy, S. R., Saritha, V., Karnena, M. K., & Dwarapureddi, B. K. (2017). Combined SBR and RO pilot scale treatment for pharmaceutical wastewater. *Desalination and Water Treatment*, *98*, 45–51. doi:10.5004/dwt.2017.21579

Roucoux, A., Schulz, J., & Patin, H. (2002). Reduced transition metal colloids: A novel family of reusable catalysts? *Chemical Reviews*, *102*(10), 3757–3778. doi:10.1021/cr010350j PMID:12371901

Samsudin, E. M., Goh, S. N., Wu, T. Y., Ling, T. T., Hamid, S. B. A., & Juan, J. C. (2015). Evaluation on the photocatalytic degradation activity of reactive blue 4 using pure anatase nano-TiO2. *Sains Malaysiana*, *44*(7), 1011–1019. doi:10.17576/jsm-2015-4407-13

Sathishkumar, P., Sweena, R., Wu, J. J., & Anandan, S. (2011). Synthesis of CuO-ZnO nanophotocatalyst for visible light assisted degradation of a textile dye in aqueous solution. *Chemical Engineering Journal*, *171*(1), 136–140. doi:10.1016/j.cej.2011.03.074

Savadogo, O., Lee, K., Oishi, K., Mitsushima, S., Kamiya, N., & Ota, K. I. (2004). New palladium alloys catalyst for the oxygen reduction reaction in an acid medium. *Electrochemistry Communications*, *6*(2), 105–109. doi:10.1016/j.elecom.2003.10.020

Schätz, A., Long, T. R., Grass, R. N., Stark, W. J., Hanson, P. R., & Reiser, O. (2010). Immobilization on a nanomagnetic Co/C surface using ROM polymerization: Generation of a hybrid material as support for a recyclable palladium catalyst. *Advanced Functional Materials*, *20*(24), 4323–4328. doi:10.1002/adfm.201000959 PMID:22308111

Schmidt, F. (2001). New catalyst preparation technologies—Observed from an industrial viewpoint. *Applied Catalysis A, General*, *221*(1-2), 15–21. doi:10.1016/S0926-860X(01)00802-X

Schwarzenbach, R. P., Escher, B. I., Fenner, K., Hofstetter, T. B., Johnson, C. A., Von Gunten, U., & Wehrli, B. (2006). The challenge of micropollutants in aquatic systems. *Science*, *313*(5790), 1072–1077. doi:10.1126cience.1127291 PMID:16931750

Shamsizadeh, A., Ghaedi, M., Ansari, A., Azizian, S., & Purkait, M. K. (2014). Tin oxide nanoparticle loaded on activated carbon as new adsorbent for efficient removal of malachite green-oxalate: Non-linear kinetics and isotherm study. *Journal of Molecular Liquids*, *195*, 212–218. doi:10.1016/j.molliq.2014.02.035

Somorjai, G. A., & Borodko, Y. G. (2001). Research in nanosciences–great opportunity for catalysis science. *Catalysis Letters*, *76*(1-2), 1–5. doi:10.1023/A:1016711323302

Stowell, C. A., & Korgel, B. A. (2005). Iridium nanocrystal synthesis and surface coating-dependent catalytic activity. *Nano Letters*, *5*(7), 1203–1207. doi:10.1021/nl050648f PMID:16178211

Tang, H., Yu, C. H., Oduoro, W., He, H., & Tsang, S. C. (2008). Engineering of a monodisperse core–shell magnetic Ti– O– Si oxidation nanocatalyst. *Langmuir, 24*(5), 1587–1590. doi:10.1021/la702489s PMID:18225927

Tang, X., Zhang, Q., Liu, Z., Pan, K., Dong, Y., & Li, Y. (2014). Removal of Cu (II) by loofah fibers as a natural and low-cost adsorbent from aqueous solutions. *Journal of Molecular Liquids, 199*, 401–407. doi:10.1016/j.molliq.2014.09.033

Teunissen, W., Bol, A. A., & Geus, J. W. (1999). Magnetic catalyst bodies. *Catalysis Today, 48*(1-4), 329–336. doi:10.1016/S0920-5861(98)00389-7

Toshima, N., & Yonezawa, T. (1998). Bimetallic nanoparticles—Novel materials for chemical and physical applications. *New Journal of Chemistry, 22*(11), 1179–1201. doi:10.1039/a805753b

Tristao, J. C., Magalhães, F., Corio, P., & Sansiviero, M. T. C. (2006). Electronic characterization and photocatalytic properties of CdS/TiO2 semiconductor composite. *Journal of Photochemistry and Photobiology A Chemistry, 181*(2-3), 152–157. doi:10.1016/j.jphotochem.2005.11.018

Van Leeuwen, P. W. (2006). *Homogeneous catalysis: understanding the art*. Springer Science & Business Media.

Weng, X., Huang, L., Chen, Z., Megharaj, M., & Naidu, R. (2013). Synthesis of iron-based nanoparticles by green tea extract and their degradation of malachite. *Industrial Crops and Products, 51*, 342–347. doi:10.1016/j.indcrop.2013.09.024

WHO. (2015). *Drinking-water: Fact sheet No. 391*. Retrieved from https://www.who.int/mediacentre/factsheets/fs391/en/

Widegren, J. A., & Finke, R. G. (2003). A review of the problem of distinguishing true homogeneous catalysis from soluble or other metal-particle heterogeneous catalysis under reducing conditions. *Journal of Molecular Catalysis A Chemical, 198*(1-2), 317–341. doi:10.1016/S1381-1169(02)00728-8

Xia, M., Chen, C., Long, M., Chen, C., Cai, W., & Zhou, B. (2011). Magnetically separable mesoporous silica nanocomposite and its application in Fenton catalysis. *Microporous and Mesoporous Materials, 145*(1-3), 217–223. doi:10.1016/j.micromeso.2011.05.017

Xia, Y., Xu, L., Peng, J., Han, J., Guo, S., Zhang, L., ... Komarneni, S. (2019). TiO2@ g-C3N4 core/shell spheres with uniform mesoporous structures for high performance visible-light photocatalytic application. *Ceramics International, 45*(15), 18844–18851. doi:10.1016/j.ceramint.2019.06.118

Yan, N., Yuan, Y., & Dyson, P. J. (2011). Rhodium nanoparticle catalysts stabilized with a polymer that enhances stability without compromising activity. *Chemical Communications, 47*(9), 2529–2531. doi:10.1039/c0cc04641h PMID:21225042

Yao, Y., Lu, F., Zhu, Y., Wei, F., Liu, X., Lian, C., & Wang, S. (2015). Magnetic core–shell CuFe2O4@ C3N4 hybrids for visible light photocatalysis of Orange II. *Journal of Hazardous Materials, 297*, 224–233. doi:10.1016/j.jhazmat.2015.04.046 PMID:25974659

Yilmaz, B., & Müller, U. (2009). Catalytic applications of zeolites in chemical industry. *Topics in Catalysis, 52*(6-7), 888–895. doi:10.100711244-009-9226-0

Yoon, T. J., Lee, W., Oh, Y. S., & Lee, J. K. (2003). Magnetic nanoparticles as a catalyst vehicle for simple and easy recycling. *New Journal of Chemistry*, *27*(2), 227–229. doi:10.1039/b209391j

Zäch, M., Hägglund, C., Chakarov, D., & Kasemo, B. (2006). Nanoscience and nanotechnology for advanced energy systems. *Current Opinion in Solid State and Materials Science*, *10*(3-4), 132–143. doi:10.1016/j.cossms.2007.04.004

Zelmanov, G., & Semiat, R. (2008). Phenol oxidation kinetics in water solution using iron (3)-oxide-based nano-catalysts. *Water Research, 42*(14), 3848-3856.

Zhang, J., Worley, J., Dénommée, S., Kingston, C., Jakubek, Z. J., Deslandes, Y., ... Botton, G. A. (2003). Synthesis of metal alloy nanoparticles in solution by laser irradiation of a metal powder suspension. *The Journal of Physical Chemistry B*, *107*(29), 6920–6923. doi:10.1021/jp027269k

Zhang, Q., Xu, R., Xu, P., Chen, R., He, Q., Zhong, J., & Gu, X. (2014). a. Performance study of ZrO2 ceramic micro-filtration membranes used in pretreatment of DMF wastewater. *Desalination*, *346*, 1–8. doi:10.1016/j.desal.2014.05.006

Zhang, Y., Yan, L., Xu, W., Guo, X., Cui, L., Gao, L., ... Du, B. (2014). b. Adsorption of Pb (II) and Hg (II) from aqueous solution using magnetic CoFe2O4-reduced graphene oxide. *Journal of Molecular Liquids*, *191*, 177–182. doi:10.1016/j.molliq.2013.12.015

Zhou, B., Balee, R., & Groenendaal, R. (2005). Nanoparticle and nanostructure catalysts: Technologies and markets. *Nanotech. L. & Bus.*, *2*, 222.

Zhou, W., Zhou, Z., Song, S., Li, W., Sun, G., Tsiakaras, P., & Xin, Q. (2003). Pt based anode catalysts for direct ethanol fuel cells. *Applied Catalysis B: Environmental*, *46*(2), 273–285. doi:10.1016/S0926-3373(03)00218-2

Zhu, H., Jiang, R., Xiao, L., Chang, Y., Guan, Y., Li, X., & Zeng, G. (2009). Photocatalytic decolorization and degradation of Congo Red on innovative crosslinked chitosan/nano-CdS composite catalyst under visible light irradiation. *Journal of Hazardous Materials*, *169*(1-3), 933–940. doi:10.1016/j.jhazmat.2009.04.037 PMID:19477069

Chapter 12
Synthesis and Characterisation of Nano-Biosorbents and Their Applications for Waste Water Treatment

Virendra Kumar Yadav
Sankalchand Patel University, India

Nisha Choudhary
https://orcid.org/0000-0002-1847-6520
Central University of Gujarat, India

Samreen Heena Khan
Central University of Gujarat, India

Parth Malik
Central University of Gujarat, India

Gajendra Kumar Inwati
Kadi University, India

R. Suriyaprabha
Central University of Gujarat, India

Raman Kumar Ravi
Central University of Gujarat, India

ABSTRACT

Nanotechnology is one of the most reliable techniques for the remediation of heavy metals. As nanoparticles have a higher surface area to volume, ratio, and high surface energies, so nano-based absorbents are very efficient. Adsorption technique is the most preferred for the remediation of wastewater pollutants. In the current study, a comparative study was done between bio sorbents, nanosorbents and bio nanosorbents. The chapter studies with the synthesis and characterization of the bio sorbents, bionanosorbents, their mechanism of sorption, their synthesis, in addition, application for the remediation of heavy metals from wastewater. The fly ash is an industrial byproduct. Biosorbents have immense applications in the field of bioremediation of heavy metals. Further, their components have also enhanced removal efficiency from the wastewater.

DOI: 10.4018/978-1-7998-1241-8.ch012

BACKGROUND

Nanoscience and Nanotechnology is a developing field and due to recent significant progress, the field has increasing number of applications in practically all areas. The adsorption based on nanoparticles (NPs) is the most efficient and reliable technique for the remediation of wastewater pollutants. The recovery of natural nanosorbents for the remediation of heavy metals makes the process eco-friendly and cost effective while the cost of nanoadsorbents also is reduced. The utilization of natural nanosorbents and bio sorbents reduces the pollution present in the form of solid waste. The waste materials, which have high potential as a biosorbents or nanosorbents, are sugarcane bagasse, rice husk, fly ash, and other industrial products like spent microbes etc. These bio and nanosorbents finds applications in the field of wastewater treatment for the removal of heavy metals and organic contaminants. But prior to that there application as an adsorbent depends on their morphology and chemical properties which can be revealed by the analysis by instruments. Here, different agro-industrial waste has been reported for the synthesis of bio and nanosorbents as an economical and eco-friendly source material.

INTRODUCTION

The major reason for the contamination of water is rapid urbanization and industrialization by the anthropogenic behaviour due to population explosion. (Khatri & Tyagi, 2015). For such a massive population there is need of more and more industries like metal plating (Venkateswaran, Vellaichamy, & Palanivelu, 2007), mining operations (Panayotova, 2016), surface finishing industry, tanneries (Qureshi, 2015), paper and pulp industries (Pöykiö, Nurmesniemi, & Keiski, 2007), color-alkali, fertilizer and pesticide industry (Marković, Radmanović, Đurović-Pejčev, Milinovic, & Kljajić, 2009), radiator manufacturing, smelting, energy and fuel production, aerospace and atomic energy installation, alloy industries, electroplating, and battery industries to fulfill their necessities (s. k. Das, Grewal, & Banerjee, 2011; Sato, Shimosato, & Klinman, 2018). Most of these industries produces heavy metals like zinc (Zn), copper (Cu), nickel (Ni), mercury (Hg), cadmium (Cd), lead (Pb), and chromium (Cr) (Jaishankar, Tseten, Anbalagan, Mathew, & Beeregowda, 2014) which leads to the water pollution. The continuous discharge of heavy metals into the water bodies has led to the abandon of several industries in various parts of the globe. The presence of these impurities in water results in an increased color, turbidity, chemical oxygen demand and several other parameters (Saritha, Srinivas, & Srikanth Vuppala, 2017). Therefore, there is an immediate requirement of cost effective, eco-friendly, efficient technique for the remediation of pollutants from such contaminated water. However, there are numerous conventional approaches for the remediation of pollutants from wastewater like precipitation (by precipitating down the heavy metals), coagulation (eliminates the heavy metals by forming a coagulation), membrane filtration (heavy metals are removed by using filters having specific pose size), ion exchange (eliminates heavy metals by exchanging ion), evaporation, absorption (heavy metals are eliminated by absorbing them), and adsorption (heavy metals are removed by adsorbing them on the surface of adsorbents) (Krishna, Kumar Chintalpudi, & Muddada, 2018). Out of which absorption is the most preferred techniques because of their simplicity of operation, handiness, less area requirement and tendency to work with diverse pollutants (Hanida, 2018). The absorption process is being applied in sensors, for the detection of gases and other molecules, in chromatography, in optical devices, in catalytic reactions, in industries, filters in wastewater treatment (Vanrolleghem & Lee, 2003). If the absorbents are recovered from the waste products of industries then it is more advantageous as it will not only reduce the cost of the absorbent and absorption mechanism but also minimize the pollution. Moreover, waste derived absorbents make the process economical and nonhazardous, as the absorbents are biodegradable. The absorbents obtained from the waste products have large number of functional groups on their surface, which further makes them more effective and efficient for the remediation of the pollutants (Güngen, Aydemir, Çoban, Düzenli, & Tasdemir, 2016); (Barakat, 2011). If the product is derived from the agricultural waste, then it has ketones, alcohols, carboxyl, carbonyl, aldehydes, amines, phenols, ethers and several other groups on their surface (Tojo & I. Fernandez, 2006). All these different groups have different mechanism of binding of pollutants and ultimately higher removal of pollutants from the wastewater at one go.

However, process of adsorption and ion exchange that comes under technique of sorption is possible substitute for wastewater purification. Adsorption is a process where a substance (adsorbate) attaches (adsorbs) to another substance (adsorbent) by physical and chemical interactions (Cerofolini, Meda, & Bandosz, 1999). The sorbate must diffuse from water or gas phases into the sorbent surface and frequently into the internal pores of sorbent. Removal of heavy metal ions from water with the help of wide use of chelating resins is possible due to their properties of high adsorption capacities, selectivity, and durability (Sartore & Dey, 2019). It combines ion exchange and complex formation in its separation process and shows great advances in extraction when compared to conventional resins.

While the other above-mentioned techniques have drawbacks of expensiveness, high maintenance and operational cost, generation of toxic products like sludge and complex mechanism. The chemical precipitation and electrochemical treatment are ineffective, produce a large quantity of sludge, and require further treatment (K. Wang, A. Vaccari, Li, & Shammas, 2007). The ion exchange, membrane technologies, and activated carbon adsorption process are extremely expensive, which are utilized in the treatment of water pollution. For the past few years, numerous novel methods have been examined for heavy metal ion removal successfully (Bolisetty, Peydayesh, & Mezzenga, 2019).

With the passage of time, there is always a need of more reliable, efficient, economical and eco-friendly techniques in order to prevent our environment from the risk associated with it. So, now the researchers have switched on to the biosorbents which are bio degradable, eco-friendly and biocompatible. Therefore, there is increase in the biosorbent-based remediation of pollutants from the wastewater in the last few decades (Dragan, Dinu, & Shankar, 2015). The major drawback of biosorbents are that they are larger i.e. in microns, which have low adsorption efficiency. However, nowadays nanotechnology has played a drastic changes in this field due to their high surface area, high surface energies etc. Nanotechnology is applied in every field of sciences whether it is electronics, drug-delivery and environmental cleanup (Khan, Saeed, & Khan, 2017). Because of these properties, they have gained huge attention in the recent years and is being used as a potential candidate for the remediation of pollutants from the wastewater. The research is continuously focusing on the development of novel and smart materials based on the cheaper materials especially derived from the waste materials.

BIO SORBENTS, BIOSORPTION AND ITS MECHANISM

Biosorbents (El-Wakeel, Moghazy, Labena, & Husien, 2019) are the material, which are derived from the biological particles like plants, agricultural wastes, animal waste etc. and are applied for the absorption of pollutants from the aqueous medium. The biosorbents contains huge number of functional groups and binding sites on their surface. Some of the common groups, which are present, are imidazole, carboxyl, imidazole, sulphydryl, amino, phosphate, sulfate, thioether, phenol, carbonyl, amide and hydroxyl moieties (Yu et al., 2016). Broadly, based on their origin, the biosorbents can be classified into three major categories: plant based, animal based, or microorganisms based or industrial waste (Sulaymon, 2014). The plant originated materials like dried powders of leaves, fruits, peels, seeds, bark, stem, roots, husk, hull, skin, shell, bran, and stone could be used as biosorbents (P. Das, Chowdhury, Gupta, Kumar, & Kumar, 2010). Besides this, plants fluids like resins, gums and other adhesive can also be used as a biosorbents. All these waste products or agriculturally waste products (AWBs) have high content of plant polysaccharides like chitin, lignin, cellulose, hemicellulose, and others. All these materials have high hydroxyl groups in their structure, which have an important role in the sorption process. Moreover, various plants extracts can be used for the coating or capping material for the conventional adsorbents and can be considered as biosorbents. Some of the common plants products or waste materials are lemon peel, coconut shell, almond shell, groundnut shell, banana peel; tea leaves waste, which are frequently used as a biosorbents (Ahmad Khan, Ibrahim, & Subramaniam, 2004). Besides this, there are various agricultural waste materials that are used, as biosorbents are rice husk ash, sugarcane bagasse, corn hub ash etc. Every year a million tonnes of such waste are produced either in the kitchen, or in agricultural fields. These materials are generally disposed of into the environment, but if these are processed properly then it could prove to be a valuable source of biosorbents.

The biosorption is a simple metabolically passive physicochemical process (physical adsorption, ion exchange, chelation, complexation, and micro-precipitation) that comprises a solid phase (sorbent or biosorbent; biological material) and an aqueous phase (solvent; usually water) containing dissolved species to be sorbed (sorbate, metal ions) (Naja, Murphy, & Volesky, 2010); (Patel, Sv, Shah, & Verma, 2018). More specifically, an adsorption technique involves the biologically derived adsorbate (commonly called biosorbents). It is also called as bioremediation, which involve the detoxification of hazardous substances instead of transferring them from one medium to another by means of microbes and plants. It is useful for the removal of most of the pollutants in their diluted form from the aqueous media. It utilizes, agro-waste or plant derived products, microbes, industrial waste and biopolymers. It is a reversible rapid mechanism involved in binding of ions onto the functional groups present on the surface of the biosorbents in liquid medium by means of numerous interactions (Javanbakht, Amir Alavi, & Zilouei, 2014).

For the development of basic scientific understanding of the biosorption, process researchers have proposed various biosorption mechanisms. Yet, large-scale commercial applications of biosorption method need to be explored widely. There are numerous factors that affects the biosorption are affinity of sorbent for sorbate, intrinsic adsorption and columbic interaction (Dostálek, 1970). The intrinsic adsorption is surface phenomenon. Therefore, the surface area is a major factor while in columbic there is consideration charges on the surface by the various molecules (Abia, 2014). The biosorption mechanism is a complex process that involves the binding of sorbate onto the biosorbent. Various naturally derived biosorbents materials could be utilized as biosorbents which involve the binding of metal ions by physical (electrostatic interaction or van der Waals forces) or chemical (displacement of either bound metal cations (ion exchange) or protons) binding, chelation, reduction, precipitation, and complexation (refer Figure 1). Biosorbents have different functional groups like amine, amide, imidazole, thioether, sulfonate, carbonyl, sulfhydryl, carboxyl, phosphodiester, phenolic, imine, and phosphate groups that can attract and sequester metal ions (Park, Yun, & Park, 2010).

The key factors controlling and characterizing these mechanisms are

Figure 1. Different hypothesis for the biosorption of heavy metals
Source: *adapted from Sri Lakshmi Ramya Krishna et al., 2018*

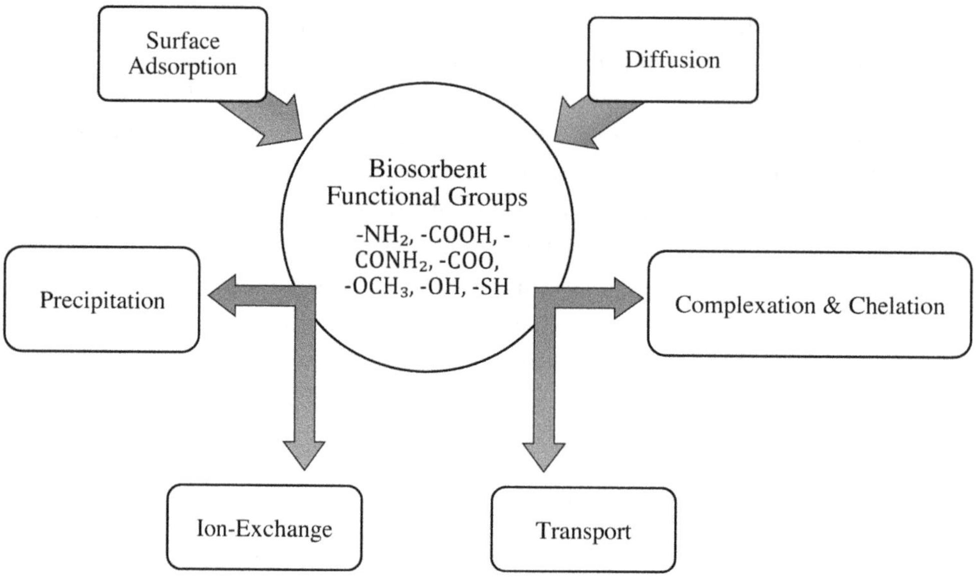

- The chemical, stereochemical, and coordination features of metal ions like molecular weight, ionic radius, and oxidation state of the targeted metal species;
- Features of biosorbent, for instance, the structure and nature (in case of microorganism—living/non-living);
- Sorts of binding sites available on the surface (biological ligand)
- The process parameters like pH, temperature, strength of sorbate and sorbent, and other competing metal ions; and
- Availability of the binding sites. The combined effects of the above parameters influence the metal speciation (the formation of new forms of metal as a result of biosorption) (Debnath, Chowdhury, & Ghosh, 2018).

Types of Biosorbents

The sources of biosorbents could be variable and it is also preferable to have a biosorbent from a wide spectrum of available materials. However, there are no compulsory criteria to have a biosorbent but the plants, animal, microbes, agricultural waste or industrial waste and various by products of industries can be considered as biosorbents. The desired characteristics of an ideal biosorbent are: that it should have higher for metals (biosorption capacity), cost effective, should be available in large quantities, should get easily desorped from the biosorbents surface and should be regenerative (Kotrba, Mackova, & Macek, 2011).

It could be derived and used directly or might me processed by acids and alkali for the surface modification and ultimately the efficiency of metal removal (J. Wang & Chen, 2009). Though the biosorbents could be synthesized in the laboratory but they are not only expensive and energy intensive, but also

Figure 2. Major classes of biosorbents

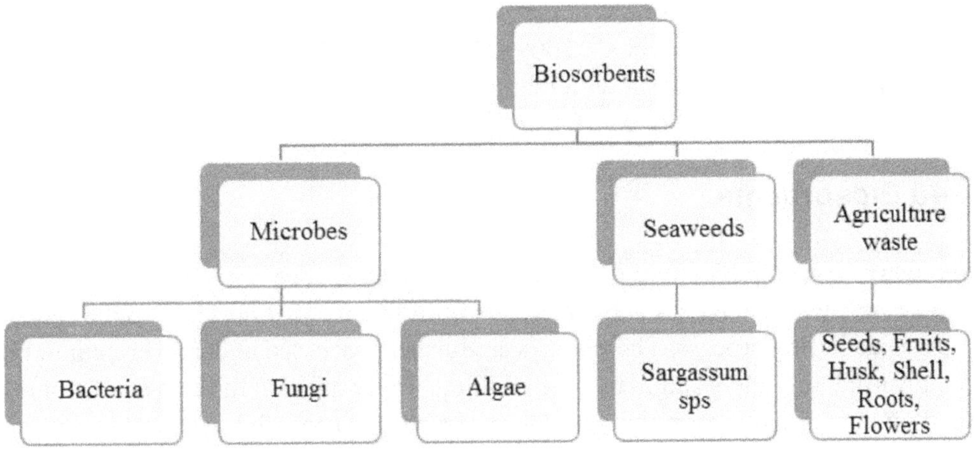

have issue of biodegradability and biocompatibility. The synthesized nanosorbents have to be chemically functionalize for the targeted absorption of pollutants. The surface functionalization of the synthesized makes it a tedious source as a biosorbents. Therefore, the biosorbents derived from the natural waste materials like agro wastes overcomes all the drawbacks of synthesized biosorbents. The major types of biosorbents are given below in the diagram.

Agricultural wastes especially those with high percentage of cellulose and lignin contains polar functional groups like amino, carbonyl, alcoholic, phenolic, and ether groups having high potential for metal binding (J. Wang & Chen, 2009). These groups donate a lone pair of electrons and form complexes with metal ions in the solution. Due to their unique chemical composition (the presence of hemicellulose, lipids, lignin, water hydrocarbons, simple sugars, and starch having a variety of functional groups) and availability, the use of agro-wastes seems to be a viable option for heavy metal remediation. Here the use of different agro waste based materials as biosorbents is explained in detail.

Shell as a Biosorbents

The fruits like coconut, almond, cashew nut, peacan nut, groundnut, etc. have a very hard covering, which are meant for the protection of the fruits. However, these have organic compositions but their composition may vary from plant to plant. Earlier different workers have reported the utilization of biosorbents from such fruits and their application in the remediation of pollutants from the wastewater. One such example is the utilization of biochar from the pyrolysis of almond shell by Kilic et al. for the removal of Ni (II) and Co (II) heavy metal ions (Pagnanelli, Mainelli, Veglio, & Toro, 2003);(Kilic, Kırbıyık, Çepelioğullar Mutlu, & E. Pütün, 2013). In another experiment conducted by Ho and Ofomaja, used the coconut shell as biosorbents for the removal of cadmium ions from aqueous solution (huamán, Maria Souza de Mesquita, Torem, & Adolfo Saavedra Pinto, 2006). One more experiment where Vaghetti et al used the peacan nut (Caryaillinoensis) shell as biosorbnets for the removal of Cr(III), Fe(III), and Zn(II) metal ions from aqueous solutions (C. P. Vaghetti et al., 2009). Further, they tried to study the effect of various parameters like absorbent dosage, pH, and time on the removal efficiency of the considered heavy metal ions. For the removal of Cr(III), Fe(III), and Zn(II) metal ions, the maximum biosorption capacity of pecan nutshell was 561 93.01, 76.59, and 107.9 mg/g, respectively. Further a team of workers lead by Senthil., treated the cashew nut shell by acid, for the removal of Cu(II), Cd(II), Zn(II), and Ni(II) ions from aqueous solution (C. P. Vaghetti et al., 2009); (Darweesh, 2018). The adsorption isotherm studies followed Langmuir adsorption isotherm with maximum adsorption capacity of 406.6 mg/g for Cu (II), 436.7 mg/g for Cd (II), 455.7 mg/g for Zn (II), and 456.3 mg/g for Ni (II).

Husk Based Biosorbents

Husks are the outer covering of some of the fruits like coconut, corn, or it may be also present in the cereal grains like rice, wheat and barley. Every year a million tonnes of husks from the cereal grains are obtained from the dehulling process. These husks are further processed and converted into fine powder, which are mainly used as fodders and in fuel for cooking. These are also the rich source of organic molecules like carbohydrates and other polymers. These can be also used directly as a biosorbent or may be further functionalized with chemical for the enhanced efficiency. Again, the developed biosorbents are biocompatible, biodegradable and eco-friendly. Some of the most widely used husk based biosorbents for the remediation of the wastewater are rice husk, wheat husk, barley husk. Rice husk is also a rich source of silica, which is about 70-90% so it is also used for the synthesis of silica nanoparticles. The husk is firstly treated with phosphoric acid to remove the most of organic contents and then further calcinated at higher temperature to obtain the ash, which are treated with strong alkali and finally silica, is obtained by sol-gel technique. While few workers have also reported, the Deodar saw dust as a biosorbents for the removal of pollutants from the wastewater. Malik et al. utilized coconut husk, i.e., unmodified coconut (*Cocos nucifera L.*) husk for the removal of Ni (II), Pb (II), Cu (II), and Zn (II) (Malik, Dahiya, & Lata, 2017). The experimental studies presented maximum adsorption at 443.0 mg/g (88.6%) for Cu (II), for Ni (II) with 404.5 mg/g (80.9%), 362.2 mg/g (72.4%) for Pb (II), and 338.0 mg/g (67.6%) for Zn (II) ion. Quantum-based theoretical study presented the rich electron donation sites of coconut husk as the main sites for the metal ions uptake. In another study, Krishnani et al. reported biosorption mechanism of nine different heavy metals onto bio-matrix from rice husk (Krishnani, Meng, Christodoulatos, & Boddu, 2008). In this study, rice husk was subjected to 1.5% alkali treatment and used for zinc, copper, cadmium, nickel, lead, manganese, cobalt, mercury, and chromium removal from single ion and mixed solutions. The order of biosorption efficiency was found as Ni (II) < Zn (II) < Cd (II) < Mn(II) < Co(II) < Cu(II) < Hg(II) < Pb(II) at 32 $\pm$ 0.5 between pH range of 5.5 and 6.0 $\pm$ 0.1. However, in other study, the pretreatment of sugarcane bagasse using NaOH and HCl had no significant effect on mercury biosorption capacity (Khoramzadeh, Nasernejad, & Halladj, 2013).

Wheat/Barley Saw Straw Dust

Every year a million tonnes of saw and straw dust are generated during the processing of the rice, wheat, barley and many others from the agriculture in the whole world. In most of the countries, these are mainly used for the burning as a fuel in low-income countries or as a feed material for the domestic animals. Earlier severe workers have reported the utilization of sawdusts from these materials as a biosorbnets for the remediation of pollutants from the wastewater. Pehlivan et al. reported the adsorption of copper, using esterified barley straw which was thermo chemically modified with citric acid (Khoramzadeh et al., 2013). Increasing the temperature, improved the reaction efficiency, but led to lower carboxyl content and increased crosslinking of modified barley straw. Increasing citric acid concentration enhanced free carboxyl groups on the biomass matrix. Miretzky and Cirelli reported that alkali treatments in comparison with acidic ones at the same conditions were more effective on metal ion removal by solving the cell wall matrix (Miretzky & Cirelli, 2009). Therefore, alkali treatments could result in better diffusion through walls and make the functional groups densest and thermodynamically more stable. The increase in Cd (II) and Cu (II) uptake on wheat straw with temperature was attributed to the increased available active sites on the surface of the adsorbent as a result of the opening of the cellulose fibers of wheat straw in a warmer solution. Asadi et al., Bulut and Tez, and Prado et al. have reported the adsorption study on sawdust (K Akunwa, Muhammad, & Akunna, 2014). It has been identified that phenolic, hydroxyl, and carboxylic functional groups of sawdust are responsible for heavy metal uptake, as heavy metal ions could accumulate in the secondary septum of wood in which the amount of lignin is very low. Palumbo et al. found that metal adsorption onto sorbent is possibly a passive binding. They highlighted the effect of pH and natural organic matter on Zn removal (Palumbo, J Daughney, Slade, & N Glover, 2013).

Šciban et al., investigated heavy metal biosorption capacity of sawdust in synthetic and real cable factory wastewater treatment (Fu & Wang, 2011). Heavy metal ion adsorption was inñuenced by the existence of other ions and organic materials in real wastewater through three phenomena, namely, synergism, antagonism, and non-interaction. This study showed that copper was better adsorbed from wastewater with multiple heavy metal ions from wastewater containing single Cu (II) ion. Whereas, Cd (III) adsorption was inhibited by other metal ions and Zn (II) removal were unaffected. As expected, some metal ions have better affinity than other ions towards lignocellulosic biosorbents due to the selectivity potential of functional groups. Bulut and Tez reported that Ni, Cd, and Pb biosorption on walnut tree sawdust was favorable at higher temperatures (Bulut & Tez, 2003). The randomness at solid–solution interface increased by temperature leading to enhanced adsorption at higher temperatures. This may be related to adsorption surface activation and/or pore size enlargement. Reverse trend was obtained for Pb biosorption on different types of sawdust as an exothermic process.

Farooq et al. utilized an easily available biosorbent material, i.e., powdered straw from *Triticum aestivum* for Cd (II) ions uptake (Farooq, Khan, & Athar, 2007). The authors, modified the straw with urea by using microwave radiation, which increased the surface area of the adsorbent. Hydroxyl, carboxyl, and amino functional groups were found to be involved in the adsorption of Cd (II) ions uptake. Mishra et al. carried out the biosorption of Zn (II) using Cedrus deodara sawdust (CDS) (Venkateswarlu, Vijaya Durga, Chitti Babu, & Venkateswara Rao, 2008). The adsorbent had more porous, non-crystalline, and heterogeneous surface prior to the sorption of Zn (II) ions. The adsorbent presented maximum uptake adsorption capacity of 97.39 mg/g at pH 5, temperature 45 °C, the initial concentration of Zn (II) ion 100 mg/L, contact time of 150 min, and agitation rate 160 rpm. Equilibrium kinetic studies favored chemosorptive forces of attraction for the sorption of Zn (II) ion on the surface of CDS. Bangham's equation and film diffusivity showed comparatively higher sorption rate of Zn(II) ion in the early phase of contact time whereas, at the later stage of contact time intraparticle or pore diffusion of Zn(II) ion inside the pores of CDS was found to be the rate limiting step. Sawdust can be used as a low-cost adsorbent to remove heavy metal ions from wastewater. For the removal of heavy metals such as Cr, Ni, and Cd, the adsorption potential of agricultural residues such as rice straw, wheat straw, and Salvinia plant biomass have been studied by performing batch adsorption experiments (Dhir & Kumar, 2010). The removal efficiency was increased at low metal concentration (35 mg/L) of toxic metal ions. Salvinia biomass showed advanced efficiency among the combination of three materials, i.e., rice straw, wheat straw, and Salvinia biomass, which were used for the removal of heavy metals such as Cr, Ni and Cd. El-Said et al. demonstrated the adsorption of Zn (II) and Se (IV) from their aqueous solution using rice husk ash, an agricultural waste, as an adsorbent (El-Shafey, 2009). The Zn (II) ions showed higher affinity than that for Se (IV) ions by rice husk ash biosorbent. Natural bamboo sawdust can also be an efficient Cu (II) ion adsorbent.

Fruit Based Biosorbents

There are several reports by the workers where the fruit based biosorbents has been used for the remediation of heavy metals from wastewater. Daneshfozoun et al. loaded the magnetic nanoparticles on oil palm fruit bunch fibres, cellulose and Ceiba pentandra for the removal of Pb(II), Cu(II), Zn(II), Mn(II), and Ni(II) from the aqueous solutions (Daneshfozoun, Abdullah, & Abdullah, 2017). In another work reported by Schiewer and Patil utilized the pectin-rich fruit wastes as biosorbents for the removal of Cd ions from the wastewater and concluded that the uptake of Cd ions decreases by decreasing the pH due to the competition of H^+ ions for binding to acidic sites (Pathak, Mandavgane, & Kulkarni, 2015). In another study, the removal of Cd (II) metal was investigated by using pectin by Schiewer and Iqbal in citrus peels, native orange peels, protonated peels, depectinated peels, and extracted pectic acid and the equilibrium was accomplished within 1 hour only (Schiewer & Iqbal, 2010). It was found that –COOH of pectin were a major contributor to the surface charge of the negative surface charge and the optimum pH for metal binding experiments was at 5. The orange peels are the major waste of fruit juices industry and the further chemical treatment makes them a suitable candidate for the heavy metal due to negative zeta potential and high physical stability and surface activity. Feng et al modified the surface of orange peel by NaOH and $CaCl_2$ for the removal of Cu (II), Pb(II), and Zn(II) where the maximum adsorption capacities for Cu (II), Pb (II), and Zn (II) of 70.73, 209.8, and 56.18 mg/g, respectively (Feng, Guo, & Liang, 2010). The column based absorption study revealed that the removal of metal takes place by means of ion exchange process, which involves neutralization of -COOH of the pectin by Ca (II) ions. Shakoor and Nasar utilized *cucumis sativus* peel for the removal of methylene blue from simulated wastewater (Shakoor & Nasar, 2016).

Flower as Biosorbents

The applications of biosorbents for the removal of heavy metal ions and other organic and inorganic micro pollutants are known all around the world. Haq et al. removed the Pb (II) and Co (II) from aqueous solutions by using flower of Rosa grussan teplitz (Red Rose) as a biosorbents (Bhatti, Khadim, & Hanif, 2019). Further Shrma et al did slight modification in it and obtained a better result than the Haq et al. Similarly, Manzoor et al. treated the Rosa bourbonia waste with organic acids, used their potential for the uptake of Pb (II) and Cu (II) from aqueous media, and concluded that the organic acid based pretreatment caused reduction in the efficiency of metal absorption (Manzoor, Nadeem, Iqbal, Saeed, & Ansari, 2013).

Bark Waste Biosorbents

Like seeds barks of almost all the above-mentioned plants have been used by different workers for the removal of heavy metals from the wastewater. In a study, pure and thioglycolic acid-treated cassava tuber bark wastes (CTBW) were investigated for the kinetics of Cd (II), Cu (II), and Zn (II) adsorption (Horsfall Jnr, Abia, & Spiff, 2006).

Seed Based Biosorbents

There are numerous examples where seeds of various plants have been used for the biosorption of various pollutants from the wastewater especially heavy metals. The seeds of cottonseed, linseed, soya been, litchi, mango seed, palm seed, carob seeds, papaya seed, Indian jujube seed, neem seed, Chinese chest nut, tamarind seed, sunflower seed powders have used most widely as a biosorbents. These seeds are rich in nutrients which are organic polymers having various functional groups which provides variation in the sort of active sites for absorption. Most of the above-mentioned seeds are actually the byproducts of oil industries, which are left as residue after extraction of oil. More over such types of seed powders have oil particles also on their surface, which provides them additional properties in terms of absorption efficiency and reactivity with the pollutants. One such example is litchi seed based biosorbents for the removal of Ni (II) ions from the wastewater in a batch sorption studies by the team of Flores–Garnica et al (Ghasemi et al., 2014). They further studied the remediation of Ni (II) ions in terms of kinetics, equilibrium, and thermodynamics. The maximum removal efficiency of Ni (II) was seen at 7.5, which was 66.62 mg/g. Ejila et al.,(2011) used Carica papaya seeds for the uptake of Mn (II) and Pb (II) which have maximum efficiency of 4.1 mg/100 cm^3 and 3.63 mg/100 cm^3 respectively after 90 minutes at 33 ^oC (Udoh, Udoh, & Umoh, 2005).

Gupte et al.,(2012) utilized the soaked seeds of *Ocimum basilicum* (sweet basil), for the uptake of Cu(II), where the removal percentage was 90% within 2.5 hour (Nazir, Wani, & Masoodi, 2017). Mancy et al.,(2013) reported the removal of Cr (VI)- 94.6% and Fe (III)-70% at pH 4 and 37.85 ^0C by using seeds of tamarind (*Tamarindus indica*) (Babu & Gupta, 2008). Mancy et al.,(2013) also reported the uptake of about 70% and 58% for Cr(IV) and Fe(III) respectively at pH 4 and temperature of 37.85 ^0C by using seeds of *Azadiractha indica* (Babu & Gupta, 2008).

MICROBIAL BIOSORBENTS

Microorganisms like fungus, yeast and bacteria are mostly used as biosorbents. These microbes have potential to tolerate high load of heavy metals, aiding in wastewater remediation. The advantage with these entities is their suitability of usage in living as well as dead form. Moreover, these are also used to immobilize the particles like silica NPs or enzymes. These microbes are rich in different biomolecules, like polysaccharides, sugars, and amino acids (either as biochemical constituents or through engineered metabolism), having a role in the heavy metal detoxification. The major chemical functionalities contributing in such activities include carbonyl, carboxyl, hydroxyl and amide linkages (Sulaymon, 2014), that bind to the metal ions through chemisorption ion exchange, physico-adsorption or electrostatic interactions. These microbes transform the toxic forms of metals ions into their nontoxic textures, thereby facilitating their easier discard. The microbially assisted remediation of heavy metals is based on the principle that when these microbes are a challenged with high concentrations of heavy metals, they could metabolize them faster (based on feedback control through their constitutive genetic mechanism). Tables 2-4 enlist the different microbial species that have been used for heavy metal remediation from the wastewater.

Table 1. Agro or plant based different biosorbents for the removal of toxic pollutants from the wastewater

Biosorbent types	Target pollutant	Adsorption efficiency (mg/g)	Functional groups
Red rose waste	Pb(II)	99.72	
Rose petals waste	Pb(II), Cu(II)	119.92, 124.21	
Orange peel	Pb(II), Zn(II)	209.8, 56.18	
Cassava tuber bank waste	Zn(II)	83.30	
Cedrus deodara sawdust	Zn(II)	97.39	
Cashew nut shell	Zn(II), Cu(II), Ni(II), Cd(II)	455.7, 406.6, 456.30, 436	
Powdered wheat straw	Cd(II)	39.22	
Rice bran	Ni(II)	46.51	
Litchi chinensis seeds	Ni(II)	66.62	
Cocoa pod husk	Ni(II)	20.10	
Pine cone powder	Pb(II), Cu(II)	32.26, 26.32	
Pineapple peel fiber	Pb(II), Cd(II)	70.29, 34.18	
Cortex orange waste	Pb(II), Cu(II)	76.80, 67.20	
Watermelon rind	Pb(II)	98.06	
Onion skins	Cd(II)	200.00	
Coffee grounds	Cd(II)	15.65	
Water lily flower, Mangrove leaves.	Cr(VI)	8.44, 8.87	
Cupressus lusitanica bark	Cr(VI)	87.5	
Garden grass	Cu(II)	58.34	
Palm oil fruits shell	Ni(II)	60.00	
Moringa oleifera bark	Ni(II)	30.38	
Caesalpinia bonducella seed	Ni(II)	188.67	
Cucurbita moschata	Cu(II), Ni(II)	12.5, 15	
Mannich base biosorbent	Pb(II)	60.5	
Watermelon shell	Cu(II)	111.1	
Organosolv lignin	Methylene blue	40.02	
Lentinus edodes	Pb(II)	21.5	
Citrus maxima peel, passion fruit shell, sugarcane bagasse	Pb(II), Cd(II), Cu(II), Ni(II)	169, 132, 84.0, 60.7 98.4, 48.6, 40.8, 25.8, 49.8, 26.3, 23.0, 16.1	
Modified *Punica granatum* L. peels.	Pb(II)	371.36	
Solanum melongena leaf powder	Pb(II)	71.42	
Pomegranate peel	Cu(II)	30.12	
Saccharum officinarum bagasse	As(III), As(V)	28.57, 34.48	
Spent-grain	Cu(II)	10.47	

continued on following page

Table 1. Continued

Biosorbent types	Target pollutant	Adsorption efficiency (mg/g)	Functional groups
Rice husk	As(V), Au, Cr(IV), Pb(II), Fe(III), Mn(II), Zn(II), Cd(II) malachite green and acid yellow		
Zirconium-loaded okara	Phosphates	44.13	
Stevia leaves	Malachite green	284.45	
Vigna radiata waste	Uranium	230	
Cabbage cauliflower waste	Pb(II)	60.57, 47.63	-CH,-OH, -C=O
Rice husk	Ni (II)	51.8%	-OH, -C=O
Sugarcane baggase	Ni (II)	2 mg/g2	Ion exchange
Neem bark	Pb	86.7%	-CH,-OH, -C=O, N-H, S=O, C-N
Iris pit	Cu (II), Ni (II)	17.6, 14.5 mg/g2	-C=N, -C=N
Date pit	Cu (II), Cd (II)	35.9, 39.5 mg/g2	
Cassava peelings	Cu (II), Cd (II)	127.3,119.6 mg/g2	

Figure 3. Microbes and their products used for the heavy metal removal

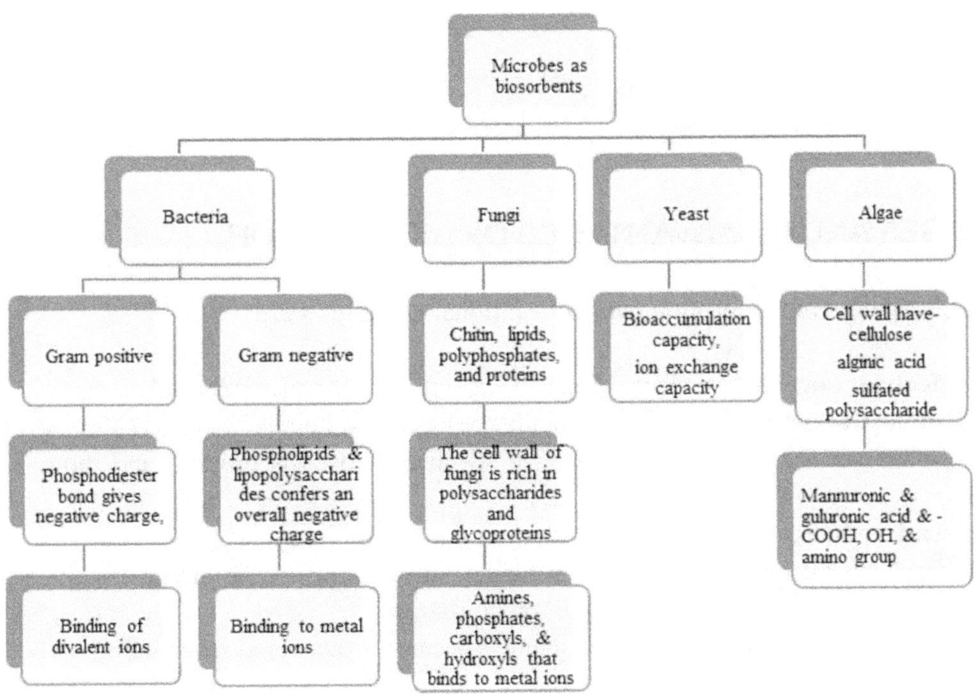

Table 2. Algal biomass as biosorbents for heavy metal removal

Biosorbent type	Metal ion	Biosorption capacity (mg/g^2 or %)	Functional groups	Mechanism
Stoechospermum marginatum	Cr (VI)	32.63		Ion exchange
Ulva lactuca sps	Cd (II)	35.72	Amido, hydroxyl, C=O, C-O	Chemisorption
Spirulina platensis	Cu (II)	90.6%		
Oedogonium hatei	Ni (II)	40.9	-OH, -CH, C=O, -CN, =C-N	
Maugcotia genuflexa	Ar (II)	57.48	Carboxyl, hydroxyl, amide	Ion exchange
Spirulina platensis	Cu	67.93		
Palmaria palmate	Cr (VI)	33.8	-NH, C=O, C-O, -S=O	Ion exchange and complexation
Fucus vesiculosus		42.6		
Enterobacter spps	Pb (II), Cu (II), Cd (II),	50, 32.5, 46.2		
Cladophora spp.	Pb (II), Cu (II)	46.51, 14.71		Physical adsorption or ion exchange
Laminiaria japonica	Zn (II)	91.5		
Spirogyra sp.	Pb (II)	140	Carboxyl, amino, amide, hydroxyl	
Ecklomia sps	Cr (VI)	60%	Amino, carboxyl	Ion exchange, Chemisorption
Saccharomyces cerevisiae loaded nano fibrous mats	Pb(II)	238.0		

ADVANTAGES AND DISADVANTAGES OF BIOSORPTION

Some basic advantages of biosorption over conventional techniques are

- **Cost Effective:** Low cost as derived from the by-product or waste material.
- **High Uptake Capacity:** Have potential to uptake higher pollutants/heavy metals from the solution
- **Metal Selective:** Specific biosorbents have specificity towards binding and removal of heavy metals
- **Regenerative:** Biosorbents can be reused, after the metal is recycled.
- **No Sludge Generation:** No secondary problems with sludge occur with biosorption, as is the case with many other techniques, for example, precipitation.
- **Metal Recovery Possible:** In case of metals, it can be recovered after being sorbed from the solution.
- Easy to operate, high efficiency and no increase in chemical oxygen demand (COD) of water, no additional nutritional requirement.
- Competitive performance: biosorption is capable of a performance comparable to the most similar technique, ion exchange treatment.

Table 3. Bacteria as biosorbents for heavy metal removal

Biosorbent type	Metal ion	Biosorption capacity (mg/g^2 or %)	Functional groups	Mechanism
Bacillus cereus	Zn (II)	66.6	Amino, carboxyl, hydroxyl, carbonyl,	Physio-chemical adsorption and ion exchange
Bacilus pumillus	Pb (II)	28.06		
Trametes versicolor	Cu (II)	140.9	$-NH_2$, -OH -C-O	Chemisorption
Lactobacillus delbrukii	Fe (ii) Zn (II)	100% 90%	carboxyl, hydroxyl,	
Lactobacilus bulagricus			carboxyl, hydroxyl,	
Streptococcus thermophilus			carboxyl, hydroxyl,	
Bacillus coagulans	Cr (II)	39.9%		
Bacilus thuringensis	Ni (II)	15.7%		
Bacillus thioparans	Cu (II) Pb (II)	27.3, 210.1 mg/g2		
E. coli	Ni (II)	6.9	C-H	Ion exchange
Pseudomonas ptuida	Zn	17.7		
Arthrobacter sps	Cu (II)	32.64		
Bacillus licheniformis	Cr (II), Fe (ii) Cu (ii)	95% 52% 32%		
Rhizobium spps	Cd (II) Co (II)	135.3, 167.5		
Streptomyces rimosus	Pb (II)		Carboxyl, hydroxyl	
Bacillus subtilis	Au (II)		Carboxyl, hydroxyl	

Disadvantages

The disadvantages of biosorption include:

- Early saturation i.e. when metal interactive sites are occupied, metal desorption is necessary prior to further use.
- The potential for biological process improvement (e.g. through genetic engineering of cells) is limited because cells are not metabolizing. There is no potential for biologically altering the metal valency state.

Table 4. Fungal biomass for biosorption of heavy metals

Biosorbent type	Metal ion	Biosorption capacity (mg/g² or %)	Functional groups	Mechanism
Penicillum canescens	As (III), Hg (II), Cd (II), Pb (II)	26.4, 54.8, 102.7, 213.2 mg/g2		
Penicillum chrysogenum	Ni	82.5		
Penicillum purpurogenum	As (III), Hg (II), Cd (II), Pb (II)	35.6, 70.4, 110.4, 252.8		
Aspergillus niger	Cu (II)	9.53		
Penicillum simpliccium	Cd (II), Zn (II), Pb (II)	52.5, 65.6, 76.9		Chemical ion exchange
Saccharomyces cerevisie	Pb (II), Ni (II), Cr (VI)	270.3, 46.3, 32.6		Physical adsorption
Lentinus sajor	Cr (VI)	18.9	C-O, N-H, C-H	Physio-chemical adsorption
Pleurotous ostreatus	Cr (VI), Cu (II), Ni (II), Zn (II), Cr (VI)	20.71%	Carboxyl Amine	Ion exchange, surface complexation and electrostatic interaction
Aspergilus terreus	Cu (II)	180	Carboxyl Amine	
Phanerochaete Chrysoporium	Ni (II), Pb (II)	55.9 53.6		Ion exchange
Trametes versicolor	Ni (II)	212.5	Carboxyl, Hydroxyl, Amine	
Mucor rouxiii	Cu (II)		Amino, carboxyl, phosphate	
Rhizopus chonii	Cd (II)		Carboxyl	

BIOPOLYMERS BASED BIO NANOSORBENTS FROM THE WASTE PRODUCTS

Nanosorbents

These are nanosized particles or nanocomposites which have very high tendency to absorb the substances from the aqueous media because high surface area to volume ratio. Further, these can be designed in such a way that it have numerous nanostructure components for the elimination of specific contaminants and propose the opportunity of even greater sorption capacities (Amin, Alazba, & Manzoor, 2014). These materials finds tremendous applications in the field of environmental clean-up, water purification etc. They have high surface area to volume ratio, which provides the pollutants larger surface area to absorb, more actives site, low intraparticle resistance. They also have numerous pores and provides high permeability. Moreover, the suitability of the best method and material for the water purification depends on the four features: - (1) How far the treatment is ñexible and efficient, (2) reusability after adsorption, (3) environmental friendliness, and (4) economic viability.

Sugarcanes Bagasse-As Lignocellulose

Lignocellulose are basically plant material which are made up of carbohydrate and aromatic polymers. These are dried after collecting from the plant and considered as lignocellulosic biomass. Sugarcane bagasse is one such agro-waste, which are rich source of lignocellulose and are widely used for the synthesis of the same throughout the world. Due to the different laboratory conditions (e.g., pH, temperature, adsorbent dose, particle size), materials and methodologies, it is not easy to conclude which biosorbent would be suitable for metal ion. For example, sugarcane bagasse, an agro-waste from sugar industries has been extensively studied because of its low price and high availability all over the world. This biosorbent exhibited very high potential in heavy metal uptake during wastewater treatment (J. Wang & Chen, 2009). The biosorption capacity of sugarcane bagasse could be noticeably improved by introducing carboxylic, amine and other functional groups into the surface materials or by removing soluble organic compounds, and increasing efficiency of metal sorption (Deng & Ting, 2005). The main factors determining the adsorption of As (V) on sugarcane bagasse modified by Fe (III) oxy-hydroxide (HFO) were electrostatic interactions, ligand exchange, and chelation between positively charged surface groups of FeOH (II), and negatively charged As (V) ions (Arcibar-Orozco, Delgado, Rios-Hurtado, & Rangel-Mendez, 2014).

Chitosan Nanoparticles

Chitosan is a linear polysaccharides made up of randomly distributed beta-linked-D-Glucosamine and N-acetyl-D-glucosamine. Chitosan is the partially deacetylated chitin prepared by the alkaline deacetylation (Y. Li & Ju, 2017). It is recovered from the shells of the crustaceans like shrimps, oysters by treating the shell with strong alkali. While chitosan nanoparticles are synthesized by the interaction of the oppositely charged macromolecules. The most commonly used material for the synthesis of chitiosan anopartciles is trisphosphate [TPP] (Tran, Duri, & Harkins, 2013). It is most widely used in the wastewater treatment, medicine as drug-delivery due to their biocompatible, biodegradable and non-toxic nature. Chitosan among other biosorbents is one of the most promising alternative adsorbents for the recovery of heavy metals from wastewater.

Hejri studied the adsorption of lead metal ions from aqueous solutions by chitosan nanoparticles. Nanoparticles were (Ahmad, Ahmed, Lal Swami, & Ikram, 2015) synthesized by maleic acid from chitosan cross-linking and then analyzing their particle size by the method of dynamic light scattering (DLS) which was in the range of 70-350 nm. At room temperature (20 ± 0.5 °C), the experiments were performed to observe the effects of the variables like pH of initial solution, absorbent dosage, and initial concentration of lead (II). Comparing the analysis of nanoparticles with chitosan polymer particles represents an increase in the effective surface of nanoparticles which increases the efficiency of absorption process, partly, and decrease the costs and the duration of the process, greatly.

Wu et al. carried out fabrication of carboxymethyl chitosan-hemicellulose resin (CMCH) by a thermal crosslinking process for adsorptive removal of Ni (II), Cd (II), Cu (II), Hg (II), Mn (VII), and Cr (VI) metals from wastewater (Wu, Dai, Kan, Shilong, & Zhu, 2016).

Figure 4. Steps involved in the synthesis of nanocellulose from the orange peels

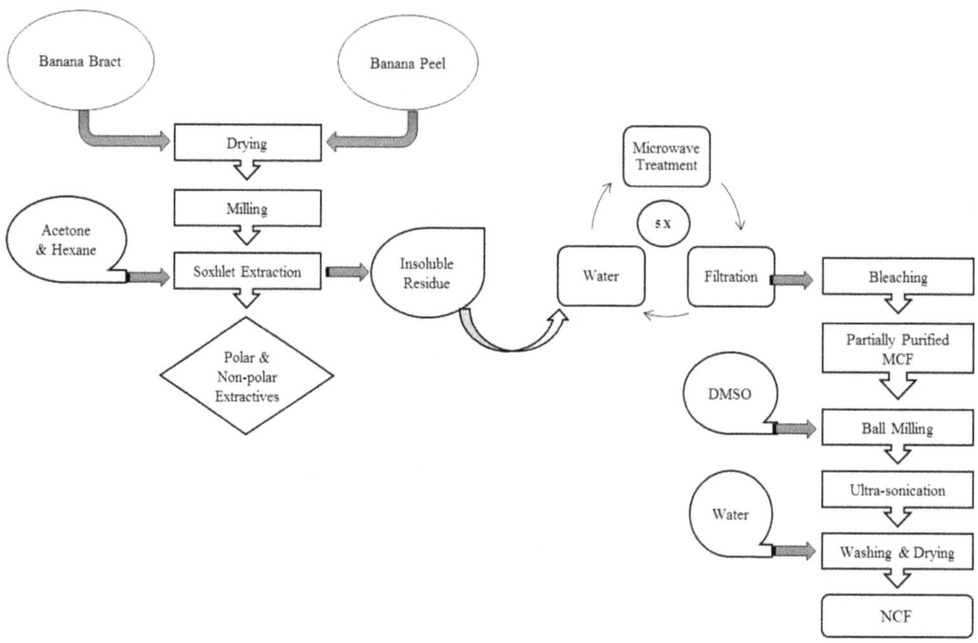

Figure 5. SEM image of nanocellulose from citrus peel waste, adapted from (Mariño, Lopes da Silva, Durán, & Tasic, 2015)

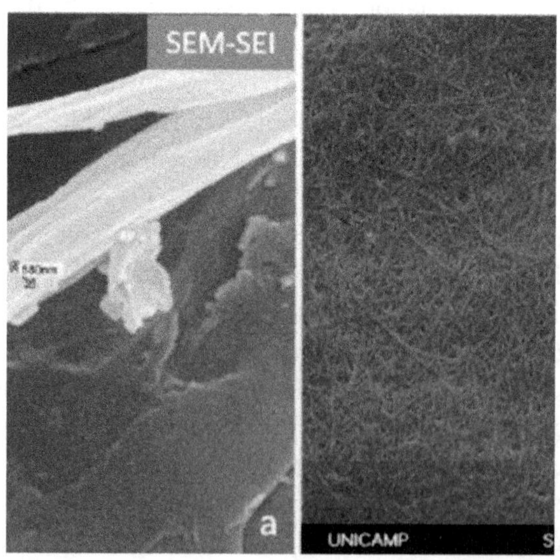

Table 5.Industrial byproducts as a source of biosorbents

Biosorbent type	Metal ion	Biosorption capacity (mg/g[2] or %)	Functional groups	Mechanism
Tea industry waste	Cr (VI)	54.65	-OH, -SO, C-O, -CN	
Baggase from sugar industry	Cd (II), Fe (II)	96.4%, 93.8%		
Peach and apricot stones	Pb (II)	97.64%, 93%		
Antibiotic waste	Cationic dye (Basic blue 41)	111		Ion exchange or complexation
Sludge	Ni (II), Cu (II), Pb (II), Cd (II)	137, 139, 14.1, 14.8		Ion exchange and physico-chemical adsorption
Waste green sands	Zn (II)	10.0		
Fly ash	Pb	22		Precipitation

Nanocellulose

Cellulose are the polysaccahridic organic molecule of plant having linear repeating units of D-glucose. These are mainly present in the primary cell wall of plants, bacteria and oomycetes. These are present in high amount in plants, which provides strength and rigidity. There are several reports where cellulose fibers of either micron sized or nanosized has been used for the application of wastewater treatment. The nanocellulose are fibrous molecule whose diameter is in nanometers whereas as the length may extend up to few microns. This nanocellulose are extracted from the banana peels, orange peels, etc. and has been used for the fabrication of textiles. The extraction of nanocellulose from the waste products makes the process eco-friendly, cost effective and biocompatible. Steps involved in the synthesis of nanocellulose from the agro waste are given below.

Nanocellulose in Remediation

Earlier nanocellulose has been applied for the treatment of sewage water, along with combination of coagulation-ñocculation process. It has potential to reduce the turbidly and COD in the suspension. It was used for the removal of Hg (II) ions, where it was recycled subsequently for the removal of metals.

The removal of heavy metals such as zinc (II), nickel (II), copper (II), cobalt (II), and cadmium (II) ions from an aqueous solutions was determined by using succinic anhydride modified mercerized nanocellulose (Nhandeyara do Carmo Ramos, Xavier, Teodoro, Frédéric Gil, & Gurgel, 2016). The pH effect, contact time, regeneration, and the concentration of metals were studied in batch mode. The range of maximum uptake of metal from 0.72 to 1.95 mmol/g and its order are the following: Cd > Cu > Zn > Co > Ni.

Cellulose-based nanosorbents of magnetic nature were fabricated and employed for the removal of Hg(II), Cu(II), and Ag(I) ions (Novi et al., 2016). The process of removal of metallic toxicants was found to be highly efficient, and the uptake capacity was found to be 2, 1.5, 1.2 mmol/g for Hg (II), Cu (II), and Ag (I), respectively. The adsorption was quite fast and the process followed pseudo-second-order kinetics. It was observed that the equilibrium adsorption reached within 5 min. From kinetic consider-

Figure 6. FESEM image of fly ash particles

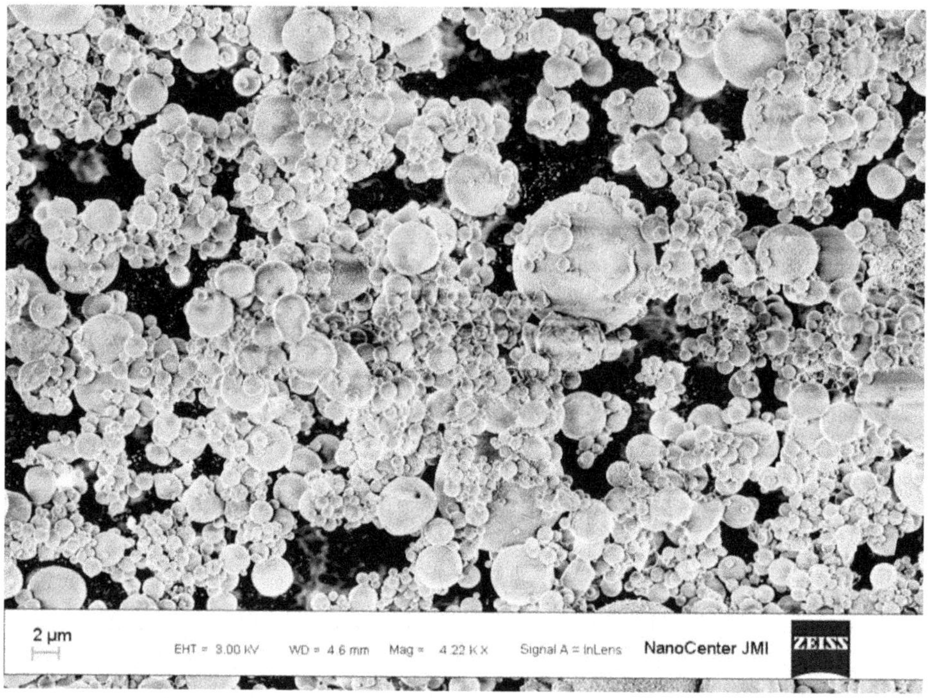

ations, the process was quite fast and exothermic in nature. The adsorbent was successfully regenerated after reaction with acidified thiourea.

INDUSTRIAL BYPRODUCTS: A LOW-COST BIOSORBENT MATERIALS

There are numerous types of waste materials that are being produced from the industries like sugar, paper mill, textile, municipal waste, sludge tea etc., most of which are the by-products (of major operational mechanism) whose utilization as biosorbents can significantly diminish the pollution extents (Ramachandra, N, & Rd, 2005). The production of such wastes is really challenging in terms of disposal, so their utilization provides an ecofriendly and cost-effective source of biosorbents. Table 5 comprises the most common biosorbents, used for heavy metal remediation from the wastewater.

Out of all the above-mentioned industrial wastes based biosorbents, fly ash is the most significant as it is a hazardous material, in comparison to others. Therefore, here we have discussed fly ash and its constituents thereof in detail for heavy metal biosorption.

Fly Ash Suitability for Being Used as Bionanosorbents

Fly ash is the by-product of thermal power plants, produced from the combustion of pulverized coal in the thermal power plants during electricity generation (Marinoni & Broekmans, 2013). Every year a million tonnes of fly ash is produced throughout the whole world. A large fraction of fly ash is disposed of in the near proximity of the thermal power plants, which encroaches the fertile and agricultural

Figure 7. FESEM image of fly ash ferrospheres

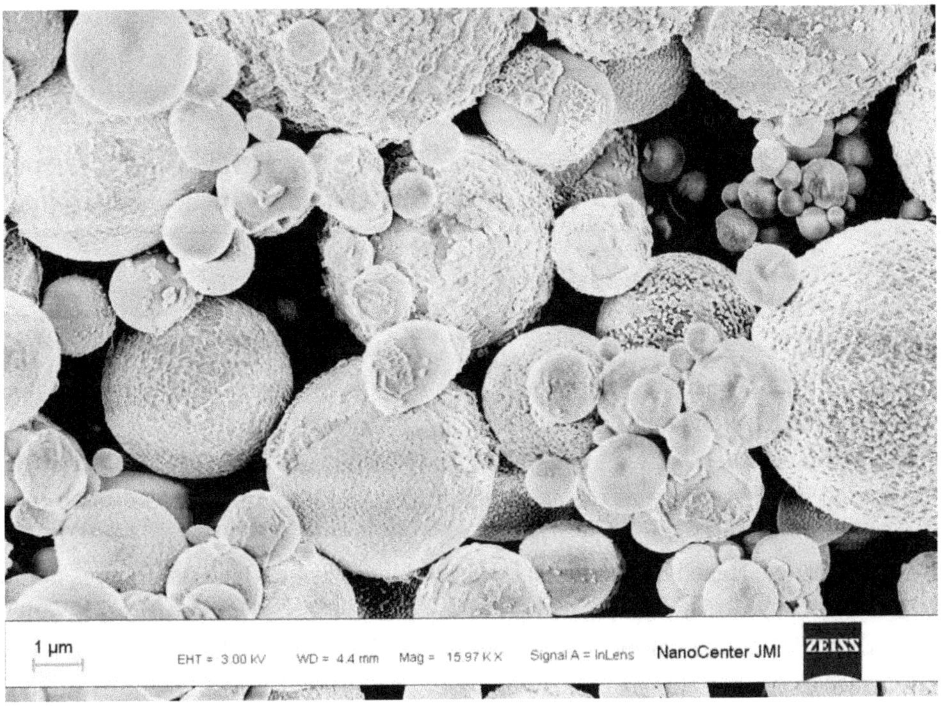

land and forests. Besides, this fly ash have chemically, silica, alumina and ferrous particles as major constituents. Fly ash have structurally different types of particles like cenospheres, ferrospheres (Hui, Chao Cy Fau - Kot, & Kot, 2005). These particles are naturally engineered entities (in the nanoscale), with mostly a spherical shape. These moieties have various surface modifications, such as presence of sulfate, N, C, and several other molecules, playing key role in the remediation of organic and inorganic pollutants. Several biomaterials could be used to enhance the working efficiency of fly ash constituents through textural modification.

Properties of Fly Ash Particles

The most ubiquitous entity of fly-ash are the spherical glassy particles, with varying size limits from 0.1 to several microns. They have mainly Si, Ai and Fe as major oxides whose composition varies on the surface of the particles. Fly-ash constituents confer cementing properties, due to their high surface adherence and carbon comprising nature.

Fly ash can be used directly or indirectly (synthesized as zeolites) for the heavy water removal from the polluted water. Several studies have reported the Cu, Pb, Zn, Mn, Cd, Cr, and Ni removal from the contaminated water. One interesting observation related to fly ash based heavy metal removal is the varying chemical inclination of fly ash, from acidic to alkaline, based on the minerals and compositions

Ahmad Ashfaq and Mohd Kaifiyan (2016) used fly ash as a cost-effective, adsorbent for Pb (II) and Fe (II) ion removal from the aqueous solutions (Ashfaq & Kaifiyan, 2016). *Bayat* compared the removal of Cr (VI) and Cd (II) from the aqueous solutions using the Turkish fly ash (Ashfaq & Kaifiyan, 2016). *Grover and Narayanaswamy* studied the effects of Cr (VI) concentrations, fly ash dosage, contact time, and pH on the removal of Cr (VI) and found that removal was effective at lower pH. *Dasmahapatra et al* used fly ash as a biosorbent for the removal of Cr(VI) ions and found the removal percentage as critically dependent on the fly ash concentration apart from its size, medium pH and the experimental temperature. *Bhattacharya et al* also reported the Cr (VI) ions remediation using fly ash as a biosorbent. *Liu et al* developed the fly ash-based adsorbent material-supported zero-valent iron for Pb (II) and Cr (VI) removal from aqueous solutions. The Fourier transform infrared (FTIR) analysis demonstrated the presence of various surface functional groups as responsible for heavy metal uptake, with a maximum removal capacity for Pb (II) (78.13 mg/g^2).

Fly Ash Ferrous Particles as Bionanosorbents

The ferrous contribution of fly ash ranges between (5-15)%, depending on the respective source. Higher grades of coal have higher ferrous contents, wherein ferrous particles are formed by ferrous sulfate and other Fe containing minerals in coal, which on decomposition within a furnace forms iron oxides. These iron oxides get settled down on the preformed Al-Si matrix surface, during cooling and condensation (Sathyamurthy, Dharmaraj, & Thadhani J, 2015). The surface deposition is extensively in the form of crystals, angular, spherules. Based on the types of ferrous deposition, the ferrospheres could be smooth or rough surfaced. The rough surfaced ferrospheres have greater ferrous depositions in either angular or spherules textures contrary to the smooth ferrospheres, bearing an even distribution of ferrous particles throughout the surface alongside being weakly magnetic. These ferrospheres can be extracted from the fly ash using magnetic separation. It could be extracted through dry or wet chemical methods, wherein the recovered fractions are reported to have various impurities i.e. C, Si, S, Mg, Ca, Ti. These particles can be used as an adsorbent for the removal of dyes, pesticides and inorganic pollutants from the wastewater. One major advantage with ferrospheres is that it can be easily recovered after the process completion, reducing the adsorption cost. Besides this, the presence of impurities and other elements enhances the working efficacy for the removal of different pollutants without any further modification.

Raw fly ash generally has low adsorption capacity so the fly ash modified with acid or alkali either by physical and chemical treatment would enhance the adsorption capacity (J. Li, Gan, Wu, & Yi, 2016), conferring an enhanced workable value. The modified fly ash or their derived products are discussed ahead, for the remediation of wastewater heavy metals.

Fly Ash: Zeolites as Biosorbents

Zeolites are the hydrated alumino-silicate minerals, connected with Al-Si and O atoms. They could be either natural or synthetic. Even though these prevail naturally but due to the supposedly large number of impurities, cannot be used in the specific tasks requiring high purity. Therefore, nowadays several zeolites are being synthesized for the industrial applications, especially using petroleum cracking, hydrogenation etc. However, the synthesis of zeolites at commercial level is expensive due to elaborate chemical approaches and toxicity concerns. Therefore, fly ash is one of the major alternative material for the synthesis of zeolites. As we have already seen that fly ash are rich source of silica so it could be used for the synthesis of different types of zeolites by varying the synthesis parameters. The synthesis of zeolites is done by alkali dissolution method, like NaOH and KOH at higher temperature i.e. 90-95 ^{0}C for 1-24 hours. The synthesis of zeolites are given below.

Figure 8. Steps involved in the synthesis of zeolites from fly ash

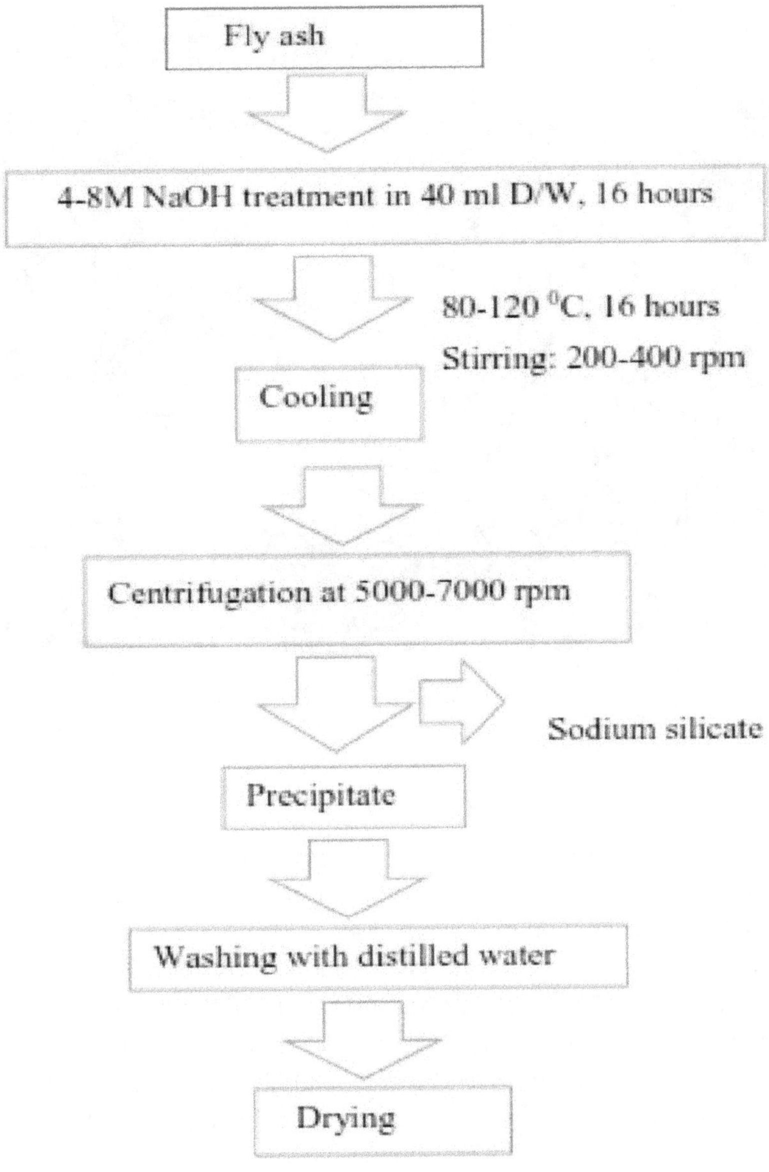

Zeolites are crystalline hydrated aluminosilicates associated with either group I or II elements as counter ions, which have a three-dimensional lattice, crumpled by an inner network of pores and channels (Thuadaija and Nuntiya 2012). Porous nature, higher surface area and cavities identical to basket like casing are some of the remarkable properties of zeolites (Y. Wang et al., 2016). The $[SiO_4]^4$ and $[AlO_4]^5$ tetrahedra, linked together by sharing centrally positioned oxygen. This tetrahedral 3-D network has a lot of voids and spaces in them, which are responsible for providing several additional properties and adsorption of molecules in huge internal channels is one of them (Sudarsanam, Peeters, Makshina, Parvulescu, & Sels, 2019). Due to these features, features, zeolites have been consistently used as promising

Figure 9. FESEM image of fly ash zeolites

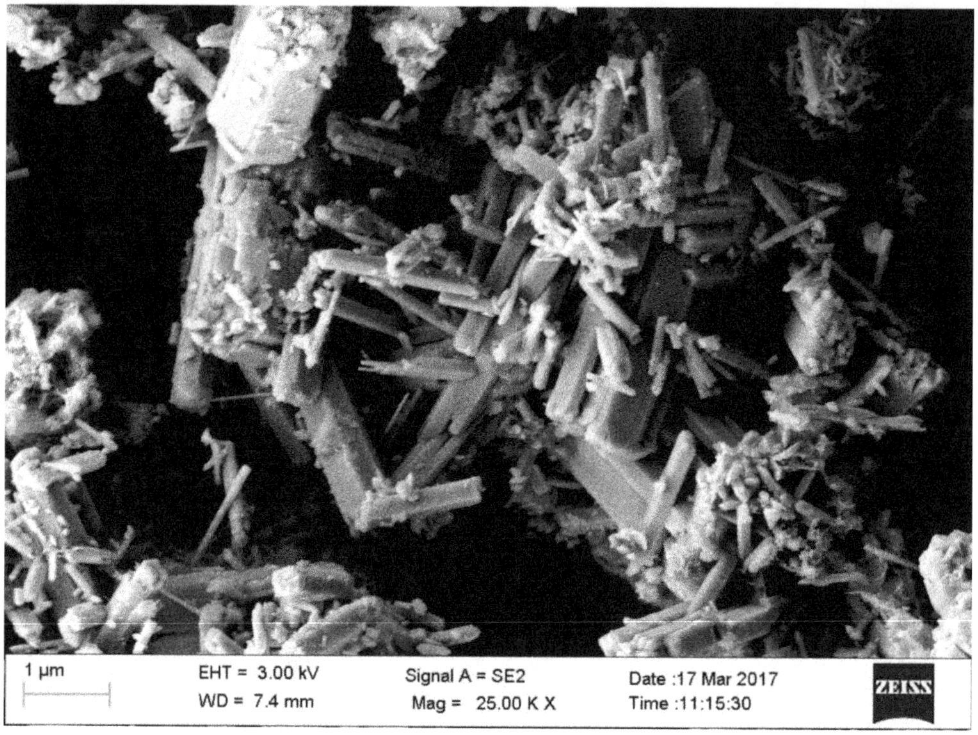

adsorbents, wastewater treatment (recovery of useful byproducts) and ion exchange driven separation methodologies (Penilla, Guerrero Bustos, & Goñi Elizalde, 2006).

The transformation of fly ash into zeolite is accomplished through treatment with NaOH or KOH using the hydrothermal alkaline method (Somerset et al., 2005). Coal fly ash (CFA) is one of the most suitable materials for the zeolite synthesis due to its richness in oxides of Al, silica alongwith traces of Ca, Mg and Mn (Přech, Pizarro, Serrano, & Čejka, 2018). The high percentage of Al and Si (60-80%) makes the CFA as highly suitable and cheaper alternative materials for the synthesis of Al-Si based zeolites (scott, Guang, Naeramitmarnsuk, Thabuot, & Amal, 2002). Previously, numerous studies have reported zeolite synthesis by transforming fly ash, concluding the impossibility to achieve entire fly ash zeolite processing and until now, only 50% conversion has been reported (Prince et al., 1999). Lower Si/Al containing fly ash gives rise to lower silica containing zeolites, having large pore volume, high affinity towards polar molecule and high CEC (cation exchange capacity) (Eliche-Quesada, Sandalio-Pérez, Martínez-Martínez, Pérez-Villarejo, & Sánchez-Soto, 2017). Fly ash generally gives rise to three classes of zeolites based on silica content, respectively being, low, intermediate and high-grade silica. Here we have focused mainly on sodalites, Na-X and faujasite. Sodalites are a class of zeolites that are produced naturally or synthetically from the Al and Si-rich minerals, for instance, fly ash alkali mediated hydrothermal synthesis (Franus, Wdowin, & Franus, 2014). While faujasite belongs to a class of zeolites having silicate minerals, which include group of elements, consisting of faujasite-sodium, calcium and magnesium. These attributes enable a similar essential formula with varying Na, Ca and Mg amounts (Amrhein et al., 1996). The structural framework of these materials is made up of sodalite cages connected *via* hexagonal prism (Błoński, Birczyński, Lalowicz, Datka, & Łodziana, 2015). The

Figure 10. FESEM micrograph of cenospheres

pores are arranged perpendicular to each other. This material either exists naturally or can be synthesized in industries and finds rigorous applications in catalytic conversion of crude petroleum products into gasoline, diesel and others.

Cenospheres

These are the alumino silicate spheres, comprising major fly ash constituents. These act as good adsorbents due to the presence of different elements on their surface. These particles are highly stable, heat and fire resistant alongside having superior mechanical properties. Therefore, such materials can be extraordinary candidates for the remediation of organic and inorganic pollutants. In general, these moieties can be extracted from the fly ash through wet slurry based buyout floatation method. The cenospheres being lighter in weight, float above and could be easily separated. Subsequently, these are dried before being surface modified for the remediation purposes.

Recovery of Cenospheres

Figure 11. Steps involved in the recovery of cenospheres from fly ash

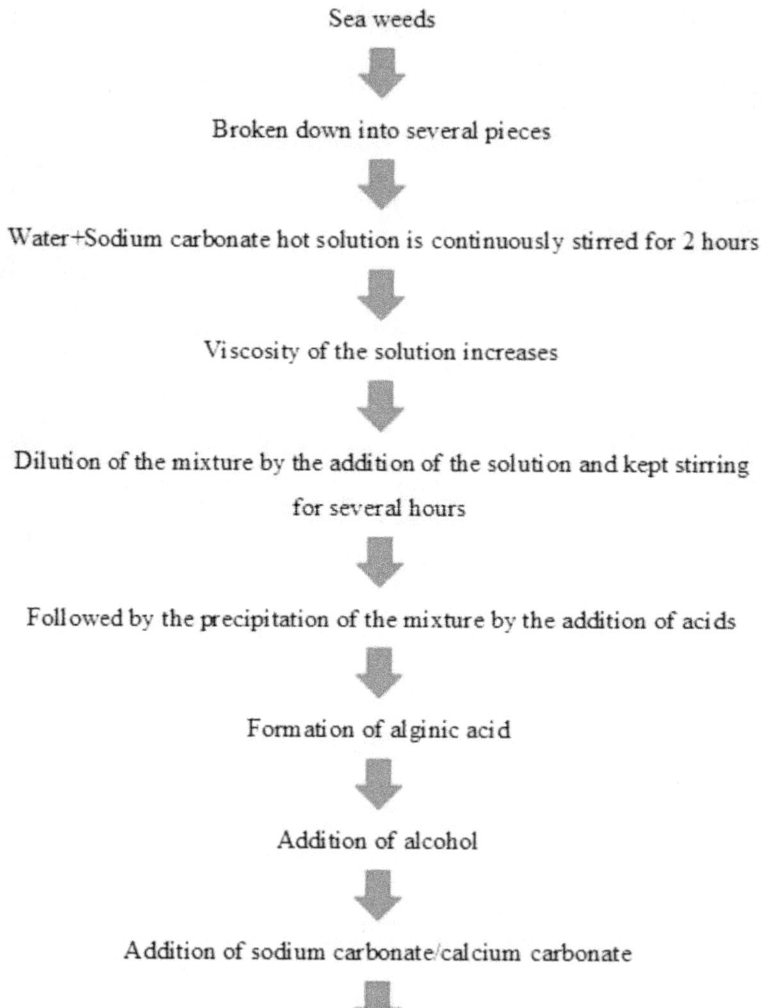

Fly Ash: Silica bio Nanosorbents

The major fraction of fly ash chemically constitutes silica (50-60)%, prevailing as silimanite, quartz with glassy amorphous texture. The extraction of silica from the crystalline form i.e. quartz and silimanite is quite difficult as both sources are inert, owing to which yield efficiency is very low. So, silica can be extracted from the fly ash using the amorphous form, which reacts much more easily. As silica is insoluble in acids, so it can be extracted by alkali treatment at (90-100)^{0}C for (60-120) minutes. The synthesis of silica nanoparticles is described ahead. The use of silica has been rare for the wastewater

Figure 12. TEM image of silica nanoparticles

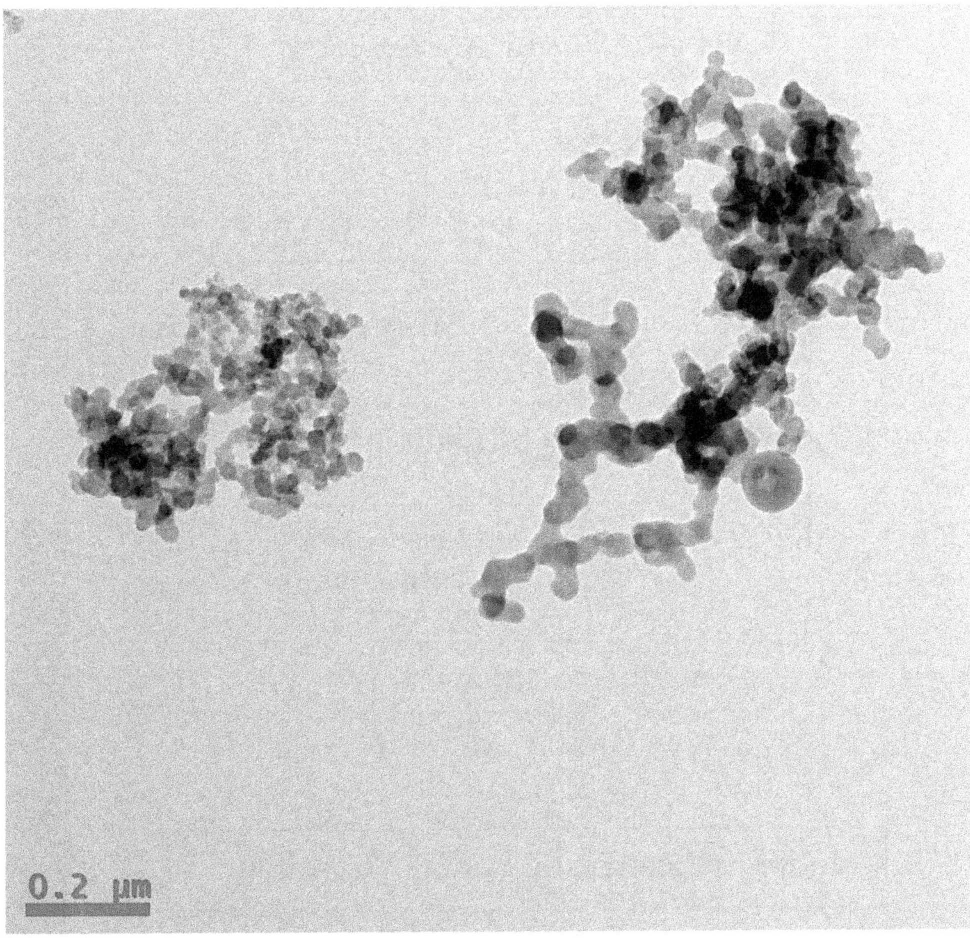

remediation and most frequently used form of it is the one obtained after surface modification or as composite material. The synthesized silica NPs are also mesoporous in nature.

Bulk form of silica and silica gels are being used as conventional adsorbents from ancient times and now are commonly used in the field of material science, medicine, chemistry and environment (Vallet-Regí & Balas, 2008). Likewise iron oxide NPs, and nanoporous silica play critical role as an adsorbent because nanoporous silica also have valuable role as an adsorbent because of wider potential for the adsorption and desorption of heavy metals ions (Yantasee et al., 2010). The mesoporous silica has advantageous property due to the well-ordered porosity enables more precise release and capture of molecules or ions (Slowing, Vivero-Escoto, Trewyn, & Lin, 2010). While the functionalized silica nanoparticles have more features that are superior then the naked one's because of better uptake and improved detoxification of heavy metals. The synthesis of nanosilica from the elemental silicon and oxygen and more specifically, from the waste like fly ash, is a green synthesis (Heleno Alves, Vitória da Silva Reis, Rovani, & Fungaro, 2017). The well-ordered porosity and desired size of obtained nanomaterials facilitates a faster and organized capture or removal of heavy metals ions (L. Lagadic, K. Mitchell, & D. Payne, 2001).

Figure 13. Steps involved in the synthesis of calcium oxide nanoparticles from fly ash

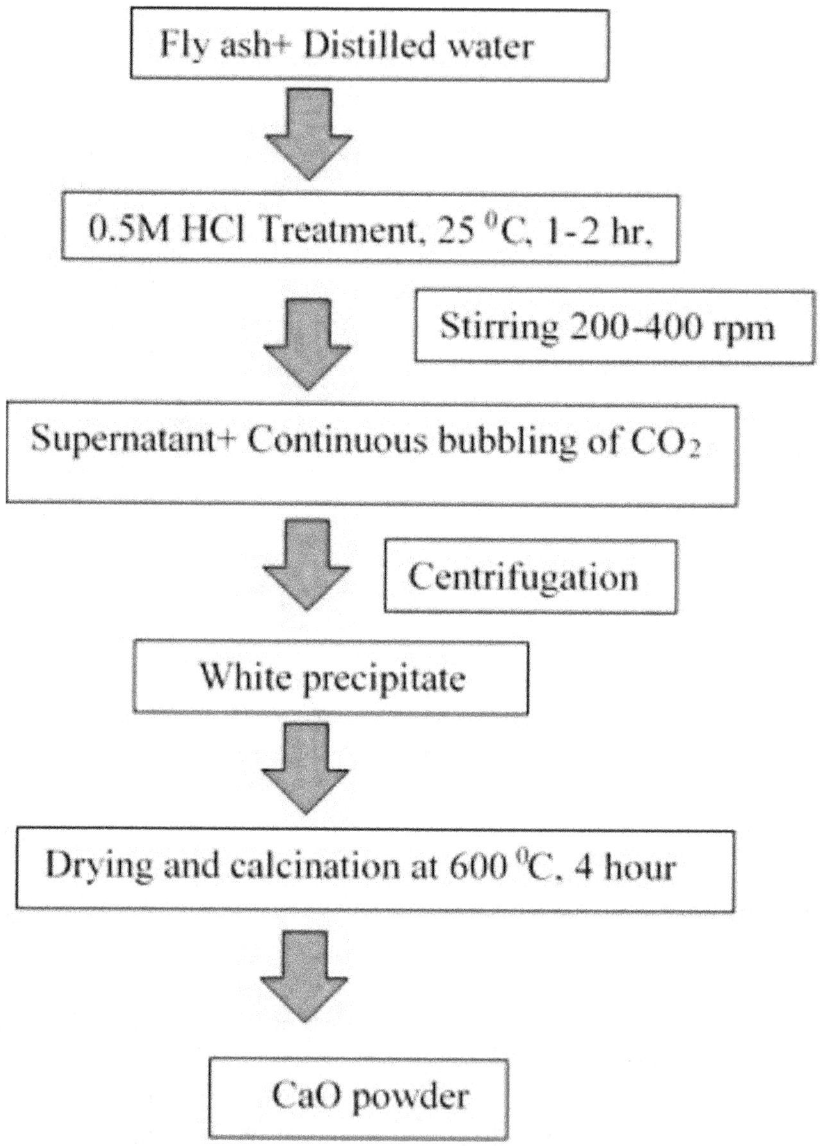

Fly ash Calcium Oxide Based bio Nanosorbents

The CaO composition in fly ash varies from (0.5-10) % based on the coal source. In general, higher grades confer class F fly ash, having < 5% CaO while the lower grades attribute to preferably more than 5% CaO. These CaO contents can be recovered from the fly ash by acidic treatment followed by precipitation and carbonation. Such particles can be surface modified for the wastewater pollutant remediation. The synthesis of CaO nanoparticles (NPs) from fly ash is described ahead.

Figure 14. FESEM image of calcium oxide particles

Carbon Nanomaterials and Their Derivatives From fly ash

Carbon and carbon materials are also one of the potential candidates to be used as nanosorbents. They are unique in their chemistry, fit to form chemical bonds with most of the elements. The different forms of carbon are graphene, graphite, bio char and carbon nanotubes. Fly ash has (5-20)% carbon in the form of soots, char, graphene, carbon nanotubes (CNTs), fullerene and organic carbon like PAH (Polyaromatic hydrocarbon), PCHs (Polycyclic hydrocarbon) (Burlakova & Novikova, 2017). This carbon fraction can be extracted using mineral oil. Upon air supply from external mode, the carbon agglomerates with the oil in the fly ash slurry. Thereafter, separated carbon can be collected from the top layer followed by drying and then subsequent application in the heavy metal remediation from wastewater. For long, CNTs have already been in use for the wastewater treatment through surface functionalization. This has been possible substantially due to their high mechanical properties, lightweight, low density, structural robustness and chemical resistance. The high surface area to volume ratio along with high surface energies, encompass ample application potential of CNTs. Apart from this, the chemical nature of CNTs allows comprehensive surface engineering (through covalent and non-covalent functionalization), enabling significant improvement of their native state properties (Guo, Wang, Weei Tjiu, Pan, &

Liu, 2012). Several studies have depicted improved adsorption of carbon nanomaterials through their manifold surface engineering (for desired interactive mode). Thereby, carbon nanomaterials are highly befitting materials for the development of efficient and sensitive adsorbents accomplishing removal of wastewater organic content (OC).

Figure 15. TEM image of fly ash char and carbon nanoparticles

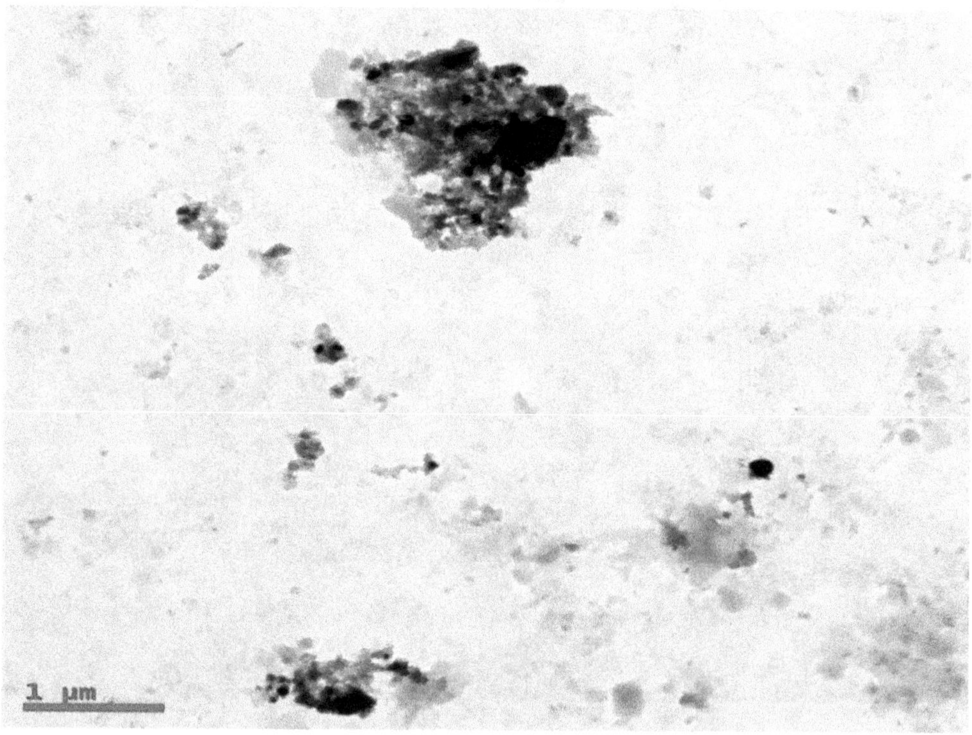

SILICA BASED REMOVAL OF HEAVY METALS

Imtithal Sheeta *et al* in 2014, used nanostructured graphite oxide (GO), silica/graphite oxide composites and silica NPs for heavy metal ions removal (from aqueous solutions), using a batch adsorption method. The adsorption of heavy metals by nanostructured GO was observed in the following order: nickel >zinc >lead > cadmium >chromium. Mona Karnib *et al* in 2014, reported the comparative study of heavy metal removal using activated carbon, silica NPs and silica activated carbon composites (Karnib, Kabbani, Holail, & Olama, 2014). A batch experiment was carried out for Cd, Pb, Zn, Cr and Ni removal from the wastewater. Ni showed the highest removal extent by the activated carbon at all concentrations and its percentage removal decreased with increase in heavy metal concentration (Mehdizadeh, Sadjadi, Ahmadi, & Outokesh, 2014). Kong Xiang-Feng *et al* in 2014, reported the selective removal of heavy metals (Pb^{2+}, Cu^{2+}, Hg^{2+}, Cd^{2+} and Zn^{2+}) from aqueous solutions using surface modified silica NPs having distinct functional groups. The silica source were the industrial waste silica fumes, which were

Figure 16. Steps involved in the extraction of alginate from algae

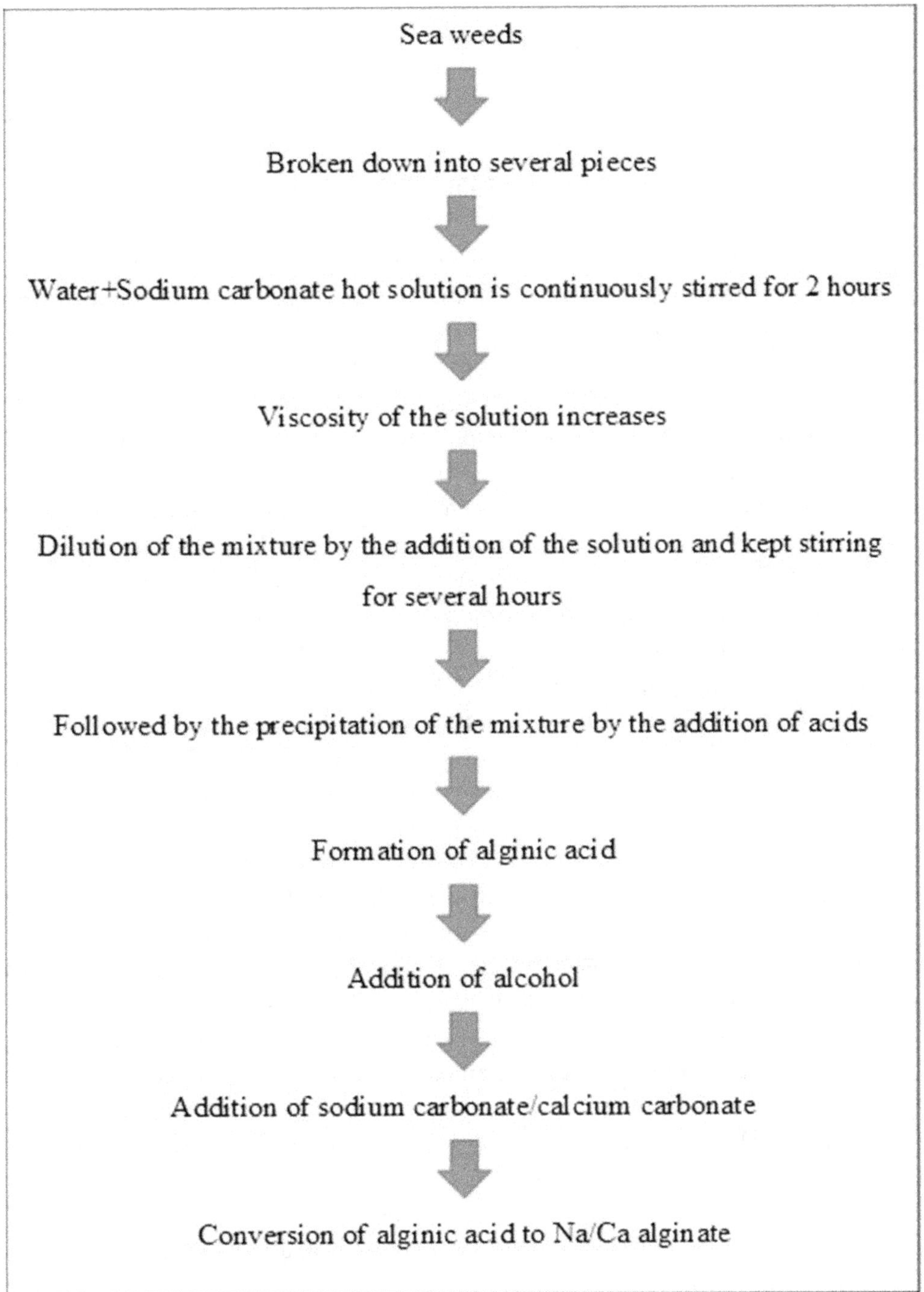

pretreated at 80°C in HNO_3 environment, subsequent to being surface modified by MPTES and APTES (Kong et al., 2014). The prepared adsorbent of surface functionalized silica NPs (having differential functional groups) was investigated for the Pb^{2+}, Cu^{2+}, Hg^{2+}, Cd^{2+} and Zn^{2+} ion selective adsorptions from the aqueous solutions (Kong et al., 2014).

ALGINATE IN THE FIELD OF BIOREMEDIATION

Alginate are anionic polysaccharides present in the cell wall of red or brown algae and kelp. It is also called as alginic acid and commercially extracted from *Laminaria hyperborean, Macrocystis pyrifera, L. digitata, Ascophyllum nodosum, L. japonica, Ecklonia maxima, Lessonia nigrescens, Durvillea antarctica, and Sargassum spp*. Alginic acid are nonhazardous, bio compatible, eco-friendly polysaccharides. Due to its water-soluble nature, it is widely used as stabilizing agent. Hydrated alginate develops into gel when exposed to divalent ions (Sun & Tan, 2013). The alginates are used in the form of beads, films for the remediation of pollutants from the wastewater (B. Wang et al., 2018). They are also used as superabsorbent by combining with magnetic nanoparticles, which drastically enhances their uptake of pollutants, increases their chemical resistance and stability and mechanical stability (B. Wang et al., 2018).

Alginate is the widely used biosorbent as it is an anionic polymer, which shows a great efficiency toward metal ions. The adsorption capacity of the algae, i.e., alginate is directly related to the presence of sites such as carboxyl and sulfate on the surface of alginate polymer, which makes it efficient in the field of water remediation (Papageorgiou, Kouvelos, & Katsaros, 2008).

Alka Tiwari and Prerna Kathane worked on the removal of Cu^{2+} ions from aqueous systems using superparamagnetic PVA-Alginate microspheres and concluded that the prepared microspheres came out as an effective adsorbent (Ackley, Rege, & Saxena, 2003). The maximum removal of approximately 98% ions was found at pH 4. Tiwari et al. also worked on the removal of organic pollutant phenol and obtained about 80% of removal efficiency from contaminated water (Zhang et al., 2019) .

Harikumar and Joseph worked on iron nanoparticles entrapped Ca-alginate bead for the removal of arsenic (III) ion and obtained up to 99.9% removal efficiency at a low adsorbent dose, and in very short time (B. Wang et al., 2018). Singh et al. used iron cross-linked alginate nanoparticles in fixed bed studies for the removal of arsenic ions and found that the nanoparticles worked well in the column and found effective against bacteriological contaminations (Singh, Singh, Bajpai, Bajpai, & Shrivastava, 2014).

Gomez et al. worked on Na-alginate and Ca-alginate film for the removal of Pb II and obtained 98% of removal efficiency using Ca-alginate film as compared to Na-alginate (Arica, arpa şahin, Ergene, Bayramoglu, & Genç, 2003). Fourest and Lee along with their coworkers studied about removal of heavy metal and Cr (VI) from aqueous solution using alginate concluded in their research article that alginate is an efficient adsorbent, and can be used in the field of water remediation (Boddu, Abburi, Talbott, & Smith, 2003). From the above discussion, it can be concluded that alginate can be used as an effective low-cost biosorbent for the removal of toxic metal from wastewater.

CONCLUSION

Absorption is no doubt, the most reliable technique for the remediation of wastewater pollutants, chiefly attributed to the simplicity of execution. Amongst all possible mechanisms, the one through biological agents is preferred, whereby utilization of agricultural, municipal and animal waste, providing cheaper, eco-friendly and biodegradable sorbents. Though agro wastes have various chemical active sites (such as carboxyl, alcohol, amine, ketone, and phenolic, aldehydic), thereby providing ample opportunities for a thorough single step wastewater pollutant remediation. However, this agro waste can be further surface modified *via* chemical treatment for enhanced and targeted pollutant removal. Thereby, nanotechnology has remarkably simplified the wastewater remediation through efficient and thorough cleansing mechanisms. The bio nanosorbents developed through combining biosorbents with nanomaterials, have additional improved performance. Fly ash alone is a warehouse of numerous natural nanosorbents, which can be used either directly or subsequent to treatment with suitable biomaterials. Thereby, the future prospects of bio nanosorbents confer an enormous potential in the domain of wastewater treatment.

REFERENCES

Abia, A. (2014). *A bioseparation process for removing heavy metals from waste water using biosorbents*. Academic Press.

Ackley, M., Rege, S., & Saxena, H. (2003). Application of natural zeolites in the purification and separation of gases. *Microporous and Mesoporous Materials*, *61*(1-3), 25–42. doi:10.1016/S1387-1811(03)00353-6

Ahmad, M., Ahmed, S., Lal Swami, B., & Ikram, S. (2015). *Adsorption of heavy metal ions: Role of chitosan and cellulose for water treatment* (Vol. 2). Academic Press.

Ahmad Khan, N., Ibrahim, S., & Subramaniam, P. (2004). *Elimination of Heavy Metals from Wastewater Using Agricultural Wastes as Adsorbents* (Vol. 23). Academic Press.

Amin, M. T., Alazba, A. A., & Manzoor, U. (2014). A Review of Removal of Pollutants from Water/Wastewater Using Different Types of Nanomaterials. *Advances in Materials Science and Engineering*, *24*, 1–24. doi:10.1155/2014/825910

Amrhein, C., Haghnia, G. H., Kim, T. S., Mosher, P. A., Gagajena, R. C., Amanios, T., & de la Torre, L. (1996). Synthesis and Properties of Zeolites from Coal Fly Ash. *Environmental Science & Technology*, *30*(3), 735–742. doi:10.1021/es940482c

Arcibar-Orozco, J., Delgado, J., Rios-Hurtado, J., & Rangel-Mendez, J. (2014). *Influence of iron content, surface area and charge distribution in the arsenic removal by activated carbons* (Vol. 249). Academic Press.

Arica, Y., Arpa Şahin, Ç., Ergene, A., Bayramoglu, G., & Genç, Ö. (2003). *Ca-alginate as a support for Pb(II) and Zn(II) biosorption with immobilized Phanerochaete chrysosporium* (Vol. 52). Academic Press.

Ashfaq, A., & Kaifiyan, M. (2016). Simultaneous Biosorption of Iron (II) and Zinc (II) onto Flyash from Aqueous Solutions. *International Journal of Current Microbiology and Applied Sciences*, *5*(7), 117–120. doi:10.20546/ijcmas.2016.507.009

Babu, B. V., & Gupta, S. (2008). Removal of Cr(VI) from wastewater using activated tamarind seeds as an adsorbent. *Journal of Environmental Engineering and Science*, *7*(5), 553–557. doi:10.1139/S08-025

Barakat, M. A. (2011). New trends in removing heavy metals from industrial wastewater. *Arabian Journal of Chemistry*, *4*(4), 361–377. doi:10.1016/j.arabjc.2010.07.019

Bhatti, H., Khadim, R., & Hanif, M. (2019). *Biosorption of Pb(II) and Co(II) on red rose waste biomass* (Vol. 30). Academic Press.

Błoński, P., Birczyński, A., Lalowicz, Z., Datka, J., & Łodziana, Z. (2015). *Structure of the Hydroxyl Groups and Adsorbed D 2 O Sites in the DX Zeolite: DFT and Experimental NMR Data* (Vol. 119). Academic Press.

Boddu, V. M., Abburi, K., Talbott, J. L., & Smith, E. D. (2003). Removal of Hexavalent Chromium from Wastewater Using a New Composite Chitosan Biosorbent. *Environmental Science & Technology*, *37*(19), 4449–4456. doi:10.1021/es021013a PMID:14572099

Bolisetty, S., Peydayesh, M., & Mezzenga, R. (2019). Sustainable technologies for water purification from heavy metals: Review and analysis. *Chemical Society Reviews*, *48*(2), 463–487. doi:10.1039/C8CS00493E PMID:30603760

Bulut, Y., & Tez, Z. (2003). *Removal of heavy metal ions by modified sawdust of walnut* (Vol. 12). Academic Press.

Burlakova, V. E., & Novikova, A. A. (2017). Extracting of fullerene-like nanoparticles from environmental soots. *Fullerenes, Nanotubes, and Carbon Nanostructures*, *25*(8), 483–487. doi:10.1080/1536383X.2017.1339274

Cerofolini, G. F., Meda, L., & Bandosz, T. J. (1999). Adsorption with soft adsorbents or adsorbates. Theory and practice. In A. Dąbrowski (Ed.), *Studies in Surface Science and Catalysis* (Vol. 120, pp. 227–272). Elsevier.

Daneshfozoun, S., Abdullah, M., & Abdullah, B. (2017). *Preparation and characterization of magnetic biosorbent based on oil palm empty fruit bunch fibers, cellulose and Ceiba pentandra for heavy metal ions removal* (Vol. 105). Academic Press.

Darweesh, M. (2018). *Adsorption Kinetic Study for Removal of Heavy Metal from Aqueous Solution by Natural Adsorbent* (Vol. 4). Academic Press.

Das, k., Grewal, A., & Banerjee, M. (2011). A brief review: Heavy metal and their analysis. *International Journal of Pharmaceutical Sciences Review and Research*, *11*, 13–18.

Das, P., Chowdhury, S., Gupta, S., Kumar, I., & Kumar, R. (2010). *Assessment on the Removal of Malachite Green Using Tamarind Fruit Shell as Biosorbent* (Vol. 38). Academic Press.

Debnath, B., Chowdhury, R., & Ghosh, S. K. (2018). Sustainability of metal recovery from E-waste. *Frontiers of Environmental Science & Engineering*, *12*(6), 2. doi:10.100711783-018-1044-9

Deng, S., & Ting, Y.-P. (2005). *Characterization of PEI-modified biomass and biosorption of Cu(II), Pb(II) and Ni,(II)* (Vol. 39). Academic Press.

Dhir, B., & Kumar, R. (2010). *Adsorption of heavy metals by Salvinia biomass and agricultural residues* (Vol. 4). Academic Press.

Dostálek, P. (1970). Immobilized Biosorbents for Bioreactors and Commercial Biosorbents. Academic Press.

Dragan, E. S., Dinu, M., & Shankar, G. (2015). *Recent developments in composite biosorbents and their applications for wastewater treatment.* Academic Press.

El-Shafey, E.-S. (2009). *Removal of Zn(II) and Hg(II) from Aqueous Solution on a Carbonaceous Sorbent Chemically Prepared from Rice Husk* (Vol. 175). Academic Press.

El-Wakeel, S., Moghazy, R., Labena, A., & Husien, S. (2019). Algal biosorbent as a basic tool for heavy metals removal; the first step for further applications. *Journal of Materials and Environmental Science, 10*, 75–87.

Eliche-Quesada, D., Sandalio-Pérez, J. A., Martínez-Martínez, S., Pérez-Villarejo, L., & Sánchez-Soto, P. (2017). *Investigation of use of coal fly ash in eco-friendly construction materials: Fired clay bricks and silica-calcareous non fired bricks* (Vol. 44). Academic Press.

Farooq, U., Khan, M., & Athar, M. (2007). *Triticum aestivum: A novel biosorbent for lead (II) ions* (Vol. 51). Academic Press.

Feng, N.-C., Guo, X.-Y., & Liang, S. (2010). *Enhanced Cu(II) adsorption by orange peel modified with sodium hydroxide* (Vol. 20). Academic Press.

Franus, W., Wdowin, M., & Franus, M. (2014). Synthesis and characterization of zeolites prepared from industrial fly ash. *Environmental Monitoring and Assessment, 186*(9), 5721–5729. doi:10.100710661-014-3815-5 PMID:24838802

Fu, F., & Wang, Q. (2011). Removal of heavy metal ions from wastewaters: A review. *Journal of Environmental Management, 92*(3), 407–418. doi:10.1016/j.jenvman.2010.11.011 PMID:21138785

Ghasemi, M., Ghasemi, N., Zahedi, G., Wan Alwi, S. R., Goodarzi, M., & Javadian, H. (2014). *Kinetic and equilibrium study of Ni(II) sorption from aqueous solutions onto Peganum harmala-L.* Academic Press.

Güngen, A. C., Aydemir, Y., Çoban, H., Düzenli, H., & Tasdemir, C. (2016). Lung cancer in patients diagnosed with silicosis should be investigated. *Respiratory Medicine Case Reports, 18*, 93–95. doi:10.1016/j.rmcr.2016.04.011 PMID:27330963

Guo, J., Wang, R., Weei Tjiu, W., Pan, J., & Liu, T. (2012). *Synthesis of Fe nanoparticles@graphene composites for environmental applications* (Vol. 225-226). Academic Press.

Hanida, S. (2018). High Potency Enviromerntal Physical and Biological Factors of Malaria Transmission Tendency in Regency Working Area of Pandean Health Center Trenggalek. *Jurnal Kesehatan Lingkungan, 10*(1), 82. doi:10.20473/jkl.v10i1.2018.82-91

Heleno Alves, R., Vitória da Silva Reis, T., Rovani, S., & Fungaro, D. (2017). *Green Synthesis and Characterization of Biosilica Produced from Sugarcane Waste Ash* (Vol. 2017). Academic Press.

Vaghetti, J., Lima, E., Royer, B., Cardoso, N., Martins, B., & Calvete, T. (2009). *Pecan Nutshell as Biosorbent to Remove Toxic Metals from Aqueous Solution* (Vol. 44). Academic Press.

Horsfall, M. Jr, Abia, A., & Spiff, A. (2006). *Kinetic studies on the adsorption of Cd2+, Cu2+ and Zn2+ ions from aqueous solutions by cassava (Manihot sculenta Cranz) tuber bark waste* (Vol. 97). Academic Press.

Huamán, G., Maria Souza de Mesquita, L., Torem, M. L., & Adolfo Saavedra Pinto, G. (2006). *Biosorption of cadmium by green coconut shell powder* (Vol. 19). Academic Press.

Akunwa, N., Muhammad, M., & Akunna, J. (2014). *Treatment of metal-contaminated wastewater: A comparison of low-cost biosorbents* (Vol. 146). Academic Press.

Hui, K. S., Chao, C., & Kot, S. (2005). Removal of mixed heavy metal ions in wastewater by zeolite 4A and residual products from recycled coal fly ash. *J Hazard Materrials*, *127*(1-3), 89–101. doi:10.1016/j.jhazmat.2005.06.027 PMID:16076523

Jaishankar, M., Tseten, T., Anbalagan, N., Mathew, B. B., & Beeregowda, K. N. (2014). Toxicity, mechanism and health effects of some heavy metals. *Interdisciplinary Toxicology*, *7*(2), 60–72. doi:10.2478/intox-2014-0009 PMID:26109881

Javanbakht, V., Amir Alavi, S., & Zilouei, H. (2014). *Mechanisms of heavy metal removal using microorganisms as biosorbent* (Vol. 69). Academic Press.

Karnib, M., Kabbani, A., Holail, H., & Olama, Z. (2014). *Heavy Metals Removal Using Activated Carbon* (Vol. 50). Silica and Silica Activated Carbon Composite.

Khan, I., Saeed, K., & Khan, I. (2017). Nanoparticles: Properties, applications and toxicities. *Arabian Journal of Chemistry*. doi:10.1016/j.arabjc.2017.05.011

Khatri, N., & Tyagi, S. (2015). Influences of natural and anthropogenic factors on surface and groundwater quality in rural and urban areas. *Frontiers in Life Science*, *8*(1), 23–39. doi:10.1080/21553769.2014.933716

Khoramzadeh, E., Nasernejad, B., & Halladj, R. (2013). *Mercury biosorption from aqueous solutions by Sugarcane Bagasse* (Vol. 44). Academic Press.

Kilic, M., Kırbıyık, Ç., Çepelioğullar Mutlu, Ö., & Pütün, A. (2013). *Adsorption of heavy metal ions from aqueous solutions by bio-char, a by-product of pyrolysis* (Vol. 283). Academic Press.

Kong, X., Yang, B., Xiong, H., Zhou, Y., Xue, S., Xu, B.-Q., & Wang, S.-X. (2014). *Selective removal of heavy metal ions from aqueous solutions with surface functionalized silica nanoparticles by different functional groups* (Vol. 21). Academic Press.

Kotrba, P., Mackova, M., & Macek, T. (2011). *Microbial Biosorption of Metals*. Academic Press.

Krishna, R., Kumar Chintalpudi, V., & Muddada, S. (2018). *Application of Biosorption for Removal of Heavy Metals from Wastewater*. Academic Press.

Krishnani, K., Meng, X., Christodoulatos, C., & Boddu, V. (2008). *Biosorption Mechanism of Nine Different Heavy Metals onto Biomatrix from Rice Husk* (Vol. 153). Academic Press.

Lagadic, I., Mitchell, M., & Payne, B. (2001). *Highly Effective Adsorption of Heavy Metal Ions by a Thiol-Functionalized Magnesium Phyllosilicate Clay* (Vol. 35). Academic Press.

Li, J., Gan, J., Wu, L., & Yi, J. (2016). *Preparation of Fly ash Based Adsorbents for Removal Active Red X-3B from Dying Wastewater* (Vol. 67). Academic Press.

Li, Y., & Ju, D. (2017). The Application, Neurotoxicity, and Related Mechanism of Cationic Polymers. In X. Jiang & H. Gao (Eds.), *Neurotoxicity of Nanomaterials and Nanomedicine* (pp. 285–329). Academic Press. doi:10.1016/B978-0-12-804598-5.00012-X

Malik, R., Dahiya, S., & Lata, S. (2017). *An Experimental and Quantum Chemical Study of Removal of Utmostly Quantified Heavy Metals in Wastewater Using Coconut Husk: A Novel Approach to Mechanism* (Vol. 98). Academic Press.

Manzoor, Q., Nadeem, R., Iqbal, M., Saeed, R., & Ansari, T. (2013). *Organic acids pretreatment effect on Rosa bourbonia phyto-biomass for removal of Pb(II) and Cu(II) from aqueous media* (Vol. 132). Academic Press.

Mariño, M., Lopes da Silva, L., Durán, N., & Tasic, L. (2015). Enhanced materials from nature: Nanocellulose from citrus waste. *Molecules (Basel, Switzerland)*, *20*(4), 5908–5923. doi:10.3390/molecules20045908 PMID:25854755

Marinoni, N., & Broekmans, M. A. T. M. (2013). Microstructure of selected aggregate quartz by XRD, and a critical review of the crystallinity index. *Cement and Concrete Research*, *54*, 215–225. doi:10.1016/j.cemconres.2013.08.007

Marković, M., Radmanović, S., Đurović-Pejčev, R., Milinovic, J., & Kljajić, P. (2009). Assessment of Heavy Metal and Pesticide Levels in Soil and Plant Products from Agricultural Area of Belgrade, Serbia. *Archives of Environmental Contamination and Toxicology*, *58*(2), 341–351. doi:10.100700244-009-9359-y PMID:19603130

Mehdizadeh, S., Sadjadi, S., Ahmadi, S. J., & Outokesh, M. (2014). Removal of heavy metals from aqueous solution using platinum nanopartcles/Zeolite-4A. *Journal of Environmental Health Science & Engineering*, *12*(1), 7–7. doi:10.1186/2052-336X-12-7 PMID:24397886

Miretzky, P., & Cirelli, A. (2009). Hg(II) Removal from Water by Chitosan and Chitosan Derivatives. *RE:view*, *167*. PMID:19232467

Naja, G. M., Murphy, V., & Volesky, B. (2010). Biosorption, Metals. Encyclopedia of Industrial Biotechnology. doi:10.1002/9780470054581.eib166

Nazir, S., Wani, I. A., & Masoodi, F. A. (2017). Extraction optimization of mucilage from Basil (Ocimum basilicum L.) seeds using response surface methodology. *Journal of Advanced Research*, *8*(3), 235–244. doi:10.1016/j.jare.2017.01.003 PMID:28239494

Nhandeyara do Carmo Ramos, S., Xavier, A. L., Teodoro, F., Frédéric Gil, L., & Gurgel, L. (2016). *Removal of cobalt(II), copper(II), and nickel(II) ions from aqueous solutions using phthalate-functionalized sugarcane bagasse: Mono- and multicomponent adsorption in batch mode* (Vol. 79). Academic Press.

Novi, V., Kumar, V., Charles, C., Vishnu, D., Ponnusamy, S. K., . . . Kumar Vaidyanathan, V. (2016). *Review on Nano Adsorbents: A Solution For Heavy Metal Removal From Waste Water* (Vol. 11). Academic Press.

Pagnanelli, F., Mainelli, S., Veglio, F., & Toro, L. (2003). *Heavy metal removal by olive pomace: Biosorbent characterisation and equilibrium modelling* (Vol. 58). Academic Press.

Palumbo, A., Daughney, C., Slade, A., & Glover, C. (2013). *Influence of pH and natural organic matter on zinc biosorption in a model lignocellulosic biofuel biorefinery effluent* (Vol. 146C). Academic Press.

Panayotova, M. (2016). Mining and mineral processing as heavy metals pollution source. Academic Press.

Papageorgiou, S., Kouvelos, E. P., & Katsaros, F. (2008). *Calcium alginate beads from Laminaria digitata for the removal of Cu+2 and Cd+2 from dilute aqueous metal solutions* (Vol. 224). Academic Press.

Park, D., Yun, Y.-S., & Park, J. (2010). *The Past* (Vol. 15). Present, and Future Trends of Biosorption.

Patel, A., Sv, A., Shah, N., & Verma, D. (2018). *Lactic Acid Bacteria As Metal Quenchers To Improve Food Safety And Quality* (Vol. 6). Academic Press.

Pathak, P., Mandavgane, S., & Kulkarni, B. (2015). *Fruit peel waste as a novel low-cost bio adsorbent* (Vol. 31). Academic Press.

Penilla, R. P., Guerrero Bustos, A., & Goñi Elizalde, S. (2006). *Immobilization of Cs, Cd, Pb and Cr by synthetic zeolites from Spanish low-calcium coal fly ash* (Vol. 85). Academic Press.

Pöykiö, R., Nurmesniemi, H., & Keiski, R. (2007). Environmental risk assessment of heavy metal extractability in a biosludge from the biological wastewater treatment plant of a pulp and paper mill. *Environmental Monitoring and Assessment, 128*(1-3), 153–164. doi:10.100710661-006-9301-y PMID:17057998

Přech, J., Pizarro, P., Serrano, D. P., & Čejka, J. (2018). From 3D to 2D zeolite catalytic materials. *Chemical Society Reviews, 47*(22), 8263–8306. doi:10.1039/C8CS00370J PMID:30167621

Prince, W., Perami, R., Espagne, M., Hollman, G. G., Steenbruggen, G., & Janssen-Jurkovicova, M. (1999). *A Two-step Process for the Synthesis of Zeolites from Coal Fly Ash* (Vol. 78). Academic Press.

Qureshi, M. (2015). *Effect of Heavy Metals from Tannery Effluent on the Soil and Groundwater using Multivariate Analysis in District Sheikhupura*. Academic Press.

Ramachandra, T. V., N, A., & Rd, K. (2005). *Biosorption: Techniques and Mechanisms*. Academic Press.

Saritha, V., Srinivas, N., & Srikanth Vuppala, N. V. (2017). Analysis and optimization of coagulation and flocculation process. *Applied Water Science, 7*(1), 451–460. doi:10.100713201-014-0262-y

Sartore, L., & Dey, K. (2019). Preparation and Heavy Metal Ions Chelating Properties of Multifunctional Polymer-Grafted Silica Hybrid Materials. *Advances in Materials Science and Engineering, 11*, 1–11. doi:10.1155/2019/7260851

Sathyamurthy, R., Dharmaraj, V., & Thadhani J, P. (2015). *Thermal and wear analysis on the effect of different compositions of flyash in Mg/SiC composite*. Academic Press.

Sato, T., Shimosato, T., & Klinman, D. M. (2018). Silicosis and lung cancer: Current perspectives. *Lung Cancer (Auckland, N.Z.)*, *9*, 91–101. doi:10.2147/LCTT.S156376 PMID:30498384

Wang, K. (2007). Chemical Precipitation (vol. 37). Academic Press.

Schiewer, S., & Iqbal, M. (2010). *The Role of Pectin in Cd Binding by Orange Peel Biosorbents: A Comparison of Peels, Depectinated Peels and Pectic Acid* (Vol. 177). Academic Press.

Scott, J., Guang, D., Naeramitmarnsuk, K., Thabuot, M., & Amal, R. (2002). *Zeolite synthesis from coal fly ash for the removal of lead ions from aqueous solution* (Vol. 77). Academic Press.

Shakoor, S., & Nasar, A. (2016). *Removal of methylene blue dye from artificially contaminated water using citrus limetta peel waste as a very low cost adsorbent* (Vol. 66). Academic Press.

Singh, P., Singh, S. K., Bajpai, J., Bajpai, A. K., & Shrivastava, R. B. (2014). Iron crosslinked alginate as novel nanosorbents for removal of arsenic ions and bacteriological contamination from water. *Journal of Materials Research and Technology*, *3*(3), 195–202. doi:10.1016/j.jmrt.2014.03.005

Slowing, I. I., Vivero-Escoto, J. L., Trewyn, B. G., & Lin, V. S. Y. (2010). Mesoporous silica nanoparticles: Structural design and applications. *Journal of Materials Chemistry*, *20*(37), 7924–7937. doi:10.1039/c0jm00554a

Somerset, V., Petrik, L. A. White, R., Klink, M. J., Key, D., & Iwuoha, E. (2005). *Alkaline hydrothermal zeolites synthesized from high SiO2 and Al2O3 co-disposal fly ash filtrates* (Vol. 84). Academic Press.

Sudarsanam, P., Peeters, E., Makshina, E. V., Parvulescu, V. I., & Sels, B. F. (2019). Advances in porous and nanoscale catalysts for viable biomass conversion. *Chemical Society Reviews*, *48*(8), 2366–2421. doi:10.1039/C8CS00452H PMID:30785143

Sulaymon, A. (2014). Biosorption of Heavy Metals. *RE:view*, *3*.

Sun, J., & Tan, H. (2013). Alginate-Based Biomaterials for Regenerative Medicine Applications. *Materials (Basel)*, *6*(4), 1285–1309. doi:10.3390/ma6041285 PMID:28809210

Tojo, G., & Fernandez, M. (2006). *Oxidation of Alcohols to Aldehydes and Ketones: A Guide to Current Common Practice*. Academic Press.

Tran, C. D., Duri, S., & Harkins, A. L. (2013). Recyclable synthesis, characterization, and antimicrobial activity of chitosan-based polysaccharide composite materials. *Journal of Biomedical Materials Research. Part A*, *101*(8), 2248–2257. doi:10.1002/jbm.a.34520 PMID:23349116

Udoh, F. V., Udoh, P. B., & Umoh, E. E. (2005). Activity of Alkaloid Extract of Carica papaya. Seeds on Reproductive Functions in Male Wistar Rats. *Pharmaceutical Biology*, *43*(6), 563–567. doi:10.1080/13880200500220961

Vallet-Regí, M., & Balas, F. (2008). Silica materials for medical applications. *The Open Biomedical Engineering Journal*, *2*(1), 1–9. doi:10.2174/1874120700802010001 PMID:19662110

Vanrolleghem, P., & Lee, D. (2003). *On-Line Monitoring Equipment for Wastewater Treatment Processes: State of the Art* (Vol. 47). Academic Press.

Venkateswaran, P., Vellaichamy, S., & Palanivelu, K. (2007). Speciation of heavy metals in electroplating industry sludge and wastewater residue using inductively coupled plasma. *International Journal of Environment Science and Technology, 4*(4). doi:10.1007/BF03325986

Venkateswarlu, P., Vijaya Durga, G., Chitti Babu, N., & Venkateswara Rao, M. (2008). *Biosorption of Zn(II) from an Aqueous Solution by Erythrina variegata Orientalis Leaf Powder* (Vol. 3). Academic Press.

Wang, B., Wan, Y., Zheng, Y. P., Lee, X., Liu, T., Yu, Z., . . . Gao, B. (2018). *Alginate-based composites for environmental applications: a critical review*. Academic Press.

Wang, J., & Chen, C. (2009). *Biosorbents for Heavy Metal Removal and Their Future* (Vol. 27). Academic Press.

Wang, Y., Wang, R., Xu, D., Sun, C., Ni, L., Fu, W., ... Qiu, S. (2016). Synthesis and properties of MFI zeolites with microporous, mesoporous and macroporous hierarchical structures by a gel-casting technique. *New Journal of Chemistry, 40*(5), 4398–4405. doi:10.1039/C5NJ03387J

Wu, S.-P., Dai, X.-Z., Kan, J.-R., Shilong, F.-D., & Zhu, M.-Y. (2016). *Fabrication of carboxymethyl chitosan–hemicellulose resin for adsorptive removal of heavy metals from wastewater* (Vol. 28). Academic Press.

Yantasee, W., Rutledge, R., Chouyyok, W., Sukwarotwat, V., Orr, G., Warner, C., . . . Shane Addleman, R. (2010). *Functionalized Nanoporous Silica for the Removal of Heavy Metals from Biological Systems: Adsorption and Application* (Vol. 2). Academic Press.

Yu, F., Sun, L., Zhou, Y., Gao, B., Gao, W., Bao, C., ... Li, Y. (2016). Biosorbents based on agricultural wastes for ionic liquid removal: An approach to agricultural wastes management. *Chemosphere, 165*, 94–99. doi:10.1016/j.chemosphere.2016.08.133 PMID:27639078

Zhang, Y., Zhu, C., Liu, F., Yuan, Y., Wu, H., & Li, A. (2019). Effects of ionic strength on removal of toxic pollutants from aqueous media with multifarious adsorbents: A review. *The Science of the Total Environment, 646*, 265–279. doi:10.1016/j.scitotenv.2018.07.279 PMID:30055489

Chapter 13
Intensive Biochemical Processes of Wastewater Treatment With High Caloricity Biogas Production

Gheorghe Duca
Institute of Chemistry, Moldova

Victor Covaliov
Moldova State University, Moldova

Olga Covaliova
Institute of Chemistry, Moldova

ABSTRACT

Anaerobic digestion of organic substrates in wastewaters from agro-industrial sector can be efficient method to reduce the pollutant contents, and also a source of biogas. The research is focused on developing of intensified biogas technology and integrated equipment, based on a series of original approaches. New types of new phyto-catalysts of methanogenic digestion were proposed: aescinum, squalene, and betulinol, introduced in the digested biomass in micro-concentrations of 10-3-10-5%. The proposed substances promote about 1.5-2 times acceleration of methanogenic process and increase biomethane contents in biogas close to the natural gas level – up to 93-97%. As a feedstock for biochemical digestion, post-distillery vinasse in mixture with farmyard manure was used. By combining of biochemical treatment with microfiltration an advanced treatment degree was reached, reducing COD from 28530 mg O_2/L to 187 mg O_2/L. In bioreactor proposed, biomass using coefficient reaches 85-93%, as compared to 60% in the conventional biogas processes.

DOI: 10.4018/978-1-7998-1241-8.ch013

INTRODUCTION

The problem of natural environment pollution preventing with wastes and waste waters is nowadays in the centre of attention of many researchers and practitioners in many countries. Although the large number of biochemical treatment technologies and related equipment have been proposed for the wastewaters purification containing the organic pollutants, along with the biogas production (McCarthy, 2001; Show et al., 2010; Khalid et al., 2011; Velmurugan et al., 2014; Mao et al., 2015), the experts always seek for the more efficient, cheap and valuable solutions.

The present research is aimed to intensify the existing biochemical water treatment technology by applying the original approach – introducing the biologically active substances (phyto-catalysts) into the fermenting biomass which makes it possible to essentially improve this process. This approach, involving the application of organic compounds in biogas processes, is novel and has no analogues in the world, although the application of enzymes in biochemical digestion processes has been tested (Parawira, 2011; Sutaryo et al., 2014). Our scope was to enlarge the nomenclature of vegetable raw materials containing the biologically active compounds, which can essentially promote biogas and biomethane formation processes, to perform microbiological and technological studies of their stimulating effects on methanogenesis processes, to develop the integrated bioreactor with the auxiliary equipment to intensify the studied biochemical processes. At the same time, an important issue is to provide the higher wastewaters treatment degree resulted from the digestion of the organic feedstock, as well as to enhance the biomass using coefficient to improve the biogas yield with high caloric value, due to the increased biomethane contents in it. To ensure the treated water reuse in agricultural activity or other technological scopes, it is necessary to explore the combination of treatment methods. The higher treatment efficiency of the combined processes has been already demonstrated by different authors (Kamiya et al., 1998; Campos et al., 2002; Oller et al., 2011; Ganzenko et al., 2014). Such treatments can be applied as a consecutive steps or can be combined within one and the same apparatus.

BACKGROUND

The pollution of surface water with organic compounds from leachates of wastes generated by industries of agricultural products processing has become an issue of serious environmental concern, especially considering the lack of efficient treatment technologies and facilities, their high maintenance and operational costs and need for energy resources for the conventional aerobic digestion treatment of such waste waters (Rozich et al., 2002; Suvilampi, 2003).

This problem is common for many countries worldwide. For instance, in the Republic of Moldova two large rivers Dniester and Prut and a network of their tributaries, small rivers, serve as the main sources of drinking water for the millions of people. However, due to the discharges of untreated wastes and waste waters from the industries of agro-industrial sector, the major economy sector developed in this area, such as wineries, cattle and poultry farms, fruit and vegetable processors the waters of small rivers become highly polluted with organics and suspended solids. As a result, such waters, when inflowing the Dniester and Prut, bring a variety of pollutants, among them rather toxic and persistent substances (Opopol, 2006; Nastasuic et al., 2016). Along with the environmental problems, the natural water sources polluted with the insufficiently treated effluents can be a matter of serious health concerns (Okoh et al., 2010; Verbyla et al., 2015; Xiao et al., 2018).

Along with the important issue of natural water pollution, the problem of renewable energy production and use can also be considered as one of the major global challenge. Existing ways to produce energy demonstrate their devastating effect on the environment (Zhang et al., 2013; Bilen et al., 2018). In case of Moldova, which is also relevant for other countries of the world, there are no natural fossil fuels, but instead, there are significant sources of the organic agricultural wastes and plant raw materials. They can serve as a feedstock to develop a renewable energy, specifically in biogas technology applying an anaerobic processes, which is capable to deliver biomethane and biohydrogen – a "green energy" sources, friendly to environment, being at the same time a way to reduce the pollutants contents (Kaliuzhny, 2004; Nasir et al., 2012; Sivasubramanian et al., 2014; Covaliov et al., 2017).

Generally, a variety of biomass resources can be converted to energy, which can be generally subdivided into the energy crops, forestry residues, industrial and agricultural residues and wastes (McKendry, 2002; Ni et al., 2006, Saxena et al., 2009; Long et al., 2013).

Despite the significant efforts of researchers and technology experts focused on the wastes, wastewaters and energy problem, it is still far from being completely and efficiently resolved.

Thus, the permanent heavy pollution of small and large rivers and other fresh water sources, along with the energy deficiency in the country have motivated us to the develop a totally new approach towards the liquid wastes management generated by agricultural sector – the industries producing wine, alcohol, yeasts, beer, sugar, fruit and vegetables, meat, poultry, diaries. Their wastewaters contain the organic substances such as carbohydrates (starch, glucose, pectin, mono- and oligosaccharides) and others, that can undergo fermentation under the certain conditions (Yakovlev & Kariukhina, 1980; Fiorentino, 2003; Mosse, 2013).

Conventional technologies and equipment used in biogas systems do not allow the biogas obtaining from these materials with the highest possible yield: according to the previous studies, no more than 30-60% of organics are turned into the biogas whereas the average biomethane contents is not exceeding 60-65% in the most cases (Demirel et al., 2010; Sitorus et al., 2013). In addition, to create and maintain the thermophilic conditions needed for the anaerobic fermentation (45-55 °C), rather high energy consumption is required. Therefore, a large part of the energy thus produced should be reused in the same process.

That is why to improve the biogas production systems efficiency, the special attention is to be paid to the modification of its specific flow sheets, technological parameters, equipment, as well as the development and application of novel scientific and technological approaches, considering the environmental aspects (Moletta et al., 1986; Covaliov, 2015).

The optimal conditions for microbial and biochemical processes can be reached through the innovative scientific and technical solutions, taking into account the achievements of related scientific disciplines, including the processes and apparatuses of chemical technology, environmental chemistry, electrochemistry. To ensure the effective combination of biochemical and physico-chemical processes in order to accelerate the digestion of organic substrate and increase the efficiency of biogas technology, it is necessary to fully explore the factors responsible for these processes and to develop innovative methods of fermentation in wastewater treatment, to intensify the processes promoting the increase in biomethane and biohydrogen yield, as sources of renewable energy, to improve the performance of anaerobic wastewater treatment and to increase the value of resulting sediments that may be further used as organic-mineral fertilizers, as well as food additives for cattle feeding (Covaliova et al., 2018). All these factors will enable the biogas technology to become more efficient and cost effective (Covaliov et al., 2012b; Farcas et al., 2014).

There are ten specific enzymes that are characteristic for the methanogenesis process, due to which the formation of methane proceeds from carbon dioxide and hydrogen under the anaerobic conditions, is realized with the methanogenic microorganisms (Daniels et al., 1984). According to the actual tendencies, five bacterial consortiums are involved in the production of anaerobic methane, which are capable to be incorporated in the four stage process, the stages being dependent on each other:

1. Enzymatic hydrolysis of low soluble organic substances up to the formation of substances with more simple bonds, during the development of aerobic bacterial group at the optimal pH = 6.5-7.6.
2. Acid stage effectuated by the heterotrophic group of microorganisms, for which carbon passed into the mixture and transformed organic substances with simple bonds serves as a nutrition source, which results in formation of volatile fatty acids, aminoacids, alcohols and hydrogen and carbon dioxide as well.
3. Acetogenic stage, which takes place due to two bacterial groups: the first is in the form of acetate with the emission of hydrogen from soluble substances on the stage preceding the formation of acids, and the second, acetogenic bacterial group, ensure the transformation of volatile fatty acids, which dissociate into iono-acetates, hydrogen cations and is further used for CO_2 recuperation.
4. Methanogenic stage – methane formation from acetic acid, as a result of a reaction of hydrogen and carbon dioxide with methane bacteria, which transform the methane in two ways – acetate division and carbon dioxide reducing with hydrogen. Anaerobic bacteria are very sensible to the environment at the optimal pH = 7-7.5, although they can work at pH = 9-10.

The mechanism of the methanogenesis energetic processes is not established yet, however, the general principal statements have been proposed. Energy production, at least during the oxidation of H_2, associated with the reduction of CO_2, is connected with the functioning of the electron-transport system, involving dehydrogenases, electron carriers and reductases (Shima et al., 2002). Hydrogenase and formate dehydrogenase have been identified as dehydrogenases. In this process, the methane formation from carbon dioxide and hydrogen occurs with the methanogenic microorganisms under the anaerobic conditions. The detailed review regarding the structure and functions of the enzymes involved in the methane synthesis from carbon dioxide and molecular hydrogen was performed some years ago (Shima, 2014).

The proposed new approach involves the development and testing of new type of biologically active micro- and nano-structural additives to stimulate the biochemical processes and manage them, with the scope to intensify the anaerobic processes of biomethane production and purification on the fermentation of hard degradable raw material and cellulose (Covaliov et al., 2012c). Therefore, the selection and testing of various groups of stimulating biologically active substances as phyto-catalysts of anaerobic treatment of the agro-industrial wastewaters represents a perspective research direction, and is to be based on the biological activity, simplicity of synthesis and/or extraction of identified active principles from the available raw materials.

Another idea is to test the vegetable raw materials as phyto-catalysts of biochemical digestion of organic substrate, instead of using the extracted or especially synthesized bioactive substances. This approach could make the overall treatment process cheaper. A number of enzymes and natural biostimulators (belonging to the class of plants growth stimulators) were preliminarily tested by us for the acceleration of biochemical hydrolysis of refractory compounds in the distillery grains. The better results were obtained using the biostimulators (Covaliov et al., 2015b; Bobeica et al., 2015). This work is devoted to the further search and testing of new biostimulators of methanogenic digestion of organic biomass and waste waters polluted with organic components.

ELABORATION OF NEW PHYTO-CATALYSTS AND INTEGRATED REACTOR FOR BIOCHEMICAL DIGESTION OF ORGANIC MATTER IN AGRO-INDUSTRIAL WASTES AND WASTEWATERS

Experimental

Experimental set-up

The biochemical processes and, specifically, the effect of bioactive substances on methanogenesis have been studied using the especially elaborated set of four bioreactors. The experimental set-up included a bench-scale bioreactor, inserted into the thermostatically controlled reservoir, devices to remove and control the volume of biogas produced, sludge formed and purified liquid. The total volume of each bioreactor was 5000 cm^3, the useful volume was 3500 cm^3. As an internal package, the grapevine rods have been introduced in bioreactor, with the total package surface of 8000 cm^2, in view to provide an additional surface for the development of immobilized methanogenic bacteria and prevent the washing out of microflora (Covaliov et al., 2010).

Fermented Biomass

The mixture of alcohol distillery vinasse (60%) and farmyard manure (40%) was gradually added in bioreactors, to substitute the initial nutrition mixture, containing milk and sugar. The initial vinasse contained dry substances – 8.4-8.8% with the C_{org} in the dry matter – 4.25%; C_{org} in the raw material – 49.5%. Its BOD_5 was 17600 mg O_2/L; COD was 28530 mg O_2/L and pH = 5.3-5.6. The stationary growth regime of methanogenic culture was maintained. Three concentrations of phyto-catalysts were studied at the same time, whereas one bioreactor served as a control. Measurements were carried out in dynamics, after 5, 24, 96 and 120 incubation hours.

Microbial Studies

As a source of methanogenic microorganisms, a sewage sludge was used taken from the Chisinau municipal treatment plant. Microbiological studies have shown that the sewage sludge had slightly alkaline reaction (pH=7.5), and high contents in organic matter (about 21.3%). The sludge was shown to be a biologically active material, in which bacteria are predominant among the microbial community, with negligible amount of fungi. Microbial carbon contents (MB) in the dry waste is 2.5 mg C/g. Thus, a tone of waste contains about 5.0 kg of dry microbial biomass. The sewage sludge contains methane bacteria in the amount of $350 \cdot 10^3$ CFU/g. Among the other bacteria types found, the following are to be named (colony forming units - $CFU \cdot 10^6$/g): heterotrophic bacteria – 1286.8; nocardia – 249.3; actinomycetes – 22.8. CO_2-producing capacity was 1.9 mg/100g/24 h, the dehydrogenase activity – 11.8 mg TPF (triphenylformasane)/10g/24h. The amount of fungi did not exceed $5.4 \cdot 10^3$ CFU/g.

Dehydrogenase Activity

It was measured by colorimetric technique on the base of occurring TPF from TTC (2 3 5 triphenyl-tetrazolium chloride), added to the sample of waste (Haziev, 2005), the substrate being 0.1 M glucose solution. The calculations were made per 10 mL of waste over a 24 h incubation period. The gas volume was measured by the volumetric method, the methane contents – by the gas chromatographic method (Zveagintsev, 1991).

Phyto-Catalysts

The preliminary comparative analysis of proposed additives to be introduced in biochemical fermentation process of the distillery of grains was performed, and the most efficient ones were selected for further research. Thus, a number of biologically active additives were tested, including the aescinum (extracted from the chestnut seeds), squalene (oil extract of amaranth), betulinol (extract from the birth tree cork), etc. All the studied additives were introduced in bioreactor in various concentrations, jointly with the distillery vinasse and farmyard manure as a biomass source (Covaliov et al., 2011).

In the preliminary studies, the change in biogas composition during the distillery grains fermentation process without additives was studied, using gas chromatography. It was shown that the anaerobic fermentation starts with hydrogen release, and then the release of carbon monoxide (CO) begins. Hydrogen contents in the emitted gas was gradually increasing, reaching the maximum of 7-10% in 4-6 days. Then, the decrease in both gases contents is observed. They are transformed into methane and in 18-20 days its contents in biogas increases to 60-65 vol.%.

Intensifying Effect of Phyto-Catalysts on Wastewaters Biochemical Treatment

The testing of a series of biologically active substances was undertaken out with the scope to explore their possible activating or inhibiting effect on methanogenic processes during the biogas formation. All the proposed phyto-catalysts were tested in biochemical digestion processes in the pure form and in the composition of crushed vegetable raw materials. Below, the research results related to the studies of various phyto-catalysts introduced into the digested biomass are presented.

Aescinum (Chestnut Seeds) as Phyto-Catalyst

The first phyto-catalyst studied in the biomass digestion processes, was aescinum. Pure aescinum, triterpene glycoside, was obtained from the chestnut seeds which were pounded coarsely and finely.

The data on the biogas emission rate in dependence of the added amounts of aescinum demonstrate that the maximal gas emission rate for anaerobic digestion in the presence of highest aescinum concentration and was reached within 2-3 days (Table 1). The results obtained clearly testify on the promising use of aescinum as a phyto-catalyst of anaerobic methanogenic digestion of post-distillery vinasse, as the 6-time increase in biogas emission rate is reached while adding the aescinum dose of 10.0 mg/L. Still, within the entire range of studied aescinum concentrations, the limit of gas emission rate was not reached.

It was also shown (Table 2) that the dose of added aescinum can modify the emitted amounts of biogas component gases.

Table 1. Change in the biogas emission rate in the methanogenic process with different doses of added aescimun

Methanogenic Process Time, h		Biogas emission rate, ml/h at the concentration of added aescinum, ml/l			
		Control, no aescinum added	1.5	5.0	10.0
2	4.2	4.2	15.1	15.2	36.1
20	8.0	8.9	33.3	34.3	49.8
40	12.1	12.2	34.8	64.3	89.0
60	12.1	12.4	20.2	38.6	65.9
80	11.5	12.6	18.2	22.4	51.4
100	4.6	11.1	12.0	30.0	30.0
120	3.1	3.2	3.1	3.1	3.1

Table 2. Composition of the emitted biogas in dependence of the amounts of added aescinum

Biogas components	Control test without additives	Dose of aescinum as a pure extract, mg/l		
		2.5	5.0	10.0
Methane, mass.%	67.4	81.8	84.7	85.2
Carbon dioxide, mass.%	21.2	7.2	4.0	3.2
Oxygen, mass.%	4.0	4.0	4.3	4.6
Nitrogen, mass.%	7.4	7.0	7.0	7.0
Carbon monoxide, ppm	2.0	4.0	14.0	13.0
Hydrogen sulfide, ppm	17.0	3.0	1.0	0.0

An increase in the aescinum dose is accompanied with the increased methane contents in the emitted biogas from 67-65 to 85-88 mass.%, which is reached due to the more complete transformation of organics into methane. This is confirmed both by the decrease in carbon dioxide contents in biogas under the higher doses of aescinum, and by the increased contents in carbon monoxide, which is an intermediate product in the methane formation in anaerobic process.

Testing of aescinum as a bio-stimulator for the process of anaerobic decomposition of the organic matter in bioreactors has demonstrated its significant impact on the dehydrogenase activity both in pure form and in the form of seeds.

Dehydrogenase activity in variant with additives was compared with the control variant after 2 and 24 h of incubation. Application of rudely shredded chestnut in the dose of 0.001% activated the dehydrogenase by 1.6 times and finely shredded chestnut in the dose of 0.001% – by 1.8 times immediately after 2 hours of incubation. A similar regularity was preserved in 24 hours after the start of the experiment. Variant with the pure aescinum had the maximum dehydrogenase activity after 24 hours of anaerobic fermentation. The values of dehydrogenase index exceeded the control variant by 1.6 times. The use of finely shredded chestnut demonstrated the stimulating effect after 72 hours of incubation and was statistically significant during the whole experiment. Perhaps this is due to the presence of other biologically active additives, except of aescinum, in the finely shredded part of chestnuts.

It could be suggested that aescinum increases the dehydrogenase activity, reduces the lag-phase and activates the exponential growth of the microbial community in bioreactors. In addition, the finely shredded chestnuts also significantly prolong the stationary growth phase. This promotes the acceleration of biochemical methanogenic processes running and the increase the biogas yield, makes it possible to utilize the waste more completely. As the distillery vinasse contains the aerobic heterotrophic microorganisms, the use of aescinum and chestnuts, stimulating aerobic bacteria growth and reproduction, can activate the aerobic stage of wastes treatment and improve their preparation for methanogenesis. These results suggest indirectly the existence of the mechanism of the biologically active impact of these substances on microorganisms.

It was shown that the use of aescinum obtained from chestnuts in the concentration of 0.001% is the effective method for the biochemical methanogenic processes acceleration. The crushed chestnuts also significantly prolong the stationary growth phase, that which is probably due with the presence of another bioactive substances in the finely shredded fraction. This makes it possible to utilize the waste more completely. The phenomenon of aerobic microorganisms activation by aescinum makes it possible to propose its using at the aerobic stage of the distillery vinasse biochemical treatment, improving the biomass preparation for the anaerobic stage.

Squalene (Amaranth Seeds) as Phyto-Catalyst

Another substance proposed as a phyto-catalyst was squalene. Two series of experiments with squalene as phyto-catalyst were carried – using the pure extracted squalene and a crushed amaranth seeds containing 10% of squalene. The oil extract of amaranth (pure squalene) with concentration of 0.0005% induced an increased biomethane yield in biogas, reaching 93-97%, as compared to the control which was 68-70%. At the same time, the additive of squalene did not render the significant effect on the biochemical process rate.

The experiments with crushed amaranth seeds, instead of pure squalene, has shown that addition of the amaranth seeds in dose of 50 mg/L has induced the 1.6-2.4 times increase in the methanogenesis efficiency after 5, 24 and 96 hours of the anaerobic fermentation of the mixture of distillery grains and farmyard manure. The highest values of gas and methane emissions in the variants with the addition of amaranth seeds have been recorded in 24 hours after the beginning of experiment. The control variant has demonstrated a gradual increase in the amount of gases. The methane contents in the total amount of biogas increases by average of 22.1% at the dose of amaranth seeds of 50 mg/L, accounting for 65.3-71.3%, while on the control remains approximately the same – 52.5-58.4%. The methane contents in the total volume of the biogas increases towards the end of the experiment from 65.3% to 71.3% on the variant with the dose of 50 mg/L of amaranths' seeds, from 54.0% to 70.3% with 100 mg/L and 53.8% to 68.2% on the variant with the dose of 200 mg/L.

It was found that the increase in amaranth seeds concentration higher than 50 mg/L is not rational, as it is not accompanied any more with the gas release rate acceleration and CH_4 emissions.

Testing the amaranth seeds as phyto-catalyst for the process of anaerobic decomposition of the organic matter in bioreactors has demonstrated its significant impact on the dehydrogenase activity at all doses. Thus, dehydrogenase activity in bioreactors with amaranth was significantly lower compared to the control test. After 5 hours from the start of the experiment, the 10.8-21 times decrease in the activity was observed. The differences were smoothed out through the 96-120 hours, but they continued to be statistically significant. Differences between the dose of amaranth of 50 mg/L and doses of 100 and 200 mg/L were statistically significant during the whole experiment. Differences between doses of amaranth of 100 and 200 mg/L were not significant.

These results suggest that the dehydrogenase index reliably reflects the decline in the contents of organic substances in fermenting wastes, as the result of their anaerobic fermentation, accompanying with biogas release. Dehydrogenase index is thus in the close negative connection with the amount of resulting gases, including biomethane.

Therefore, it was demonstrated that the adding of amaranth seeds to the methanogenic bioreactor has a clear inhibitory effect on the dehydrogenase activity and stimulates the process of methanogenesis. It may be suggested that both squalene and betulinol, the other tested phyto-additive, stimulate the aerobic microorganisms' multiplication, which can activate the aerobic stage of wastes treatment and improve their preparation for the methanogenesis process.

Betulin Derivatives as Phyto-Catalysts

In addition to the previously proposed phyto-catalysts, using of betulin in the biomass digestion has also been proposed.

In was shown that iIntroduction of isoprenoids (betulin derivatives) in concentration 0.0005%, caused more than two times increase in the biogas emission rate, reaching 12.5-15.0 mL/h, and induced 3-4 times increase in the methanogenesis efficiency. Both amaranth and betulinol were found to stimulate the growth of heterotrophic aerobic bacteria.

The biological activity of substances in the methanogenic process depends on their availability to microorganisms. As the amaranth oil in the seeds is in the homogeneous state in the inter-cellular space, it is possible to avoid the squalene extraction operation from the seeds. Thus, the seeds can be used in the crashed state in the biochemical fermentation processes. This would make the process cheaper. If the phyto-catalyst within the vegetable cells is made of the dense cellulose, as in the case of betulinol, they are not easily available for microorganisms. Therefore, betulinol extraction is needed.

Application of a Mixture of Phyto-Catalysts

Initially, pure substances were studied as phyto-catalysts in the biogas formation processes. However, it was interesting also to explorey the behavior of the mixture of the active additives on the methanogenic process.

The initial suspension of biologically active additives contained 1 mg of betuline and 1 mg of amaranth oil per mL of active additives contained (0.2% with regard to the active substances). The following concentrations of biologically active additives have been selected for the research: 0.00025%, 0.0005% and 0.001%, which was equivalent to the introduction in bioreactor of each 2.5, 5.0 and 10 mg of active substance per litre of the distillery grains introduced.

The results of the study of various concentrations of biologically active additives on the formation rate, yield and general volume of biogas, resulting from the methanogenic process, have shown that the joint application of two phyto-catalysts essentially increases the rate of gas release (3.5-4 times) (Covaliov et al., 2012a).

The increase in phyto-catalysts concentration higher than 5 mg/L is not rational, as it is not accompanied by further gas emission rate acceleration. The analysis of qualitative and quantitative composition of biogas has shown that the methane content made 58-60% in the control sample, whereas in the gas produced with introduction of phyto-catalysts, was 90-92% of methane.

The data obtained have demonstrated that the additives in the methanogenic process make it possible to obtain practically the same biogas volume during 25 hours, compared to the 93 hours in the control test, the biogas emission rate being thus about 4 times higher. The explanation of this phenomenon can be connected with the molecular structure of these compounds. Intensification of anaerobic fermentation seems to be ensured due to the antioxidant, antihypoxant and amnimutagenic activity of the additives of betulinol, which extract stabilizes the cellular membranes of microorganisms, enhances the stability of their cells, increases the stability of their cells with regard to the hypoxia (oxygen lack), decreases the peroxide oxidation of lipids and prevents the damage of cellular membranes. This makes it possible to accelerate the running of biochemical methanogenic processes and increase the biogas yield.

Squalene is a biochemically active substance, which has a property to capture oxygen and saturate the microorganism's tissues with it during the simple biochemical interaction with water. Introduction of squalene in biochemical digestion process ensures the more complete running of methanogenic process. In its turn, this promotes the higher biogas yield and higher contents in biomethane in it. In the presence of the mixture of squalene and betulinol or its derivatives, the synergetic effect appears, that promotes the more complete running of methanogenic process and therefore, more efficient anaerobic fermentation of organic substrate.

As a result of a combined biochemical treatment in the presence of proposed phyto-additives and the following microfiltration treatment, the initial concentration of organics in feedstock (COD=28530 mg O_2/l), has been essentially decreased to about 187 mg O_2/L.

An increase in the dose of biologically active substance from 5 to 10 mg/L increased the maximal rate of gas emission within one to four days. In the case of regular introduction of amaranth, it becomes possible to keep the maximal gas emission rate during the entire process. It is to be mentioned that the introduction of bioactive agent containing betulinol, inhibits the methanogenic process, however, adding of betulinol and squalene jointly, gives the positive effect, improving the efficiency of biochemical process. This different effect of additives may be explained by the different mechanisms of their influence on the organic substrate fermentation. At the same time, it was shown that the introduction of microadditives at extremely low concentrations into the biochemical fermentation process has a number of advantages. Thus, the microorganisms do not cause a damage to the environment, and help to re-establish and maintain the ecological balance. They do not show any unfavorable effects on microorganisms, present at the treatment plants, and do not bring any harm to them.

INTEGRATED ANAEROBIC REACTOR FOR THE ADVANCED MICROFILTRATION TREATMENT OF WASTEWATER

To provide practical application of the intensified anaerobic process elaborated and reach the advanced water purification from the organic pollutants, the integrated bioreactor has been developed. It involves several built-in components, including the colloidal mill and membrane microfiltration module. Automatic control of the system can be provided with the help of control panel, according to the sensor readings of biomethane contents in biogas composition.

An important intensifying factor, enhancing the anaerobic digestion efficiency, is introduction of phyto-stimulating bioactive micro-additives and their and uniform distribution within biomass, specifically, betulinol, squalene, etc., affecting the microorganisms' living activity, rising the energy and growth rate of the methanogenic bacterial consortium. Intensification of methanogenic process, rate increase of the biochemical reactions improve the technological indicators, including the higher biogas yield and biomethane contents in it.

The liquid biomass and phyto-catalyst are introduced into the colloidal mill for grinding and homogenization. Application of colloidal allows to use the crushed vegetable feedstock (such as squalene-containing amaranth seeds) as phyto-catalyst, instead of the extracted pure bioactive substances, which makes the technological process cheaper. Due to the destruction of large molecular structures and fine dispersion of mechanical suspensions in biomass, up to the particles size of less than 1-0.05 mkm, the microorganisms' availability rises, with regard to the biochemical transformations. In this way, the efficiency of the biomass anaerobic digestion is increased.

To provide an intensive mass-exchange and mass-transfer, the reciprocating motion of the reverse funnel is used, with the digested biomass recirculation within the reactor, which is the principally novel and original approach. Unlike the other agitation methods, this approach allows preventing the microorganisms of biomass ruin because of the mechanical action (Covaliov et al., 2015a). In addition, the uniform temperature distribution within the reactor and optimal digestion conditions are created.

Biochemically treated liquid is a suspension which is settling in the conic part of the reactor. The clarified portion of liquid from the upper zone of reactor is conveyed into the microfiltration module, which makes it possible to separate the low-concentrated water fraction (permeate) from the undigested portion of active sludge (retentate), the latter fraction being re-introduced in biochemical process and repeatedly digested. An advantage of this process is advanced water treatment, which makes it possible to avoid using of the aerobic treatment technology for this scope, which rather costly.

Due to the microfiltration separation of phases, not only the clarified liquid is produced, but the biomass using coefficient is rising, along with the increased amount of biomethane emitted per biomass unit. Under the standard conditions up to 60% of biomass is consumed for gas formation during the anaerobic digestion, and 40% of biomass remains unused. Unlike this, the proposed microfiltration system and undigested biomass reuse in methanogenic process increase the biomass decay coefficient up to 85-95%.

Membrane phase separation in microfiltration unit also ensures practically complete absence or retention of microbial cultures inside the reactor, including the pathogenic agents. Therefore, the high microflora concentration in bioreactors is maintained with the simultaneous decontamination of treated wastes. Another specifics of bioreactor elaborated is using of package for microflora fixation, which prevents its washing away and intensifies the gas formation. In this way, immobilization of microorganisms cells on package surface allows to maintain the high density and concentration of active microflora with multiple reuse of microbial cells, promoting their adaptation towards the unforeseen discharges of high pollutants concentrations.

Based on the proposed approach and technological operations, the generalized process scheme was elaborated (Figure 1) comprising the specific operations to reach the most efficient indicators of waste water treatment from the organic pollutants, with the production of purified water and biogas.

Figure 1. Flow sheet of the intensified treatment of waste waters biomass, with the production of high caloricity biogas

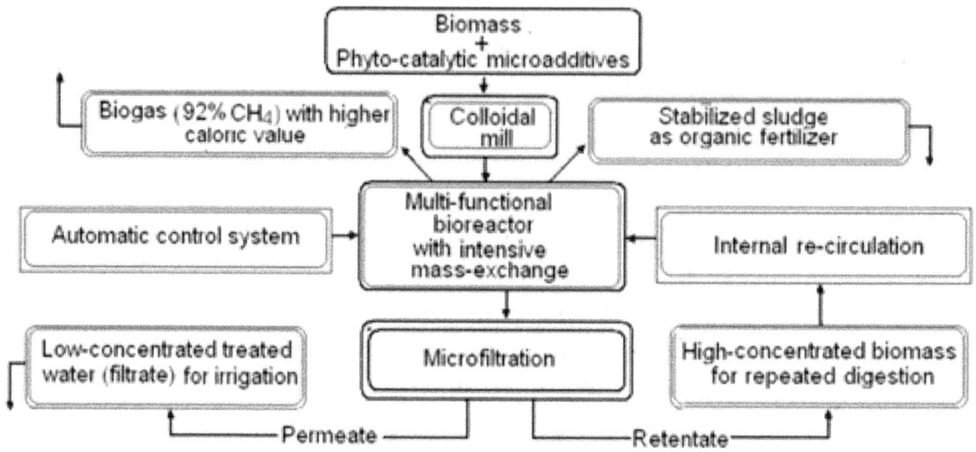

CONCLUSION

New promising phyto-catalysts of biochemical digestion processes of organic matter in the agro-industrial wastes and wastewaters were tested, specifically: aescinum, squalene and betulinol. Introducing of aescinum makes it possible to increase the gas emission rate, compared to the control test: 2.5 times increase is reached at the aescinum dose of 2.5 mg/l; 4-times increase – at the aescinum dose of 5.0 mg/l; 6-times increase – at the aescinum dose of 10.0 mg/l. The phenomenon of aerobic microorganisms activation by aescinum makes it possible to propose its using at the aerobic stage of the distillery vinasse biochemical treatment.

The pure squalene with concentration of 0.0005% induced an increased biomethane yield in biogas, reaching 93-97%, as compared to the control which was 68-70%. Using of amaranth seeds as a squalene source has shown that the methane contents in the total amount of biogas emitted was increased by about 22% at the dose of amaranth seeds of 50 mg/L, accounting for about 71.3%.

Using of betulinol in concentration 0.0005%, provoked more than 2-times rising of biogas emission rate, reaching 12.5-15.0 mL/h, and induced 3-4 times increase in the methanogenesis efficiency. Both amaranth and betulinol stimulate the growth of heterotrophic aerobic bacteria. The possibility of using the raw materials as phyto-catalysts – aescinum-containing chestnut seeds, squalene-containing crushed seeds of amaranth, betulinol-containing birch-tree cortex - in the biochemical methanogenic technology has shown their lower efficiency as compared to the pure extracted products, but provides the possibility to make the biochemical treatment process cheaper.

A combined biochemical-microfiltration treatment (the latter process is used instead of traditional anaerobic process), in the presence of proposed phyto-additives, allows to essentially reduce COD contents in treated water – from 28530 mg O_2/L to about 187 mg O_2/L.

To ensure practical using of elaborated intensified anaerobic process, while reaching the advanced water purification from the organic pollutants, an integrated bioreactor has been elaborated, based on some original technical approaches: colloidal mill for biomass efficient homogenization and membrane microfiltration module, applied instead of aerobic treatment stage. Automatic control system is envisaged. Both microfiltration system and biomass reuse in methanogenic digestion rise the biomass decay coefficient up to 85-95%.

An integrated scheme of intensified treatment of waste waters biomass was elaborated, comprising the series of steps and proposed approaches, aimed at the with the production of high caloricity biogas and high quality treated water. Intensive biochemical treatment technology, along with the developed integrated bioreactor make it possible to reach an enhanced degree of biomass using and its digestion efficiency, along with the advanced water treatment for its discharges in natural water bodies.

Acknowledgment

This research was supported by the STCU (American Science and Technology Centre in Ukraine) Projects No. 5832, No. 5998, as well as Institutional Project Nr. 15.817.02.35A financed by the Government of the Republic of Moldova. The authors wish to thank the team of researchers who were involved in experimental work and theoretical justification of the results obtained – Dr. Hab. Valentin Bobeica, Dr. Dumitru Ungureanu, Dr. Vladimir Nenno, Dr. Irina Senicovscaia.

REFERENCES

Bilen, K., Ozyurt, O., Bakırcı, K., Karslı, S., Erdogan, S., Yılmaz, M., & Comaklı, O. (2018). Energy production, consumption, and environmental pollution for sustainable development: A case study in Turkey. *Renewable & Sustainable Energy Reviews*, *12*(6), 1529–1561. doi:10.1016/j.rser.2007.03.003

Bobeica, V., Covaliov, V., Nenno, V., & Covaliova, O. (2015). Phytochemical compounds and plant products containing active principles as stimulators of the methanogenic processes of the anaerobic microorganisms. In *Proceedings of the SIMI 2015 International Symposium "The Environment & The Industry"* (pp. 57-65). INCD ECOIND.

Campos, J. C., Borges, R. M. H., Oliveira Filho, A. M., Nobrega, R., & Sant'Anna Jr, G. L. (2002). Oilfield wastewater treatment by combined microfiltration and biological processes. *Water Research*, *36*(1), 95–104. doi:10.1016/S0043-1354(01)00203-2 PMID:11767743

Covaliov, V. (2015). Improving the biogas production systems (Rus.). *Academos*, *4*, 38–44.

Covaliov, V., Bobeica, V., Ungureanu, D., Covaliova, O., Duca, Gh., Nenno, V. & Senicovschaia, I. (2012). *Method of anaerobic treatment of liquid organic wastes*. Moldova Patent, No. 4189.

Covaliov, V., Covaliova, O. & Ungureanu, D. (2010). *Process for the obtaining of a volume charge for microflora fixation and anaerobic reactor with its use for biochemical treatment of waste waters*. Moldova Patent, No. 187Y.

Covaliov, V., Covaliova, O. & Ungureanu, D. (2015). *Combined reactor for biogas production*. Moldova Patent, No. 4382.

Covaliov, V., Covaliova, O., Ungureanu, D. & Nenno, V. (2011). *Reactor for biogas production*. Moldova Patent, No. 4130.

Covaliov, V., Duca, Gh., & Covaliova, O. (2017). "Green Energy": Innovative Ecobiotechnologies and Integrated Reactors. Anthology of Inventions. Chisinau, Moldova: CEP USM.

Covaliov, V., Nenno, V., Bobeica, V., Senicovscaia, I., Covaliova, O., & Ungureanu, D. (2012). Stimulation of the methanogenic fermentation processes of liquid organic wastes. *Studia Universitatis Moldaviae. Real and Natural Sciences Series.*, *1*(51), 160–168.

Covaliov, V., Senicovscaia, I., Nenno, V., Bobeica, V., & Covaliova, O. (2015). Effects of Amaranths' Seeds on Dehydrogenase Activity and Gases Emission in Methanogenic Bioreactors (Rus.). *Studia Universitatis Moldaviae. Real and Natural Sciences Series*, *1*(81), 230–235.

Covaliov, V., Ungureanu, D., & Covaliova, O. (2012). Theoretical and applied aspects of improving the biogas technology. In *Problems of the Regional Power Engineering* (pp. 3–17). Chisinau, Moldova: Institute of Power Engineering ASM.

Covaliova, O., Duca, Gh., Covaliov, V., & Bobeica, V. (2018). Stimulation of Biochemical production of Cyanocobalamin (Vitamin B12) with Phyto-Additives and Electrolytic Hydrogen. In Compounds and materials for drugs development and biomedical applications (pp. 159-180). Bucuresti: Ed. Academiei Romane; Braila: Istros Publi. House.

Daniels, L., Sparling, R., & Sprott, G. (1984). The bioenergetics of methanogenesis. *Biochimica et Biophysica Acta*, *6*(2), 113–163. doi:10.1016/0304-4173(84)90002-8 PMID:6236847

Demirel, B., Scherer, P., Yenigun, O., & Onay, T. (2010). Production of Methane and Hydrogen from Biomass through Conventional and High-Rate Anaerobic Digestion Processes. *Critical Reviews in Environmental Science and Technology*, *40*(2), 116–146. doi:10.1080/10643380802013415

Farcas, N., & Matei, F. (2014). Main factors affecting biogas production - an overview. *Romanian Biotechnological Letters*, *19*(3), 9283–9296.

Fiorentino, A., Gentili, A., Isidori, M., Monaco, P., Nardelli, A., Parrella, A., & Temussi, F. (2003). Environmental Effects Caused by Olive Mill Wastewaters: Toxicity Comparison of Low-Molecular-Weight Phenol Components. *Journal of Agricultural and Food Chemistry*, *51*(4), 1005–1009. doi:10.1021/jf020887d PMID:12568563

Ganzenko, O., Huguenot, D., Esposito, G., & Oturan, M. A. (2014). Electrochemical advanced oxidation and biological processes for wastewater treatment: A review of the combined approaches. *Environmental Science and Pollution Research International*, *21*(14), 8493–8524. doi:10.100711356-014-2770-6 PMID:24965093

Haziev, F. (2005). Methods of soil enzymology. Moscow: Nauka.

Kaliuzhny, S. (2004). High-intensive anaerobic industrial wastewaters treatment technologies. *Cataysis in Industry*, *6*, 42-50.

Kamiya, T., & Hirotsuji, J. (1998). New combined system of biological process and intermittent ozonation for advanced wastewater treatment. *Water Science and Technology*, *38*(8-9), 145–153, 147–153. doi:10.2166/wst.1998.0801

Khalid, A., Arshad, M., Anjum, M., Mahmood, T., & Dawson, L. (2011). The anaerobic digestion of solid organic waste. *Waste Management (New York, N.Y.)*, *31*(8), 1737–1744. doi:10.1016/j.wasman.2011.03.021 PMID:21530224

Long, H., Li, X., Wang, H., & Jia, J. (2013). Biomass resources and their bioenergy potential estimation: A review. *Renewable & Sustainable Energy Reviews*, *26*, 344–352. doi:10.1016/j.rser.2013.05.035

Mao, C., Feng, Y., Wang, X., & Ren, G. (2015). Review on research achievements of biogas from anaerobic digestion. *Renewable & Sustainable Energy Reviews*, *45*, 540–555. doi:10.1016/j.rser.2015.02.032

McCarthy, P. L. (2001). The development of anaerobic treatment and its future. *Water Science and Technology*, *44*(8), 149–156. doi:10.2166/wst.2001.0487 PMID:11730130

Moletta, R., Verrier, D., & Albagnac, G. (1986). Dynamic modeling of anaerobic digestion. *Water Research*, *20*(4), 427–434. doi:10.1016/0043-1354(86)90189-2

Mosse, K. P., Verheyen, T. V., Cruickshank, A. J., Patti, A. F., & Cavagnaro, T. R. (2013). Soluble organic components of winery wastewater and implications for reuse. *Agricultural Water Management*, *120*, 5–10. doi:10.1016/j.agwat.2012.05.011

Nasir, M. I., Mohd Ghazi, T. I., & Omar, R. (2012). Production of biogas from solid organic wastes through anaerobic digestion: A review. *Applied Microbiology and Biotechnology*, *95*(2), 321–329. doi:10.100700253-012-4152-7 PMID:22622840

Nastasiuc, L., Bogdevici, O., Overcenco, A., Culighin, E., Sidorenko, A., & Vaseashta, A. (2016). Monitoring Water Contaminants: A Case Study for the Republic of Moldova. *Polish Journal of Environmental Studies*, *25*(1), 1–11. doi:10.15244/pjoes/58888

Ni, M., Leung, D., Leung, M., & Sumathy, K. (2006). An overview of hydrogen production from biomass. *Fuel Processing Technology*, *87*(5), 461–472. doi:10.1016/j.fuproc.2005.11.003

Okoh, A. I., Sibanda, T., & Gusha, S. S. (2010). Inadequately treated wastewater as a source of human enteric viruses in the environment. *International Journal of Environmental Research and Public Health*, *7*(6), 2620–2637. doi:10.3390/ijerph7062620 PMID:20644692

Oller, I., Malato, S., & Sánchez-Pérez, J. A. (2011). Combination of Advanced Oxidation Processes and biological treatments for wastewater decontamination - A review. *The Science of the Total Environment*, *409*(20), 4141–4166. doi:10.1016/j.scitotenv.2010.08.061 PMID:20956012

Opopol, N. (2006). Water management in Republic of Moldova. Principal Threats. In Management of Intentional and Accidental Water Pollution (pp. 167-176). Springer.

Parawira, W. (2011). Enzyme research and applications in biotechnological intensification of biogas production. *Critical Reviews in Biotechnology*, *32*(2), 172–186. doi:10.3109/07388551.2011.595384 PMID:21851320

Rozich, A., & Bordacs, K. (2002). Use of thermophilic biological aerobic technology for industrial wastewater treatment. *Water Science and Technology*, *46*(4-5), 83–89. doi:10.2166/wst.2002.0557 PMID:12361053

Saxena, R. C., Adhikari, D. K., & Goyal, H. B. (2009). Biomass-based energy fuel through biochemical routes: A review. *Renewable & Sustainable Energy Reviews*, *13*(1), 167–178. doi:10.1016/j.rser.2007.07.011

Shima, S. (2014). Enzyme chemistry of methanogenesis and anaerobic oxidation of methane. *Proceedings of Max Plank Institute for Terrestrial Microbiology*, *52*, 307–312.

Shima, S., Warkentin, E., Thauer, R., & Ermler, U. (2002). Structure and function of enzymes involved in the methanogenic pathway utilizing carbon dioxide and molecular hydrogen. *Journal of Bioscience and Bioengineering*, *93*(6), 519–530. doi:10.1016/S1389-1723(02)80232-8 PMID:16233244

Show, K. Y., Tay, J. H., & Hung, Y. T. (2010). Global Perspective of Anaerobic Treatment of Industrial Wastewater. In L. Wang, J. H. Tay, S. Tay, & Y. T. Hung (Eds.), *Environmental Bioengineering. Handbook of Environmental Engineering* (Vol. 11, pp. 773–807). Totowa, NJ: Humana Press. doi:10.1007/978-1-60327-031-1_23

Sitorus, B., Sukandar, & Panjaitan, S. D. (2013). Biogas Recovery from Anaerobic Digestion Process of Mixed Fruit -Vegetable Wastes. *Energy Procedia*, *32*, 176–182. doi:10.1016/j.egypro.2013.05.023

Sivasubramanian, V., & Jayaraj, S. (2014). Biogas Generation through Anaerobic Digestion Process. *Research Journal of Chemistry and Environment*, *18*(5), 80–93.

Sutaryo, S., Ward, A. J., & Moller, H. B. (2014). The effect of mixed-enzyme addition in anaerobic digestion on methane yield of dairy cattle manure. *Environmental Technology*, *35*(17-20), 2476–2482. doi:10.1080/09593330.2014.911356 PMID:25145202

Suvilampi, J., & Rintala, J. (2003). Thermophilic aerobic wastewater treatment, process performance, biomass characteristics, and effluent quality. *Reviews in Environmental Science and Biotechnology*, *2*(1), 35–51. doi:10.1023/B:RESB.0000022959.46025.9a

Velmurugan, S., Deepanraj, B., & Jayaraj, S. (2014). Biogas Generation through Anaerobic Digestion Process – An Overview. *Research Journal of Chemistry and Environment*, *18*(5), 80–94.

Verbyla, M. E., Cairns, M. R., Gonzalez, P. A., Whiteford, L. M., & Mihelcic, J. R. (2015). Emerging challenges and synergies for pathogen control and resource recovery in natural wastewater treatment systems. *WIREs. Water*, *2*(6), 701–714. doi:10.1002/wat2.1101

Xiao, S., Hu, S., Zhang, Y., Zhao, X., & Pan, W. (2018). Influence of sewage treatment plant effluent discharge into multipurpose river on its water quality: A quantitative health risk assessment of Cryptosporidium and Giardia. *Environmental Pollution*, *233*, 797–805. doi:10.1016/j.envpol.2017.11.010 PMID:29141237

Yakovlev, S., & Kariukhina, T. (1980). Biochemical processes in the biochemical treatment of wastewaters. Moscow: Academic Press.

Zhang, C., & Diaz Anadon, L. (2013). Life Cycle Water Use of Energy Production and Its Environmental Impacts in China. *Environmental Science & Technology*, *47*(24), 14459–14467. doi:10.1021/es402556x PMID:24125477

Zveagintsev, D. (1991). Methods of soil microbiology and biochemistry. Moscow: MGU.

Chapter 14
Current Approaches of Nanotechnology for Potential Drinking Water Purification

Rohini Singh
https://orcid.org/0000-0003-2967-3439
Sitarambhai Naranji Patel Institute of Technology and Research Centre, India

Suman Dutta
https://orcid.org/0000-0002-2695-8108
Indian Institute of Technology, Dhanbad, India

ABSTRACT

Uninterrupted supply of adequate and clean drinking water is a fundamental human necessity, and essential to maintain the healthy surroundings. It also influences the development and economic growth, specifically in developing countries. It provides an excellent opportunity for treatment of surface water, groundwater, and wastewater contaminated by toxic metal ions, organic and inorganic solutes, and microorganisms. It facilitates an opportunity to optimize the current conventional techniques and to offer novel and innovative methods of water purification. Various nanomaterials on which purification of contaminated drinking water is based includes nanostructured catalytic membranes, nanosorbents, nanocatalysts, and bioactive nanoparticles. Current and potential applications of nanotechnology in water purification are adsorption, membrane separation processes, photocatalysis, disinfection, and microbial control. This chapter reviews the recent research efforts accomplished in the area of nanotechnology for drinking water purification.

INTRODUCTION

Affordable and clean drinking water became a major global challenge in 21st century as water is the origin of human civilization. The continuous increase in population, urbanization and industrialization are the main reason for the contamination of water (Qu, Alvarez and Li 2013). Water is an essential fac-

DOI: 10.4018/978-1-7998-1241-8.ch014

Table 1. Major drinking water contaminants and their permissible limit (WHO 2011)

Major pollutant	Origin	Permissible Limits(mg/l)	Affected countries	Health effects
Fluoride	- Geological origin, mineral weathering, coal mining	0.5- 1	India, China, Central Africa and South America	Dental and skeletal fluorosis, Muscle fiber regeneration, nervous system malfunction
Cadmium	Fertilizers, impurities in the zinc of galvanized pipes and solders and some metal fittings	0.003		Kidney damage
Selenium	Earth's crust	0.04		gastrointestinal disturbances, discoloration of the skin, decayed teeth, hair or nail loss
Arsenic	- Geological origin	0.01	Bangladesh, India, China, Pakistan, Nepal, Myanmar, Vietnam	High blood pressure, hyperpigmentation, ,cancer
Mercury	- Industrial pollution, dental filling, Food (fish)	0.006	Indonesia, China, Africa,Philippines, Japan,Kazakhstan, USA,Brazil,Australia, Taiwan	kidney damage, respiratory failure, gastrointestinal failures.
Lead	- Old piping lines, mineral weathering, paint	0.01	Egypt, EU, USA, Thailand, China, Cambodia	Delays in physical/ mental development, Kidney problem, high blood pressure

tor to sustain life, and an adequate, safe and accessible supply must be available to all. Improving access to safe drinking water can result in tangible benefits to health. Every effort should be made to achieve drinking water that is as safe as practicable. The era of water purification had begun since the first governmental act was passed in 1852 entitled; Metropolis Water Act of 1852 (Pradeep and Anshup 2009). The principal objective of wastewater purification is generally to dispose of human and industrial effluents without affecting human health or causing damage to the natural environment. Water is considered as contaminated if the level of introduced pollutants exceeds a certain permissible limit, making it unsuitable for human consumption. WHO produces international norms on water quality and human health in the form of guidelines that are used as the basis for regulation and standard setting, in developing and developed countries worldwide. Table 1 lists some of the major drinking water contaminants and their permissible limits according to WHO guidelines for drinking-water quality (WHO 2011) designed for the protection of public health and the management of the risk from hazards.

Major organic pollutants include hazardous organic dyes such as methylene blue, rhodamine 6G, Eosin Y and many more.

Table 2 lists various sources and example of chemical contaminants. The seriousness of pure drinking water must be looked at from two perspectives: first, the amount of accessible water and second, the quality of drinking water.

Table 2. Categorization of chemical contaminants source

Sources of chemical contaminants	Examples of sources
Natural	Rocks, soils and sewage inputs
Industrial	Mining, manufacturing and processing industries, solid wastes, urban runoff, fuel leakages
Agricultural	Manures, fertilizers and pesticides

Table 3. Major conventional technologies utilized for drinking water purification

	Reverse osmosis	Ultrafiltration	Electro-dialysis	Chemical Precipitation
Description	Flow of solvent through membrane on the application of osmotic pressure	Hydrostatic pressure forces a liquid against a semipermeable membrane	Ions are transported through semi permeable membrane, under the influence of an electric potential	Suspended particles form larger aggregates in the solution
Advantages	Low energy and space requirement	Simple and compact, Low energy consumption	Low energy consumption,	Relatively inexpensive, minimal maintenance
Disadvantages	High pressure requirement, susceptible to fouling	Cannot separate dissolved salts or low molecular weight (MW) species	Organic matter, colloids and SiO_2 are not removed	Corrosive chemicals may pose safety concerns

Increasing urbanization and industrialization have a significant contribution in the economic growth. However, it has also been trade off with many environment related issues such as contaminated water resources. Natural wastewater treatment systems include constructed wetlands, infiltration field and many more. Advantage of the natural treatment of wastewater is its simple technology but possess low efficiency. Ways to overcome these issues involve the development of novel and facile strategies for producing clean drinking water. Table 3 describes the conventional technologies utilized for drinking water purification. Nanotechnology enabled water and wastewater treatment promise to overcome major challenges faced by existing treatment technologies. It helps to provide new treatment capabilities that could allow economic utilization of unconventional water sources to expand the water supply.

Other significant properties of nanomaterials include- superparamagnetism, localized surface plasmon resonance, and quantum confinement effect (Zhang et al. 2016; Adeyemi et al. 2016). The major objective of this chapter is to provide an integrated perspective of the research efforts in the area of drinking water purification utilizing nanotechnology.

ROLE OF NANOTECHNOLOGY IN WASTEWATER TREATMENT

Nanotechnology is the act of manipulation of matter at very tiny scales or nanoscale. The size and shape directly affect the intrinsic properties of such materials. Nanomaterials can be broadly grouped into four categories depending upon the confinement: (a) zero-dimensional (0-D), (b) one-dimensional (1-D), (c) two-dimensional (2-D), and (d) three-dimensional (3-D) nanomaterials. According to the geometry, they can be classified as isotropic and anisotropic nanoparticles. The general rules of physics and chemistry often fail when materials have one or more of their dimensions less than 100 nanometres and displays unique and dramatically improved properties. Low cost and highly efficient nano-based technology in the water treatment sector are capable of providing clean drinking water in very remote regions. Nanotechnology can significantly contribute in the effective removal of the contaminants such as heavy metals (e.g. mercury, arsenic, and hexavalent chromium) and biological toxins, organic and inorganic contaminants (Brame et al. 2011; Hua et al. 2012; Woo and Park 2016).

Eventually, nanotechnology has the capability to-

Enhance Potable Water Supplies: By developing the economically feasible portable purifiers, filters and other techniques, clean drinking water can be provided in the rural areas close to industrial belts, where the accessible water is heavily contaminated.

Desalinate Seawater: By the effective separation of salt from seawater, another huge source of consumable water can be offered which would be more economically than existing methods.

Safeguard Industrial Effluent: To protect the environment from industrial practices many new, effective and cost efficient techniques could be applied. For example, mining industries utilizes nanotechnology to prevent the contamination of groundwater from inactive mines by cleaning of acid mine drainage sources.

NANOMATERIALS AND WATER PURIFICATION

Nanomaterials are commonly characterized as materials smaller than 100 nm in no less than one dimension. At this scale, materials frequently have novel size-dependent properties many of which have been investigated for applications in water and wastewater treatment. Some of the properties of nanomaterials are high specific surface area, which is responsible for fast dissolution, high reactivity, strong sorption as well as discontinuous properties, such as superparamagnetism, localized surface plasmon resonance, and quantum confinement effect. Broadly, nanoscale materials utilized for water purification can be classified as: (1) zeolites (2) dendrimers (3) carbonaceous nanomaterials and (4) metal-containing nanoparticles (5) semiconductor nano-photocatalysts . The above mentioned nano materials possess wide range of physical and chemical properties that make them highly suitable for the purpose of water purification. There are two noteworthy research requirements for full-scale uses of nanotechnology in water/wastewater treatment. To start with, the execution of different nanotechnologies in treating genuine regular and wastewaters should be tried. Future studies should be done under more reasonable conditions to evaluate the pertinence and proficiency of distinctive nanotechnologies. Secondly, the long-term efficiency evaluation of these nanotechnologies is in great need. Despite the efficient performance, the acceptance of innovative technologies strongly depends on the cost effectiveness and the potential risk involved. The current cost of nanomaterials is high with few exceptions such as nano-TiO_2, nanoscale iron oxide, and polymeric nanofibers (Liu et al. 2016).

Dendrimers for Water Purification

Dendrimers are nano-sized hyperbranched molecules with radial symmetry. The invention of dendritic polymers as the result of macromolecular chemistry advancement can be considered as an excellent platform to develop efficient UF processes for the treatment of water contaminated with toxic metal ions, organic and inorganic pollutants and microorganisms such as- virus and bacteria (Lancelot et al. 2017; Beraa et al. 2017). Magnetic nanoparticles modified by third-generation dendrimers (MNP-G3) and MNP-G3 and further treated by ethylenediaminetetraacetic acid (EDTA) (MNP-G3-EDTA) were investigated for their ability to recovery Au(III), Pd(IV), Ag(I) and Pd(II) from water (Yen et al. 2017).

Metal Nanoparticle

Nanoparticles are considered as effective adsorbents due to their large surface area as compared to that of corresponding bulk particles and can be functionalized with various chemical groups to increase their affinity towards target contaminants. The catalytic performances of metal nanoparticles are based on the factors such as particle size, shape, metal dispersion and metal-support interactions (Campisi et al. 2016). Copper nanoparticles have been investigated to eliminate a wide variety of microorganisms, such as Vibrio cholerae, Shigella, Escherichia coli, Salmonella, fungi and viruses (Dankovich and Smith 2014). Zinc oxide nanoparticles have been used for both cationic and anionic dye removal due to the high adsorption capability and selectivity (Zhang et al. 2016). Bimetallic nanoparticles show novel electronic, optical, catalytic or photocatalytic properties due to a synergy between two metals (Zaleska-Medynska et al. 2016; Amendola et al. 2017). Calcium alginate encapsulated Ni/Fe bimetallic nanoparticles beads (CA–Ni/Fe beads) showed improved response in the simultaneously removal of Cu (II) and monochlorobenzene (MCB) from aqueous solution (Kuang et al. 2015). Highly effective removal of chromium (VI) was proposed via active Fe/Ni bimetallic nanoparticles (Zhou et al. 2016). In addition, Fe/Ni has been investigated for the acetamiprid degradation in aqueous systems (Marcelo et al. 2016). Biochar supported Ni/Fe bimetallic nanoparticles have been successfully studied to remove 1,1,1-trichloroethane under various reaction conditions (Li et al. 2017). Iron-based nanoparticles (Fe NPs) were prepared by Euphorbia cochinchensis and Syzygium jambos (L.) Alston leaves extracts and utilized for the degradation of 2,4-dichlorophenol and Cr(VI) respectively. (Guo et al. 2017; Xiao et al. 2017).

Nanocatalysts, Magnetic and Noble Metal Nanoparticles

Nanocatalysts are particles with catalytic properties that can chemically degrade pollutants. It has the potential for treating contaminants at extremely low levels, especially where the current treatment techniques are ineffective or very expensive to use. Magnetic nanoparticles have large surface areas relative to their mass and have the capability to bind easily with the contaminants. They are preferred for the purpose of water treatment due to their ability to bind with certain contaminants, and microbial communities that can be conveniently removed using a magnet. However, effective modification on magnetic nanoparticles is required to balance their reactivity and reusability (Tang and Lo 2013). Noble metals such as- gold, silver, platinum and palladium are resistant to corrosion and oxidation in moist air. Gold (Au) and silver (Ag) nanoparticles have been intensely researched by different scientific community due to their physical and chemical properties (Vajtai 2013; Lin et al. 2013; Qian et al. 2013; Wu and Feng 2016; Wu et al. 2017; Shirkhanloo et al. 2017). Gold (Au) nanoparticles offers cost effective nanotechnology based water treatment solution due to its enhanced catalytic activity, visible surface plasmon resonance color changes, and chemical stability. Silver nanoparticles (AgNPs, nanosilver) are used as an excellent antimicrobial agent in wastewater treatment (He et al. 2017). Nano sized Pt and Au exhibits multifaceted properties in catalysis, optical, biosensing and magnetic applications (Zhao et al. 2017). Au, Ag and Pd noble metals were separately doped on ZnO nano photocatalysts by borohydride reduction method for effective Congo red degradation (Guy and Ozacar 2016). Fe/Pd nanoparticles prepared by a simple successive reduction method were utilized for the degradation of Reactive Black 5 (RB5) in aqueous solution (Samiee et al. 2016). The major reason for the significant application of noble metals is the minimal reactivity at the bulk scale, which can be illustrated by numerous concepts such as electrochemical potential, relativistic contraction, molecular orbital theory, etc. Realizing the molecular nature of contamination in drinking water, significant progress has been made to utilize the chemistry of these noble metal nanoparticles for wastewater treatment

Zeolite

Zeolites are considered as efficient sorbent material and ion exchange media for metal ions (Meng et al. 2017). Natural zeolites are abundant and cost effective crystalline hydrated aluminosilicates with a honeycomb-like structure having pores occupied by water, alkali and alkaline earth cations. Natural zeolites have been widely used as adsorbents in separation and water treatment processes in the past decades due to their high cation-exchange and molecular sieve properties. Various natural zeolites around the world show different ion-exchange capacity for cations such as ammonium and heavy metal ions. However, some zeolites also show adsorption of anions and organics from aqueous solution. Modification of natural zeolites can be done in several methods such as acid treatment, ion exchange, and surfactant functionalization for achieving higher adsorption capacity for organics and anions. Sintered filter materials were prepared from the mixtures of zeolite, bentonite, kaolinite, and charcoal powder with controlled porosity for water purification (Mopoung, Sriprang and Namahoot 2014). Sorption characteristics of salt ions onto natural zeolite (clinoptilolite) were effectively investigated and showed favorable response (Wibiwo et al. 2017).Natural zeolite membranes have been experimented to deionize water and remove toluene from a mixture of toluene and saline water at high temperature and required further modification of the membranes to improve water fluxes (Swenson et al. 2012). A novel hydrophilic hydrophobic zeolite 3A incorporated polyether-block-amide composite membrane had been prepared for the efficient desalination of sea water (Nigiz, Veli and Hilmioglu 2017). The preparation of Zeolite/Zinc Oxide Nanocomposite (Zeolite/ZnO NCs) by using a co-precipitation method has been tested for adsorption of Lead Pb (II) and Arsenic As (V) from aqueous solution under the room pressure and temperature (Alswata et al. 2017).

Carbonaceous Nanoparticle

Carbonaceous nanomaterials possess high adsorption capacity for the effective removal of organic solutes from aqueous solutions (Mukhtar and See 2016; Thines et al. 2017). Carbonaceous nanomaterials, such as carbon nanotubes and graphene exhibits enhanced antimicrobial as well as adsorptive characteristics. These nanomaterials have been efficiently used in the removal of chemical and biological contaminants. However, significant improvements in terms of cost effectiveness and production capacity of these materials must be required for the effective commercialization of such technology (Smith and Rodrigues 2015). Development of nitrogen-enriched carbonaceous material coated titania nanotubes array, carbonaceous char-titania composite hollow fibers and uniform carbonaceous nanofibers (CNFs) have been done for the efficient wastewater treatment (Ma et al. 2017; Wang et al. 2016; Cheng et al. 2016).

Semiconductor Nano-Photocatalysts

Nano sized semiconductor photocatalyst such as- TiO_2 and ZnO possess a huge potential in the effective elimination of complex organic pollutants in a sustainable and environmental friendly manner. TiO_2 and ZnO are considered as highly efficient photocatalysts due to low cost, nontoxicity and high photosensitivity.

NANOMATERIAL BASED APPROACHES FOR WATER PURIFICATION

Nanotechnology can impact water purification process in several ways:

Photocatalysis

Photocatalytic oxidation is an important pre-treatment process for hazardous and non-biodegradable contaminants to enhance their biodegradability. It can be defined as the advanced oxidation process for the elimination of trace contaminants and microbial pathogens from drinking water and can also be utilized as a polishing step to treat stubborn organic compounds (Matilainen and Sillanpää 2010). The major hindrance for its wide application is the slow kinetics due to limited light influence and photocatalytic activity. Photocatalytic nanomaterials utilize ultraviolet light to destroy pesticides, industrial solvents and germs. TiO_2 is the most commonly used semiconductor photocatalyst in water/wastewater treatment owing to its low toxicity, chemical stability, low cost, and abundance as raw material. It generates an electron/hole pair upon absorbing a UV photon, which later either migrate to the surface or undergo undesired recombination (Chong et al. 2010). The photoactivity of nano-TiO_2 can be improved by optimizing particle size and shape, reducing electron/hole recombination rate by noble metal doping, maximizing reactive sites, and surface treatment to enhance adsorption of the contaminants. The overall efficiency of a photocatalytic water purification process strongly depends on the design parameters and operating condition of the photo-reactor. Two configurations are commonly used for the purification of drinking water: (1) slurry reactors and (2) immobilized reactors. Various techniques of dispersion, recovery and immobilization of catalysts are being researched in lab and pilot scale system to enhance the reactor efficiency. According to the Pilot tests conducted, photocatalyst system showed highly efficiency in removing organics without producing waste streams and consumes very less energy for its operation. Nano-TiO_2 facilitated solar disinfection (SODIS) has been extensively tested and appears to be a justified option to provide pure drinking water in remote areas of developing countries. The SODIS system can be used as small scale or for one person or can be scaled up to medium size solar compound parabolic collectors. Photocatalysis is a sustainable, economic and environmental friendly water purification technology. However, there are various technical hurdles for its large-scale acceptance, including 1) catalyst optimization for the synthesis of visible light active photocatalyst 2) effective design of photocatalytic reactor and 3) development of efficient catalyst recovery/ immobilization techniques. Metal oxide nanomaterials such as TiO_2 and ZnO nanoparticles have been studied as photocatalysts in heterogeneous catalysis for rapid and comparatively complete degradation of organic contaminants due to their large specific surface area. Hierarchically porous anatase TiO_2 microspheres composited

Figure 1. Schematic illustration of organic pollutants abatement via photocatalysis

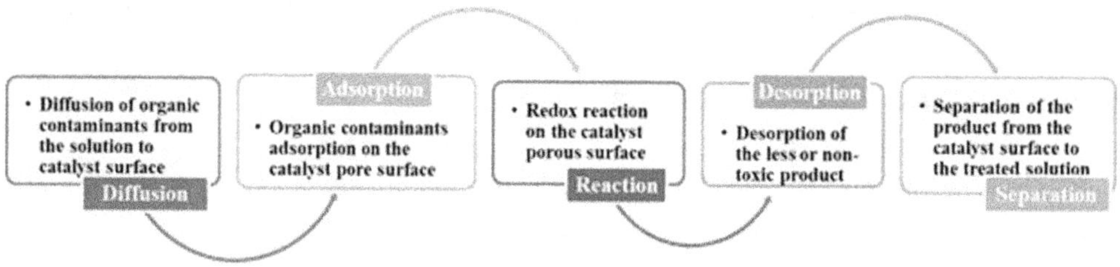

with carbonaceous species (TCS) have been successfully fabricated as a visible light photocatalyst with remarkable dye degradation capability (Liu et al. 2017). Figure.1 illustrates the general procedure of organic pollutants degradation via photocatalysis.

Disinfection and Microbial Control

Nanoantimicrobial nanomaterials are being researched for the successful disinfection of wastewater systems against infectious microorganisms (Santhosh et al. 2016). Several nanoparticles possess anti-microbial properties such as- TiO_2, MgO, ZnO, Ag and CNT (Hossain et al. 2014). Ag nanoparticles have good potential for application in wastewater treatment against waterborne pathogens for vulnerable population. The antimicrobial properties, fibrous shape, and high conductivity of CNTs enable novel CNT filters for both bacteria and virus removal (Li et al. 2008). Chitosan possess antimicrobial properties against Gram positive, Gram negative and fungal strains that was enhanced by copper, silver nanoparticles and CNTs (Morsi et al. 2017). However, disadvantage of disinfection approaches enabled by nanomaterial involves lack of disinfection residue, which is important for microbial growth control during storage and distribution of water.

Nanofiltration Membranes

These membranes selectively filter out particles smaller than their pores size, and thus remove hazardous contaminants and retain useful nutrients present in water. The filters and membranes are made from a variety of nanomaterials including carbon nanotubes, nonporous ceramics (clays), dendrimers, zeolites, nanofibers and nanosponges. Multi–tasking filtration systems should be developed to detect and remove/detoxify the contaminants. A major challenge of the membrane technology is the inherent trade-off between membrane selectivity and permeability. The high energy consumption is a major drawback for the wide application of pressure driven membrane processes. Membrane fouling reduces the lifetime of membranes and membrane modules. Membrane material is the major factor that decides the efficiency of membrane systems. Multifunctional hybrid composites of zeolite/palladium (Ze/Pd) on polymer nanofiber membranes were fabricated and explored for sustainable contaminant removal (Choi et al. 2016). Functional nanomaterials into membranes can improve the fouling resistance, membrane permeability, mechanical and thermal stability, as well as offers an opportunity for contaminant degradation and self-cleaning (Ying et al. 2017).

Adsorption

Adsorption is generally considered as a significant method to remove organic and inorganic contaminants during wastewater treatment. Extremely high surface area and active sites of the nano-adsorbents plays a vital role in the enhancement of overall efficiency of adsorption of hazardous pollutant from wastewater. Few examples of nanoadsorbant used for water purification process are-carbon based nanosorbents, metal based nanosorbents and polymeric nano-adsorbents. Among all, metal based nano-adsorbents can be produced at relatively low cost. The high adsorption capacity, low cost, easy separation and regeneration make metal based nanosorbents technologically and economically advantageous (Li et al. 2017). Nano-adsorbents can be readily integrated into existing water treatment processes such as in slurry reactors or as adsorbers. Nitim and Mitra (2012) successfully investigated multiwall carbon nanotube–zirconia nanohybrid as a potential adsorbent for arsenic removal from drinking water and found that sorption capacity was independent of pH over a certain range. A co-precipitation method was adopted to synthesize bimetallic Fe–Zn nanoparticle for the adsorptive removal of carcinogenic dye such as- Congo red and Malachite green from contaminated water (Gautam et al. 2015). Ag nanoparticles-modified activated carbon prepared from brewer's spent grains can be considered as a promising adsorbent for the desulphurization of liquid fuels (Olajire et al. 2017).

Sensing and Detection

Small and portable sensors are being developed via nanotechnology with enhanced sensing and detection capability. These can detect extremely low concentration levels of biological and chemical contaminants present in water. The sensitive and selective detection of copper (II) in various drinking water sample have been done via Graphene quantum dots decorated graphene as an enhanced sensing platform (Wang et al. 2017). Waste derivitized blue luminescent carbon quantum dots have been prepared for selenite sensing in water (Devi et al. 2017). A simple and versatile strategy was adopted by the researchers in the fabrication of bright orange-fluorescent gold–silver nanoclusters and graphene quantum dots (GQD) for the utilization in cyanide and Cr (VI) sensing respectively (Tian et al. 2017; Carrasco et al. 2016). A water-stable metal-organic framework (MOF) EuNDC has been synthesized for selective detection of the well-known contaminant and toxicant picric acid (PA) in aqueous solution (Xia et al. 2017).

BENEFITS OF USING NANOTECHNOLOGY

1. **Increased Contaminant Detection and Removal Efficiency**: Nanotechnology could more efficiently detect and remove the contaminants even at a very low concentration level.
2. **Removal of new Contaminants**: Contaminants or pollutants that were previously unable to remove effectively could now be treated with the assistance of nanoscale adsorbents due to their increased surface area.
3. **Simplification:** Nanotechnology could simplify the overall water purification/treatment process due to the reduction in the number of steps and resources needed that make its implementation suitable even in rural areas.
4. **Reduced Cost:** Nanotechnology based water treatments requires initial capital investment but once established, it could be beneficial for the lower cost water treatment in the long run.

RECOVERY AND RECYCLE OF NANOMATERIALS

Recovery and recycle of nanomaterials are the major factors to be considered for the cost efficient usage of nanotechnology-enabled water treatment processes and public health concerns. It can be achieved generally by immobilizing nanomaterials in the water purification system or by applying a separation device.

Immobilizing

Different supports are being used to immobilize nanomaterials such as polymer surface layer (Siegel et al. 2017), microchannel walls (Zhang et al. 2017) and membranes to avoid the need for further separation and for the recycled usage of nanoparticles. Therefore, a sincere effort is required to develop simple and economically tuned methods to immobilize nanomaterials without significantly affecting its performance.

Applying Membrane Separation Device

Membrane filtration is a promising separation process that allows a continuous operation to retain nanoparticles from the water after the treatment process to recycle them effectively. Ceramic membranes are more suitable than polymeric membranes in photocatalytic applications as they are more resistant to ultraviolet (UV) radiation and chemical oxidants. The suspended particles in the untreated water are detrimental to reactor-membrane hybrid systems as they can be captured by the membrane and reduces the reaction efficiency. Therefore, pretreatment of contaminated water is usually required to reduce the turbidity and enhance the life of the reaction system.

DRAWBACKS IN USING NANOPARTICLES FOR WATER PURIFICATION

There are few considerations of potential threats that need to be considered using nanoparticles. Lack of information and methods of characterizing nanomaterials make existing technology extremely difficult to detect the nanoparticles in air or water for environmental protection. The chemical structure is a critical factor in determining the toxicity of the material. Complete risk assessment regarding the safety of human health and environmental impact of nanotechnology needs to be evaluated at all stages. Nanoparticles can cause various respiratory problems. The risks that should be considered in the assessment are the exposure risk, toxicity and transport risk, persistence risk, transformation risk and recycling risk. Lack of trained engineers and workers can cause further concerns.

CONCLUSION

In conclusion, Nanotechnology for water purification is gradually accepted worldwide because of its increased effectiveness, unique properties of nanoparticles and simplicity. Despite all the benefits, the future development and commercialization of nanotechnology has to face a variety of challenges including cost effectiveness and potential environmental and human risk. Although many nanotechnologies are still in the laboratory research stage, some of them are being tested in pilot plants whereas some are even commercialized. However, some major concern to be considered related to water and nanotechnology include:

Better understanding of nanotechnology – Public awareness and understanding plays a vital role in the successful recognition and establishment of any developing technology. Credible information should be communicated to the general public and other target audiences to ensure community preparedness for this technology.

Risk assessment of nanotechnology – For the long term industrial applications and acceptance of nanotechnology–based products by the society, it is necessary to investigate the risk concerns associated with the technology along with the proper management and communication channel to the public.

ACKNOWLEDGMENT

This work was supported by the Department of Chemical Engineering, Indian Institute of Technology (ISM), Dhanbad.

REFERENCES

Adeleye, A. S., Conway, J. R., Garner, K., Huang, Y., Su, Y., & Keller, A. A. (2016). Engineered nanomaterials for water treatment and remediation: Costs, benefits, and applicability. *Chemical Engineering Journal*, *286*, 640–662. doi:10.1016/j.cej.2015.10.105

Akir, S., Barras, A., Coffinier, Y., Bououdina, M., Boukherroub, R., & Omrani, A. D. (2016). Eco-friendly synthesis of ZnO nanoparticles with different morphologies and their visible light photocatalytic performance for the degradation of Rhodamine B. *Ceramics International*, *42*(8), 10259–10265. doi:10.1016/j.ceramint.2016.03.153

Ali, L. I., El-Molla, S. A., Ibrahim, M. M., Mahmoud, H. R., & Naghmash, M. A. (2016). Effect of preparation methods and optical band gap of ZnO nanomaterials on photodegradation studies. *Optical Materials*, *58*, 484–490. doi:10.1016/j.optmat.2016.05.034

Alswata, A. A., Ahmad, M. B., Al-Hada, N. M., Kamari, H. M., Hussein, M. Z. B., & Ibrahim, N. A. (2017). Preparation of Zeolite/Zinc Oxide Nanocomposites for toxic metals removal from water. *Results in Physics*, *7*, 723–731. doi:10.1016/j.rinp.2017.01.036

Amendola, V., Scaramuzza, S., Carraro, F., & Cattaruzza, E. (2017). Formation of alloy nanoparticles by laser ablation of Au/Fe multilayer films in liquid environment. *Journal of Colloid and Interface Science*, *489*, 18–27. doi:10.1016/j.jcis.2016.10.023 PMID:27770998

Bai, X., Sun, C., Liu, D., Luo, X., Li, D., Wang, J., ... Zhu, Y. (2017). Photocatalytic degradation of deoxynivalenol using graphene/ZnO hybrids in aqueous suspension. *Applied Catalysis B: Environmental*, *204*, 11–20. doi:10.1016/j.apcatb.2016.11.010

Beraa, A., Hajjaji, M., Laurent, R., & Caminade, A. M. (2017). Dendrimers-containing organoclays: Characterization and interaction with methylene blue. *Applied Clay Science*, *136*, 142–151. doi:10.1016/j.clay.2016.11.017

Brame, J., Li, Q., & Alvarez, P. J. (2011). Nanotechnology-enabled water treatment and reuse: Emerging opportunities and challenges for developing countries. *Trends in Food Science & Technology*, *22*(11), 618–624. doi:10.1016/j.tifs.2011.01.004

Campisi, S., Chan-Thaw, C. E., Wang, D., Villa, A., & Prati, L. (2016). Metal nanoparticles on carbon based supports: The effect of the protective agent removal. *Catalysis Today*, *278*, 91–96. doi:10.1016/j.cattod.2016.04.026

Carrasco, P. M., García, I., Yate, L., Zaera, R. T., Cabañero, G., Grande, H. J., & Ruiz, V. (2016). Graphene quantum dot membranes as fluorescent sensing platforms for Cr (VI) detection. *Carbon*, *109*, 658–665. doi:10.1016/j.carbon.2016.08.038

Cheng, W., Ding, C., Wang, X., Wu, Z., Sun, Y., Yu, S., ... Wang, X. (2016). Competitive sorption of As (V) and Cr (VI) on carbonaceous nanofibers. *Chemical Engineering Journal*, *293*, 311–318. doi:10.1016/j.cej.2016.02.073

Choi, J., Chan, S., Yip, G., Joo, H., Yang, H., & Ko, F. K. (2016). Palladium-Zeolite nanofiber as an effective recyclable catalyst membrane for water treatment. *Water Research*, *101*, 46–54. doi:10.1016/j.watres.2016.05.051 PMID:27253639

Chong, M. N., Jin, B., Chow, C. W., & Saint, C. (2010). Recent developments in photocatalytic water treatment technology: A review. *Water Research*, *44*(10), 2997–3027. doi:10.1016/j.watres.2010.02.039 PMID:20378145

Dankovich, T. A., & Smith, J. A. (2014). Incorporation of copper nanoparticles into paper for point-of-use water purification. *Water Research*, *63*, 245–251. doi:10.1016/j.watres.2014.06.022 PMID:25014431

Devi, P., Kaur, G., Thakur, A., Kaur, N., Grewal, A., & Kumar, P. (2017). Waste derivitized blue luminescent carbon quantum dots for selenite sensing in water. *Talanta*, *170*, 49–55. doi:10.1016/j.talanta.2017.03.069 PMID:28501201

El-Gendy, N. S., El-Salamony, R. A., & Younis, S. A. (2016). Green synthesis of fluorapatite from waste animal bones and the photo-catalytic degradation activity of a new ZnO/green biocatalyst nanocomposite for removal of chlorophenols. *Journal of Water Process Engineering*, *12*, 8–19. doi:10.1016/j.jwpe.2016.05.007

Gautam, R. K., Rawat, V., Banerjee, S., Sanroman, M. A., Soni, S., Singh, S. K., & Chattopadhyaya, M. C. (2015). Synthesis of bimetallic Fe–Zn nanoparticles and its application towards adsorptive removal of carcinogenic dye malachite green and Congo red in water. *Journal of Molecular Liquids*, *212*, 227–236. doi:10.1016/j.molliq.2015.09.006

Guo, M., Weng, X., Wang, T., & Chen, Z. (2017). Biosynthesized iron-based nanoparticles used as a heterogeneous catalyst for the removal of 2, 4-dichlorophenol. *Separation and Purification Technology*, *175*, 222–228. doi:10.1016/j.seppur.2016.11.042

Güy, N., & Özacar, M. (2016). The influence of noble metals on photocatalytic activity of ZnO for Congo red degradation. *International Journal of Hydrogen Energy*, *41*(44), 20100–20112. doi:10.1016/j.ijhydene.2016.07.063

Hao, C., Wang, J., Cheng, Q., Bai, Y., Wang, X., & Yang, Y. (2017). Anionic surfactants-assisted solution-phase synthesis of ZnO with improved photocatalytic performance. *Journal of Photochemistry and Photobiology A Chemistry*, *332*, 384–390. doi:10.1016/j.jphotochem.2016.09.013

He, X., Wang, F., Liu, H., Li, J., & Niu, L. (2017). Synthesis of quartz crystals supporting Ag nanoparticle powder with enhanced antibacterial properties. *Surfaces and Interfaces*, *6*, 122–126. doi:10.1016/j.surfin.2017.01.002

Hossain, F., Perales-Perez, O. J., Hwang, S., & Roman, F. (2014). Antimicrobial nanomaterials as water disinfectant: Applications, limitations and future perspectives. *The Science of the Total Environment*, *466*, 1047–1059. doi:10.1016/j.scitotenv.2013.08.009 PMID:23994736

Hou, X. (2015). ZnO/Ag heterostructured nanoassemblies: Wet-chemical preparation and improved visible-light photocatalytic performance. *Materials Letters*, *139*, 201–204. doi:10.1016/j.matlet.2014.10.053

Hua, M., Zhang, S., Pan, B., Zhang, W., Lv, L., & Zhang, Q. (2012). Heavy metal removal from water/wastewater by nanosized metal oxides: A review. *Journal of Hazardous Materials*, *211*, 317–331. doi:10.1016/j.jhazmat.2011.10.016 PMID:22018872

Kuang, Y., Du, J., Zhou, R., Chen, Z., Megharaj, M., & Naidu, R. (2015). Calcium alginate encapsulated Ni/Fe nanoparticles beads for simultaneous removal of Cu (II) and monochlorobenzene. *Journal of Colloid and Interface Science*, *447*, 85–91. doi:10.1016/j.jcis.2015.01.080 PMID:25700214

Kumar, A., Subash, B., Krishnakumar, B., Sobral, A. J. F. N., & Sankaran, K. R. (2016). Synthesis, characterization and excellent catalytic activity of modified ZnO photocatalyst for RR 120 dye degradation under UV-A and solar light illumination. *Journal of Water Process Engineering*, *13*, 6–15. doi:10.1016/j.jwpe.2016.07.009

Lancelot, A., Clavería-Gimeno, R., Velázquez-Campoy, A., Abian, O., Serrano, J. L., & Sierra, T. (2017). Nanostructures based on ammonium-terminated amphiphilic Janus dendrimers as camptothecin carriers with antiviral activity. *European Polymer Journal*, *90*, 136–149. doi:10.1016/j.eurpolymj.2017.03.012

Li, H., Qiu, Y. F., Wang, X. L., Yang, J., Yu, Y. J., Chen, Y. Q., & Liu, Y. (2017). Biochar supported Ni/Fe bimetallic nanoparticles to remove 1, 1, 1-trichloroethane under various reaction conditions. *Chemosphere*, *169*, 534–541. doi:10.1016/j.chemosphere.2016.11.117 PMID:27898326

Li, J., Jin, P., Dai, W., Wang, C., Li, R., Wu, T., & Tang, C. (2017). Excellent performance for water purification achieved by activated porous boron nitride nanosheets. *Materials Chemistry and Physics*, *196*, 186–193. doi:10.1016/j.matchemphys.2017.02.049

Li, Q., Mahendra, S., Lyon, D. Y., Brunet, L., Liga, M. V., Li, D., & Alvarez, P. J. (2008). Antimicrobial nanomaterials for water disinfection and microbial control: Potential applications and implications. *Water Research*, *42*(18), 4591–4602. doi:10.1016/j.watres.2008.08.015 PMID:18804836

Lin, S., Huang, R., Cheng, Y., Liu, J., Lau, B. L., & Wiesner, M. R. (2013). Silver nanoparticle-alginate composite beads for point-of-use drinking water disinfection. *Water Research*, *47*(12), 3959–3965. doi:10.1016/j.watres.2012.09.005 PMID:23036278

Liu, W., Xu, Y., Zhou, W., Zhang, X., Cheng, X., Zhao, H., ... Huo, L. (2017). A Facile Synthesis of Hierarchically Porous TiO_2 Microspheres with Carbonaceous Species for Visible-light Photocatalysis. *Journal of Materials Science and Technology*, *33*(1), 39–46. doi:10.1016/j.jmst.2016.04.007

Liu, Y., Deng, J., Xie, S., Wang, Z., & Dai, H. (2016). Catalytic removal of volatile organic compounds using ordered porous transition metal oxide and supported noble metal catalysts. *Chinese Journal of Catalysis*, *37*(8), 1193–1205. doi:10.1016/S1872-2067(16)62457-9

Lu, J., Li, L., Wang, Z., Wen, B., & Cao, J. (2013). Synthesis of visible-light-active TiO_2-based photocatalysts by a modified sol–gel method. *Materials Letters*, *94*, 147–149. doi:10.1016/j.matlet.2012.11.137

Ma, M., Wang, H., Zhen, Q., Zhang, M., & Du, X. (2017). Development of nitrogen-enriched carbonaceous material coated titania nanotubes array as a fiber coating for solid-phase microextraction of ultraviolet filters in environmental water. *Talanta*, *167*, 118–125. doi:10.1016/j.talanta.2017.02.002 PMID:28340701

Marcelo, C. R., Lopes, R. P., Cruz, J. C., Nascimento, M. A., Silva, A. A., & Lima, C. F. (2016). Evaluation of different parameters on the acetamiprid degradation by bimetallic Fe/Ni nanoparticles. *Separation and Purification Technology*, *171*, 256–262. doi:10.1016/j.seppur.2016.07.032

Matilainen, A., & Sillanpää, M. (2010). Removal of natural organic matter from drinking water by advanced oxidation processes. *Chemosphere*, *80*(4), 351–365. doi:10.1016/j.chemosphere.2010.04.067 PMID:20494399

Meng, Q., Chen, H., Lin, J., Lin, Z., & Sun, J. (2017). Zeolite A synthesized from alkaline assisted pre-activated halloysite for efficient heavy metal removal in polluted river water and industrial wastewater. *Journal of Environmental Sciences (China)*, *56*, 254–262. doi:10.1016/j.jes.2016.10.010 PMID:28571862

Molea, A., Popescu, V., Rowson, N. A., & Dinescu, A. M. (2014). Influence of pH on the formulation of TiO_2 nano-crystalline powders with high photocatalytic activity. *Powder Technology*, *253*, 22–28. doi:10.1016/j.powtec.2013.10.040

Mopoung, S., Sriprang, N., & Namahoot, J. (2014). Sintered filter materials with controlled porosity for water purification prepared from mixtures with optimal ratio of zeolite, bentonite, kaolinite, and charcoal. *Applied Clay Science*, *88*, 123–128. doi:10.1016/j.clay.2013.11.035

Morsi, R. E., Alsabagh, A. M., Nasr, S. A., & Zaki, M. M. (2017). Multifunctional nanocomposites of chitosan, silver nanoparticles, copper nanoparticles and carbon nanotubes for water treatment: Antimicrobial characteristics. *International Journal of Biological Macromolecules*, *97*, 264–269. doi:10.1016/j.ijbiomac.2017.01.032 PMID:28082228

Mukhtar, N. H., & See, H. H. (2016). Carbonaceous nanomaterials immobilised mixed matrix membrane microextraction for the determination of polycyclic aromatic hydrocarbons in sewage pond water samples. *Analytica Chimica Acta*, *931*, 57–63. doi:10.1016/j.aca.2016.04.032 PMID:27282751

Muñoz-Fernandez, L., Sierra-Fernandez, A., Milošević, O., & Rabanal, M. E. (2016). Solvothermal synthesis of Ag/ZnO and Pt/ZnO nanocomposites and comparison of their photocatalytic behaviors on dyes degradation. *Advanced Powder Technology*, *27*(3), 983–993. doi:10.1016/j.apt.2016.03.021

Nigiz, F. U., Veli, S., & Hilmioglu, N. D. (2017). Deep purification of seawater using a novel zeolite 3A incorporated polyether-block-amide composite membrane. *Separation and Purification Technology*, *188*, 90–97. doi:10.1016/j.seppur.2017.07.017

Ntim, S. A., & Mitra, S. (2012). Adsorption of arsenic on multiwall carbon nanotube–zirconia nanohybrid for potential drinking water purification. *Journal of Colloid and Interface Science*, *375*(1), 154–159. doi:10.1016/j.jcis.2012.01.063 PMID:22424815

Nuengmatcha, P., Chanthai, S., Mahachai, R., & Oh, W. C. (2016). Visible light-driven photocatalytic degradation of rhodamine B and industrial dyes (texbrite BAC-L and texbrite NFW-L) by ZnO-graphene-TiO_2 composite. *Journal of Environmental Chemical Engineering*, *4*(2), 2170–2177. doi:10.1016/j.jece.2016.03.045

Olajire, A. A., Abidemi, J. J., Lateef, A., & Benson, N. U. (2017). Adsorptive desulphurization of model oil by Ag nanoparticles-modified activated carbon prepared from brewer's spent grains. *Journal of Environmental Chemical Engineering*, *5*(1), 147–159. doi:10.1016/j.jece.2016.11.033

Pradeep, T., & Anshup. (2009). Noble metal nanoparticles for water purification: A critical review. *Thin Solid Films*, *517*(24), 6441–6478. doi:10.1016/j.tsf.2009.03.195

Qian, H., Pretzer, L. A., Velazquez, J. C., Zhao, Z., & Wong, M. S. (2013). Gold nanoparticles for cleaning contaminated water. *Journal of Chemical Technology and Biotechnology (Oxford, Oxfordshire)*, *88*(5), 735–741. doi:10.1002/jctb.4030

Qu, X., Alvarez, P. J., & Li, Q. (2013). Applications of nanotechnology in water and wastewater treatment. *Water Research*, *47*(12), 3931–3946. doi:10.1016/j.watres.2012.09.058 PMID:23571110

Rajeswari, A., Vismaiya, S., & Pius, A. (2017). Preparation, characterization of nano ZnO-blended cellulose acetate-polyurethane membrane for photocatalytic degradation of dyes from water. *Chemical Engineering Journal*, *313*, 928–937. doi:10.1016/j.cej.2016.10.124

Sabbaghan, M., Anaraki Firooz, A., & Jan Ahmadi, V. (2012). The effect of template on morphology, optical and photocatalytic properties of ZnO nanostructures. *Journal of Molecular Liquids*, *175*, 135–140. doi:10.1016/j.molliq.2012.08.019

Saleh, R., & Djaja, N. F. (2014). Transition-metal-doped ZnO nanoparticles: Synthesis, characterization and photocatalytic activity under UV light. *Spectrochimica Acta. Part A: Molecular and Biomolecular Spectroscopy*, *130*, 581–590. doi:10.1016/j.saa.2014.03.089 PMID:24813289

Samiee, S., Goharshadi, E. K., & Nancarrow, P. (2016). Successful degradation of Reactive Black 5 by engineered Fe/Pd nanoparticles: Mechanism and kinetics aspects. *Journal of the Taiwan Institute of Chemical Engineers*, *67*, 406–417. doi:10.1016/j.jtice.2016.07.012

Santhosh, C., Velmurugan, V., Jacob, G., Jeong, S. K., Grace, A. N., & Bhatnagar, A. (2016). Role of nanomaterials in water treatment applications: A review. *Chemical Engineering Journal*, *306*, 1116–1137. doi:10.1016/j.cej.2016.08.053

Selvam, N. C. S., Narayanan, S., Kennedy, L. J., & Vijaya, J. J. (2013). Pure and Mg-doped self-assembled ZnO nano-particles for the enhanced photocatalytic degradation of 4-chlorophenol. *Journal of Environmental Sciences (China)*, *25*(10), 2157–2167. doi:10.1016/S1001-0742(12)60277-0 PMID:24494504

Shavisi, Y., Sharifnia, S., & Mohamadi, Z. (2016). Solar-light-harvesting degradation of aqueous ammonia by CuO/ZnO immobilized on pottery plate: Linear kinetic modeling for adsorption and photocatalysis process. *Journal of Environmental Chemical Engineering*, *4*(3), 2736–2744. doi:10.1016/j.jece.2016.04.035

Shirkhanloo, H., Osanloo, M., Ghazaghi, M., & Hassani, H. (2017). Validation of a new and cost-effective method for mercury vapor removal based on silver nanoparticles coating on micro glassy balls. *Atmospheric Pollution Research*, *8*(2), 359–365. doi:10.1016/j.apr.2016.10.004

Siegel, J., Lyutakov, O., Polívková, M., Staszek, M., Hubáček, T., & Švorčík, V. (2017). Laser-assisted immobilization of colloid silver nanoparticles on polyethyleneterephthalate. *Applied Surface Science*, *420*, 661–668. doi:10.1016/j.apsusc.2017.05.151

Silva, I. M. P., Byzynski, G., Ribeiro, C., & Longo, E. (2016). Different dye degradation mechanisms for ZnO and ZnO doped with N (ZnO:N). *Journal of Molecular Catalysis A Chemical*, *417*, 89–100. doi:10.1016/j.molcata.2016.02.027

Smith, S. C., & Rodrigues, D. F. (2015). Carbon-based nanomaterials for removal of chemical and biological contaminants from water: A review of mechanisms and applications. *Carbon*, *91*, 122–143. doi:10.1016/j.carbon.2015.04.043

Swenson, P., Tanchuk, B., Bastida, E., An, W., & Kuznicki, S. M. (2012). Water desalination and de-oiling with natural zeolite membranes—Potential application for purification of SAGD process water. *Desalination*, *286*, 442–446. doi:10.1016/j.desal.2011.12.008

Tang, S. C., & Lo, I. M. (2013). Magnetic nanoparticles: Essential factors for sustainable environmental applications. *Water Research*, *47*(8), 2613–2632. doi:10.1016/j.watres.2013.02.039 PMID:23515106

Thines, R. K., Mubarak, N. M., Nizamuddin, S., Sahu, J. N., Abdullah, E. C., & Ganesan, P. (2017). Application potential of carbon nanomaterials in water and wastewater treatment: A review. *Journal of the Taiwan Institute of Chemical Engineers*, *72*, 116–133. doi:10.1016/j.jtice.2017.01.018

Tian, L., Li, Y., Ren, T., Tong, Y., Yang, B., & Li, Y. (2017). Novel bimetallic gold– silver nanoclusters with "Synergy"-enhanced fluorescence for cyanide sensing, cell imaging and temperature sensing. *Talanta*, *170*, 530–539. doi:10.1016/j.talanta.2017.03.107 PMID:28501207

Vajtai, R. (2013). *Springer handbook of nanomaterials*. Springer Science & Business Media. doi:10.1007/978-3-642-20595-8

Wang, D. K., Elma, M., Motuzas, J., Hou, W. C., Schmeda-Lopez, D. R., Zhang, T., & Zhang, X. (2016). Physicochemical and photocatalytic properties of carbonaceous char and titania composite hollow fibers for wastewater treatment. *Carbon*, *109*, 182–191. doi:10.1016/j.carbon.2016.08.001

Wang, L., Wu, Y., Chen, F., & Yang, X. (2014). Photocatalytic enhancement of Mg-doped znO nanocrystals hybridized with reduced graphene oxide sheets. *Progress in Natural Science: Materials International, 24*(1), 6–12. doi:10.1016/j.pnsc.2014.01.002

Wang, Y., Yang, J., Li, Y., Jiang, T., Chen, J., & Wang, J. (2015). Controllable preparation of ZnO nanostructure using hydrothermal-electrodeposited method and its properties. *Materials Chemistry and Physics, 153*, 266–273. doi:10.1016/j.matchemphys.2015.01.013

Wang, Y., Yang, Y., Xi, L., Zhang, X., Jia, M., Xu, H., & Wu, H. (2016). A simple hydrothermal synthesis of flower-like ZnO microspheres and their improved photocatalytic activity. *Materials Letters, 180*, 55–58. doi:10.1016/j.matlet.2016.05.107

Wang, Y., Zhao, S., Li, M., Li, W., Zhao, Y., Qi, J., & Cui, X. (2017). Graphene quantum dots decorated graphene as an enhanced sensing platform for sensitive and selective detection of copper (II). *Journal of Electroanalytical Chemistry (Lausanne, Switzerland), 797*, 113–120. doi:10.1016/j.jelechem.2017.05.031

Welderfael, T., Pattabi, M., Pattabi, R. M., & G, A. K. T. (2016). Photocatalytic activity of Ag-N co-doped ZnO nanorods under visible and solar light irradiations for MB degradation. *Journal of Water Process Engineering, 14*, 117–123. doi:10.1016/j.jwpe.2016.11.001

WHO. (2011). Guidelines for drinking-water quality. *WHO Chronicle, 38*, 104–108. PMID:6485306

Wibowo, E., Rokhmat, M., & Abdullah, M. (2017). Reduction of seawater salinity by natural zeolite (Clinoptilolite): Adsorption isotherms, thermodynamics and kinetics. *Desalination, 409*, 146–156. doi:10.1016/j.desal.2017.01.026

Woo, H., & Park, K. H. (2016). Recent developments in hybrid iron oxide–noble metal nanocatalysts for organic reactions. *Catalysis Today, 278*, 209–226. doi:10.1016/j.cattod.2016.01.030

Wu, H., & Feng, Q. (2017). Fabrication of bimetallic Ag/Fe immobilized on modified biochar for removal of carbon tetrachloride. *Journal of Environmental Sciences (China), 54*, 346–357. doi:10.1016/j.jes.2016.11.017 PMID:28391946

Wu, H., Feng, Q., Yang, H., Alam, E., Gao, B., & Gu, D. (2017). Modified biochar supported Ag/Fe nanoparticles used for removal of cephalexin in solution: Characterization, kinetics and mechanisms. *Colloids and Surfaces. A, Physicochemical and Engineering Aspects, 517*, 63–71. doi:10.1016/j.colsurfa.2017.01.005

Xia, T., Zhu, F., Cui, Y., Yang, Y., Wang, Z., & Qian, G. (2017). Highly selective luminescent sensing of picric acid based on a water-stable europium metal-organic framework. *Journal of Solid State Chemistry, 245*, 127–131. doi:10.1016/j.jssc.2016.10.016

Xiao, Z., Zhang, H., Xu, Y., Yuan, M., Jing, X., Huang, J., ... Sun, D. (2017). Ultra-efficient removal of chromium from aqueous medium by biogenic iron based nanoparticles. *Separation and Purification Technology, 174*, 466–473. doi:10.1016/j.seppur.2016.10.047

Yang, Y., Li, H., Hou, F., Hu, J., Zhang, X., & Wang, Y. (2016). Facile synthesis of ZnO/Ag nanocomposites with enhanced photocatalytic properties under visible light. *Materials Letters, 180*, 97–100. doi:10.1016/j.matlet.2016.05.117

Yen, C. H., Lien, H. L., Chung, J. S., & Yeh, H. D. (2017). Adsorption of precious metals in water by dendrimer modified magnetic nanoparticles. *Journal of Hazardous Materials*, *322*, 215–222. doi:10.1016/j.jhazmat.2016.02.029 PMID:26921983

Ying, Y., Ying, W., Li, Q., Meng, D., Ren, G., Yan, R., & Peng, X. (2017). Recent advances of nanomaterial-based membrane for water purification. *Applied Materials Today*, *7*, 144–158. doi:10.1016/j.apmt.2017.02.010

Zaleska-Medynska, A., Marchelek, M., Diak, M., & Grabowska, E. (2016). Noble metal-based bimetallic nanoparticles: The effect of the structure on the optical, catalytic and photocatalytic properties. *Advances in Colloid and Interface Science*, *229*, 80–107. doi:10.1016/j.cis.2015.12.008 PMID:26805520

Zhang, F., Chen, X., Wu, F., & Ji, Y. (2016). High adsorption capability and selectivity of ZnO nanoparticles for dye removal. *Colloids and Surfaces. A, Physicochemical and Engineering Aspects*, *509*, 474–483. doi:10.1016/j.colsurfa.2016.09.059

Zhang, L., Liu, Z., Wang, Y., Xie, R., Ju, X. J., Wang, W., ... Chu, L. Y. (2017). Facile immobilization of Ag nanoparticles on microchannel walls in microreactors for catalytic applications. *Chemical Engineering Journal*, *309*, 691–699. doi:10.1016/j.cej.2016.10.038

Zhang, Y., Wu, B., Xu, H., Liu, H., Wang, M., He, Y., & Pan, B. (2016). Nanomaterials-enabled water and wastewater treatment. *NanoImpact*, *3*, 22–39. doi:10.1016/j.impact.2016.09.004

Zhao, M., Zhong, N., & Ji, Y. (2017). Ultra-stable colloidal porous Pt-Au-Ag nanoparticles. *Materials Letters*, *191*, 38–41. doi:10.1016/j.matlet.2017.01.003

Zhou, X., Jing, G., Lv, B., Zhou, Z., & Zhu, R. (2016). Highly efficient removal of chromium (VI) by Fe/Ni bimetallic nanoparticles in an ultrasound-assisted system. *Chemosphere*, *160*, 332–341. doi:10.1016/j.chemosphere.2016.06.103 PMID:27393969

Chapter 15

Remote Detection of Environmentally Vulnerable Terrain in Terms of Chemical Pollution of Soil and Groundwater

Inna Romanciuc
https://orcid.org/0000-0002-2891-4686
Scientific Centre for Aerospace Research of the Earth of the Institute of Geological Science of the National Academy of Sciences of Ukraine, Ukraine

Natalia Pazynych
Scientific Centre for Aerospace Research of the Earth of the Institute of Geological Science of the National Academy of Sciences of Ukraine, Ukraine

Anton I. Manilov
https://orcid.org/0000-0003-2225-2754
Taras Shevchenko National University of Kyiv, Ukraine

ABSTRACT

Remote sensing is an efficient way to detect environmentally vulnerable terrain, where waste chemicals and excess fertilizers must result to considerable pollution of soil and groundwater. The indicators of such vulnerable formations were analyzed and applied for investigation of ground microdepressions. First indicator was the humidity level that was well recognized from the satellite images by calculation of normalized water index (NWI). Another active indicator used for the detection was the relief. The method of relief plastics joint with morphodynamic analysis was suggested for the data processing. The scheme of water basins and the map of environmentally fragile areas were created by combination of the indicators of soil humidity and topography. On the basis of the created maps, the restrictions for used chemicals can be developed in order to preserve considerable environmental impact of the pollution.

DOI: 10.4018/978-1-7998-1241-8.ch015

INTRODUCTION

The problems of ecological monitoring and remote analysis of chemical pollution are closely related to environmental chemistry (Bahadir & Duca, 2009). It is important to develop the methods of remote sensing that can be efficiently applied for detection of environmentally fragile areas, where waste chemicals and excess fertilizers must result to considerable pollution of soil and groundwater. Such vulnerable terrestrial formations like temporary ponds or ground microdepressions were not widely investigated. However, these formations greatly influence the environment, because pollutants tend to concentrate here and then propagate to groundwater (Keesstra et al., 2012).

There are different sources of contaminants including agrochemicals, natural gas, petroleum hydrocarbons and potentially toxic elements (Gholizadeh et al., 2018). The pollutants can be detected using extensive sampling and conventional techniques of analytical chemistry. This requires a complex infrastructure to support the entire sampling, laboratory and field based analysis, as well as data processing activities (McLean et al., 2019). So, it is important to develop more simple, rapid and cost-effective sensing technologies that can be used for monitoring of large areas, where the analytical networks have not been constructed.

Remote sensing approach is based on characterization of investigated object without direct contact. There are a lot of sensor systems that are developed for this purpose: object presence and motion detectors, directional microphones and thermometers, variety of photodetectors, radar stations etc. In contrary, widely used chemical sensors are not appropriate for remote detection. Analyzed substance interacts with a sensor by way of physical or chemical processes on its surface (Manilov et al., 2019). Thus, a contact of the analyte and the detector is necessary. It can be used to analyze the sample at the monitoring site. However, the ability to determine certain chemical compounds without contact is very important for applications in aerospace sensor systems.

Remote sensing data for characterization of the Earth's surface can be registered by terrestrial stations, aircraft (especially pilotless vehicles), space stations or satellites. Passive or active sensing methods are used (Lillesand, Kiefer, & Chipman, 2015). Main way of the passive data collection is detection of electromagnetic waves received from investigated objects. Measurements of spectral reflectance are made to separate areas with different types of soil, water and vegetation. The most valuable results are obtained in visible, near infrared and thermal, long wavelength infrared range. The multichannel detectors working in corresponding spectral bands are installed on research aircraft and satellites. The active methods of remote sensing are mainly released by radar systems working in microwave band. Main focus of the radar measurements is exploration of land and water surface in order to produce maps and digital elevation models. Monitoring of an atmosphere and meteorological processes are also provided (Chen et al., 2005; Clevers et al., 2008; Maeda et al., 2015).

Most of the remote sensing systems are devoted to the problems of mapping, weather and climate control, natural resources and agriculture, as well as disaster risk management. However the problems of ecological monitoring and remote analysis of chemical pollution are also important. The data obtained by remote sensing are used for detection of heavy metals and excessive fertilizers in soil, crops and forests. Analysis of atmospheric pollution, emission from urban zones and vehicles is also supported by satellite data, as well as search of oil spills in water. The specialized detectors are developed that help to improve the quality of remote sensing data from the standpoint of environmental chemistry (Keesstra et al., 2012; Kostyuchenko et al., 2013).

With all applications of remote sensing for analysis of present chemical pollution and its environmental impact, it is also necessary to determine the areas that are characterized with high danger of chemical pollution. The aim of this chapter is to present the methods of remote data analysis that can be efficiently applied for exploration of terrain areas, where waste chemicals and excess fertilizers must result to considerable pollution of soil and groundwater. The proposed methods have been successfully tested by State institution "Scientific Centre for Aerospace Research of the Earth Institute of Geological Science National Academy of Science of Ukraine" (Lyalko et al., 2016; Lyalko & Sakhatsky, 2013; Kostuychenko et al., 2003; Kharytonov et al., 2013).

BACKGROUND

Modern agriculture leads to considerable violation of environment. Using mineral fertilizers, herbicides and pesticides is followed by their accumulation in soil above the permissible limits, contamination of plants and agricultural products, as well as propagation in groundwater. Not all chemical elements of the soil solutions are absorbed by plants. They often exist in excessive quantities under the influence of downward flows of water that results in migration flows. Thus, the soil cover is the initial chain of migration of chemical elements. The main way of the migration is water. The removal of chemical elements from soil profile and weathering crust occurs due to atmospheric precipitation and surface runoff. Then the dissolved chemicals move under surface layer and appear in springs. So, the chemical composition of water in rivers ultimately depends on the elements that are present in soil solutions of the territories from which the rivers flow (Kazeev & Kolesnikov, 2018; Keesta et al, 2012).

The migration of chemical elements is governed by geochemical regularities. It is influenced by the peculiarities of climate, relief, soil formation and geodynamic processes. The migration occurs both in vertical and horizontal profile. The assessment of water exchange between surface and groundwater identifies the areas, where this process is the most intensive. Microdepressions of relief are considered as zones of accelerated migration in the processes of moisture exchange (Bublyas & Shestopalov, 2001; Appels et al., 2011; Sycheva et al., 2019; Whigham & Jordan, 2003).

Microdepressions include genetically heterogeneous, closed lowering of upper layer of the Earth's crust, characterized with relatively small size. Such formations were found on the territories of Eastern Europe, Brazil, Argentina, Australia and Canada. They exist as pods, steppe saucers, ponds, microcavities and other negative closed forms at the level of micro and mesorelief (Figure 1). Such lowered forms of relief can be characterized by size from a few meters to 1.5 km, as well as by morphological type including linear, linear-rounded and rounded forms (Azimov & Bublyas, 2005; Pischulov, 2013).

A wide range of hypotheses have been advanced regarding to the origin of microdepressions. Taking into account morphological features the following dominant factors were considered (Azimov & Bublyas, 2005):

- Degradation of sedimentary rocks within the cracks inside the crystalline massif (tectonic factor);
- Weathering of particles from rocks (deflationary factor);
- Subsidence of soil as a result of uneven wetting in some areas (subsidence);
- Selective removal of easy soluble elements (suffusion);
- Deep processes of suffusion and leaching in easily dissolving rocks (karst);
- Thermal karst cracking and freezing (thermokarst);

Figure 1. Microdepressions of relief: examples on space image and in situ

- Glacial erosion with formation of valleys and cirques;
- Residual fragments of old streambed (oxbow);
- Polygenetic factor in case of multicomponent origin.

Microdepressions of relief were studied both as natural processes and in scope of human activity (Bublyas, 2006; Wigham & Jordan, 2003). There are a lot of methods concerning to search and analysis of these formations. Multispectral imaging, geomorphological analysis, digital models of relief, as well as in situ field observations were mainly used for this purpose (Helming et al., 1993; Ghergina el al., 2008; Mostaquir Rahman, 2019).

Remote sensing systems placed on space satellites are the most convenient way to explore the Earth's surface including search of the lowered land formations. A remote sensing system receives electromagnetic radiation reflected or emitted by investigated object. The received radiation is described by two variables: how much reflection was received, and when it was received (Voloshin & Korchinsky, 2007).

The temporal characteristics are applicable only for active systems, in which the time of electromagnetic emission is controlled. In such systems, it is possible to determine the distance from the sensor to the investigated object. This principle is a basis of active remote sensing systems like laser surveyors, optical locators (lidars), radar altimeters and other devices. Passive systems give information only about the amount of radiation that has arrived at the sensor. If the received radiation is a result of thermal emission, its value characterizes temperature and emittance of the source.

When both passive and active sensing methods are used, the amount of received energy is determined by the sum of emitted and reflected radiation of the investigated object. Thus, the information received directly by remote sensing is significantly limited: only distance to the object, its reflectivity and combination of temperature and reflectivity can be measured. However, radiation can be registered at different times, wavelengths and polarizations. It contributes to increasing information capability of the data.

The data obtained from a remote sensing system must be further processed and analyzed. In most cases the data are presented in a form of two-dimensional images showing the distribution of radiation intensity within different spectral ranges or scanning signal parameters. Processing of the images passes through several steps (Swarnalatha & Sevugan, 2018). First, the correction of measured data is made preventing random and systematic errors caused by atmospheric disturbances, drift of the sensor parameters, geometrical distortions, etc. Then the optimal interrelation of spectral sources is determined in order to obtain the best view of the image. The chosen algorithm of the image processing depends on the objects that are expected for recognition. The recognition is made by classification methods that make possible to determine and separate the areas of certain type.

Analysis of the obtained images can be based on visual interpretation or mathematical treatment using pattern recognition algorithms. When multispectral imaging is done in n spectral ranges, each analyzed element of the Earth's surface is characterized by the set of measured values of spectral brightness: vector or point in n-dimensional space. The recognition procedure is intended to classify the feature vector to one of the classes of natural entities. Supervised learning approach is mostly used. Experimental results of the field measurements are taken as the referent points for this purpose. Also, unsupervised learning can be applied for automated qualitative characterization of the investigated area. Depending on the used pattern recognition method there are different approaches for a further quantitative characterization of the dedicated objects.

So, the use of remote sensing data and image processing algorithms in order to investigate the selected class of objects, namely microdepressions of relief, require retrieval and development of assisting methods that consider their determinative properties (Romanciuc, 2018).

MAIN FOCUS OF THE CHAPTER

The described mechanism of water migration clearly indicates the correlation between chemical substances in soil and chemical composition of groundwater. The microdepressions represent environmentally fragile areas that tend to accumulate chemicals from surrounding territory. These areas considerably influence an environment, because pollutants tend to concentrate here and then propagate to groundwater. However, such vulnerable terrestrial formations were not widely investigated (Bublias & Shestopalov, 2006; Keesstra et al., 2012; Pristaer et al., 2016; Shestopalov & Bublyas, 2016).

Using hydrological and hydrophysical methods it was previously determined that much of surface runoff and infiltration into groundwater lies within the area of microdepressions. The amount of water infiltrated within the vulnerable territory was 2-4 times higher than the values corresponding to referent areas, where no lowered forms of relief were existed (Shestopalov & Bublyas, 2016). Research teams of the Institute of Geological Sciences of the National Academy of Sciences of Ukraine consider microdepressions as the anomalous zones, where land degassing, as well as mass exchange and vertical migration processes occur. Preliminary calculations indicated that moisture overflow in the anomalous zone could be 4-6 times higher than the corresponding values in the referent zone. The studies have shown that the areas of rapid migration are characterized with a higher level of mass exchange and more active geochemical processes (Bublias & Shestopalov, 2006; Lyalko & Sakhatsky, 2013).

The methods of remote sensing provide new possibilities for visual observation and determination of the anomalous zones. The shapes of ground microdepressions are well recognized on agricultural lands, when using satellite images. The territory of observation should not be hidden under canopy of trees and their foliage in the mature state should not overlap the soil. The characteristic parameters of microdepressions are specific composition of soil and water content by which they can be remotely determined. The bottom of these formations is usually characterized by soil disposed to gleization, waterlogging and salinization, as well as increased values of humidity (Romanciuc et al, 2018; Sakhatsky, 2007; Starodubtsev & Basarab, 2019).

This chapter is focused on remote detection of microdepressions and similar forms of relief that are vulnerable for chemical pollution. The map of environmentally fragile areas can be created by processing the data obtained from different remote sensing systems taking into account the specific features of microdepressions. On the basis of the created maps the restrictions for used chemicals can be developed in order to preserve considerable environmental impact of the pollution.

SOLUTIONS AND RECOMMENDATIONS

Efficient detection of the environmentally vulnerable terrain can be realized by combination of the remote sensing methods based on different indicators supplied with in situ verification (Figure 2). The first indicator used for identification of microdepressions was the humidity level. It was good recognized from the satellite images by calculation of normalized water index (Romanciuc, 2018; Sakhatsky, 2007). Another active indicator used for the detection was the relief. The method of relief plasticity joint with morphodynamic analysis was suggested for the data processing. The method of relief plasticity included the data of digital elevation model and the topographic map. The morphodynamic analysis allowed to identify the direction of runoff and to detect drainage basins at the microrelief level.

Classification of Space Images

Classification is the process dividing the volume of data into several classes (groups). Each pixel of the analyzed space image is attributed to one of the selected classes. It results in automatic separation of the objects depicted in the images and drawing up a map of the examined area.

The authors used supervised classification algorithms, where the specific standards for each class were set. The brightness value of each pixel was compared with the standards that had been previously set by the expert. The brightness values of pixels within the reference areas were automatically measured. Then the computer compared all other pixels of the image with the reference ones, and each of them was attributed to the closest class. As a result the classification map was created. The classification of space images allowed determination of the most problematic negative forms of relief. The applications of maximum likelihood method, Mahalanobis distance and minimum distance for the classification of space images were described in (Romanciuc, 2018). The best results were obtained using the method of minimum distance. The results of the classification made it possible to identify the most humid areas, as well as the clearly separated microdepressions (Figure 3).

Figure 2. The methods used for detection of environmentally vulnerable formations

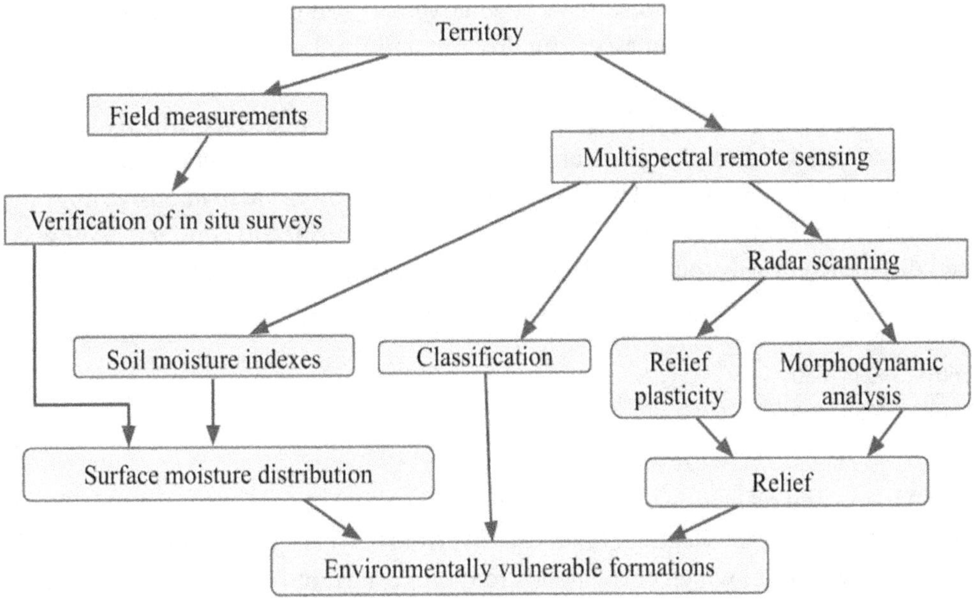

Figure 3. Microdepressions of relief on the space image detected by minimum distance method of classification

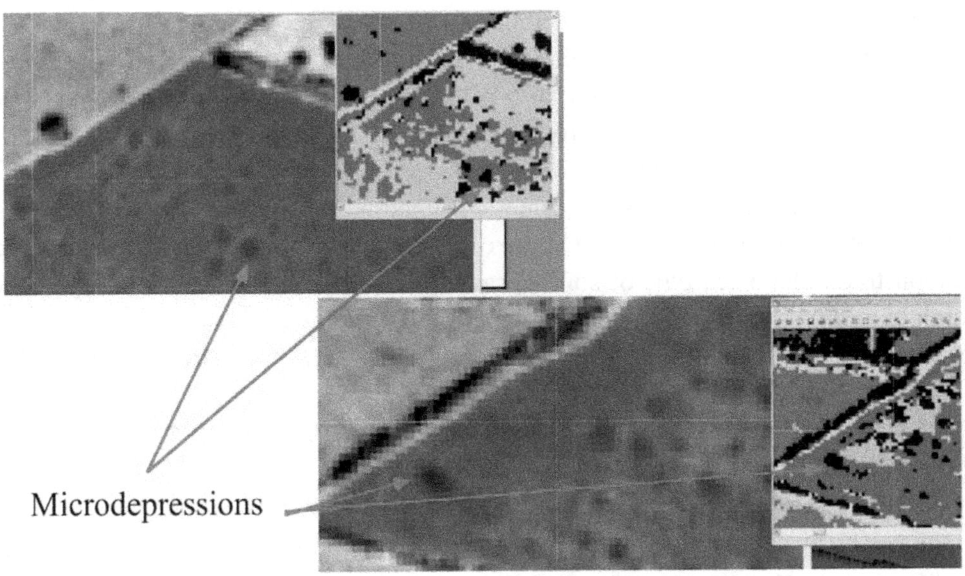

Calculation of Water Indexes

Water indexes were designed to estimate the content of moisture in vegetation. The presence of water is an important indicator. The high content of water characterizes healthy vegetation. Such vegetation grows rapidly and it is more resistant to fire. Near and middle infrared ranges were used to calculate the indexes estimating the presence of moisture in vegetative and soil cover (Lyalko & Sakhatsky, 2013). These indexes were also applied when assessing drought and fire hazards in analyzed territories, along with other indexes describing vegetation status.

The water index named Normalized Water Index (NWI) was proposed for effective measurements of the moisture both in vegetation and soil cover (Sakhatsky, 2007). This method indicated the areas with microdepressions of relief characterized with high values of humidity, where intensive migration of water is possible, as well as movement of chemicals from the Earth's surface to aquifers. NWI was calculated as normalized difference of spectral reflection in green and middle infrared bands of the spectrum.

When using images from Sentinel-2 satellite, the formula is:

$$NWI = \frac{GR(560) - SWIR(1650)}{GR(560) + SWIR(1650)} \tag{1}$$

where GR (560) is spectral reflection within the range of 540-580 nm wavelength, which corresponds to the green spectral channel of the Sentinel-2 survey; SWIR (1650) is spectral reflection within the range of 1570-1660 nm wavelength, which corresponds to the mid-infrared channel of the Sentinel-2 survey.

The space images of Sentinel series are open to access on the official website of the European Space Agency (http://sentinels.copernicus.eu). The authors used the multi-spectral satellite images with resolution of 10, 20 and 60 meters. The dates of satellite imagery were chosen according to the time of the corresponding field measurements providing verification of the obtained data. The samples collected during the field measurements included different levels of soil with varying humidity. The percentage of water in the samples was detected by the method of thermostatic weighting.

A large number of the relief microdepressions were observed on the agricultural fields filled with water after prolonged precipitation. The calculation of NWI in each pixel of the corresponding space images was used for successful detection of the microdepressions due to the presence of excessive moisture in vegetation cover and upper layer of open soil.

A good correlation between the moisture values obtained with NWI and in situ field measurements was estimated. As a result, the map of soil moisture of the analyzed surface layer was elaborated (Figure 4). It was demonstrated that such type of soil degradation as microdepressions are represented by the areas with high humidity.

The microdepressions are represented by the areas with high humidity (Romanciuc et al., 2018).

The Method of Relief Plasticity and Morphodynamic Analysis

The method of relief plasticity is widely used to study soil cover and soil degradation processes (Stepanov, 2006; Dmitruk & Struzhuk, 2014; Pazynych & Romanciuc, 2018; Voloshin et al, 2013). It makes possible to fix even a slight increase or decrease in the structure of terrains of the Earth's surface. The relief plasticity map is based on digital elevation model (DEM) with elements of morphodynamic analysis and simulates the movements of water and soil over observed surface (Lastochkin, 2016).

To build the basic relief plasticity map, the digital elevation model of the Shuttle Radar Topography Mission was used supplemented by subsequent processing with the Global Mapper program. The information content of DEM was determined by the distance between the survey points at which the height

Figure 4. The map of soil humidity (%) obtained by calculation of NW

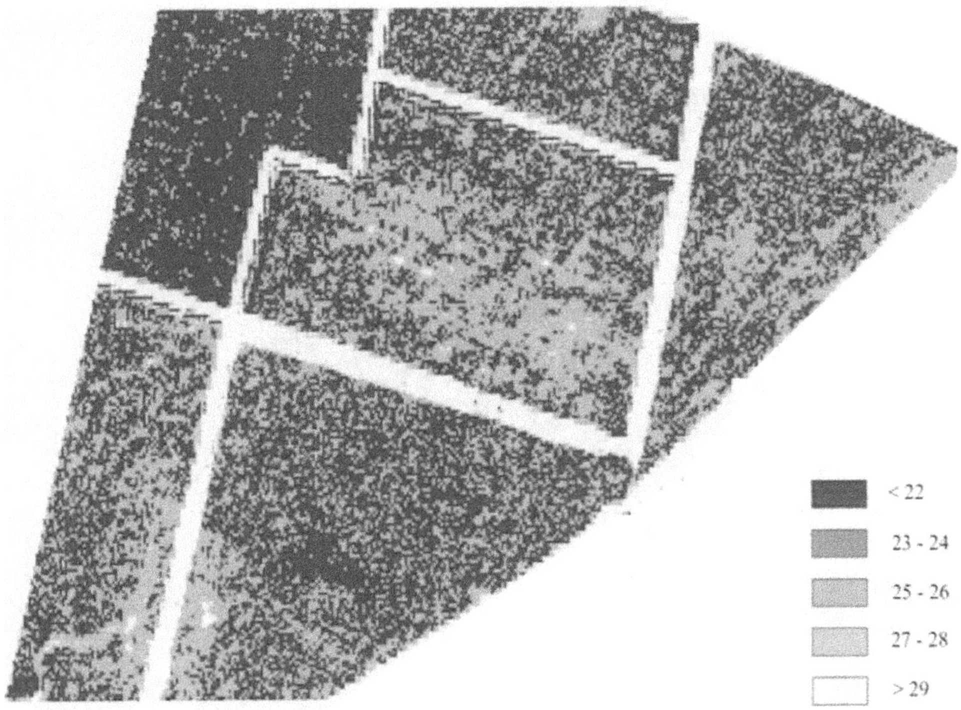

was set. The digital elevation model used in the study had a spatial resolution of 90 m (3 arc-second). The topography contours were generated through the program menu. Considering the flat terrain, the contour line tracing was set to 1 m increments. DEM and the horizontals were transformed with a mapping program MapInfo in order to create the relief plasticity map supplemented by morphodynamic analysis. The spatial resolution of DEM allowed reliable detection of microdepressions and other landforms of 180 meters size or more.

The surface of the Earth was represented by a plane where the movement of substances occurred as a result of kinetic energy received from showers, melting snow, solar thermal energy and anthropogenic influence. To determine the patterns of these movements according to the principles of morphodynamic analysis, the Earth's surface was divided by conditional lines representing the framework of relief's margins. The lines were drawn according to absolute values of the surface obtained from the contour lines of topographic maps or DEM. A computer processing of DEM data was carried out with aim to build the framework of relief making possible to obtain drainage lines and ridge lines. After editing, this basic framework structure was used in the morphodynamic analysis.

The selected frame lines formed the basis of the relief plasticity map. The terrain plasticity map was created by tracing zero morphoisograph line, which integrated the visually defined points of zero curvature located at the border of concave and convex parts of the contour lines obtained after processing of DEM. The visualization of relief plasticity map for the examined area is illustrated in Figure 5.

Yellow color reflects the top layer of the map, namely elevated watershed slope area, where removal and movement of soil masses, surface waters and fertilizers take place to the lower area of the relief. The bottom, background layer of the map directly reflects the image of DEM. These accumulation areas

Figure 5. The relief plasticity map with elements of morphodynamic analysis
Legend: 1 - relief plasticity map; 2 - microdepressions; 3 - cavities; 4 - drainage lines: 5 - ridge lines

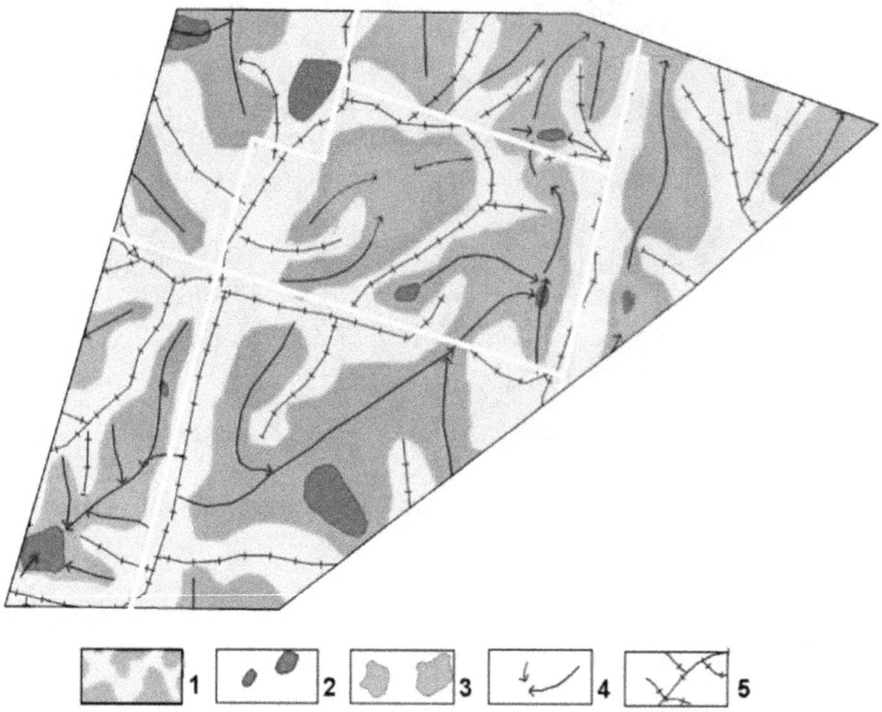

are more wet and characterized by increased power and humus content. The vectors of circular lines determine the directions of lateral material flows. The map indicates the presence of microdepressions and lowlands. These negative forms of relief differ from above-ground levels of soil and fertilizers and appear as a result of accumulation of excess moisture.

Computer processing of DEM makes possible to characterize the surface runoff by separation of drainage network and individual basins. Thus the network of drainage lines and ridges lines was obtained (Figure 6). The drainage network has a fairly complex structure. Various colors show the fragments of valleys depending on their order. The flow directions are determined by the structure of drain lines laid down on the lowest parts of the surface. Respectively, the surrounding areas have an increased humidity of soil. This model should be taken into account when determining the flow direction of surface and groundwater moving particles of soil and chemicals.

Recommendations for Precise Land use

The relief plasticity map, as well as the maps of soil humidity and the maps of morphodynamic analysis were integrated in geographic information system (GIS). This system is the instrument to highlight the areas that are vulnerable for pollution of groundwater and soil. The results are presented in the form of a scheme, which is possible to apply when maintaining precise land use. The scheme consists of three layers defining areas of demolition, accumulation and infiltration (Figure 7).

Figure 6. The computer model indicating basin structure of the analyzed area
Symbols: 1 - separate basins; 2 - ridge lines; 3. - drainage lines

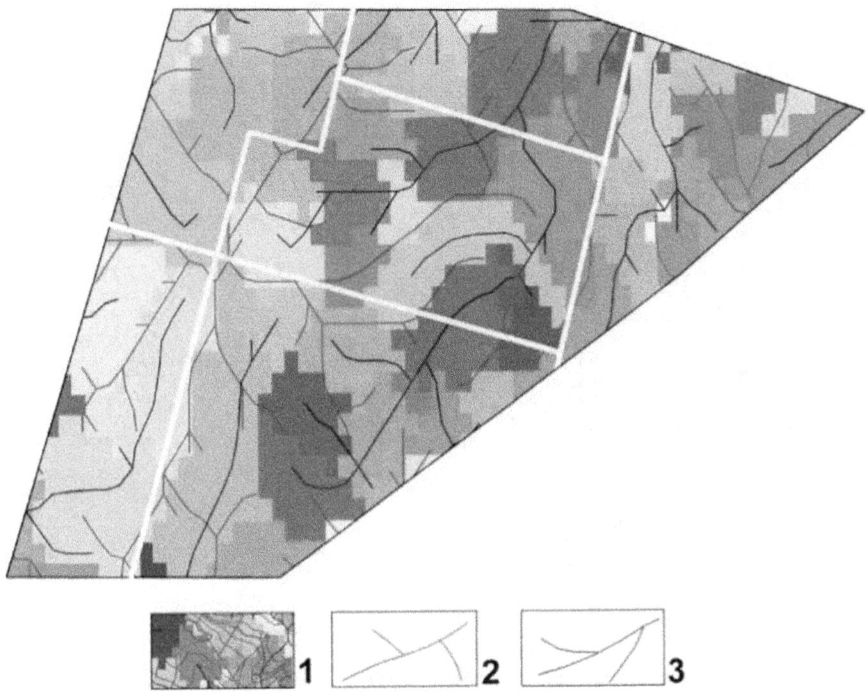

The elevated areas of the surface with relatively low humidity represent the sites from which a permanent removal and movement of soil particles, groundwater, fertilizers and other chemicals occur. The lowered areas reflected on the scheme obtain organic and mineral compounds from the higher levels. The areas of microdepressions, highlighted by white color on the scheme, are the areas of accumulation. They are characterized by excessive moisture, flushing of soil particles and solutions from the higher levels and active infiltration processes. The introduction of chemicals and fertilizers within these areas should be limited in order to prevent pollution of groundwater. The density of crops should depend on soil fertility, moisture reserves and surface topography.

The elaborated scheme of recommendations for precise land use can preserve soil and groundwater from hazardous chemical pollution that should have considerable environmental impact within the area of microdepressions. Also, it can be used in order to increase agricultural potential of cultivated fields through differentiation of the rate of fertilization, which will lead to save resources and to improve productivity. The described methods are able to determine the environmentally vulnerable areas and the corresponding regulations for new chemical additives. The management of the already polluted microdepressions using the proposed approach will be the scope of a further research.

Figure 7. The scheme of recommendations for precise land use
Symbols: 1 - elevated areas, areas of wear; 2 - lowered areas, areas of accumulation; 3 - microdepressions and bottomless lowlands, areas of infiltration.

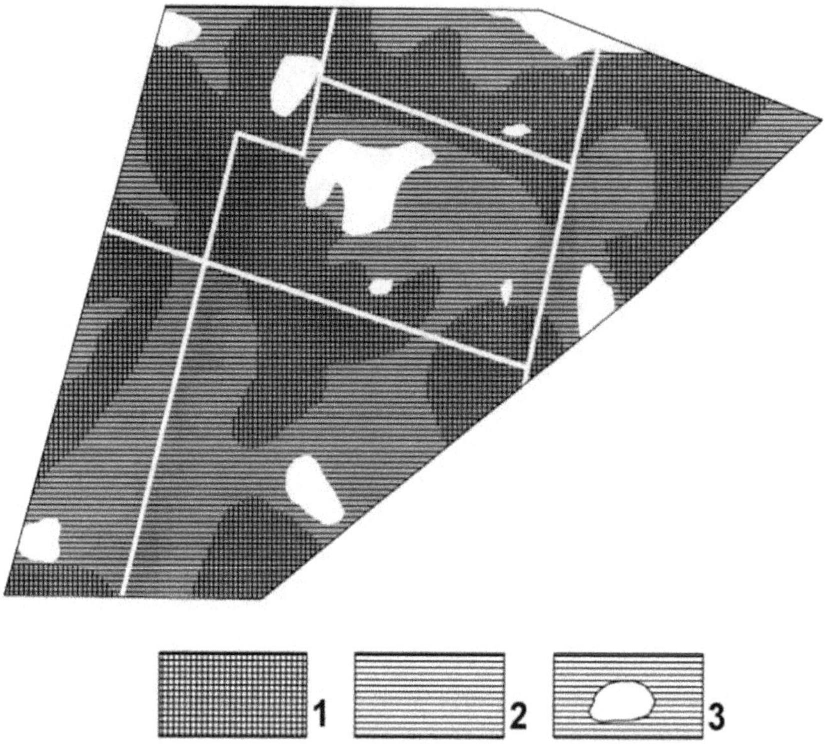

FUTURE RESEARCH DIRECTIONS

The ecological aspect of remote analysis contributes to both increasing productivity of agriculture and enhancing protection of environment, especially groundwater, from fertilizer contaminants or other chemicals. Precision farming in the future should become a common technology that will ensure sustainable land use, high yields of crops, while preserving natural resources from excessive chemical processing. Application of the proposed methods for zoning of used territories tends to decrease environmental impact and to reduce production costs and human efforts.

The methods of remote sensing as the basis for detection of environmentally vulnerable terrain can be further improved. It belongs to the theory and practice of mapping soil structure, as well as for development of precision farming and waste management. Optimization of the calculation algorithms and pattern recognition techniques will ensure development of the automated system producing maps of microdepressions and schemes of land use.

The proposed schemes of recommendations are prospective for applications in modern agricultural machinery. Advanced technology can be developed, which enables the installation of the corresponding GIS data on board computers of agricultural machinery with further automatic differentiation of using fertilizers at various sites within a cultivated field.

CONCLUSION

The territories containing microdepressions and other lowering forms of relief considerably influence an environment because of significant capabilities to accumulate and transfer chemical substances by water flows. Presence of hazardous chemicals within the vulnerable area should cause their intensive propagation in soil, rivers and groundwater.

The data of remote sensing systems obtained from satellites can be used for effective detection of the vulnerable formations of relief. The methods of image classification, calculation of water indexes, relief plasticity and morphodynamic analysis are proposed to solve this task. The classification of space images separates the investigated area from other objects on the Earth's surface. The calculation of normalized water index allows detection of microdepressions due to the high values of humidity. The relief plasticity map based on digital elevation model supported with morphodynamic analysis highlights the lowered forms of relief and displays the movement of substances over the observed surface. The proposed methods of remote sensing correlate with in situ field measurements.

The schemes of land use are composed using the created maps and the processed data of remote sensing. On the basis of the schemes the recommendations for precise agriculture and restrictions for used chemicals can be developed in order to preserve considerable environmental impact of the pollution.

REFERENCES

Appels, W. M., Bogaart, P. W., & Sjoerd, E. A. T. M. (2011). Influence of spatial variations of microtopography and infiltration on surface runoff and field scale hydrological connectivity. *Advances in Water Resources*, *34*(2), 303–313. doi:10.1016/j.advwatres.2010.12.003

Azimov, O. T., & Bublyas, V. M. (2005). Investigation of geodynamic processes in anomaly transfer zones. *Visnyk KNU. Geology*, *34-35*, 97–102.

Bahadir, M., & Duca, G. (Eds.). (2009). *The role of ecological chemistry in pollution research and sustainable development: Proceedings of the NATO Advanced Research Workshop on the role of ecological chemistry in pollution research and sustainable development, Chisinau, Moldova, 8-11 October 2008.* Dordrecht, The Netherlands: Springer. 10.1007/978-90-481-2903-4

Bublias, V. N., & Shestopalov, V. M. (2001). Anomalous zones and their role in the redistribution of radionuclides from the soil surface to groundwater. In Water exchange in hydrageological structure and the Chernobyl disaster. Part 1. Distribution of Chernobyl radionuclides in hydrogeological structure. (pp. 251-356) Kiev: Geological Institute.

Chen, D., Huang, J., & Jackson, T. J. (2005). Vegetation water content estimation for corn and soybeans using spectral indices derived from MODIS near- and short- wave infrared bands. *Remote Sensing of Environment*, *98*(2-3), 225–236. doi:10.1016/j.rse.2005.07.008

Clevers, J. G. P. W., Kooistra, L., & Schaepman, M. E. (2008). Using spectral information from the NIR water absorption features for the retrieval of canopy water content. *International Journal of Applied Earth Observation and Geoinformation*, *10*(3), 388–397. doi:10.1016/j.jag.2008.03.003

Dmitruk, Y. M., & Stuzhuk, O. V. (2014). Cartographic modeling and soil erosion by method of plastic relief. *Geopolitics and Ecogeodynamics of Regions, 10*(1), 41-43.

Gherghina, A., Grecu, F., & Molin, P. (2008). Morphometrical Analysis of Microdepressions in the Central Baragan Plain (Romania). *Revista de Geomorfologie, 10*, 31-38.

Gholizadeh, A., Mohammadmehdi, S., Ben-Dor, E., & Borůvka, L. (2018). Monitoring of selected soil contaminants using proximal and remote sensing techniques: Background, state-of-the-art and future perspectives. *Critical Reviews in Environmental Science and Technology*, *48*(3), 1–36. doi:10.1080/10643389.2018.1447717

Helming, K., Roth, Ch. H., Wolf, R., & Diestel, H. (1993). Characterization of rainfall-microrelief interactions within runoff using parameters derived from digital elevation models (DEMs). *Soil Technology*, *6*(3), 273–286. doi:10.1016/0933-3630(93)90016-8

Keesstra, S., Geissen, V., Mosse, K., Piiranen, S., Scudiero, E., Leistra, M., & van Schaik, L. (2012). Soil as a filter for groundwater quality. *Current Opinion in Environmental Sustainability*, *4*(5), 507–516. doi:10.1016/j.cosust.2012.10.007

Kharytonov, M. M., Mitsik, O., Stankevich, S. A., & Titarenko, O. V. (2013). Geomining Sate ecological assessment and reclamation along costal line of the Kerch peninsula. In A. Veziroglu & Tsitskishvili (Ed.), Black Sea Energy Resource Development and Hydrogen Energy Problems (pp. 325-336). Dordrecht: Springer.

Kostuychanko, Y. V., Bilous, Yu., Movchan, D., Laszlo, M., & Kopachevsky, I. (2013). Toward methodology of satellite observation utilization for agricultural production risk assessment. IERI Procedia, 1, 1-6.

Kostyuchenko, Y. V., Movchan, D., Artemenko, I., & Kopachevsky, I. (2016). Stochastic Approach to Uncertainty Control in Multiphysics Systems: Modeling of Carbon Balance and Analysis of GHG Emission Using Satellite Tools. In R. Mangey & J. Paulo Davim (Ed.), Mathematical Contepts and Applications in Mechanical Engineering and Mechatronics (pp. 350-378). IGI Global. DOI: 10.4018/978-1-5225-1639-2.ch017

Kovac, L., Luptacik, P., Miklisova, D., & Mati, R. (2001). Soil Oribatida and Collembola communitias across a land depression in an arable field. *European Journal of Soil Biology*, *37*(4), 285–289. doi:10.1016/S1164-5563(01)01106-2

Lastochkin, A. N. (2016a). *Foundations of the general theory of geosystems: a textbook in 2 parts. Part 1.* Russia, St. Petersburg: SPbGU.

Lastochkin, A. N. (2016b). *Foundations of the general theory of geosystems: a textbook in 2 parts. Part 2.* Russia, St. Petersburg: SPbGU.

Lillesand, T. M., Kiefer, R. W., & Chipman, J. W. (2015). *Remote sensing and image interpretation* (7th ed.). Hoboken, NJ: John Wiley & Sons, Inc.

Lyalko, V. I., & Sakhatsky, A. I. (2013). The Use of multispectral satellite images for the hydrogeological model certain to evaluate the water balance and groundwater resources. In A. G. Zagorodny & Yu. M. Yermoliev (Ed.), Integrated modeling of food, energy and water security management for sustainable social, economic and environmental developments (pp. 308–319). Akademperiodyka.

Lyalko, V. I., Sakhatsky, A. I., Zholobak, G. M., & Apostolov, A. A. (2016). The research of the dynamics of the annual average of gross primary productivity in Ukraine from 2000-2010. *Ukrainian Journal of Remote Sensing*, *9*, 30–35. https://ujrs.org.ua/ujrs/article/view/77/96

Maeda, K., Kondo, A., & Muir Wood, D. (2015). Micro and Macro Modeling of Internal Erosion. In *Geomechanics from Micro to Macro* (pp. 871–882). London: Taylor & Francis Group.

Manilov, A. I., Kozinetz, A. V., Litvinenko, S. V., Skryshevsky, V. A., Al Araimi, M., & Rozhin, A. (2019). Effect of nano-carbon dispersions on signal in silicon-based sensor structure with photoelectrical transducer principle. *Current Applied Physics*, *19*(3), 308–313. doi:10.1016/j.cap.2018.12.012

McLean, M. I., Evers, L., Bowman, A. W., Bonte, M., & Jones, W. R. (2019). Statistical modelling of groundwater contamination monitoring data: A comparison of spatial and spatiotemporal methods. *The Science of the Total Environment*, *652*, 1339–1346. doi:10.1016/j.scitotenv.2018.10.231 PMID:30586819

Mostaquimur Rahman, A. S. M., Rosolem, R., Kollet, S. J., & Wagener, T. (2019). Towards a computationally efficient free-surface groundwater flow boundary condition for large-scale hydrological modelling. *Advances in Water Resources*, *123*, 225–233. doi:10.1016/j.advwatres.2018.11.015

Pazynych, N., & Romanciuc, I. (2018). Characteristic of the soil moisture state based on cartographic methods and satellite survey data. Odessa National University, Geography and Geology, 23(33), 43-54. doi:10.18524/2303-9914.2018.2(33).146631

Pischulov, S. A. (2013). Sors: An example of arid landforms. *Geomorfologiya*, *3*, 88–98. doi:10.15356/0435-4281-2013-3-

Pristaer, B. S., Kliuchnykov, A. A., Baryakhtar, V. G., Shestopalov, V. M., & Kuhar, V. P. (Eds.). (2016). The safety problems of the nuclear power the lessons of chernobyl. NAS of Ukraine.

Romanciuc, I. F. (2018). Investigation the influence of soils moisture regime on their degradation using the remote sensing and ground field verification. *Ukrainian Journal of Remote Sensing*, *17*(17), 26–30. doi:10.36023/ujrs.2018.17.126

Romanciuc, I. F., Sakhatsky A. I., & Apostolov A. A. (2018). The estimation of soil humidity by the satellite Sentinel-2 imageries (object of study is the Baryshevs'kyi district of the Kiev region. Dopov. Nac. akad. nauk Ukr, 1, 60-66. doi:10.15407/dopovidi2018.01.060

Sakhatsky, A. I. (2007). Improved hydrological forecasts using water index multispectral satellite filming. *Geological Journal*, *4*, 42–47.

Shestopalov, V. M., & Bublyas, V. N. (2016). On formation of depression-channel migration structures. *Geological Journal*, *3*(356), 73–88. doi:10.30836/igs.1025-6814.2016.3.97248

Staodubtsev, V. M., & Basarab, R. M. (2019). Ecologic and economic estimation of land productivity spatial heterogeneity in forest-steppe zone. *Resour Environ Econ.*, *1*(1), 39–42. doi:10.25082/REE.2019.01.005

Stepanov, I. N. (2006). Theory of relief plasticity and new thematic maps. Moscow: Nauka.

Swarnalatha, P., & Sevugan, P. (Eds.). (2018). *Big Data Analytics for Satellite Image Processing and Remote Sensing*. IGI Global. doi:10.4018/978-1-5225-3643-7

Swarnalatha, P., & Sevugan, P. (Eds.). (2018). *Big Data Analytics for Satellite Image Processing and Remote Sensing*. IGI Global. doi:10.4018/978-1-5225-3643-7

Sycheva, S., Khokhlova, O., Pushkina, P., & Ukrainsky, P. (2019). Interrelations of the Bryansk paleosol (end of MIS 3) with the Holocene surface soils in microdepressions of the central forest-steppe within the Russian Upland. *Catena, 172*, 619–633. doi:10.1016/j.catena.2018.09.021

Voloshin, V., Korol, P., & Rudyk, O. (2013). Mathematical and cartographic support for monitoring studies of erosion-prone land using a method of relief plastics. *Lviv Polytechnic National University, 1*(25), 96-99. http://ena.lp.edu.ua/handle/ntb/18863

Voloshin, V. I., & Korchinsky, V. M. (Eds.). (2007). *Distant sounding of the Earth Information Technology, collection, and the maintenance of the Earth's Earth Watch*. Prospectus.

Whigham, D. F., & Jordan, T. E. (2003). Isolated wetlands and water quality. *Wetlands. Springer, 23*(3), 541–549. doi:10.1672/0277-5212(2003)023[0541:IWAWQ]2.0.CO;2

Chapter 16
Impacts of Remediation of Halogenated Organic Compounds in Soils and Sediments

Hale Demirtepe
Izmir Institute of Technology, Turkey

ABSTRACT

Halogenated hydrophobic organic compounds (HOCs) have been used in various industrial applications and are present in many commercial products. Due to their emissions during manufacturing and discharges as wastes, halogenated HOCs such as polychlorinated biphenyls and polybrominated diphenyl ethers are ubiquitously found in the environment and create contaminated sites. To remove the contamination from these sites, various remediation techniques have been useful. The purpose of this chapter is to investigate the impacts of traditional and emerging remediation techniques on ecosystem. One of the traditional remediation techniques is dredging and the mostly studied emerging remediation techniques are bioaugmentation and biostimulation. The efficiency of these techniques is also evaluated regarding reduction in contaminant mass. Overall, this chapter presents the efficiency and possible impacts of dredging, bioaugmentation and biostimulation of soils and sediments, and the implications include the evaluation of most feasible remediation techniques by using life cycle assessment.

INTRODUCTION

Halogenated hydrophobic organic compounds (HOCs) have been produced and used in various industrial applications due to their appealing properties such as chemical and physical stability. They were released into the environment due to their emissions during manufacturing and from products, discharges as wastes and unintentional production. They tend to partition in the organic-rich solid phase, i.e. soils and sediments, and constitute the contaminated sites. After the recognition of the contaminated sites problem, remediation techniques have been deemed as promising solutions. Predominantly used remedia-

DOI: 10.4018/978-1-7998-1241-8.ch016

tion techniques are environmental dredging, capping and monitored natural attenuation (Perelo, 2010; Sowers & May, 2013).

Environmental dredging is the mechanical removal of sediments, hence it may create resuspension of sediments and release of contaminants into water, which can affect the ecosystem. Realization of adverse impacts of dredging directed the remediation efforts towards in situ bioremediation, such as bioaugmentation and biostimulation. Both techniques rely on microbial transformation of pollutants in soils and sediments. They are applied to enhance the intrinsic biotransformation of contaminants. Enhancement of biotransformation can be achieved via addition of microorganisms (Payne, Ghosh, May, Marshall, & Sowers, 2019), nutrients (Chen, Zhou, Wang, Zhu, & Tam, 2015), electron donors (Qiu et al., 2012), surfactants (Huang, Chang, & Lee, 2014), and zerovalent iron (Keum & Li, 2005). It is essential to investigate the effects of these amendments in the environment regarding both the efficiency of bioremediation and the impacts on the natural status of soils and sediments. Plausible solutions should be evaluated before application of these techniques in contaminated sites.

The aim of this chapter is to investigate the efficiency and possible impacts of traditional and emerging remediation techniques applied to soils and sediments contaminated with halogenated HOCs. Halogenated HOCs, polychlorinated biphenyls (PCBs) and polybrominated diphenyl ethers (PBDEs) are selected for this evaluation since they are widely studied in the literature and they have similar molecular structure. The remediation techniques evaluated in this chapter are dredging, biostimulation and bioaugmentation. Previous review studies discussed the advantages and disadvantages of traditional and emerging remediation techniques (e.g. Perelo, 2010). Overall, the dredging has been applied in many sites, and the data collected has been evaluated in terms of effectiveness and impacts (Bridges et al., 2008; National Research Council, 2007). However, since biostimulation and bioaugmentation have not been extensively applied in field, their possible impacts have not been studied in detail. Hence, this chapter, as a new contribution to the literature, compiles the findings of the recent studies on impacts of bioremediation techniques to evaluate them with that of traditional techniques. Furthermore, as the implication of this evaluation, the use of life cycle assessment is discussed regarding the selection of the most efficient remediation technique with the least impact to be implemented in the contaminated site.

BACKGROUND

Hydrophobic organic compounds (HOCs) are released into the environment due to the anthropogenic activities such as emissions during the manufacturing and industrial discharges. They have low water solubility and high octanol-water partitioning coefficient (Kow), which leads to their strong sorption on solid phase. They partition on organic-rich particulate matter, can be transported on suspended particulates and accumulate in soil and sediment particles. When HOCs enter aquatic environment, the ecology of the aquatic environment is affected negatively (Jaffé, 1991). After they reach to bottom sediments, particle-bound HOCs are preserved in sediments and may continue to be accumulated there for decades (Jaffé, 1991), as reflected by core sediment samples (de Souza et al., 2018; Marvin, Waltho, Jia, & Burniston, 2013). Due to the high Kow of HOCs, they are bioaccumulated in the aquatic organisms and hence pose a direct threat to human health. Therefore, the presence of HOCs in the environment has significant impacts on the ecosystem and human health.

Halogenated HOCs were manufactured for industrial purposes to provide chemical and physical stability to products. They had high production volumes and were widely used in industry. Some examples include organochloride pesticides to be used in agriculture for pest control, polychlorinated biphenyls (PCBs) in industry for thermal stability, and polybrominated diphenyl ethers (PBDEs) for flame retardancy. After the recognition of their persistence and toxicity, their production and use were regulated by the countries and international treaties (Stockholm Convention, 2019). However, the environmental contamination with these compounds persists and their emissions may continue from the products which are still in use. Hence, the contamination by halogenated HOCs is still of great concern. Within the scope of this chapter, the focus is on PCBs and PBDEs since they are widely studied in laboratory and in field for remediation of soils and sediments and they are structurally similar compounds which have similar degradation pattern.

PCBs were used as the dielectric fluids in transformers and capacitors, hydraulic fluids and additives in paints and adhesives (Erickson, 2001). They have two aromatic rings connected, and differing number of chlorines attached to the carbon atoms from various positions (Figure 1). This structure constitutes 209 possible PCB congeners. PCBs were manufactured as mixtures of congeners with trade names as Aroclors in USA, Clophen in Germany, Kanechlors in Japan, and Delor in Czechoslovakia (Breivik, Sweetman, Pacyna, & Jones, 2002). Total global production was estimated to be 1.3 million tons (Breivik, Sweetman, Pacyna, & Jones, 2007). This high production volume resulted in the release of PCBs into the environment by emissions from industries and products and waste discharges.

The degree of chlorination affects the physicochemical properties of PCB congeners in such a way that while the water solubility (range from 0.000761 to 2 mg/L) decreases with chlorination degree, log Kow (4.3 – 8.26) increases (Mackay, Shiu, Ma, & Lee, 2006). Hence, when they enter the environment, they partition to air, water, soil and sediments according to their physicochemical properties. The lower chlorinated PCBs tend to partition into water column and pore water, while higher chlorinated ones tend to accumulate in sediments and soil. They are bioaccumulated in the aquatic organisms and are biomagnified in the food chain (Erickson, 2001). The adverse effects of PCBs on human health have been identified as being potential neurotoxins, endocrine disruptors (Safe, 1994) and carcinogenic to humans (Group 1) (IARC, 2016). Additionally, twelve congeners show dioxin-like toxicity owing to their coplanar structure (Van den Berg et al., 2006).

PBDEs were used as flame retardants in the electronic equipment, plastics and insulation materials (Alaee, Arias, Sjödin, & Bergman, 2003; Rahman, Langford, Scrimshaw, & Lester, 2001). They have two aromatic rings attached with oxygen atom, and bromine atoms attached to the carbon atoms (Figure 1). Similar to PCBs, 209 PBDE congeners are theoretically possible. PBDEs were produced as mixtures of congeners with similar degree of bromination, e.g. penta-BDE mixture mainly consists of PBDEs with four, five and six bromines (La Guardia, Hale, & Harvey, 2006). The total global PBDE production was approximately 1.5 million tons (UNEP, 2010).

Water solubility of PBDE congeners ranges from $1.3*10^{-8}$ to 0.088 mg/L and log Kow ranges from 5.03 to 11.15 (Palm et al., 2002). Hence, they mostly partition in the particulates and solid phase in the environment. The bioaccumulative congeners are mostly ones having three to five bromines (USEPA, 2010), which may also biomagnify in aquatic food web (Shao, Tao, Wang, Jia, & Li, 2016). PBDEs have been shown to affect nervous, reproductive, endocrine and immune systems in humans (ATSDR, 2017).

Figure 1. The molecular structures of polychlorinated biphenyls (PCBs) in the upper panel, and polybrominated diphenyl ethers (PBDEs) in the lower panel

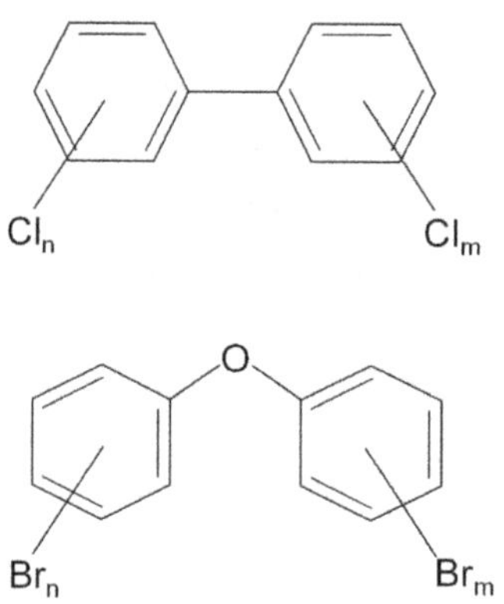

Remediation of Soils and Sediments

The accumulation of HOCs in soils and sediments require attention regarding the human health. The soils and sediments having high concentrations of hazardous pollutants which pose risk to human health and the environment are called contaminated sites. Remediation of contaminated sites involve the activities to control, reduce and remove the contamination, and hence to reduce the risk to human health and the environment. Traditional remediation techniques are dredging, capping and monitored natural attenuation, while emerging techniques include biostimulation, bioaugmentation and phytoremediation (Perelo, 2010). The most frequently applied traditional technique, dredging, and the mostly studied emerging techniques, biostimulation and bioaugmentation, are explained below.

Environmental Dredging

Environmental dredging is the mechanical removal of sediments having exceeded levels of contamination from a water body to remediate the sediment and minimize the adverse effects of contamination on the aquatic species (Palermo & Hays, 2014; Palermo, Schroeder, Estes, & Francingues, 2008). The objectives of environmental dredging are to:

- Supply the sufficient accuracy for the removal of contaminated sediments so that the desired cleanup levels are achieved without destroying the clean sediments,
- Provide the reasonable time period for dredging operation and subsequent transport for the treatment or disposal of sediments,
- Reduce or if possible eliminate the resuspension of sediments, downstream transport of resuspended sediments and contaminants release to water and air (Palermo et al., 2008).

Environmental dredging with subsequent treatment or disposal is the most common approach for remediation of contaminated sediments. It has been reported that dredging was applied in more than 100 contaminated sites in North America, 15 – 20% of which were also followed by capping or monitored natural attenuation as remediation techniques (Patmont, LaRosa, Narayanan, & Forrest, 2018; US EPA,

2005). When environmental dredging is the first component of a contaminated site remediation project, it should be fully integrated with the following steps such as transportation, treatment and/or disposal of the dredged sediment and with the subsequent application of monitored natural attenuation or capping of the sediments to achieve the desired cleanup level (Palermo & Hays, 2014).

Biostimulation and Bioaugmentation

Another method to remove/reduce the contamination is bioremediation. It involves biotransformation of contaminants in sediments or soil by microbial activity. In order for biotransformation to occur in the environment, the suitable microbial consortia and the proper environmental conditions are required (Himmelheber & Hughes, 2014). The environmental conditions such as redox potential, presence of electron acceptors/donors, pH, nutrients, and temperature are important factors influencing the biotransformation, as well as the bioavailability of contaminants. The presence and number of microbial consortia and the need for co-culturing species determine the biotransformation efficiency. In some cases, the microbial population present in the contaminated site, and the environmental conditions of the site are suitable so that the contaminant biotransformation, i.e. intrinsic biotransformation, can occur without any interference, hence, the natural attenuation can be used as a bioremediation technique (Himmelheber & Hughes, 2014). However, in most cases the intrinsic microbial activity is limited or absent, hence, the interference is needed to enhance the biotransformation processes. Biostimulation is the addition of stimulants such as nutrients, electron acceptors/donors to the contaminated site so that biotransformation by indigenous microbial population is enhanced (Himmelheber & Hughes, 2014). Bioaugmentation is the addition of microorganisms able to biotransform the contaminant of concern to the contaminated site (Himmelheber & Hughes, 2014).

Biotransformation of halogenated HOCs, i.e. PCBs and PBDEs, can occur as aerobic degradation and anaerobic dehalogenation processes. Anaerobic dehalogenation is the main process since the anaerobic conditions dominate in soil and sediment systems. Anaerobic dehalogenation of PCBs and PBDEs is achieved by organohalide respiring bacteria (He, Robrock, & Alvarez-Cohen, 2006; Liang et al., 2018; Wiegel & Wu, 2000). The process includes replacement of halogen atoms with hydrogen on the aromatic ring, transforming higher halogenated congeners to lower halogenated ones (Wiegel & Wu, 2000). Dehalogenating bacteria use halogenated compounds as terminal electron acceptors for reductive dehalogenation (Borja, Taleon, Auresenia, & Gallardo, 2005; Liang et al., 2018).

When compared to traditional remediation techniques such as dredging, biostimulation and bioaugmentation are less frequently applied in soil and sediments in field due to the difficulties in even distribution of amendments throughout the site. These bioremediation methods are mostly tested in laboratory or pilot scale. Bioaugmentation is generally applied for the groundwater remediation, and sometimes is applied together with biostimulation to ensure the continuous activity of the amended microorganisms (Himmelheber & Hughes, 2014). Bioaugmentation of PCB contaminated sediments has recently been applied in field as a pilot study by incorporating the microorganisms to activated carbon (Payne et al., 2019).

EFFECTIVENESS AND IMPACTS OF REMEDIATION TECHNIQUES

Effectiveness and Impacts of Environmental Dredging

Environmental dredging is found to be an effective remediation method regarding the removal of contaminant mass. Especially in the contaminated sites, where mobilization and transport of contaminants are possible under the high-flow conditions, the removal of contaminant mass by the dredging may reduce the risk. It was shown to result in a significant reduction in total contaminant mass. For example, the environmental dredging achieved up to 96% reduction in total PCB mass in sediments of Fox River (National Research Council, 2007). However, considering all the ecological impacts of contaminants in an ecosystem, the mass removal may not reduce risk. Removal of contaminant mass does not directly result in the decrease in water column contaminant concentrations. For instance, in Grasse River, no reduction in water column or fish PCB concentrations was observed although 98% of PCB mass was removed from sediments (National Research Council, 2007). Additionally, the removal of contaminant mass may not be effective to meet the cleanup goal in the contaminated sediments. About one quarter of 20 contaminated sites evaluated by the National Research Council (2007) could meet the remediation goals, the remaining either could not meet the goals or did not have sufficient monitoring data to evaluate.

Environmental dredging is mainly the disturbance of sediments, which further results in the changes in chemical properties of the sediment such as redox potential and pH. The disturbance in anoxic sediments creates an influx of dissolved oxygen leading to an increase in the redox potential and accordingly an increase in microbial activity, and a decrease in pH of sediment. The change in the chemical properties of sediments would then cause the contaminants mobilization and transfer in the sediments (Eggleton & Thomas, 2004). Therefore, the alterations in the natural status of sediments due to the dredging might trigger the other problems.

Environmental dredging results in the sediments resuspension by the particles detaching from the sediment bed during the dredging operation (Bridges et al., 2008). These detached particles are dispersed in the water column, create turbidity, and may cause several water quality and ecological problems (Bridges et al., 2008; Goossens & Zwolsman, 1996). The turbidity can be continued for a short period of time after dredging operation finishes (Eggleton & Thomas, 2004). The turbid plume formed in the dredging area could affect the behavior of fish and other mobile receptors, and could have an impact on the health of less mobile aquatic vertebrates and invertebrates. Furthermore, the resettling of suspended sediments might affect the bottom-dwelling organisms (Bridges et al., 2008). The benthic organisms are also affected by the reduction in available light due to the turbidity created during the dredging (Fraser et al., 2017). Hence, the turbidity is one of the outcomes of dredging that affects the aquatic ecosystem.

The resuspension rates reported from various dredging operations range from less than 0.1% to over 5% of the total volume of sediment removed (Bridges et al., 2008). The resuspension rate varies depending on the sediment and site characteristics and dredging operation. The sediment properties such as bulk density, particle size distribution and mineralogy influence the rate and extent of resuspension. Also, the site characteristics, e.g. water depth, current, waves and salinity affect the resuspension rate. The rate is also affected by the dredging equipment used, mechanical dredges, cutter suction heads, horizontal augers, etc (Bridges et al., 2008). It was also reported that erosion of generated residuals can also contribute to the resuspension of sediments (Palermo & Hays, 2014).

During the environmental dredging operation, the contaminants within the sediment particles and pore water can be released into the water column or air depending on the physical and biological characteristics of the ecosystem (Bridges et al., 2008; Eggleton & Thomas, 2004). HOCs can be desorbed from the resuspended sediments depending on the desorption rate. The desorption rate of varies with the sediment particle size and physical-chemical properties of the compounds (Eggleton & Thomas, 2004). For example, degree of chlorination is an important factor in rate of desorption of PCBs, highly chlorinated ones having the slower desorption rates (Borglin, Wilke, Jepsen, & Lick, 1996). Hence, the lower halogenated congeners can be desorbed more easily and be dissolved in the water column. Once the contaminants adsorbed on the resuspended particles partition to the water column and the ones dissolved in the pore water are released, they may be transported for long distances in the dissolved form, even many months after the dredging (Eggleton & Thomas, 2004). PCB levels in the water of the Grasse River increased right after the dredging was started, and fluctuated until the end of dredging operation, then decreased again to the baseline levels after the dredging has been completed (National Research Council, 2007). A laboratory simulation of physical resuspension of sediments also reported the release of the dissolved phase PCBs more than that of particulate phase PCBs after resuspension (Hedman, Tocca, & Gunnarsson, 2009). Another laboratory study has observed the increased PCB concentrations in both dissolved and particulate forms, enriched with the lower and higher chlorinated congeners, respectively, after resuspension of sediments (Di Leo et al., 2016). Additionally, the dissolved contaminants may become readily bioavailable (Bridges et al., 2008). The bioavailability of contaminants depends on the site characteristics and level of contamination, as well as the organisms present in the ecosystem (Eggleton & Thomas, 2004). Exposure studies have showed the accumulation of PCBs in fish exposed to resuspended contaminated sediments under the different conditions, such as aerated and anoxic sediments (Seelye, Hesselberg, & Mac, 1982). Previous field studies showed that PCB uptake by fish was influenced by water column concentration moving downstream of a dredging site and increased for six months after dredging (Eggleton & Thomas, 2004; Rice & White, 1987). The long-term monitoring of a group of HOCs with similar water solubility and Kow as PCBs, polychlorinated naphthalene (PCN) levels in fish and gulls in the Great Lakes has also demonstrated an increase in PCN levels in biota after 1993, when the remediation projects including dredging had been implemented in the Detroit River (McGoldrick, Pelletier, de Solla, Marvin, & Martin, 2018). The authors also discussed that the sediment resuspension after the dredging might cause the changes in the congener profiles of bioavailable PCNs (McGoldrick et al., 2018). Therefore, the contaminant release due to resuspension is another outcome of dredging that influences the aquatic ecosystem.

The residual formation at the post-dredging surface of sediment is another limitation of environmental dredging operation. When the resuspended sediment particles settle, they become part of the dredging residuals. Contaminants in the residuals may also be released to the water by densification, diffusion and bioturbation (Bridges et al., 2008). The previous applications indicated that the post-dredging residual sediments had high contamination levels that could not meet the desired cleanup level (Patmont et al., 2018). Additionally, the post-dredging residuals can result in continuous exposure of biota for long period of time after completion of dredging operation (Patmont et al., 2018). For example, PCB concentrations in fish were found to be increased during and many years after the dredging of bay sediments (Patmont et al., 2018; West, O'Neill, & Ylitalo, 2017). Hence, even after the settling of resuspended sediments, they are still a contamination source in the aquatic environment.

The final disposal of dredged sediments may create another challenge in a remediation project. To protect the human health and the environment from contamination, the dredged sediments are required to be placed in a temporary or permanent structure such as landfills, confined disposal facilities and contained aquatic disposal pits. Sanitary or hazardous waste landfills are alternative disposal facilities for dredged sediments depending on the type of wastes they are designed to contain and the content of the sediments. Hazardous waste landfills are more appropriate compared to the sanitary ones if the sediments have the high levels of contamination. Regardless the landfill type, the contaminant release should be minimized via the caps, covers and leachate collection systems. The contaminants may become mobilized and transported via the physical, chemical and biological processes. The release rates are dependent on characteristics of dredged material, method of dredging and disposal, location of disposal facility and their design, operation and management (US EPA, 1996). The loss of sediments during the filling, surface runoff due to precipitation, seepage from bottom or dike walls, volatilization of contaminants to air and contaminant uptake by plants and animals are the possible mechanisms that result in the potential risks (US EPA, 2005). The effluent generated due to the surface runoff should be controlled by accurate design and be treated properly. Also, to reduce the plant and animal uptake of sediments, the disposal facility must be covered with proper materials (US EPA, 1996). On the other hand, the dredged sediments may also have a beneficial use such as additives in construction material production. The use of sediments in such applications depends on the physical properties of the sediments, type and amount of the contaminants (US EPA, 2005). Therefore, the proper disposal or beneficial use of the dredged sediments is an important part of a remediation project that needs to be designed appropriately not to result in any environmental contamination.

Effectiveness and Impacts of Biostimulation and Bioaugmentation

The disturbance of sediments by dredging has lead researchers to investigate the effectiveness of bioremediation techniques. Biostimulation and bioaugmentation are shown to be effective in reducing contamination and toxicity of halogenated organic compounds in soils and sediments. Biostimulation of PCBs and PBDEs was achieved by carbon source, nutrients, electron donors and other halogenated organic compounds addition to improve microbial activity in soils and sediments. The commonly used biostimulants also include surfactants to increase the bioavailability of HOCs (Perelo, 2010) and zerovalent iron (Keum & Li, 2005). The effectiveness and impacts of amendment with nutrients, electron donors, carbon sources and halogenated organic compounds are evaluated below.

When compared to intrinsic biotransformation, i.e. natural attenuation processes, biostimulation with electron donors showed an enhancement of the reduction percentages from 11% to 44% in sediment microcosms (Demirtepe & Imamoglu, 2019; Huang et al., 2014), and bioaugmentation enhanced the reduction by 23% (Demirtepe & Imamoglu, 2019). The examples of laboratory studies on enhancing anaerobic dehalogenation of PCBs and PBDEs with various amendments are presented in Table 1. Percent reduction of PCBs and PBDEs had a wide range from nearly zero percent to nearly 99%. This range is related to the soil and sediment properties, contaminant types, incubation conditions and time. The organic matter contents in the soil and sediment determines the bioavailability of pollutants and acts on the maintenance of inoculated microbial strains, hence it is an important parameter in the effectiveness of bioaugmentation (Lebeau, 2011). The intrinsic nutrient contents in the soils and sediments and supplied nutrients influence the effectiveness of bioremediation, regarding partial or complete microbial degradation of pollutants (Lebeau, 2011). Hence, these bioremediation techniques have the low predictability

Table 1. Laboratory studies on enhanced anaerobic dehalogenation of PCBs and PBDEs

Soil/ Sediment	Contaminant	Amendments	% Reduction	Incubation time (days)	Reference
PCBs					
Sediment mesocosm	Hepta-CBs Hexa-CBs Penta-CBs	Bioaugmentation with anaerobic methanogenic granules with poplar wood powder as nutrient source	65% 50% 90%	224	(Natarajan et al., 1997)
Sediment microcosm	PCB-61 PCB-23 PCB-65 PCB-24	Estuarine salt medium with sodium acetate, propionate and butyrate	89% 100% 99% 97%	128	(Berkaw, Sowers, & May, 1996)
Sediment microcosms	PCBs	Iron addition to produce H_2 + after 11 months Biaougmentation with enriched indigenous bacteria	56% (based on Cl per biphenyl)	400	(Winchell & Novak, 2008)
Sediment mesocosm	PCBs (penta and higher chlorinated CBs)	Bioaugmentation with *Dehalobium chlorocoercia* strain DF-1	56%	120	(Payne, May, & Sowers, 2011)
Sediment microcosms	PCB-151	Bioaugmentation with Bacterium o-17, *Dehalobium chlorocoercia* DF-1, and phylotypes DEH10 & SF1 and biostimulation with acetate, propionate and butyrate	~80%	300	(Fagervold, Watts, May, & Sowers, 2011)
Sediment microcosms	PCBs	Sodium acetate	26.9%	12 months	(Di Gregorio, Azaizeh, & Lorenzi, 2013)
Sediment microcosms	PCBs	Bioaugmentation with *Dehalococcoides mccartyi*	19.9%	320	(Matturro, Ubaldi, Grenni, Caracciolo, & Rossetti, 2016)
Soil microcosms	PCBs-179, 183, 189, 198, 209	Lactate	>50%	90	(Song et al., 2015)
PBDEs					
Sediment microcosms	BDE-47 BDE-99 BDE-209	Phosphate buffer, methanol, dextrose	30% 3% 3.8%	240 240 300	(Tokarz III, Ahn, Leng, Filley, & Nies, 2008)
Sediment microcosms	BDE-209	Yeast and various electron donors	5 – 12%	90	(Qiu et al., 2012)
Sediment slurry	BDE-209	Lactate/acetate/pyruvate Sulfate/bicarbonate/nitrate	82 – 88% 75 – 81%	56	(Huang et al., 2014)
Sediment microcosms (with various type of sediments)	BDE-47 BDE-153 BDE-209	Mineral salts medium	62 – 82.4% 31.4 – 98.8% ~0%	90	(Zhu, Wang, Wang, Luan, & Tam, 2014)
Soil microcosms	BDE-209 BDE-47	Lactate	39.7% 29.5%	90	(Song et al., 2015)
Sediment microcosms	BDE-47	Biochar	64.3%	140	(Chen, Wang, Pan, Farzana, & Tam, 2017)
Sediment microcosms	BDE-209	Ethanol and sodium formate	55.3%	180	(Demirtepe & Imamoglu, 2019)
Sediment microcosms	BDE-209	Bioaugmentation with *Dehalobium chlorocoercia* strain DF-1	40.2%	180	(Demirtepe & Imamoglu, 2019)
Sediment mesocosms	BDE-209 BDE-99	Ethanol and sodium formate	32.3% 34.2%	250	(Demirtepe, 2017)

in terms of the final contaminant reduction due to the biological processes involved (Perelo, 2010). It has been also shown that a biostimulant which was found to be effective in the sediments from one site may not be effective for the sediments from another site, indicating an insufficient number of dechlorinating bacteria in the latter sediments (Winchell & Novak, 2008). Hence, in order for biostimulation to be effective, the sufficient population of dehalogenating bacteria should be present in the contaminated site. Furthermore, as can be seen from Table 1, the incubation times for laboratory studies reach up to more than one year indicating that even under the favorable anaerobic conditions, the contaminant reduction percentages are not very high. Field applications of these bioremediation techniques would require longer times due to the harsh environmental conditions such as lower temperatures and presence of co-contaminants. Therefore, one of the main disadvantages of bioremediation is the long period of time that might be needed to achieve the desired level or to meet the remediation goal.

Biostimulation of dehalogenating bacteria with the halogenated organic compounds, i.e. haloprimers, has been studied for PCBs and PBDEs. The addition of haloprimers may increase dehalogenating biomass, induce genes required for dehalogenation and support cometabolism of PCBs and PBDEs (Sowers & May, 2013). For example, to stimulate dechlorination of weathered Aroclor 1260 in sediments, a specific PCB congener was added, which resulted in the enrichment of the dechlorinating bacteria in sediments (Bedard, Bunnell, & Smullen, 1996). The researchers have also demonstrated that polybrominated biphenyls are successful in stimulation of PCB dechlorination (Bedard, Van Dort, & Deweerd, 1998). However, the addition of these compounds in the environment, especially in high concentrations (Bedard et al., 1998), would mean introducing toxic compounds, which are listed as Persistent Organic Pollutants under Stockholm Convention (Stockholm Convention, 2019). The other haloprimers for PCB biostimulation were halogenated benzoates and pentachloronitrobenzene (Krumins et al., 2009; Sowers & May, 2013). Similarly, PBDE debromination was shown to be more extensive when trichloroethene (TCE) was present in the sediment microcosms amended with pyruvate as a carbon source for dehalogenating bacteria (Lee & He, 2010). TCE is known as causing adverse effects on the central nervous system, liver, kidneys, gastrointestinal system and skin, and even can lead to kidney and liver cancer and lymphoma (ATSDR, 2016; WHO, 2005). A very recent study investigated the effects of tetrachloroethene (PCE) addition to enhance PCB dechlorination in sediments (Xu, Lu, Yu, & Wang, 2019). PCE addition enhanced dechlorination of PCB-180 and Aroclor 1260 in sediments, while PCE concentrations higher than 0.2 mM were shown to inhibit PCB-180 dechlorination in microcosms in the absence of sediments (Xu et al., 2019). Although the authors have observed the complete dechlorination of PCE to non-toxic ethene, PCE itself is identified as probably carcinogenic to humans and as toxic to aquatic species (IARC, 2014; WHO, 2006). Considering the main aim of bioremediation as reducing the risk posed by contaminants at those sites, introducing new toxic compounds in the environment, even if they are completely degraded at the end of operation, might have significant impacts on the ecosystem and human health.

Biostimulation with nutrients, carbon sources and electron donors is also commonly studied for PCB and PBDE biotransformation in soils and sediments. For instance, nitrogen addition was found not to significantly affect BDE-47 reduction but to stimulate BDE-47 phytoremediation in mangrove soil (Chen et al., 2015). Nitrogen is a key element in microbial activity, and determines the abundance and diversity of microorganisms in soils (Chen et al., 2015). The authors have also discussed that despite the higher abundance of total bacteria in nitrogen amended unplanted soil, the growth of dehalogenating bacteria was inhibited (Chen et al., 2015). The effects of more complex nutrient mixture addition on the removal of PBDEs were tested by adding various types of synthetic wastewater to lake sediments (Chen,

Wang, Wang, Miao, & Wang, 2018). BDE-47 debromination was inhibited by the high nitrite and nitrate concentrations, while it was enhanced at low nutrient concentrations (Chen et al., 2018). The similar effect of high nutrient concentrations was observed for the abundance of dehalogenating bacteria, i.e. high ammonium and dissolved organic carbon resulted in lower abundance of dehalogenating bacteria, possibly due to a competition for available electrons with carbon decomposition (Chen et al., 2018). These studies investigated the short-term effects of nitrogen addition, but the long-term effects are still unknown (Chen et al., 2015). Nevertheless, these studies demonstrated the possibility of inhibition of dehalogenating bacteria under the certain conditions.

Addition of biostimulant consisting of high amount of carbon source and electron donor may also lead to increase in the chemical oxygen demand (COD) in the receiving aquatic environment. COD is a water quality parameter that measures the amount of oxygen needed to oxidize organics in a water body, i.e. a measure of the amount of organics in water. It is also used to determine the effect of a discharge to a receiving water body. Hence, water quality regulations set limits on the allowable COD for industrial effluents. Accordingly, a study measured the COD of a biostimulant solution to be 10 times higher than the allowable industrial COD discharge and 35 to 100 times higher than allowable COD of surface waters (Demirtepe, 2017). Hence, the addition of high COD biostimulant solution to sediments might result in the drastic changes in nutrient balance of the aquatic ecosystem, and might cause eutrophication in the water body.

Bioaugmentation with organohalide respiring bacteria may have little effects on the indigenous population since the organohalide respiring bacteria use hydrogen and acetate which are common fermentation products by the anaerobic bacteria (Liang et al., 2018). However, the niche competition can also occur in the aged contaminated soil communities during the bioaugmentation (Lebeau, 2011). Other than alterations of genetic structures of microbial populations, bioaugmentation can also result in the impairment of metabolic activities of more than one biological compartments in a soil system (Lebeau, 2011). Bioaugmentation of the aged PCB contaminated sediments with Dehalococcoides sp. did not enhance the biodegradation of PCBs, resulting in 19.9% of total PCB reduction compared to 26.7% reduction in non-bioaugmented sediments (Matturro et al., 2016). The researchers have detected the microbial community change during the incubation under the favorable anaerobic conditions in microcosms with and without bioaugmentation, and have reported the similar microbial community distribution for both. They have concluded that the PCB dechlorinating organisms were already adapted to the long-term contamination by the aged sediments, were active in microcosms, while the bioaugmented species were not able to adapt to high PCB concentrations (Matturro et al., 2016). On the other hand, the pilot scale bioaugmentation study for PCB contaminated sediments showed that the total indigenous microbial diversity did not significantly change in different sites and depth of the sediments (Payne et al., 2019). The dechlorinating species used in the study was previously grown with PCB congeners and tested in smaller scale laboratory studies (Payne, Fagervold, May, & Sowers, 2013; Payne et al., 2011). Hence, it is important to acclimatize the dehalogenating species to environmental contamination levels and conditions before bioaugmentation for the effective bioremediation in contaminated sites.

A challenge for the effectiveness of bioaugmentation is the survival and activity of microorganisms introduced in the environment (Perelo, 2010). The activity and survival of microorganisms depends on the pH and redox of the system, concentration and bioavailability of the contaminants, and the presence of necessary substrates (Perelo, 2010). For an effective bioaugmentation, the adequate number of viable cells should be introduced and maintained (Singh, Parmar, Kuhad, & Ward, 2011). Additionally, an appropriate bacterial strain should be selected for bioaugmentation to proceed successfully (Perelo, 2010).

Dehalogenation products of biostimulation and bioaugmentation can create another challenge if these products are more bioaccumulative or toxic than the original contaminant. A study investigating the effect of BDE-153 spike and debromination on microbial community compositions of four different types of sediments have revealed that the microbial community changes were significantly related to the formation of BDE-52 as the debromination product, salinity and total organic carbon of the sediments (Pan, Chen, Zhou, & Tam, 2018). Hence, depending on the generation of dehalogenation products, the accumulating congeners might affect the microbial composition of the environment and accordingly the ecosystem. Formation and accumulation of toxic congeners is a significant concern. As mentioned before, twelve PCB congeners exhibit dioxin-like toxicity. Previous studies have identified the biotransformation of toxic PCB congeners to non-toxic ones (Demirtepe, Kjellerup, Sowers, & Imamoglu, 2015). On the other hand, debromination of higher brominated PBDE congeners leads to the formation of congeners having three or four bromines (Demirtepe & Imamoglu, 2019; He et al., 2006; Huang et al., 2014; Robrock, Korytár, & Alvarez-Cohen, 2008; Tokarz III et al., 2008), which are more bioaccumulative than parent congeners. Therefore, accumulation of the toxic or bioaccumulative congeners should be monitored during a bioremediation project.

SOLUTIONS AND RECOMMENDATIONS

In order to reduce the possible impacts of remediation techniques, certain actions can be taken during the project implementation. For example, to limit the particles dispersion during the dredging, the special dredging equipments or silt screens can be used (Goossens & Zwolsman, 1996), or non-toxic biostimulants at acceptable concentrations can be used in biostimulation. These actions can be determined via prior laboratory or pilot scale studies. The studies investigating the use of various techniques to minimize resuspension and contaminants release into the water column should include testing of contaminants bioaccumulation in the indigenous species. More studies can be conducted to identify the most effective and environmentally friendly biostimulants at the appropriate concentrations for the biotransformation of halogenated HOCs. However, to make a comprehensive assessment of the effectiveness and impacts of remediation techniques, the tools like a life cycle assessment (LCA) can be useful during the decision making process.

LCA is a tool to analyze and evaluate the burdens of a product or activity covering its all life cycle (Diamond, Page, Campbell, McKenna, & Lall, 1999; Søndergaard & Owsianiak, 2018; Visentin, da Silva Trentin, Braun, & Thomé, 2019). It can be applied for the evaluation of environmental activities, including also their economic efficiency. LCA application for remediation includes all the raw material and energy inputs, the outputs such as waste and air emissions, site processing, i.e. efficiency/result of the remediation activity, and post-site processing, i.e. monitoring and maintenance after remediation, along with the associated costs (Diamond et al., 1999). LCA can be conducted either qualitatively or quantitatively. Suer and collegues (2004) suggested that LCA to be limited in space, time and inclusion of secondary processes to obtain manageable results. In terms of spatial scale, global, regional or local boundaries should be set and the time scale should be determined based on the period of possible impacts of evaluated remediation techniques (Suèr, Nilsson-Påledal, & Norrman, 2004). LCA can be applied for remediation projects in terms of i) selection of the remediation technique having minor impacts, and ii) optimization of remediation technique regarding the aspects considered to lower the possible impacts of the technique (Visentin et al., 2019). An example for optimization of the remediation technique by LCA can be comparing of two biostimulants to be used in the biostimulation of contaminated site to select the one causing the smallest impacts.

The studies reviewing the LCA application on soils and sediments remediation have revealed that the techniques compared mainly included the capping, excavation and landfill, ex-situ treatments, in-situ bioremediation and no action (Lemming, Hauschild, & Bjerg, 2010; Suèr et al., 2004; Visentin et al., 2019). In general, the excavation based remediation techniques resulted in the higher impacts considering the land use and the heavy equipments used, which cause the air emissions (Visentin et al., 2019). Bioremediation using electron acceptors was found as the best technique unless the specific electron acceptors, generating very high impacts, were used (Suèr et al., 2004). A couple of studies have applied LCA for remediation of PCBs. The first study compared incineration of PCB contaminated soils with the base catalyzed decomposition process, and found incineration to result in higher impacts (Hu, Zhu, & Ding, 2011). The second study has compared the aerobic bioremediation techniques, i.e. mechanical aeration and electric aeration, with incineration of PCB contaminated soils, and has determined that the bioremediation was the lower impact generating technique (Busset, Sangely, Montrejaud-Vignoles, Thannberger, & Sablayrolles, 2012). The future studies may be focused on the evaluation of remediation techniques to be used for halogenated HOCs both to help decision makers to select the most appropriate remediation technique and to optimize the remediation techniques which can result in lower environmental impacts.

FUTURE RESEARCH DIRECTIONS

The remediation of halogenated HOCs is an ever-growing research area. Although PCBs and PBDEs were restricted, the sites contaminated with these compounds still require attention since the compounds are persistent and remediation has not been applied to all the contaminated sites yet. Traditional and emerging remediation techniques need to be improved so that the remediation application would not pose risk to human health and ecosystem. Since the dredging was shown to have high environmental impacts both by field observations and LCA analysis, more sustainable and cost-effective remediation techniques should be developed further.

The emerging remediation techniques need to be optimized in terms of types of amendments and their effective application. Investigation of the impacts of biostimulant addition appears as an essential research direction. Especially, the studies aiming to determine the alterations in water quality and fate of biostimulant in the contaminated environment can be conducted. Additionally, the bioaugmentation effectiveness in terms of the niche competition, species activity under the extreme conditions such as high contaminant concentration, presence of co-contaminants, and low temperatures, which are normally observed in the subsurface environment, should be investigated further. Lastly, the distribution of biostimulants and bioaugmented species in the contaminated site should be studied. Bioaugmentation using activated carbon proved to be useful in previous pilot studies (e.g. Payne et al., 2019). The similar attempts should be made for the biostimulation of soils and sediments.

In terms of the field studies, the monitoring before, during and after remediation would determine the effectiveness of the specific remediation project and could help to identify the precautions to eliminate possible risks related to the exposure to contaminants. Especially, the long-term monitoring of each environmental compartment including the biota will result in the effective management of remediation projects. Incorporating LCA and similar impact assessment tools for the decision making process of remediation projects emerges as a promising research direction. Additionally, conducting laboratory or pilot scale studies for the purpose of providing information on LCA as much as possible can facilitate use of these assessment tools. For LCA to cover all the environmental impacts of remediation techniques, the stages should be optimized to include formation of possible toxic/bioaccumulative compounds, fate of biostimulants and microbial community change.

CONCLUSION

This chapter evaluates the traditional and emerging remediation techniques for soils and sediments contaminated with PCBs and PBDEs. Environmental dredging, a frequently applied technique, was shown to generate resuspension in the sediments, release the contaminants in water and generate residuals as a continuous source of contamination. Biostimulation and bioaugmentation, despite being efficient techniques, require long time periods to achieve the remediation goals. The amendments used in biostimulation might result in introduction of toxic compounds, inhibition of microbial activity with the increased biostimulant concentration, generation of toxic or bioaccumulative compounds and water quality destruction in the aquatic environment. Bioaugmentation might lead to the niche competition and change in the microbial diversity. To evaluate the most efficient remediation technique with the least impacts, the tools such as LCA can be used before field applications. This chapter adds to the existing literature on the observed impacts of the emerging remediation techniques on the whole ecosystem, while emphasizing the importance of LCA in remediation applications.

ACKNOWLEDGMENT

This research received no specific grant from any funding agency in the public, commercial, or not-for-profit sectors.

REFERENCES

Alaee, M., Arias, P., Sjödin, A., & Bergman, A. (2003). An overview of commercially used brominated flame retardants, their applications, their use patterns in different countries/regions and possible modes of release. *Environment International*, *29*(6), 683–689. doi:10.1016/S0160-4120(03)00121-1 PMID:12850087

ATSDR. (2016). *Public Health Statement for Trichloroethylene*. Author.

ATSDR. (2017). *Toxicological profile for polybrominated biphenyl and polybrominated diphenyl ethers*. Atlanta, GA: Author.

Bedard, D. L., Bunnell, S. C., & Smullen, L. A. (1996). Stimulation of Microbial para-Dechlorination of Polychlorinated Biphenyls That Have Persisted in Housatonic River Sediment for Decades. *Environmental Science & Technology*, *30*(2), 687–694. doi:10.1021/es950463i

Bedard, D. L., Van Dort, H., & Deweerd, K. A. (1998). Brominated biphenyls prime extensive microbial reductive dehalogenation of Aroclor 1260 in housatonic river sediment. *Applied and Environmental Microbiology*, *64*(5), 1786–1795. doi:10.1128/AEM.64.5.1786-1795.1998 PMID:9572952

Berkaw, M., Sowers, K. R., & May, H. D. (1996). Anaerobic ortho dechlorination of polychlorinated biphenyls by estuarine sediments from Baltimore Harbor. *Applied and Environmental Microbiology*, *62*(7), 2534–2539. doi:10.1128/AEM.62.7.2534-2539.1996 PMID:16535360

Borglin, S., Wilke, A., Jepsen, R., & Lick, W. (1996). Parameters affecting the desorption of hydrophobic organic chemicals from suspended sediments. *Environmental Toxicology and Chemistry*, *15*(12), 2254–2262. doi:10.1002/etc.5620151219

Borja, J., Taleon, D. M., Auresenia, J., & Gallardo, S. (2005). Polychlorinated biphenyls and their biodegradation. *Process Biochemistry*, *40*(6), 1999–2013. doi:10.1016/j.procbio.2004.08.006

Breivik, K., Sweetman, A., Pacyna, J. M., & Jones, K. C. (2002). Towards a global historical emission inventory for selected PCB congeners - A mass balance approach: 1. Global production and consumption. *The Science of the Total Environment*, *290*(1–3), 181–198. doi:10.1016/S0048-9697(01)01075-0 PMID:12083709

Breivik, K., Sweetman, A., Pacyna, J. M., & Jones, K. C. (2007). Towards a global historical emission inventory for selected PCB congeners--a mass balance approach 3. An update. *The Science of the Total Environment*, *377*(2–3), 296–307. doi:10.1016/j.scitotenv.2007.02.026 PMID:17395248

Bridges, T. S., Ells, S., Hayes, D., Mount, D., Nadeau, S. C., Palermo, M. R., … Schroeder, P. (2008). *The Four Rs of Environmental Dredging : Resuspension, Release, Residual, and Risk (Report No. ERDC/EL TR-08-4)*. Washington, DC: US Army Corps of Engineers.

Busset, G., Sangely, M., Montrejaud-Vignoles, M., Thannberger, L., & Sablayrolles, C. (2012). Life cycle assessment of polychlorinated biphenyl contaminated soil remediation processes. *The International Journal of Life Cycle Assessment*, *17*(3), 325–336. doi:10.100711367-011-0366-7

Chen, J., Wang, C., Pan, Y., Farzana, S. S., & Tam, N. F.-Y. (2017). Biochar accelerates microbial reductive debromination of 2,2¢,4,4¢-tetrabromodiphenyl ether (BDE-47) in anaerobic mangrove sediments. *Journal of Hazardous Materials*, *341*, 177–186. doi:10.1016/j.jhazmat.2017.07.063 PMID:28777963

Chen, J., Wang, P. F., Wang, C., Miao, H. C., & Wang, X. (2018). How wastewater with different nutrient levels influences microbial degradation of 2,2',4,4'-tetrabromodiphenyl ether (BDE-47) in anaerobic sediments. *Chemosphere*, *211*(1), 128–138. doi:10.1016/j.chemosphere.2018.07.122 PMID:30071424

Chen, J., Zhou, H. C., Wang, C., Zhu, C. Q., & Tam, N. F. Y. (2015). Short-term enhancement effect of nitrogen addition on microbial degradation and plant uptake of polybrominated diphenyl ethers (PBDEs) in contaminated mangrove soil. *Journal of Hazardous Materials*, *300*, 84–92. doi:10.1016/j.jhazmat.2015.06.053 PMID:26156523

de Souza, A. C., Taniguchi, S., Lopes Figueira, R. C., Montone, R. C., Caruso Bícego, M., & Martins, C. C. (2018). Historical records and spatial distribution of high hazard PCBs levels in sediments around a large South American industrial coastal area (Santos Estuary, Brazil). *Journal of Hazardous Materials*, *360*(April), 428–435. doi:10.1016/j.jhazmat.2018.08.041 PMID:30138903

Demirtepe, H. (2017). *Sustainable Remediation of Aquatic Sediments Contaminated with Polybrominated Diphenyl Ethers and Hexabromocyclododecane*. Ankara, Turkey: Middle East Technical University.

Demirtepe, H., & Imamoglu, I. (2019). Degradation of decabromodiphenyl ether (BDE-209) in microcosms mimicking sediment environment subjected to comparative bioremediation strategies. *Journal of Environmental Management*, *233*, 120–130. doi:10.1016/j.jenvman.2018.11.144 PMID:30576959

Demirtepe, H., Kjellerup, B., Sowers, K. R., & Imamoglu, I. (2015). Evaluation of PCB dechlorination pathways in anaerobic sediment microcosms using an anaerobic dechlorination model. *Journal of Hazardous Materials*, *296*, 120–127. doi:10.1016/j.jhazmat.2015.04.033 PMID:25913678

Di Gregorio, S., Azaizeh, H., & Lorenzi, R. (2013). Biostimulation of the autochthonous microbial community for the depletion of polychlorinated biphenyls (PCBs) in contaminated sediments. *Environmental Science and Pollution Research International*, *20*(6), 3989–3999. doi:10.100711356-012-1350-x PMID:23208754

Di Leo, A., Annicchiarico, C., Cardellicchio, N., Cibic, T., Comici, C., Giandomenico, S., & Spada, L. (2016). Mobilization of trace metals and PCBs from contaminated marine sediments of the Mar Piccolo in Taranto during simulated resuspension experiment. *Environmental Science and Pollution Research International*, *23*(13), 12777–12790. doi:10.100711356-015-5472-9 PMID:26438371

Diamond, M. L., Page, C. A., Campbell, M., McKenna, S., & Lall, R. (1999). Life-cycle framework for assessment of site remediation options: Method and generic survey. *Environmental Toxicology and Chemistry*, *18*(4), 788–800. doi:10.1002/etc.5620180427

Eggleton, J., & Thomas, K. V. (2004). A review of factors affecting the release and bioavailability of contaminants during sediment disturbance events. *Environment International*, *30*(7), 973–980. doi:10.1016/j.envint.2004.03.001 PMID:15196845

Erickson, M. D. (2001). Introduction: PCB properties, uses, occurrence, and regulatory history. In L. W. Robertson & L. G. Hansen (Eds.), *PCBs: Recent Advances in Environmental Toxicology and Health Effects* (pp. xi–xxx). The University Press of Kentucky.

Fagervold, S. K., Watts, J. E. M., May, H. D., & Sowers, K. R. (2011). Effects of bioaugmentation on indigenous PCB dechlorinating activity in sediment microcosms. *Water Research*, *45*(13), 3899–3907. doi:10.1016/j.watres.2011.04.048 PMID:21601905

Fraser, M. W., Short, J., Kendrick, G., McLean, D., Keesing, J., Byrne, M., ... Wu, P. (2017). Effects of dredging on critical ecological processes for marine invertebrates, seagrasses and macroalgae, and the potential for management with environmental windows using Western Australia as a case study. *Ecological Indicators*, *78*, 229–242. doi:10.1016/j.ecolind.2017.03.026

Goossens, H., & Zwolsman, J. J. G. (1996). Evaluation of the Behaviour of Pollutants During Dredging Activities. *Terra et Aqua*, *62*(March), 20–27.

He, J., Robrock, K. R., & Alvarez-Cohen, L. (2006). Microbial Reductive Debromination of Polybrominated Diphenyl Ethers (PBDEs). *Environmental Science & Technology*, *40*(14), 4429–4434. doi:10.1021/es052508d PMID:16903281

Hedman, J. E., Tocca, J. S., & Gunnarsson, J. S. (2009). Remobilization of polychlorinated biphenyl from baltic sea sediment: Comparing the roles of bioturbation and physical resuspension. *Environmental Toxicology and Chemistry, 28*(11), 2241–2249. doi:10.1897/08-576.1 PMID:19499969

Himmelheber, D. W., & Hughes, J. (2014). In situ biotransformation of contaminants in sediments. In D. D. Reible (Ed.), *Processes, Assessment and Remediation of Contaminated Sediments*. New York: Springer. doi:10.1007/978-1-4614-6726-7_10

Hu, X., Zhu, J., & Ding, Q. (2011). Environmental life-cycle comparisons of two polychlorinated biphenyl remediation technologies: Incineration and base catalyzed decomposition. *Journal of Hazardous Materials, 191*(1–3), 258–268. doi:10.1016/j.jhazmat.2011.04.073 PMID:21571422

Huang, H. W., Chang, B. V., & Lee, C. C. (2014). Reductive debromination of decabromodiphenyl ether by anaerobic microbes from river sediment. *International Biodeterioration & Biodegradation, 87*, 60–65. doi:10.1016/j.ibiod.2013.10.011

IARC. (2014). *IARC monographs on the evaluation of carcinogenic risks to humans: Trichloroethylene, tetrachloroethylene and some other chlorinated agents*. Lyon, France: World Health Organization.

IARC. (2016). IARC Monographs on the Evaluation of Carcinogenic Risks to Humans: Polychlorinated biphenyls and polybrominated biphenyls (Vol. 107). Lyon, France: IARC.

Jaffé, R. (1991). Fate of hydrophobic organic pollutants in the aquatic environment: A review. *Environmental Pollution, 69*(2–3), 237–257. doi:10.1016/0269-7491(91)90147-O PMID:15092165

Keum, Y.-S., & Li, Q. X. (2005). Reductive debromination of polybrominated diphenyl ethers by zerovalent iron. *Environmental Science & Technology, 39*(7), 2280–2286. doi:10.1021/es048846g PMID:15871265

Krumins, V., Park, J. W., Son, E. K., Rodenburg, L. A., Kerkhof, L. J., Häggblom, M. M., & Fennell, D. E. (2009). PCB dechlorination enhancement in Anacostia River sediment microcosms. *Water Research, 43*(18), 4549–4558. doi:10.1016/j.watres.2009.08.003 PMID:19744693

La Guardia, M. J., Hale, R. C., & Harvey, E. (2006). Detailed polybrominated diphenyl ether (PBDE) congener composition of the widely used penta-, octa-, and deca-PBDE technical flame-retardant mixtures. *Environmental Science & Technology, 40*(20), 6247–6254. doi:10.1021/es060630m PMID:17120549

Lebeau, T. (2011). Bioaugmentation for in situ soil remediation: How to ensure the success of such a process. In A. Singh, N. Parmar, & R. C. Kuhad (Eds.), *Bioaugmentation, Biostimulation, and Biocontrol* (pp. 129–186). Heidelberg, Germany: Springer Berlin Heidelberg. doi:10.1007/978-3-642-19769-7_7

Lee, L. K., & He, J. (2010). Reductive debromination of polybrominated diphenyl ethers by anaerobic bacteria from soils and sediments. *Applied and Environmental Microbiology, 76*(3), 794–802. doi:10.1128/AEM.01872-09 PMID:20008168

Lemming, G., Hauschild, M. Z., & Bjerg, P. L. (2010). Life cycle assessment of soil and groundwater remediation technologies: Literature review. *The International Journal of Life Cycle Assessment, 15*(1), 115–127. doi:10.100711367-009-0129-x

Liang, D., Siegert, M., Liu, X., Lu, Q., Xu, G., Qiu, L., ... Yu, L. (2018). Electron transport chains in organohalide-respiring bacteria and bioremediation implications. *Biotechnology Advances*, *36*(4), 1194–1206. doi:10.1016/j.biotechadv.2018.03.018 PMID:29631017

Mackay, D., Shiu, W. Y., Ma, K. C., & Lee, S. C. (2006). Handbook of Physical- Chemical Properties and Environmental Fate for Organic Chemicals.: Vol. II. *Halogenated Hydrocarbons*. Boca Raton, FL: CRC Press Taylor & Francis Group.

Marvin, C., Waltho, J., Jia, J., & Burniston, D. (2013). Spatial distributions and temporal trends in polybrominated diphenyl ethers in Detroit River suspended sediments. *Chemosphere*, *91*(6), 778–783. doi:10.1016/j.chemosphere.2013.02.009 PMID:23478126

Matturro, B., Ubaldi, C., Grenni, P., Caracciolo, A. B., & Rossetti, S. (2016). Polychlorinated biphenyl (PCB) anaerobic degradation in marine sediments: Microcosm study and role of autochthonous microbial communities. *Environmental Science and Pollution Research International*, *23*(13), 12613–12623. doi:10.100711356-015-4960-2 PMID:26162439

McGoldrick, D. J., Pelletier, M., de Solla, S. R., Marvin, C. H., & Martin, P. A. (2018). Legacy of legacies: Chlorinated naphthalenes in Lake Trout, Walleye, Herring Gull eggs and sediments from the Laurentian Great Lakes indicate possible resuspension during contaminated sediment remediation. *The Science of the Total Environment*, *634*, 1424–1434. doi:10.1016/j.scitotenv.2018.04.077 PMID:29710642

Natarajan, M. R., Nye, J., Wu, W., Wang, H., Jain, M. K., & International, M. B. I. (1997). Reductive Dechlorination of PCB- Contaminated Raisin River Sediments by Anaerobic Microbial Granules. *Biotechnology and Bioengineering*, *55*(1), 182–190. doi:10.1002/(SICI)1097-0290(19970705)55:1<182::AID-BIT19>3.0.CO;2-O PMID:18636456

National Research Council. (2007). *Sediment Dredging at Superfund Megasites*. Washington, DC: Author.

Palermo, M., & Hays, D. F. (2014). Sediment Dredging, Treatment and Disposal. In D. D. Reible (Ed.), *Processes, Assessment and Remediation of Contaminated Sediments*. New York: Springer. doi:10.1007/978-1-4614-6726-7_13

Palermo, M. R., Schroeder, P. R., Estes, T. J., & Francingues, N. R. (2008). *Technical Guidelines for Environmental Dredging of Contaminated Sediments (Report No. ERDC/EL TR-08-29). US Army Corps of Engineers*. Vicksburg, MS: US Army Corps of Engineers Engineer Research and Development Center.

Palm, A., Cousins, I. T., Mackay, D., Tysklind, M., Metcalfe, C., & Alaee, M. (2002). Assessing the environmental fate of chemicals of emerging concern: a case study of the polybrominated diphenyl ethers. *Environmental Pollution, 117*(2), 195–213. Retrieved from https://www.ncbi.nlm.nih.gov/pubmed/11916035

Pan, Y., Chen, J., Zhou, H., & Tam, N. F. Y. (2018). Changes in microbial community during removal of BDE-153 in four types of aquatic sediments. *The Science of the Total Environment*, *613–614*, 644–652. doi:10.1016/j.scitotenv.2017.09.130 PMID:28934686

Patmont, C., LaRosa, P., Narayanan, R., & Forrest, C. (2018). Environmental dredging residual generation and management. *Integrated Environmental Assessment and Management*, *14*(3), 335–343. doi:10.1002/ieam.4032 PMID:29400910

Payne, R. B., Fagervold, S. K., May, H. D., & Sowers, K. R. (2013). Remediation of polychlorinated biphenyl impacted sediment by concurrent bioaugmentation with anaerobic halorespiring and aerobic degrading bacteria. *Environmental Science & Technology*, *47*(8), 3807–3815. doi:10.1021/es304372t PMID:23463900

Payne, R. B., Ghosh, U., May, H. D., Marshall, C. W., & Sowers, K. R. (2019). A Pilot-Scale Field Study: In Situ Treatment of PCB-Impacted Sediments with Bioamended Activated Carbon. *Environmental Science & Technology*, *53*(5), 2626–2634. doi:10.1021/acs.est.8b05019

Payne, R. B., May, H. D., & Sowers, K. R. (2011). Enhanced Reductive Dechlorination of Polychlorinated Biphenyl Impacted Sediment by Bioaugmentation with a Dehalorespiring Bacterium. *Environmental Science & Technology*, *45*(20), 8772–8779. doi:10.1021/es201553c PMID:21902247

Perelo, L. W. (2010). Review: In situ and bioremediation of organic pollutants in aquatic sediments. *Journal of Hazardous Materials*, *177*(1–3), 81–89. doi:10.1016/j.jhazmat.2009.12.090 PMID:20138425

Qiu, M., Chen, X., Deng, D., Guo, J., Sun, G., Mai, B., & Xu, M. (2012). Effects of electron donors on anaerobic microbial debromination of polybrominated diphenyl ethers (PBDEs). *Biodegradation*, *23*(3), 351–361. doi:10.100710532-011-9514-9 PMID:21910024

Rahman, F., Langford, K. H., Scrimshaw, M. D., & Lester, J. N. (2001). Polybrominated diphenyl ether (PBDE) flame retardants. *The Science of the Total Environment*, *275*(1-3), 1–17. doi:10.1016/S0048-9697(01)00852-X PMID:11482396

Rice, C. P., & White, D. S. (1987). PCB availability assessment of river dredging using caged clams and fish. *Environmental Toxicology and Chemistry*, *6*(4), 259–274. doi:10.1002/etc.5620060404

Robrock, K. R., Korytár, P., & Alvarez-Cohen, L. (2008). Pathways for the anaerobic microbial debromination of polybrominated diphenyl ethers. *Environmental Science & Technology*, *42*(8), 2845–2852. doi:10.1021/es0720917 PMID:18497133

Safe, S. H. (1994). Polychlorinated biphenyls (PCBs): Environmental impact, biochemical and toxic responses, and implications for risk assessment. *Critical Reviews in Toxicology*, *24*(2), 87–149. doi:10.3109/10408449409049308 PMID:8037844

Seelye, J. G., Hesselberg, R. J., & Mac, M. J. (1982). Accumulation by Fish of Contaminants Released from Dredged Sediments. *Environmental Science & Technology*, *16*(8), 459–464. doi:10.1021/es00102a006 PMID:22236268

Shao, M., Tao, P., Wang, M., Jia, H., & Li, Y. F. (2016). Trophic magnification of polybrominated diphenyl ethers in the marine food web from coastal area of Bohai Bay, North China. *Environmental Pollution*, *213*, 379–385. doi:10.1016/j.envpol.2016.02.034 PMID:26942685

Singh, A., Parmar, N., Kuhad, R. C., & Ward, O. P. (2011). Bioaugmentation, Biostimulation, and Biocontrol in Soil Biology. In A. Singh, N. Parmar, & R. C. Kuhad (Eds.), *Bioaugmentation, Biostimulation, and Biocontrol* (pp. 1–23). Berlin: Springer. doi:10.1007/978-3-642-19769-7_1

Søndergaard, G. L., & Owsianiak, M. (2018). LCA of soil and groundwater remediation. In M. Z. Hauschild, S. I. Olsen, & R. K. Rosenbaum (Eds.), *Life Cycle Assessment Theory and Practice*. Springer. doi:10.1007/978-3-319-56475-3_36

Song, M., Luo, C., Li, F., Jiang, L., Wang, Y., Zhang, D., & Zhang, G. (2015). Anaerobic degradation of Polychlorinated Biphenyls (PCBs) and Polychlorinated Biphenyls Ethers (PBDEs), and microbial community dynamics of electronic waste-contaminated soil. *The Science of the Total Environment*, *502*, 426–433. doi:10.1016/j.scitotenv.2014.09.045 PMID:25268572

Sowers, K. R., & May, H. D. (2013). In situ treatment of PCBs by anaerobic microbial dechlorination in aquatic sediment: Are we there yet? *Current Opinion in Biotechnology*, *24*(3), 482–488. doi:10.1016/j.copbio.2012.10.004 PMID:23102490

Stockholm Convention. (2019). *Stockholm Convention Website*. Retrieved from http://pops.int/

Suèr, P., Nilsson-Påledal, S., & Norrman, J. (2004). LCA for site remediation: A literature review. *Soil & Sediment Contamination*, *13*(4), 415–425. doi:10.1080/10588330490471304

Tokarz, J. A. III, Ahn, M.-Y., Leng, J., Filley, T. R., & Nies, L. (2008). Reductive debromination of polybrominated diphenyl ethers in anaerobic sediment and a biomimetic system. *Environmental Science & Technology*, *42*(4), 1157–1164. doi:10.1021/es071989t PMID:18351087

UNEP. (2010). *Technical Review of the Implications of Recycling Commercial Pentabromodiphenyl Ether and Commercial Octabromodiphenyl Ether*. UNEP.

US EPA. (1996). *Assessment and remediation of contaminated sediments (ARCS) program*. Author.

US EPA. (2005). *Contaminated sediment remediation guidance for hazardous waste sites (Report No. EPA-540-R-05-012)*. Retrieved from https://nepis.epa.gov/

USEPA. (2010). *An Exposure Assessment of Polybrominated Diphenyl Ethers (Report No. EPA/600/R-08-086F)*. Washington, DC: National Center for Environmental Assessment Office of Research and Development.

Van den Berg, M., Birnbaum, L. S., Denison, M., De Vito, M., Farland, W., Feeley, M., ... Peterson, R. E. (2006). The 2005 World Health Organization reevaluation of human and Mammalian toxic equivalency factors for dioxins and dioxin-like compounds. *Toxicological Sciences*, *93*(2), 223–241. doi:10.1093/toxsci/kfl055 PMID:16829543

Visentin, C., da Silva Trentin, A. W., Braun, A. B., & Thomé, A. (2019). Application of life cycle assessment as a tool for evaluating the sustainability of contaminated sites remediation: A systematic and bibliographic analysis. *The Science of the Total Environment*, *672*, 893–905. doi:10.1016/j.scitotenv.2019.04.034 PMID:30978551

West, J. E., O'Neill, S. M., & Ylitalo, G. M. (2017). Time Trends of Persistent Organic Pollutants in Benthic and Pelagic Indicator Fishes from Puget Sound, Washington, USA. *Archives of Environmental Contamination and Toxicology*, *73*(2), 207–229. doi:10.100700244-017-0383-z PMID:28528416

WHO. (2005). *Trichloroethylene in drinking-water: Background document for development of WHO guidelines for drinking-water quality (Report No. WHO/SDE/WSH/05.08/22)*. WHO.

WHO. (2006). *Concise international chemical assessment document: Tetrachloroethene*. Geneva: World Health Organization.

Wiegel, J., & Wu, Q. (2000). Microbial reductive dehalogenation of polychlorinated biphenyls. *FEMS Microbiology Ecology*, *32*(1), 1–15. doi:10.1111/j.1574-6941.2000.tb00693.x PMID:10779614

Winchell, L. J., & Novak, P. J. (2008). Enhancing polychlorinated biphenyl dechlorination in fresh water sediment with biostimulation and bioaugmentation. *Chemosphere*, *71*(1), 176–182. doi:10.1016/j.chemosphere.2007.10.021 PMID:18037157

Xu, G., Lu, Q., Yu, L., & Wang, S. (2019). Tetrachloroethene primes reductive dechlorination of polychlorinated biphenyls in a river sediment microcosm. *Water Research*, *152*, 87–95. doi:10.1016/j.watres.2018.12.061 PMID:30665163

Zhu, H., Wang, Y., Wang, X., Luan, T., & Tam, N. F. Y. (2014). Intrinsic debromination potential of polybrominated diphenyl ethers in different sediment slurries. *Environmental Science & Technology*, *48*(9), 4724–4731. doi:10.1021/es4053818 PMID:24679249

ADDITIONAL READING

Bridges, T. S., Ells, S., Hayes, D., Mount, D., Nadeau, S. C., & Palermo, M. R. … Schroeder, P. (2008). *The Four Rs of Environmental Dredging : Resuspension, Release, Residual, and Risk (Report No. ERDC/EL TR-08-4)*. Washington, DC: US Army Corps of Engineers.

Chen, J., Wang, P. F., Wang, C., Miao, H. C., & Wang, X. (2018). How wastewater with different nutrient levels influences microbial degradation of 2,2¢,4,4¢-tetrabromodiphenyl ether (BDE-47) in anaerobic sediments. *Chemosphere*, *211*(1), 128–138. doi:10.1016/j.chemosphere.2018.07.122 PMID:30071424

Chen, J., Zhou, H. C., Wang, C., Zhu, C. Q., & Tam, N. F. Y. (2015). Short-term enhancement effect of nitrogen addition on microbial degradation and plant uptake of polybrominated diphenyl ethers (PBDEs) in contaminated mangrove soil. *Journal of Hazardous Materials*, *300*, 84–92. doi:10.1016/j.jhazmat.2015.06.053 PMID:26156523

Diamond, M. L., Page, C. A., Campbell, M., McKenna, S., & Lall, R. (1999). Life-cycle framework for assessment of site remediation options: Method and generic survey. *Environmental Toxicology and Chemistry*, *18*(4), 788–800. doi:10.1002/etc.5620180427

Himmelheber, D. W., & Hughes, J. (2014). In situ biotransformation of contaminants in sediments. In D. D. Reible (Ed.), *Processes, Assessment and Remediation of Contaminated Sediments*. New York: Springer. doi:10.1007/978-1-4614-6726-7_10

Lebeau, T. (2011). Bioaugmentation for in situ soil remediation: How to ensure the success of such a process. In A. Singh, N. Parmar, & R. C. Kuhad (Eds.), *Bioaugmentation, Biostimulation, and Biocontrol* (pp. 129–186). Heidelberg: Springer Berlin Heidelberg. doi:10.1007/978-3-642-19769-7_7

Payne, R. B., Ghosh, U., May, H. D., Marshall, C. W., & Sowers, K. R. (2019). A Pilot-Scale Field Study: In Situ Treatment of PCB-Impacted Sediments with Bioamended Activated Carbon. Environmental Science and Technology, 53(5), 2626–2634. research-article. doi:10.1021/acs.est.8b05019

Visentin, C., da Silva Trentin, A. W., Braun, A. B., & Thomé, A. (2019). Application of life cycle assessment as a tool for evaluating the sustainability of contaminated sites remediation: A systematic and bibliographic analysis. *The Science of the Total Environment*, *672*, 893–905. doi:10.1016/j.scitotenv.2019.04.034 PMID:30978551

KEY TERMS AND DEFINITIONS

Bioaugmentation: Addition of microorganisms to contaminated sites where microbial activity is limited or absent.

Biostimulation: Addition of materials to enhance the growth and activity of indigenous microorganisms in a contaminated site.

Biotransformation: The process of microbial degradation of organic chemicals.

Contaminated Site: Sites having high concentrations of hazardous pollutants which pose risks to human health and the environment.

Environmental Dredging: Mechanical removal of soils and sediments to remove the contaminated parts.

Hydrophobicity: The tendency of an organic compound to partition in organic-rich media rather than water phase.

Life Cycle Assessment: A tool to analyze and evaluate the burdens of a product or an activity covering its all life cycle.

Remediation: The activities to control, reduce and remove the contamination, aiming at reducing the risk to human health and the environment.

Section 3
Interaction of Nanoparticles

Chapter 17
Ionic Liquid-Mediated Synthesis of Metal Nanoparticles

Vivek Srivastava
NIIT University, India

ABSTRACT

In this chapter, the authors discussed the recent as well as significant development in the synthesis of ionic liquid-mediated nanoparticles (considered a less toxic way to make nanoparticles). Apart from that, they also disused a small background of ionic liquid considering types, properties, and effect of ionic liquid's counterparts on nanoparticles synthesis. There are promising consequences proposing that the ionic liquid route can lead the design and fabrication of nanoparticles through the variety of size, nature, configuration, and functionality. All discussed examples in this chapter clearly depicted that ionic liquids add great value to the area of nanoparticle synthesis.

INTRODUCTION

Ever since the ending of the 1990s, the continuous growth of nanosciences and nanocatalysis has evidently turned up as an interesteing reach area, which interface between homogeneous as well as heterogeneous catalysis (Khan, Saeed, & Khan, 2017; Polshettiwar & Varma, 2010; Renna, Boyle, Gehan, & Venkataraman, 2015). The fundamental target is to develop efficient and prominent catalysts, that consist of either metal nanoparticles (MNPs) or a nanomaterial in the form of support. This type nanocatalysts have the ability to present the compelling advantages of homogeneous and heterogeneous catalysts, mainly superior efficiency, selectivity, durability and simple recovery or recycling. Precise reactivity in nanao-catalytic system is typically anticipated as a result of their nano dimensions configured to afford definite characteristics, which will be never attained with normal, non-nano components (Dhakshinamoorthy, Navalon, Alvaro, & Garcia, 2012; Fan & Gao, 2006).

Nanoparticles (NPs) or nanoscale materials (nanomaterials) are any solid which has size ranges between 1-100 nanometre (10^{-9} meter) with at least one nanometer dimension. In the nanoscale approach, neither quantum chemistry nor the traditional rules of physics confine. In materials, where, powerful chemical bonding occurs and delocalization of electrons are typically considerable, therefore the intensity

DOI: 10.4018/978-1-7998-1241-8.ch017

of delocalization varies with the dimension of the structure (Astruc, 2008; Bahadory, 2008; Kalidindi & Jagirdar, 2012a; Scholten, Leal, & Dupont, 2012). This result, combined with structural variations, can cause several chemical and even physical characteristics with respect to size. When it comes to additional characteristics, surface reactivity of nanoscale particles is thereby extremely size-dependent. The surface energies and surface morphologies are too size-dependent have importance in nanochemistry, which can leads to improved intrinsic surface reactivity.

Currently, the list of consumer goods in the nano-field is steadily growing, including many complex products including, the nanomaterials, such as sunscreens, toothbrushes, medicines, paints, textiles, plastic wrap, waterless car wash, corrosion resistance, tennis rackets, solar batteries, catalysts, and microelectronic devices (Kalidindi & Jagirdar, 2012b; Roldan Cuenya & Behafarid, 2015; Sheldon, 2012; Zhu & Hosmane, 2015). Engineered nanoparticles are also examined for *in situ* applications in environmental remediation (Narayan. However, current investigations have shown the obvious toxicity of nanoparticles to living organisms, as well as its potentially negative impact on environmental ecosystems (ecotoxicity); several features of this topic are actively addressed in scientific journals (Singh et al., 2019b; Walters, Pool, & Somerset, 2016). A review of industrial organizations that either production, handle, research, or use nanomaterials has shown that most of the organizations do not have nano-specific health and safety practices at the workplace. Nanotoxicology developed as a new subspecialty of particle toxicology, which discusses the toxicology of nanoparticles that seem to have significant toxicity forms in connection with larger particles. Because of their tiny sizes, nanoparticles can readily pass the biological barriers and penetrate cells inducing many toxic outcomes following in enhanced oxidative stress, inflammation, and cell death (Bopp et al., 2019; Lamon, Aschberger, Asturiol, Richarz, & Worth, 2019; Lewis, Bertsch, & McNear, 2019; Singh et al., 2019a; Wigger & Nowack, 2019). Nanoparticles can get into touch with and insert a human body via the skin, eyes, respiratory system, and gastrointestinal (GI) system. Once inside a human organism, nanoparticles can be acknowledged by the innate immunity and be internalized by phagocytic cells, accumulate in the body (if not biodegradable), and may interfere with body regulatory mechanisms, even causing changes in protein structure (Bopp et al., 2019; Lamon et al., 2019; Lewis et al., 2019; Singh et al., 2019a, 2019b; Walters et al., 2016; Wigger & Nowack, 2019). Therefore, concerns are raised because the toxicity and environmental impact of the newly developed nanomaterials are seldom hidden and hard to predict.

In concern with the greenhouse effect, researchers are looking to create or search fewer volatile compounds which can replace the conventional, toxic and volatile organic solvents. The curiosity of the researcher had increased over the last few decades in the area of Ionic Liquids (ILs) (Prechtl, 2016; Sowmiah, I. Cheng, & Chu, 2012; Villa et al., 2019). The chemical and physical properties of ILs make them more interesting as an area of research. Mainly researches are looking to apply ILs as a suitable solvent or as catalyst in different type's organic transformations as they show non-detectable vapor pressure along with other unique properties like high stability, high density, large liquid range (*i.e.* more than 3000C), less vapor pressure, high viscosity, and good solubility makes them a unique and fascinating solvent system over conventional solvents (Gordon, 2001; Tom Welton, 2018). The less volatile nature under ambient conditions like high temperature and pressure decreases the chances of harmful organic substance emission in the environment.[3-a,b] ILs entirely made up of poorly coordinated ions (cations/anions). Such chemical nature of ionic liquid makes them facilitate various fascinating properties which make them a new class of material that can be modified as per the desired requirement.

In recent years, the synthesis of metal nanoparticles (NPs) with controlled size and shape is one of the highly important areas of study due to their multiple applications in different segments like material science, organic transformations, and pharmaceutical production (Chaturvedi, Dave, & Shah, 2012; Polshettiwar & Varma, 2010; Shenhar, Norsten, & Rotello, 2005). The unique properties of metal NPs can be judged by their composition, structure, shape, size, and stability. In the area of metal nanoparticle synthesis, the major concern of the researcher is to create stable, well disperse and agglomeration free metal NPs with narrow size distribution (Fan & Gao, 2006; Kalidindi & Jagirdar, 2012a). The bottom-up, the wet chemistry approach is one of the common ways to synthesize metal NPs. Although in most of the synthesis, the metals NPs are synthesized by the wet chemistry route in aqueous or organic solvent systems, their properties like density, polarity, solubility, high toxicity, low boiling point, and flammability make them unattractive. In the present scenario researchers are looking ILs as a promising reaction medium for the synthesis of metal NPs, due to their unique properties of ILs which overcome the demerits of conventional solvent systems in the synthesis metal NPs (Chaturvedi et al., 2012).

Various reports have been published, concerning the synthesis of ionic liquid mediated nanoparticles (NPs) (Kalidindi & Jagirdar, 2012b; Prechtl, 2016). The cationic and anionic species present in ILs are electrostatic in nature, hence they act as electrostatic stabilizers, therefore in ILs the stable suspension of metal NPs can be synthesized easily. The catalytic rate and selectivity of metal NPs in major organic transformations is higher under ionic liquid medium with respect to conventional solvent system (Olveira, Forster, & Seeger, 2014). In the ILs, the cations and anions create an unique microenvironment; due to this the desired selective transformation was observed in many reactions.

Hence, we can say that reactions catalyzed by metal NPs are task-specific and the recycling of immobilized NPs is easily possible due to tunable miscibility of ILs. In recent advancement, the ionic liquid mediated metal NPs have been employed to various catalytic reactions like, for hydrogenation, oxidation, coupling reactions, etc (He & Alexandridis, 2015; H. Zhang & Cui, 2009). The good dispersion, high catalytic activity or selectivity and less agglomeration is the properties of an ideal nano catalytic system. Some of the metal NPs still show negligible dispersion in ILs, poor catalytic activity, and agglomeration in the recycling experiment (Janiak, 2015; Verma, Ebenso, & Quraishi, 2019; Wegner & Janiak, n.d.; Zvereva, Katsyuba, Dyson, & Aleksandrov, 2016). To overcome these challenges, researchers are working day and night to overcome these problems. In this chapter, we are discussing the use of IL's for the synthesis of Nanomaterials under different reaction condition, as not only the nanomaterials but also the way of nanomaterial synthesis also create the toxicity in the environment.

IONIC LIQUIDS

In general, the ILs defined as salt, which is liquid at room temperature or melts at or below 1000C, having poor coordination of ions *i.e.* organic cations and inorganic anions. At ambient temperature, ionic liquids are freely flowing liquid (Austen Angell, Ansari, & Zhao, 2012; Eyckens & Henderson, 2019b; Mandai, Yoshida, Ueno, Dokko, & Watanabe, 2014; Wasserscheid & Keim, 2000). The ILs are unstable crystal lattice because they have one or more than one ion with delocalized charge and having one organic component in its total structure. Among different definitions, four of them are most acceptable.

Definition 1

The term ionic liquid implies a material that is fluid at (or close to) ambient temperature, is colorless, has a low viscosity and is easily handled and Room temperature ionic liquids [RTILs] are generally salts of organic cations, e.g. tetraalkylammonium, tetraalkylphosphonium, N-alkylpyridinium, 1,3-dialkylimidazolium and trialkylsulfonium cations. (Austen Angell et al., 2012; Wasserscheid & Keim, 2000)

Definition 2

The most basic definition of a room temperature ionic liquid is a salt that has a melting point at or near room temperature. (Austen Angell et al., 2012; Wasserscheid & Keim, 2000).

Definition 3

Organic salts with melting points below ambient or reaction temperature. (Austen Angell et al., 2012; Wasserscheid & Keim, 2000).

Definition 4

Ionic liquid is a salt with a melting temperature below the boiling point of water. (Austen Angell et al., 2012; Wasserscheid & Keim, 2000).

There are many synonyms used like room temperature ionic liquid (RTIL), molten salt, room temperature molten salt, ambient temperature molten salt or ionic liquid (IL), task-specific ionic liquid (TSIL), liquid organic salt, fused salt, neoteric solvent, etc to represent ionic liquid. In the total structure of ionic liquids, the low symmetry, weak intermolecular interaction and low charge density make cations bulky for example asymmetric ammonium, phosphonium salts, and heteroaromatic (Eyckens & Henderson, 2019a; Plechkova & Seddon, 2008; Slattery, Daguenet, Dyson, Schubert, & Krossing, 2007; Ueno, Tokuda, & Watanabe, 2010). In the latest advancement, there are some reports, where inorganic anions have been replaced with organic anions for the synthesis of ILs. The nature of cations and anions decides the chemical and physical properties of ILs (Angell, Byrne, & Belieres, 2007; Austen Angell et al., 2012; Eyckens & Henderson, 2019b; Hallett & Welton, 2011; Wasserscheid & Keim, 2000). The properties of ILs are unique, which makes ILs act as a solvent for the reactions, electrocatalysis, as a catalyst, as co-catalyst, as ligand source and as supporting materials. The properties like relatively low viscosity, low vapor pressure under ambient condition show its non-volatile nature, immiscibility with numerous organic solvents which makes ILs more promising solvent for chemical transformations (Ahrens, Peritz, & Strassner, 2009; Disasa Irge, 2016; Dong, Liu, Dong, Zhang, & Zhang, 2017; Lei, Chen, Koo, & MacFarlane, 2017; Tom Welton, 2018).

As compared to conventional solvents, the ILs are quite viscous (but least viscous at room temperature). The viscosities of ILs are decided by the length of the alkyl chain or functionalized alkyl chain. The shorter alkyl chain length makes low viscous ILs, while a large alkyl chain increases the viscosity of the ionic liquid. Also, the long-chain alkyl group provides a low density while increasing the halogen group, it makes high-density ILs. The density of ILs is least affected by impurities or temperature. The unsymmetrical cations give the ionic liquids at room temperature, while the low melting temperature of ILs was obtained by more weakly coordinating anions. As compared to low boiling point solvents

Figure 1. The most common synthetic route of formation of ILs

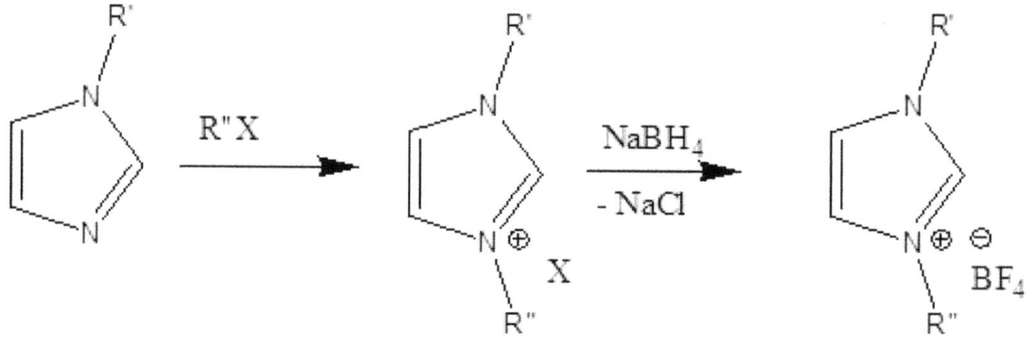

offers much lower toxicity because of negligible volatility. The synthesis and purification of ILs involve organic solvents, which contradicts the greenness of ILs, as the impacts of ILs preparation on the environment are more severe than organic solvent preparation. The recent research study on the greenness of ILs like bioaccumulation, toxicity, and degradability was shown that they are not as eco-friendly to the environment as initially predicted. The ILs can be said to be green if the following points can be achieved successfully a. improvement in the synthesis route of ILs, b. increasing degradability, c. easy recyclability and recovery process, d. the suitable purification process of ILs (Dong et al., 2017; Lei et al., 2017; Tom Welton, 2018).

The most common synthetic process for IL synthesis is divided in two parts a.) Alkylation and b.) Metathesis (Figure 1) ("A review of ionic liquids: Applications towards catalytic organic transformations - ScienceDirect," n.d.; Mandai et al., 2014). Imidazolium, pyrollidinium, pyridinium, ammonium, phosphonium and sulphonium ions are the most commonly used typical cations also known as heteroatomic cations, while trifluoromethanesulfonate, *Bis* (trifluoromethanesulfonyl) amide, hexafluorophosphate, tetrafluoroburate and dicyanimide are the most commonly used typical anions.

TYPES OF IONIC LIQUID

The nature of ILs decided by the type of cations and anions are used during its synthesis. Depending upon the group attached to the cations or anions, the ionic liquid behaves as an acidic, basic, or neutral. The anionic nature ILs, determines the coordination properties and the acidity of ILs but at some level the cations of ILs also influence the acidity of IL system. The concentrations of cations or anions present in ILs or the combining strength of cations and anions give acidic or basic properties to ILs. The neutral ionic liquid (NILs) can be obtained by matching a wide of anions with the cation's components. In Table. 1 some examples of typical cations and anions had given which are commonly used for preparation of ILs (Austen Angell et al., 2012; Dong et al., 2017; Lei et al., 2017; Mandai et al., 2014; Tom Welton, 2018).

The efficiency or productivity of any reaction process substantially influences by the acidity or basicity of the reaction medium, due to this a special attention has been given to acidic or basic ILs as they show a great tendency to increase the efficiency of many reactions processes (Eyckens & Henderson, 2019b; Wasserscheid & Keim, 2000). The basic ionic liquids (BILs) are the combination of ILs and inorganic bases show the properties like stability to water & air, high catalytic efficiency & recyclability,

Table 1. Typical cations and anions for ILs (Austen Angell et al., 2012)

CATIONS	ANIONS
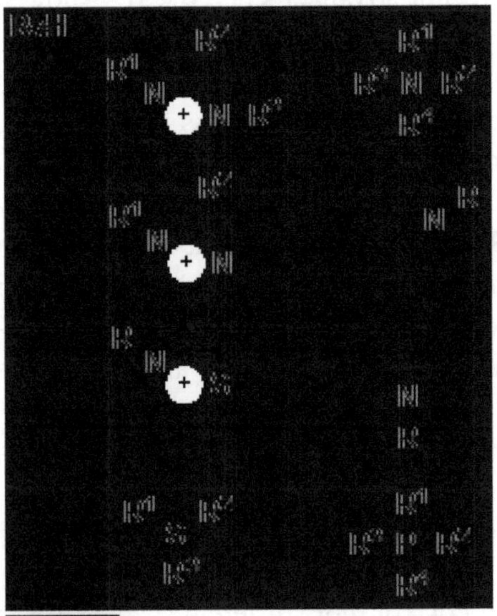 	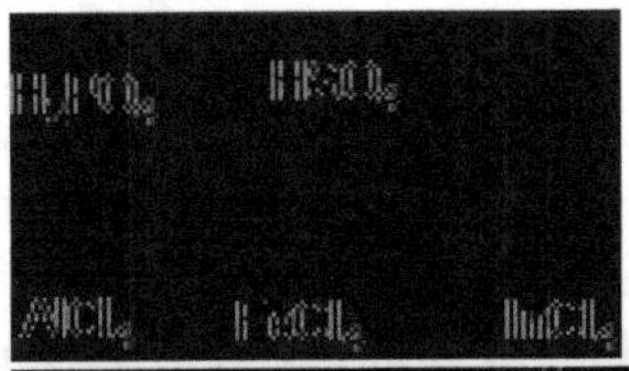  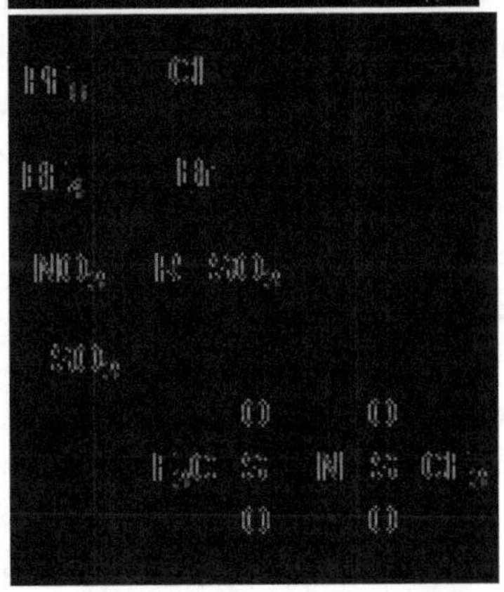

and easy separation it makes them a good option to replace conventional catalysts and solvent system. The high selectivity & activity, also recyclability of catalysts made BILs replace traditional bases like KOH, NaOH, $NaHCO_3$, NaOAc, trimethylamine, etc. BILs are immiscible with many different organic solvents; also, non-volatile and non-corrosive in nature (Austen Angell et al., 2012; Eyckens & Henderson, 2019b; Hallett & Welton, 2011; Tokuda, Tsuzuki, Susan, Hayamizu, & Watanabe, 2006; Tom Welton, 2018). There are many base-catalyzed reactions in which BILs are used, like Knoevenagel condensation, Marknovnikov addition, Michael addition, Mannich reaction, etc. The Bronsted acidic ionic liquids and Lewis acidic ionic liquids are the types of acidic ionic liquid (AILs). In general, Bronsted acidic ILs can be prepared by attaching of alkane sulfonic acid or carboxylate acid group to the side chain of the cations, while Lewis acidic ILs are prepared from metal halides and organic halides. The problem like corrosion, residual waste, and environmental issues arise, by using traditional acids and bases, which generated the need for an environmentally friendly catalyst. In many organic reactions, the acid ionic liquid (AILs) and neutral ionic liquid (NILs) have been used successfully while the report related to the BILs shows very rare cases of organic reaction. The Neutral ionic liquid [BMIm]BF_4 promoted high selective esterification of tertiary alcohols has been achieved by acetic anhydride, while the acidic ILs and basic ILs show less selectivity under similar conditions (Gadilohar & Shankarling, 2017; Mandai et al., 2014; Tokuda et al., 2006; Vekariya, 2017; Tom Welton, 2018).

APPLICATIONS OF IONIC LIQUIDS

ILs show great potential for chemical analysis as a solvent. In recent advancements, the ILs showed various analytical applications in chromatography, extraction, electrochemistry and sensor technology.[34] There are various applications of ILs as a solvent, catalyst and as support for several important organic reactions like Hydrogenation, Hydroformylation, Alkoxycarbonylation, Cross-coupling like Heck, Suzuki, Negishi, Stille reactions, Allylic substitution, Friedel-Crafts alkylation, Diels-Alder, Diol or carbonyl protection, Epoxidation/Epoxide opening, Cyanosilyation of aldehydes, esterification, ring-closing metathesis, Knoevenagel condensation, Baylis-Hillman, Robinson annulation, Alcohol oxidation Friedlander reaction, Nitration of phenols, bromination of alkynes, biocatalysis and chiral solvent for asymmetric synthesis (Table 2) (Ghandi, 2014; Ghernaout, Ghernaout, & Naceur, 2011; Janus, Goc-Maciejewska, Łozyński, & Pernak, 2006; Johnson, Pagni, & Bartmess, 2007). ILs was also utilized as a storage medium for toxic gases, lubricant additives in pigments and as propellants.

NANOPARTICLES

The nanoparticles are composed of identical atoms, molecules, and two or more different species. It is a solid particle with size ranges from 1 nm to 100 nm and having specific physical and chemical properties. The nanoparticles have very small size means more defects *i.e.* more edge and corners compared to large particles. The nanoparticles have high catalytic activity in various reactions due to their larger surface area to volume ratio and residual energy. The nanocatalyst requires the stabilization concerning agglomeration to the bulk metal, as they are kinetically quite unstable. The catalyst modulation and catalyst recyclability both are the advantages of homogeneous and heterogeneous catalysis respectively. While combining these two properties the more efficient catalytic system can be achieved. To obtain such a catalytic system, modern nanochemistry is very helpful as it is possible to make soluble analogs of the heterogeneous catalyst using this technology (Ahrens et al., 2009; Disasa Irge, 2016; Dong et al., 2017; Ghandi, 2014; Janus et al., 2006; Lei et al., 2017; Tom Welton, 2018).

It is well documented that ILs not only stabilizes the metal nanoparticles by providing electro-static stabilization support, but also gives more surface area for the reaction to take place.[35] Therefore, the stability and activity of nanoparticles are controlled by ILs. Thus, ILs mediated nanoparticles provide easy recovery of nanoparticles and also improve the total turnover number during the catalytic run. The common transition metals which were studied as NPs are like Ag, Au, Sn, Pd, Pt, Ru, Rh, Ni and Cu. There are mainly two types of metal NPs such as monometallic and bimetallic (Capello, Fischer, & Hungerbühler, 2007; Gordon, 2001; Greaves & Drummond, 2008; Prechtl, 2016; Sheldon, 2012) . The most used transition metals, which are used for preparation of monometallic NPs catalyst includes mostly Pd, Pt, Cu, Ni and Rh etc.

Two universal methods are usually applied for the synthesis of metal nanoparticles are as under below (Ahmad et al., 2019; Corrêa, Bizeto, & Camilo, 2016; Dupont & Suarez, 2006; Ealias & Saravanakumar, 2017; Heera & Shanmugam, 2015; Jeevanandam, Barhoum, Chan, Dufresne, & Danquah, 2018; Sharma, Kanchi, & Bisetty, 2019):

Method-1: Subdivision of metallic aggregates (physical method)

Method-2: Nucleation and growth of metallic nanoparticles (chemical method).

Table 2. Chemical application of ionic liquid (Ghandi, 2014; Gordon, 2001; Greaves & Drummond, 2008; Lei et al., 2017; Sheldon, 2012; Sowmiah et al., 2012; Sowmiah, Srinivasadesikan, Tseng, & Chu, 2009; Thomas Welton, 1999; Tom Welton, 2018)

Entry	Reaction Name	Ionic Liquid / Solvent / catalyst	Substrate 1	Substrate 2	Product	Yield (%)
1.	Diol Protection					100
2.	Friedlander Synthesis		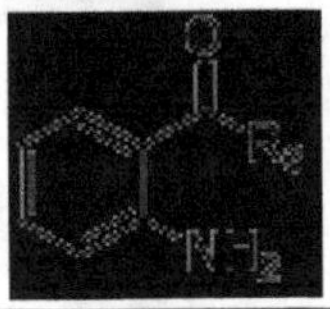	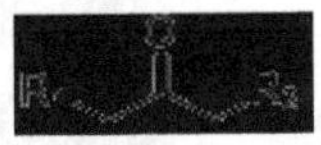	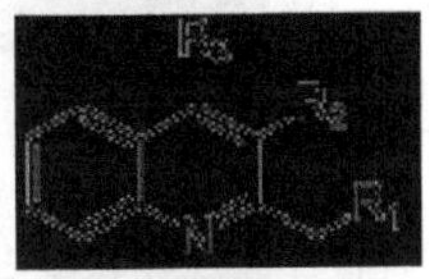	96
3.	Robinson Annulation	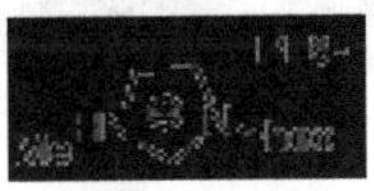 + glycine		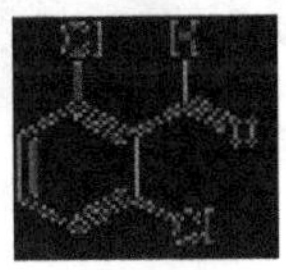	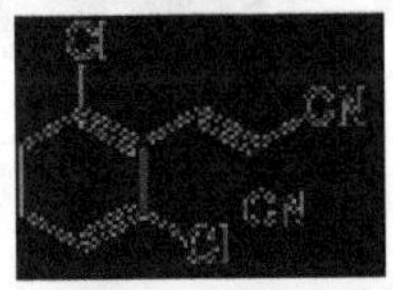	86
4.	Hydrogenation	 + [RuCl$_2$- (S)- BINAP]$_2$•NEt$_3$		i –PrOH	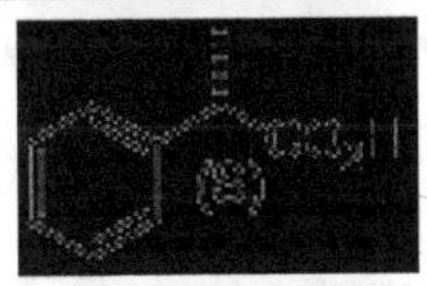	100
5.	Diels-Alder reaction	 + Sc(OTf)$_3$	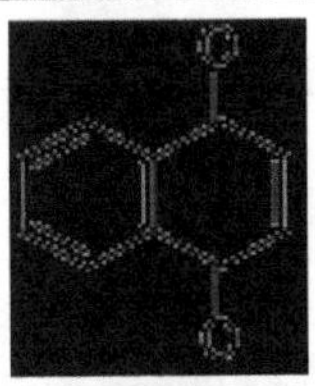	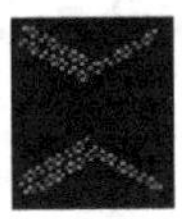	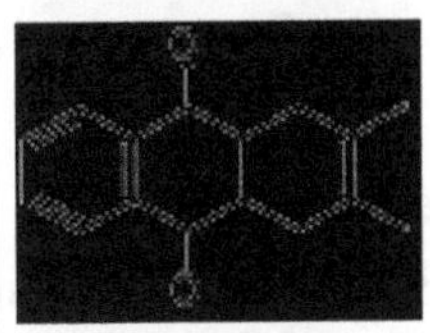	> 99
6.	Olefin Epoxidation	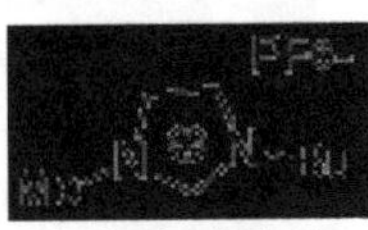 + (R, R)-Jacobsen's Catalyst	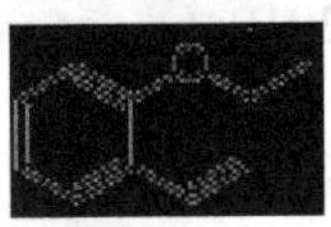	-	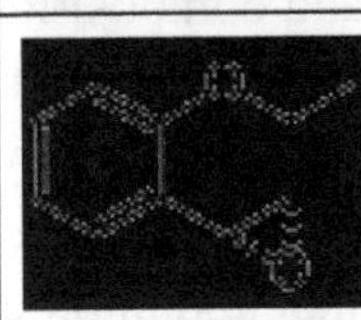	86
7.	Heck Reaction	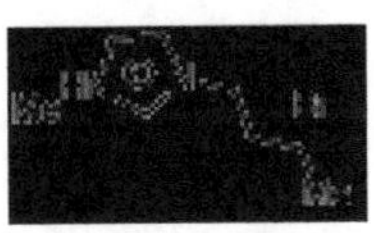			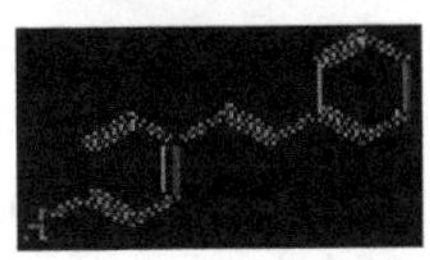	99
8.	Wittig reaction	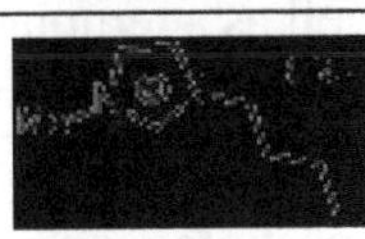		PhCHO		82

continued on following page

Table 2. Continued

Entry	Reaction Name	Ionic Liquid / Solvent / catalyst	Substrate 1	Substrate 2	Product	Yield (%)
9.	Suzuki-Miyaura Coupling reaction	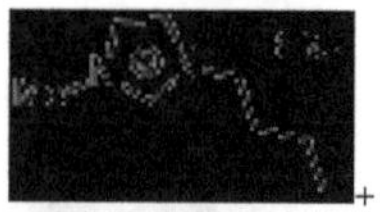 + $Pd(PPh_3)_4$				81
10.	Stille reaction	 + $PdCl_2(PhCN)_2$	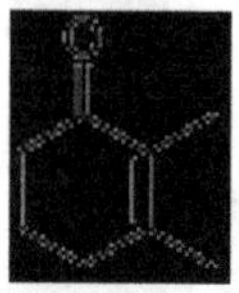		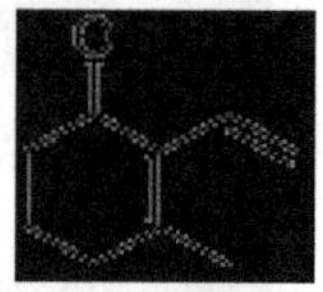	82
11.	Friedel – Craft reaction	 a. + $Cu(OTF)_2$		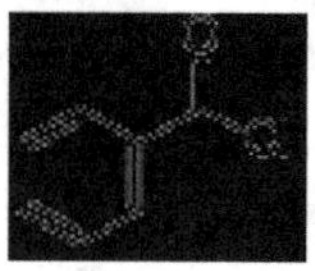		p/o ratio 96/4

Figure 2. Common structure of Imidazolium ILs used for preparation of metal NPs

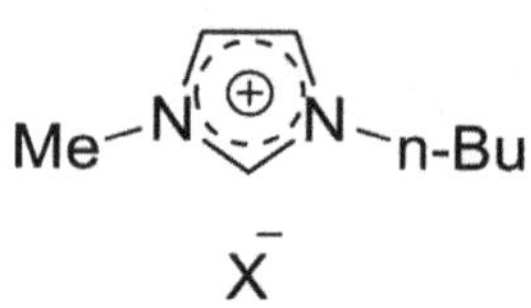

Where X= BF_4 or PF_6 or $N(SO_2CF_3)_2$ or CF_3SO_3

The chemical process is commonly used to design NPs, as it provides better control over particle size, aggregation and composition of nano material. Three major routs can be applied for synthesizing NP followed by chemical method (Ahmad et al., 2019; Ghoranneviss, Soni, Talebitaher, & Aslan, 2015; Heera & Shanmugam, 2015; Jeevanandam et al., 2018):

Route-1: Chemical reduction of metallic salts.

Route-2: Thermal and photochemical decomposition of metal complexes.

Route-3: Electrochemical reduction of metallic salts.

In the chemical reduction process, many reducing agents were used such as molecular hydrogen, organic and inorganic chemicals, but among all the reducing agents, NaBH4 and molecular hydrogen are the most common to obtain agglomeration free, highly stable, well dispersed and active NPs. Whereas templates for the fabrication of NPs, ionic liquids have attracted a lot of interest in recent years. There are many reports, where ILs have been applied as reaction medium as well as a template for the synthesis of various metallic nanoparticles such as inorganic oxide materials including zeolite materials, mesoporous Titania and Titania nanoparticles. The high nucleation rates in many ILs are due to the low surface tension and in consequence, to small particles. ILs mediated metal nanoparticles can be synthesized through chemical reduction, where metal atoms are in a positive oxidation state in their metal salts.

APPLICATIONS OF IONIC LIQUIDS IN METAL NANOPARTICLE SYNTHESIS

Ionic liquids mostly applied to obtain metal nanoparticles, metal oxide nanoparticles, and alloy nanoparticles. The physicochemical properties of metal nanoparticles, their surface properties, size, shape, and their molecular arrangement make them attentive for various applications such as medicine, chemicals, and electronics. The surface properties of one metal NPs are mainly different from another because of their different physiochemical properties (as developed after synthesis). In ILs mediated metal nanoparticle synthesis, ILs act as *in situ* stabilizing agent and offers a variety of surface properties and applications. The preparation and stabilization of metal nanoparticles make them more efficient in terms of catalyst loading, catalytic selectivity and catalyst recycling to ionic liquid mediated organometallic or metal salt catalysts (Dupont & Suarez, 2006; Janiak, 2013). The imidazolium-based ILs are the most studied and recommended class of ionic liquid, which was successfully used to synthesize a series of NPs (Figure 1).

In IL mediated NPs synthesis, both anions and cations plays an important role not only to stabilize the NPs of metal but also offers the interactions of anion with metal surface of NPs which is more stronger than cations and hence anion are more active for stabilization of NPs (Dupont, 2004; Li & Liu, 2011; Olveira et al., 2014; Prechtl, n.d.; Scholten et al., 2012; Wasserscheid & Keim, 2000) . However, the high catalytic activity can be achieved due to weak interaction between metal NPs and non-functionalized ILs but the problem of poor stability and recyclability also occurs with NPs during the reaction. Such glitches can be resolved by using functionalized ILs, as it is well documented that -OH group functionalized ILs generate more stable NPs (B. Zhang & Yan, 2013).

IONIC LIQUID MEDIATED METAL NANOPARTICLES OF RU, PD, PT AND TIO_2

Synthesis of Ionic liquid mediated Ru-NPs

The reduction of metal salts is the most applicable method to generate metal nanoparticles (MNPs) in ILs. MNPs of metals like Ru, Pd, Pt, Au, Ag, Rh and Ir with positive oxidation state (M^{n+}) have been prepared in ILs. In general, the reducing agents like $NaBH_4$ and molecular hydrogen (4 bar) were commonly used for the synthesis of Ru NPs using a series of metal precursors such as RuO_2•$3H_2O$, RuO_2, [Ru(C_8H_{12})(2-methyallyl)$_2$], [Ru(C_8H_{12})(C_8H_{10})]. The Ru metal precursors found highly dispersed in different ILs like 1-*n*-butyl-3-methylimidazolium hexafluorophosphate [BMIM][PF_6], or 1-*n*-butyl-3-methylimidazolium trifluoromethanesulphonate [BMIM][CF_3SO_3], or 1-butyl-3-methylimidazolium tetrafluoroborate [BMIM][BF_4], or in different cations of imidazolium 1-n-decyl-3-methylimidazolium [DMI] and 1-butyronitilre-3-methylimidazolium[(BCN)MI] with anions PF_6, NTf_2, BF_4 (Ghoranneviss et al., 2015; Kalarikkal, Thomas, & Koshy, 2018; Kaur & Singh, 2017; Ratti, 2014; Rodriguez-Liviano, Nuñez, Rivera-Fernández, de la Fuente, & Ocaña, 2013; Praveenkumar Upadhyay & Srivastava, 2019).

The chemical reduction method offers very fine, stable and agglomeration free Ru(0) NPs in the ionic liquid medium.The H_2 reduction is one of the commercially available reducing agents to Ru metal precursor $RuO_2 \cdot 3H_2O$, which easily offers Ru(0) nanoparticles. Also, this process does not require any organometallic precursor. The Ru(0) nanoparticle was successfully applied to benzene hydrogenation and synthesis of hydrogenated benzene derivatives as desired requirements. Biphasic experiments were also utilized to synthesize Ru nanoparticles, where prior to the addition of the substrate, RuO_2 was dispersed in ionic liquid (1-*n*-butyl-3-methylimidazolium hexafluorophosphate [BMIM][PF_6], or 1-n-butyl-3-

Figure 3. Reduction of Pd(II) species through intermediate followed by subsequent decomplexation and reduction (Prechtl et al., 2010)

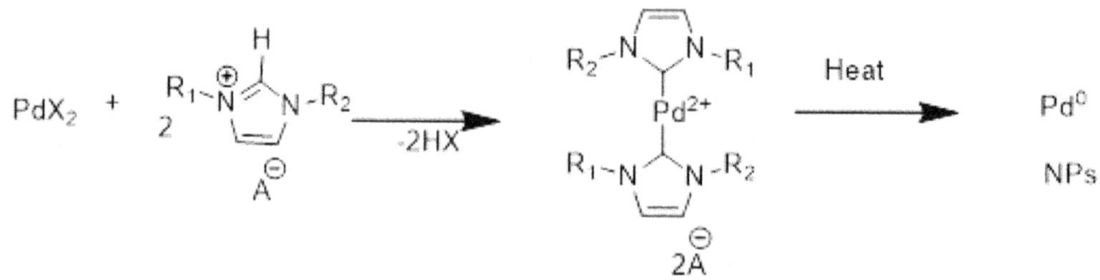

methylimidazolium trifluoromethanesulphonate [BMIM][CF_3SO_3], or 1-butyl-3-methylimidazolium tetrafluoroborate [BMIM][BF_4]). Later, the vacuum was created in a reactor at 75ºC under constant stirring. The reactor was connected to the hydrogen gas source and bypassing H_2 gas (4 atm) the reaction was initiated. The pressure drop of H_2 gas confirms the progress of the reaction (P. R. Upadhyay & Srivastava, 2016d, 2016b, 2016a, 2017; P. Upadhyay & Srivastava, 2015, 2016). The ILs containing *N*-bis(trifluoromethanesulfonyl)imidates gives NPs in lesser size to the less coordinated tetrafluoroborate analogs.

[Ru(COD)(COT)] (COD = 1,5-cyclooctadiene, COT=1,3,5-cyclooctatriene) is zero-valent organometallic compounds, which were used as a significant precursor to obtaining Ru(0) metal atom followed by decomposition of COD and COT ligand under hydrogen pressure. The concentration of [Ru(COD)(COT)] probably controls the crystal growth of NPs and hence it limits the size of the non-polar domains. The degree of self-organization, as well as the structural morphology of the Imidazolium based ionic liquid, oversees the size of Ru NPs. The reduction of [Ru(COD)(2-methylallyl)$_2$] (COD = 1,5-cyclooctadiene) associated with the *N*-bis(trifluoromethanesulfonyl)imidates (NTf_2) or with the tetrafluoroborates (BF_4), offers well distributed and ultra-narrow size Ru NPs at 50ºC in hydrogen pressure (4 bar) (Kalidindi & Jagirdar, 2012a; P. R. Upadhyay & Srivastava, 2016c, 2017).

The $RuCl_3$ reduction in ILs (1-n-butyl-3-methylimidazolium hexafluorophosphate [BMIM][PF_6] and 1-n-butyl-3-methylimidazolium tetrafluoroborate [BMIM][BF_4]) with reducing agents $NaBH_4$, is a simple and reproducible method to obtain ruthenium oxide NPs. This well-accepted method offers stable Ru NPs with a narrow size distribution (2 - 3 nm). Chemists achieved high catalytic activity of ruthenium oxide NPs in the biphasic hydrogenation of olefins and arenes under mild reaction conditions with the added advantage of 10 times catalysts recycling for model reaction (hydrogenation of 1-hexene yielding a total turnover number for exposed Ru(0) atoms of 175,000) (Rossi et al., 2004).The trinuclear Ru metal carbonyls ($Ru_3(CO)_{12}$) in *n*-butylmethylimidazolium tetrafluoroborate, [BMIM][BF_4], under an argon atmosphere gets deoxygenated. The dispersed metal carbonyl in [BMIM][BF_4] ionic liquid was thermally decomposed 250°C for several hours. The brown Ru NPs of size 1.5-2.5 nm obtained, which was stable for several months under argon (Kim, Demberelnyamba, & Lee, 2004; Krämer, Redel, Thomann, & Janiak, 2008).

Figure 4. Guan ILs cattion, $[Guan]^+$ = 1,1-dibutyl-2,2,3,3-tetramethyl-guanidinium

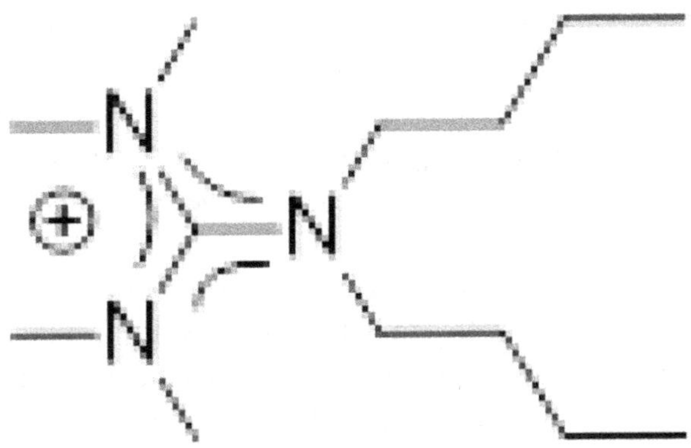

Synthesis of ILs Mediated Pd-NPs

Imidazolium-based IL were found highly promising for the synthesis of Pd-NPs from palladium (II) salt without using any additional reducing agent. It is reported that the formation of Pd-NPs takes place after the development of Pd-N-heterocyclic carbine complexes as intermediates [Figure 3] (Deshmukh, Rajagopal, & Srinivasan, 2001; Prechtl, Scholten, & Dupont, 2010).

As a precursor, the Imidazolium salts are also counted as mild reducing agents and applied for the synthesis of Pd-NPs. Thermal decomposition of $Pd(OAc)_2$ is well-established protocol and it was well studied in various conventional solvents. Simply heating of $Pd(OAc)_2$ in [BMIm][Tf_2N] at 80^0C in the presence of triphenyl phosphine (PPh_3). Pd-NPs were obtained with a diameter of nearly 1 nm were formed [59], while in the absence of an additional reducing agent, Pd nanoparticles of 5 and 10 nm were obtained by heating the proper mixture of $Pd(acac)_2$ and [HOBMIm][Tf_2N] (Anderson, Cortiñas Fernández, Hardacre, & Marr, 2004; Bradley, Hill, Klein, Chaudret, & Duteil, 1993; Ruta, Laurenczy, Dyson, & Kiwi-Minsker, 2008; Scholten, Ebeling, & Dupont, 2006; Tano, Esumi, & Meguro, 1989).

Heating of well-dispersed $Pd(OAc)_2$ in 1-butyronitrile-3-methyl-imidazolium *N*-bis(trifluoromethane sulfonyl)amide [NCBMIm][Tf_2N], under reduced pressure, leads to the formation of stable and small-sized Pd-NPs. In a series of hydroxy-functionalized ILs with the $[HOEMIm]^+$ cation, the effect of the anions on the decomposition rate of $Pd(OAc)_2$ was found in the given order $[Tf_2N]^-$, $[PF_6]^-$ >$[BF_4]^-$ >$[OTf]^-$ >$[TFA]^-$, based on a determination of the percentage of $Pd(OAc)_2$ remaining in the sample (Venkatesan et al., 2011). Thermal treatment in hydroxyl functionalized ILs with the 1-(20-hydroxyethyl)-3-methyl-imidazolium $[HOEMIm]^+$ cation and non-functionalized control IL, Pd-NPs were synthesized from $Pd(OAc)_2$. The OH-functionalized ILs [HOEMIm][Tf_2N] gave smaller Pd-NPs with diameters 4.0 (± 0.6 nm) as it is accepted that the alcohol group in the $[HOEMIm]^+$ cation is not working as the reducing agent (this is also confirmed by ^1H NMR spectra recorded before and after reduction of $Pd(OAc)_2$ showed no difference) (Venkatesan et al., 2011; Yuan et al., 2012).

The Pd(0) stable NPs with a size of 4.9 (±0.8) nm were obtained by reduction of $Pd(acac)_2$(acac=acetylacetonate) under high- pressure hydrogen atmosphere (4 atm) at 75 °C. Before, going to reduction, the $Pd(acac)_2$(acac=acetylacetonate) was dissolved in ionic liquid like (1-*n*-butyl-3-methylimidazolium hexafluorophosphate [BMIM][PF_6] or 1-n-butyl-3-methylimidazolium tetrafluoroborate [BMIM][BF_4]). These developed Pd NPs were further tested for the selective partial hydrogenation 1,3-butadiene (Umpierre, Machado, Fecher, Morais, & Dupont, 2005). *In situ* laser radiation process was also trying to break large Pd(0) nanoparticles fragmentation into smaller particles with narrow size distribution in 1-n-butyl-3-methylimidazolium hexafluorophosphate ionic liquid. In another approach, simple hydrogen reduction of $PdCl_2$ dispersed in 1-*n*-butyl-3-methylimidazolium hexafluorophosphate at 75^0C, the formation of Pd(0) nanoparticles indicated by a dark solution which was obtained after 15 minutes (Umpierre et al., 2005). Guan-ILs (Fig. 2) as a mediator for the nucleation and growth process uniform Pd nanoparticles supported on Vulcan XC-72 carbon were synthesized from H_2PdCl_4 and $NaBH_4$ (Umpierre et al., 2005).

The ultrasonic approach was also applied to irradiate $Pd(OAc)_2$ or $PdCl_2$ salt solution in imidazolium-based ILs such as [BBIm][Br] or [BBIm][BF_4] to obtain nearly spherical Pd-NPs (20 nm). The succeeding sonolytic conversion of Pd biscarbene complexes as intermediates to Pd NPs have been recognized by NMR/MS and TEM analysis, respectively (Prechtl et al., 2010). Electrochemical reduction of palladium foil as the anode and a platinum foil as the cathode were used to prepare stabilized Pd NPs in morpholinium based ionic liquid, [BMMor][BF_4]. At Pd anode, Pd ions migrate towards Pt cathode and nanoparticles forms due to the reduction of Pd ions at the cathode. An increase in temperature as well as electrolysis duration along with a decrease in the current density, the surprising increase in particle size of Pd, was observed. The particles had a crystalline structure was confirmed by the electron diffraction patterns of the resulting NPs (Umpierre et al., 2005).

Synthesis of Ionic Liquid Mediated Pt-NPs

The well crystalline, distinct, small and longtime stable Pt-nanocrystal synthesis was carried out by thermal, photolytic or microwave- assisted decomposition of the air/ moisture stable organometallic Pt(IV) precursor (MeCp)$PtMe_3$ in following ionic liquid reaction medium [BMIm][BF_4] and [BtMA][Tf_2N] (Full name of IL should be entered). Ionic liquid mediated Pt NPs (0:0125 mol:-%) offered the biphasic hydrosilylation of phenylacetylene with triethylsilane, which leads to giving the distal and proximal products triethyl (2- and 1-phenylvinyl) silane with TOF 96 000 h^{-1} (Fernández-Lodeiro et al., 2019; Huang et al., 2019)(Marquardt, Barthel, Braun, Ganter, & Janiak, 2012). The catalytically sound and active Pt (0) (2.3nm size) was obtained by controlling thermal decomposition of the organometallic precursor $Pt_2(dba)_3$ (dba = dibenzylideneacetone) in reported [BMIM][BF_4] and [BMIM][PF_6], ionic liquid medium. Pt NPs also synthesized from $H_2Pt(OH)_6$ and formic acid in [BMIm][FEP] (FEP=tris(pentafluoroethyl) trifluorophosphate) which offers chemo selective hydrogenation of 3-nitrostyrene (yields exceeding 90%). Recyclability of Pt NPs found while immobilizing on solid SiO_2 or supported carbon nanotube (CNTs) . Pt sputtering without stabilizing agents onto the IL trimethyl-n-propylammonium bis((trifluoromethyl) sulfonyl)amide [Me_3PrN][Tf_2N] produce Pt NPs of mean particle diameters of ca. 2.3 – 2.4 nm but they are independent of sputtering time. Without any pretreatment of SWCNTs or any chemical reagent, the Pt-NP-SWCNT composites were achieved by using the Pt sputtered ILs and the Pt-NP immobilization onto single- walled carbon nanotubes (Tsuda, Yoshii, Torimoto, & Kuwabata, 2010; Yang et al., 2011).

Synthesis of Ionic liquid mediated TiO_2-NPs

Titanium oxide (TiO_2) is one of the most important nanomaterial, which has attracted great attention due to its unique properties like optical, dielectric, and catalytic properties(Dahl, Liu, & Yin, 2014; Ding et al., 2007; Haider, Jameel, & Al-Hussaini, 2019; Yoshii et al., 2012). Such properties of TiO_2 nanomaterials Ti NPs make them nearer to following industrial applications such as pigments, fillers, catalyst supports, and photo-catalysts. It is reported that the overall physiochemical properties of this material may vary with size, morphology and crystalline phase of Ti NPs (Dahl et al., 2014; Haider et al., 2019). Various significant methods such as liquid process (sol–gel, solvothermal, hydrothermal), solid state processing routes (mechanical alloying/milling, mechanochemical, RF thermal plasma and other routes such as laser ablation have been developed over the last decade. Among a series of ionic liquids, 1-butyl-3-methylimidazolium bis(triflylmethylsulfonyl)imide [BMIM][Tf_2N is used several times in the synthesis of titanium oxide nanoparticles. The anatase, brookite, rutile is the polymorph of titanium (S. & M., 2016; Zheng, Liu, Yan, & Zhu, 2009). The synthesis of nano size rutile phase of titanium is considered as most promising material for the development of photocatalyst as well as electrode material. The under mild condition, IL mediated synthesis of titanium dioxide nanocrystal of 2-3 nm size was done due to their self-assembly toward mesoporous TiO_2 spheres. The resulting larger spheres with a considerable high surface area and narrow pore size distribution had shown a great potential for catalysis, solar energy conversion and optics – electronic devices. 1-butyl-3-methylimidazoliumtetrafluoroborate was utilized as solvent for the synthesis of TiO_2 NPs. The hydrolysis of titanium tetrachloride in hydrochloric acid with the support of 1-ethyl-3-methyl-imidazolium bromide successfully synthesized the pure rutile and rutile-anatase nanocomposites. Simple optimization of 1-ethyl-3-methyl-imidazolium bromide ionic liquid added to the reaction mixture may control the content of rutile in the composites. As per the approach, the reaction can be performed under atmospheric pressure in a glass vessel and with high yield. The composites with arbitrary content of rutile have been rationally prepared and the contents are controllable with high crystallinity. This method can be improved in a general way to synthesize other metal oxides NPs on a huge scale (Kaur & Singh, 2017; Zheng et al., 2009) . An alkoxide sol-gel method was found applicable for the synthesis of nanocrystalline particles of TiO_2 in the presence of water-immiscible ionic liquids as a solvent medium. Also, later non-ionic surfactant was used as pore-templating material for the modified synthesis of TiO_2.

CONCLUSION

In this book chapter, we framed the most significant developments in the area of ionic liquid mediated important metal nanoparticles. The basic aim of this article is to provide comprehensive information concisely to new researchers by taking the support of the best scientific reports. Here, we are not only covering the impact of imidazolium-based ionic liquid, but we also discussed some other important group of ionic liquids to make the article most versatile in the area of ionic liquid mediated metal nanoparticle synthesis. The fascinating properties of ionic liquid not only control the size- and shape-of the nanoparticle, but also provide a stable environment to make them more reactive towards their applications. As per the reported literature, the ionic liquid provides an opportunity to achieve the synthesis of well-dispersed, small size, metallic nanoparticles with different chemical compositions. Anions of ionic liquid play a more noteworthy role in stabilizing NPs, as they interact more strongly and effectively with the metal surface other than the cationic part of the ionic liquid. However, it is worth noted here that, NPs are only kinetically stable and neither anion nor cation could prevent the process of agglomeration.

ACKNOWLEDGMENT

This work is financially supported by DST Fast Track (SB/FT/CS-124/2012). Author is also thankful for the library of NIIT University for providing useful journals and books for the compilation of this review article.

REFERENCES

Ahmad, S., Munir, S., Zeb, N., Ullah, A., Khan, B., Ali, J., … Ali, S. (2019). Green nanotechnology: A review on green synthesis of silver nanoparticles — An ecofriendly approach. *International Journal of Nanomedicine*.

Ahrens, S., Peritz, A., & Strassner, T. (2009). Tunable aryl alkyl ionic liquids (TAAILs): The next generation of ionic liquids. *Angewandte Chemie International Edition*, *48*(42), 7908–7910. doi:10.1002/anie.200903399 PMID:19760688

Anderson, K., Cortiñas Fernández, S., Hardacre, C., & Marr, P. C. (2004). Preparation of nanoparticulate metal catalysts in porous supports using an ionic liquid route; hydrogenation and C-C coupling. *Inorganic Chemistry Communications*, *7*(1), 73–76. doi:10.1016/j.inoche.2003.10.008

Angell, C. A., Byrne, N., & Belieres, J. P. (2007). Parallel developments in aprotic and protic ionic liquids: Physical chemistry and applications. *Accounts of Chemical Research*, *40*(11), 1228–1236. doi:10.1021/ar7001842 PMID:17979250

Astruc, D. (2008). Transition-metal Nanoparticles in Catalysis: From Historical Background to the State-of-the Art. *Nanoparticles and Catalysis*.

Austen Angell, C., Ansari, Y., & Zhao, Z. (2012). Ionic Liquids: Past, present and future. *Faraday Discussions*, *154*, 9–27. doi:10.1039/C1FD00112D PMID:22455011

Bahadory, M. (2008). Synthesis of noble metal nanoparticles. *Materials Science and Engineering C*, *22*(August), 204.

Bopp, S. K., Kienzler, A., Richarz, A. N., van der Linden, S. C., Paini, A., Parissis, N., & Worth, A. P. (2019, February 7). Regulatory assessment and risk management of chemical mixtures: challenges and ways forward. *Critical Reviews in Toxicology*.

Bradley, J. S., Hill, E. W., Klein, C., Chaudret, B., & Duteil, A. (1993). Synthesis of monodispersed bimetallic palladium-copper nanoscale colloids. *Chemistry of Materials*, *5*(3), 254–256. doi:10.1021/cm00027a005

Capello, C., Fischer, U., & Hungerbühler, K. (2007). What is a green solvent? A comprehensive framework for the environmental assessment of solvents. *Green Chemistry*, *9*(9), 927–934. doi:10.1039/b617536h

Chaturvedi, S., Dave, P. N., & Shah, N. K. (2012). Applications of nano-catalyst in new era. *Journal of Saudi Chemical Society*, *16*(3), 307–325. doi:10.1016/j.jscs.2011.01.015

Corrêa, C. M., Bizeto, M. A., & Camilo, F. F. (2016). Direct synthesis of silver nanoparticles in ionic liquid. *Journal of Nanoparticle Research, 18*(5), 1–10. doi:10.100711051-016-3436-8

Dahl, M., Liu, Y., & Yin, Y. (2014, October 8). Composite titanium dioxide nanomaterials. *Chemical Reviews.*

Deshmukh, R. R., Rajagopal, R., & Srinivasan, K. V. (2001). Ultrasound promoted C-C bond formation: Heck reaction at ambient conditions in room temperature ionic liquids. *Chemical Communications, 1*(17), 1544–1545. doi:10.1039/b104532f PMID:12240373

Dhakshinamoorthy, A., Navalon, S., Alvaro, M., & Garcia, H. (2012). Metal Nanoparticles as Heterogeneous Fenton Catalysts. *ChemSusChem, 5*(1), 46–64. doi:10.1002/cssc.201100517 PMID:22250135

Ding, K., Miao, Z., Liu, Z., Zhang, Z., Han, B., An, G., ... Xie, Y. (2007). Facile Synthesis of High Quality TiO_2 Nanocrystals in Ionic Liquid via a Microwave-Assisted Process. *Journal of the American Chemical Society, 129*(20), 6362–6363. doi:10.1021/ja070809c PMID:17469828

Disasa Irge, D. (2016). Ionic Liquids: A Review on Greener Chemistry Applications, Quality Ionic Liquid Synthesis and Economical Viability in a Chemical Processes. *American Journal of Physical Chemistry, 5*(3), 74. doi:10.11648/j.ajpc.20160503.14

Dong, K., Liu, X., Dong, H., Zhang, X., & Zhang, S. (2017, May 24). Multiscale Studies on Ionic Liquids. *Chemical Reviews.*

Dupont, J. (2004). On the solid, liquid and solution structural organization of imidazolium ionic liquids. *Journal of the Brazilian Chemical Society.*

Dupont, J., & Suarez, P. A. Z. (2006). Physico-chemical processes in imidazolium ionic liquids. *Physical Chemistry Chemical Physics, 8*(21), 2441–2452. doi:10.1039/b602046a PMID:16721427

Ealias, A. M., & Saravanakumar, M. P. (2017). A review on the classification, characterisation, synthesis of nanoparticles and their application. *IOP Conference Series. Materials Science and Engineering, 263*, 032019. doi:10.1088/1757-899X/263/3/032019

Eyckens, D. J., & Henderson, L. C. (2019a). *A review of solvate ionic liquids: Physical parameters and synthetic applications. Frontiers in Chemistry* .

Eyckens, D. J., & Henderson, L. C. (2019b, April 18). A review of solvate ionic liquids: Physical parameters and synthetic applications. *Frontiers in Chemistry.*

Fan, J., & Gao, Y. (2006). Nanoparticle-supported catalysts and catalytic reactions – a mini-review. *Journal of Experimental Nanoscience, 1*(4), 457–475. doi:10.1080/17458080601067708

Fernández-Lodeiro, A., Djafari, J., Lopez-Tejedor, D., Perez-Rizquez, C., Rodríguez-González, B., Capelo, J. L., ... Fernández-Lodeiro, J. (2019). Highly accessible aqueous synthesis of well-dispersed dendrimer type platinum nanoparticles and their catalytic applications. *Nano Research, 12*(5), 1083–1092. doi:10.100712274-019-2350-7

Gadilohar, B. L., & Shankarling, G. S. (2017, February 1). Choline based ionic liquids and their applications in organic transformation. *Journal of Molecular Liquids.*

Ghandi, K. (2014). A Review of Ionic Liquids, Their Limits and Applications. *Green and Sustainable Chemistry*, *04*(01), 44–53. doi:10.4236/gsc.2014.41008

Ghernaout, D., Ghernaout, B., & Naceur, M. W. (2011, April 15). Embodying the chemical water treatment in the green chemistry-A review. *Desalination*, *271*(1-3), 1–10. doi:10.1016/j.desal.2011.01.032

Ghoranneviss, M., Soni, A., Talebitaher, A., & Aslan, N. (2015). Nanomaterial synthesis, characterization, and application. *Journal of Nanomaterials*.

Gordon, C. M. (2001). New developments in catalysis using ionic liquids. *Applied Catalysis A, General*, *222*(1–2), 101–117. doi:10.1016/S0926-860X(01)00834-1

Greaves, T. L., & Drummond, C. J. (2008, January). Protic ionic liquids: Properties and applications. *Chemical Reviews*, *108*(1), 206–237. doi:10.1021/cr068040u PMID:18095716

Haider, A. J., Jameel, Z. N., & Al-Hussaini, I. H. M. (2019). Review on: Titanium dioxide applications. *Energy Procedia*, *157*, 17–29. doi:10.1016/j.egypro.2018.11.159

Hallett, J. P., & Welton, T. (2011, May 11). Room-temperature ionic liquids: Solvents for synthesis and catalysis. *Chemical Reviews*.

He, Z., & Alexandridis, P. (2015, July 28). Nanoparticles in ionic liquids: Interactions and organization. *Physical Chemistry Chemical Physics*.

Heera, P., & Shanmugam, S. (2015). Review Article Nanoparticle Characterization and Application: An Overview. *Int. J. Curr. Microbiol. App. Sci*, *4*.

Huang, X., Li, Z., Yu, Z., Deng, X., Xin, Y., & Jesionowski, T. (2019). Recent Advances in the Synthesis, Properties, and Biological Applications of Platinum Nanoclusters. *Journal of Nanomaterials*.

Janiak, C. (2013). Ionic Liquids for the Synthesis and Stabilization of Metal Nanoparticles. *Z. Naturforsch*, *68*(10), 1059–1089. doi:10.5560/znb.2013-3140

Janiak, C. (2015). Metal nanoparticle synthesis in ionic liquids. *Topics in Organometallic Chemistry*, *51*, 17–54. doi:10.1007/3418_2013_70

Janus, E., Goc-Maciejewska, I., Łozyński, M., & Pernak, J. (2006). Diels-Alder reaction in protic ionic liquids. *Tetrahedron Letters*, *47*(24), 4079–4083. doi:10.1016/j.tetlet.2006.03.172

Jeevanandam, J., Barhoum, A., Chan, Y. S., Dufresne, A., & Danquah, M. K. (2018, April 3). Review on nanoparticles and nanostructured materials: History, sources, toxicity and regulations. *Beilstein Journal of Nanotechnology*.

Johnson, K. E., Pagni, R. M., & Bartmess, J. (2007). Brønsted acids in ionic liquids: Fundamentals, organic reactions, and comparisons. *Monatshefte für Chemie*, *138*(11), 1077–1101. doi:10.100700706-007-0755-6

Kalarikkal, N., Thomas, S., & Koshy, O. (Eds.). (2018). Nanomaterials. Oakville, Canada: Apple Academic Press.

Kalidindi, S. B., & Jagirdar, B. R. (2012). Nanocatalysis and Prospects of Green Chemistry. *ChemSusChem*, *5*(1), 65–75. doi:10.1002/cssc.201100377 PMID:22190344

Kaur, N., & Singh, V. (2017). Current status and future challenges in ionic liquids, functionalized ionic liquids and deep eutectic solvent-mediated synthesis of nanostructured TiO2: A review. *New Journal of Chemistry*, *41*(8), 2844–2868. doi:10.1039/C6NJ04073J

Khan, I., Saeed, K., & Khan, I. (2017). Nanoparticles: Properties, applications and toxicities. *Arabian Journal of Chemistry*.

Kim, K.-S., Demberelnyamba, D., & Lee, H. (2004). Size-Selective Synthesis of Gold and Platinum Nanoparticles Using Novel Thiol-Functionalized Ionic Liquids. *Langmuir*, *20*(3), 556–560. doi:10.1021/la0355848 PMID:15773074

Krämer, J., Redel, E., Thomann, R., & Janiak, C. (2008). Use of Ionic Liquids for the Synthesis of Iron, Ruthenium, and Osmium Nanoparticles from Their Metal Carbonyl Precursors. *Organometallics*, *27*(9), 1976–1978. doi:10.1021/om800056z

Lamon, L., Aschberger, K., Asturiol, D., Richarz, A., & Worth, A. (2019, January 2). Grouping of nanomaterials to read-across hazard endpoints: a review. *Nanotoxicology*.

Lei, Z., Chen, B., Koo, Y.-M., & MacFarlane, D. R. (2017). Introduction: Ionic Liquids. *Chemical Reviews*, *117*(10), 6633–6635. doi:10.1021/acs.chemrev.7b00246 PMID:28535681

Lewis, R. W., Bertsch, P. M., & McNear, D. H. (2019, March 16). Nanotoxicity of engineered nanomaterials (ENMs) to environmentally relevant beneficial soil bacteria–a critical review. *Nanotoxicology*.

Li, Y. F., & Liu, Z. P. (2011). Particle size, shape and activity for photocatalysis on titania anatase nanoparticles in aqueous surroundings. *Journal of the American Chemical Society*, *133*(39), 15743–15752. doi:10.1021/ja206153v PMID:21879719

Mandai, T., Yoshida, K., Ueno, K., Dokko, K., & Watanabe, M. (2014, May 21). Criteria for solvate ionic liquids. *Physical Chemistry Chemical Physics*.

Marquardt, D., Barthel, J., Braun, M., Ganter, C., & Janiak, C. (2012). Weakly-coordinated stable platinum nanocrystals. *CrystEngComm*, *14*(22), 7607–7615. doi:10.1039/c2ce25904d

Olveira, S., Forster, S. P., & Seeger, S. (2014). *Nanocatalysis: Academic Discipline and Industrial Realities*. Academic Press.

Plechkova, N. V., & Seddon, K. R. (2008). Applications of ionic liquids in the chemical industry. *Chemical Society Reviews*, *37*(1), 123–150. doi:10.1039/B006677J PMID:18197338

Polshettiwar, V., & Varma, R. S. (2010). Green chemistry by nano-catalysis. *Green Chemistry*, *12*(5), 743–754. doi:10.1039/b921171c

Prechtl, M. H. G. (2016). *Nanocatalysis in Ionic Liquids. Nanocatalysis in Ionic Liquids*. Wiley.

Prechtl, M. H. G., Scholten, J. D., & Dupont, J. (2010). Carbon-carbon cross coupling reactions in ionic liquids catalysed by palladium metal nanoparticles. *Molecules (Basel, Switzerland)*, *15*(5), 3441–3461. doi:10.3390/molecules15053441 PMID:20657493

Ratti, R. (2014). Ionic Liquids: Synthesis and Applications in Catalysis. *Advances in Chemistry, 2014*, 1–16. doi:10.1155/2014/729842

Renna, L. A., Boyle, C. J., Gehan, T. S., & Venkataraman, D. (2015). Polymer Nanoparticle Assemblies: A Versatile Route to Functional Mesostructures. *Macromolecules, 48*(18), 6353–6368. doi:10.1021/acs.macromol.5b00375

Rodriguez-Liviano, S., Nuñez, N. O., Rivera-Fernández, S., de la Fuente, J. M., & Ocaña, M. (2013). Ionic Liquid Mediated Synthesis and Surface Modification of Multifunctional Mesoporous Eu:GdF $_3$ Nanoparticles for Biomedical Applications. *Langmuir, 29*(10), 3411–3418. doi:10.1021/la4001076 PMID:23402647

Roldan Cuenya, B., & Behafarid, F. (2015). *Nanocatalysis: Size- and shape-dependent chemisorption and catalytic reactivity. Surface Science Reports*. Elsevier B.V.

Rossi, L. M., Dupont, J., Machado, G., Fichtner, P. F. P., Radtke, C., Baumvol, I. J. R., & Teixeira, S. R. (2004). Ruthenium dioxide nanoparticles in ionic liquids: Synthesis, characterization and catalytic properties in hydrogenation of olefins and arenes. *Journal of the Brazilian Chemical Society, 15*(6), 904–910. doi:10.1590/S0103-50532004000600018

Ruta, M., Laurenczy, G., Dyson, P. J., & Kiwi-Minsker, L. (2008). Pd Nanoparticles in a Supported Ionic Liquid Phase: Highly Stable Catalysts for Selective Acetylene Hydrogenation under Continuous-Flow Conditions. *The Journal of Physical Chemistry C, 112*(46), 17814–17819. doi:10.1021/jp806603f

S., A., & M., S. (2016). [EMIM] BF4 ionic liquid-mediated synthesis of TiO2 nanoparticles using Vitex negundo Linn extract and its antibacterial activity. *Journal of Molecular Liquids, 221*, 986–992.

Scholten, J. D., Ebeling, G., & Dupont, J. (2006). On the involvement of NHC carbenes in catalytic reactions by iridium complexes, nanoparticle and bulk metal dispersed in imidazolium ionic liquids. *Journal of the Chemical Society, Dalton Transactions: Inorganic Chemistry*, (47): 5554–5560. PMID:18043817

Scholten, J. D., Leal, B. C., & Dupont, J. (2012). Transition Metal Nanoparticle Catalysis in Ionic Liquids. *ACS Catalysis, 2*(1), 184–200. doi:10.1021/cs200525e

Sharma, D., Kanchi, S., & Bisetty, K. (2019, December 1). Biogenic synthesis of nanoparticles: A review. *Arabian Journal of Chemistry*.

Sheldon, R. A. (2012, February 21). Fundamentals of green chemistry: Efficiency in reaction design. *Chemical Society Reviews, 41*(4), 1437–1451. doi:10.1039/C1CS15219J PMID:22033698

Shenhar, R., Norsten, T. B., & Rotello, V. M. (2005). Polymer-Mediated Nanoparticle Assembly: Structural Control and Applications. *Advanced Materials, 17*(6), 657–669. doi:10.1002/adma.200401291

Singh, A. V., Laux, P., Luch, A., Sudrik, C., Wiehr, S., Wild, A. M., … Sitti, M. (2019b, June 13). Review of emerging concepts in nanotoxicology: opportunities and challenges for safer nanomaterial design. *Toxicology Mechanisms and Methods*.

Singh, A. V., Laux, P., Luch, A., Sudrik, C., Wiehr, S., Wild, A.-M., ... Sitti, M. (2019a). Review of emerging concepts in nanotoxicology: Opportunities and challenges for safer nanomaterial design. *Toxicology Mechanisms and Methods, 29*(5), 378–387. doi:10.1080/15376516.2019.1566425 PMID:30636497

Slattery, J. M., Daguenet, C., Dyson, P. J., Schubert, T. J. S., & Krossing, I. (2007). How to predict the physical properties of ionic liquids: A volume-based approach. *Angewandte Chemie International Edition, 46*(28), 5384–5388. doi:10.1002/anie.200700941 PMID:17568467

Sowmiah, S., & Cheng, I, C., & Chu, Y.-H. (2012). Ionic Liquids for Green Organic Synthesis. *Current Organic Synthesis, 9*(1), 74–95. doi:10.2174/157017912798889116

Sowmiah, S., Srinivasadesikan, V., Tseng, M. C., & Chu, Y. H. (2009, September). On the chemical stabilities of ionic liquids. *Molecules (Basel, Switzerland), 14*(9), 3780–3813. doi:10.3390/molecules14093780 PMID:19783957

Tano, T., Esumi, K., & Meguro, K. (1989). Preparation of organopalladium sols by thermal decomposition of palladium acetate. *Journal of Colloid and Interface Science, 133*(2), 530–533. doi:10.1016/S0021-9797(89)80069-4

Tokuda, H., Tsuzuki, S., Susan, M. A. B. H., Hayamizu, K., & Watanabe, M. (2006). How ionic are room-temperature ionic liquids? An indicator of the physicochemical properties. *The Journal of Physical Chemistry B, 110*(39), 19593–19600. doi:10.1021/jp064159v PMID:17004825

Tsuda, T., Yoshii, K., Torimoto, T., & Kuwabata, S. (2010). Oxygen reduction catalytic ability of platinum nanoparticles prepared by room-temperature ionic liquid-sputtering method. *Journal of Power Sources, 195*(18), 5980–5985. doi:10.1016/j.jpowsour.2009.11.027

Ueno, K., Tokuda, H., & Watanabe, M. (2010). Ionicity in ionic liquids: Correlation with ionic structure and physicochemical properties. *Physical Chemistry Chemical Physics, 12*(8), 1649–1658. doi:10.1039/b921462n PMID:20145829

Umpierre, A. P., Machado, G., Fecher, G. H., Morais, J., & Dupont, J. (2005). Selective Hydrogenation of 1,3-Butadiene to 1-Butene by Pd(0) Nanoparticles Embedded in Imidazolium Ionic Liquids. *Advanced Synthesis & Catalysis, 347*(10), 1404–1412. doi:10.1002/adsc.200404313

Upadhyay, P., & Srivastava, V. (2019). Synthesis of Ionic Liquid Mediated Nanoparticle Synthesis. In *Nanomaterials* (pp. 67–89). Apple Academic Press.

Upadhyay, P., & Srivastava, V. (2015). Synthesis of supported Ru-nanoparticles for selective hydrogenation of carbonyl compounds. *Letters in Organic Chemistry, 12*(8), 528–533. doi:10.2174/15701786 1208150826112807

Upadhyay, P., & Srivastava, V. (2016). Synthesis of Monometallic Ru/TiO2 Catalysts and Selective Hydrogenation of CO2 to Formic Acid in Ionic Liquid. *Catalysis Letters, 146*(1), 12–21. doi:10.100710562-015-1654-9

Upadhyay, P. R., & Srivastava, V. (2016a). Heterogeneous Silica Tethered Ruthenium Catalysts for Carbon Sequestration Reaction. *Catalysis Letters, 146*(8), 1478–1486. doi:10.100710562-016-1772-z

Upadhyay, P. R., & Srivastava, V. (2016b). Recyclable graphene-supported palladium nanocomposites for Suzuki coupling reaction. *Green Processing and Synthesis, 5*(2), 123–129. doi:10.1515/gps-2015-0112

Upadhyay, P. R., & Srivastava, V. (2016c). Selective Hydrogenation of CO2 Using Ruthenium Nanoparticles Intercalated Montmorillonite Clay. *Letters in Organic Chemistry*, *13*(6), 459–465. doi:10.2174/1570178613666160815095733

Upadhyay, P. R., & Srivastava, V. (2016d). Silica Tethered Ruthenium Catalyst for the Hydrogenation of CO2 Gas. *Letters in Organic Chemistry*, *13*(5), 380–387. doi:10.2174/1570178613666160303002129

Upadhyay, P. R., & Srivastava, V. (2017). Ionic Liquid Mediated In Situ Synthesis of Ru Nanoparticles for CO2 Hydrogenation Reaction. *Catalysis Letters*, *147*(4), 1051–1060. doi:10.100710562-017-1995-7

Vekariya, R. L. (2017, February 1). A review of ionic liquids: Applications towards catalytic organic transformations. *Journal of Molecular Liquids*.

Venkatesan, R., Prechtl, M. H. G., Scholten, J. D., Pezzi, R. P., MacHado, G., & Dupont, J. (2011). Palladium nanoparticle catalysts in ionic liquids: Synthesis, characterisation and selective partial hydrogenation of alkynes to Z-alkenes. *Journal of Materials Chemistry*, *21*(9), 3030–3036. doi:10.1039/c0jm03557b

Verma, C., Ebenso, E. E., & Quraishi, M. A. (2019, February 15). Transition metal nanoparticles in ionic liquids: Synthesis and stabilization. *Journal of Molecular Liquids*.

Villa, R., Alvarez, E., Porcar, R., Garcia-Verdugo, E., Luis, S. V., & Lozano, P. (2019, December 10). Ionic liquids as an enabling tool to integrate reaction and separation processes. *Green Chemistry*.

Walters, C., Pool, E., & Somerset, V. (2016). Nanotoxicology: A Review. In Toxicology - New Aspects to This Scientific Conundrum. InTech. doi:10.5772/64754

Wasserscheid, P., & Keim, W. (2000). Ionic Liquids—New "Solutions" for Transition Metal Catalysis. *Angewandte Chemie*, *39*(21), 3772–3789. doi:10.1002/1521-3773(20001103)39:21<3772::AID-ANIE3772>3.0.CO;2-5 PMID:11091453

Wegner, S., & Janiak, C. (n.d.). Metal Nanoparticles in Ionic Liquids. *Topics in Current Chemistry*. PMID:28589266

Welton, T. (1999). Room-Temperature Ionic Liquids. Solvents for Synthesis and Catalysis. *Chemical Reviews*, *99*(8), 2071–2083. doi:10.1021/cr980032t PMID:11849019

Welton, T. (2018, June 1). Ionic liquids: a brief history. *Biophysical Reviews*.

Wigger, H., & Nowack, B. (2019). Material-specific properties applied to an environmental risk assessment of engineered nanomaterials–implications on grouping and read-across concepts. *Nanotoxicology*, *13*(5), 623–643. doi:10.1080/17435390.2019.1568604 PMID:30727799

Yang, J., Qiu, L., Liu, B., Peng, Y., Yan, F., & Shang, S. (2011). Synthesis of polymeric ionic liquid microsphere/Pt nanoparticle hybrids for electrocatalytic oxidation of methanol and catalytic oxidation of benzyl alcohol. *Journal of Polymer Science. Part A, Polymer Chemistry*, *49*(21), 4531–4538. doi:10.1002/pola.24895

Yoshii, K., Tsuda, T., Arimura, T., Imanishi, A., Torimoto, T., & Kuwabata, S. (2012). Platinum nanoparticle immobilization onto carbon nanotubes using Pt-sputtered room-temperature ionic liquid. *RSC Advances*, *2*(22), 8262. doi:10.1039/c2ra21243a

Yuan, X., Yan, N., Katsyuba, S. A., Zvereva, E. E., Kou, Y., & Dyson, P. J. (2012). A remarkable anion effect on palladium nanoparticle formation and stabilization in hydroxyl-functionalized ionic liquids. *Physical Chemistry Chemical Physics*, *14*(17), 6026–6033. doi:10.1039/c2cp23931k PMID:22446618

Zhang, B., & Yan, N. (2013). Towards Rational Design of Nanoparticle Catalysis in Ionic Liquids. *Catalysts*, *3*(2), 543–562. doi:10.3390/catal3020543

Zhang, H., & Cui, H. (2009). Synthesis and Characterization of Functionalized Ionic Liquid-Stabilized Metal (Gold and Platinum) Nanoparticles and Metal Nanoparticle/Carbon Nanotube Hybrids. *Langmuir*, *25*(5), 2604–2612. doi:10.1021/la803347h PMID:19437685

Zheng, W., Liu, X., Yan, Z., & Zhu, L. (2009). Ionic liquid-assisted synthesis of large-scale TiO2 nanoparticles with controllable phase by hydrolysis of TiCl4. *ACS Nano*, *3*(1), 115–122. doi:10.1021/nn800713w PMID:19206257

Zhu, Y., & Hosmane, N. S. (2015, June 15). Nanocatalysis: Recent advances and applications in boron chemistry. *Coordination Chemistry Reviews*.

Zvereva, E. E., Katsyuba, S. A., Dyson, P. J., & Aleksandrov, A. V. (2016). Solvation of Palladium Clusters in an Ionic Liquid: A QM/MM Molecular Dynamics Study. *The Journal of Physical Chemistry C*, *120*(8), 4596–4604. doi:10.1021/acs.jpcc.5b11229

Chapter 18
Nanotechnology-Based Nano-Biosorbents

Mintu Maan Dutta
Gauhati University, India

Anushmita Charingia
https://orcid.org/0000-0003-4607-3419
North Eastern Hill University, India

ABSTRACT

Most countries worldwide face the problem related to clean and safe water. The major source of contamination of water bodies includes the discharges of wastes from various domestic and industrial sources (heavy metal ions, dyes, and pharmaceuticals). To overcome those problems, various water purification methods such as coagulation, flocculation, adsorption, membrane separation, biological, and electrochemical methods have been engaged. Adsorption using nano-biosorbents based on chitin, chitosan and modified chitosan, cellulose, alginate, micro algal has emerged as a better alternative for the removal of contaminants from drinking and wastewater treatment. The main advantages of these nano-biosorbents include its biodegradability and eco-friendliness, which have attracted researchers to this field.

INTRODUCTION

Nanotechnology is a newly emerging branch of science that deals with the study of matter that is in the range of 1-100 nm. Materials in nanometer scale possess specific physical, chemical and biological properties that differs from properties of bulk. (Ealias & Saravanakumar., 2017) Potential application of nanotechnology includes its use in catalysis, scientific tools, electronics, biomedical application, environmental remediation and many industrial applications. (Pangiet al., 2003) (Pal et al., 2011) (Jeevanandam et al., 2018) Recently, reports have been found which explains the use of nanotechnology in solving various social problems that are persisting in various developing countries. Clean and safe water is one of the major problems faced by many countries worldwide. (Fleischer & Grunwald, 2008) Release of effluents (heavy metal ions, dyes and pharmaceuticals) from industries is the prime source

DOI: 10.4018/978-1-7998-1241-8.ch018

of contamination of water bodies. (Mallampati et al., 2015) These pollutants are often very stable and non-biodegradable. They often gets accumulated in the living tissues. (Gokila et al., 2017) Once it is accumulated, its removal becomes a cumbersome process. [Castro et al., 2014) To overcome this problem various water purification methods such as coagulation, flocculation, adsorption, membrane separation, biological, and electrochemical methods have been employed, out of which adsorption using bio adsorbents has emerged a better alternative. (Ouyang et al., 2018) (Crini, 2006)

Advantages of Bio Sorbents: (Crini, 2006)

- Cost effective,
- Easy availability,
- Environmental friendly and highly efficient.

MATERIALS AND METHODS

Different bio sorbents such as rice husk, banana peels, sugarcane bagasse, tomato peels, orange peels, eggshells, tea waste, kiwi peels, pine bark, and other peels have been used for the removal of heavy metals, dyes and organic molecular pollutants. (Mallampati et al., 2015)Most of the household vegetable wastes contains polar functional groups such as -OH, $-NH_2$, -COOH which helps in the process of adsorption. (Feng et al., 2011) (Negi et al., 2012) After initial reports on bio sorbents, to make the process of adsorption more effective and efficient it has been replaced by nano adsorbents. Nano adsorbents possess unique physical and chemical properties, which take account of its high chemical activity and adsorption capacity at its surface. (Kyzas & Matis (2015) These nano adsorbents are widely used to solve the problems related to water. Nano adsorbents remove the toxic substances via ion exchange, ion precipitation and adsorption. (Lata & Samadder., 2016)

Some of the commonly used nano adsorbents are as follows: (Lata & Samadder., 2016)

- Iron oxides/hydroxides nanoparticles
- Alumina nanoparticles
- Copper oxide nanoparticles
- Titanium dioxide nanoparticles
- Zinc oxide nanoparticles
- Mixed metal oxide (MMO's) nanoparticles
- Carbon nano tubes
- Nano-biosorbents

In this chapter, our interest lies with the use of nano-biosorbents for the removal of toxic substances form water.

Types of Nano-Biosorbents

Biosorption has emerged as an important technique for the removal of contaminants from drinking and wastewater. The biodegradability and eco-friendliness of biocomposites has attracted the researchers to this field and sudden rise in its use have been observed. (Rahim et al., 2015) Chitin, chitosan & modified

Figure 1. Structure of chitin

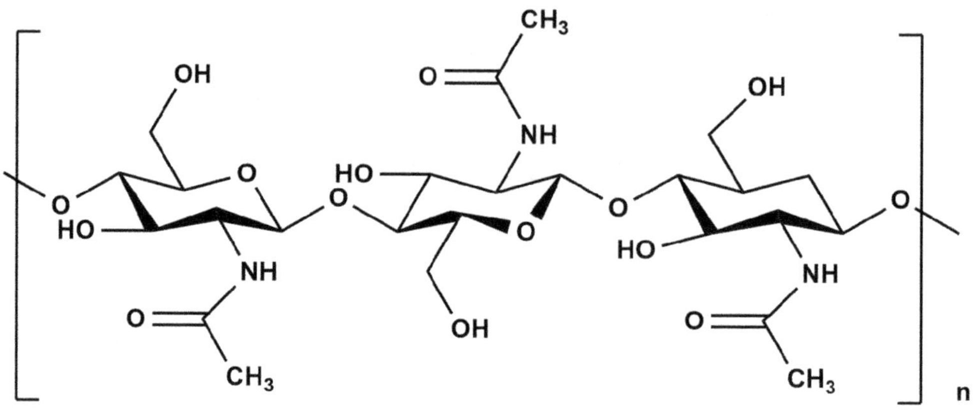

chitosan (lanthanum incorporated chitosan, chitosan coated silica, chelated chitosan biosorbent beads etc.), cellulose, alginate, micro algal have been used in biosorption for the effective removal of heavy metals. (Zhang et al., 2013)

Chitin

Chitin is a linear long-chain polymer that has *N*-Acetyl group in its C-2 position. It is a derivative of glucose. (Rahim et al., 2015) It has several uses in medicinal and food processing industries. Chitosan has been derived from chitin.

Chitosan

Chitosan is a naturally available, nontoxic, biodegradable polymer. It is the second most abundant polymer after cellulose. (Rahim et al., 2015) Chitosan is composed of β (1 ®4)-linked D-glucopyranosamine basic units. Commercial sources of chitosan include the shells of crustaceans such as prawns, crabs, shrimps etc. Chitosan possess chelating amino group at C-2 position. Continuous efforts have been made by various research groups to modify chitosan to increase its surface area, porosity, chemical stability and adsorption sites. It has found application as adsorbents for the removal of heavy metals from water.

Alginate (AL)

Alginate is a linear biodegradable and biocompatible polysaccharide. It is made up of α-L-guluronic (G) and β-D-mannuronic acid (M) repeating units linked via 1-4 glycosidic bond. (Rahim et al., 2015) (Gombotz & Wee, 1998) (Rowley et al., 1999) It is found in brown algae, intracellular matrix and some bacteria. Alginate possesses carboxyl and hydroxyl functional group that can be easily modified. It has potential for pharmaceutical and biomedical applications.

Figure 2. Structure of CHITOSAN

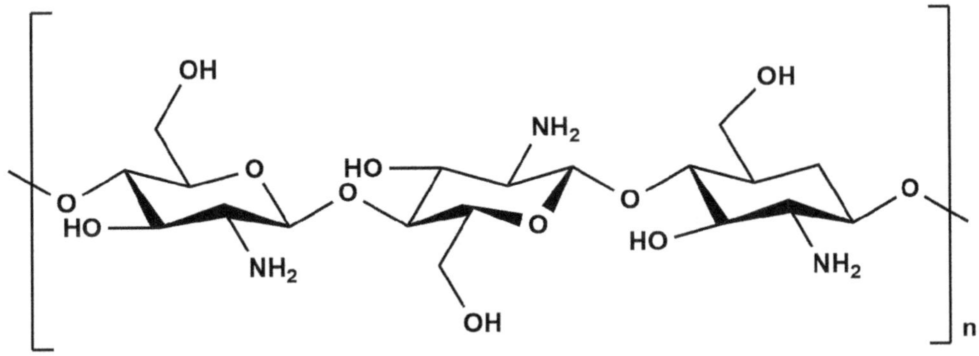

Dextrin (Dex)

It is a low cost polysaccharides formed by hydrolysis of starch or glycogen. (Lotfy, 2009) (Zare, et al., 2018) It is composed of D-glucose repeating units, which are connected by α-(1→4) or α-(1→6) glycosidic bonds. It has potential application in agro and textile industries.

Figure 3. Structure of Alginate

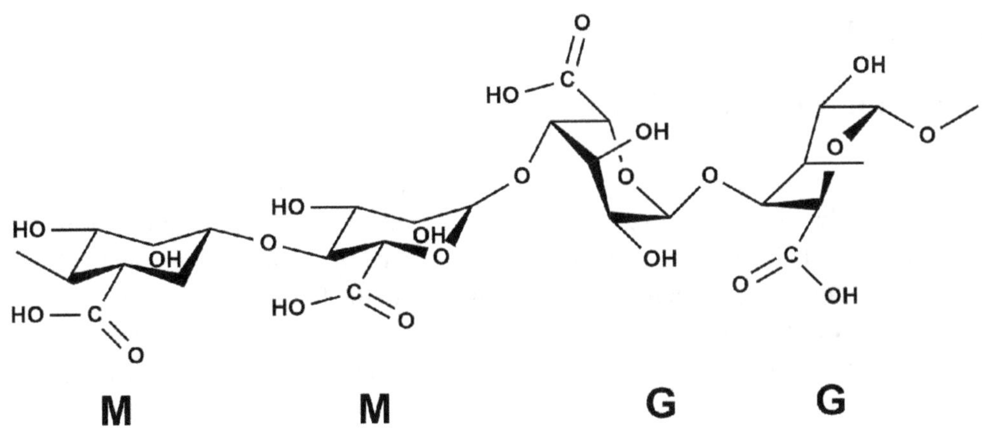

Application of Nano-Biosorbents

1. Removal of organic molecules, dyes (Congo red, methylene blue, Rhodamine B, Reactive Black 5, Acid Blue 29 and Victoria Blue B etc.)
2. Removal of heavy metals (Such as As, Cu, Hg, Cr, Pb, Ni, Zn, Cd, Fe, Ag etc.)

Figure 4. Structure of Dextrin

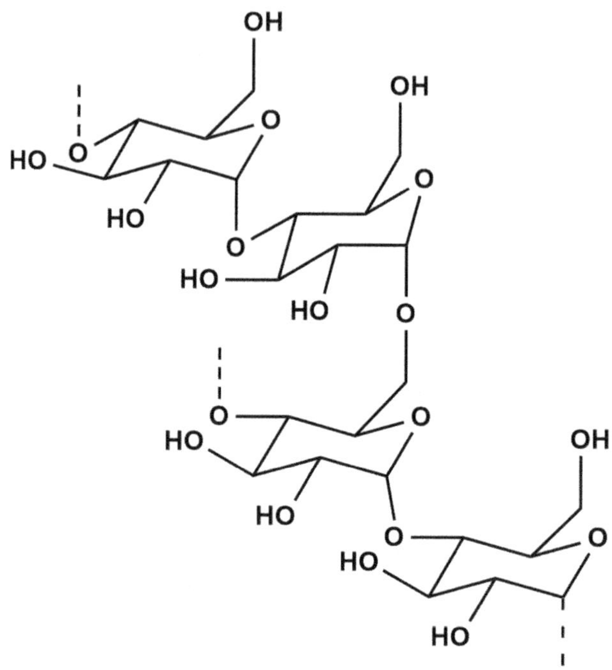

RESULTS AND DISCUSSIONS

Nano-biosorbent: Removal of Organic Molecules from Wastewater

Zare et al developed a novel nano-biosorbent based on dextrin-g-poly m-phenylenediamine (DgPmPDA) by using the solution blending technique and utilized it as adsorbent for the removal of methylene blue (MB) from water. (Zare et al., 2018) Maximum adsorption capacity of methylene blue (MB) from water was found to be 76.33 mg/g. Experimental results fits well with the Langmuir adsorption isotherm model and the kinetics of MB followed the pseudo-second order models. Zhao et al reported the synthesis of sodium alginate beads via post-crosslinking method. Sodium alginate beads was utilized for the removal of cationic dye i.e. methylene blue (MB) from wastewater. (Zhao et al., 2015) Maximum adsorption capacity of methylene blue (MB) from water was found to be 572.7 mg/g. Thus, the adsorption process fits well with the Freundlich adsorption isotherm model and pseudo-second order kinetic model.

Arotiba et al designed a novel Titania grafted sodium alginate cross linked polyacrylic acid nanocomposite (SA-cl-poly (AA)-TiO_2) via copolymerization of acrylic acid (AA) onto sodium alginate (SA). (Arotiba et al., 2016) The nano-biosorbents was utilized for the removal of cationic dyes i.e. methylene blue (MB) from wastewater. Maximum adsorption capacity of methylene blue (MB) from water was found to be 2257.36 mg/g. Thus, the experimental data fits well with the Langmuir adsorption isotherm and the kinetics data fits well with the pseudo-second order kinetic model. Ouyang et al synthesized carboxy methylcellulose (CMC)/k-carrageenan (kC)/activated montmorillonite (AMMT) and utilized the nano biosorbents for the removal of cationic dyes (MB) from aqueous solution. (Ouyang et al., 2018) Adsorption data fits well with the Langmuir isotherm model and kinetic data fits with the pseudo-second order kinetic model. Maximum adsorption capacity of methylene blue (MB) was found to be 12.25 mg/g.

Figure 5. Structure of reactive blue B-RN

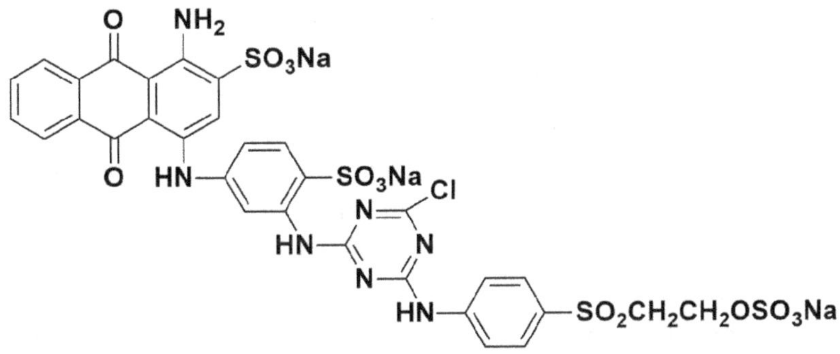

Figure 6. Structure of reactive yellow B-4RFN

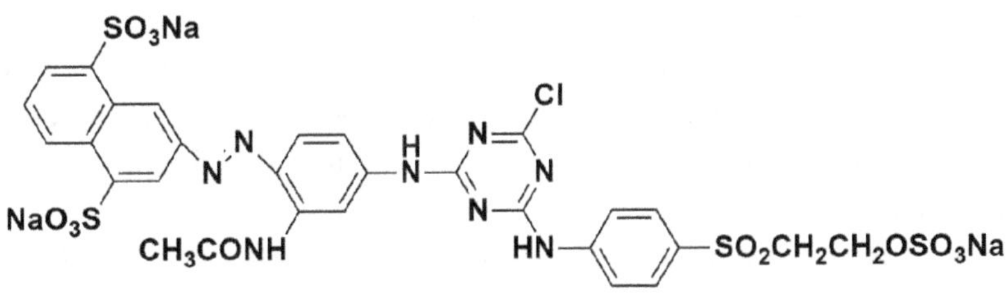

Chan and his coworkers reported the use of amyloid fibrils obtained from hen egg white lysozyme and used it as nano biosorbent for the removal of both cationic and anionic dyes (Victoria Blue B, Reactive Black 5 and Acid Blue 29) form wastewater. (Chan et al., 2015) Maximum adsorption capacity of different dyes using amyloid fibrils was found to be 159.2, 103.1, 100.2 mg/g for Reactive Black 5, Acid Blue 29 and Victoria Blue B respectively. The experimental data fits more precisely with the Langmuir adsorption isotherm than the Freundlich adsorption isotherm. The group also tried to study the adsorption kinetics using both pseudo-first-order and pseudo second-order kinetics model, but the rapid adsorption of all the dyes by lysozyme nanofibers constrains them to collect enough data during initial phase of dye adsorption.

Xie et al synthesized nano-cellulose hybrid containing polyhedral oligomeric silsesquioxane (POSS) and utilized it for the removal of Reactive Yellow B-4RFN and Reactive Blue B-RN from aqueous solutions. (Xie et al., 2011) Maximum adsorption capacity of 14.40 mg/g, 16.61 mg/g was obtained for reactive blue B-RN and Reactive Yellow B-4RFN at 293 K. Adsorption capacity of nano-cellulose hybrid decreases with increase in temperature to 323 K. Adsorption data fits well with the Langmuir adsorption isotherm and the adsorption kinetics data fits with the pseudo second-order kinetics model.

Sohni and his coworkers modified the surface of chitosan with magnetite nanoparticles and graphene oxide to produce a new nano-biosorbent i.e. Fe_3O_4 supported chitosan-graphene oxide composite (Fe_3O_4©-GO). (Sohni et al., 2016) This biosorbent was used for the removal of both cationic and anionic dyes [Methyl violet (MV) and Alizarin yellow R (AY)], from aqueous solution. Adsorption kinetics modelling data fits well with pseudo second-order kinetics model. Marcovich et al reported the synthesis of Chitosan/iron oxide nanocomposite films via two different pathways: one of method involves the in situ generation of iron oxide nanoparticles and the other method involves the dispersion of previously synthesized iron oxide nanoparticles over the film via sonication. (Marcovich et al., 2019) They have utilized it as nano biosorbent for the removal of anionic Congo red from wastewater. Adsorption capacity of the nanocomposite synthesized by both the pathways was compared and nanocomposite synthesized by second pathway exhibited more adsorption capacity up to 700 mg/g. The experimental data of nanocomposites synthesized by first method (in situ) follows the Langmuir adsorption isotherm whereas most of the experimental data of nanocomposites synthesized by second method follows the Freundlich adsorption isotherm.

Liu et al fabricated cellulose nanocrystal (CNC) via its extraction from Carexmeyeriana Kunth (CMK) and used it as nano-biosorbent for the removal of methylene blue (MB) from aqueous solutions.(Liu et al., 2017) The maximum adsorption capacity of the nano-biosorbent was found to be 217.4 mg/g. The experimental data fits with the Langmuir adsorption isotherm and the adsorption kinetics follows the pseudo-second order kinetic model. Jin and his coworkers synthesized micro gel based on nanocellulose and amphoteric polyvinylamine (PVAm) by following a two-step method. (Jin et al., 2015)The nanocomposite micro gel was used as adsorbent for the removal of anionic dye (Congo red 4BS, acid red GR and reactive light yellow K-4G) under acidic conditions. The maximum adsorption capacities of different anionic dyes are 869.1 mg/g, 1469.7 mg/g and 1250.9 mg/g respectively. The experimental data for adsorption of three different dyes fits the Sips model and the kinetics data follows the pseudo second order kinetics model.

Zhou et al initially synthesized quaternized cellulose Nano fibrils (Q-NFC) by treating wood pulp with glycidyltrimethylammonium chloride. (Zhou et al., 2013) Thereafter they have prepared cellulose nanopapers from Q-NFC. These nanopapers possess high tensile strength, Young's modulus and high porosity. These Q-NFC Nano fibrils were utilized for the removal of anionic dyes (Congo red and acid green 25) from aqueous waste streams. The adsorption capacity of Q-NFC Nano fibrils is greatly dependent on trimethylammonium chloride, i.e., it increases with increase in trimethylammonium chloride content. The maximum adsorption capacity of anionic dyes i.e. Congo red and acid green 25 are 664 mg/g and 683 mg/g from aqueous solution. Wu et al reported the synthesis of hydrolyzed polyacrylamide/cellulose nanocrystal (HPAM/CNC) nanocomposite hydrogels and utilized it as adsorbent for the removal of methylene blue (MB) from wastewater. (Wu et al., 2014) The maximum adsorption capacity of methylene blue (MB) dye from wastewater was found to be 326 mg/g. Adsorption data fits well with the Langmuir adsorption isotherm and kinetics data follows Pseudo-second-order and Elovich models.

Mathew et al utilized low cost cellulose obtained from various industries and fabricated cross-linked composite membrane nanoporous membranes with cellulose nanocrystals (CNCs) as functional entity in chitosan. (Mathew et al., 2014) The membrane was used for the removal of Victoria Blue 2B, Methyl Violet 2B and Rhodamine 6G dye from water. The mechanism involved in the removal of dyes from wastewater comprises the electrostatic interaction between negatively charged CNCs and the positively charged dyes. In another report, Mathew et al reported the synthesis of cellulose acetate (CA) membranes coated with chitin nanocrystals (ChNC) and used it for purification of water. (Mathew et al., 2016)

Mallakpour and Tabesh synthesized hydrogel nanocomposites (HNCs) based on Tragacanth gum (TG), $CaCO_3$ nanoparticles (CC NPs) and EDTA to produce TG/CC@EDTA ef HNC. (Mallakpour & Tabesh, 2019) They have used the hydrogel nanocomposites for the removal of methylene blue dye from aqueous solution. The maximum adsorption capacity of MB dye from wastewater using HNCs was found to be 468.62 mg/g. The experimental data fitted well with the Langmuir adsorption isotherm and kinetics data follows the pseudo-first-order kinetic model. Tang et al developed cellulose Nano fibril (CNF) based aerogels and used as adsorbent for the removal of methyl orange (MO) from wastewater. (Tang et al., 2019) The sorbent was highly efficient for the adsorption of MO with maximum removal capacity of 265.9 mg/g. The adsorption data matches with the Langmuir adsorption isotherm and the kinetics data follows the pseudo-second-order kinetic model. Further intra-particle diffusion model was used for describing the mechanism of adsorption.

Tan and his coworkers isolated nanocrystalline cellulose (NCC) from oil palm biomass waste and utilized it for the removal of methylene blue from aqueous solution. (Tan et al., 2019) This adsorbent shows a maximum adsorption of 144.93 mg/g of MB dye from textile effluents. Thus, the adsorption data and kinetics data fits well with Langmuir isotherm and pseudo-second-order kinetic model respectively. Ma et al reported a synthetic pathway for the synthesis of β-cyclodextrin modified magnetic graphene oxide nanohybrids (β-CD/MGO). (Ma et al., 2018) The synthesis was carried via one-step inverse phase co-precipitation method using β-cyclodextrin (β-CD), ferrous sulfate hepta hydrate ($FeSO_4 \cdot 7H_2O$) ferric chloride hexahydrate ($FeCl_3 \cdot 6H_2O$) as iron sources, NaOH as precipitant and graphene oxide as carrier. β-CD/MGO nanohybrids was used as adsorbent for elimination of methylene blue from wastewater and showed an adsorption maximum of 93.97 mg/g. The adsorption process followed Langmuir isotherm and pseudo-second-order kinetic model.

Pandey and his coworkers synthesized MNP-biomass-alginate (MBCaAb) beads using iron oxide nanoparticles (MNPs), Agrobacterium fabrum strain SLAJ731 present in biomass and calcium alginate. (Pandey et al., 2018) The synthesized composite was used as nano-biosorbent for the elimination of methylene blue (MB) dye with maximum adsorption capacity of 91 mg/g. Thermodynamic studies of methylene blue adsorption shows that process is exothermic and spontaneous in nature. Further adsorption data fits with Freundlich adsorption isotherm and kinetics data follows pseudo-second-order kinetic model. Shoueir et al deposited Sol-gel titanium dioxide (TiO_2) nanoparticles over the surface of calcined eggshell (CES) via insitu precipitation method for the synthesis of hybrid nano-biosorbent (TiO_2-CES). (Shoueir et al., 2018) This hybrid nano-biosorbent was utilized as adsorbent for the removal of acid red nylon 57 (AN57) dye with maximum adsorption capacity of 222.20 mg/g. Thus, the experimental data best agrees with the Langmuir adsorption isotherm and the kinetics data is best described by pseudo-second-order kinetic model.

Zhou and his coworkers synthesized magnetic peach gum bead (MPGB) via cross linking of natural peach gum polysaccharide and magnetic nanoparticles. (Zhou et al., 2018) The composite was used as magnetic biosorbent for the removal of cationic dyes from wastewater with maximum adsorption capacity of 231.5 mg/g. The adsorption experimental data best agrees with the Langmuir adsorption isotherm and the kinetics data is best described by pseudo-second-order kinetic model. Further thermodynamic studies of adsorption shows the adsorption process is exothermic and spontaneous in nature. Sheshmani and Mashhadi developed magnetite reduced graphene oxide (rGO)/chitosan nanocomposite (M/Rgo/Ch). (Sheshmani & Mashhadi, 2017) They have used the nanocomposite as biosorbent for the removal of Remazol Black B (RBB) and Acid red 22 (AR22) from wastewater. The experimental data best agrees with both Langmuir and Freundlich equation and the kinetics data is best described by Elovich and Blanchard models. Further the magnetic property of the nanocomposites enable the easy removal of RBB and AR22 dye pollutant from water using an external magnet and reused in multiple cycles.

Tural et al reported the synthesis of magnetic biosorbent via immobilization of *Bacillus Subtilis* (*B. subtilis*) over magnetic nanoparticles. (Tural et al., 2017) The magnetic biosorbent was used for elimination of methylene blue dye with maximum adsorption capacity of 59±0.6 mg/g. The experimental data best agrees with Freundlich model and kinetic data follows the pseudo-second-order kinetic model. As this particular biosorbent is magnetic in nature, it could be reused for at least five successive runs with high biosorption capacity. Zhou et al synthesized lignin/silica nanocomposites using cellulosic ethanol residue (CER) for the generation of lignin and sulfuric acid as precipitating reagent. (Zhou & Guo, 2017) The lignin/silica nanocomposites possess high adsorption and dispersive properties. It was used as nano-biosorbent for the adsorption of methylene blue (MB) dye from wastewater.

In the year 2018, Ali fabricated Cellulose/SiO_2 nanocomposite (CSNC), using a readily available source of silica i.e. banana peel waste. (Ali, 2018) The composite was used as nano-biosorbent for the elimination of methylene blue from dye sample and real samples with maximum adsorption capacity of 78.75 mg/g and 58.81 mg/g. The experimental data best agrees with Langmuir adsorption isotherm and the kinetics data is best described by pseudo-second order kinetics. In addition, the thermodynamic studies reveal the adsorption process to be exothermic and spontaneous in nature. Sivasubramanian et al synthesized zero-valent iron-Sargassumswartzii (nZVI-SS) nano biocomposites using $FeCl_3$ and macro algae (*Sargassumswartzii*). (Sivasubramanian et al., 2017) The biocomposites was used for the elimination of crystal violet (CV) from aqueous solution with maximum adsorption capacity of 200 mg/g at pH 8. The adsorption data best fits with Langmuir adsorption isotherm and the kinetics data is best described by pseudo-second order kinetics. In addition to this thermodynamic study reveals adsorption process is endothermic and spontaneous in nature.

Satish Kumar and his coworkers reported the synthesis of silver nanocomposite beads (SNC) using *Mukiamaderaspatna* plant extract and utilized it as biosorbent for the removal of methylene blue from wastewater. Maximum adsorption capacity of SNC was found to be 213.7 mg/g. The adsorption experimental data fits well with Freundlich and Sips adsorption isotherm models whereas the kinetics data follows the pseudo-first order kinetics model. (Sathish Kumar et al., 2016).

Thus, several nano-biosorbent have been used for the removal of various cationic and anionic dyes from water. In addition to these, few other sorbents have been used as adsorbents for the purification of water that are listed in **table 1.**

The presence of organic molecules in aqueous solution is another environmental problem. Organic molecules can be categorized into organic dye, solvent, drug, etc. It is also important to remove the organic molecules from aqueous solution. Other than organic dye, some other organic molecules removal by nano-biosorbents is described here.

Tural et al reported the synthesis of chitosan based magnetic nanoparticles through co-precipitation method. The as-synthesized magnetic chitosan was used as adsorbents for removal of phenolic compounds such as phenol, 2-chlorophenol and 4-chlorophenol from aqueous solution. The adsorption process is best described by Freundlich isotherm model and the kinetic study of this adsorbent obeys the pseudo-second order reaction model. The maximum adsorption capacities were obtained at pH 6-7. (Tural et al., 2016)

Li et al synthesized a novel magnetic nano-scale biosorbent, i.e. Fe_3O_4/MFX via cross-linking of extracellular polymeric substances (MFX) from *Klebsiella sp. J1* and Fe_3O_4. (Li et al., 2017) This synthesized nano biosorbent was used for adsorption of tetracycline. The experimental data fits with Freundlich and Langmuir isotherm. The kinetic study follows pseudo second order model.

Table 1. Adsorption capacity of various nano-biosorbent for removal of dye from wastewater

Adsorbent	Dye	qe (mg/g)		Reference
Quaternized cellulose Nano fibrils nanocellulose (Q-NFC)	Acid green 25 Congo red	683 664		Pei et al., 2013
Partially hydrolyzed polyacrylamide/ cellulose nanocrystal hydrogel	Methylene blue	326	Langmuir equation; pseudo-second-order and Elovich models.	Zhou et al., 2014
Nanoporous membranes with cellulose nanocrystals as functional entity in chitosan	Victoria Blue 2B, Methyl Violet 2B Rhodamine 6G			Karim et al.,2014
ZM/Fe-Cu nanoparticles Fe-Cu stalk composite Fe-Cu bagasse composite	Malachite Green	119.46 60.95	Langmuir-Freundlich isotherm model; pseudo-second order kinetic	Lara-Vásquez et al., 2016
Nano-biosorbent of rose watering waste	Reactive Yellow			Rabbani et al., 2016
Magnetic polydopamine-chitosan nanoparticles PDA/CS/Fe_3O_4 Bio-Nano sorbents	Methylene Blue Malachite Green	201 61	Langmuir isotherm model; pseudo-second order kinetic	Wang et al.,2016
Graphene Nano sheets (GNS) rGO-N_3-Glc-Fe_3O_4 rGO-diazide-Glc-Fe_3O_4	Methylene Blue	111 105		Namvari et al., 2016
Carbon–silica composite	Methylene Blue	340.1	Langmuir isotherm model; pseudo-second order kinetic	Lv et al., 2016
Moringaoleifera-encapsulated alginate beads (MEA)	Anionic blue-FFS (AB-FFS)	76.92	Langmuir and Freundlich isotherm models; pseudo-first order and intra-particle diffusion model.	Kannan et al.,2016
Spirulina platensis micro and nanoparticles	FD&C red no. 40 and Acid blue 9	295 and 1450	Pseudo-first order model-biosorption on micro particles and Elovich model for biosorption onto Nanoparticles	Dotto et al., 2012
Surfactant-modified sawdust	Methylene Blue (MB)	129.68	Langmuir and Freundlich isotherm models, Thomas model	Ansari et al., 2012
Magnetic-cyclodextrin–chitosan nanoparticles (CDC)	Methylene Blue(MB)	2780	Langmuir isotherm model; pseudo-second-order kinetic model; spontaneous and exothermic	Fan et al., 2012
Amino-functionalized nanocrystalline cellulose	Acid red GR Congo red 4BS Reactive light yellow K-4G	555.6	Langmuir model; pseudo second order kinetics	Jin et al., 2015
Carboxylate-functionalized cellulose nanocrystals	Crystal Violet, Methylene Blue, Malachite Green and Basic fuchsin	243.9	Langmuir isotherm model; pseudo-second-order kinetic model	Qiao et al., 2015
Xanthan gum/silica hybrid nanocomposite	Congo red	209.2	Langmuir isotherm model; pseudo-second-order kinetic model	Ghorai et al., 2013
Chitosan/Montmorillonite nanocomposite	Congo red	54.5	Langmuir isotherm model; pseudo-second-order kinetic model	Wang et al., 2007
Nano-mycocomposite	Congo red	127.2	Langmuir isotherm model; pseudo-second-order kinetic model	Azina et al., 2017

Figure 7. Structure of phenolic compounds

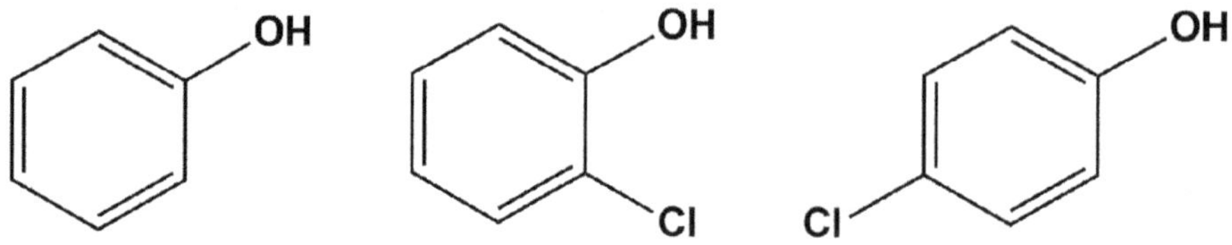

Ge et al synthesized Fe_3O_4@CTS nanoparticles through two- steps wet chemical process. (Ge et al., 2017) The first step involves the preparation of Fe_3O_4 nanoparticles via co-precipitation method and in the second step; the surface of chitosan was laminated with magnetic nanoparticles via suspension crosslinking method. The as-synthesized Fe_3O_4@CTS nanoparticles was coated with inactivated CICC1769 cells to produce a new nano-biosorbent and it used for adsorption of patulin in orange juice. The adsorption study of that new nano-biosorbent was found to be reducing 90% of patulin in the orange juice.

Figure 8. Structure of tetracycline

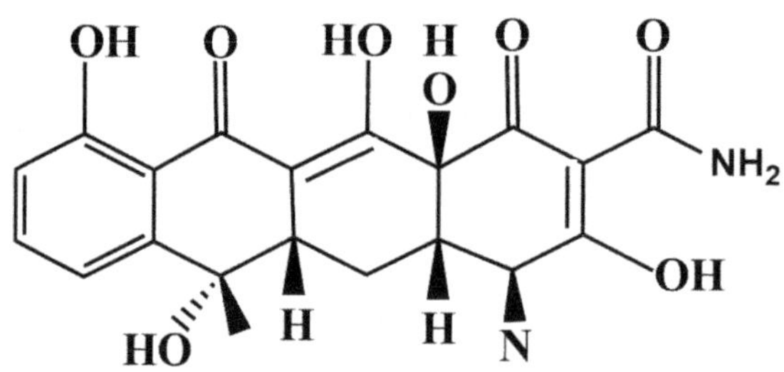

Figure 9. Structure of patulin

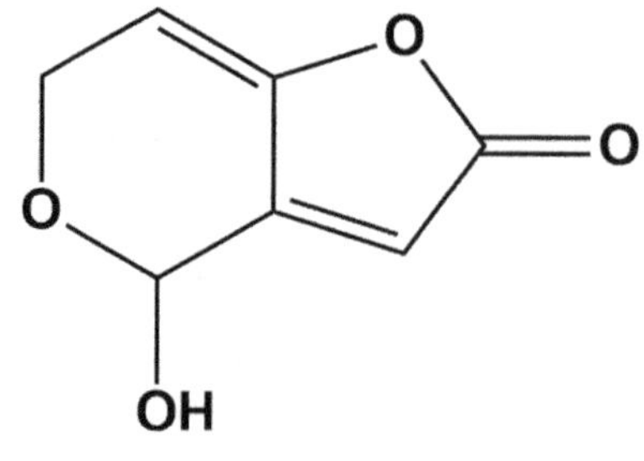

Nano-Biosorbents: Removal of Heavy Metals from Wastewater

Heavy metals, due to non- biodegradable and stability properties, find ubiquitous pollutant in wastewater system. They lean to gather in animal and human body through food chain system, which causes severe problems in nervous system, kidney and liver infection, skin infection, deafness, vision loss, dementia etc. It is necessary to remove heavy metals and prevent water free from pollutant. We have described some reported nano-biosorbents that have been used as adsorbents for removal of heavy metals from water. Alkaff and his team modified the surface of Nano chitosan with *Eichhornia crassipes* to produce a new nano-biosorbent. (Alkaf et al, 2018) This nano biosorbent was used for removal of lead (II), cadmium (II) and copper (II) ions from aqueous solution. They have better adsorption capacity at neutral p^H. Kazemi and his research team developed chitosan based sodium alginate multilayer membrane by layer by layer method and denoted as PES/$(CHI\text{-}ALG)_{3.5}$/Fe^0@WO_3. The application of this novel photocatalytic membrane includes removal of hexavalent chromium ion in batch experiment as well as filtration system. (Kazemi et al, 2018) Under batch experiment, the highest removal percentage of chromium (VI) was 84.9 at 10mg/l feed concentration. Daneshvar et al have synthesized dead, alkaline activated biomass of Aspergillusniger based magnetic nano biosorbent, and it used for removal of hexavalent chromium (VI) ion. The maximum chromium removal was 92% at pH 5.8, 23.4 mg/L Cr concentration, 3.72 g/L adsorbent dose, 300-rpm agitation rate, and 11 min time duration. The adsorption capacity was found to be 5.81 mg/g. The adsorption results are well accordance with Temkin model and follow physical adsorption. (Daneshvar et al, 2018)

Nath and coworkers have synthesized bio nanocomposite (ZnO:CeO_2:nanocellulose:PANI) using a modified polymerization process. (Nath et al., 2016) That Synthesized bio nanocomposite was used as adsorbent for arsenic removal. The adsorption process of arsenic removal by nano biosorbent confirms the Freundlich and Dubinin-Radushkevich isotherms and follows pseudo-second order kinetics. Shao and his research team have synthesized sodium alginate-based cross-linking beads, denoted as SA/PAMPS10 (Shao et al., 2018) and used it as adsorbent for removal of copper (II), lead (II), nickel (II) and cadmium (II). The maximum adsorption capacities were found to be 2042, 254, 50 and 843 mg/g for lead (II), copper (II), nickel (II) and cadmium (II), respectively. Esmaeili & Khoshnevisan reported the synthesis of Cu and Co doped biomass alginate coated chitosan nanoparticles (Alg-CS-NPs) by adding aqueous solution of $CuCl_2.H_2O$ in Alg-CS-NPs nanoparticles. This nanoparticle was utilized as adsorbents for removal of nickel (II). The maximum percentage of removal of nickel was found to be 94.48 at pH 3; 70 mg/L nickel concentration; 0.3 g biomass dose and 30 min contact time. The experimental data follows pseudo second-order model and fits well with Freundlich isotherm models. (Esmaeili et al., 2016) Plohla et al reported the synthesis of a novel adsorbent as covalent linkage between carboxymethyl chitosan and silica coated core-shell magnetic nanoparticles. (Plohla et al., 2019)

These nanoparticles were utilized as adsorbent for removal of copper (II). The maximum adsorption capacity was found to be 350 mg/g. Jose et al have synthesized yeast/Fe_3O_4 nano biocomposites by adding yeast biomass into magnetite nanoparticles. (Jose et al., 2019) The advantage of that composite is use for removal of copper (II) from aqueous solutions. The experimental value of copper adsorbed at equilibrium was 8.33 mg/g. The adsorption process is best described by Dubinin-Radushkevich (D-R) isotherm model. The efficiency of biosorbent was found to be 77.27% under first batch experiment. Mirzajani et al have reported the synthesis of chitosan based SnO_2 nanoparticle and utilized it as green biosorbent for removal of iron (II) ions. The better adsorption capacity for the Fe (II) was 17.24 mg/g. The adsorption studies describes well by Freundlich isotherm model. (Mirzajani et al., 2017)

Fan et al have developed a new nano-biosorbent, i.e. magnetic chitosan/graphene oxide composite (MCGO) by grafting of magnetic chitosan with activated grapheme oxide. (Fan et al., 2013) This MCGO nanoparticle was used as adsorbents for removal of lead (II) from aqueous solutions. The adsorption data of lead (II) follows the Langmuir model. The maximum adsorption capacity for lead(II) was determined and its value found to be 76.94 mg/g. Li et al modified the cellulose nanocrystal with cysteamine to develop a new nano-biosorbent (Li et al., 2019) and used it as absorbent for removal of mercury (II) from water. The adsorption data follows Langmuir monolayer model and the maximum adsorption capacity for mercury (II) was estimated to be 849 mg/g. The kinetics of mercury (II) adsorption follows pseudo-second order kinetics. The removal process by this nano-biosorbent was very fast; i.e. 99% of mercury ions removal within 10 min. Hong et al modified the nanocellulose with polyethyleneimine to investigate the different novel Nano adsorbents. (Hong et al., 2019) The different cellulose sources are tunicate (T-CNF) and hardwood pulp (P-CNF) and nanocellulose type P-CNF (cellulose Nano fibril) and P-CNC (cellulose nanocrystal). These nano-bio adsorbents were used for removal of platinum ions from water streams. The polyethyleneimine modified T-CNF had exhibited highest adsorption capacity, i.e. 600 mg/g and the experimental adsorption data fits with Langmuir isotherm model. Li and his coworkers synthesized a new biosorbent by mixing thiol functionalized graphene oxide with chitosan, denoted as CS/GO-SH. (Li et al., 2015) That composite was used for adsorption of copper (II), lead (II) and cadmium (II). The adsorption kinetics obeys pseudo second order equation and the experimental data fits well with Freundlich isotherm. The adsorption capacities of copper (II) and lead (II) are similar while cadmium(II) is lowest at pH 5.Wang et al reported the synthesis of ferrite nanoparticle coated *agrocybecylindracea* (Wang et al., 2018) and used it for removal of chromium (VI). Maximum adsorption capacity was estimated to be 73.88%. The adsorption data is well accordance with Langmuir isotherm. Lian et al modified carbonized pomelo peel biosorbent (MCPP) with nanoscale zero-valent iron (NZVI) and cetyl- trimethyl ammonium bromide (CTAB) to produce new nano bio adsorbent (Lian et al., 2017*)* and used it for removal of molybdenum (VI) from aqueous solutions. The adsorption data follows pseudo-second order kinetics. The data fits well with Langmuir model. Zare and his coworkers synthesized a novel nano-biosorbent based on dextrin-g-poly m-phenylenediamine (DgPmPDA) and graphene oxide (GO) via solution blending technique and employed it as effective adsorbent for removal of lead (II). (Zare et al., 2018) Maximum adsorption capacity was found to be 80 mg/g. The adsorption data best fits with Langmuir isotherm. The kinetic studies show both pseudo-first order and pseudo-second order model. Li and his research team fabricated β-cyclodextrin (β-CD) modified magnetic graphene oxide (β-CD / MGO) nano hybrid via one step inverse co-precipitation method and used it for removal of chromium (VI) and lead (II). (Li et al., 2018) Maximum adsorption capacities for removal of chromium (VI) and lead (II) were estimated to be 279.21 and 51.29 mg/g, respectively. Langmuir isotherm describes well for removal of chromium (VI) and lead (II) on β-CD /MGO. The kinetic study shows that the adsorption process follows pseudo-first order kinetics. Sreeram and his coworkers prepared Nano chitosan emulsion by mixing tripolyphosphate in to a chitosan solution and obtained it at pH 5. (Sreeram et al., 2017) The synthesized Nano chitosan was utilized for adsorption of arsenate. The experimental data is well accordance with sips isotherm model. The kinetic study for removal of arsenate into Nano chitosan follows pseudo-second order model. The optimized mass and contact time for removal of arsenate was found to be around 1-3 kg and 100 min, respectively. Narwade and his research team reported the synthesis of hydroxyapatite-cellulose Nano fibrils (CNF/TCNF-HApHybrids) via in situ method involves addition of CNF or TCNF to the solution of HAp precursors. (Narwade et al., 2018) The synthesized adsorbent was utilized for removal of cobalt (II) from aqueous solutions. The experimental adsorption data fits with Freundlich isotherm. Maximum adsorption capacity was found to be 25 mg/g. The adsorption process follows pseudo-first order and Elovich kinetic model. Gokila et al synthesized chitosan-Alginate nanocomposites by mixing chitosan nanoparticles with alginate nanoparticles (Gokila et al, 2017) and used it as nano biosorbent for removal of chromium (VI).The adsorption data fits well with Freundlich isotherm model and pseudo-second order kinetics. The sorption capacity of this nanocomposite was very high. Zhou et al synthesized a novel magnetic biosorbent (MCIT) via functional modification and used

Table 2. Different useful nano-biosorbents for removal of heavy metals from wastewater and their adsorption parameters:

Adsorbents	Metal	q_m (mg/g)	Adsorption parameters	Reference
Chitosan/PVA nano-fiber membrane	Ni(II)	8.13	Langmuir isotherm model; pseudo-first order kinetic model	Esmaeili et al., 2018
Chitosan/ Fe_2O_3 nanocomposite with shrimp shell	Ni(II) Pb(II) Cd(II)	58 12 202		Keshvardoostchoka et al., 2017
Nano chitosan	As(III) AS(V)	6.1 13		Kwok et al., 2018
Walnut skin nanoparticles	Pb(II)	17.39	Langmuir isotherm model	Karami et al., 2017
Sawdust nano biosorbent (SNBS)	Pb(II)	10	Langmuir isotherm model	Karami et al., 2017
Amino-thiourea functionalized magnetic chitosan (AT-MCS)	Hg(II)	246		Hou et al., 2018
Hydrogel nanocomposite (HNC)	Pb(II)	192	Dubinin-Radushkevich isotherm model; pseudo-second order kinetic model	Mallakpour et al., 2018
TiO_2-cellulose composite	Pb(II)			Zhang et al., 2018
Coriolus versicolor Immobilized γ-Fe_2O_3 nanoparticles	Ag(I)	31.7		Serkan Yalcin et al., 2019
Oak shell modified magnetic nanoparticles (OSMMN)	Ni(II)	454.54	Langmuir isotherm model	Mousavi et al., 2018
orange peel residue carbon nanoparticles	Cu(II)		Temkin model	Safari et al., 2019
Gelatin-based nanocomposite	Cu(II) Cd(II)	192.4 172.3	Langmuir model; pseudo-second order kinetic model; endothermic	Nematidil et al., 2018
Pleurotus eryngii loaded Fe_2O_3 magnetic nanoparticles	Co(II) Hg(II)	25.4 30.3		Ozdemir et al., 2018
Nanochitosan/ sodium alginate/microcryst. cellulose bead (NCS/SA/MC)	Pb(II)	188.96	Freundlich adsorption isotherm model: pseudo-second order kinetic model	Vijayalakshmi et al., 2017
Nano-TiO_2/feldspar embedded chitosan	As		Sips isotherm; pseudo-second order kinetic model	Yazdani et ai., 2017
S. cerevisiae loaded nanofibrous mats	Pb(II)	238		Xin et al., 2017
magnetic chitosan-4-((pyridin-2-ylimino)methyl) benzaldehyde Schiff's Base (m-CSPIB)	Pb(II)	104.16 (323 K)	Langmuir isotherm model; pseudo-second order kinetic model ; spontaneous; endothermic	Gutha et al., 2016
nSe^0 - Phanerochaete chrysosporium pellets	Zn	3.07	Sips model; pseudo-second order kinetic model	Espinosa-Ortiz et al., 2016
chitosan-g-poly(butyl acrylate)/silica gel nanocomposite (Cs-g-PBA/SG)	Cr(VI)	55.71	Langmuir isotherm model; pseudo-second-order kinetic model	Nithya et al., 2016
nanostructured cedar leaf ash	Pb(II) Zn(II)	7.23 4.79	Sips model; pseudo-second order kinetic model	Hafshejani et al., 2015
Pleurotus ostreatus Nanoparticles (PONP)	Mn(II)	130.625 (298 K)	Langmuir isotherm model; pseudo-second order kinetic model	Ma et al., 2013

it for adsorption of mercury (II). (Zhou et al., 2017) The adsorption process for removal of mercury (II) on MCIT is well accordance with complexation isotherm model, which is similar to Langmuir model. Maximum adsorption capacities calculated from Langmuir model were found to be 290.8, 314.6, 351.1 and 385.3 mg/g at 303 K, 308 K, 313 K and 318 K, respectively. Zhang et al synthesized Bamboo cellulose nanofibers-graft-poly (acrylic acid) (BCN-g-PAA) and bamboo cellulose nanofibers-graft-poly (acrylic acid)/sodium humate (BCN-g-PAA/SH) and utilized for removal of copper (II). (Zhang et al., 2014) The adsorption experiments fit well with Freundlich isotherm model and follows pseudo-second order kinetics. The adsorption capacity of BCN-g-PAA/SH is higher than BCN-g-PAA. Hong et al modified the amyloid fibrils of hen lysozyme with ethylene diamine and used it as nano biosorbent for removal of chromium (VI) in water. (Hong et al., 2016) Rathinam and his coworker synthesized a novel nano biosorbent by modifying the surface of chitosan with polyaniline and used it for removal of cadmium (II) and lead (II) from aqueous solutions. The experimental value fits precisely with Freundlich isotherm model and follows pseudo-second order kinetic. The removal of cadmium (II) and lead (II) by this adsorbent is directly proportional to pH. (Karthik et al., 2014)

A few different nano-biosorbents other than these are given in **table 2,** utilized as adsorbents for elimination of heavy metals from wastewater system.

CONCLUSION

Nowadays, decontamination of water is one of the major tasks for materials science. Life cannot exist without water. It is imperative to prevent water free from pollutants. Adsorption is one of the most effective process for decomposition and reduction of pollutants from wastewater. In this chapter, we have described the use of different nano-biosorbents for removal of hazardous pollutants from wastewater. Adsorption using nano-biosorbents based on chitin, chitosan & modified chitosan, cellulose, alginate, micro algal has emerged better alternative for the removal of contaminants from drinking and wastewater treatment. Wastewater treatment includes the removal of organic molecules, dyes (such as Congo red, methylene blue, Rhodamine B, Reactive Black 5, Acid Blue 29 etc.) heavy metals (Such as As, Cu, Hg, Cr, Pb, Ni, Zn, etc.) The main advantages of these nano-biosorbents include its cost effectiveness, biodegradability and eco-friendliness that has attracted the researchers to this field.

REFERENCES

Ali, S. M. (2018). Fabrication of a nanocomposite from an agricultural waste and its application as a biosorbent for organic pollutants. *International Journal of Environmental Science and Technology*, *15*(6), 1169–1178. doi:10.100713762-017-1477-x

Alkaff, A. H., Hendri, H., Farozy, I. H., Annisa, M., & Aritonang, R. P. (2018). Article. *Materials Science and Engineering*, *299*, 012003.

Ansari, R., Seyghali, B., Mohammad-khah, A., & Zanjanchi, M. A. (2012). Application of Nano Surfactant Modified Biosorbent as an Efficient Adsorbent for Dye Removal. *Separation Science and Technology*, *47*(12), 1802–1812. doi:10.1080/01496395.2012.658485

Azina, E., Moghimia, H., & Taheri, R. A. (2017). Article. *Desalination and Water Treatment*, *94*, 222–230. doi:10.5004/dwt.2017.21586

Castro, E., Avellaneda, A., & Marco, P. (2014). Combination of advanced oxidation processes and biological treatment for the removal of benzidine-derived dyes. *Environmental Progress & Sustainable Energy*, *33*(3), 873–885. doi:10.1002/ep.11865

Crini, G. (2006). Non-conventional low-cost adsorbents for dye removal: A review. *Bioresource Technology*, *97*(9), 1061–1085. doi:10.1016/j.biortech.2005.05.001 PMID:15993052

Daneshvar, M., & Raouf Hosseini, M. (2018). Kinetics, isotherm, and optimization of the hexavalent chromium removal from aqueous solution by a magnetic nanobiosorbent. *Environmental Science and Pollution Research International*, *25*(28), 28654–28666. doi:10.100711356-018-2878-1 PMID:30094673

Devi, G. K., Kumar, P. S., & Kumar, K. S. (2016). Article. *Desalination and Water Treatment*, 1–14. do i:10.1080/19443994.2016.1178178

Dotto, G. L., Cadaval, T. R. S., & Pinto, L. A. A. (2012). Use of Spirulina platensis micro and nanoparticles for the removal synthetic dyes from aqueous solutions by biosorption. *Process Biochemistry*, *47*(9), 1335–1343. doi:10.1016/j.procbio.2012.04.029

Ealias, A. M., & Saravanakumar, M. P. (2017). Article. *IOP Conf. Series: Materials Science and Engineering*, *263*, 1-14.

El-Kemary, M. A., El-mehasseb, I. M., Shoueir, K. R., El-Shafey, S. E., El-Shafey, O. I., Aljohani, H. A., & Fouad, R. R. (2018). Sol-gel TiO_2 decorated on eggshell nanocrystal as engineered adsorbents for removal of acid dye. *Journal of Dispersion Science and Technology*, *39*(6), 911–921. doi:10.1080/01932691.2017.1410829

Esmaeili, A., & Aghababai Beni, A. (2018). Optimization and design of a continuous biosorption process using brown algae and chitosan/PVA nano-fiber membrane for removal of nickel by a new biosorbent. *International Journal of Environmental Science and Technology*, *15*(4), 765–778. doi:10.100713762-017-1409-9

Esmaeili, A., & Khoshnevisan, N. (2016). Optimization of process parameters for removal of heavy metals by biomass of Cu and Co-doped alginate-coated chitosan nanoparticles. *Bioresource Technology*, *21*, 650–658. doi:10.1016/j.biortech.2016.07.005 PMID:27416515

Espinosa-Ortiz, E. J., Shakya, M., Jain, R., Rene, E. R., Van Hullebusch, E. D., & Lens, P. N. L. (2016). Sorption of zinc onto elemental selenium nanoparticles immobilized in Phanerochaete chrysosporium pellets. *Environmental Science and Pollution Research International*, *23*(21), 21619–21630. doi:10.100711356-016-7333-6 PMID:27519902

Fan, L., Luo, C., Sun, M., Li, X., & Qiu, H. (2013). Highly selective adsorption of lead ions by water-dispersible magnetic chitosan/graphene oxide composites. *Colloids and Surfaces. B, Biointerfaces*, *103*, 523–529. doi:10.1016/j.colsurfb.2012.11.006 PMID:23261576

Fan, L., Zhang, Y., Luo, C., Lu, F., Qiu, H., & Sun, M. (2012). Synthesis and characterization of magnetic β-cyclodextrin–chitosan nanoparticles as nano-adsorbents for removal of methyl blue. *International Journal of Biological Macromolecules*, *50*(2), 444–450. doi:10.1016/j.ijbiomac.2011.12.016 PMID:22227306

Feng, Y., Yang, F., Wang, Y., Ma, L., Wu, Y., Kerr, P. G., & Yang, L. (2011). Basic dye adsorption onto an agro-based waste material – Sesame hull (Sesamum indicum L.). *Bioresource Technology*, *102*(22), 10280–10285. doi:10.1016/j.biortech.2011.08.090 PMID:21962534

Fleischer, T., & Grunwald, A. (2008). Making nanotechnology developments sustainable. A role for technology assessment? *Journal of Cleaner Production*, *16*(8-9), 889–898. doi:10.1016/j.jclepro.2007.04.018

Ge, N., Xu, J., Li, F., Peng, B., & Pan, S. (2017). Immobilization of inactivated microbial cells on magnetic Fe3O4@CTS nanoparticles for constructing a new biosorbent for removal of patulin in fruit juice. *Food Control*, *82*, 83–90. doi:10.1016/j.foodcont.2017.06.027

Ghorai, S., Sarkar, A. K., Panda, A. B., & Pal, S. (2013). Effective removal of Congo red dye from aqueous solution using modified xanthan gum/silica hybrid nanocomposite as adsorbent. *Bioresource Technology*, *144*, 485–491. doi:10.1016/j.biortech.2013.06.108 PMID:23896441

Goetz, L. A., Jalvo, B., Rosal, R., & Mathew, A. P. (2016). Superhydrophilic anti-fouling electrospun cellulose acetate membranes coated with chitin nanocrystals for water filtration. *Journal of Membrane Science*, *510*, 238–248. doi:10.1016/j.memsci.2016.02.069

Gokila, S., Gomathi, T., Sudha, P. N., & Anil, S. (2017). Removal of the heavy metal ion chromiuim(VI) using Chitosan and Alginate nanocomposites. *International Journal of Biological Macromolecules*, *104*, 1459–1468. doi:10.1016/j.ijbiomac.2017.05.117 PMID:28551438

Gombotz, W. R., & Wee, S. F. (1998). Protein release from alginate matrices. *Advanced Drug Delivery Reviews*, *31*(3), 267–285. doi:10.1016/S0169-409X(97)00124-5 PMID:10837629

Gul, K., Sohni, S., Waqar, M., Ahmada, F., Norulaini, N. A. N., & Mohd. Omar, A. K. (2016).. . *Carbohydrate Polymers*, *152*, 520–531. doi:10.1016/j.carbpol.2016.06.045 PMID:27516300

Gutha, Y., Munagapati, V. S. (2016). Article. *International Journal of Biological Macromolecules*, *93*(A), 408-417.

Hafshejani, L. D., Nasab, S. B., Gholami, R. M., Moradzadeh, M., Izadpanah, Z., Hafshejani, S. B., & Bhatnagar, A. (2015). Removal of zinc and lead from aqueous solution by nanostructured cedar leaf ash as biosorbent. *Journal of Molecular Liquids*, *211*, 448–456. doi:10.1016/j.molliq.2015.07.044

Hong, H. J., Yu, H., Park, M., & Jeong, H. S. (2019). Recovery of platinum from waste effluent using polyethyleneimine-modified nanocelluloses: Effects of the cellulose source and type. *Carbohydrate Polymers*, *210*, 167–174. doi:10.1016/j.carbpol.2019.01.079 PMID:30732749

Hou, C., Zhao, D., Zhang, S., & Wang, Y. (2018). Highly selective adsorption of Hg(II) by the monodisperse magnetic functional chitosan nano-biosorbent. *Colloid & Polymer Science*, *296*(3), 547–555. doi:10.100700396-017-4253-z

Jeevanandam, J., Barhoum, A., Chan, Y. S., Dufresne, A., & Danquah, M. K. (2018). Review on nanoparticles and nanostructured materials: History, sources, toxicity and regulations. *Beilstein Journal of Nanotechnology*, *9*, 1050–1074. doi:10.3762/bjnano.9.98 PMID:29719757

Jerold, M., Vasantharaj, K., Joseph, D., & Sivasubramanian, V. (2017). Fabrication of hybrid biosorbent nanoscale zero-valent iron- *Sargassum swartzii* biocomposite for the removal of crystal violet from aqueous solution. *International Journal of Phytoremediation*, *19*(3), 214–224. doi:10.1080/15226514.2016.1207607 PMID:27420340

Jin, L., Li, W., Xu, Q., & Sun, Q. (2015). Amino-functionalized nanocrystalline cellulose as an adsorbent for anionic dyes. *Cellulose (London, England)*, *22*(4), 2443–2456. doi:10.100710570-015-0649-4

Jin, L., Sun, Q., Xu, Q., & Xu, Y. (2015). Adsorptive removal of anionic dyes from aqueous solutions using microgel based on nanocellulose and polyvinylamine. *Bioresource Technology*, *197*, 348–355. doi:10.1016/j.biortech.2015.08.093 PMID:26344242

Jose, J. C., Debs, K. B., Carvalh, J. T. T., Labuto, G., & Carrilho, E. N. V. M. (2019). Article. *Chemical Engineering Communications*.

Kannan, R., Lakshmi, S., Aparna, N., Prabhakar, S., Thilagaraj, W. R. (2016). Article. *International Journal of Industrial Chemistry*. DOI doi:10.100740090-016-0088-6

Karami, H., Moosakhani, Z., & Goli, F. (2017). Walnut skin nanoparticles as a novel and fast nano-biosorbent for the fast removal of lead ions from polluted water. *Desalination and Water Treatment*, *60*, 291–300. doi:10.5004/dwt.2017.0088

Karami, H., Yaghoobi, F., & Goli, F. (2017). Application of sawdust nanoparticles as a cheap and non-toxic biosorbent for the removal of lead ions from polluted waters. *Desalination and Water Treatment*, *58*, 120–128. doi:10.5004/dwt.2017.1754

Karima, Z., Mathewa, A. P., Grahn, M., Mouzon, J., & Oksman, K. (2014). Article. *Carbohydrate Polymers*, *112*, 668–676. doi:10.1016/j.carbpol.2014.06.048 PMID:25129796

Kazemi, M., Jahanshahi, M., & Peyravi, M. (2018). Chitosan-sodium alginate multilayer membrane developed by Fe0@WO3 nanoparticles: Photocatalytic removal of hexavalent chromium. *Carbohydrate Polymers*, *198*, 164–174. doi:10.1016/j.carbpol.2018.06.069 PMID:30092986

Keshvardoostchokami, M., Babaei, L., Zamani, A. A., Parizanganeh, A. H., & Piri, F. (2017). Global. *Journal of Environmental Science and Management*, *3*, 267–278.

Kloster, G. A., Mosiewicki, M. A., & Marcovich, N. E. (2019). Chitosan/iron oxide nanocomposite films: Effect of the composition and preparation methods on the adsorption of congo red. *Carbohydrate Polymers*, *221*, 186–194. doi:10.1016/j.carbpol.2019.05.089 PMID:31227157

Kwok, K. C. M., Koong, L. F., Ansari, T. A., & McKay, G. (2018). Adsorption/desorption of arsenite and arsenate on chitosan and nanochitosan. *Environmental Science and Pollution Research International*, *25*(15), 14734–14742. doi:10.100711356-018-1501-9 PMID:29536422

Kyzas, G. Z., & Matis, K. A. (2015). Nanoadsorbents for pollutants removal: A review. *Journal of Molecular Liquids*, *203*, 159–168. doi:10.1016/j.molliq.2015.01.004

Lara-Vásquez, E. J., Solache-Ríos, M., & Gutiérrez-Segura, E. (2016). Malachite green dye behaviors in the presence of biosorbents from maize (Zea mays L.), their Fe-Cu nanoparticles composites and Fe-Cu nanoparticles. *Journal of Environmental Chemical Engineering*, *4*(2), 1594–1603. doi:10.1016/j.jece.2016.02.008

Lata, S., & Samadder, S. R. (2016). Removal of arsenic from water using nano adsorbents and challenges: A review. *Journal of Environmental Management*, *166*, 387–406. doi:10.1016/j.jenvman.2015.10.039 PMID:26546885

Leung, W.-H., Lo, W.-H., & Chan, P.-H. (2015). Amyloid fibrils as rapid and efficient nano-biosorbents for removal of dye pollutants. *RSC Advances*, *5*(109), 90022–90030. doi:10.1039/C5RA17182B

Li, C., Wang, X., Meng, D., Zhou, L. (2018). Article. *International Journal of Biological Macromolecules, 107*(Part B), 1871-1878.

Li, W., Ju, B., & Zhang, S. (2019). Preparation of cysteamine-modified cellulose nanocrystal adsorbent for removal of mercury ions from aqueous solutions. *Cellulose (London, England)*, *26*(8), 4971–4985. doi:10.100710570-019-02420-1

Li, X., Zhou, H., Wu, W., Wei, S., Xu, Y., & Kuang, Y. (2015). Studies of heavy metal ion adsorption on Chitosan/Sulfydryl-functionalized graphene oxide composites. *Journal of Colloid and Interface Science*, *448*, 389–397. doi:10.1016/j.jcis.2015.02.039 PMID:25746192

Lian, J. J., Huang, Y. G., Chen, B., Wang, S. S., Wang, P., Niu, S. P., & Liu, Z. L. (2017). Article. *Water Science and Technology*, *3*, 859–868.

Liu, C., Omer, A. M., & Ouyang, X.-K. (2018). Adsorptive removal of cationic methylene blue dye using carboxymethyl cellulose/k-carrageenan/activated montmorillonite composite beads: Isotherm and kinetic studies. *International Journal of Biological Macromolecules*, *106*, 823–833. doi:10.1016/j.ijbiomac.2017.08.084 PMID:28834705

Lotfy, S. (2009). Controlling degradation of low-molecular-weight natural polymer "dextrin" using gamma irradiation. *International Journal of Biological Macromolecules*, *44*(1), 57–63. doi:10.1016/j.ijbiomac.2008.10.002 PMID:18996409

Lu, T., Xiang, T., Huang, X.-L., Li, C., Zhao, W.-F., Zhang, Q., & Zhao, C.-S. (2015). Post-crosslinking towards stimuli-responsive sodium alginate beads for the removal of dye and heavy metals. *Carbohydrate Polymers*, *133*, 587–595. doi:10.1016/j.carbpol.2015.07.048 PMID:26344317

Lv, J.-L., Zhai, S.-R., Wang, Z.-Z., Lei, Z.-M., & An, Q.-D. (2016). Carbon–silica composite bio-sorbents with a high density of oxygen-containing sites for efficient methylene blue adsorption. *Research on Chemical Intermediates*, *42*(2), 839–854. doi:10.100711164-015-2058-6

Ma, L., Peng, Y., Wu, B., Lei, D., & Xu, H. (2013). Article. *Chemical Engineering Journal (Amsterdam, Netherlands)*, *225*, 59–67.

Ma, Y.-X., Shao, W.-J., Sun, W., Kou, Y.-L., Li, X., & Yang, H.-P. (2018). One-step fabrication of β-cyclodextrin modified magnetic graphene oxide nanohybrids for adsorption of Pb(II), Cu(II) and methylene blue in aqueous solutions. *Applied Surface Science*, *459*, 544–553. doi:10.1016/j.apsusc.2018.08.025

Mallakpour, S., Abdolmaleki, A., & Tabesh, F. (2018). Ultrasonic-assisted manufacturing of new hydrogel nanocomposite biosorbent containing calcium carbonate nanoparticles and tragacanth gum for removal of heavy metal. *Ultrasonics Sonochemistry*, *41*, 572–581. doi:10.1016/j.ultsonch.2017.10.022 PMID:29137788

Mallakpour, S., & Tabesh, F. (2019). Tragacanth gum based hydrogel nanocomposites for the adsorption of methylene blue: Comparison of linear and non-linear forms of different adsorption isotherm and kinetics models. *International Journal of Biological Macromolecules*, *133*, 754–766. doi:10.1016/j.ijbiomac.2019.04.129 PMID:31004645

Mallampati, R., Xuanjun, L., Adin, A., & Valiyaveettil, S. (2015). Fruit Peels as Efficient Renewable Adsorbents for Removal of Dissolved Heavy Metals and Dyes from Water. *ACS Sustainable Chemistry & Engineering*, *3*(6), 1117–1124. doi:10.1021/acssuschemeng.5b00207

Mirzajani, R., Pourreza, N., & Burromandpiroze, J. (2017). Synthesis of chitosan-SnO2 nanoparticles biocomposite for dispersive micro-solid phase extraction of iron in environmental and wastewater samples and its determination by microvolume spectrophotometic detection. *Desalination and Water Treatment*, *97*, 213–224. doi:10.5004/dwt.2017.21536

Mousavi, S. M., Hashemi, S. A., Esmaeili, H., Amani, A. M., & Mojoudi, F. (2018). Synthesis of Fe3O4 Nanoparticles Modified by Oak Shell for Treatment of Wastewater Containing Ni(II). *Acta Chimica Slovenica*, *65*(3), 750–756. doi:10.17344/acsi.2018.4536

Namvari, M., & Namazi, H. (2016). Magnetic sweet graphene nanosheets: Preparation, characterization and application in removal of methylene blue. *International Journal of Environmental Science and Technology*, *13*(2), 599–606. doi:10.100713762-015-0885-z

Narwade, V. N., Khairnar, R. S., & Kokol, V. (2018). In Situ Synthesized Hydroxyapatite—Cellulose Nanofibrils as Biosorbents for Heavy Metal Ions Removal. *Journal of Polymers and the Environment*, *26*(5), 2130–2141. doi:10.100710924-017-1101-7

Nath, B. K., Chaliha, C., Kalita, E., & Kalita, M. C. (2016). Synthesis and characterization of ZnO:CeO2:nanocellulose:PANI bionanocomposite. A bimodal agent for arsenic adsorption and antibacterial action. *Carbohydrate Polymers*, *148*, 397–405. doi:10.1016/j.carbpol.2016.03.091 PMID:27185154

Negi, R., Satpathy, G., Tyagi, Y. K., & Gupta, R. K. (2012). Biosorption of heavy metals by utilising onion and garlic wastes. *International Journal of Environment and Pollution*, *49*(3/4), 179–196. doi:10.1504/IJEP.2012.050898

Nematidil, N., & Sadeghi, M. (2018). Article. *Polymer Bulletin*.

Nithya, R., Gomathi, T., Sudha, P. N., Venkatesan, J., Anil, S., & Kim, S. (2016). Removal of Cr(VI) from aqueous solution using chitosan-g-poly(butyl acrylate)/silica gel nanocomposite. *International Journal of Biological Macromolecules*, *87*, 545–554. doi:10.1016/j.ijbiomac.2016.02.076 PMID:26952703

Ozdemir, S., Mohamedsaid, S. A., Kilinc, E., Yildirim, A., & Soylak, M. (2018). Application of magnetized fungal solid phase extractor with Fe2O3 nanoparticle for determination and preconcentration of Co(II) and Hg(II) from natural water samples. *Microchemical Journal*, *143*, 198–204. doi:10.1016/j.microc.2018.07.032

Pal, S. L., Jana, U., Manna, P. K., Mohanta, G. P., & Manavalan, R. (2011). Article. *Journal of Applied Pharmaceutical Science, 01*(06), 228–234.

Pangi, Z., Beletsi, A., & Evangelatos, K. (2003). Article. *Advanced Drug Delivery Reviews, 24*, 403–419.

Pei, A., Butchosa, N., Berglund, L. A., & Zhou, Q. (2013). Surface quaternized cellulose nanofibrils with high water absorbency and adsorption capacity for anionic dyes. *Soft Matter, 9*(6), 2047–2055. doi:10.1039/c2sm27344f

Pi, S., Li, A., Wei, W., Feng, L., Zhang, G., Chen, T., ... Ma, F. (2017). Synthesis of a novel magnetic nano-scale biosorbent using extracellular polymeric substances from Klebsiella sp. J1 for tetracycline adsorption. *Bioresource Technology, 24*, 471–476. doi:10.1016/j.biortech.2017.08.190 PMID:28898846

Plohl, O., Ajdnik, U., Ban, I., Gyergyek, S., Vesel, A., Glaser, T. K., & Zemljic, L. (2019). Superior stability and high biosorbent efficiency of carboxymethylchitosan covalently linked to silica-coated core-shell magnetic nanoparticles for application in copper removal. *Journal of Environmental Chemical Engineering, 7*(1), 102913. doi:10.1016/j.jece.2019.102913

Qiao, H., Zhou, Y., Yu, F., Wang, E., Min, Y., Huang, Q., ... Ma, T. (2015). Effective removal of cationic dyes using carboxylate-functionalized cellulose nanocrystals. *Chemosphere, 141*, 297–303. doi:10.1016/j.chemosphere.2015.07.078 PMID:26298027

Rabbani, D., Mahmoudkashi, N., Mehdizad, F., & Shaterian, M. (2016).. . *Journal of Environmental Science and Technology, 9*(1), 121–130. doi:10.3923/jest.2016.121.130

Rahim, M., & Haris, M. R. H. M. (2015). Application of biopolymer composites in arsenic removal from aqueous medium: A review. *Journal of Radiation Research and Applied Sciences, 8*(2), 255–263. doi:10.1016/j.jrras.2015.03.001

Rathinam, K., & Sankaran, M. (2014). Article. *Chemical Engineering Journal.*

Rowley, J. A., Madlambayan, G., & Mooney, D. J. (1999). Alginate hydrogels as synthetic extracellular matrix materials. *Biomaterials, 20*(1), 45–53. doi:10.1016/S0142-9612(98)00107-0 PMID:9916770

Safari, E., Rahemi, N., Kahforoushan, D., & Allahyari, S. (2019). Article. *Journal of Environmental Chemical Engineering.*

Shanmugarajah, B., Chew, I. M., Mubarak, N. M., Choong, T. S. Y., Yoo, C. K., & Tan, K. W. (2019). Valorization of palm oil agro-waste into cellulose biosorbents for highly effective textile effluent remediation. *Journal of Cleaner Production, 210*, 697–709. doi:10.1016/j.jclepro.2018.10.342

Shao, Z., Huang, X., Yang, F., Zhao, W., Zhou, X., & Zhao, C. (2018). Engineering sodium alginate-based cross-linked beads with high removal ability of toxic metal ions and cationic dyes. *Carbohydrate Polymers, 187*, 85–93. doi:10.1016/j.carbpol.2018.01.092 PMID:29486848

Sharma, S., Hasan, A., Kumar, N., & Pandey, L. M. (2018). Removal of methylene blue dye from aqueous solution using immobilized Agrobacterium fabrum biomass along with iron oxide nanoparticles as biosorbent. *Environmental Science and Pollution Research International, 25*(22), 21605–21615. doi:10.100711356-018-2280-z PMID:29785597

Sheshmani, S., & Mashhadi, S. (2017). Article. *Polymer Composites*. doi:10.1002/pc.24608

Sreeram, A., Hadi, P., Hui, C. W., Ansari, T. A., & McKay, G. (2017). Optimisation of the removal of arsenate from water using nanochitosan. *Desalination and Water Treatment*, *70*, 235–243. doi:10.5004/dwt.2017.20561

Tang, J., Song, Y., Zhao, F., Spinney, S., Bernardes, J., & Tam, K. C. (2019). Compressible cellulose nanofibril (CNF) based aerogels produced via a bio-inspired strategy for heavy metal ion and dye removal. *Carbohydrate Polymers*, *208*, 404–412. doi:10.1016/j.carbpol.2018.12.079 PMID:30658817

Thakur, S., Pandey, S., & Arotiba, O. A. (2016). Development of a sodium alginate-based organic/inorganic superabsorbent composite hydrogel for adsorption of methylene blue. *Carbohydrate Polymers*, *153*, 34–46. doi:10.1016/j.carbpol.2016.06.104 PMID:27561469

Tian, W., Li, H., Zhou, J., & Guo, Y. (2017). Preparation, characterization and the adsorption characteristics of lignin/silica nanocomposites from cellulosic ethanol residue. *RSC Advances*, *7*(65), 41176–41181. doi:10.1039/C7RA06322A

Tural, B., Ertas, E., Enez, B., Fincan, S. A., & Tural, S. (2017). Preparation and characterization of a novel magnetic biosorbent functionalized with biomass of Bacillus Subtilis: Kinetic and isotherm studies of biosorption processes in the removal of Methylene Blue. *Journal of Environmental Chemical Engineering*, *5*(5), 4795–4802. doi:10.1016/j.jece.2017.09.019

Tural, B., Ertas, E., & Tural, S. (2016). Removal of phenolic pollutants from aqueous solutions by a simple magnetic separation. *Desalination and Water Treatment*, *54*(54), 26153–26164. doi:10.1080/19443994.2016.1162202

Vijayalakshmi, K., Devi, B. M., Latha, S., Gomathi, T., Sudha, P. N., Venkatesan, J., & Anil, S. (2017). Batch adsorption and desorption studies on the removal of lead (II) from aqueous solution using nanochitosan/sodium alginate/microcrystalline cellulose beads. *International Journal of Biological Macromolecules B.*, *104*, 1483–1494. doi:10.1016/j.ijbiomac.2017.04.120

Wang, C., Liu, H., Liu, Z., Gao, Y., Wu, B., & Xursos, H. (2018). Fe_3O_4 nanoparticle-coated mushroom source biomaterial for Cr(VI) polluted liquid treatment and mechanism research. *Royal Society Open Science*, *5*(5), 171776–171790. doi:10.1098/rsos.171776 PMID:29892371

Wang, L., & Wang, A. (2007). Adsorption characteristics of Congo Red onto the chitosan/montmorillonite nanocomposite. *Journal of Hazardous Materials*, *147*(3), 979–985. doi:10.1016/j.jhazmat.2007.01.145 PMID:17349744

Wang, Y., Zhang, Y., Hou, C., & Liu, M. (2016). Mussel-inspired synthesis of magnetic polydopamine–chitosan nanoparticles as biosorbent for dyes and metals removal. *Journal of the Taiwan Institute of Chemical Engineers*, *61*, 292–298. doi:10.1016/j.jtice.2016.01.008

Xie, K., Zhao, W., & He, X. (2011). Adsorption properties of nano-cellulose hybrid containing polyhedral oligomeric silsesquioxane and removal of reactive dyes from aqueous solution. *Carbohydrate Polymers*, *83*(4), 1516–1520. doi:10.1016/j.carbpol.2010.09.064

Xin, S., Zeng, Z., Zhou, X., Luo, W., Shi, X., Wang, Q., Deng, H., Du, Y. (2017). Article. *Journal of Hazardous Materials, 324*(B), 365-372.

Yanga, X., Liua, H., Hana, F., Jiang, S., Liu, L., & Xia, Z. (2017). Article. *Carbohydrate Polymers, 175*, 464–472. doi:10.1016/j.carbpol.2017.08.007 PMID:28917889

Yazdani, M., Bhatnagar, A., & Vahala, R. (2017). Article. *Chemical Engineering Journal (Amsterdam, Netherlands), 316*, 370–382.

Zare, E. N., Lakouraj, M. M., & Kasirian, N. (2018). Development of effective nano-biosorbent based on poly m-phenylenediamine grafted dextrin for removal of Pb (II) and methylene blue from water. *Carbohydrate Polymers, 201*, 539–548. doi:10.1016/j.carbpol.2018.08.091 PMID:30241851

Zhang, J., Li, L., Li, Y., & Yang, C. (2018). Adsorptive filtration of low-concentration lead with wet-laid filter media made from nano-TiO_2 decorated cellulose fibers. *Journal of Chemical Technology and Biotechnology (Oxford, Oxfordshire), 93*(2), 432–439. doi:10.1002/jctb.5373

Zhang, L., Xia, W., Teng, B., Liu, X., & Zhang, W. (2013). Zirconium cross-linked chitosan composite: Preparation, characterization and application in adsorption of Cr(VI). *Chemical Engineering Journal, 229*, 1–8. doi:10.1016/j.cej.2013.05.102

Zhang, X., Zhao, J., Cheng, L., Lu, C., Wang, Y., He, X., & Zhang, W. (2014). Acrylic acid grafted and acrylic acid/sodium humate grafted bamboo cellulose nanofibers for Cu^{2+} adsorption. *RSC Advances, 4*(98), 55195–55201. doi:10.1039/C4RA08307E

Zhou, C., Wu, Q., Lei, T., & Negulescu, I. I. (2014). Adsorption kinetic and equilibrium studies for methylene blue dye by partially hydrolyzed polyacrylamide/cellulose nanocrystal nanocomposite hydrogels. *Chemical Engineering Journal, 251*, 17–24. doi:10.1016/j.cej.2014.04.034

Zhou, J., Liu, Y., Zhou, X., Ren, J., & Zhong, C. (2017). Removal of mercury ions from aqueous solution by thiourea-functionalized magnetic biosorbent: Preparation and mechanism study. *Journal of Colloid and Interface Science, 507*, 107–118. doi:10.1016/j.jcis.2017.07.110 PMID:28780330

Chapter 19
Photocatalytic Activity of ZnO Nanoparticles on the Degradation of Organic Dyes Under Solar Light

Sivakumar Krishnamoorthy
Department of Chemistry, Sri Chandrasekharendra Saraswathi Viswa Mahavidyalaya (Deemed), India

Dharani M.
Department of Chemistry, Sri Chandrasekharendra Saraswathi Viswa Mahavidyalaya (Deemed), India

ABSTRACT

Zinc oxide (ZnO) nanoparticles prepared using simple co-precipitation method are characterized and photocatalytic activity is tested on the degradation of methylene blue and rhodamine B organic pollutants. Morphological and structural properties of synthesized nanomaterial have been characterized using FESEM, EDAX spectroscopy, and XRD, while UV-visible DRS spectroscopy and photoluminescence have been used to understand their optical properties. The photocatalytic behaviour of synthesized nanoparticles was evaluated on the degradation of methylene blue (MB) and rhodamine B (RhB) organic pollutants under solar light irradiation. The highest degradation was achieved for MB (100%) over RhB (96%). Preliminary investigation shows the effective degradation of organic pollutants by ZnO nanoparticles.

INTRODUCTION

Dyes are important organic pollutants from various industries like textile, rubber, plastics, paper etc. Dyes released through industrial effluents pollute the natural water resources, causes eutrophication by consuming dissolved oxygen and destroy the aquatic life (Rahman et al., 2013). 10-15% of dyes produced are released into the natural environment during their synthesis and manufacturing process (Chaibakhsh

DOI: 10.4018/978-1-7998-1241-8.ch019

et al., 2015). Among various dyes methylene blue (MB) and rhodamine B (RhB) dyes are widely used coloring agent in textile industry (Ahamed et al., 2014; Saravanan et al., 2013). RhB has moderate wash and light fastness properties on wool (Chakravertty et al., 1979). It is a well-known water tracer fluorescent (Mahasin et al., 2011). It is also a useful analytical reagent for the detection and determination of metals (Shekari et al., 2017). It is generally used as a coloring agent in textiles and food stuffs. Use of RhB as food colorant has been restricted for a number of years due to its carcinogenic nature (Gupta et al., 2004). It causes irritation to eyes, skin and respiratory tract (Jain et al., 2003). It causes neurotoxicity, chronic toxicity; reproductive and developmental toxicity and carcinogenicity towards humans and animals have been experimentally proven (Jian Deng et al., 2014). Methylene blue is an important basic dye widely used by industries as coloring agent in silk, cotton, wool, leather (Baeissa, 2016) and as indicator in oxidation-reduction process (Xu et al., 2017). Methylene blue is not much hazardous as rhodamine blue but it can cause some harmful effects. Intense degree of methylene blue exposure can lead to increased pulse rate, vomiting, shock, Heinz body formation, carbon monoxide poison, jaundice, and death of body tissues in humans (Hameed et al., 2007).

PHOTO CATALYSIS

Photocatalysis is a "green" technology for the treatment of environmental pollutants with the use of solar energy. Photo catalysis has been a widely studied technology since 1970s. Environmental pollution and mineralization of organic pollutant in a global scale have drawn attention to the vital important need for totally new environmental friendly clean chemical technologies and process, the most important challenge for the chemical scientist in the field of green chemistry. Strong contenders as environmentally harmonious catalysts are photo catalyst that operates at room temperature and in a clean manner.

Semi Conductor Photo Catalyst

Photocatalysis using semiconductors is an emerging technique for sustained developments which uses conducting nanoparticles as catalytic agent to enhance the rate of reaction for the degradation of pollutants (Chang et al., 2015; Zhuang et al., 2010). Various semiconductor substances such as TiO_2, ZnO, Fe_2O_3, CdS, CeO_2 and ZnS are used as photocatalyst to degrade organic pollutants. The role of semiconductor in photocatalytic degradation of organic pollutant (dyes) is sensitizers for light induced redox processes because of their electronic structure, which is characterized by filled valence band and an empty conduction band (Hua et al., 1997; Sangave et al., 2004). Semiconductor metal oxide (and sulfide) catalysts, such as TiO_2, ZnO, CdS, and WO_3 etc., have much attention (Velusamy et al., 2012). Among these ZnO and TiO_2 has grown primarily in the area of photo catalysis because of their physical and chemical stability, high oxidative capacity, low cost and ease of availability with wide band gap (3.2 eV) (Wan et al., 2005). TiO_2 absorption falls in ultra-violet region is one of the disadvantages for photocatalytic application (Zhao et al., 2008). Zinc oxide (ZnO) is widely used than TiO_2 due to the utilization of only a small portion of UV radiation in the solar light (<5%) in photocatalytic reactions (Wang et al., 2007). In the presence of light, semiconductor generates the electron-hole pairs on crystal surface which are responsible for mineralization of dyes (Modirshahla et al., 2011). Photocatalysis occurs on the surface of semiconductor and its efficiency is strongly influenced by morphology and surface modification of the oxide. Hence, catalytic ability of different nanostructures are evaluated and reported in the literature; nanoflowers (Long et al., 2017), nanodisks (Zeng et al., 2009), nanowire (Huang, et al., 2008), nanorod (Guo et al., 2002), nanosheet (Kuo-Chuan Ho et al., 2010) and nanonail (Ren et al., 2003). Reports on the application of different types of nanostructures in various fields such as solar cells, photodetection, gas sensor and medicine are available. A heterogeneous photocatalytic system consists of semiconductor particles (photo catalyst) which are in close contact with a liquid or gaseous reaction medium, which increase the rate of a thermodynamically allowed reaction by reducing their activation energy in the presence of light with creation of some new reaction pathways involving photo generated species (Pang et al., 2010; Rahman et al., 2016). Generally, there are four essential key steps in the mechanism of heterogeneous photocatalysis on the surface of semiconductor (ZnO), namely, (a) charge-carrier generation, (b) charge-carrier trapping, (c) charge-carrier recombination and, (d) photocatalytic degradation of organic pollutants (Chen et al., 2015; Changa et al., 2015; Zhuang et al., 2010; Velusamy et al, 2014; Rehman et al., 2009; Hernandez et al., 2009; Malato et al., 2009; Sakthivel et al., 2003; Chen et al., 2007; Kansal et al., 2007).

Charge-Carrier Generation

The role of semiconductor in photocatalytic degradation of organic pollutant (dyes) is sensitizers for light induced redox processes because of their wide band gap, which is characterized by filled valence band and an empty conduction band. The band gap is defined as the energy difference between the valence and conduction bands. The semiconductor photocatalyst absorbs incident photons with energies equal to or higher than its band gap or threshold energy. Upon irradiation of semiconductor (ZnO) with UV light energy of equivalent or greater than its band gap energy, the electron is excited from the valence band to the conduction band.

Charge-Carrier Trapping

The ($e^- h^+$) pair is trapped by electron and hole scavengers and inhibited from recombination. The positive hole is a strong oxidant which can either directly oxidize pollutants or react with electron donors such as water or hydroxyl ions (OH^-) to form hydroxyl radical ($^{-}$•OH), which is also a potent oxidizing agent. One of the efficient electron acceptors is molecular oxygen (O^{2-}). Through the reduction of O_2 with electron, reactive superoxide radical anions (O^{2-}) are produced. Protonated form of Oxidizing species such as hydroperoxyl radicals (HO^-_2) and hydrogen peroxide (H_2O_2) are also subsequently formed.

Charge-Carrier Recombination

In competition with charge transfer to adsorbed pollutants, there is the opportunity that both e^-h^+ pair recombination and trapped carrier recombination happen. This recombination can occur either in the volume of the photo catalyst or on the surface of photo catalyst with a by-product of heat liberation. The characteristic times for various steps in the photo catalysis mechanism have been reported in literatures (Pare et al., 2008; Wan et al., 2005; Zha et al., 2008). It has been reviewed that the charge-carrier trapping takes place about a nanosecond (100ps to 10ns). Charge-carrier recombination exhibits a characteristic time of 10–l00ns. Interfacial charge transfer of holes is slow (~100ns) but the slowest step is the interfacial charge transfer of electrons to the electron acceptor (ms) degradation.

Photo Catalytic Degradation

The primary photoreaction indicated the important role of e^- h^+ pairs in photo catalytic degradation. Essentially, $^{\cdot}OH$, $HO^{\cdot}_2$ and $O_2^{\cdot}$ radicals as well as photo generated hole (h^+) are highly reactive intermediates that will attack repeatedly in the reacting system and ultimately lead to complete mineralization of the organic pollutants.

Modification of ZnO Nanostructures

The degradation efficiency and rate of the catalyst are surface area dependent since electron–hole transfer takes place on the surface (Wan et al., 2007; Zhang et al., 2011; Fang et al., 2011). Thus, the catalyst morphology is an important property in determining the effectiveness of the catalyst. When the surface-to-volume ratio increases, the rate of electron–hole transfer from the semiconductor to the adsorbed molecule also increases due to the ability of the semiconductor to absorb more molecules on its surface. For these reasons, many researchers have recently attempted and succeeded in creating a number of different morphologies of different sizes of ZnO such as nanopellets, nanorods, nanosheets, nanoflowers, nanoplates, nanopieces, nanoneedles, nanocups, nanocombs, nanowires and nanoplatelets (Fang et al., 2011; 49-61). ZnO is an II-VI semi conductor compound, which exist as crystals in wurtzite, zincblende and rocksalt. ZnO prefer the hexagonal wurtzite structure with lattice parameters a = 3.25Å and c = 5.20Å at ordinary pressure and temperature. ZnO hexagonal wurtzite structure with number of alternating planes contain tetrahedrally surrounded by $O^{2-}$and Zn^{2+} ions placed alternately along the c-axis. The tetrahedral coordination in ZnO results in piezoelectric and pyroelectric due to the non central symmetric structure. Another important characteristic of ZnO is polar surfaces, which are the basal planes (0001 and 0 000). One end of the basal polar plane (0001) is terminated by Zn atoms and the other end is terminated by oxygen atoms. Dipole moment and spontaneous polarization along the c-axis as well as a divergence in surface energy as a result of the oppositely charged ions produce positively charged Zn (0001) and negatively charged O-(0000) surfaces. Thus, to maintain electrical neutrality, the polar surfaces typically have facet or exhibit massive surface reconstructions, but not in ZnO ± (0 0 01) surfaces that they are atomically flat, stable and with no reconstruction (Porter 1991; Wang 2004). Although the wurtzite structure of ZnO is the thermodynamically stable phase, two other structures of ZnO exist, which are known as the zincblende and the rocksalt. ZnO has a strong natural tendency to crystallize in the wurtzite structure and thus nearly all photocatalytic studies are focused on this structure. Depending on the optical absorption properties, the refractive index of ZnO (2.0) is smaller than that of TiO2 (2.5–2.7). So, ZnO hardly scatters light, thereby making it colorless and enhancing the transparency. In addition, ZnO has a considerably high thermal conductivity of 54W/mK increases the appeal of ZnO as a substrate for homoepitaxy or heteroepitaxy (Baruah 2009; Zgu 2005). The basic investigation of the electrical properties of ZnO nanostructures is important to extend their future applications in nanoelectronics. The advantages of a wide band gap (3.27 eV at room temperature) are having higher values of breakdown voltages, sustaining high-power operations, high-temperature and large electric fields. ZnO is an n-type semi conductor. Non-stoichiometry defects are usually responsible for n-type character. Due to defects such as zinc interstitials and oxygen vacancies ZnO nanoparticle is previously reported to have semiconductor type behavior. The main application of ZnO in electronics and photonics is as p-n junction diodes and light emitting diodes (Jacak et al., 1998). Zinc oxide is highly transparent to visible light but strongly absorbs ultra violet light below 375nm. Therefore it is used in white pigment. Zinc oxide appears in white colour, but rutile and anatase titanium dioxide have a higher refractive index and thus has a superior opacity. The band gap energy (between valence and conducting bands) is 3.2 eV, this corresponds to the energy of 375 nm wave length of photons zinc oxide is photoconductive under ultra violet light. The combination of optical and semiconductor properties gives application to gas sensors, varistors, and generators of surface acoustic waves (Harper 2001). Photocatalytic reaction using ZnO nano particles occurs in neutral solutions and hence it makes the ZnO as superior material for environmental remediation over its competitors.

Removal of Dyes

Removal of dyes is necessary before they were mixed with the natural environment due to its harmful effects for humans and animals in the environment. The removal of dye pollutants, various conventional methods such as adsorption on activated carbon, ultra filtration, reverse osmosis, ion exchange chromatography coagulation by chemical agent etc can generally be used efficiently (Pitchaimuthu et al., 2015; Rajalakshmi et al., 2017). But these treating methods are ineffective since they just transfer organic compounds from waste water into secondary pollutant (Mukit et al., 2016). In recent years, ultrasonic degradation as one of the advanced oxidation process has been studied and extensively used to treat hazardous organic pollutants in effluents before polluting the water resources. Thus, due to, (i) disadvantages of various methods, (ii) and limitation in their large-scale implementation of reported methods, developing more efficient and economically affordable degradation methods is unavoidable (Dai et al., 2015).

In this report, ZnO nanoparticle prepared through a simple and cost effective method using $ZnNO_3.6H_2O$ and NaOH through co-precipitation process, standard analyzed using techniques such as, FESEM, XRD, EDAX, FTIR, and DRS to characterize the synthesized ZnO nanoparticles. The photocatalytic ability of ZnO nanoparticles were tested on MB and RhB as standard reference dyes using UV–visible spectrophotometer.

MATERIALS AND METHODS

Zinc nitrate hexahydrate was purchased from Molychem. Sodium hydroxide, Methylene blue and rhodamine B were purchased from Sigma Aldrich. Double distilled water was used for aqueous solutions and washings. The ZnO nanoparticles were prepared by drop wise addition of aqueous solution of Zn $(NO_3)_2$ $.6H_2O$ (3 g in 80 mL) into aqueous NaOH solution (1.6 g in 80 mL) to form white slurry and kept at $80\,^{\circ}C$ on a hotplate for 6 hour without stirring. The white precipitate thus obtained was centrifuged at 6000 rpm and washed with water and ethanol several times to remove the soluble impurities. Finally, the precipitate was vacuum dried at 60° for 12 hours and the resultant white powder ZnO nanoparticles was used for further studies. Diffuse reflectance spectra (DRS) of ZnO nanoparticles were obtained using Jasco UV–Vis spectrophotometer Model V-670. Photoluminescence (PL) spectra was recorded using a Perkin Elmer spectrophotometer, model 204 A. X-ray diffraction (XRD) pattern of the samples was recorded on XPERT analytical diffractometer with Cu Ka as radiation source ($\lambda = 1.5418$Å). Scanning electron microscope (SEM) images were obtained using JEOL JSM 5610 LV at accelerated voltage 20.0kV. The photocatalytic activity of prepared nano particles were tested on MB and RhB as standard reference dyes using UV–visible spectrophotometer. In a typical process, 10 mg of photo catalyst was dispersed into the 80 ml of dye solutions (10^{-4}M) under constant stirring. Before irradiation, the suspension was kept in the dark under 30 minutes stirring to attain adsorption-desorption equilibrium. After 30 minutes, the reaction mixture was irradiated under natural sun light as radiation source. All the experiments were carried out under intense sunny days (9 am and 2 pm.) At regular time of interval, the 5 mL of sample aliquots was withdrawn from the reaction mixture and tested for degradation of dye using UV–visible spectrophotometer (Analytic Jena, Germany) by measuring the absorption intensity at 664 nm for MB and 554 nm for RhB. The dye degradation percentage was calculated using the following equation:.

$$\text{Dye degradation }\% = \frac{(A_0 - A)}{A_0} * 100$$

where, 'A 'is the absorbance value at maximum wavelength of solution (664 for MB and 554 nm for RhB). 'A_0' is the value before treatment and 'A' is the value after treatment process.

RESULT AND DISCUSSIONS

The UV-diffuse reflectance spectra of synthesized ZnO nanoparticles Figure 1(a) were measured in the range of 200-600 nm. It was obtained in terms of F(R) by applying the Kubelka-Munk algorithm. The spectrum shows absorbance below 400 nm corresponding to the intrinsic band-gap of ZnO which is related to transition of electron from valence band to the conduction band (Pandikumar et al., 2012). There is no absorption observed above 400 nm which is characteristic of ZnO nano structure. The band-gap energy Figure 1(b) of synthesized ZnO nanoparticles is estimated using Tauc's plot relation (photon energy $h\nu$ versus $(\alpha h\nu)\ 1^{/2)}$ (Dhanalakshmi et al., 2013).

$$\alpha h\nu = A\left(h\nu - Ebg\right)^{1/2}$$

Where, 'h' is Planck's constant, 'ν' is a frequency of vibration, α is an absorption coefficient, 'Ebg' is band-gap energy and 'A' is the absorption constant and 'α' is directly proportional to F(R). A linear portion of the plot is extrapolated to intersect the *X*-axis and this band gap energy value is 3.07 eV, which is lower than band gap ~3.3 eV of ZnO bulk. (Rokesh et al., 2014). Reduced band gap is more beneficial for visible light utilization of photo electro catalytic activity of ZnO. Photoluminescence (PL) is directly related to the photochemical and optical properties and gives detailed information on structural defects. Bulk and low dimensional semi conductors have different electronic behavior due to the difference in electronic density of states. It is generally known that the room temperature PL spectra of ZnO exhibit three major peaks. The UV peak is assigned to the near band-edge emission, while the two broad bands in visible region are characteristic to deep-level defects of zinc and oxygen in ZnO crystal (Zou B., 2011; Rahman et al., 2014; Suresh et. al., 2012). Figure 2 shows the PL spectra of ZnO nanoparticles. Peak at 380 nm is attributed to the band-edge emission whose energy is 3.22 eV, while the broad visible bands (540 nm) are due to the deep-level defects in the ZnO crystal. (Murugesan et al., 2011).

Figure 1. (a) and (b)

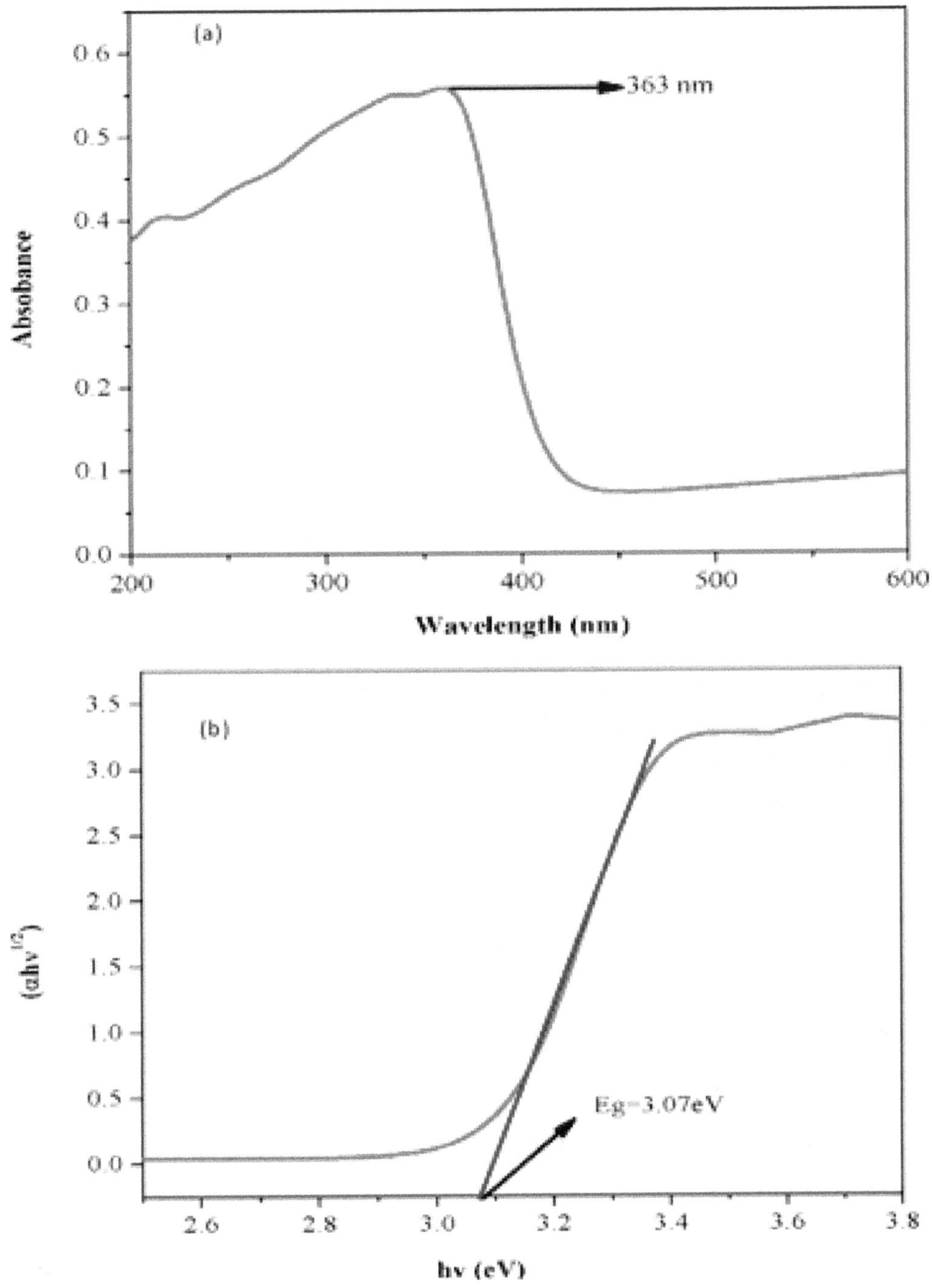

Figure 2.

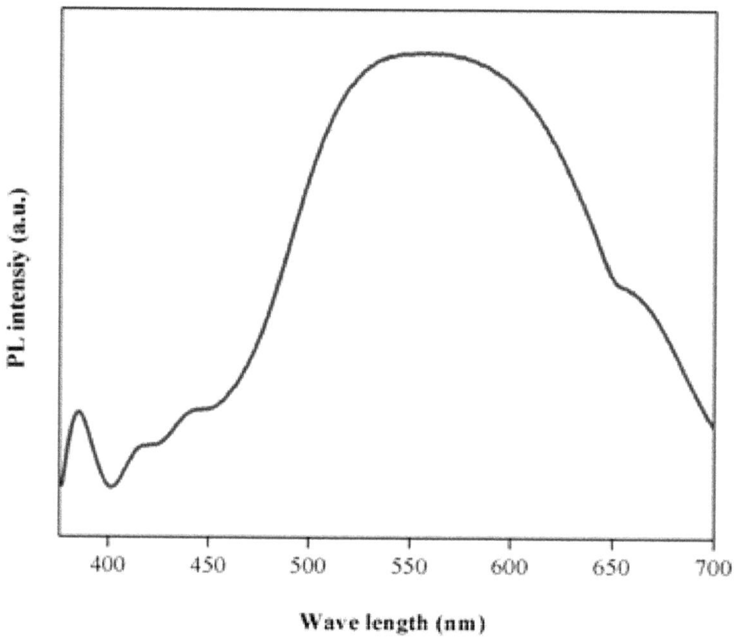

The peaks obtained for XRD pattern (Figure 3) of ZnO are 31.93°, 34.38°, 36.33°, 47.43°, 56.74°, 66.83°, and 71.80° corresponded to planes (100), (002), (101), (102), (110), (112) and (004) and indentified to hexagonal wurtzite structure with the dominant peak (101) (JCPDS no. 36-1451). The crystallite sizes were calculated using Debye- Scherer's equation,

Figure 3.

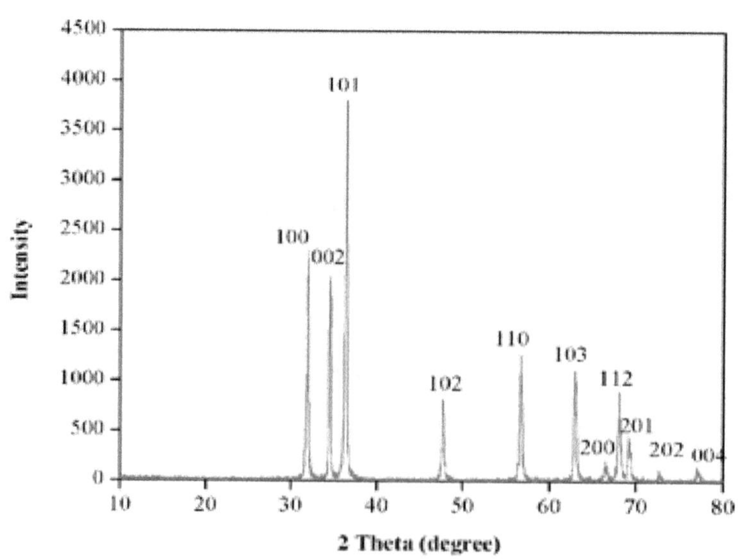

$$D = 0.89\lambda / \beta \cos\theta$$

Where, D is the crystallite size or particle size, λ is x-ray wavelength (1.5406 Å), is full width at half maximum (FWHM) of the observed peak and is the diffraction angle. The average crystallite size is found to be 36nm by resolving three intense peaks. Furthermore, the hexagonal wurtzite structure of ZnO can be reconfirmed in terms of lattice constant 'a' and 'c' values (Chan et al., 2013; Murugesan et al., 2011).

$$a = \lambda / \sin\theta\sqrt{3}$$

$$c = \lambda / \sin\theta$$

Moreover, dislocation density δ which represents the number of defects in thin films was determined by the equation,

$$\delta = 1 / D^2$$

ZnO bond length, L can also be calculated given by,

$$u = a^{(2)} / (3c^2) + 0.25$$

$$L = \left(\sqrt{\frac{a^2}{3} + 0.5 - u^2}\right) c^2$$

Data relevant to XRD parameters are listed in Table 1.

Table 1. Data obtained from X-ray diffraction pattern analysis

Diffraction angle 2Θ	Lattice parameter (Å)		Inter spacing d (hkl) (Å)	FWHW (deg)	Crystallite size (nm)	Positional parameter u	Bond length (Å)	$\delta.10^{15}$ Lines/ meter²
	a	c						
31.8 (100)	3.249	5.622	2.811	0.22093	39.07	0.3613	2.02	6.54
34.39 (002)	2.952	5.1078	2.5538	0.22093	39.03	0.3613	1.84	6.56
36.3 (101)	2.8585	4.944	2.466	0.22093	39.55	0.3614	1.78	6.39
Average	3.013	5.2246	2.610	0.2293	36.21	0.36135	1.88	6.49

The FTIR spectra of ZnO nanoparticles are shown in Figure 4. Existence of Zn-O bond is confirmed by the appearance of absorption band at 433cm^{-1} (Noor Azmy et al., 2017). The band at 490 cm^{-1} may be related with oxygen vacancy or oxygen deficiency in ZnO. The peaks at 1739 cm^{-1} correspond to the C-O absorption of ZnO surface (Becheri et al., 2008; Hedayati et al., 2015). The broad peaks around 3419 cm^{-1} is attributed to the O-H stretching and 1635 cm^{-1} (bending) are due to asymmetrical stretching of the zinc carboxylate. In these results identified the impurities which exist near ZnO surface. The weak absorption band at 2928 cm^{-1} stands for presence of C-H species (George kutty et al., 2008). No other peak was not observed in the spectra indicated that the synthetic route is high pure. SEM displays the detailed morphology of the synthesized ZnO nanomaterial. SEM images infer the formation of bricks shaped ZnO particles in Figure (5a). ZnO nanoparticle size varies from10-1000 nm Figure (5b) and the average size is found to be 47 nm. The energy dispersive X-ray diffractive (EDAX) study was used to define the elemental composition. EDAX analysis confirms the presence of zinc and oxygen signals of zinc oxide nanoparticle as shown in Figure 6. The emission peaks corresponding to the elements O and Zn were observed at 0.53 KeV (O) and 1.12 KeV, 8.60 KeV, 9.63 KeV (Zn), respectively and this analysis showed the peaks that are corresponded to the optical absorption of the produced ZnO nanoparticle. The elemental analysis of the synthesized nanoparticle yielded 66% of zinc and 34% of oxygen which proves that no other impurities present in it. (Snehal et al., 2016; Kumar et al., 2013).

Figure 4.

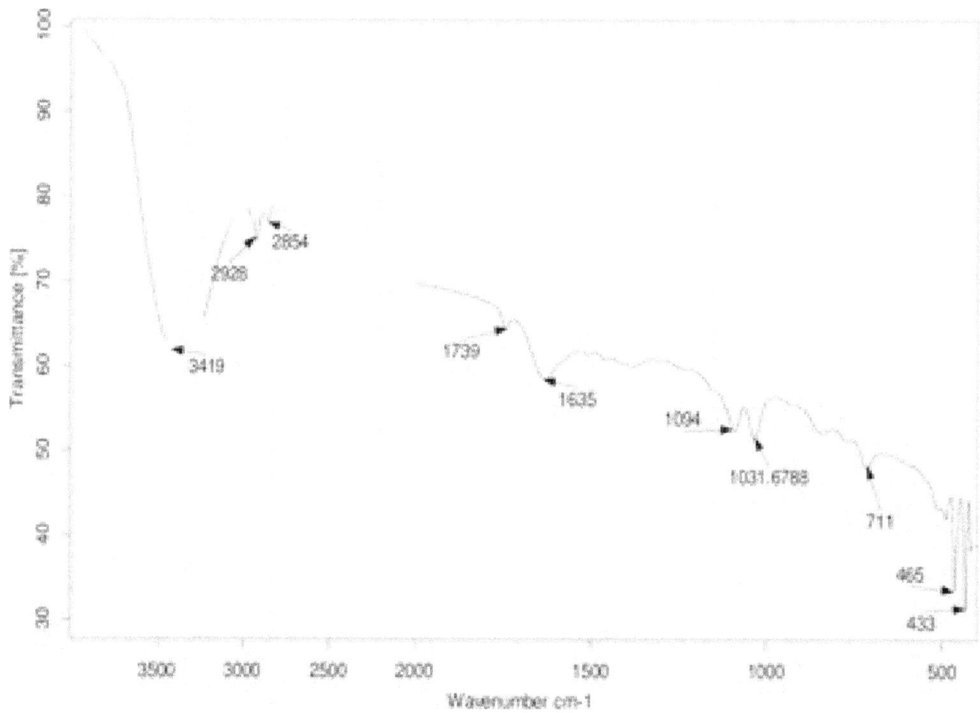

The photocatalytic ability of ZnO semiconductor nano particles was tested by evaluating the degradation of MB and RhB dyes under solar light irradiation. The reaction mixture RhB or MB with ZnO nanoparticles in water medium was exposed to sun light. The color of the dyes solution faded slowly after six hours under sun light; Figure 7 (a) and 8 (a). The Beer-Lambert law stated that the absorption is directly proportional to the concentration of the absorbing species. Under solar light irradiation of sufficient energy (hν), equal to or greater than the band gap energy of ZnO (hν ≥ 3.2) the electrons (e^-) move from valence band (VB) to the conduction band (e^-CB). Meanwhile, holes (h^+VB) will be left at the valence band. A characteristic feature of semiconducting metal oxides is the highly reactive electrons and holes at the surface of the photo catalyst as demonstrated in Figure 9. They can react with water to

Figure 5. (a) and (b)

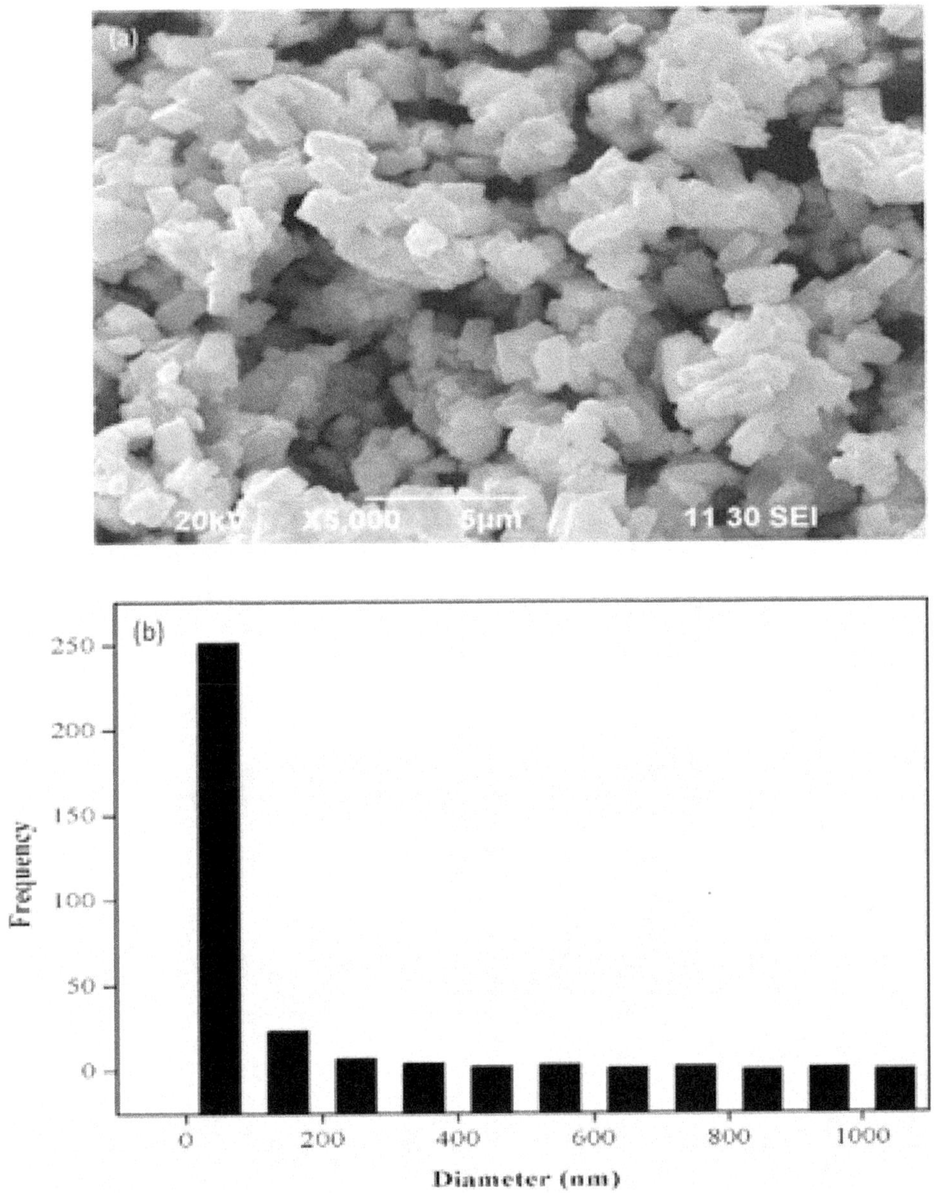

produce reduction and oxidation reactions to generate the highly reactive hydroxyl radical (•OH) and superoxide anion radicals (O_2^-), respectively.

The photo catalytic mechanism for degradation of organic pollutants can be explained through the following reactions. (Lam et al., 2012).

$ZnO + h\nu \longrightarrow ZnO\ (h^+VB^+\ e^-CB)$

$h^+\ VB^+\ dye \longrightarrow dye^+ \longrightarrow$ oxidation of the dye

Figure 6.

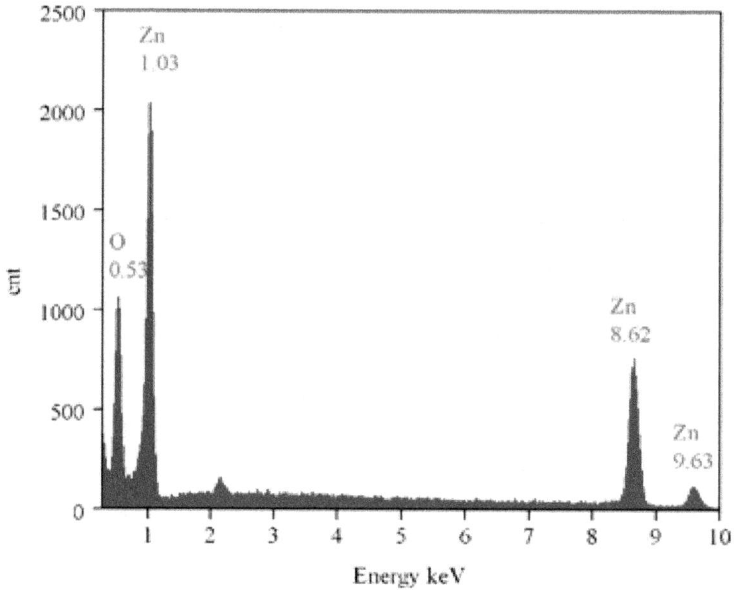

Figure 7. (a) and (b)

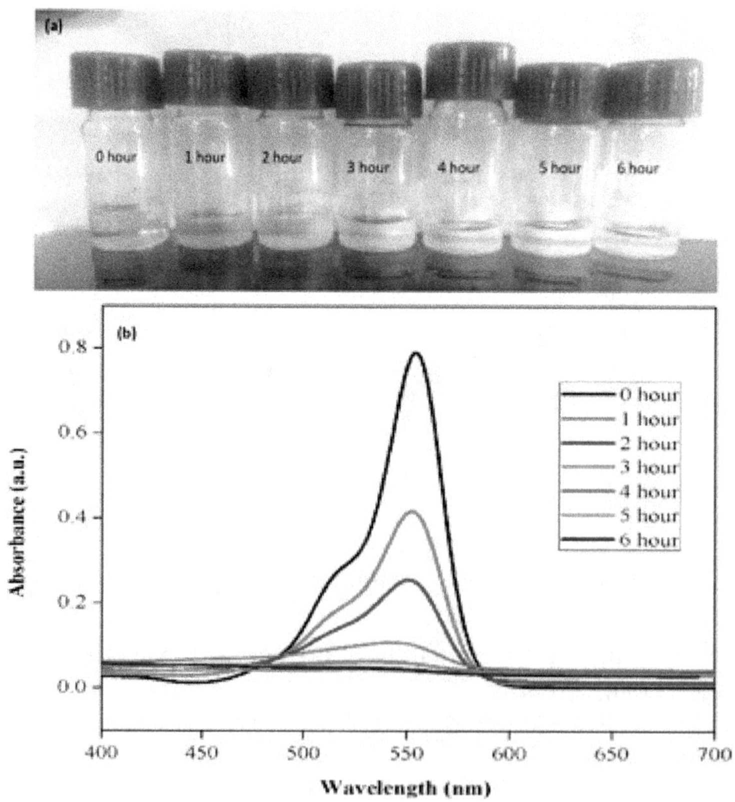

Figure 8. (a) and (b)

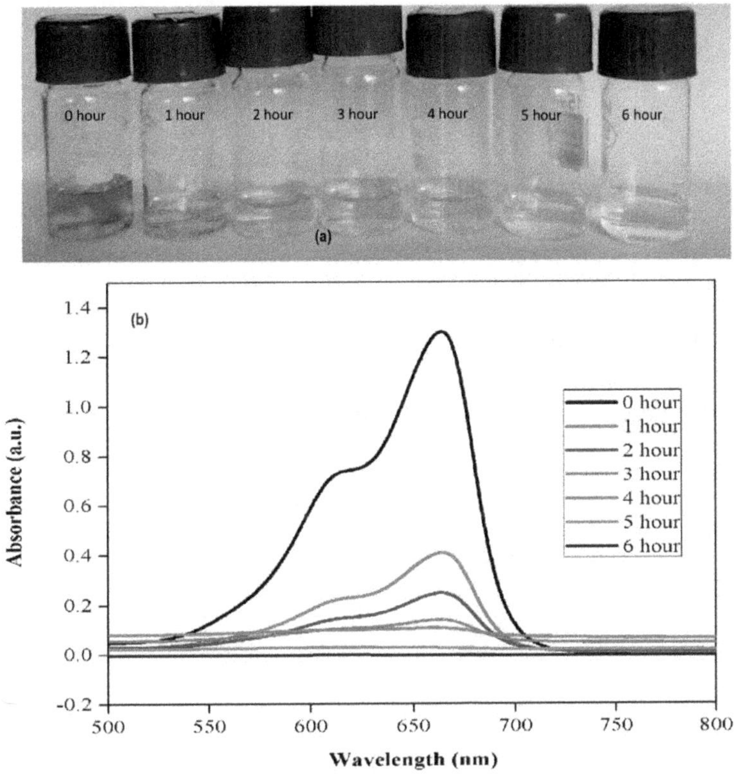

Figure 9.

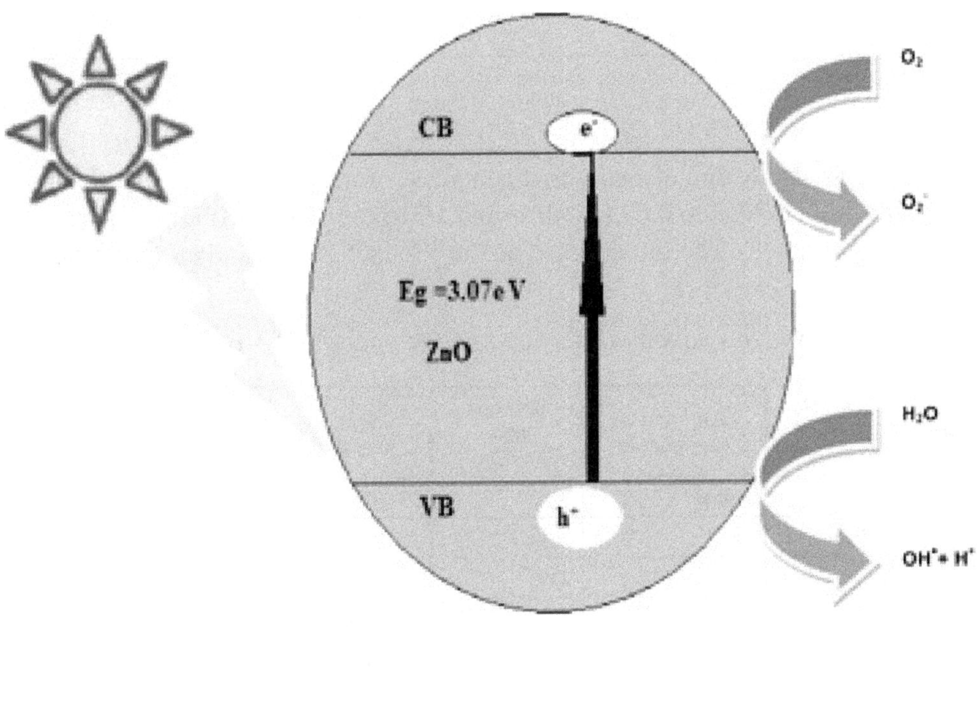

$H_2O+h^+VB \longrightarrow \bullet OH+H^+$

$O_2+ e^-CB \longrightarrow O_2^-$

$O_2^-+H^+ \longrightarrow HO_2^-$

$H^++HO_2^- \longrightarrow H_2O_2$

$H_2O_2+h\nu \longrightarrow 2\ \bullet OH$

$\bullet OH+$ dye $\longrightarrow$ degradation of the dye.

The percentage of dye degradation can be determined with respect to time as shown in Table 2.

Table 2. Percentage of degradation of methylene blue and rhodamine B dyes

Rhodamine B*		Methylene Blue*	
Time (in hrs)	*% of Degradation*	*Time (in hrs)*	*% of Degradation*
0	0	0	0
1	47	1	69
2	54	2	81
3	86	3	89
4	92	4	92
5	94	5	98
6	96	6	100

**Initial concentration: 10^{-4}M*

The colour of dye gradually diminishes with the increase of irradiation time. The absorbance of dyes decreased gradually from 0.7918 to 0.0357 (RhB) and 1.315 to 0.003(MB) after six hours of sun light exposure Figure 7 (b) and 8 (b). Therefore, based on the obtained results and on comparing the obtained

Table 3. Performance of various nanomaterials on the degradation of MB and RhB.

Nanomaterial	Particle size (nm)	Light used for degradation	Reaction time (in hrs)	% of dye degradation	Reference
ZnO	200	visible	3	29 (RhB)	(XuanGuo et al., 2014)
Sb doped ZnO	150	uv	5	95 (MB)	(Anukorn et al., 2010)
Co doped ZnO	-	visible	24	100 (RhB)	(Xiaoqing Qiu et al., 2008)
ZnO/TiO_2	25	visible	19	100 (Rhb)	(Gang Liu et al., 2009)
ZnO	47	solar light	6 6	96 (RhB) 100 (MB)	This work

degradation percentage of MB and RhB (Table 3), the performance of prepared ZnO nanoparticles are significant and it is recommended for real-time applications after optimization.

CONCLUSION

The ZnO nanoparticle was synthesized through co-precipitation method. This is a cheap, reproducible and suitable for large scale production of ZnO materials. UV-DRS and PL spectroscopy was used to investigate their optical property. The band gap energy is 3.07eV was calculated from Tacu's plot and decrease in band gap energy is beneficial for solar-driven photocatalytic activity. FT-IR explains molecular species present in the material. XRD pattern reveals that the phase properties of synthesized materials exhibit wurtzite hexagonal structure and its crystallite size is 47 nm. EDX spectrum defined the elemental composition and material purity of ZnO. The morphology of ZnO was confirmed by FE-SEM and exhibits the brick like morphology as nanostructure. The prepared material exhibited efficient solar-light active photocatalytic activity and good reusability examined by rhodamine B and methylene blue.

REFERENCES

Ahmad, M., Ahmed, E., Hong, E., Ahmed, Z. L., Elhissi, A., & Khalid, N. R. (2014). Photocatalytic, sonocatalytic and sonophotocatalytic degradation of rhodamine B using ZnO/CNTs composites photocatalysts. *Ultrasonics Sonochemistry*, *21*(2), 761–773. doi:10.1016/j.ultsonch.2013.08.014 PMID:24055646

Baeissa, E. S. (2016). Photocatalytic degradation of methylene blue dye under visible light irradiation using In/ZnO nanocomposite. *Front Nanosci Nanotech*, *2*(5), 1–5. doi:10.15761/FNN.1000134

Baruah, S., & Dutta, J. (2009). Hydrothermal growth of ZnO nanostructures. *Science and Technology of Advanced Materials*, *10*(1), 013001. doi:10.1088/1468-6996/10/1/013001 PMID:27877250

Baruah, S., Thanachayanont, C., & Dutta, J. (2008). Growth of ZnO nanowires on nonwoven polyethylene fibers. *Science and Technology of Advanced Materials*, *9*(2), 1–8. doi:10.1088/1468-6996/9/2/025009 PMID:27877984

Becheri, A., Durr, M., Lo Nostro, P., & Baglioni, P. (2008). Synthesis and characterization of zinc oxide nanoparticles: Application to textiles as UV-absorbers. *Journal of Nanoparticle Research*, *10*(4), 679–689. doi:10.100711051-007-9318-3

Chaibakhsh, N., Ahmadi, N., & Zanjanchi, M. (2016). Optimization of photo catalytic degradation of neutral red dye using TiO2 nanocatalyst via Box-Behnken design. *Desalination and Water Treatment*, *57*(20), 1–11. doi:10.1080/19443994.2015.1030705

Chakravertty, R. R., & Trivedi, S. S. (1979). *Technology of Bleaching & Dyeing of Textile Fibers*. Ahmedabad, Gujarat, India: Mahajan Brothers.

Chan, Y. L., Pung, S. Y., Hussain, N. S., Sreekantan, S., & Yeoh, F. Y. (2013). Photocatalytic degradation of Rhodamine B using MnO_2 and ZnO nanoparticles. *Materials Science Forum*, *756*, 379–382. doi:10.4028/www.scientific.net/MSF.756.167

Chang, B.Y.Z., Mehmood, M.S., Pandikumar, A., Huanga, N.M., Limb, H.N., & Marlinda, Ab. R., Yusoff, N., & Chiu, W. (2015). Hydrothermally prepared graphene-titania nanocomposite for the solar photo catalytic degradation of methylene blue. *Desalination and Water Treatment*, *57*, 238–245.

Chen, C. (2007). Degradation pathways of ethyl violet by photocatalytic reaction with ZnO dispersions. *Journal of Molecular Catalysis A Chemical*, *264*(1-2), 82–92. doi:10.1016/j.molcata.2006.09.013

Chen, X., Dai, J., Shi, G., Li, L., Wang, G., & Yang, Y. (2015). Sonocatalytic degradation of Rhodamine B catalyzed by β-$Bi2O_3$ particles under ultrasonic irradiation. *Ultrasonics Sonochemistry*, *27*, 172–177. PMID:26584995

Chiu, W. S., Khiew, P. S., Cloke, M., Isa, D., Tan, T. K., Radiman, S., ... Chia, C. H. (2010). Photocatalytic study of two dimensional ZnO nanopellets in the decomposition of methylene blue. *Chemical Engineering Journal*, *158*(2), 345–352. doi:10.1016/j.cej.2010.01.052

Dai, J., Chen, X., Shi, G., Li, L., Wang, G., & Yang, W. (2015). Sonocatalytic degradation of Rhodamine B catalyzed by β-Bi_2O_3 particles under ultrasonic irradiation. *Ultrasonics Sonochemistry*, *27*, 172–177. PMID:26584995

Deng, J., Huang, K., Hu, C., Chen, D., Xiao, N., & Ju, S. (2014). Mixed micelle-mediated extraction for simultaneous trace determination of rhodamine B and rhodamine 6Gby spectrophotometry coupled with the partial least squares regression. *ACAIJ*, *14*, 150–159.

Dhanalakshmi, R., Pandikumar, A., Sujatha, K., & Gunasekaran, P. (2011). Photo catalytic and antimicrobial activities of functionalized silicate sol–gel embedded ZnO–TiO_2 nanocomposite materials. *Materials Express*, *307*, 291–298.

Fang, X. S., Zhai, V., Gautam, U. K., Li, L., Wu, L. M., Bando, Y., & Golberg, D. (2011). ZnS nanostructures: From synthesis to applications. *Progress in Materials Science*, *56*(2), 175–287. doi:10.1016/j.pmatsci.2010.10.001

Georgekutty, R., Seery, M. K., & Pillai, S. C. (2008). A highly efficient Ag-ZnO photocatalyst: Synthesis, properties, and mechanism. *Journal of Physical Chemistry*, *112*, 13563–13570.

Guo, L., Liang, L. H. B., Xu, H., Simon, P., & Wu, Z. Y. (2002). Regularly shaped, single crystalline ZnO nanorods with wurtzite structure. *Journal of the American Chemical Society*, *124*(50), 14864–14865. doi:10.1021/ja027947g PMID:12475325

Guo, X., Zhu, H., & Li, Q. (2014). Visible-light-driven photo catalytic properties of ZnO/ZnFe2O4 core/shell nanocablearrays. *Applied Catalysis B: Environmental*, *160*, 408–414. doi:10.1016/j.apcatb.2014.05.047

Guo, X., Zhu, H., & Li, Q. (2014). Visible-light-driven photocatalytic properties of ZnO/$ZnFe_2O4$ core/shell nanocablearrays. *Applied Catalysis B: Environmental*, *160–161*, 408–414. doi:10.1016/j.apcatb.2014.05.047

Gupta, V. K., Suhas, Ali, I., & Saini, V. K. (2004). Removal of rhodamine B, fast green, and methylene blue from wastewater using red mud, an aluminum industry waste. *Industrial & Engineering Chemistry Research*, *43*(7), 1740–1747. doi:10.1021/ie034218g

Hameed, B. H., Ahmad, A. L., & Latiff, K. N. A. (2007). Adsorption of basic dye (methylene blue) onto activated carbon prepared from rattan sawdust. *Dyes and Pigments*, *75*(1), 143–149. doi:10.1016/j.dyepig.2006.05.039

Harper, C. A. (2001). *Guide to Ceramic Materials. Handbook of Ceramics, Glasses, and Diamonds*. New York: McGraw-Hill.

Hedayati, K. (2015). Fabrication and optical characterization of zinc oxide nanoparticles prepared via a simple sol-gel method. *J Nanostruct*, *5*, 395–401.

Hernandez, M. D. (2009). Development of alternative photocatalysts to TiO_2: Challenges and opportunities. *Energy & Environmental Science*, *2*(12), 1231–1257. doi:10.1039/b907933e

Hua, I., & Hoffmann, M. R. (1997). Optimization of ultrasonic irradiation as an advanced oxidation technology. *Environmental Science & Technology*, *31*(8), 2237–2243. doi:10.1021/es960717f

Huang, M. H. (2001). Room-temperature ultraviolet nanowire nanolasers. *Science*, *292*(5523), 1897–1899. doi:10.1126cience.1060367 PMID:11397941

Jacak, L., Hawrylak, P., & Wojs, A. (1998). Quantum Dots. Springer-Verlag.

Jain, R., Sharma, N., & Bhargava, M. (2003). Electrochemical degradation of RhB in textile and paper industry effluents. *Journal of Scientific and Industrial Research (New Delhi, India)*, *62*, 1138–1144.

Kansal, S. K., Singh, M., & Sud, D. (2007). Studies on photodegradation of two commercial dyes in aqueous phase using different photocatalysts. *Journal of Hazardous Materials*, *141*(3), 581–590. doi:10.1016/j.jhazmat.2006.07.035 PMID:16919871

Kumar, S. S., Venkateswarlu, P., Ranga Rao, V., & Nageswara Rao, G. (2013). Synthesis, characterization and optical properties of zinc oxide nanoparticles. *International Nano Letters*, *30*, 1–6.

Kuo-Chuan, H., Lin, C.-Y., Lai, Y.-H., & Chen, H.-W. (2011). Highly efficient dye-sensitized solar cell with a ZnO nanosheet-based photoanode. *Energy & Environment*, *4*(9), 3448–3455. doi:10.1039/c0ee00587h

Lam, S. M., Sin, J. C., Abdullah, A. Z., & Mohamed, A. R. (2012). Degradation of wastewaters containing organic dyes photocatalysed by zinc oxide: A review. *Desalination and Water Treatment*, *41*(1-3), 131–169. doi:10.1080/19443994.2012.664698

Liu, G., Li, G., Qiu, X., & Li, L. (2009). Synthesis of ZnO/titanate nanocomposites with highly photocatalytic activity under visible light irradiation. *Journal of Alloys and Compounds*, *481*(1-2), 492–497. doi:10.1016/j.jallcom.2009.03.021

Liu, G., Li, G., Qiu, X., & Li, L. (2009). Synthesis of ZnO/titanate nanocomposites with highly photo catalytic activity under visible light irradiation. *Journal of Alloys and Compounds*, *481*(1-2), 492–497. doi:10.1016/j.jallcom.2009.03.021

Long, W., & Li, F. (2011). Hydrothermal synthesis of ZnO flower-reduced graphene oxide composite for electrochemical determination of ascorbic acid. *Energy & Environment*, *4*, 3448–3455.

Lu, Y. C., Wang, L. L., Wang, D. J., Xie, T. F., Chen, L. P., & Lin, Y. H. (2011). A comparative study on plate-like and flower-like ZnO nanocrystals surface photovoltage property and photocatalytic activity. *Materials Chemistry and Physics*, *129*(1-2), 281–287. doi:10.1016/j.matchemphys.2011.04.004

Ma, S. S., Li, R., Lv, C. P., Xu, W., & Gou, X. L. (2011). Facile synthesis of ZnO nanorod arrays and hierarchical nanostructures for photocatalysis and gas sensor applications. *Journal of Hazardous Materials*, *192*(2), 730–740. doi:10.1016/j.jhazmat.2011.05.082 PMID:21684076

Mahasin, F., Al-Kadhemy, Israa, F., Alsharue, Ali Abid, D., & Al-Zuky. (2011). Analysis of the effect of the concentration of rhodamine B in ethanol on the fluorescence spectrum using the "gauss mod" function. *Journal of Physical Chemistry*, *22*(2), 77–86.

Malato, S., Fernandez-Ibanez, P., Maldonado, M. I., Blanco, J., & Gernjak, W. (2009). Decontamination and disinfection of water by solar photocatalysis: Recent overview and trends. *Catalysis Today*, *147*(1), 1–59. doi:10.1016/j.cattod.2009.06.018

Marlinda, A., Pandikumar, A., Yusoff, N., Huang, N., & Lim, H. N. (2015). Electrochemical sensing of nitrite using a glassy carbon electrode modified with reduced functionalized graphene oxide decorated with flower-like zinc oxide. *Mikrochimica Acta*, *182*(5-6), 1113–1123. doi:10.100700604-014-1436-x

Modirshahla, N., Hassani, A., & Mohammad, A. (2011). Effect of operational parameters on decolorization of acid yellow 23 from wastewater by UV irradiation using ZnO and ZnO/SnO_2 photocatalysts. *Desalination*, *271*(1-3), 187–192. doi:10.1016/j.desal.2010.12.027

Mohajerani, M. S., Lak, A., & Simchi, A. (2009). Effect of morphology on the solar photocatalytic behavior of ZnO nanostructures. *Journal of Alloys and Compounds*, *485*(1-2), 616–620. doi:10.1016/j.jallcom.2009.06.054

Mukit, M. S. H., Rahman, M., & Roy, D. (2016). Investigation on the relative degradation of methylene blue (MB) and rhodamine-B (RB) dyes under UV-visible light using thermally treated commercial and doped TiO_2/ZnO photocatalysts. *Int. J. Integ. Sci. Tech*, *2*, 14–18.

Murugesan, S., Suresh, S., Pandikumar, A., Ramaraj, R., & Paul Raj, S. (2011). Metal-free low-cost organic dye-sensitized zno-nanorod photoanode for solid-state solar cell. *Materials Express*, *1*(4), 307–314. doi:10.1166/mex.2011.1035

Noor Azwen, N., Ahmad Ashrif, B., Norhana, A., Sarada, I., & Abdul Rahman, A. (2017). Enhancement of ZnO-rGO nanocomposite thin films by gamma radiation for E.coli sensor. *Applied Surface Science*, *392*, 1134–1143. doi:10.1016/j.apsusc.2016.09.144

Pandikumar, A., Saranya, K. M., & Ramaraj, R. (2012). Sheaf-like-ZnO@Ag nanocomposite materials modified photoanode for low-cost metal-free organic dye-sensitized solid-state solar cell. *Applied Physics Letters*, *101*(9), 187–192. doi:10.1063/1.4748287

Pang, Y. L., Abdullah, A. Z., & Bhatia, S. (2010). Effect of annealing temperature on the characteristics, sonocatalytic activity and reusability of nanotubes TiO_2 in the degradation of Rhodamine B. *Applied Catalysis B: Environmental*, *100*(1-2), 393–402. doi:10.1016/j.apcatb.2010.08.016

Pare, B., Jonnalagadda, S. B., Tomar, H., Singh, P., & Bhagwat, V. M. (2008). ZnO assisted photocatalytic degradation of acridineorange in aqueous solution using visible irradiation. *Desalination*, *232*(1-3), 80–90. doi:10.1016/j.desal.2008.01.007

Phuruangrat, A., Kongpet, W., Yayapao, O., Kuntalue, B., Thongtem, S., & Thongtem, T. (2014). Ultrasonic-assisted synthesis, characterization, and optical properties of Sb doped ZnO and their photocatalytic activities. *Journal of Nanomaterials*, 1–10.

Pitchaimuthu, S., Lakshmi, G., & Velusamy, P. (2015). Enhanced photo catalytic activity of TiO_2 using β-cyclodextrin on solar light assisted decoloration of azocarmine G dye. *J. Adv. Chem. Sci*, *1*, 9–14.

Porter, F. (1991). *Zinc dust and compounds. Zinc handbook properties, processing, and use in design.* New York: Marcel Dekker. doi:10.1201/9781482276947

Qiu, X., Li, G., Sun, X., Li, L., & Fu, X. (2008). Doping effects of Co^{2+} ions on ZnO nanorods and their photo catalytic properties. *Journal of Nanotechnology*, *19*(21), 215703. doi:10.1088/0957-4484/19/21/215703 PMID:21730582

Qiu, X., Li, G., Sun, X., Li, L., & Fu, X. (2008). Doping effects of Co^{2+} ions on ZnO nanorods and their photocatalytic properties. *Nanotechnology*, *19*(21), 5703. doi:10.1088/0957-4484/19/21/215703 PMID:21730582

Rahman, N. M., Roy, D., & Mukit, M. S. H. (2016). Investigation on the relative degradation of Methylene Blue (MB) and Rhodamine-B (RB) dyes under UV-visible light using thermally treated commercial and doped TiO2/ZnO photocatalysts. *J Mat Let*, *2*, 14–18.

Rahman, Q. I., Ahmad, M., Misra, S., & Lohani, M. (2013). Effective photocatalytic degradation of rhodamine B dye by ZnO nanoparticles. *J Mat Let*, *91*, 170–174. doi:10.1016/j.matlet.2012.09.044

Rajalakshmi, S., Kannan, N., & Velusamy, P. (2017). Enhanced photocatalytic activity of metal oxidesβ-cyclodextrin nanocomposites for decoloration of Rhodamine B dye under solar light irradiation. *Applied Water Science*, *7*(1), 115–1274. doi:10.100713201-014-0223-5

Rehman, S., Ullah, R., Butt, A. M., & Gohar, N. D. (2009). Strategies of making TiO_2 and ZnO visible light active. *Journal of Hazardous Materials*, *170*(2-3), 560–569. doi:10.1016/j.jhazmat.2009.05.064 PMID:19540666

Ren, Z. F., Lao, J. Y., Huang, J. Y., & Wang, D. Z. (2003). ZnO nanobridges and nanonails. *Nano Letters*, *3*(2), 235–238. doi:10.1021/nl025884u

Rokesh, K., Pandikumar, A., & Jothivenkatachalam, K. (2014). ZincOxide nanopillar: Preparation, characterization and its photoelectron catalytic activity. *Mater. Focus*, *3*(5), 1–5. doi:10.1166/mat.2014.1172

Sakthivel, S., Neppolian, B., Shankar, M. V., Arabindoo, B., Palanichamy, M., & Murugesan, V. (2003). Solar photocatalytic degradationof azo dye: Comparison of photocatalytic efficiency of ZnO and TiO_2. *Solar Energy Materials and Solar Cells*, *77*(1), 65–82. doi:10.1016/S0927-0248(02)00255-6

Sangave, P. C., & Pandit, A. B. (2004). Ultrasound pre-treatment for enhanced biodegradability of the distillery wastewater. *Ultrasonics Sonochemistry*, *11*(3-4), 197–203. doi:10.1016/j.ultsonch.2004.01.026 PMID:15081981

Saravanan, R., Karthikeyan, S., Gupta, V. K., Sekaran, G., Narayanan, V., & Stephen, A. (2013). Enhanced photocatalytic activity of ZnO/CuO nanocomposite for the degradation of textile dye on visible light illumination. *Materials Science and Engineering C, 33*(1), 91–98. doi:10.1016/j.msec.2012.08.011 PMID:25428048

Shekari, Z., Younesi, H., Heydari, A., Tajbakhsh, M., Chaichi, M., Shahbazi, A., & Saberi, D. (2017). Fluorescence chemosensory determination of Cu^{2+} using a new rhodamine–morpholine conjugate. *Chemosensors, 5*(3), 26–40. doi:10.3390/chemosensors5030026

Snehal, Y., Chandra, M., & Prakash, M. (2016). Biosynthesis of zinc oxide nanoparticles using ixora coccinea leaf extract-a green approach. *Open Journal of Synthesis Theory and Applications, 5*(01), 1–14. doi:10.4236/ojsta.2016.51001

Suresh, S., Pandikumar, A., Murugesan, S., Ramaraj, R., & Paul Raj, S. (2011). Photovoltaic performance of solid-state solar cells based on ZnO nanosheets sensitized with low-cost metal-free organic dye. *Solar Energy, 85*(9), 1787–1793. doi:10.1016/j.solener.2011.04.016

Umar, A. (2009). Growth of comb-like ZnO nanostructures for dyesensitized solar cells applications. *Nanoscale Research Letters, 4*(9), 1004–1008. doi:10.100711671-009-9353-3 PMID:20596445

Umar, A., Chauhan, M. S., Chauhan, S., Kumar, R., Kumar, G., Al-Sayari, S. A., ... Al-Hajry, A. (2011). Large-scale synthesis of ZnO balls made of fluffy thin nanosheets by simple solution process: Structural, optical and photocatalytic properties. *Journal of Colloid and Interface Science, 363*(2), 521–528. doi:10.1016/j.jcis.2011.07.058 PMID:21862032

Velusamy, P., Rajalakshmi, S., Pitchaimuthu, S., & Kannan, N. (2014). Modification of the photocatalytic activity of TiO_2 by β-Cyclodextrin in decoloration of ethyl violet dye. *Journal of Advanced Research, 5*(1), 19–25. doi:10.1016/j.jare.2012.10.001 PMID:25685468

Wan, Q., Wang, T. H., & Zhao, J. C. (2005). Enhanced photocatalytic activity of ZnO nanotetrapods. *Applied Physics Letters, 87*(8), 083105–3. doi:10.1063/1.2034092

Wang, H., Xie, C., Zhang, W., Cai, S., Yang, Z., & Gui, Y. (2007). Comparison of dye degradation efficiency using ZnO powders with various size scales. *Journal of Hazardous Materials, 141*(3), 645–652. doi:10.1016/j.jhazmat.2006.07.021 PMID:16930825

Wang, Z. L. (2004). Zinc oxide nanostructures: Growth, properties and applications. *Journal of Physics Condensed Matter, 16*(25), R829–R858. doi:10.1088/0953-8984/16/25/R01

Warule, S. S., Chaudhari, N. S., Kale, M. A., & More, M. A. (2009). Novel sonochemical assisted hydrothermal approach towards the controllable synthesis of ZnO nanorods, nanocups and nanoneedles and their photocatalytic study. *CrystEngComm, 11*(12), 2776–2783. doi:10.1039/b909916f

Xie, J., Li, Y. T., Zhao, W., Bian, L., & Wei, Y. (2011). Simple fabrication and photocatalytic activity of ZnO particles with different morphologies. *Powder Technology, 207*(1-3), 140–144. doi:10.1016/j.powtec.2010.10.019

Xu, A. N., Liu, Y. N., Zhou, X., Wang, X., Liang, K., Yang, Z. K., ... Sahar, S. (2017). Hydrogenation/oxidation induced efficient reversible color switching between methylene blue and leuco-methylene blue. *RSC Advances*, *7*(48), 30080–30085. doi:10.1039/C7RA04498D

Yan, W. H., Hou, J. B., Fu, Z. P., Yang, B. F., Yang, P. H., Liu, K. P., ... Chen, Y. J. (2009). Growth and photocatalytic properties of one-dimensional ZnO nanostructures prepared by thermal evaporation. *Materials Research Bulletin*, *44*(10), 1954–1958. doi:10.1016/j.materresbull.2009.06.014

Ye, C. H., Bando, Y., Shen, G. Z., & Golberg, D. (2006). Thickness-dependent photocatalytic performance of ZnO nanoplatel. *The Journal of Physical Chemistry B*, *110*(31), 15146–15151. doi:10.1021/jp061874w PMID:16884228

Zeng, J. H., Jin, B., & Wang, F. (2009). Facet enhanced photocatalytic effect with uniform single-crystalline zinc oxide nanodisks. *Chemical Physics Letters*, *472*(1-3), 90–95. doi:10.1016/j.cplett.2009.02.082

Zgu, R., Alivov, Y. Z., Liu, C., Teke, A., Reshchikov, C. A., Dogan, S., ... Morkoc, H¸. A. (2005). Comprehensive review of ZnO materials and devices. *Journal of Applied Physics*, *98*, 1–103.

Zha, Y., Li, C., Liu, X., Gu, F., Du, H. L., & Shi, L. (2008). Zn-doped TiO_2 nanoparticles with high photocatalytic activity synthesized by hydrogen–oxygen diffusion flame. *Applied Catalysis B: Environmental*, *79*(3), 208–215. doi:10.1016/j.apcatb.2007.09.044

Zhang, L., Yang, H. Q., Ma, J. H., Li, L., Wang, X. W., Zhang, L. H., ... Wang, X. Y. (2010). Controllable synthesis and shape-dependent photocatalytic activity of ZnO nanorods with a cone and different aspect ratios and of short-and-fat ZnO microrods by varying the reaction temperature and time. *Appl. Phys. Astronomy*, *100*, 106–1067.

Zhang, Q. F., & Cao, G. Z. (2011). Nanostructured photoelectrodes for dye-sensitized solar cells. *Nano Today*, *6*(1), 91–109. doi:10.1016/j.nantod.2010.12.007

Zhang, W., Zhou, C., Zhou, W., Lei, A., Zhang, Q., Wan, Q., & Zou, B. (2011). Fast and considerable adsorption of methylene blue dye onto graphene. *Bulletin of Environmental Contamination and Toxicology*, *87*(1), 86–90. doi:10.100700128-011-0304-1 PMID:21567134

Zhu, Q. Y., Chen, J., Zhu, Q., Cui, Y. M., Liu, L., Li, B., & Zhou, X. F. (2010). Monodispersed hollow microsphere of ZnO mesoporous nanopieces: Preparation, growth mechanism and photocatalytic performance. *Materials Research Bulletin*, *45*(12), 2024–2030. doi:10.1016/j.materresbull.2010.06.016

Zhuang, J., Dai, W., Tian, Q., Li, Z., Xie, J., Wang, J., ... Wang, D. (2010). Photocatalytic degradation of RhB over TiO_2 bilayer films: Effect of defects and their location. *Langmuir*, *26*(12), 9686–9694. doi:10.1021/la100302m PMID:20334418

Zou, B., Wang, X., Zhang, Q., Wan, Q., Dai, G., & Zhou, C. (2011). Controllable ZnO architectures by ethanolamine-assisted hydrothermal reaction for enhanced photocatalytic activity. *The Journal of Physical Chemistry C*, *115*(6), 2769–2775. doi:10.1021/jp1096822

ADDITIONAL READING

Bahnemann, D. (2004). Photocatalytic water treatment: Solar energy applications. *Solar Energy*, *77*(5), 445–459. doi:10.1016/j.solener.2004.03.031

Curcó, D., Giménez, J., Addardak, A., Cervera-March, S., & Esplugas, S. (2002). Effects of radiation absorption and catalyst concentration on the photocatalytic degradation of pollutants. *Catalysis Today*, *76*(2-4), 177–188. doi:10.1016/S0920-5861(02)00217-1

Field, M. S., Wilhelm, R. G., Quinlan, J. F., & Aley, T. J. (1995). An assessment of the potential adverse properties of fluorescent tracer dyes used for groundwater tracing. *Environmental Monitoring and Assessment*, *38*(1), 75–96. doi:10.1007/BF00547128 PMID:24197914

Herrmann, J. M. (1999). Heterogeneous photocatalysis: Fundamentals and applications to the removal of various types of aqueous pollutants. *Catalysis Today*, *53*(1), 115–129. doi:10.1016/S0920-5861(99)00107-8

Ong, C. B., Ng, L. Y., & Mohammad, A. W. (2018). A review of ZnO nanoparticles as solar photocatalysts: Synthesis, mechanisms and applications. *Renewable & Sustainable Energy Reviews*, *81*, 536–551. doi:10.1016/j.rser.2017.08.020

Somorjai, G. A., & Borodko, Y. G. (2001). Research in nanosciences-great opportunity for catalysis science. *Catalysis Letters*, *76*(1/2), 1–5. doi:10.1023/A:1016711323302

Vutskits, L., Briner, A., Klauser, P., Gascon, E., Dayer, A. G., Kiss, J. Z., ... Morel, D. R. (2008). Adverse effects of methylene blue on the central nervous system. *Anesthesiology*, *108*(4), 684–692. doi:10.1097/ALN.0b013e3181684be4 PMID:18362601

Zeltner, W. A. (2005). Tompkin, D.T. Shedding light on photocatalysis. *Ashrae Trans*, *111*, 532–534.

Zhao, J., & Yang, X. (2003). Photocatalytic oxidation for indoor air purification: A literature review. *Building and Environment*, *38*(5), 645–654. doi:10.1016/S0360-1323(02)00212-3

KEY TERMS AND DEFINITIONS

Carcinogenic: Potential of any chemical substance or radiation to cause cancer in cells. Exposure to through absorption, ingestion, and inhalation causes perilous changes at the living tissues at cellular level.

Chronic Toxicity: Adverse effects due to long term exposure to toxic chemicals.

Eutrophication: Enrichment of nutrient salts in water resources due to run-off from the land that affects the ecosystem such as increased growth of aquatic plants, production of algae, lowering of fish species, deterioration of water quality, etc.

Hazardous: Anything (example: material or radiation) with the potential to cause harm.

Heinz Body Formation: Condition of destruction of red blood cells may be due to certain medications or food.

Neurotoxicity: Toxicity in which causes adverse effect on the structure or function of the nervous system probably by a biological, chemical, or physical agent.

Photocatalysis: Process of change (increase, generally) in rate of a chemical reaction due to the enhanced activity of catalyst under any type of radiation like ultraviolet, visible, infrared.

Photocatalyst: Chemical compounds with the property to absorb radiation and utilize for chemical transformation.

Photodegradation: Destruction of a molecule from higher molecular weight to lower molecular weight fragments.

Chapter 20
Silver Nanoparticles:
Synthesis, Characteristics, and Application

Boguslaw Buszewski
Centre for Modern Interdisciplinary Technologies, Nicolaus Copernicus University, Poland

Viorica Railean Plugaru
Centre for Modern Interdisciplinary Technologies, Nicolaus Copernicus University, Poland

Pawel Pomastowski
Centre for Modern Interdisciplinary Technologies, Nicolaus Copernicus University, Poland

Anatoli Sidorenco
Institute of Electronic Engineering and Nanotechnologies "D.GHITU", Technical University of Moldova, Moldova

ABSTRACT

Silver nanoparticles (AgNPs), as a material with attractive properties, has received interest worldwide by the implementation of synthesis particles at nano-size. AgNPs has triggered intense research on the production of nanoparticles using different synthesis methods due actual and future application. The increasing attention of AgNPs resulted in the growth and development of nanoparticles synthesis methods. Recently, various approaches, including chemical and biological, have been used for them preparation. In order to obtain information about size, shape, or optical properties of nanoparticles, a broad range of analytical methods—microscopy, spectroscopy, and spectrometry techniques—have been used. AgNPs exhibit attractive antimicrobial properties against bacteria (Gram-positive and Gram-negative), fungi, and show also selective toxicity toward normal and cancerous cells. This chapter reviewed the methods of AgNPs synthesis, their characteristics, antimicrobial activity, and cytotoxicity against normal and tumor cells.

DOI: 10.4018/978-1-7998-1241-8.ch020

INTRODUCTION

Nanotechnology has rapidly gained importance in modern research in the last few decades. The synthesis and manipulation of particles structure at nanoscale allow nanotechnology to provide innovative solutions in the field of biomedicine, chemical industries, and electronics.

Many types of nanoparticles have been synthesized and are now being used in different fields for various applications. Silver nanoparticles have received big interest worldwide particularly because of their significant and broad bioactivity. Due to the surface area to volume ratio silver nanoparticles have been used as potential antimicrobial agent in many biomedical applications. Besides physical and chemical methods, the biological systems are also used for the synthesis of metal nanoparticles. Biological synthesis plays a crucial role due to its eco-friendly nature. Bacteria, fungi, algae and plants are used for the intra and extracellular synthesis of silver nanoparticles (Dar et al. 2013; Sadhasivam et al., 2010; Gaikwad.et al, 2013).

Antimicrobial properties of silver have been known from ancient time, similar to other heavy metals, silver is toxic to microorganisms. Along with discovery of several antibiotics, the use of silver compounds has been remarkably declined. Nowadays, there is an increasing concern about the emergence and re-emergence of drug resistance of many pathogens such as bacterial and fungal strains (Shahverdi, 2007). Therefore, the development of new antimicrobial compounds or the modification of available ones to improve their antimicrobial activity became high priority research. In relation to its broad-spectrum activity, efficacy, and lower cost, nanosilver-based antimicrobial agents are excellent candidates for therapeutic use. It seems that a good alternative for chemical, physical and biological methods may be a search not only for new antimicrobial agent but also for new possible inexpensive, convenient sources able to synthesize safe silver composites for human cells without affecting environment and future generation.

In this fact, the present chapter reviewed the methods of AgNPs synthesis, the knowledge about interdisciplinary approaches regarding the physico-chemical characterization of silver nanoparticles using various combination techniques and their antimicrobial and cytotoxic effect. Moreover, it has summarized the current state of understanding on a several topics regarding the antibiotics development, action mechanisms, as well as resistance ability.

ANTIBIOTICS STRATEGY

The "Father of microbiology", Antonie van Leeuwenhoek was one of the first people (26 December 1676) who observed cells of bacteria with the microscope constructed by himself. As later has been noticed, microorganisms populate the entire planet, and due to the property of mutation, they are adapted even for extreme conditions such as high temperature and low acidity (Stewart, 2002). Many of the bacteria prove profitable activity for human - bacteria occupying a gastrointestinal tract, participating in the correct processes of digestion (microflora), holding appropriate pH on the surface of mucous membranes, and determining barrier protective on the surface of skin (Michael et al, 2010). They are also used for different types of fermentation in food industry. However, only after many years of intensive research on the microorganisms their pathogenic activity has been proven as a result of which millions of people died. This problem encouraged mankind to search medicine against the threatening microbes. Many scientists took research in this direction, but only in 1929 year became isolated the first antibiotic. Alexander Fleming observed the inhibitory influence of fungus *Penicillium notatum* on the *Staphylococcus strain.*

Name of antibiotic comes from Greek language, from words: *anti* - against, *bios* - life and their height means substances which they can destroy bacteria or suppress, but are out of order to healthy cells of the organism. The basis of antibiotic therapy it is considered the Ehrlich rule, based on which the antibiotic is a chemical connection, without showing high toxicity to eukaryotic cells, causes the damage or death of microorganisms. Nevertheless, it is important to note that, firstly, the antibiotic was produced by living organisms through insulation and was not product of the chemical synthesis (Kumar et al, 2005).

The antibiotics can act as bacteriostatic or bactericidal. The bacteriostatic action consists on the inhibition of bacterial cell division without impairment of her activity. The non-reproductive microorganism is destroyed by the body's defense forces - phagocytes and antibodies. The main factor in the selection of antibiotics is the type of bacteria fighting mechanism. According to mechanism action, the antibiotics can be divided by different action: damaging the structure and functions of a cell wall (e.g. β lactams, cephalosporins), provoking disorders of the function and biosynthesis of a cell wall (e.g. antifungal drugs), hindering the intracellular biosynthesis of proteins and nucleic acids (e.g. chloramphenicol, tetracyclines, aminoglycosides, macrolides) and hindering the synthesis of nucleic acids (e.g. klotrimazol, quinolone antibiotics, metronidazol) (Kumar et al, 2005). Since the discovery of antibiotics, despite all of Fleming's warnings against the abuse of this type of medication, which can lead to the creation of bacterial resistance, the doctors have prescribed repeatedly for patients in order to fight infection. Mechanisms of different kind of the resistance are embracing changes or alterations of the destination of the effect of the medicine, the decline of the molecule of the antibiotic or reducing the effective concentration of the antibiotic in the unit of bacterium. The genes responsible for the resistance are being handed over by chromosomes, or other elements so as plasmids or jumping genes (the conjugation, the transduction and the transformation) (Gould, 2005). Methicillin-Resistance *Staphylococcus aureus* (MRSA) created the chromosomal resistance by producing enzyme spreading the penicillin (penicillinases). MRSA is also manufacturing proteins binding penicillin (PBP). These bacteria are often only sensitive on vancomycin antibiotic of much lower productivity, in comparison for example to β-lactam. Unfortunately some strains of the *Staphylococcus* were also detected as resistant strains to vancomycin (Vancomycin Resistant *Staphylococcus Aureus*, VRSA). Because of the acquired wide refractoriness the MRSA constitutes the serious problem of hospital infection, therefore it is important from a clinical point of view (Jarmuła et al, 2011). Efficiency of various antibiotics (e.g. fluorochinols) depends on their intracellular concentration. In the bacteria cell wall are generated two mechanisms: IP (Influx Pump) – responsible for the movement of a substance between compartments, in this case from the extracellular to the intracellular space and EP mechanism (Efflux Pump) – that consist in pumping out unwanted toxic substances, from the intracellular to the extracellular space. Obstinacy of bacteria can consist in triggering of IP or activating of EP. EP is the system of medications deflation for help conveyors with the use of a source energy to perform their function, but without degradation or change of structure of medications. In connection with the differences in construction of the Gram (+) and Gram (-) cells, also the systems of the medicines removal are different. Despite Gram (-) bacteria step out the systems of the medicines removal belonging to all mentioned earlier families, the most significant and characteristic for this group are the transporters belonging to the family RND (Resistance - Nodulation - Cell Division) usually coded by chromosomal genes. In case of *Escherichia coli,* step out seven known transporters RND, from which five are well characterized, however *Serratia marcescens* has three mechanisms RND (Michael et al, 2010).

Exists still many undiscovered drug resistance mechanisms. This is a problem on a large scale, because the chances of the new antibiotics discovery are the minimal, but creating unusually quickly refractoriness on the antibiotics causes the decrease of chances on the chance of bacterial infection treatment. This situation, obliges scientists to the search of a new generation of antiseptics, on which microorganisms will not generate the drug resistance mechanisms.

SILVER SYSTEM

One of experience strategies is the use of different forms of metals as antiseptics. Silver is one of noble metals that has got known the earliest, and his healing remarkable properties were known 4000 years BC. Egyptians knew that wounds were healing more quickly after using silver coins, however ancient Greeks, to the purpose of the protection against illnesses, covered everyday articles with silver as well as stored water, wine and other liquids in dishes made of silver in order to extend their period of the usefulness for the consumption (Railen-Plugaru et al, 2014 A). During the World War I silver was used in order to prevent infections, before commercial use of antibiotics, silver appeared as solutions of the silver nitrate, and then of, saline silicon with silver and sulfadiazine which were applied mainly to burns wound. Rapidly after that, antiseptic action of silver was used against infection caused by destroyed germs of typhus, malaria, cholera and dysenteries (Kędziora et al, 2013). Nowadays, silver has been used in medicine in many antiseptic formulations in the form of colloidal silver, silver nanoparticles (AgNPs), for prevention, exudation, psoriasis or acne, but also in the prevention of post-operative infections of the genitourinary system, most commonly caused by *Escherichia* or *Enterococcus* species, and even in the treatment of wounds. Recent developments in the pharmaceutical industry also include preparations containing AgNPs in the form of drops for external use in otitis or rhinitis. AgNPs are also added in acne and psoriasis cosmetics. However, the overuse of silver solutions in oral form can lead to a disease called silver (blue skin color), which is a consequence of excessive metal accumulation in cells and to necrosis of liver tissues (Żelazowska et al, 2017).

Moreover, silver is a well-known catalyst for the oxidation of methanol to formaldehyde and ethylene to ethylene oxide (Nagy et al, 1999). In the United States only over 4mnl tones of silver were used in the United States in 2000. Silver is currently of particular interest due to its characteristic properties such as good conductivity, chemical and catalytic stability, and antibacterial activity (Frattini, 2005). Colloidal silver is used as a substrate for Raman spectroscopy, since it requires a partially conductive surface (Tessier et al, 2000).

The supposed mechanism of action of silver nanoparticles is based on the secondary release of silver cations (Trojan Horse Mechanism, THM). Ag^+ ions have a high affinity for functional groups such as phosphate, carboxy, amino, sulfhydryl, and imidazole. These groups are part of the proteins and nucleic acids contained in both prokaryotic and eukaryotic cells. It is suggested that silver ions penetrate the outer layers of the bacterial cell, binding to the phospholipid membrane of the cytoplasmic membrane. This is followed by the accumulation of silver in the wall and the cell membrane, creating free spaces in them. This causes their destabilization and high permeability and as a result uncontrolled transport of ions and outflow of metabolites outside the cell. Even low concentrations of Ag^+ ions induce proton leakage and lead to the death of the microorganism. In addition, contact of silver ions with cell surface leads to protein denaturation and inactivation of many enzymes. Ag^+ ions also bind to the thiol groups (-SH) of the proteins, causing them to denature. In addition, they break the disulfide bridges of proteins, preventing them from adopting a proper conformation for proper functioning (Pomastowski at al, 2016). The affinity of Ag^+ ions to functional groups of the ribosome active site results in the inhibition of translational processes and the blocking of the oxidative enzymes involved in the respiratory chain. This leads to excessive production of free radicals and blocking of translation processes. Ag^+ ion interacts with the nitrogenous bases in the DNA which leads to the formation of a condensed nucleic acid form. Due to the fact that replication can only undergo DNA in relaxed form, the presence of Ag^+ ions blocks bacterial and eukaryotic cell division. This results in the cytotoxic properties of silver ions for the human body.

Unfortunately, all the antiseptic mechanisms of silver nanoparticles have not been proven in sufficient ways. One of the variables influencing silver antisepticism is to generate nanoparticle synthesis.

SILVER NANOPARTICLES SYNTHESIS

The most commonly used method for obtaining silver nanoparticles (AgNPs) as a stable colloidal suspension in water or organic solvents is chemical reduction (Tao et al., 2006). Commonly used reducing agents are hydrogen bromide, citrate, ascorbate and elemental hydrogen (Lee et al, 1982). The reduction of silver ions (Ag^+) in aqueous solution generally yields colloidal silver with particle diameters of a few nanometers (Wiley et al, 2015). Initially, the reduction of various compounds with Ag^+ ions leads to the formation of silver atoms (Ag^0), which then, by agglomeration, form oligomeric clusters. These clusters often lead to the formation of colloidal silver particles (Kapoor et al, 1994). The mechanism of silver nanoparticles creation is three-steps. The first is the mentioned reduction of silver cations, which leads to the formation of metallic silver. This is due to the redox reaction where electrons from the reducer (reduction agent) go to metal ions. In the following step, the reduced silver (Ag^0) plays a role of nucleation sites where silver cations (Ag^+) are reduced to atomic silver by an electrochemical process that lead to the formation of silver nanoclusters (Ag^0_n). These nanoclusters are subject to coalescence and creation of stabilized nanoparticles performed by additional dispersion stabilizer such as citrate (Buszewski et al, 2015; Pomastowski et al, 2016).

When the colloidal particles are significantly smaller than the visible wavelength, the solutions are characterized by a yellow color with an intense band of 380-400 nm, and other less intense bands at higher wavelengths in the absorption spectrum (Cao et al, 2002). The use of a strong reducer, such as borohydride, resulted in the formation of small particles that were monodisperse, while the formation of larger particles was difficult to control. The use of a weaker reducing agent, such as citrate, resulted in a slower reduction rate, but the size distribution was far from narrow (Lee et al, 1982). Silver nanoparticle syntheses can be controlled on a two-step reduction process. In this technique, a strong reducing agent is used to produce small Ag particles that are magnified in the second stage using a weaker reducing agent. In the second stage, the particle size increases from 20-45 nm to 120-170 nm (Schneider et al, 1999). Unfortunately, the method is poorly reproducible and requires specialized equipment. Synthesis of silver nanoparticles by chemical reduction is often carried out in the presence of stabilizers to prevent undesirable colloidal agglomeration.

The popular method of AgNPs chemical synthesis is the use of microemulsions. Microemulsions are called a liquid mixture of water with a surfactant, oil (ternary mixture) and co-surfactant (four-component mixture). Microcapsules have the ability to form oil-in-water (O/W) or water-in-oil (W/O) micelles. The microcrotus form is dependent on the proportion of micellar constituents and hydrophilic-lipid balance. W/O micelles are used for the synthesis of metal nanoparticles. This type of microemulsion is transparent and consists of water droplets (of manometric size) that are suspended in the diffusing oil phase and are stabilized at the interface by the surfactant molecules (Zhang et al, 2007A). Metal nanoparticle synthesis reactions occur in water droplets surrounded by a surfactant, which are specific micro-reactors reaction.

Referring to the three-stage nanoparticle synthesis by the redox method, the first stage of the nucleation reaction and nanoparticle growth takes place in the micelle. The stabilizer in this case is a surfactant, in addition to inhibiting the nanoparticle aggregation process, it must protect the nanoparticles from excessive growth because the size of the microcapsules is limited (Zhang et al, 2007B). Synthesis of metal nanoparticles in the reverse micelle system takes place by mixing two microemulsions, one containing metal salt and the other reducing agent. Reducing agents used in the microemulsion method are identical to those used in the chemical reduction method, it is a matter of identical redox reactions occurring in the micelle as described in the previous subsection. At the moment of mixing two microemulsions (one with metal salt and the other with reducing agent), microcapsules collide resulting in the exchange of reagents between micelles. The general process of metal nanoparticle synthesis by microemulsion is based on micelle collisions and reagent exchange, followed by nucleation, aggregation and growth of AgNPs (Xu et al, 2006).

Analyzing the nowadays increasing ecological problems, although new applications appear on the horizon, the future will be focused more on the idea of sustainable synthesis development of silver nanoparticles and their use in food and pharmaceuticals field. Certainly, searching for new safe source for the silver composites synthesis, taking into consideration the present needs, without compromising the future generations, will be of the increased interest. One of this solution is using of microorganisms as a source of the nanoparticles synthesis.

The “green synthesis” of silver nanoparticles is becoming increasingly popular and comprises three main steps that must be based on the principles of green chemistry, including 1) solvent selection, 2) choice of mild reducing agent, 3) selection of non-toxic AgNPs stabilizing substances (Raveendran et al, 2003).

Microorganisms such as bacteria, yeasts, fungi and algae are capable of synthesizing metal nanoparticles according to the “green synthesis” principle. The main principle of the synthesis is to reduce the metal ion to nanoparticle. Many different microorganisms have been recorded for the synthesis of nanomaterials. Attention was focused mainly on nanoparticles of silver and gold, which is conditioned by their properties. Metal-derived nanoparticles are mainly characterized by a lack of toxicity, which is why they are used in medicine and the pharmaceutical and cosmetic industries (Buszewski et al, 2016). Although there are many publications based on biological research of nanoparticle synthesis, only a very small part focuses on the exact mechanism of this process.

Microorganisms produce inorganic compounds both inside and extracellular. The mechanisms of the external and intracellular synthesis of nanoparticles differ from each other by the biological agents present. The intracellular method involves the characteristic transport of ions into the microbial cell. In this method, the cell wall of microorganisms is very important. The mechanism involves electrostatic interactions between the positive-charged metal ion and the negatively charged cell wall. Enzymes contained in the cell wall reduce metal ions to nanoparticles that then diffuse through the cell wall. The mechanism of extracellular synthesis of metal nanoparticles using microorganisms is dependent on the presence of nitrate reductase. For example, the nitrate-reductase-induced fungus helps in metal ion and nanoparticle bioreduction. The share of enzymes showing redox potential and being able to act as electron transfer plays a key role in reducing metal ions. (Buszewski et al, 2016, Railen-Plugaru et al, 2016B). Among environmental resources and prokaryotic cells, the use of nanoparticles in the synthesis of metals has found bacteria that are considered promising microbial nanoparticle microbes. One reason for interest in bacteria is simple manipulation of their culture and metabolism. Bacteria produce both internal and extracellular synthesis. Various parameters, that are used to synthesize nanoparticles by

Figure 1. Scheme of chemical and biological approach of silver nanoparticles

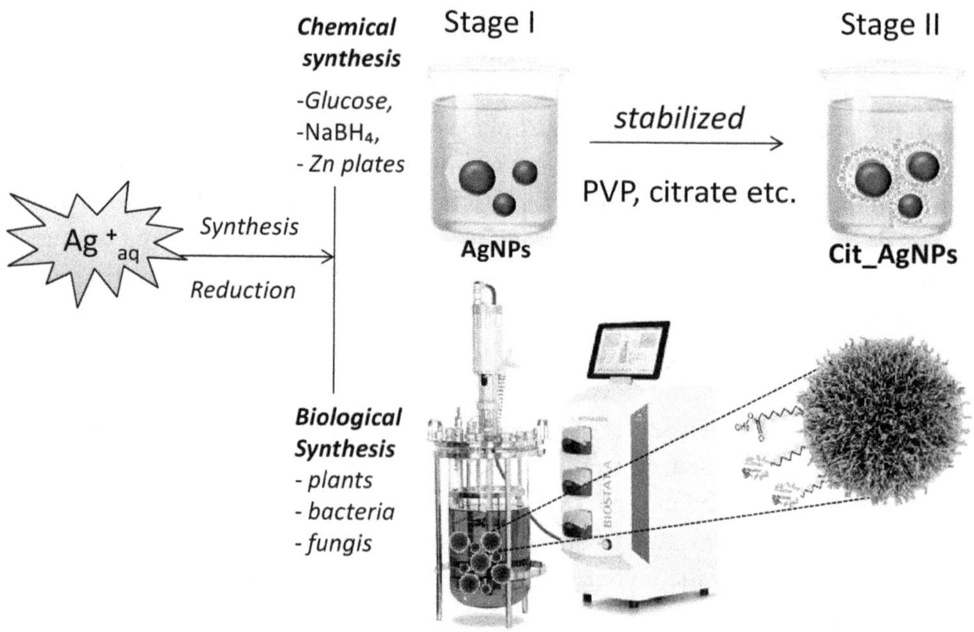

microorganisms include alteration of solubility and toxicity by reduction or oxidation, lack of a specific metal transport system, biosorption, extracellular complexation or precipitation of metals, bioaccumulation and effluent (Husseiny et al, 2007). AgNPs, which are synthesized by microorganisms, achieve sizes and crystallic properties as obtained by the chemical process. The difference is the one-step process of biological synthesis, spontaneous reduction and biological stabilization by microbial enzymes, secreted metabolites **(Figure 1)**.

With the help of various microorganisms we are also able to manipulate the size, shape and composition of nanoparticles (Nasreen et al, 2014). This example compares the synthesis of AgNPs with bacteria isolated from *Streptacidiphilus durhamensis* strain and probiotic bacteria (e.g. *Lactococcus lactis*). In all cases metallic nanoparticles of silver were obtained up to 50 nm, but the different organic deposit naturally coated on the surface of the nanoparticle was determined by various properties, eg spectral or cytotoxic (Buszewski et al, 2016; Railen-Plugaru et al, 2016 B, Railean-Plugaru et al, 2017).

Biosynthesis of silver nanoparticles using fungi is widely used because of the possibility of obtaining monodisperse nanoparticles with well-defined dimensions, different structures and sizes. Due to the presence of various enzymes in the cells and simple synthesis and eukaryotic metabolism, mushrooms are considered a good biological precursor. Mushrooms, compared to bacteria, synthesize larger amounts of nanoparticles. The productivity of nanoparticle synthesis is higher than that of bacteria due to the production of more proteins by fungi (Mohanpuria et al, 2007).

Table 1. Biologically synthetized silver nanoparticles

	Precursor	Size and shape	References
bacteria	*Streptacidiphilus durhamensis* strain	5-50 nm, spherical	[B.Buszewski et al. 2016] [V.Railean-Plugaru et al. 2016]
	Pseudomonas meridian (MTCC 4993)	2.2–21.5 nm, spherical	[S. Shivaji et al, 2011]
yeast	*Saccharomycetes cereviasae* strain MKY3	2–5 nm, hexagonal	[M. Kowshik et al. 2003]
fungy	*Fusarium semitectum* Berk. & Ravenel	10–60 nm, spherical	[S.S. Basavaraja et al. 2008]

Other advantages of mushrooms are:

- Intracellular metal absorption capacity
- Specific enzymes, such as fungal reductases, facilitate the biosynthesis of metal nanoparticles of different chemical compositions.
- Cells culture on the surface of the inorganic substrate, leading to the distribution of metal as a catalyst.
- Large amounts of enzymes are produced per biomass unit.

Monodisperse nanoparticles of 5-25 nm silver were produced in 10 minutes with *Aspergillus fumigatus* (Bhainsa, 2006). The production of nanoparticles using these fungi is faster than the chemical and physical methods of nanoparticle synthesis. As such, *A. fumigatus* can be used to produce large scale nanoparticles FTIR studies have shown that nanoparticle-limiting amine groups determine the stability of nanoparticles and proteins act as a stabilizer and blocking agent. Another example is the synthesis of silver nanoparticles using the fungus *Penicillium brevicompactum* (Shaligram et al, 2009). It has been shown that the synthesis of NADH-dependent reductase nitrate was responsible for the reduction of silver ions to metallic silver. NADH-reductase-dependent could act as an electron carrier from NADH in the reduction process. FTIR studies show that carbonyl peptides containing amino acid residues have a strong ability to bind and form a layer of silver nanoparticles to prevent agglomeration. The precise mechanism of silver nanoparticle synthesis using these fungi is not yet known.

Numerous publications show that all types of yeast are able to reduce heavy metals. Yeast species have developed many mechanisms that aim to protect them against toxic properties of heavy metals, including but not limited to: enzymatic oxidation or reduction, cell wall sorption and, in some cases, chelation of extracellular peptides or polysaccharides or their active outflow (Breierova et al, 2002). The mechanism of detoxication in yeast cells is caused by GSH (glutathione) and two metal groups associated with metallothionein and phytochelatin (PC). In most yeast species studied, these molecules determine the mechanism of nanoparticle formation and stabilization of complexes.

The ability of yeast cells to change the bonding of metal ions by complex nontoxic polymers, eg by popular baking yeast (*Saccharomyces cerevisiae*), is responsible for the reduction of ionic silver and the formation of biologically active silver nanoparticles. These nanocolloids have three important properties: crystallinity, narrow size distribution and good water solubility. Due to the presence of the peptide coating, the nanoparticles do not stick together and do not cause Ostwald's maturation, resulting in greater stability than with synthetic nanoparticles. In addition, the presence of the peptide coating provides nanoparticles with a hydrophilic surface without additional preparatory steps. In Table 1 it was presented example of biologically synthetysied silver nanoparticles **(Table 1)**.

Figure 2. Physicochemical characteristics of AgNPs

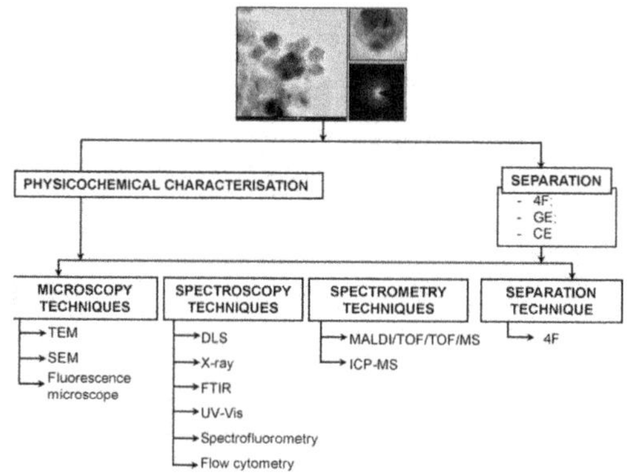

PHYSICOCHEMICAL CHARACTERISTICS OF AgNPs

One of the most stimulating aspect of fizico-chemical characteristic of silver nanoparticles is to design modern interdisciplinary technologies. Interdisciplinary approach allows to highlight the complementary aspect of the nanoparticles characterization and evaluation of their particularities. In **Figure 2** was presented the physicochemical characteristics strategy of AgNPs.

Dynamic light scattering (DLS) and spectroscopic techniques are the most popular methods of physicochemical characterization of silver nanoparticles. Dynamic light scattering (DLS) is a method of measuring the size of nanoparticles dispersed in a liquid. This technique consists of illuminating the laser beam dispersed in the nanoparticle solution, then the scattered laser light is captured by the detector and converted into an electrical signal. The signal on the detector is changing by the continuous changes in nanoparticles caused by Brownian motion. This method is based on the Stokes-Einstein equation by converting the particle size distribution based on the measurement of particle movement velocity as a result of Brownian motion.

Zeta Potential (ZP) is a physicochemical term referring to biocoid systems, is a method used to study surface charge, surface condition of nanoparticles and predictability of nanoparticles. Nanoparticles have a surface electrical charge that attracts a thin layer of ions with opposing charges on the surface of the nanoparticles.

The electrical potential that occurs on the double boundary of the layer is known as the zeta potential of the biocolide. The zeta potential is determined by the electrophoretic mobility (μe) according to the Smoluchowski's equation **(Figure 3)**. Nanoparticles with zeta potential values between -30 mV and +30 mV typically have a low degree of stability. Deviations from these values indicate greater stability of the biocolloids (Buszewski et al, 2015).

Figure 3. Double layer, zeta potential and Smoluchowski's equation

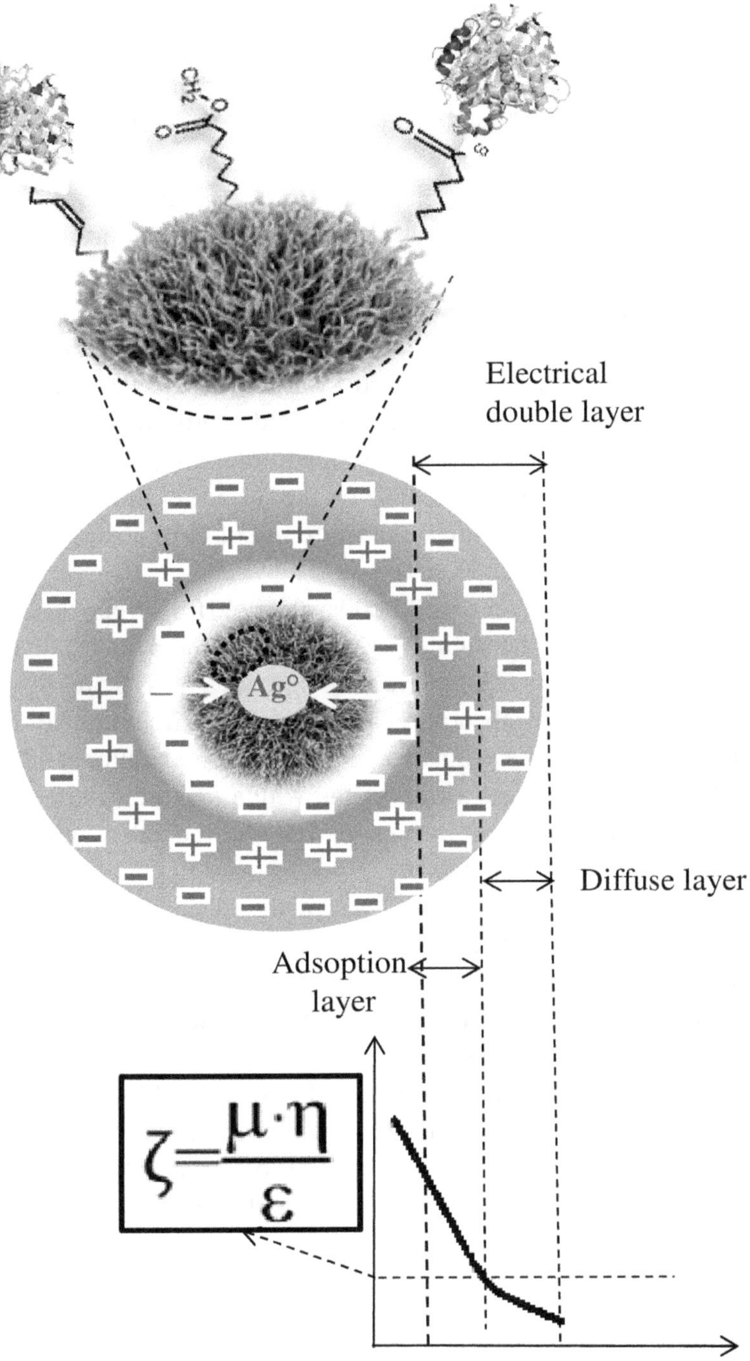

The main problem encountered during the nanoparticle study by DLS and ZP is their aggregation. This problem is usually eliminated by the use of ultrasonification, thanks to which the nanoparticle aggregates are broken up. The zeta potential and size of the hydrodynamic bio colloids also affect pH, ionic strength and temperature of the solvent. The hydrodynamic size of biologically synthesized silver nanoparticles ranges between 100 ± 1.5 nm. At a pH of 2 to 4, the size of silver nanoparticles decreased from 700 to 200 nm. Nanoparticles of size <100 nm were obtained in the pH range of 5-10 (Railean-Plugaru et al, 2016 B).

Fourier transform infrared spectroscopy (FTIR) is useful for obtaining information about the composition of nanomaterials by identifying functional groups occurring on the surface of nanoparticles. Most popular method of FTIR spectra registration is based on potassium bromide (KBr) by the tablet (disks) and calcium fluorite (CaF_2) by dry drop method. Spectra range are mostly register at middle infrared (MIR) range in $\nu = 400 - 4000cm^{-1}$. The bands observed in 3427±5/ cm, 3421±5/ cm are characteristic for hydroxyl and amino groups. Another common feature is the band in the range of 1384.4±5/ cm corresponding to N-O stretching vibrations. For AgNPs, this signal is less intense compared to $AgNO_3$ may be the result of the oxidation of amines located on the surface of the biocolloids. The new signal for 1623±5/cm may be due to the presence of tertiary amide groups that are derived from the carbonyl stretching (C = O). The 1480±5/cm band is also new and is attributed to the C-N stretching and NH bending vibrations, indicating the presence of a second-order amide. The resulting FTIR spectrum in MIR suggests the presence of a protein or peptide on the surface of the synthesized silver nanoparticle. Signals in the range of 800-875±5/cm indicate the presence of C-Br absorption bands that are the result of the KBr disc (Railean-Plugaru et al, 2016B).

The UV-Vis spectroscopy technique (λ = 200-800nm) is used for the optical characteristics of nanoparticles. Thanks to it we can monitor the stability of nanoparticles and the fact that we received them. For silver the characteristic band occurs at a wavelength of about 415-470 nm. Silver nanoparticles with a diameter of 7 ± 2 nm have a narrow absorption band with a maximum of 410 nm, silver nanoparticles with a diameter of 29 ± 5 nm also have a narrow band with a maximum at 425 nm, and nanoparticles with a diameter of 89 ± 5 nm represent a wider band with a maximum 490 nm. Thus, the application along with the increase in the diameter of the nanoparticles of their absorption bands is shifted towards longer waves (Martínez-Castañòn et al, 2008).

To determine the structure and crystalline form of the metal nanoparticles, the X-ray diffraction spectroscopy (XRD) is used. X-ray diffraction is also operated to evaluate the degree of metal oxidation associated with biomass (Mittal et al, 2014).

To visualize silver nanoparticles, microscopic techniques such as transmission electron microscope (TEM), scanning electron microscope (SEM), scanning transmission electron microscopy (STEM) and atomic forces microscopy (AFM) are commonly used. Using the TEM and SEM techniques, it is possible to determine the shape and size of the received nanoparticles of the metal. SEM and AFM technique allow to study the morphology of the obtained nanocomposites of metals, check the presence of nanoparticle aggregates and the effects of nanoparticle-bacteria compartment (Buszewski et al, 2016). Electron microscopy techniques together with Energy-dispersive X-ray detectors (EDX) enable it to determine the atomic composition of the obtained nanoparticles. Silver nanoparticles are mainly polydisperse shaped spherical, triangular and cubic.

The inductively coupled plasma mass spectrometry (ICP-MS) method is a technique for determining elements at the trace level (at ppb level) of transition metals including silver. The procedure is based on the ionization of the sample in an inductively coupled plasma and then the mass spectrometer is determined using a mass spectrometer. Another spectrometric analysis such as matrix-assisted laser desorption ionization – time of flight (MALDI TOF MS) analysis is performed to study metal nanoparticels biocolloids. This method is focused on biological materials, mainly on proteins, peptides and other biologically active molecules. The principle of operation is based on the precise transfer of the laser energy assisted by the matrix when approached to the analyte, resulting in the phenomenon of desorption of large ions. The result is a measurement of the mass of the test sample with high accuracy

(Pomastowski et al, 2015; Railean-Plugaru et al, 2018). The advantage of this technique is that there is no damage to the sample. MALDI spectra illustrate the present of silver isotopes, silver-organics connections and clusters of silver in analyzed sample of bionanoparticels of silver. MALDI spectra illustrate molecular fingerprint of synthetized silver biocolloids, registered in refection and positive mode.

BIOLOGICAL ACTIVITY OF SILVER NANOPARTICLES

Silver is known for its antimicrobial properties and has been used for many years in medicine against various gram positive and gram negative microorganisms.

The most commonly used methods for determination of the antiseptic and antimicrobial properties of AgNPs are Minimum inhibitory concentration (MIC), Disc diffusion and Well diffusion methods. MIC is the method that determines the lowest concentration of antimicrobial agent that inhibits visible growth of the microorganism at a certain incubation time. This method is primarily used in diagnostic laboratories to confirm the drug resistance properties or as a research tool investigating in vitro activity of new antibacterial agents. Disc and well diffusion method are another commonly used method for investigation of drug resistance properties of microorganisms. This method is based on the diffusion of the drug from the disk/well, to the substrate (medium). When the antimicrobial agent is applied to a medium inoculated with microorganisms (bacteria or fungus), the size of the growth inhibition zone is proportional to the sensitivity of the microorganisms to the drug. It is considered that the microorganisms are more susceptible to the given drug when the area of inhibition zone is larger.

During the last years the fluorescence microscope and flow cytometry techniques have become well-established methodology used as a complementary tools to assess viability of the bacterial cells under silver nanoparticles treatment (Railean-Plugaru et al, 2016C). Proper selection of the methods for the antimicrobial investigation is crucial.

According to the literature, silver nanoparticles are efficient antimicrobial agent against dangerous environmental strain of bacteria e.g. *Bacillus subtilis* and clinical pathogens e.g. MRSA, MSSA, *Pseudomonas aeruginosa or Klebsiella pneumonia* (Railean-Plugaru et al, 2016C; Buszewski et al, 2017; Railean-Plugaru et al, 2017). However, it is well known that size and source of silver biocolloids influence the physicochemical properties, mechanisms of toxic action and in consequence the bioavailability.

Moreover, silver nanoparticles have demonstrate also the ability to prevent HIV (human immunodeficiency virus) by their binding to host cells. In addition, silver increase popularity also as a disinfectant for water and surface disinfection.

The detailed mechanism of antibacterial activity of silver nanoparticles is still incomprehensible. Many study have confirmed the ability of silver nanoparticles to anchor the bacterial cell walls and cause the structural changes in the cell membrane which in consequence increase its permeability. Furthermore, it is suggested that antibacterial activity can also occur through direct physical contact between nanoparticle and microorganisms, causing structural damage to the cell wall. Small silver nanoparticles having a larger surface area of silver ions release increase the antimicrobial activity in comparison to the larger particles. On the other hand, secondary oxidized silver ions based on Trojan-Horse Mechanism description can exerts toxicity to bacterial cells by damaging cellular membrane. Silver nanoparticles are considered to be a source of slowly released silver ions that react with the thiol group of proteins to

Figure 4. Biological activity of AgNPs

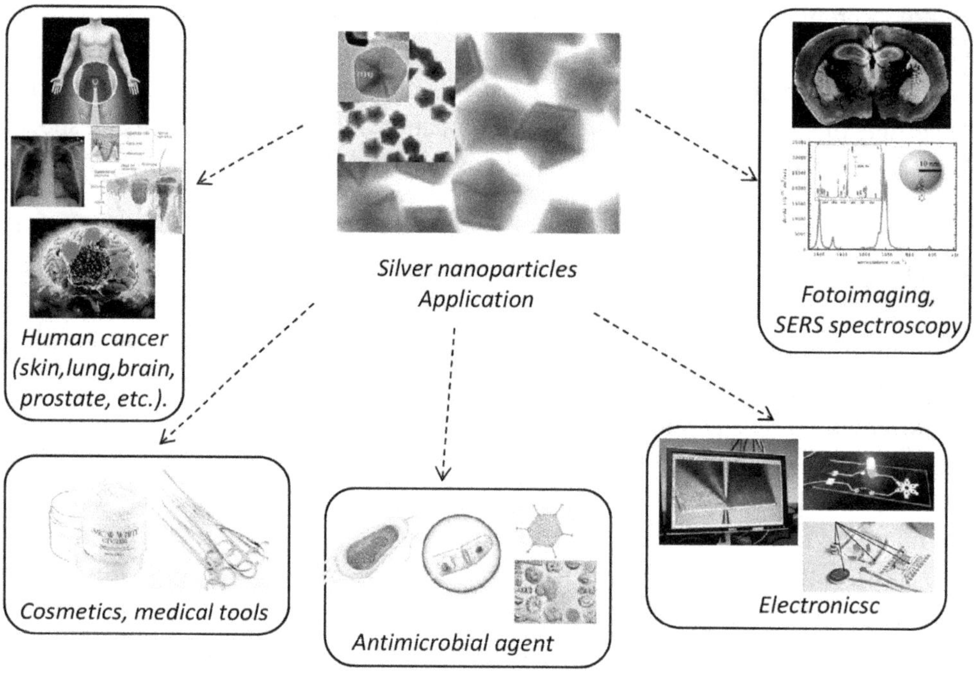

interfere with DNA replication. As it is known, silver also produces free radicals that damage the bacterial membrane (Buszewski et al, 2016; Railean-Plugaru et al, 2016C).

In recent years, severe fungal infections have significantly contributed to the increasing morbidity and mortality of immunocompromised patients who need intensive antibiotic treatment. Candida spp. Represent one of the most common groups of pathogens responsible for fungal infections, often causing hospitalal infections including sepsis, which are associated with mortality up to 40% (Patterson et al. 2007). Currently most of the available and effective antifungal agents are based on polyunsaturated (Amphotericin B), triazole (fluconazole, itraconazole, voriconazole, posaconazole) or echinocandins (Kaspofungin, mycafungin and anidulafungin). However, the use of these agents is often accompanied by various complications and toxicity in combination with other drugs. Drug resistance is also increasingly common. Silver nanoparticles are an interesting alternative especially for the treatment of candidoses or fingernail or fingernail fungus. Silver nanoparticles exhibit antifungal activity against clinical isolates *Trichophyton mentagrophytes* i *Candida spp.* (Kim et al, 2008).

In Figure 4 it is presented the application schema of AgNPs.

Nanoparticles are currently used in bio-applications as therapeutic agents, antimicrobial agents, transfection vectors, and fluorescence. The rapid progress of nanotechnology and the early acceptance of received materials increase the potential for adverse health effects. There are no established effects of long-term exposure at different concentrations in humans and the environment. However, the significant impact of nanomaterials on the environment is expected in the future. In particular, the behavior of nanoparticles inside the cells is still a mystery because no immune reactions induced by these particles have been observed up to now. This problem is addressed by nanotoxicology by setting regulations for bioaccumulation and nanoparticle toxicity at the cellular level. Silver nanoparticles (AgNPs) are gaining in popularity due to the wide spectrum of antimicrobial activity. Silver nanoparticles are now impregnated with catheters and dressings used in hospitals. Despite the extensive use of AgNPs in dressings, which provides easy access to cells, there are very few reports of silver nanoparticle toxicity. Mechanisms involved in nanoparticle toxicity on microorganisms may also be active in humans. It is suspected that the size of the surface of the nanoparticles affects the activity *in vivo*. Recent reports show the contribution of mitochondria in the toxic pathway of AgNPs. In vivo experiments in rats, changes in lung function and inflammation have been observed. It is also reported that starch sterilized silver nanoparticles and BSA (bovine serum albumin) induce various developmental defects in embryos. The nanoparticle nanocatalyst, which is measured as the rate of absorption of nanoparticles, intracellular distribution and exocytosis, contributes to toxicity. The size of the nanoparticles, their surface and functionalization of the surface are the main factors influencing the biocomics and therefore the toxicity. Silver nanoparticles coated by organics deposit (e.g. proteins, peptides) are less cytotoxic to eukaryotic cells in comparison with chemically synthetized AgNPs. Moreover, AgNPs functionalized by antibiotics (e.g. tetracycline) exhibit synergistic effect against pathogens (Buszewski et al, 2016).

Environmental Impact: The increasing ecological problems encourages the researchers to be focused more on searching of safe sources for the synthesis of the silver nanoparticles, taking into account the nowadays needs without compromising the future generation. Therefore, in this fact, the present chapter is coming to summarized/supplement the current knowledge concerning the synthesis, phisico-chemical characteristics and antimicrobial efficiency of silver nanoparticles.

ACKNOWLEDGMENT

This work was financially supported by National Science Centre in frame of Opus 14 project No. 2017/27/B/ST4/02628 (2018-2021)

REFERENCES

Basavaraja, S. S., Balaji, S. D., Lagashetty, A. K., Rajasab, A. H., & Venkataraman, A. (2008). Extracellular bio-synthesis of silver nanoparticles using the fungus Fugarium semitectum. *Materials Research Bulletin*, *43*(5), 1164–1170. doi:10.1016/j.materresbull.2007.06.020

Bhainsa, K. C., & Souza, S. F. D. (2006). Extracellular biosynthesis of silver nanoparticles using the fungus Aspergillusfumigatus. *Colloids and Surfaces. B, Biointerfaces*, *47*(2), 160–164. doi:10.1016/j.colsurfb.2005.11.026 PMID:16420977

Breierova, E., Vajczikova, I., Sasinkova, V., Stratilova, E., Fisera, M., Gregor, T., & Sajbidor, J. (2002). Biosorption of cadmium ions by different yeast species. *Z Naturforsch*, *57*(7-8), 634–639. doi:10.1515/znc-2002-7-815 PMID:12240989

Buszewski, B., & Pomastowski, P. (2015). Influence of heterogeneity biocolloids surface on their electrophoretic separation. *Wiadomości chemiczne*, *69*, 9–10.

Buszewski, B., Rafińska, K., Pomastowski, P., Walczak, J., & Rogowska, A. (2016). Corrigendum to Novel aspects of silver nanoparticles functionalization. *Colloids and Surfaces. A, Physicochemical and Engineering Aspects*, *506*, 170–178. doi:10.1016/j.colsurfa.2016.05.058

Buszewski, B., Railean-Plugaru, V., Pomastowski, P., Rafinska, K., Szultka-Mlynska, M., Golinska, P., ... Dahm, H. (2018). Antimicrobial activity of biosilver nanoparticles produced by a novel Streptacidiphilus durhamensis strain. *Journal of Microbiology, Immunology, and Infection*, *51*(1), 45–54. doi:10.1016/j.jmii.2016.03.002 PMID:27103501

Buszewski, B., Railean-Plugarua, V., Pomastowski, P., Rafińska, K., Szultka-Mlynska, M., & Kowalkowski, T. (2017). Antimicrobial Effectiveness of Bioactive Silver Nanoparticles Synthesised by Actinomycetes HGG16n Strain. *Current Pharmaceutical Biotechnology*, *18*(15), 1–9. PMID:28056728

Dar, M. A., Ingle, A., & Rai, M. (2013). Enhanced antimicrobial activity of silver nanoparticles synthesized by Cryphonectria sp. evaluated singly and in combination with antibiotics. *Nanotech. Biology and Medicine*, *9*, 105–110. PMID:22633901

Frattini, A., Pellegri, N., Nicastro, D., & Sanctis, O. (2005). Effect of amine groups in the synthesis of Ag nanoparticles using aminosilanes. *Materials Chemistry and Physics*, *94*(1), 148–152. doi:10.1016/j.matchemphys.2005.04.023

Gaikwad, S., Birla, S., Ingle, A., Gade, A., Marcato, P., Rai, M., & Duran, N. (2013). Screening of different *Fusarium* species to select potential species for the synthesis of silver nanoparticles. *Journal of the Brazilian Chemical Society*, *24*(2), 1974–1982.

Gould, I. M. (2007). MRSA bacteraemia. *International Journal of Antimicrobial Agents*, *30*(1), 66–70. doi:10.1016/j.ijantimicag.2007.06.023 PMID:17875386

Husseiny, M. I., Ei-Aziz, M. A., Bad, R. Y., & Mahmoud, M. A. (2007). Biosynthesis of gold nanoparticles using Pseudomonas aeruginosa. *Spectrochimica Acta. Part A: Molecular and Biomolecular Spectroscopy*, *673–674*(3-4), 1003–1006. doi:10.1016/j.saa.2006.09.028 PMID:17084659

Jarmuła, A., Obłąk, E., Wawrzycka, D., & Gutowicz, J. (2011). Oporność wielolekowa związana z aktywnym usuwaniem leków z komórek drobnoustrojów. *Postepy Higieny i Medycyny Doswiadczalnej*, *65*, 216–227. doi:10.5604/17322693.937011 PMID:21502698

Kapoor, S., Lawless, D., Kennepohl, P., Meisel, D., & Serpone, N. (1994). Reduction and Aggregation of Silver Ions in Aqueous Gelatin Solutions. *Langmuir*, *10*(9), 3018–3022. doi:10.1021/la00021a026

Kędziora, A., & Sobik, K. (2013). Oporność bakterii na srebro- problem stary czy nowy? *Kosmos (Stockholm)*, *62*(4), 557–570.

Kim, K. J., Sung, W. S., Moon, S. K., Choi, J. S., Kim, J. G., & Lee, D. G. (2008). Antifungal effect of silver nanoparticles on dermatophytes. *Journal of Microbiology and Biotechnology*, *18*, 1482–1484. PMID:18756112

Kowshik, M., Ashtaputre, S., Kharrazi, S., Vogel, W., Urban, J., Kulkarni, S. K., & Paknikar, K. M. (2003). Extracellular synthesis of silver nanoparticles by a silver-tolerant yeast strain MKY3. *Nanotechnology*, *14*(1), 95–100. doi:10.1088/0957-4484/14/1/321

Kumar, A., & Schweizer, H. (2005). Bacterial resistance to antibiotics: Active efflux and reduced uptake. *Advanced Drug Delivery Reviews*, *57*(10), 1486–1513. doi:10.1016/j.addr.2005.04.004 PMID:15939505

Lee, P. C., & Meisel, D. (1982). Adsorption and surface-enhanced Raman of dyes on silver and gold sols. *The Journal of Chemical Physics*, *86*(17), 3391–3395. doi:10.1021/j100214a025

Lee, P. C., & Meisel, D. (1982). Adsorption and surface-enhanced Raman of dyes on silver and gold sols. *Journal of Physical Chemistry*, *86*(17), 3391–3395. doi:10.1021/j100214a025

Martínez-Castañòn, G. A., Niño-Martínez, N., Martínez-Gutierrez, F., Martínez-Mendoza, J. R., & Ruiz, F. J. (2008). Synthesis and antibacterial activity of silver nanoparticles with different sizes. *Journal of Nanoparticle Research*, *10*(8), 1343–1348. doi:10.100711051-008-9428-6

Michael, A., Kohanski, D. J., Dwyer, J., & Collins, J. (2010). How antibiotics kill bacteria: From targets to networks. *Nature Reviews. Microbiology*, *8*(6), 423–433. doi:10.1038/nrmicro2333 PMID:20440275

Mittal, A. K., Bhaumik, J., Kumar, S., & Banerjee, U. C. (2014). Biosynthesis of silver nanoparticles: Elucidation of prospective mechanism and therapeutic potential. *Journal of Colloid and Interface Science*, *415*, 39–47. doi:10.1016/j.jcis.2013.10.018 PMID:24267328

Mohanpuria, P., Rana, N. K., & Yadav, S. K. (2007). Biosynthesis of nanoparticles: Technological concepts and future applications. *Journal of Nanoparticle Research*, *10*(3), 507–517. doi:10.100711051-007-9275-x

Nagy, A., & Mestl, G. (1999). High temperature partial oxidation reactions over silver catalysts. *Applied Catalysis A*, *188*(1-2), 337–353. doi:10.1016/S0926-860X(99)00246-X

Patterson, T. F. (2007). Treatment and prevention of fungal infections. Focus on candidemia. New York. *Applied Clinical Education*, *2007*, VII–VIII.

Pomastowski, P., Sprynskyy, M., Žuvela, P., Rafińska, K., Milanowski, M., Liu, J. J., ... Buszewski, B. (2016). Silver-Lactoferrin Nanocomplexes as a Potent Antimicrobial Agent. *Journal of the American Chemical Society*, *138*(25), 7899–7909. doi:10.1021/jacs.6b02699 PMID:27263865

Pomastowski, P., Szultka-Młyńska, M., Kupczyk, W., Jackowski, M., & Buszewski, B. (2015). Evaluation of Intact Cell Matrix-Assisted Laser Desorption/Ionization Time-of-Flight Mass Spectrometry for Capillary Electrophoresis Detection of Controlled Bacterial Clumping. *Journal of Analytical & Bioanalytical Techniques, S13,008,* 1–7.

Railean-Plugaru, V., Pomastowski, P., Kowalkowski, T., Sprynskyy, M., & Buszewski, B. (2018). Physicochemical study of natural fractionated biocolloid by asymmetric flow field-flow fractionation in tandem with various complementary techniques using biologically synthesized silver nanocomposites. *Analytical and Bioanalytical Chemistry*, *410*(11), 2837–2847. doi:10.100700216-018-0967-0 PMID:29616293

Railean-Plugaru, V., Pomastowski, P., Meller, K., Zloch, M., Rafinska, K., & Buszewski, B. (2017). Lactococcus lactis as a safe and inexpensive source of bioactive silver composites. *Applied Microbiology and Biotechnology*, *101*(19), 7141–7153. doi:10.100700253-017-8443-x PMID:28842740

Railean-Plugaru, V., Pomastowski, P., Rafińska, K., Kupczyk, W., Jackowski, M., & Buszewski, B. (2016). Antimicrobial properties of biosynthesized silver nanoparticles studied by flow cytometry and related techniques. *Electrophoresis*, *37*(5–6), 752–761C. doi:10.1002/elps.201500507 PMID:26763104

Railean-Plugaru, V., Pomastowski, P., Szultka, M., & Buszewski, B. (2014). Bionanokompleksy srebra w walce z patogenami. *Analityka*, *3*, 8–17.

Railean-Plugaru, V., Pomastowski, P., Wypij, M., Szultka-Młyńska, M., Rafińska, K., Golińska, P., ... Buszewski, B. (2016). Study of silver nanoparticles synthesized by acidophilic strain of Actinobacteria isolated from the of Picea sitchensis forest soil. *Journal of Applied Microbiology*, *120*(5), 1250–1263. doi:10.1111/jam.13093 PMID:26864807

Rani, A. (2009). Cytotoxicity and genotoxicity of silver nanoparticles in human cells. *ACS Nano*, *3*(2), 279–290. doi:10.1021/nn800596w PMID:19236062

Raveendran, P., Fu, J., & Wallen, S. L. (2003). Completely "Green" Synthesis and Stabilization of Metal Nanoparticles. *Journal of the American Chemical Society*, *125*(46), 13940–13941. doi:10.1021/ja029267j PMID:14611213

Sadhasivam, S., Shanmugam, P., & Yun, K.-S. (2010). Biosynthesis of silver nanoparticles by Streptpmyces hygroscopicus and antimicrobial activity against medically important pathogenic microorganism. *Colloids and Surfaces. B, Biointerfaces*, *81*(1), 358–362. doi:10.1016/j.colsurfb.2010.07.036 PMID:20705438

Schneider, S., Halbig, P., Grau, H., & Nickel, U. (1994). Reproducible preparation of silver sols with uniform particle size for application in surface-enhanced raman spectroscopy. *Photochemistry and Photobiology*, *60*(6), 605–610. doi:10.1111/j.1751-1097.1994.tb05156.x

Shahverdi, A. R., Minaian, S., Shahverdi, H. R., Jamalifar, H., & Nohi, A. A. (2007). Rapid synthesis of silver nanoparticles using culture supernatatnt of *Enterobacteria*: A novel biological approach. *Process Biochemistry (Barking, London, England)*, *42*(5), 919–923. doi:10.1016/j.procbio.2007.02.005

Shaligram, N. S., Bule, M., Bhambure, R., Singhal, R. S., Singh, S. K., Zakacs, G., & Pandey, A. (2009). Biosynthesis of silver nanoparticles using aqueous extract from the compactin producing fungal strain. *Process Biochemistry (Barking, London, England)*, *44*(8), 939–943. doi:10.1016/j.procbio.2009.04.009

Shivaji, S., Madhu, S., & Singh, S. (2011). Extracellular synthesis of antibacterial silver nanoparticles using psychrophilic bacteria. *Process Biochemistry (Barking, London, England)*, *46*(9), 1800–1807. doi:10.1016/j.procbio.2011.06.008

Stewart, P. S. (2002). Mechanisms of antibiotic resistance in bacterial biofilms. *International Journal of Medical Microbiology*, *292*(2), 110–113. doi:10.1078/1438-4221-00196 PMID:12195733

Tao, A., Sinsermsuksaku, P., & Yang, P. (2006). Polyhedral Silver Nanocrystals with Distinct Scattering Signatures. *Angewandte Chemie International Edition*, *45*(28), 4597–4601. doi:10.1002/anie.200601277 PMID:16791902

Velev, P. M., Kalambur, O. D., Rabolt, A. T., Lenhoff, J. F., Kaler, A. M., & Kaler, E. W. (2000). Assembly of Gold Nanostructured Films Templated by Colloidal Crystals and Use in Surface-Enhanced Raman Spectroscopy. *Journal of the American Chemical Society*, *122*(39), 9554–9555. doi:10.1021/ja0022831

Wiley, B., Sun, Y., Mayers, B., & Xi, Y. (2005). Shape-Controlled Synthesis of Metal Nanostructures: The Case of Silver. *Chemistry (Weinheim an der Bergstrasse, Germany)*, *11*(2), 454–463. doi:10.1002/chem.200400927 PMID:15565727

Xu, J., Han, X., Liu, H., & Hu, Y. (2006). Synthesis of silver nanoparticles in revers micelles stabilized by natural biosurfactant. *Colloids and Surfaces. A, Physicochemical and Engineering Aspects*, *273*, 179–183. doi:10.1016/j.colsurfa.2005.08.019

Żelazowska, R., & Pasternak, K. (2007). Matale szlachetne: Srebro (Au), złoto (Au), platyna w biologii i medycynie. *Bromatologia i Chemia Toksykologiczna*, *2*, 205–209.

Zhang, W., Qiao, H., & Chen, J. (2007b). Synthesis of silver nanoparticles—Effects of concerned parameters in water/oil microemulsion. *Materials Science and Engineering B*, *142*(1), 1–15. doi:10.1016/j.mseb.2007.06.014

Zhang, W., Qiao, X., & Chen, J. (2007a). Synthesis of Nanosilver Colloidal Particles in Water/Oil Microemulsion. *Colloids and Surfaces. A, Physicochemical and Engineering Aspects*, *299*(1-3), 22–28. doi:10.1016/j.colsurfa.2006.11.012

Chapter 21
Recovery of Natural Nanostructured Minerals:
Ferrospheres, Plerospheres, Cenospheres, and Carbonaceous Particles From Fly Ash

Nisha Choudhary
https://orcid.org/0000-0002-1847-6520
Central University of Gujarat, India

Virendra Kumar Yadav
Sankalchand Patel University, India

Parth Malik
Central University of Gujarat, India

Samreen Heena Khan
Central University of Gujarat, India

Gajendra Kumar Inwati
Department of Chemistry, Kadi University, India

Suriyaprabha R.
Central University of Gujarat, India

Bijendra Singh
Central University of Gujarat, India

A. K. Yadav
Central University of Gujarat, India

Raman Kumar Ravi
Central University of Gujarat, India

ABSTRACT

Nanotechnology is playing an important role in every field of science, medicine, electronics, catalysis, and environmental cleanup. Due to the high demand of nanoparticles, it is synthesized at commercial level by chemical and physical methods, which are expensive and energy intensive. The recovery of natural nanoparticles from the waste materials makes the process cost-effective and simultaneously reduces pollution. Fly ash or coal fly ash (CFA), a by-product of thermal power plants in-houses numerous natural nanoparticles like ferrospheres, plerospheres, cenospheres, and carbon nanoparticles, which finds applications in ceramics, wastewater treatment, lightweight materials, fillers, and composite manufacturing. Every year a million tons of fly ash are produced around the globe. In the chapter, all these fly ash natural nanoparticles have been discussed in detail including their properties, applications, and possible future applications.

DOI: 10.4018/978-1-7998-1241-8.ch021

BACKGROUND

Nanosciences and nanotechnology has played an important role in the electronics, medicine, wastewater treatment and catalysis. Nanomaterials being smaller in size have high surface area to volume ratio and high surface energy which makes them suitable material for adsorbents, catalysis, drug delivery etc. However, their synthesis by chemical methods not only makes the process hazardous but also energy intensive and expensive due to utilization of chemical precursors. Therefore, the recovery of nanomaterial from the industrial waste product like fly ash reduces the cost of the product and overall expenditure of the process. Fly ash naturally have nanostructured materials like cenospheres, ferrospheres, plerospheres and carbonaceous nanomaterials which can be recovered and applied for various applications where purity is not a priority.

INTRODUCTION

Nanotechnology is the branch of science that deals with the fabrication of objects with dimensions of roughly 1–100 nm (Mathew & Juang, 2007). Nanotechnology is the science and art of matter that is manipulated at the atomic and molecular scale. According to the National Nanotechnology Initiative (NNI), nanotechnology was defined as 'anything smaller than 100 nm with novel properties' (Medical Advisory, 2006). The important features of nanotechnology and nanoparticles are attributed to the shape, size surface characteristics and surface energy. As the nanoparticles have infinite size (1-100 nm), so the surface area is relatively large and consequently have high reactivity and sorption to several heavy metals (Al-Saad et al., 2012). These nanotechnological features allows the nanoparticles to become chemically more reactive by changing its surface activity and other properties. Hence, concerning the fortification of the environment, nanotechnology embraces the assurance of providing new and inimitable improvements (Martins & Mata, 2015). Today nanoparticles finds applications in all the fields of sciences whether it is energy, electronics, medicine, environmental cleanup etc. (Khan, Saeed, & Khan, 2017). The nanoparticles can be available in nature or it could be synthesized in the laboratory under optimized conditions. Nanoparticles are not new to the nature or human society; they are present from ancient times only in our surroundings (Ivanova et al., 2019). Nowadays, due to the advances in the instrumentation and technology we have identified their detailed information in the nature. For instance, clay, volcanic ash, different ashes, fly ash and several biological entities are the source of natural nanoparticles (Griffin et al., 2017). Today, these nanoparticles have gained huge importance in the industry and other application purpose. One such source of natural nanoparticles are clay (layered silicates) which are a member of phyllosilicates, which occurs naturally due to the continuous chemical weathering of silicate rich minerals like rocks, volcanic ash or soil (Gräfe, McFarlane, & Klauber, 2017); (Calabi-Floody, Theng, & Mora, 2009). From last decade, due to advancement in the nanotechnology, interests of scientist in clay minerals has accelerated and consequently a drastic growth in the clay-based research has observed. Clay minerals are generally considered nanomaterials, as they possess layers with thickness and inter- layer spaces both within the nanoscale range (Zhu & Njuguna, 2014), so the term 'clay minerals' is often used interchangeably with 'nanoclay' (Rytwo, 2012). This nanoclay consists of sheets of tetrahedron (SiO_4) and octahedron silicate (comprising Mg, Al and Fe). Natural clay minerals are built of layered structural units with a layer thickness of approximately one to a few nanometers and lateral dimensions varying from 30 nm to several microns, which gives a ratio of length to thickness greater than 1000 (Fu, Sun, Huang, Li, & Hu, 2019). Based on the chemical composition, net charge and structure characteristics and net charge, clay minerals are majorly classified into seven groups: (i) chlorite, (ii) pyrophyllite (iii) smectite, (iv) mica, (v) vermiculite, (vi) kaolin and (vii) interstratified clay minerals (Wakeel, Gul, & Sanaullah, 2013). Other sources of natural nanoparticles are volcanic ashes yet the particle size distribution below 1000 nm is only rarely documented. Volcanic ashes results from the explosive volcanic eruptions and contain several natural nanoparticles in them (Hegde, Goswami, Sarma, Dasu, & Brar, 2015). Fly ash are also considered as one of major reservoir of natural nanoparticles. Fly ash are byproducts of thermal power plants produced during generation of electricity from the burning of pulverized coal (Yadav & Fulekar, 2018). They are considered as a hazardous materials and every year produced in million tonnes around the globe. They are enriched with several natural nanoparticles. Some of the scientists have considered living organisms and their organelles as natural nanoparticles like viruses, viral capsids and microbial cell organelles. These biological nanoparticles can be used as whole or their parts. Viruses are used for gene delivery in molecular biology, while their capsids are also used for the numerous biological applications. Viruses like mycophages, bacteriophages, picornaviruses and several other virus like particles i.e. prions falls in the nano regime. Viruses have been used as template for the synthesis of bionanomolecules (Willner & Willner, 2010), biomimetic materials (Yannas & Burke, 1980) from a decade.

In contrary to the natural nanoparticles there are anthropogenically synthesized nanoparticles also which could be of two types: incidental and engineered nanoparticles (Morales, 2018). The incidental nanoparticles are less ordered in size, no controlled growth while engineered one are well controlled size, crystals and other parameters. Synthetically all the types of nanoparticles could be synthesized in laboratory conditions (Patra & Baek, 2014). Nevertheless, they require a precursor material, energy and utilizes harmful chemicals and such methods are energy intensive, expensive and chemical products are challenging threat for the environment. Therefore, we need to switch on to the renewable sources of nanoparticles or nanoparticles synthesis from the waste materials. Synthesis of nanoparticles from waste materials not only reduces the pollution but also provides an alternative material for the synthesis of specific nanoparticles (Das, Chakraborty, Chatterjee, & Kumar, 2018). There are several types of waste like agricultural waste (rice husk, coconut shell, almond shell, sugarcane bagasse), industrial waste (fly ash, red mud, iron scrap, alumina waste) etc. (Ali, Asim, & Khan, 2012). All these waste products are major problem in the form of solid waste and need to be managed properly. Utilization of such industrial and agricultural waste will solve the environmental pollution and energy demand. Moreover, synthesis of nanoparticles from such waste products are green chemistry approach. Besides these, there are clay, volcanic ash that could also acts as a source of natural nanoparticles.

FLY ASH: A SOURCE OF NATURAL NANOPARTICLES

Properties of fly ash

Fly ash is a very fine coal ash particles produced as a byproduct during production of electricity in thermal power plants. Fly ash has obtained its name "fly" ash since it is being transported from the combustion chamber through exhaust gases (Fuller et al., 2018). Fly ash is the fine powder formed from the mineral matter in coal, comprising of the noncombustible matter in coal in addition to a small amount of carbon that remaining parts due to incomplete combustion. Fly ash are generally light in color and comprises of mainly silt-sized and clay-sized glassy spheres, which gives fly ash a consistency fairly like talcum powder (Sahu, 2019). The major components of bituminous coal fly ash are oxides of aluminum, iron, silica, and calcium, with varying amounts of carbon (J. Mccarthy, K. Solem, E. Manz, & J. Hassett, 1989), as estimated by the loss on ignition (LOI). Lignite and sub bituminous coal fly ash have higher convergences of oxides of Ca and Mg and diminished percentages of silica and iron oxide (Wardhono, 2018). They have also lower carbon content, compared with bituminous coal fly ash. Lignite and sub bituminous coal fly ash may likewise have a higher concentration of sulfate compounds than bituminous coal fly ashes (Fuller et al., 2018). The size of fly ash varies from 0.1 microns to 7.0 microns. Their major elemental compositions are alumina silica and ferrous besides they have traces of Ti, Na, Mg, Na and other heavy metals. Based on the source of coal used for burning, fly ash are of two type's class F and class C. Where the class C-fly ash are derived from the lower grades of coal (lignite, peat) while class F are derived from the higher grades of coal (anthracite, bituminous, sub-bituminous). The primary difference between them are the total composition of ferro alumino silicate (FAS) and CaO percentage, i.e. FAS content in class C is below 70% and CaO more than 5% while class F have more than 70% FAS composition and CaO less than 5% (Dwivedi & Jain, 2014). Out of both, class F are more preferable for the extraction of natural nanomaterials as their elemental composition is higher than the class C. They have higher silica, ferrous and alumina content. Besides this higher grades of coal are also rich in carbon so it could also acts as a source of carbon nanomaterial (carbon nanotubes, fullerenes) and hydrocarbons (polyaromatic and polycyclic).

One of the fundamental clarifications that coal fly ash nanoparticles are so reactive is that, as the size of a molecule diminishes, the ratio of its surface area to volume increases drastically, subsequently increasing the amount of surface available for reactions (Adeyemo, Adeoye, & Bello, 2017). Fly ash have silica nanominerals, alumina, ferrous, carbonaceous nanominerals and clay as major constituents. Fly ash based nanoparticles can be either extracted or it can be synthesized by using it as a precursor material. The major natural nanoparticles present in fly ash are ferrospheres, aluminosilicates particles, cenospheres, plerospheres and carbonaceous nanoparticles. While, the products like alumina, zeolites and silica can be synthesized from fly ash after prior activation by chemical method. Morphologically, fly ash spherical particles (plerospheres, ferrospheres, cenospheres, and aluminosilicates spheres) and irregular shaped unburned carbon also called as char, porous carbon or carbonaceous nanoparticles are present naturally in the coal fly ash (CFA). These nanoparticles are formed during the combustion of coal, vaporization-condensation of the minerals of coal at higher temperature in the thermal power plants. There is recrystallization, reformation of various minerals present in coal, and there is reduction and oxidation of minerals and elements during combustion of coal in the thermal power plant chamber. The temperature of the burning chamber is around 1200-1800 ^{0}C, which are accountable for formation of such natural nanoparticles. The size of most of the fly ash nanoparticles varies from 60 nm to several microns depending on the operating conditions, design and types of TPP and ESP. The silica and alumina nanoparticles which are synthesized from fly ash have smaller size while natural nanoparticles like ferrospheres (Sharonova, Anshits, Fedorchak, Zhizhaev, & Anshits, 2015) plerospheres (Goodarzi & Sanei, 2009) cenospheres (Beddu et al., 2018); (Żyrkowski, Neto, Santos, & Witkowski, 2016) and carbon particles (Salah et al., 2016) have comparatively larger size. As they are grown under uncontrolled conditions, where different minerals come in contact and coalesce together in the molten state. The spherical particles form the major fraction of fly ash particles formed at the time of combustion process (Goodarzi & Sanei, 2009). All these natural nanoparticles are discussed below according to their extraction, properties and applications.

Table 1. Normal range of chemical composition for fly ash produced from different coal types (expressed as percent by weight).

Components	Bituminous	Subbituminous	Lignite
SiO_2	20-60	40-60	15-45
Al_2O_3	5-35	20-30	10-25
Fe_2O_3	10-40	4-10	4-15
CaO	1-12	5-30	15-40
MgO	0-5	1-6	3-10
SO_3	0-4	0-2	0-10
Na_2O	0-4	0-2	0-6
K_2O	0-3	0-4	0-4
LOI	0-15	0-3	0-5

FERROSPHERES

These are spherical fly ash particles, having depositions of ferrous particles on their external surface, which have magnetic property. Their size varies from 90 nm to several microns. Their major elemental compositions are Fe and O along with Ca, Mg, Na, K, Ti and C. The pattern of ferrous depositions on their surface particles provides them morphological properties. Based on the type of ferrous depositions on their surface ferrospheres can be categorized into two type smooth surface and rough surface. Rough surface ferrospheres have mainly numerous angular or spherule shaped ferrous depositions on their surface. Due to these, its texture is rough and termed as rough surfaced ferrospheres (Aaron Fuller et al., 2018). While the other type of ferrospheres is smooth surfaced ferrospheres, where ferrous particles are not deposited on their surface rather evenly distributed on the surface and have smooth texture. They are generally weakly magnetized in comparison to the rough surfaced ferrospheres. Besides these there are dendritic shaped (Norton, Markuszewski, & Shanks, 1986), where the ferrous particles are deposited in the form of dendrites. The deposited ferrous particles have mixed phases of iron oxides viz: magnetite, hematite and maghemite, which are deposited during vaporization and condensation on the external surface of CFA particles. These iron particles comes from the combustion of coal, which on reduction and oxidation of iron pyrites, iron sulfides converts into oxides. The ferrous fractions are about 5-16% based on the source of fly ash, coal, operating conditions of thermal power plants and other parameters. As ferrospheres, have depositions of Fe on the surface of Al-Si rich spherical particles. Therefore, ferrospheres have majorly ferrous, aluminum and silica content. Besides this, they also have traces of heavy metals like Hg, Mo, Ni, Co, Cu and As. The ferrous present in fly ash are always incorporated with the impurities in the form of Na, Mg, K, P, Mn, Ti, Ca etc. in trace amount (Gomes et al., 1999). Ferrospheres of fly ashes could be recovered economically by magnetic separation techniques (Xue & Lu, 2008). Besides ferrospheres and plerospheres there is another class also called ferroplerospheres, which are plerospheres having depositions of ferrous particles on them, and acquires magnetic properties.

Figure 1. FESEM micrograph of fly ash ferrospheres nanoparticles

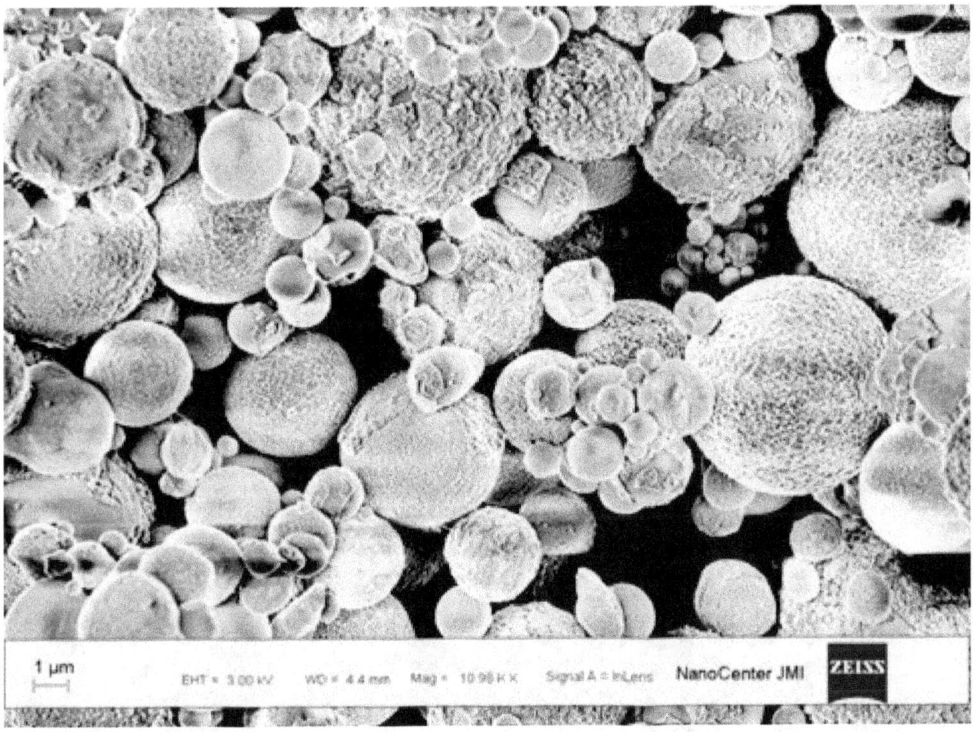

Extraction of Ferrous Particles from fly ash

Extraction of ferrous particles is possible by the magnetic separation methods, which will also segregate the ferrous from the non-ferrous. Based on the presence of impurities, ferrospheres can be processed further for the purification and development of iron based products. Several researchers have isolated the ferrous particles from fly ash by using magnetic separation method either by wet method or by dry method. Dry method generally uses the machines with conveyer belts. The conveyer belts are fitted with magnet at one end, which separates the ferrous from the nonferrous. While the wet method separates the ferrous particles from the slurry. These ferrospheres are further dried and used for future applications. The FESEM micrograph of ferrospheres in fig.1 reveals ferrospheres with different morphology. Few ferrospheres have polygonal deposits of ferrous on their surface while few have ferrous depositions in the form of small spherules. As such, particles have texture due to deposition of ferrous on their surface they are called as rough surfaced ferrospheres. These ferrous depositions provide them magnetic properties.

These deposited ferrous particles are crystalline in nature, and have mixtures of magnetite, maghemite and hematite. Size of these deposited ferrous particles varies from are 50-200 nm. While few particles are smooth surfaced, which have uniform depositions of ferrous on their surface which provides them magnetic properties and are weakly magnetic in nature. Such ferrospheres are called smooth surfaced ferrospheres. TEM image in fig.2 reveals a group of ferrospheres particles of size 300 nm to several microns. The fly ash ferrospheres are generally polycrystalline in nature as evident from literatures.

Figure 2. TEM micrograph of ferrospheres nanoparticles from fly ash

Properties and Applications

Ferrospheres are spherical shaped, smooth to rough surface having magnetic properties. The deposited ferrous particles could be of polygonal, angular, spherules shaped. They are rich in ferrous and have associations of Al, Si, Ca, Ti, and O. The matrix of ferrospheres is made up of Al and Si while the surface consists of Fe. The most common applications of fly ash extracted ferrospheres is as a fillers for composite materials (Fuller et al., 2018), raw material in steel and other iron based industries.

CENOSPHERES

The name 'cenosphere' originates from two Greek words, which are kenos (empty or hollow) and sphaira (circle or sphere). Cenospheres are hollow ceramic microspheres that establish a little however vital part of by-products from coal combustion process (Ranjbar & Künzel, 2017). The cenospheres are thin-walled, ultra-lightweight, hollow and inert ceramic microspheres which are loaded up with air or gases and cenospheres having the density lesser than water produced as by-products in fly ash. They are also called as aluminosilicates spheres or ash microspheres. The fig.3 shows SEM micrograph of a broken

Figure 3. SEM micrograph of a broken cenosphere from fly ash, adapted from (Shao, Jia, & Liu, 2009)

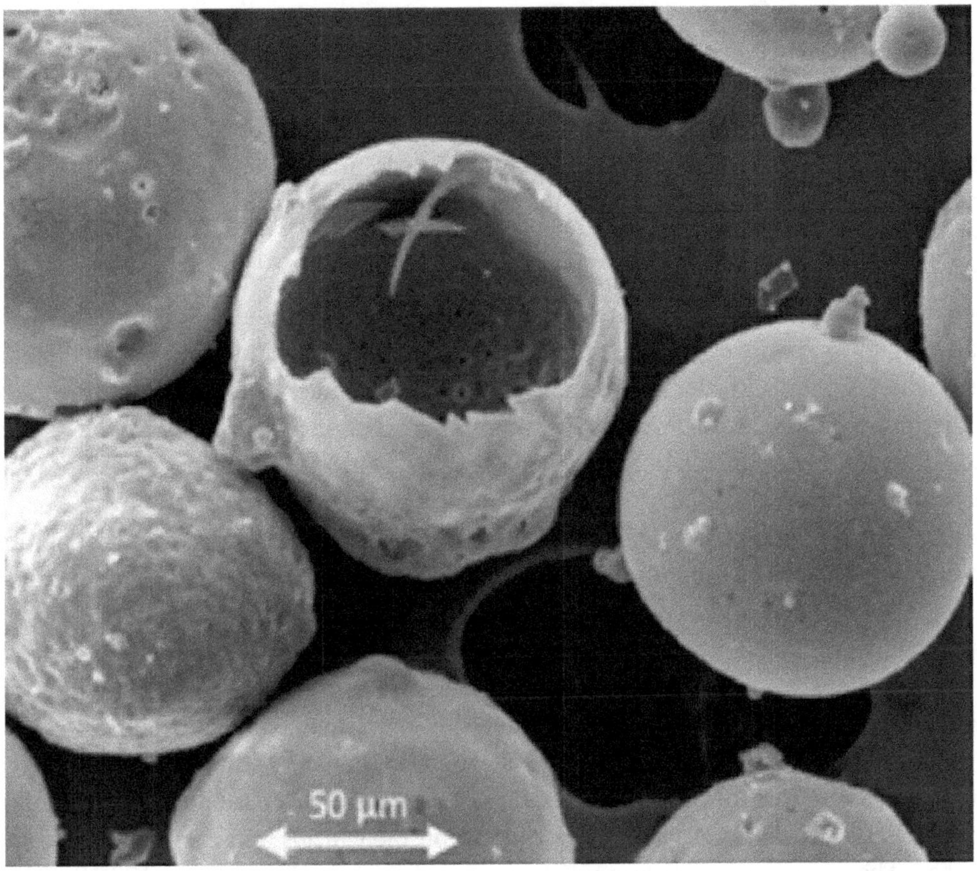

cenospheres from the fly ash by Shao et al., 2009. While there are rough and smooth surface particles, which have depositions of white color small carbon particles on their surface.

The cenospheres are made up of glassy surface majorly and matrix of crystalline like mullites, which have covering of nano-film of 30-50 nm (Katiyar & Balasubramanian, 2014). They have shell like appearance whose thickness varies 2-30 microns. They are most wanted by-products of coal combustion process nowadays, whose size generally varies from 10 – 1000 μm diameter and constitute around 1-2% of the fly ash obtained from the coal combustion process. They have gained huge consideration an essential subject of coal TPPs, because of their remarkable properties, namely their low density and very high mechanical strength (S. Pandian, Rajasekhar, & Sridharan, 1998). The properties of cenospheres rely upon the nature of the coal utilized, combustion, grinding operations and withdrawal in the process of electricity generation (Hirajima et al., 2010). Aluminosilicates microspheres are basically solidified melts made up of oxides of Si, Al and vitreous substance as the major portion. At high aluminum concentrations (higher than 34–36% on an oxide basis), cenospheres are occasionally pierced by mullite or, more rarely, silimanite crystals (Ngu, Wu, & Zhang, 2007). The amount of cenospheres or plerospheres in an ash normally depends on its carbon and iron contents.

Formation of Cenospheres in fly ash

The process of formation of cenospheres from the coal combustion is a really an overwhelming mechanism. This process depends on the properties of the fuel and combustion environmental conditions. This process is almost similar to the glass blowing process (Ghosal & Self, 1995). In this way, it is helpful to have a profound knowledge of the principle of glass development. Cenospheres are formed from the coal combustion ash amid the molten state in the thermal power plants. Streaming with the ignition gas stream, the temperature of the liquid particles is quickly extinguished, consequently 'solidifying in' a circular shape. Any gas bubbles within the molten particles are likewise caught inside the spheres. These bubbles cause the production of cenospheres; bubbles may form in numerous structures inside the 'solidified' particles, or as single, concentric structures that are almost as extraordinary as the diameter of the particles (Kolay & Singh, 2001). Because of wettability parameters and short duration of interaction, individual dissolved smaller scale drops converge into bigger drops by keeping up the silicate and ferrous melted component relationship between the two. At present two theory have been proposed for the development of cenospheres. The most acknowledged one is that dense gases are formed because of the disintegration of organics, carbonates, sulfides, sulfates, hydro silicates, nitrogen and evaporation of pore water in coal and are caught in the liquid microspheres. These gases are equipped to expand the liquid drop up to a diameter of 500 μ, producing spherical and hollow particles. The other proposed course is that in the high temperature zone of a boiler, hot flue gas can be infiltrated into the molten ash droplets and when the particles are cooled, it stays inside. Nonetheless, after binding, these hollow spheres are called cenosphere. The variation in the liquid properties of molten ash droplets and gas pressure will obviously prompts distinctive development procedures and arrangement and subsequently, cenospheres with extensive range wall thicknesses and molecule sizes are formed. The stability of the hollow spheres is reliant on the equilibrium of surface tension forces on melt drops and the intrinsic pressure of the internal hot gas. The average gas pressure inside a cenosphere is under 1 atm (0.172– 0.227) which decreases in bigger particles since more gases would be required for the expansion. The low pressure is a consequence of the compression of high-temperature gas bolted inside the cenosphere upon fast cooling and causes a noteworthy pressure force to compel applying on the cenospheres wall. On the off-chance that the surface tension capitulates to gas pressure, the hollow sphere explodes into numerous dense or little hollow spheres (10– 25μ). The drop fragments regularly adhere to the surfaces of coarser cooled particles or blended with the melts of other particles.

Based on the viscosity of liquid composition, there are two kinds of cenospheres - high consistency or nonmagnetic particles while other is low consistency or magnetic particles. The non- magnetic particles have oxides of silica and aluminum while inadequate in Fe, Na and Ca, though magnetic fractions have higher contents of oxides of Fe, Ca and Na. Highly viscous particles tend to catch gases and extend subsequently they have larger diameter, thinner shells, yet there are no pores. In spite of the low viscosity type where the hot gas can flow through the materials because of the decreased viscosity and lesser possibility to capture the gas; in this way, the diameter/shell thickness reduced to around 3–4 μ with a porous shell structure. The formation of magnetic globules occurs with the investment of two discrete melts, to be specific, the Al-Si and Fe-Al-Si melts (Ranjbar & Künzel, 2017).

Recovery of Cenospheres from fly ash

Cenospheres can be extracted from fly ash either by dry separation method or by wet separation method. Therefore, the traditional cenosphere separation depends on wet processes, namely simple sedimentation and flotation (Sahoo, Kim, Powell, & Equeenuddin, 2016). Separation of cenosphere from fly ash degrees of separation in the recovery of cenospheres from coal fly ash have been estimated using float and sink method. Different liquids viz. water (1g/cc) and acetone (0.789 g/cc) were used for separation of cenosphere from fly ash. Fly ash is taken in a vessel to which is added water. Complete mass is stirred for four hours and is then allowed to settle for ten hours. All cenospheres having density less than 1 g/cc float up and are separated (Manocha, Ram, & Manocha, 2011).

Wet Separation

To separate the different components of coal fly ash the most widely recognized technique is froth floatation pursued by gravity separation. During the gravity separation, cenospheres can be removed in the form of a hydraulic separation method utilizing water as a medium. The cenospheres floats on the surface of a fly ash suspension and are obtained by pond skimming. This technique is widely applied due to its basic accessibility and simplicity. However, the effectiveness of the gravity separation techniques depends upon:

- Differences between the density of solid particles and used liquid and used fluid,
- Buoyancy forces,
- Concentration of feeding particles,
- Porosity and texture of particles (which means particles with a smooth surface floats quicker compared to a rough surface because of lower associations between surface and liquid) and
- Refinement cycles.

An additionally preferred standpoint of the wet separation process is that it tends to be utilized to separate intact from broken cenospheres by varying the process properties.

Dry Separation

Dry separation is an alternative method, which has been utilized in the most recent years to surmount the disadvantages of wet separation (Wheelock & Markuszewski, 1984). The benefits of this method are that the chemical composition stays unaltered, water pollution and energy consumption because of drying would be avoided, and lesser area is required for the technical instruments in comparison to the wet separation. One of the applied technology is the air classifier, which is a unit operation to separate solid particles based on their difference in particle geometric or hydrodynamic/aerodynamic equivalent diameter and density (Shapiro & Galperin, 2005). There are five sorts of air classifiers: gravitational; cascade; fluidized bed; inertial and centrifugal air classifiers. Among these, only inertial and centrifugal air classifiers are intended to separate particles with a micrometer size while the rest can be applied for particles with a size of millimeters. The separation principle depends on the fact that the particles suspended in a streaming gas amid which gravity and drag forces acting on the particle in inverse directions. Heavy particles with terminal settling velocity larger than the velocity of the air move downwards against the air stream, while the light particles whose terminal settling velocity is smaller than the velocity of air ascend alongside the air stream to the top of the column. However, a noteworthy disadvantage of this separation technique is the multifaceted nature of the preliminary setup and their mechanisms compare to wet processing.

Properties and Applications of Cenospheres

Their color ranges from white to dark gray, which is also referred as glass beads, hollow ceramic sphere, or microspheres hollow ceramic microspheres and micro-balloons. Their content in fly ash varies from 0.01-4.80 wt. %. Chemically they have oxides of Ca, Na, Ti, Al, Si, Fe, Mg and K. Concentrates of cenospheres are represented by the hollow spherical particles whose size varies from 5-500 um. There is a covering of a nanosized thin film whose thickness varies from 30-50 nm on the outer and inner surface of cenospheres (Anshits et al., 2005). Outer shell formed from the alumino-silicate glass, while the skeleton is made up of mullite, magnetite, quartz etc. The outer and inner surface of cenospheres forms a shell, which is made up of glass-crystalline phase whose thickness varies from 2-30 microns and glass phase 50-90% (Borisovich Fenelonov, S Melgunov, & Parmon, 2010). The various crystalline phases in cenospheres are magnetite, mullite, cristobolite, calcite, feldspar and hematite. The difference in chemical and phase-mineralogical compositions give rise to differently shaped globules. The phase mineral-compositions of fly ash have, glassy-70%, mineral matter-22% and remaining 2% char (Vassilev, Menendez, Díaz-Somoano, & Martínez-Tarazona, 2004). Cenospheres could be of two types: Thin-walled sphere with a continuous shell, spheres with porous shells, perforated spheres, spheres with having numerous silicates globules in their volumes, network shaped structure. They are a fraction of fly ash, with super hydrophobic surface low thermal conductivity, high fusion temperatures, ultra-low density, and protection against electromagnetic interference high mechanical strength, high porosity and resistant to oxidizing and corrosive environments. High porosity enhances their water absorption capacity. They have ~18 times more water absorption capacity than the sand due to the porous nature (Bose & Shukla, 2019). They have thermal shock resistant nature due to the high mullite content in them (Urunkar et al., 2018). Their hollow spherical shape is because of the cooling and cementing of fly ash particles around trapped gases, which are for the most part CO_2 and N_2, and traces of CO, O_2 and H_2O.

The most common application of cenospheres includes in wastewater-purification (Blissett & Rowson, 2012), in electronics and radar applications by coating them with various metals of interest whereby it acts as a magnetic shield (Blissett & Rowson, 2012). It can be used in refractory materials due to their low thermal conductivity and high temperature resistance. They are also referring as a material for hydrogen storage. Due to their lightweight and high strength, they are most suitable material for the design of modern lightweight composite materials, which are demanded, in automotive and aerospace industry (Blissett & Rowson, 2012). These cenospheres can be used as reinforcements for polymer matrix composites and can be added in paints, ceramics, plastics, construction, recreation, automotive, fillers, energy and technology because of enhanced properties and lessened expense of the final product (Kolay & Singh, 2001); (Vassilev et al., 2004).

Figure 4. SEM micrographs of thin walled fly ash plerosphere, adapted from (Kolay & Singh, 2001)

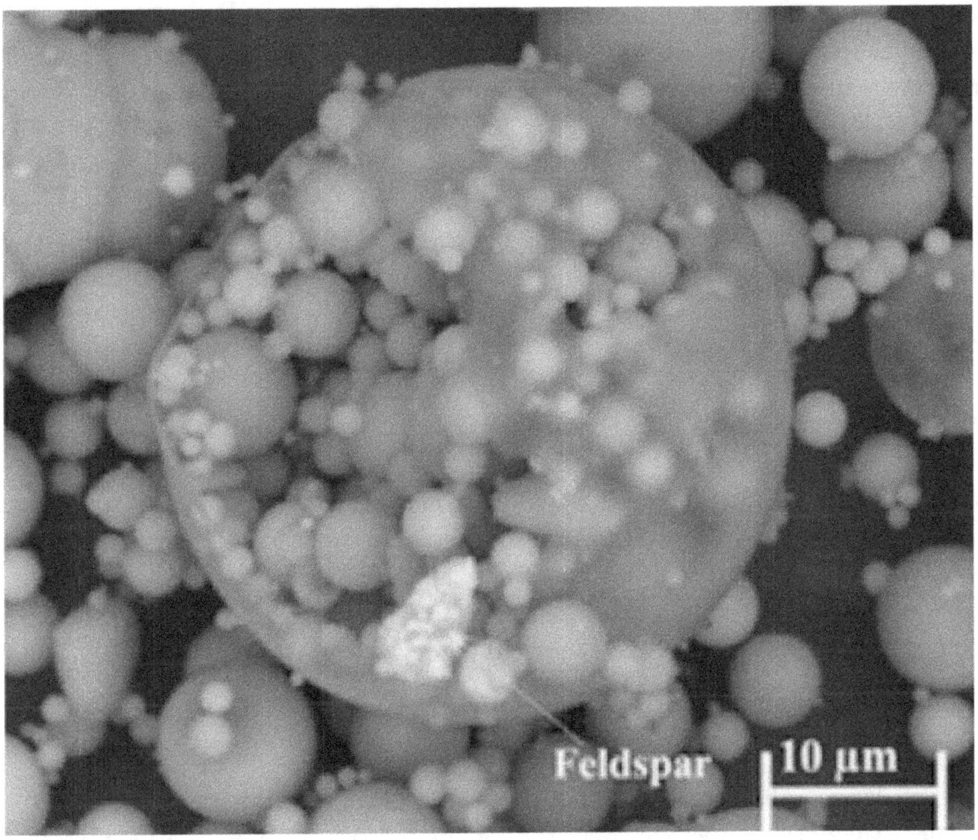

PLEROSPHERES

The plerosphere (or spheres within spheres) is basically a spherical shaped hollow, and for the most part, thin-walled particle which encapsulates some of the smaller fly ash particles of various sizes (mainly <10 µm). The formation of plerospheres is because of differential melting of particles and the inñuence of gases produced by: (i) dehydration of clay minerals (ii) decomposition of $CaCO_3$ and carbon present on the surface or inside the ash slag particles (Ngu et al., 2007). Raask, states "plerospheres originate from the coal particles with high dispersed mineral matter during the combustion and a fused silicate envelope, inside which the carbon residue prevents coalescence of encapsulated ash particles". This is perceived as an essential process of plerosphere formation, recommending that the microspheres inside the plerosphere are in certainty indigenous and formed within the plerosphere (primary plerosphere) formed due to the capture of microspheres by a cenosphere amid combustion.

Formation Mechanism of Plerospheres

At the point when both cenospheres and plerospheres are available in a similar fly ash sample then there is possibility that just a portion of the cenosphere may transform to secondary plerospheres conceivably through the following mechanism: Due to evolution of gas from the disintegration of carbon, carbonates or carbides or by dehydration of clay, there is an expansion of fly ash particles to a stable cenospheres and subsequently there is increase in the internal pressure of the cenosphere. This is followed by the advancement of an opening in the over-expanded cenospheres, which discharges the gases in the wake of puncturing the walls of cenospheres (this additionally can occur by the collision of particles) and forming apertures of framing different sizes. The cracked cenospheres are impregnated with the microspheres, which are conðned inside the cenosphere's wall and forms a secondary plerosphere. The real forces that triggers the entering of microspheres into the cenospheres is at present not clear. This demonstrates that the ðner particles trapped by the plerospheres are held by particle control devices inside the coal-ðred control plants and therefore diminish the emission of ðne particles into the atmosphere. The plerospheres are heavier than hollow cenospheres of comparative size, because of the extra weight of trapped microspheres.

Recovery of Plerospheres

The plerospheres are heavier than the cenospheres due to encapsulation of numerous smaller cenospheres by the plerospheres in them. So, the two can be easily separated by mean of density gradation. Therefore, the best way to recover the plerospheres is by centrifugal force, or centrifugation. Cenospheres will settle down early while plerospheres will settle down later on. Hence, plerospheres have a superior probability of being removed by the centrifugal force generated in the cyclone.

Properties and Applications

Plerospheres are also morphologically similar to the cenospheres and co-exist with cenospheres in the fly ash particles. They have numerous small (hollow or solid) spheres within them (Kolay & Singh, 2001). They are of two types: primary plerospheres and secondary plerospheres. They are used in civil engineering works, construction of road. They are used as an adsorbent for the removal of pollutants. They are used as cementing materials, as ceramics and in glass industries.

CARBON NANOPARTICLES

The carbonaceous particles of fly ash are generally named as unburned carbon or chars, which are blends of porous, char particles and aggregates of sub-micron particles. The carbon, which is available in fly ash, are generally, those, which remained unburned amid the burning of coal in the thermal power plants. In this way, it is better to articulate as unburnt carbon. A portion of the unburnt carbon residue is collected with the fly ash in the precipitators (Barranco et al., 2007). Bituminous-coal-derived but, evidently not subbituminous derived (J. Hower et al., 2017) fly ash additionally have amorphous and crystalline carbons attached to and binding together other CFA particles (Diamond, 1986). These carbons require high-resolution transmission electron microscopy (HRTEM) for proper identification. Significant fractions of carbon particles in fly ash are the unburnt carbon, which can be utilized as a source of natural nanostructured carbon materials. The level of unburnt carbon in fly ash, generally, fluctuates from 2-12%, while sometimes more than 20% and few have reported up to 57%. Fly ash having higher unburnt carbon is mostly called as "high carbon fly ash" (HCFA). In any case, a few different parameters influence the carbon content in the fly ash. The parameters influencing unburnt carbon levels in fly ash fall inside

Figure 5. TEM micrograph of carbon nanoparticles from fly ash

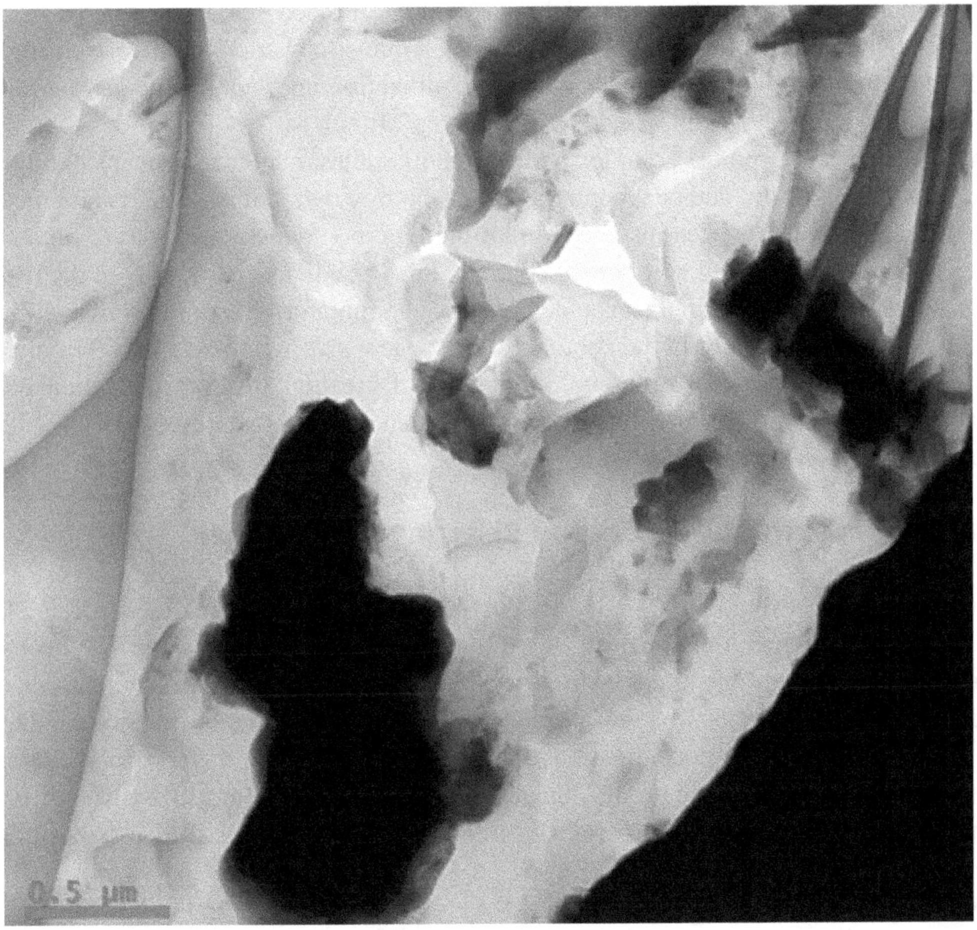

two major groups -the effect of coal qualities (e.g., coal rank, moisture, particle size, maceral composition, calorific value, mineral matter and volatile matter) and the effect of combustion system design and working conditions (e.g., residence time accessible for combustion in furnace, boiler temperature and pressure, oxygen feed, and coal/air ratio, furnace heating and loading, flame pattern and heat flow rate) (Suárez-Ruiz & Parra, 2007); (Taggart, Hower, S Dwyer, & Hsu-Kim, 2016).

Generally, carbonaceous particles in fly ash are present either in the form of char, which are smaller in size or in the form large sized particles, aggregates of smaller particles. The surface area of unburnt carbon varies depending on the parent coal and combustion residence time. Fundamentally as a micropore structures, fly ash carbon seldom surpass 10 m^2/gm with the large majority of surface area related with macropores and mesopores. The unburnt carbon can likewise be considered as an important precursor for the production of premium carbon products.

The unburnt carbon in the fly ash formed from the low-NOx pulverized coal combustion was discovered and it is a blend of soot and coal char. The soot, as generally recognized by electron microscope, are available as chains or aggregates of 10–50 nm diameter. As smaller particles have high surface energies and tends to come closer and forms an aggregate. Essentially particles of soot, ordinarily 10–100 nm diameter, can coagulate to form macroscopic aggregates. Such particles will merge with other similar particles and will form a larger particles whose size falls mainly in macrons. The incomplete burnout of soot aggregate between the end of the fuel-rich zone and the furnace exit shows that there is deficient time at elevated temperature to complete soot oxidation. Amid combustion of pulverized coal particles, they are changed into porous, irregular char by devolatilization and heterogeneous partial oxidation (Torresi, Fornarelli, Fortunato, Camporeale, & Saponaro, 2017).

Carbons in the soot incorporate complex carbon nanoparticles, nanotubes encapsulating Hg, onion like structures with polyhedral and quasi-spherical morphology with hollow centers and metal-bearing multi-walled nanotubes. It has been discovered that these carbon nanoparticles have both amorphous and inorganic phases and traces of components like Pb, Se. The enhancement of N in the residue is because of the fractionation of NH_3 in the arrangement of Sal ammoniac. Selenium, Pb, and Zn are found in moderately high concentrations in the soot. Until now a few investigators from different parts of the globe has reported the presence of fullerenes, which are of great enthusiasm since their revelation in natural systems (Kroto, 1993). Several investigators have also reported the presence of CNTs both single walled and multi-walled in the fly ash.

Extraction of Nanostructured Carbon Materials from fly ash

Fly ash carbon can be recovered by both dry and wet based techniques which incorporate electrostatic separation, gravitational separation, or froth flotation, density separation (Walker & Wheelock, 2006). Notwithstanding, considering the undeniable differentiation between the interface features of unburnt carbon and ash particles, froth flotation is as yet the most reliable and preferable method for the extraction of carbon from fly ash. Different examinations have been completed on the recovery of combustibles from fly ash materials by froth floatation (Mandavgane, Tajane, & P. Jatav, 2018). The different methods, which are used, for the recovery of carbon or unburnt carbon from fly ash, are discussed concisely.

Froth Flotation

Froth floatation is a standout amongst the most broadly used strategy for the recovery of carbonaceous nanoparticles, soots and unburned carbon from the high carbon fly ash. The use of froth and column floatation for the recovery of carbon from fly ash has been the subject of investigations among the scientist (Li et al., 2015) and patents (Hurst and Styron, 1978; Hwang, 1991); (Gray, Champagne, Soong, & Finseth, 2001), and it is among the most reliable techniques for recovery of unburned carbon from fly ash. The froth floatation technique depends on the capacity of air bubbles to specifically adhere to the surface of a specific particle surface. In this way, carbon particles are specifically recovered from an agitated tank by attachment to air bubbles, while hydrophilic ash particles stay in suspension. Kerosene and orfom, pine oil, methyl isobutyl carbinol and sodium metasilicate and butanol are the common reagents used for the column floatation method. Here kerosene or orfam (acts as collector), methyl isobutyl carbinol/pine oil (as frother), sodium meta silicate as dispersant/depressant and butanol as a promoter to improve floatability of carbonaceous particles in the tank (Uçurum, Toraman, Depci, & Yoğurtcuoğlu, 2011). Froth floatation has additionally been marketed in the US, with carbon grade recovery as high as 70% (Baltrus, Wells, Fauth, Diehl, & White, 2001).

Oil Aggregation

The oil aggregation methods also works on the principle of froth floatation. The oil agglomeration procedure depends on the preferential wetting of oleophilic/hydrophobic particles by oil that is added to an aqueous slurry. Such procedure allows the unburned carbon particles to be wetted by the oil and the mineral particles of ash to stay in suspension. Amid agitation, the unburned carbon particles covered with oil would collide into one another and forms agglomerates, which have a lesser density than the mineral particles present in the suspension. Oil/unburned carbon agglomerates would ascend to the top of the aqueous slurry system and recovery of unburned carbon from the fly ash would be accomplished (Yao, Xia, Sarker, & Chen, 2014). The use of oil agglomeration ensures the recovery of high-purity unburned carbon (66-71% C; (Eisele & Kawatra, 2002); (Özer, Basha, & Morsi, 2017).

Properties and Applications of Carbon Nanoparticles/ Unburnt Carbons of fly ash

They are basically unburned carbons, soot and hydrocarbons. They are mixtures of soot, fullerenes and carbon nanotubes (CNTs), which are associated with trace elements like Hg, Pb etc. They are mostly present on the surface of fly ash particles. The unburned carbon recovered from fly ash can be used directly or after purification for numerous applications like. A low-cost adsorbents replacing activated carbon for flue gas treatment. Unburnt carbon may also be an important as a reasonable source of activated carbon (Mittal, Sharma, & Singh, 2012). The carbon present in fly ash can be a precursor of activated carbons since it has experienced a devolatilization amid the combustion in the furnace of the power station and, in this manner; it just requires a procedure of activation (J. C. Hower et al., 2010). As an adsorbent for treating wastewater from dyes industry. As an adsorbent for mercury, or, which is a function of the temperature of the flue gas, in other words of the temperature of the vent gas, the extent of unburned carbon in the fly ash and the kind of unburned carbon (Sakulpitakphon, Hower, S. Trimble, H. Schram, & Thomas, 2000). As an impetus in low temperature, NO decrease with smelling salts. A product of beneficiation processes (e.g., size classification, electrostatic separation, or froth flotation), or a source of carbon to feedstock into the boiler. For graphite-like materials, or as a source of carbon to feedback into the boiler.

CONCLUSION

Fly ash are rich source of natural nanoparticles namely ferrospheres, plerospheres, cenospheres and carbon nanoparticles. These natural nanoparticles have sizes from few nanometers to several microns. These are formed during combustion of coal and reduction and oxidation of molten minerals of the coal slag. The utilization of waste product like fly ash for the source of natural nanoparticles will reduce environmental pollution as well as also acts as an alternate material for Al, Si, Fe and carbon-based industries. Further the applications of these nanoparticles in electronics, wastewater treatment and medicine is possible either directly or after purification. Such approach for the extraction of natural nanoparticles from fly ash is a green chemistry approach. The recovery of natural nanoparticles from ash will save huge resources and will reduce burden on the environment in the form of solid waste.

REFERENCES

Adeyemo, A. A., Adeoye, I. O., & Bello, O. S. (2017). Adsorption of dyes using different types of clay: A review. *Applied Water Science*, *7*(2), 543–568. doi:10.100713201-015-0322-y

Al-Saad, K., Amr, M., Hadi, D., Arar, R., Al-Sulaiti, M., Abdulmalik, T., & Alsahamary, N. (2012). *Iron oxide nanoparticles: applicability for heavy metal removal from contaminated water* (Vol. 45). Academic Press.

Ali, I., Asim, M., & Khan, T. A. (2012). Low cost adsorbents for the removal of organic pollutants from wastewater. *Journal of Environmental Management*, *113*, 170–183. doi:10.1016/j.jenvman.2012.08.028 PMID:23023039

Anshits, N., Vereshchagina, T., Fomenko, E., Kruchek, D. M., Bayukov, O., Kyrenskii, L. V., . . . Anshits, A. G. (2005). *Coal fly ash cenospheres and their application for immobilization of liquid radioactive waste* (Vol. 2005). Academic Press.

Baltrus, J. P., Wells, A. W., Fauth, D. J., Diehl, J. R., & White, C. M. (2001). Characterization of Carbon Concentrates from Coal-Combustion Fly Ash. *Energy & Fuels*, *15*(2), 455–462. doi:10.1021/ef000201o

Barranco, R., Gong, M., Thompson, A., Cloke, M., Hanson, S., Gibb, W., & Lester, E. (2007). *The impact of fly ash resistivity and carbon content on electrostatic precipitator performance* (Vol. 86). Academic Press.

Beddu, S., Mohd Zainoodin, M., Kamal, N., Mohamad, D., Sadon, S., & Mohamed Nazri, F. (2018). *Material characterization of coal combustion product (cenosphere) generated from coal – fired power plant in Malaysia*. Academic Press.

Blissett, R. S., & Rowson, N. A. (2012). A review of the multi-component utilisation of coal fly ash. *Fuel*, *97*, 1–23. doi:10.1016/j.fuel.2012.03.024

Borisovich Fenelonov, V., Melgunov, M., & Parmon, V. (2010). *The Properties of Cenospheres and the Mechanism of Their Formation During High-Temperature Coal Combustion at Thermal Power Plans* (Vol. 28). Academic Press.

Bose, A., & Shukla, A. (2019). *Processing of cenosphere-cement/asphalt composite materials and evaluation of their mechanical and acoustic properties*. Academic Press.

Calabi-Floody, M., Theng, B., & Mora, M. L. (2009). *Natural nanoclays: Applications and future trends-A chilean perspective* (Vol. 44). Academic Press.

Das, S., Chakraborty, J., Chatterjee, S., & Kumar, H. (2018). Prospects of biosynthesized nanomaterials for the remediation of organic and inorganic environmental contaminants. *Environmental Science. Nano*, *5*(12), 2784–2808. doi:10.1039/C8EN00799C

Diamond, S. (1986). *Particle morphologies in fly ash* (Vol. 16). Academic Press.

Dwivedi, A., & Jain, M. (2014). Fly ash – waste management and overview. *RE:view*, *6*.

Eisele, T., & Kawatra, S. (2002). *Use of froth flotation to remove unburned carbon from fly ash* (Vol. 23). Academic Press.

Fu, S.-Y., Sun, Z., Huang, P., Li, Y.-Q., & Hu, N. (2019). *Some basic aspects of polymer nanocomposites: A critical review*. Nano Materials Science; doi:10.1016/j.nanoms.2019.02.006

Fuller, A., Maier, J., Karampinis, E., Kalivodova, J., Grammelis, P., Kakaras, E., & Scheffknecht, G. (2018). Fly Ash Formation and Characteristics from (co-)Combustion of an Herbaceous Biomass and a Greek Lignite (Low-Rank Coal) in a Pulverized Fuel Pilot-Scale Test Facility. *Energies*, *11*(6), 1581. doi:10.3390/en11061581

Ghosal, S., & Self, S. A. (1995). Particle size-density relation and cenosphere content of coal fly ash. *Fuel*, *74*(4), 522–529. doi:10.1016/0016-2361(95)98354-H

Gomes, S., François, M., Abdelmoula, M., Refait, P., Pellissier, C., & Evrard, O. (1999). *Characterization of magnetite in silico-aluminous fly ash by SEM, TEM, XRD, magnetic susceptibility, and Mössbauer spectroscopy* (Vol. 29). Academic Press.

Goodarzi, F., & Sanei, H. (2009). *Plerosphere and its role in reduction of emitted fine fly ash particles from pulverized coal-fired power plants* (Vol. 88). Academic Press.

Gräfe, M., McFarlane, A., & Klauber, C. (2017). Clays and the Minerals Processing Value Chain (MPVC). In A. J. McFarlane, C. Klauber, D. J. Robinson, & M. Gräfe (Eds.), *Clays in the Minerals Processing Value Chain* (pp. 1–80). Cambridge: Cambridge University Press. doi:10.1017/9781316661888.003

Gray, M., Champagne, K. J., Soong, Y., & Finseth, D. H. (2001). *Parametric study of the column oil agglomeration of fly ash* (Vol. 80). Academic Press.

Griffin, S., Masood, M., Nasim, M., Sarfraz, M., Ebokaiwe, A., Schäfer, K.-H., ... Jacob, C. (2017). Natural Nanoparticles: A Particular Matter Inspired by Nature. *Antioxidants*, *7*(1), 3. doi:10.3390/antiox7010003 PMID:29286304

Hegde, K., Goswami, R., Sarma, S., Dasu, V., & Brar, S. (2015). *Nano-Ecotoxicology of Natural and Engineered Nanomaterials for Different Ecosystems*. doi:10.1061/9780784414088.ch19

Hirajima, T., Petrus, H. T. B. M., Oosako, Y., Nonaka, M., Sasaki, K., & Ando, T. (2010). *Recovery of cenospheres from coal fly ash using a dry separation process: Separation estimation and potential application* (Vol. 95). Academic Press.

Hower, J., Groppo, J., Graham, U., Ward, C., Kostova, I., Maroto-Valer, M., & Dai, S. (2017). Coal-derived unburned carbons in fly ash. *RE:view, 179.*

Hower, J. C., Senior, C. L., Suuberg, E. M., Hurt, R. H., Wilcox, J. L., & Olson, E. S. (2010). Mercury capture by native fly ash carbons in coal-fired power plants. *Progress in Energy and Combustion Science, 36*(4). doi:10.1016/j.pecs.2009.12.003

Ivanova, N., Gugleva, V., Dobreva, M., Pehlivanov, I., Stefanov, S., & Andonova, V. (2019). Silver Nanoparticles as Multi-Functional. *Drug Delivery System*. doi:10.5772/intechopen.80238

Katiyar, N., & Balasubramanian, K. (2014). Thermal modelling of hybrid composites of nano cenosphere and polycarbonate for a thermal protection system. *RSC Advances*, *4*(88), 47529–47535. doi:10.1039/C4RA07973F

Khan, I., Saeed, K., & Khan, I. (2017). Nanoparticles: Properties, applications and toxicities. *Arabian Journal of Chemistry*. doi:10.1016/j.arabjc.2017.05.011

Kolay, P., & Singh, D. (2001). *Physical, chemical, mineralogical, and thermal properties of cenospheres from an ash lagoon* (Vol. 31). Academic Press.

Kroto, H. (1993). *The Birth of C60: Buckminsterfullerene.* Paper presented at the Electronic Properties of Fullerenes, Berlin, Germany.

Li, G., Deng, L., Liu, J., Cao, Y., Zhang, H., & Ran, J. (2015). A New Technique for Removing Unburned Carbon from Coal Fly Ash at an Industrial Scale. *International Journal of Coal Preparation and Utilization*, *35*(5), 273–279. doi:10.1080/19392699.2015.1008098

Mandavgane, S., Tajane, S., & Jatav, P. (2018). *A Process of Carbon Enrichment of Bottom Slag Ash for Value-Added Applications*. Academic Press.

Manocha, L. M., Ram, K. A., & Manocha, S. (2011). *Separation of Cenospheres from Fly Ashes by Floatation Method* (Vol. 13). Academic Press.

Martins, A., & Mata, T. (2015). Nanotechnology and Sustainability— Current Status and Future Challenges. Academic Press.

Mathew, D. S., & Juang, R.-S. (2007). An overview of the structure and magnetism of spinel ferrite nanoparticles and their synthesis in microemulsions. *Chemical Engineering Journal*, *129*(1-3), 51–65. doi:10.1016/j.cej.2006.11.001

McCarthy, G., Solem, J., Manz, O., & Hassett, D. (1989). *Use of a Database of Chemical, Mineralogical and Physical Properties of North American Fly Ash to Study the Nature of Fly Ashand Its Utilization as a Mineral Admixture in Concrete* (Vol. 178). Academic Press.

Medical Advisory, S. (2006). Nanotechnology: An evidence-based analysis. *Ontario Health Technology Assessment Series*, *6*(19), 1–43. PMID:23074489

Mittal, M., Sharma, C., & Singh, R. (2012). *Estimates of Emissions from Coal Fired Thermal Power Plants in India*. Academic Press.

Morales, R. (2018). *Nanomaterials – Do the benefits outweigh their risks*. Academic Press.

Ngu, L., Wu, H., & Zhang, D. (2007). Characterization of Ash Cenospheres in Fly Ash from Australian Power Stations. *Energy & Fuels*, *21*(6), 3437–3445. doi:10.1021/ef700340k

Norton, G. A., Markuszewski, R., & Shanks, H. R. (1986). Morphological and chemical characterization of iron-rich fly ash fractions. *Environmental Science & Technology*, *20*(4), 409–413. doi:10.1021/es00146a016 PMID:22300215

Özer, M., Basha, O. M., & Morsi, B. (2017). Coal-Agglomeration Processes: A Review. *International Journal of Coal Preparation and Utilization*, *37*(3), 131–167. doi:10.1080/19392699.2016.1142443

Pandian, N., Rajasekhar, C., & Sridharan, A. (1998). *Studies of the Specific Gravity of Some Indian Coal Ashes* (Vol. 26). Academic Press.

Patra, J. K., & Baek, K.-H. (2014). Green Nanobiotechnology: Factors Affecting Synthesis and Characterization Techniques. *Journal of Nanomaterials*, *12*. doi:10.1155/2014/417305

Ranjbar, N., & Künzel, C. (2017). Cenospheres. *RE:view*, *207*.

Rytwo, G. (2012). The use of clay-polymer nanocomposites in wastewater pretreatment. *TheScientificWorldJournal*, *2012*, 498503–498503. doi:10.1100/2012/498503 PMID:22454607

Sahoo, P., Kim, K., Powell, M., & Equeenuddin, S. (2016). *Recovery of metals and other beneficial products from coal fly ash: a sustainable approach for fly ash management* (Vol. 3). Academic Press.

Sahu, P. (2019). *Characterization of Coal Combustion By-Products (CCBs) for their Effective Management and Utilization*. Academic Press.

Sakulpitakphon, T., Hower, J., S. Trimble, A., H. Schram, W., & Thomas, G. (2000). *Mercury Capture by Fly Ash: Study of the Combustion of a High-Mercury Coal at a Utility Boiler* (Vol. 14). Academic Press.

Salah, N., Habib, S. S., Khan, Z. H., Alshahrie, A., Memic, A., & Al-ghamdi, A. A. (2016). Carbon rich fly ash and their nanostructures. *Carbon Letters, 19*, 23-31. doi:10.5714/cl.2016.19.023

Shao, Y., Jia, D., & Liu, B. (2009). *Characterization of porous silicon nitride ceramics by pressureless sintering using fly ash cenosphere as a pore-forming agent* (Vol. 29). Academic Press.

Shapiro, M., & Galperin, V. (2005). Air classification of solid particles. *RE:view*, *44*.

Sharonova, O. M., Anshits, N. N., Fedorchak, M. A., Zhizhaev, A. M., & Anshits, A. G. (2015). Characterization of Ferrospheres Recovered from High-Calcium Fly Ash. *Energy & Fuels*, *29*(8), 5404–5414. doi:10.1021/acs.energyfuels.5b01618

Suárez-Ruiz, I., & Parra, J. B. (2007). Relationship between Textural Properties, Fly Ash Carbons, and Hg Capture in Fly Ashes Derived from the Combustion of Anthracitic Pulverized Feed Blends. *Energy & Fuels*, *21*(4), 1915–1923. doi:10.1021/ef060656z

Taggart, R., Hower, J., S Dwyer, G., & Hsu-Kim, H. (2016). *Trends in the Rare Earth Element Content of U.S.-Based Coal Combustion Fly Ashes* (Vol. 50). Academic Press.

Torresi, M., Fornarelli, F., Fortunato, B., Camporeale, M. S., & Saponaro, A. (2017). Assessment against Experiments of Devolatilization and Char Burnout Models for the Simulation of an Aerodynamically Staged Swirled Low-NOx Pulverized Coal Burner. *Energies*, *10*(1), 66. doi:10.3390/en10010066

Uçurum, M., Toraman, O., Depci, T., & Yoğurtcuoğlu, E. (2011). *A Study on Characterization and Use of Flotation to Separate Unburned Carbon in Bottom Ash from Çayirhan Power Plant* (Vol. 33). Academic Press.

Urunkar, Y., Pandit, A., Bhargava, P., Joshi, J. B., Mathpati, C. S., Sudarshan, V., . . . More, V. (2018). *Light-weight thermal insulating fly ash cenosphere ceramics*. Academic Press.

Vassilev, S., Menendez, R., Díaz-Somoano, M., & Martínez-Tarazona, M. (2004). *Phase-mineral and chemical composition of coal fly ashes as a basis for their multicomponent utilization. 2. Characterization of ceramic cenosphere and salt concentrates* (Vol. 83). Academic Press.

Wakeel, A., Gul, M., & Sanaullah, M. (2013). *Potassium dynamics in three alluvial soils differing in clay contents* (Vol. 25). Academic Press.

Walker, A., & Wheelock, T. D. (2006). Separation of Carbon from Fly Ash Using Froth Flotation. *Coal Preparation, 26*(4), 235–250. doi:10.1080/07349340601104883

Wardhono, A. (2018). *Comparison Study of Class F and Class C Fly Ashes as Cement Replacement Material on Strength Development of Non-Cement Mortar* (Vol. 288). Academic Press.

Wheelock, T. D., & Markuszewski, R. (1984). Coal Preparation and Cleaning. In B. R. Cooper & W. A. Ellingson (Eds.), *The Science and Technology of Coal and Coal Utilization* (pp. 47–123). Boston, MA: Springer US. doi:10.1007/978-1-4684-4580-0_3

Willner, I., & Willner, B. (2010). Biomolecule-Based Nanomaterials and Nanostructures. *Nano Letters, 10*(10), 3805–3815. doi:10.1021/nl102083j PMID:20843088

Xue, Q., & Lu, S. (2008). Microstructure of ferrospheres in fly ashes: SEM, EDX and ESEM analysis. *Journal of Zhejiang University. Science A, 9*(11), 1595–1600. doi:10.1631/jzus.A0820051

Yadav, V. K., & Fulekar, M. H. (2018). The current scenario of thermal power plants and fly ash: Production and utilization with a focus in India. *International Journal of Advance Engineering and Research Development, 5*(4), 768–777.

Yannas, I. V., & Burke, J. F. (1980). Design of an artificial skin. I. Basic design principles. *Journal of Biomedical Materials Research, 14*(1), 65–81. doi:10.1002/jbm.820140108 PMID:6987234

Yao, Z. T., Xia, M. S., Sarker, P. K., & Chen, T. (2014). A review of the alumina recovery from coal fly ash, with a focus in China. *Fuel, 120*, 74–85. doi:10.1016/j.fuel.2013.12.003

Zhu, H., & Njuguna, J. (2014). 7 - Nanolayered silicates/clay minerals: uses and effects on health. In J. Njuguna, K. Pielichowski, & H. Zhu (Eds.), *Health and Environmental Safety of Nanomaterials* (pp. 133–146). Woodhead Publishing. doi:10.1533/9780857096678.3.133

Żyrkowski, M., Neto, R. C., Santos, L. F., & Witkowski, K. (2016). Characterization of fly-ash cenospheres from coal-fired power plant unit. *Fuel, 174*, 49–53. doi:10.1016/j.fuel.2016.01.061

Section 4

Food Chain, Agriculture, and Safety

Chapter 22
Ecological Chemistry Aspects of Food Safety

Rodica Sturza
https://orcid.org/0000-0002-2412-5874
Technical University of Moldova, Moldova

ABSTRACT

The presented results reflect the researches carried out over the last decade, having as their object the soil, water, vegetal raw materials, and wines from the Republic of Moldova. The analysis of the possible anthropogenic contamination (NAA method) demonstrated the absence of systematic soil pollution. A total of 30 elements were determined in soil samples and the soil-leaves-fruit transfer factors were calculated. Approximately 3000 samples of local wines have been analysed to determine the residual quantities of pesticides. POPs were not found in any of the wine samples. In most of the examined cases (> 60% of samples), the lack of organic pesticide residues was observed. The migration of phthalates into different solutions from polymeric materials (PVC, rubber) and the influence of the temperature on the extraction rate were investigated. It has been shown that the contamination with phthalate residues occurs predominantly at the stage of grape processing, technological treatment, and storage.

INTRODUCTION

Food is a system that is closely related to the individual and it is the fundamental biological link of man to the environment (Weiss, 2016). Thus, any environmental disturbance finds the ideal way to reach man via food. A characteristic of food products is the wide variety of factors that can influence safety and hygienic quality (Mylona, 2016). In a brief classification, they can be distributed as follows: biological pollution agents, pollution and chemical contamination agents, natural toxic substances and toxic compounds formed by food processing.

The main agents that can reduce or even annihilate food safety, with varying degrees of harmfulness and extremely random penetration, can be synthesized in the following groups (Sturza, 2017):

DOI: 10.4018/978-1-7998-1241-8.ch022

– Toxic compounds can be found naturally in foods, including toxic aminoacids in seeds, unripe fruits and vegetables. Biogenic amines, alkaloids, glycosides releasing hydrogen cyanide, toxins from fungi fall also into this category. Their effects on the body are manifested by growth retardation and renal tissue damage (produced by amino acids), cardiac lesions, effects on the adrenal gland, increased blood pressure (caused by amines), gastrointestinal disorders, vomiting, diarrhea, haemolytic lesions by alkaloids), nerve disorders and even death caused by toxins from poisonous fungi.

– Toxic compounds that are formed in foodstuffs within technological and preservation processes: nitrosamines from plants grown on nitrogen-treated lands, condensed polycyclic hydrocarbons (3 – 4 benzo[a]pyrene), azotates and nitrites from plants, meat, milk. These chemicals have irritating action on the digestive tract, causing congestion and haemorrhage, increased blood pressure, severe hepatotoxic effects, liver damage.

– Pesticides and fertilizers representing the chemicals used in agriculture and veterinary medicine to combat various pest categories: fungicides, insecticides, rodenticides; act primarily on the liver which increases its volume and intensifies enzymatic action, causing disturbances in human metabolism and leading to tumor formation.

– Industrial chemicals: organics, packaging materials, plastics, synthetic rubber, lacquers, industrial solvents, aerosols, nitrates and nitrites, polycyclic aromatic hydrocarbons, metals with toxic potential, antibiotics, hormones, detergents, radionuclides, etc.

The large number of exogenous aggression factors on food requires food control throughout the food chain. As the chemicals diffuse into the food chain, the chemicals and their metabolites accumulate at all levels of the food chain. Effects on living organisms can range from mild discomfort to serious illness, such as cancer or physical deformities. Experts recognize that the effects of pollution are quite often underestimated and that more research is needed to understand the connections between pollution and its effects on all forms of life. Conditions related to water polluted by chemicals (such as pesticides, hydrocarbons, persistent organic pollutants, phthalates, heavy metals, nitrates) can induce cancer, hormonal problems that can disrupt reproduction and development processes, damage to the nervous system, liver and kidneys, DNA damage. Soil pollution also has many effects, including cancer and leukemia. Lead from the soil can be transfer in food via fruits and vegetables; it is particularly dangerous for children, causing damage to brain development. Mercury increases the risk of kidney damage; pesticide residues can cause liver toxicity. Other effects may include neuromuscular blockage, central nervous system depression, headaches, nausea, fatigue, eye irritation and rash. The random nature and impact of different types of toxins implies the knowledge of the sources of pollution.

The purpose of this study was to elucidate the impact of same exogenous factors (environmental contamination) on food harmlessness. These aspects will be examined, such as the heavy metal accumulation in plant foods depending on the soil composition, the mineral transfer factor in the soil-leaf-fruit systems; accumulation of pesticide residues in horticultural products; the causes of contamination of the food chain with phthalate residues.

CIRCUIT OF MINERAL ELEMENTS IN THE SOIL-PLANT-FRUIT-PRODUCTS SYSTEM BACKGROUND

Analysis of published data indicates that heavy metals, such as cadmium, arsenic, chromium, lead and mercury, naturally occur. However, anthropogenic activities contribute significantly to environmental contamination. The heavy metal contamination of food products by vegetal origin comes from soil and atmosphere, especially when the crops are approximately enterprises, urban wastewaters, and intensely circulated streets. These metals are systemic toxins known to induce adverse effects on human health. Toxic elements induce cardiovascular disease, developmental abnormalities, neurological disorders, diabetes, hearing loss, hematological and immunological disorders and cancer. The severity of adverse effects on health is related to the type of heavy metals and their chemical form; it is dependent on the exposure time and the dose. An essential role in the toxic kinetics of metals have valence, solubility, bioavailability and chemical form. Their bioavailability is influenced by physical factors, such as temperature, phase association, adsorption and sequestration. It is also affected by chemical factors that influence thermodynamic balance, complexation kinetics, lipid solubility and partition coefficients. Biological factors such as species characteristics, trophic interactions and biochemical / physiological adaptation also play an important role.

All metals are toxic at higher concentrations. Excessive levels can be damaging to the organism (Reena Singh, 2011). If the acute and chronic effects are known for some elements, little is known about the impact of mixtures of toxic elements. It have pointed that toxic elements may interfere metabolically with nutritionally essential metals such as iron, calcium, copper, and zinc (López Alonso M, 2004). The literature is low regarding the combined toxicity of heavy metals. Simultaneous exposure to several heavy metals can produce a toxic effect that is additive, antagonistic or synergistic (Paul B Tchounwou, 2012). There is a potential synergism between Cu and Zn, Cu and As, Zn and As, but an antagonism between these elements and Pb. The synergistic or antagonistic effect depends on the dose and duration of application of the substances (Sturza, 2016).

Lead is used in the form of lead tetraethyl on the gasoline additive. By removing it from the exhaust gases, the air and the adjacent land of the road are polluted over a distance of 200 – 250 m. Plants such as cabbage, celery, beet, corn, peach collect a lot of lead (Botnari, 2016).

Mercury comes from the use of pesticides, from gas or industrial emissions. It can to precipitate cytoplasmic proteins and block some enzymes. Cadmium is used in stainless steel alloys, dental technology, dye enameling, etc. The use of fertilizers (with Cd residue) leads to its accumulation in the soil, where it is taken up by the plants, rapidly migrating into their organs. Rice, wheat accumulate large amounts of cadmium. Zinc has an important biological role, but in the case of high amounts, it produces toxic effects. It is used in the form of compounds (oxide, sulfide, sulfate, chloride) for various industrial uses. The use of insecticides and fungicides based on organic zinc compounds leads to contamination of agro-food products and feed. It is accumulated in beans, corn, sorghum (Veliksar, 2016).

Copper is used in phytosanitary treatments (control of green algae in ponds, vine diseases) as well as in industry. It originates from the soil, absorbed by plants, but also as a result of the antifungal treatments with copper-based preparations, copper oxychloride (Stasiev, 2016). It is accumulated mainly in root vegetables. Although copper is part of the category of essential elements, being present in excessive amount (> 1 mg / kg) can cause intoxication of the body?

Tin (Sn) is used both in industry (metal packaging for cans, cutlery) and in agriculture. Products are contaminated by the use of various fungicides and through contact with processing equipment or storage packaging. Arsenic is an element being present in organisms, having an incompletely elucidated role. It is used in therapeutics, food industry and agriculture, its residues being active for years (Balan, 2008).

At the same time, the elements are an important component of food. Sources that contribute to the elemental composition of food products can be distinguished into two groups: natural and anthropic, allowing the definition of a representative "fingerprint", which is particularly important for foods manufactured in specific regions, such as Protected Designation of Origin (Galani-Nikolakaki S. M., 2007). The natural factors that influence the endogenous metallic content are the metabolic specificity of the plants, the degree of maturity, the climatic conditions, the nature of the soil, the quality of the water. The technological factors that influence the metal content include fertilizers, inorganic pesticides, processing equipment, steel containers and pollution of the surrounding environmental industries.

The elemental analysis of food products is of great interest to consumers and producers, as it can offer guarantees of quality and harmlessness, identify the geographical origin of the product, determining criteria for their price (Frontasyeva M.V., 2011).

Issues, Controversies, Problems

Are there soil pollution with mineral elements? What is the transfer rate of soil minerals in plants, fruits and food? Can the transfer of heavy metals from soil to food affect health? Can the content of mineral elements provide information on the origin of food?

Study carried out by Epithermal Neuron Activation Analysis (NAA) was focused on the distribution of microelements in the soils of the Republic of Moldova (Zinicovscaia, 2018). The comparison of two analytical techniques – inductively coupled plasma atomic emission spectroscopy (ICP-AES) and NAA using as example eight matrix and trace elements: Al, Ba, Ca, Fe, K, Mg, Na, and Zn was previously performed (Sturza, 2015). All data, interpreted within the Upper Continental Crust model, showed first of all that within experimental uncertainties, the soils of both vineyards are identical and close to the average composition of the UCC. A more detailed analysis of the distribution of Rb, Sr, and Zr revealed a significant degree of chemical weathering as well as a prolonged process of sorting and recycling of minerals.

At the same time, the soil content of Cr, Mn, Co, Zn and As proved, exception for As, the absence of any systematic heavy element pollution. In this regard, although the As content exceeds the UCC value of 1.5 mg/kg, almost reaching the alarm threshold according to some national regulations, its increased content can be regarded, according to literature data, as a characteristic of Moldavian soils, without any traces of anthropogenic contamination (Stasiev, 2016).

In a recent study, metals Transfer Factor (TF) was analyzed from soil to plant (Zinicovscaia, 2018). The transfer factor (TF) is defined in the following equation: $TF = \frac{C_{plant}}{C_{soil}}$, where C_{plant} is the concentration of an element in the plant material (dry weight basis) and C_{soil} is the total concentration of the same element in the soil (dry weight basis) where the plant was grown. The higher the value of the TF, the more mobile/available the metal is.

The values of TF were calculated to reveal accumulation of elements the following systems: $TF_{L/S}=C_{leaves}/C_{soil}$, $TF_{F/S}=C_{fruits}/C_{soil}$, $TF_{F/L}=C_{fruits}/C_{leaves}$, $TF_{F/St}=C_{fruits}/C_{stone}$, and $TF_{St/S}=C_{stone}/C_{soil}$. The content of major and trace elements in selected varieties of fruits (apple, plum and grape) as well as in leaves and soils samples were determined using neutron activation analysis. A total of 30 elements were determined in soil samples and 25 elements in fruit samples, originals from the Republic of Moldova.

It has been found that in leaves from soils mostly there have been accumulated K ($TF_{L/S}$=3.1), Zn ($TF_{L/S}$=2.1), Sr ($TF_{L/S}$ =1.9), and Mo ($TF_{L/S}$ =1.1) and Ca ($TF_{L/S}$=0.96). In a smaller extent there have been accumulated Ca ($TF_{L/S}$=0.96) and Br ($TF_{L/S}$ =0.73). For other elements $TF_{L/S}$ values were found to be lower than 1.0, for most of them in the order of 0.1. Concerning the elements transfer from soil in fruits the highest TF values were obtained for Ca ($TF_{L/S}$ = 3.3), Sr ($TF_{L/S}$ = 2.8) and K ($TF_{F/S}$= 1.6). In system leaf- fruit (apple) TF> 1.0 was obtained for Na, Cl, and Cr. For all studied systems, the accumulation of elements (both rare earth and pollutants) was very low. It is known that plants act as a barrier to accumulation of pollutants in the food chain. The low level of uptake and the accumulation of some elements in fruits can also be explained by the low level of available forms of these elements in the soil, as well as to the slightly acid to neutral soil pH, which is unfavorable in terms of their uptake (Murtić, 2014).

It is known that the macro-and microelement composition of plants and living organisms depends on the elemental composition of the habitat (Balasubramanian, 2008). European researchers found that even 10 years after bottling, the wine still carries the "chemical signature" of the vineyard and the forest from which the barrel was made, in which the wine was aged (Amorós, 2013). Naturally, this fact can be used to confirm the origin of wines with a geographical name, as well as to detect falsification of wine.

Identification of a product by its place of origin means that it has some special characteristics associated with this territory due to the unique natural and climatic conditions and existing traditions. Also, the geographical name carries the information that there is a specific control over such production. In essence, this is recognition of the product's valorization due to its place of origin.

A large number of natural and anthropogenic factors such as soil characteristics, grape type, production area, environment conditions, fertilizers, inorganic pesticides, winemaking practices, application of additives, transport and storage could significantly influence the concentration of the major and trace elements in wine (Geana, 2012).

From a toxicological point of view, the determination of the content of some mineral elements – possible contaminants (Pb, Cd, Cu, Zn, As, Hg) is mandatory for all food products. At the same time, some countries include in the list of requested characteristics of wine and other macronutrients – potassium (normally its content in wine varies in the range of 0.5–0.8 g/l), sodium (0.02 – 0.35 g/l), calcium (0.1 – 0.5 g/l), magnesium (0.05 – 0.4 g/l). Some micro and oligoelements like F, Al, B, Mn, Se, Br differ in a larger range in which concentrations can vary – fluorine: 0.6 – 2.8 mg/l; aluminum: 1 – 5 mg/l; bromine: 0.1 – 0.7 mg/l; selenium: 0.02 – 0.8 mg/l; manganese: 0.4 – 2.0 mg/l), etc. Therefore, if in the first case the content of macronutrients is of interest mainly from a technological point of view, in the case of trace elements their content / ratio is a carrier of information on the geographical origin of the product (Gonzálvez, 2009).

The ICP-AES and NAA techniques was used to determine 35 elements in vineyard chernozem soil and 18 elements in wines from Republic of Moldova and expand our understanding of the elemental transfer from soil to wine (Zinicovscaia, 2017). Soil elemental content allowed evidencing more similarities between considered soils and the Upper Continental Crust and the World Average Soil as well as to calculate the soil-to-wine transfer factor for 18 of investigated elements. From all 28 trace elements evidenced in soil, only 13, the soluble ones, were found in all wine samples, which finally allowed determining the corresponding transfer factors whose values varied between 0.02 mg/l (U) and 38 mg/l (K). In this regard, all sorts of wines showed a significant concentration of potassium, varying from

370 to 700 mg/l. Comparison of the results obtained for Na, K, Ca, Fe, Zn, and Ba (figure 1) show a good agreement between ICP-AES and NAA techniques. In case of Mg and Al the correlation between methods was better than wines were divided by producers. Mg and Al can be defined as a representative “fingerprint” for grouping wines into distinct classes, depending on their origin (figure 1).

Figure 1. Comparison of ICP-AES and NAA data for K and Al

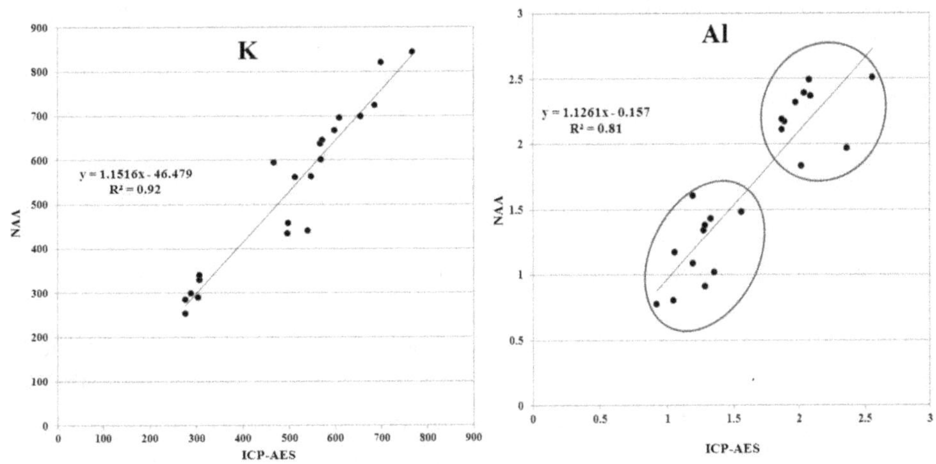

In conclusion, the activation analysis enabled to evidence the presence of all seven major elements, which are present in soil, i.e., Na, Mg, Al, K, Ca, Mn, and Fe, and only 10 soluble trace elements, i.e., Co, Ni, Zn, As, Br, Rb, Sr, Sb, Cs, Ba, and U, whose concentrations were above the detection limits. All types of wine presented a high content of K, whose average values varied between 383 mg/l (white wine) and 693 mg/l (red wine), results which are in good concordance with literature data which also point toward a significantly increased content of K in all sorts of wine.

The content of 18 elements detected in all 24 wine samples examined allowed, through discriminatory analyzes, the classification of wines into distinct categories according to sort, type (red and white) and vineyard, which proves that multielement analysis represents a valid tool for the determination of the origin of wines.

ANALYSIS OF CONTAMINATION OF FOODSTUFFS WITH RESIDUES OF PLANT PROTECTION PRODUCTS – BACKGROUND

The presence of pesticide residues in food is a major concern for consumers. A Eurobarometer about EU citizens' general fears and fears about food shows that 63%are worried about the presence of pesticide residues in food (EURACTIV, 2019).

A recent report provides an overview of the 2016 the results of control program (EUCP) and the results of national control programs (NP) about the situation of pesticide residues in food products that are the most widely consumed in the EU, carried out in the European Union (EU) Member States, Iceland and Norway. Overall, 98.3% of the samples analyzed in the 2016 EUCP fell within the legal limits (11,965 samples); 6,367 samples (52.3%) were free of quantifiable residues (residues below the limit of quantification (LOQ). The number of samples with quantified residues, but within the legally permitted level (at or above the LOQ but below the maximum residue level (MRL) was 5,598 (46.0%). In 1.7% of the samples analyzed the MRL was exceeded (203 samples). In total, 0.9% of the samples (107 samples) were considered to be not compliant with the legal limits taking into account the measurement uncertainty (EU Report, 2018). Compared with 2013, when the same commodities were analyzed, the MRL exceedance rate in 2016 is higher (1.7% in 2016 compared to 0.9% in 2013). This significant increase was mainly driven by the high number of exceedances regarding residues of chlorpyrifos, an active substance for which the MRLs were lowered for many crops during 2016 (59 exceedances were reported out of the 10,212 samples analyzed for this pesticide).

The presence of considerable levels of pesticide residues in the European food chain is the direct result of pesticide dependence on conventional agriculture. Each year, over 220 000 tons of pesticides are used in the European Union environment. The European Union annual results on the use of pesticides show that 108 000 tons of fungicides, 84 000 tons of herbicides, 21 000 tons of insecticide and 7 000 tons of plant growth regulators – representing approximately one kilo of the active substance per capita (Eurostat, 2010).

The study "*Message in a Bottle*", *2008* on the presence of pesticide residues in wine present in the European consumer network was coordinated by Pan-Europa and supported by MDRGF for France, Global 2000 for Greenpeace Austria and Germany. 40 bottles of red wine from France, Austria, Germany, Italy, Portugal, South Africa, Australia and Chile were analyzed, of which 34 were from intensive agriculture and 6 from organic farming. Wines have been selected from the least-priced brands as well as the world's most famous brands. 100% of the conventional wines tested were contaminated. It has been established that each sample from traditional agriculture contains more than 4 different pesticide residues in the environment, the most contaminated among them containing up to 10 pesticides. In some of the wines tested, the residual content of plant protection products exceeded 5800 times the maximum allowable concentrations authorized for drinking water.

These numerous residues characterize a very intensive use of pesticides in viticulture. Many of these are carcinogenic or potentially toxic for reproduction or development. Altogether, 15 different pesticide residues were detected in conventional-grape wines. Among the most dangerous substances there was procymidone, classified as a carcinogen and endocrine substance. The study shows that the intensive use of pesticides in viticulture leads to the systematic presence of many phytosanitary residues in wine. It has been established that about 1/3 of the pesticides present in grapes are systematically transferred to the finished product („Message in a Bottle", 2008).

The amount of pesticide (active substance) used in viticulture is 21.4 kg of active ingredients (AI) per hectare. Much of this refers to the application of sulfur – relatively less dangerous than the chemicals used to protect against oidium (Cabras, 2000). However, in addition to sulfur-based fungicides, grapes receive doses of synthetic fungicides averaging 4.7 kg/ha – and higher than any other crop, except for potatoes. Of the 2163 grape samples, 57% of the tests were positive for at least one pesticide, 5% of the samples contained more pesticides than the legal limits. Among the most frequently, detected residues there were many synthetic fungicides: procymidone, dicarboximide (present in 22.41% of grape samples evaluated), iprodione, imidazole present in 16.26% of evaluated grape samples (Čuš, 2010).

The results obtained in these studies are by no means able to alleviate public opinion – procymidone, incriminated as a substance with potential carcinogenic, toxic for reproduction / development and deregulator of the endocrine system, which is the leader in this chapter, was detected in 22.41% of contaminated grape seed samples. Chlorpyrifos – a substance with neurotoxic effect present in 17.33% of cases, carcinogenic iprodione – in 16.26%, and for maneb (substance with carcinogenic potential and endocrine deregulator), the incidence is 14.33%. Numerous researches have shown that people with frequent activities requiring contact with pesticides (field spraying, chemical plants, etc.) show a higher incidence of respiratory problems, chronic rhinitis, cancer and chromosomal abnormalities, alterations in neurological capacity associated with the risk of occurrence of neurodegenerative diseases.

The results of these studies show that only the issue of contact with pesticides either directly (people involved in agricultural works, for example) or indirectly – contamination with residues present in food can significantly affect the health of the population. A major issue – the cumulative effect of pesticide residues remains on the agenda and requires extensive research (Grinbaum, 2010).

Issues, Controversies, Problems

What is the rate of accumulation of organic pesticide residues in horticultural products? In wines? Transfer of pesticide residues into wine can affect health?

In the Republic of Moldova, the area of the vineyard plantations is about 150 thousand hectares. For phytosanitary protection of the vineyards, 1000 tons of pesticides (per active substance) are used annually, which is an average of 6.5 kg per hectare. The State Register includes 158 pesticides, approved for the protection of vineyards, including 112 fungicides. In 2007, it was decided to include in the State Register and to admit only plant and plant products of grade III and IV of toxicity (with a medium and low toxicity level) for plant protection (State Register of Pesticides, 2016).

Approximately 3000 samples of wine have been analyzed to determine the residual quantities of pesticides in wines produced in the Republic of Moldova. Persistent organic pollutants, POPs (DDT and its isomers / metabolites DDE, DDD), alpha-, beta-, gamma-HCH, Aldrin and Heptaclor were not found in any of the wine samples. Chlorinated organic pesticides have been banned for use in the Republic of Moldova since 1970, including DDT and its metabolites: hexachlorocyclohexane, heptachlor, aldrin, etc. banned after 1980. The lack of POPs in wine products is an extremely important thing, which proves that there is no systemic pollution of the vineyards (Gh. Duca, 2012).

In wine samples there were identified insignificant quantities of metalaxyl and mefenoxam or its isomer (metalaxyl-M). Mefenoxam and metalaxyl are active ingredients of some modern systemic fungicides, such as Ridomil Gold, Protexyl, Acedan, Metaxyl, applied on vineyards with the purpose to control diseases caused by air- and soil-borne Peronosporales.

Detected residues of metalaxyl and mefenoxam in wine were generally lower than MAL (Maximal Admitted Level) for grapes – 0.03 mg/kg and EU MAL – 0.2 mg/kg. The results obtained during 2008 – 2009 are presented in Table 1.

After the return of Moldovan wine from Russia under the pretext of containing pesticide residues, 384 raw material wine and bottled wine samples were analyzed in order to determine the organic pesticide residue content. This category does not include the lots, which according to phytosanitary certificate, there were not treated with organic pesticides, but with sulfur-based fungicides.

The distribution median for metalaxyl in the examined wines can serve as a statistical index. At the same time it must be noted that wine production originating from the same enterprises presents, in most cases, similar data about metalaxyl residual content or the lack of this (for the same crop year). The generalized data regarding the distribution of wine samples tested for pesticide residues is presented in Table 2.

Table 1. Pesticide residues in wines produced in Republic of Moldova

Found active ingredient	Corresponding pesticide	Concentration, µg/l	MAL µg/kg, grapes	Wines
metalaxyl	Profexyl 350 FS Ridomil Gold MZ 68 WG Valsalaxil720 WP Pyrenomil720WP	6 7 9 30-35	30	Cabernet[1] Merlot[1] Muscat[1] wine raw material[3]
vinclozolin	Ronilan 50 WG	13	300	Kahor[2]
metalaxyl	Profexyl 350 FS, Ridomil Gold MZ 68 WG	16 2 3-13 11 16 6	30	Muscat[4] Cabernet[5] Chardonnay[5] Feteasca[6] Portwein[7] Zemfira[8]
metalaxyl, procymidone	Sumilex, WG	13 49	30 500	Sauvignon[9] Chardonnay[9]
tebuconazole azoxystrobin metalaxyl	Falcon 460 EC, Quadris 250 EC	11 9 22	25* 50 30	Wine raw material[1o] Riesling[6] Aligote[6]
kresoxim-methyl azoxystrobin metalaxyl	Stroby, Quadris 250 EC	2 9 23	50 50 30	Sauvignon[6]
azoxystrobin mefenoxam	Quadris 250 EC, Ridomil Gold MZ 68 WG	7 6	50 30	Rkatsiteli[6]

*for grains, not allowed in grapes; [1]-[1o] –wine samples originating from different regions

Table 2. Distribution of pesticide active ingredients found in wine samples

Wine category	Metalaxyl[1] residues µg/l				
	<5	5-10	10-20	20-30	[3]30
Raw material wine: white	10	17	18	7	3
red	16	19	23	9	2
Bottled white wines (including sparkling wines)	6	2	3	-	-
Bottled red wines	5	7	3	3	-
Total	37	45	47	19	5
Median of distribution	3.77±0.89	7.98±1.02	13.66±1.56	24.21±2.07	35.2±6.28

[1]MAL for metalaxyl – 30 µg/l

The monitoring of pesticide residues content in grapes and wine has a short history in Republic of Moldova. Before 2006, when the embargo for Moldovan wines in Russian Federation was imposed, the content of pesticide residues in wines was not analyzed, and this one was not regulated (moreover, this has not changed at present). In addition, there persists the opinion that pesticide residues cannot be accumulated, because during technological treatment these are eliminated. In grapes only the residual content of persistent pesticides (chlorinated organic) - DDT and metabolites, hexachlorocyclohexane, heptachlor, aldrin was analyzed. The methods used for residual analysis of pesticides in grapes were very simple – thin layer chromatography, which had not allowed a veridical analysis.

Almost in all cases, for which traces of pesticides' active ingredients were found, their residual content in wine was not exceeding MAL (for grapes). However, if the fact that some pesticides contain 2 – 3 active ingredients is considered and every active ingredient presents a different degree of toxicity, it is obvious that the cumulative effect of these must be taken into account.

Based on the results obtained in 2008 the Interagency Council approved a decision on increasing the waiting period (time between last treatment and grape harvest) from 20 to 60 days, no later than two weeks after the flowering of vine. The Council decision had a beneficial effect – not only the range of active ingredients' residues of detected pesticides was significantly reduced, but also the contamination level of wine products.

By July 2010 the analysis of pesticide residues in wines bears a random character, dictated by the demands of importers (Belarus, Russian Federation). After the return of Moldovan wines from Russian Federation on the grounds of metalaxyl residues detection (in two cases the concentrations of 0.037 and 0.032 mg/l metalaxyl were reported) the Ministry of Agriculture and Food Industry has ordered compulsory testing of the pesticide residue content for each batch of exported wine. For organic pesticide testing, the phytosanitary certificate, which indicates the used pesticide and date of last treatment, must be taken into account.

According to the obtained results of 384 analyzed samples, in 231 cases traces of organic pesticides were not found, although according to the phytosanitary certificate, they have been used for vine treatment. Thus, if the pesticide application doses and the waiting period are met, they can be used without causing damage to the products. The lack of organic pesticide traces in 60% of examined cases evidence that most growers comply with the rules of use of organic pesticides.

With regard to wine samples in which organic pesticide residues were detected, it should be noted that in most cases metalaxyl was found, which indicates that the pesticides mainly used to treat vines were Profexyl 350 FS, Ridomil Gold MZ 68 WG, Pyrenomil 720 WP, Valsalaxil 20 WP. In terms of quantity, the distribution of samples is shown in Figure 1 (white wine) and 1b (red wine).

The data analysis from Table 2 attests that most manufacturers test raw materials before bottling, allowing them to reduce the cost of analysis and product losses related to the refusal by the importer. It appears, that in 24% of the cases, the content of metalaxyl in raw material and wine samples did not exceed 5 µg/l. In 30% of the cases it was situated within 5 to 10 µg / l; in 30.5% of the cases the residue content of metalaxyl was varying within the range 10 – 20 µg/l, which is well below the MAL established in Russian Federation for wines at 30 µg/l. In 12% of samples positively attested to the presence of metalaxyl, its concentration was located near the MAC and varied within 20 to 30 µg/l. In 3% of the cases excesses of the MAL were discovered. It is noted that excesses of the MAL were not observed in finished products (bottled wines). In addition, values close to the MAL, located within 20 to 30 µg/l were only found in a very small number of examined samples.

In red wines, the percentage of samples without organic pesticide residues was 56%. For white wines, pesticide residues were not detected in 64% of examined samples. Decreased rate of contamination of white wines compared to red wines suggests the idea that there might be a correlation between applied winemaking technology, red or white – the skin maceration could contribute to a higher degree of pesticide residues' transfer in wine.

In addition to wine samples, at the demand of importers from Russian Federation some manufacturers have tested spirits category – divin, wine and brandy distillates (a total of 42 samples). Divin is a spirit drink obtained from wine distillates, aged at least 3 years, with the addition of softened water, sugar syrup and, where appropriate, with caramel sugar and alcoholic water, up to the organoleptic and physico-chemical characteristics established in the respective regulations. Divin (from Latin *divinus*) is a product with a geographical indication from the Republic of Moldova, synonymous with the well-known Cognac.

Traces of pesticides were not found in any of the examined samples confirming the lack of pesticide residues transfer in the process of wine distillation. Obviously, this conclusion cannot be generalized, because previous data on the content of organic pesticide residues in raw materials subjected to wine distillation are lacking.

The following conclusions can be made from the generalization of the data presented:

The problem of organic pesticide residues in wines is significant for countries with intensive viticulture. In the Republic of Moldova it is noted a low level of organic pesticide residues in wines, taken in relation to the major wine-producing countries. Persistent organic pesticides (chlorinated organic) were not detected in any of the samples examined in the last three years (over 3000 samples) using advanced methods with high-resolution, GC/MS. The presence of multi residues is rarely detected. The active ingredient most commonly found in wine production is metalaxyl.

The lack of organic pesticide residues is attested in most of the examined cases (60% of samples). This is due firstly to the relatively low number of applied phytosanitary treatments, but also the prevalence of sulfur-based fungicides. The increase of the waiting period at vineyards treatment with organic pesticides also had an important role. The share of MAL (for grapes) excess cases is very low – <3%, and refers only to raw material wines. In bottled wines, MAL surpluses were not registered.

In order to reduce the contamination risks of wine production with organic pesticide residues, it is necessary to revise the list of pesticides approved for the treatment of vineyards in the State Register, taking into account the degree of toxicity, and to inform the grape producers about the effectuated changes.

FOOD CHAIN CONTAMINATION WITH PHTHALATE RESIDUES

BACKGROUND

Phthalates (esters of phthalic acid) are included in the compositions of almost all types of plastics, rubber, paints and varnishes, giving them elasticity and strength. The annual production of phthalates was estimated by the World Health Organization (WHO) to approach 8 million tons (by data on 1992) and 5 million tons by data on 2011 (Phthalates, 2011). Most phthalates produced (90%) are used as plasticizers (Wei Wang, 2018). Since February 2017, 4 phthalates (DEHP, DIBP, DBP and BBzP) are recognized by European regulations REACH on the list of substances of very high concern to humans. DEHP is classified as a carcinogen (category 2B) by the International Agency for Research on Cancer (IARC). DEHP and DBP are classified as Category 1B reproductive toxic (likely effect in human). Standards are

Figure 2. Migration of DBP (a) and DEHP (b) in various solvents from PVC.

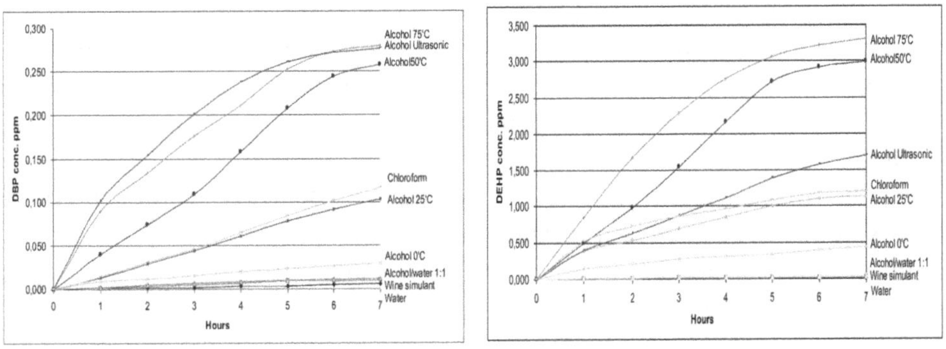

defined based on the individual data of each phthalate, but they can act according to a cocktail effect, including with other currently unregulated phthalates (eg DEP).

The main source of contamination is food, due to the ingestion of food contaminated by phthalate migration from plastic packaging (Arnold Schecter, 2013). When DEHP, a semi-volatile organic compound, is used in floor coverings, it constitutes 20 to 40% of the weight. As it is not chemically related to materials, it is slowly emitted in the indoor environment. An American study has calculated that the estimated exposures from this source range from 5µg / kg / day to 180µg/kg/day with a median at 38µg/kg/day, twice the toxic value of reference of the US EPA (Marianne Balck, 2015).

The authors (van Gelder, 2010) provided that phthalates be accumulated in the human body. Phtalates negatively affect hormones, liver and kidneys and can also cause allergies, asthma, neurodeveloped disorders and abnormalities in the development of children (Lacey Robinson, 2015). A mixture of plastic derived compounds, phthalates can promote epigenetic transgenerational inheritance of adult onset disease. The sperm provide potential epigenetic biomarkers for transgenerational disease and/or ancestral environmental exposures (Mohan Manikkam, 2013).

Most researchers from different organizations suggest that in most cases, the influence of phthalates is less than tolerable daily doses (Jung-Wei Chang, (2017). However, it is difficult to accurately determine the exposure dose as the spread of phthalates is everywhere (Xu-Liang Cao, 2010). Determination of the sources and causes of food contamination with phthalates is the key moments in the issue of product safety. A number of European laboratories investigated the migration of harmful substances from the polymer by acting on them with solutions simulating saliva, blood, milk, soft drinks, etc. Most of the procedures of the study of migration are described of existing publications and directives (Chatonnet, 2014, Yukihiro Ouchi, 2019).

Issues, Controversies, Problems

What are the factors that influence the migration of phthalates in wines and other beverages?

Migration of phthalates has been studied in samples of rubber and plasticized PVC. For each series of experiments, samples of polymers were taken with the same mass (5.0g of rubber, 1.0g of PVC) and surface area $S=32,98см^2$ and $S=15,57см^2$ respectively. The volumes of extractants are also the same for every series – 100 ml. As the solvents, there were investigated: water – as a basic component of alcoholic beverages (T=25,0°C), solution of water/alcohol (1:1) for modeling strong alcoholic beverages

(T=25,0°C), wine simulant (T=25,0°C), alcohol ethylic – as an essential component (T=0,0°C, 25,0°C, 50,0°C, 75,0°C), chloroform as alternative organic solvent (T=25,0°C), alcohol ethylic (T=25,0°C under the ultrasonic influence). All samples of polymers, completely immersed in the extractants, were kept for 7 hours at static conditions, except the case of ultrasonic extraction. Samples were taken for analysis every hour (all extractions were performed in two parallels). DBP and DEHP were detected in the investigated samples of polymers.

Figure 2 shows the kinetic curves reflecting the migration of DBP (fig. 2a) and DEHP (fig. 2b) from PVC in different solvents, at different temperatures.

The migration of DBP and DEHP in water, wine simulant (12% vol. ethanol) and water/ethanol solution (1: 1) runs slowly during the examined time (25°C). The migration rate of phthalates in chloroform and ethanol (25°C) is considerably higher. The temperature and the action of the ultrasound accelerate exponentially the migration process of phthalates into alcohol.

The study of phthalates migration from the rubber was carried out similar to the same experiment with PVC (figure 3). Although the allure of the curves is similar to that attested for PVC, the migration of phthalates in similar time intervals is considerably higher, a fact most likely caused by the texture of the material.

Figure 3. Migration of DBP (a) and DEHP (b) in various solvents from rubber

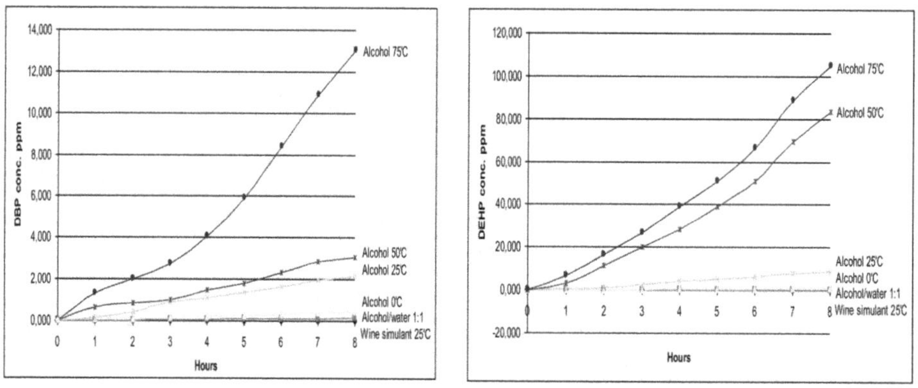

The migration of phthalates from polymeric materials proceeds extremely fast, depending on the nature and texture of the polymer, but also on the nature of the solution. In solutions with a low degree of polarity, the intensity of the transfer is considerably higher than in water and low alcohol solutions. Extraction at high temperatures increase the migration of phthalates .For every 10°C in water, there is an increase of 1.5 – 2.5 times, and for alcohol - an increase of 2.5 – 15 times. The use of the ultrasonic bath (45 kHz, 100 W) for the study of migration initially increases the migration rate up to 6 – 10 times.

The samples of polymeric materials in contact with wine /wine raw material at different stages of production were investigated. The samples for study, such as lining for containers, hoses, rubber and PVC pipe have been provided by some manufacturers, to determine the availability for use in the winery. Results of the phthalates migration in alcohol are given in tab.3. The data are expressed as mg phthalate per kg of investigated material (25°C, 1 hour).

Table 3. Migration of phthalates in alcohol from polymeric materials for 1 hour, mg/L

Polymeric material	DMP	DEP	DBP	DEHP
Paints	<LD	<LD	0 – 867.4	0 – 55.7
Lining for containers	<LD	<LD	0 – 94.6	0 – 34.44
Enamels	<LD	<LD	0 – 29.5	0 – 11.589
Varnish	<LD	<LD	63.7	13.15
Hardenger	<LD	<LD	33.2	<LD
Hoses	<LD	<LD	0 – 170.0	0 – 2478.5
Filter paper	<LD	<LD	0.052	0.042
Filter sheets	<LD	0 – 0.020	0.068 – 0.406	0.025 – 0.241
Perlite	<LD	<LD	0.000	0.000
Nutrient additions	<LD	<LD	0 – 0.019	0 – 1.016
Bentonite	<LD	<LD	0.023	0.000
Gelatin	<LD	<LD	0.000	0.000
Kieselguhr	0 – 0.082	0 – 0.104	0.05 – 4.204	0 – 0.067

Also 36 different samples of cork stoppers (for wine and brandy), 6 samples of polymer stoppers for sealing wines and more than 20 samples of caps, seals, dispensers bottle and other polymer elements, which can contact with bottled alcoholic beverages have been studied as a potential source of contamination. All the samples were crushed to accelerate the potential migration of phthalates in the model solution and better quantify the amount of phthalates present.

In some cases, migration of phthalates was determined from the surface of the products. As the model solutions there was used water-alcohol solution and acidic water-alcohol mixture simulating wine. As a result, it should be noted that DBP was detected in trace amounts only in four of the investigated samples. In these cases, the observed DBP was on the surface of cork, that is probably due to a violation of the conditions of capping material storage. No significant amounts of phthalates were found in any of the samples studied by coating materials (Duca, 2014).

Although the costs of replacing DBP and DEHP include both the higher cost of alternative materials, exposure risks can not be ruled out as a contributing factor to multiple health hazards (Sturza, 2017).

CONCLUSION

The environment influences food along the raw material-finished product pathway, at different times the food-environment relationship causes changes that concern the composition and harmlessness of the product intended for human consumption. The raw materials make the first impact with the environment. The study of the ways and dimensions in which the environment interacts with the raw materials involves assessing the transfer mechanisms, the persistence and the accumulation of various xenobiotics. The research allows us to draw the following conclusions:

- The analysis of the possible anthropogenic contamination content, made by the Neutron Activation Analysis (NAA) has demonstrated, with the exception for the As, the absence of any systematic soil pollution in the Republic of Moldova. In this respect, although the As content exceeds a few times the UCC value of 1.5 mg / kg, reaching the alarm threshold according to some national regulations, its increased content can be considered, according to the literature, as a characteristic of soils Moldovans, without traces of anthropogenic contamination.
- The TF values were calculated to show the accumulation of elements in the following systems: $TF_{L/S}=C_{leaves}/C_{soil}$, $TF_{F/S}=C_{fruits}/C_{soil}$, $TF_{F/L}=C_{fruits}/C_{leaves}$, $TF_{F/St}=C_{fruits}/C_{stone}$, and $TF_{St/S}=C_{stone}/C_{soil}$. A total amount of 30 elements were determined in soil samples and 25 elements in the samples of leaves and fruits (apples, plums and grapes), originating from the Republic of Moldova. For all studied systems, the accumulation of rare earth elements and elements, which can be considered as environmental pollutants, was very low.
- The NAA was used to determine 35 elements in vineyard chernozem soil and 18 elements in wines from Republic of Moldova. From all 28 trace elements evidenced in soil, only 13, the soluble ones, were found in all wine samples. Using discriminant analysis, the content of each element and wine sample could be classified into categories according to sort, type (red and white) and vineyard. The method can be applied to guarantee origin and quality of wines of controlled origin from the Republic of Moldova, as well as to prevent fraud.
- In the Republic of Moldova it is noted a low level of organic pesticide residues in wines, taken in relation to the major wine producing countries. Persistent organic pesticides (chlorinated organic) were not detected in any of the samples examined in the last three years (over 3000 samples) using advanced methods with high-resolution, GC/MS. The presence of multi residues is rarely detected. The active ingredient most commonly found in wine production is metalaxyl. The share of MAC (for grapes) excess cases is very low – <3%, and refers only to raw material wines. In finished products, MAC surpluses were not registered.
- Migration of phthalates in various solutions from polymeric materials was investigated. A number of materials that might be potential sources of phthalates and influence of temperature on the rate and extent of extraction was carried out. It has been shown that the contamination with phthalate residues of beverages occurs predominantly at the stage of grape processing, technological treatments and storage.

FUTURE RESEARCH DIRECTIONS

Research has shown that environmental pollution, agrobiological and technological processes applied to the industrial processing of raw materials can be the cause of food contamination. It is necessary to monitor the content of pesticide residues in plant raw materials, as the active principles used are very varied. One concern is the contamination of foodstuffs with phthalates residues, both from environmental pollution, from packaging and utensils used to produce them. Thus, a rigorous and continuous control of the degree of environmental pollution and contaminants transfer factors in raw materials and food is necessary.

ACKNOWLEDGMENT

The author expresses his gratitude to all colleagues involved in the research mentioned in this chapter: ASM, JINR-Dubna, CNVCPA.

REFERENCES

Amorós, J. A., Pérez-de-los Reyes, C., Francisco, J. G. N., Bravo, S., Chacón, J. L., Martínez, J., & Ballesta, R. J. (2013). Bioaccumulation of mineral elements in grapevine varieties cultivated in "La Mancha". *Journal of Plant Nutrition and Soil Science*, *176*(6), 843–850. doi:10.1002/jpln.201300015

Balan, V. (2008). Development of Fruit Growing In the Republic Of Moldova. *Bulletin UASVM. Horticulture*, *65*(1), 252–256.

Balasubramanian, A. (2008). *Ecosystem and its Components*. Technical Report. Retrieved from: https://www.researchgate.net/publication/314213426_ECOSYSTEM_AND_ITS_COMPONENTS

Balck, M. (2015). *Phthalates in preschool dust the relation between phthalates and parameters in the preschool environment*. Degree project in biology, Master of Science. Retrieved from https://www.diva-portal.org/smash/get/diva2:852215/FULLTEXT01.pdf

Botnari, V. (2016). Use of microelements in vegetable growing. In *Microelements in the composition of the biosphere and their application in agriculture and medicine* (pp. 164–167). Chisinau: ASM Publishing House.

Cabras, P., & Angioni, A. (2000). Pesticide Residues in Grapes, Wine, and Their Processing Products. *Journal of Agricultural and Food Chemistry*, *48*(4), 967–973. doi:10.1021/jf990727a PMID:10775335

Cao, X.-L. (2010). Phthalate Esters in Foods: Sources, Occurrence, and Analytical Methods. *Comprehensive Reviews in Food Science and Food Safety*, *9*(1), 21–43. doi:10.1111/j.1541-4337.2009.00093.x

Chang, J.-W., Lee, C.-C., Pan, W.-H., Chou, W.-C., Huang, H.-B., Chiang, H.-C., & Huang, P.-C. (2017). Estimated Daily Intake and Cumulative Risk Assessment of Phthalates in the General Taiwanese after the 2011 DEHP Food Scandal. *Scientific Reports*, *7*(1), 45009. doi:10.1038rep45009 PMID:28327585

Chatonnet, P., Boutou, S., & Plana, A. (2014). Contamination of wines and spirits by phthalates: Types of contaminants present, contamination sources and means of prevention. *Food Additives & Contaminants. Part A, Chemistry, Analysis, Control, Exposure & Risk Assessment*, *31*(9), 1–11. doi:10.1080/19440049.2014.941947 PMID:25099435

Čuš, F., Česnik, H. B., Bolta, Š. V., & Gregorčič, A. (2010). Pesticide residues in grapes and during vinification process. *Food Control*, *21*(11), 1512–1518. doi:10.1016/j.foodcont.2010.04.024

Determination of Migration Rates for Certain Phthalates. (2016). *Survey of chemical substances in consumer products No. 149*. Retrieved from https://www2.mst.dk/Udgiv/publications/2016/08/978-87-93529-01-4.pdf

Duca, G., Sturza, R., & Lazacovici, D. (2014, Dec.). Comparative study of the analysis techniques of phthalates in wine products. *Revue Roumaine de Chimie*, 11–15.

Duca, Gh., Sturza, R., & Siretanu, L. (2012). Estimation of organic pesticide residue content in wines. *CLEAN - Soil, Air, Water*, *40*(6), 661–666. doi:10.1002/clen.201100081

EURACTIV. (2012). *Pesticides are EU citizens' top food-related health concern.* Retrieved from: https://www.euractiv.com/section/health-consumers/news/pesticides-are-eu-citizens-top-food-related-health-concern/

European Union report on pesticide residues in food. Scientific Report. (2018). EFSA Journal, 16(7). doi:10.2903/j.efsa.2018.5348

Eurostat. (2010). *The use of plant protection products in the European Union.* Retrieved from: https://ec.europa.eu/eurostat/web/products-statistical-books/-/KS-76-06-669

Frontasyeva, M. V. (2011). Neutron activation analysis for the Life Sciences. A review. *Physics of Particles and Nuclei*, *42*(2), 332–378. doi:10.1134/S1063779611020043

Galani-Nikolakaki, S. M., & Kallithrakas-Kontos, N. G. (2007). Elemental Content of Wines. In P. Szefer & J. O. Nriagu (Eds.), Mineral components in foods (pp. 323–338). Academic Press.

Geana, I., Iordache, A., Ionete, R., Marinescu, A., Ranca, A., & Culeac, M. (2013). Geographical origin identification of Romanian wines by ICP-MS elemental analysis. *Food Chemistry*, *138*(2-3), 1125–1134. doi:10.1016/j.foodchem.2012.11.104 PMID:23411223

Gonzálvez, A., Llorens, A., Cervera, M. L., Armenta, S., & Guardia, M. (2009). Elemental fingerprint of wines from the protected designation of origin Valencia. *Food Chemistry*, *112*(1), 26–34. doi:10.1016/j.foodchem.2008.05.043

Grinbaum. (2014). *Le point sur les résidus des produits phytosanitaires dans les vins. Œnologies Bordeaux.* Retrieved from: https://www.canal.tv/video/universite_de_bordeaux/

Kalliopi, Petros-Achillefs, Anne-Katrin, Jan, Sandra, & Franz (2016). *Delivering on EU Food Safety and Nutrition in 2050 – Future challenges and policy preparedness.* Publications Office of the European Union.

López Alonso, M., Prieto Montaña, F., Miranda, M., Castillo, C., Hernández, J., & Luis Benedito, J. (2004). Interactions between toxic (As, Cd, Hg and Pb) and nutritional essential (Ca, Co, Cr, Cu, Fe, Mn, Mo, Ni, Se, Zn) elements in the tissues of cattle from NW Spain. *Biometals*, *17*(4), 389–397. doi:10.1023/B:BIOM.0000029434.89679.a2 PMID:15259359

Manikkam, M., Tracey, R., Guerrero-Bosagna, C., & Skinner, M. K. (2013). Plastics Derived Endocrine Disruptors (BPA, DEHP and DBP) Induce Epigenetic Transgenerational Inheritance of Obesity, Reproductive Disease and Sperm Epimutations. *PLoS One*, *8*(1), e55387. doi:10.1371/journal.pone.0055387 PMID:23359474

Message in a Bottle. (2008). *Supporting Information.* Retrieved from: https://www.pan-europe.info/old/Resources/Briefings/Message_in_a_bottle_Background.pdf

Meyts, E. R.-D. (2006). Developmental model for the pathogenesis of testicular carcinoma in situ: Genetic and environmental aspects. *Human Reproduction Update*, *12*(3), 303–323. doi:10.1093/humupd/dmk006 PMID:16540528

Murtić, S., Brkovic, D., Djuric, M., & Vujinovic, I. (2014). Heavy metal dynamics in the soil–leaf–fruit system under intensive apple cultivation. *Acta Agriculturae Serbica*, *19*(38), 123–132. doi:10.5937/AASer1438123M

Ouchi, Y., Yanagisawa, H., & Fujimaki, S. (2019). Evaluating Phthalate Contaminant Migration Using Thermal Desorption–Gas Chromatography–Mass Spectrometry (TD–GC–MS). *Polymers*, *11*(4), 683. doi:10.3390/polym11040683 PMID:30991697

Paul, B. (2012). Heavy Metals Toxicity and the Environment. *EXS*, *101*, 133–164. doi:10.1007/978-3-7643-8340-4_6 PMID:22945569

Phthalates and their alternatives health and environmental concerns. (2011). University of Massachusetts Lowell.

Robinson, L., & Miller, R. (2015). The Impact of Bisphenol A and Phthalates on Allergy, Asthma, and Immune Function: A Review of Latest Findings. *Current Environmental Health Reports*, *2*(4), 379–387. doi:10.100740572-015-0066-8 PMID:26337065

Rodica, S. (2016). Microelements in food. In Microelements in the composition of the biosphere and their application in agriculture and medicine, (pp.174-194). Chisinau: ASM Publishing House.

Rodica, S., Boris, G., Roxana, I. E., & Diana, C. (2017). The authenticity and harmlessness of uvological products. Chişinău, "MS Logo" Publishing House.

Schecter, A., Lorber, M., Guo, Y., Wu, Q., Yun, S. H., Kannan, K., ... Birnbaum, L. S. (2013). Phthalate Concentrations and Dietary Exposure from Food Purchased in New York State. *Environmental Health Perspectives*, *121*(4), 473–479. doi:10.1289/ehp.1206367 PMID:23461894

Serrano, S. E. (2014, June 2). Phthalates and diet: A review of the food monitoring and epidemiology data. *Environmental Health*, 13. PMID:24894065

Singh, R., Gautam, N., Mishra, A., & Gupta, R. (2011). Heavy metals and living systems: An overview. *Indian Journal of Pharmacology*, *43*(3), 246–253. doi:10.4103/0253-7613.81505 PMID:21713085

Stasiev, G., & Toma, S. (2016). Ecological content and evaluation of microelements in soil (2016). In *Microelements in the composition of the biosphere and their application in agriculture and medicine* (pp. 27–56). Chisinau: ASM Publishing House.

State Register of Pesticides Approved in the Republic of Moldova. (2016). Retrieved from http://www.pesticide.md/registrul-de-stat/

Sturza & Lazacovich. (2017). *Monitoring the overall phthalate burden – a public health problem*. RIVEmed no.2, "Grigore T. Popa" Publishing House Iasi.

Sturza, R., Bilici, C., Zinicovscaia, I., Culicov, O. A., Gundorina, S., & Duca, G. (2015). Moldavian wine analysis by ICP-AES and NAA techniques: Comparison study. *Revue Roumaine de Chimie*, *60*(11-12), 1065–1071.

van Gelder, M. M. H. J., van Rooij, I. A. L. M., Miller, R. K., Zielhuis, G. A., de Jong-van den Berg, L. T. W., & Roeleveld, N. (2010). Teratogenic mechanisms of medical drugs. *Human Reproduction Update*, *16*(4), 378–394. doi:10.1093/humupd/dmp052 PMID:20061329

Veliksar, S. (2016). Microelements in achieving the adaptive productivity potential and resistance in horticulture. In *Microelements in the composition of the biosphere and their application in agriculture and medicine* (pp. 142–163). Chisinau: ASM Publishing House.

Wang, W., Leung, A. O. W., Chu, L. H., & Wong, M. H. (2018). Phthalates Contamination in China: Status, Trends and Human Exposure-with an Emphasis on Oral Intake. *Environmental Pollution*, *238*, 771–782. doi:10.1016/j.envpol.2018.02.088 PMID:29625301

Weiss, J. M., Lignell, S., Darnerud, P. O., & Kotova, N. (2016). Human Biomonitoring of Environmental Contaminants - Examples Offering Tools Towards Safe Food in Sweden. *European Journal of Nutrition & Food Safety*, *6*(3), 132–147. doi:10.9734/EJNFS/2016/20212

Zinicovscaia, Duliu, Culicov, Frontasyeva, & Sturza. (2018). Major and trace elements distribution in Moldavian soils. Rom. Rep. Phys., 70(2).

Zinicovscaia, I., Duliu, O. G., Culicov, O. A., Sturza, R., Bilici, C., & Gundorina, S. (2017). Geographical Origin Identification of Moldavian Wines by Neutron Activation Analysis. *Food Analytical Methods*, *10*(11), 3523–3530. doi:10.100712161-017-0913-3

Zinicovscaia, I., Sturza, R., Gurmeza, I., Vergel, K., Gundorina, S., & Duca, G. (2018). Metal bioaccumulation in the soil-leaf-fruit system determined by neutron activation analysis. *Journal of Food Measurement and Characterization*. doi:10.100711694-018-9972-4

Chapter 23
The Study of N-Nitrosoamines Formation in Model and Real Gastric Juice Systems

Maria Vasile Gonta
State University of Moldova, Moldova

ABSTRACT

The formation of N-nitrosamines in gastric juice during the nitrosation of secondary amines by nitrite ions has been investigated. It has been established that the ratio of nitrosation of amines in gastric juice is higher than in model systems due to the presence of catalysts. On the nitrosation of the formed metabolites generated during the proteolysis of casein and albumin in gastric juice, the rate of this process is pH and enzyme activity dependent. The dihydroxyfumaric acid and sodium dihydroxy fumarate demonstrate inhibition properties during the formation of N-nitrosamines in gastric juice at the proteolysis of proteins in the presence of different proteolysis enzymes. The result of the comparative study demonstrates that the rate of NO2-consumption in the presence of DFH4 is the highest, while the constants of the reaction speed vary as follows kDFH4(0,94•102 M-1s-1) > kAAs (0,6•102 M-1s-1) > k (+) Ct (0,4•102 M-1s-1).

INTRODUCTION

The ecological chemistry of nitrogen compounds (nitrogen oxides, nitrates, and nitrites) has a direct connection with the formation of the N-nitrosamines in the atmosphere, under water, in the combustion of tobacco products (Stepanov I, et al., 2005), and in the gastrointestinal system (Porubin D. et al, 2007). The wide spread of N-nitrso compounds in the environment, the highlighted property of forming tumors in various organs, carcinogenic, mutagenic effects indicates the danger of these compounds for animals and humans.

N-nitrosocompounds (NNCs) are carcinogenic substances. About 80% of them possess carcinogenic action and lead to the appearance of malignant tumours in humans and animal'sinternal organs (Martelli A. et al., 2012).

DOI: 10.4018/978-1-7998-1241-8.ch023

N-nitrosocompounds are formed quite easily in the human body, environment, food, and cigarette smoke.The occurrence of these compounds is predominantly due to the interactionbetween secondary amines and nitrites or nitrogen oxides.

The purpose of this paper is to establish the amine nitrosation mechanisms in model and real systems and to develop inhibition procedures by using of different substances obtained from the winemaking by-products.

To achieve the aim, the following objectives have been highlighted:

- establishing the nitrosation mechanisms of the aliphatic, heterocyclic secondary amines, of the amides and the determination of the kinetic laws of nitrosation according to different parameters;
- research of the nitrate, nitrite transformation and N-nitrosamines formation in different model systems;
- establishing the enzymatic and non-enzymatic nitrosation peculiarities of some amines and peptides in the gastric juice;
- development of inhibition methods for the formation of N-nitrosocompounds by the use of natural reductions obtained from the wine products and further determination of their antioxidant / anti-radical activity.

BACKGROUND

Nitrate and Nitrite Ions

Protecting groundwater from nitrogen compounds (ammonium ions, nitrites, nitrates, etc.) contamination is one of the key tasks concern for ecological chemistry studies. Nitrate-ions are the most stable form of nitrogen compounds. But penetrating into the gastrointestinal system nitrates are converted into nitrites, and the latter are transformed and serve as nitrosating agents for various nitrosatable compounds (Simonov Ya et al, 2005). The evaluation of nitrate ion content shows that most of the countries groundwater are affected by nitrate pollution. The World Health Organization expressed concern about the growing global trends in nitrate content in groundwater.

In the Republic of Moldova, groundwater presents an important source of water supply, because for 85-90% of the rural population, groundwater remains as the main source of supply.

Groundwater is considered to be polluted if the nitrate ion content exceeds 50 mg / L. The European Commission report, for recent years, it was found that the level of nitrate pollution of waters in some EU countries exceeds the allowable limit. The same situation is for Republic of Moldova.

A study of the quality of the centralized water supply in the country, indicates that samples of water that do not meet the hygienic requirements according to the chemical composition, which reach 73,7% more than 60% of them do not comply with the requirements as normative content of the nitrate ion.

The consequences of human activity are increasingly reflected on the changes in chemical composition and physicochemical properties of groundwater. Thus, the problem of the environmental pollution with nitrogen compounds is a current matter, which has a negative impact on the health of the population.

The causes of the groundwater nitrogen pollution and its contribution of the river phreatic aquifer are multiple and have a cumulative character. In addition to the natural sources of nitrate, the majority of the nitrates in groundwater came from a wide range of anthropogenic sources: land poverty and the application of the soil fertilizer, nitrogen, irrigation, and other activities.

The main source of anthropogenic pollution of groundwater with nitrogen compounds is agriculture, which can lead to soil, groundwater and surface pollution through excessive use of fertilizers, pesticides, improper irrigation. By-products waste from agriculture, especially from large livestock farming complexes in the industrial system, is a problem for the environment.

Sources of industrial pollution with nitrogen oxides are emitted from exhaust gases and from industrial inorganic and organic plants or from other sources. Nitrogen oxides can also be formed by electrical discharges.

Moreover, the main cause of pollution is the impact of the social systems such as the lack of compliant landfills, sewerage systems and treatment plants. In rural localities the situation is quite critical, due to the fact that the treatment plants are completely missing. One of the stringentele environmental problems faced by the residents is the absence of capitalization of waste management, the unmanaged forests, both in the rural as well as urban areas, can pose a serious threat to the environmental factors and the health of the population.

Thus pollution of water, soil, atmospheric air with nitrogen compounds leads to pollution of food products.

Nitrate ions are natural components in various foods, especially in vegetable and meat products. Almost 80% of nitrates occurrence in the body are from vegetables and 19% from meat products. Approximate amount of nitrate ions entering the human body from the food per day has been calculated. The nitrate quantity in the human diet is over 100 mg per day and varies depending on the country. Thus, this quantity in Japan is 300 mg, Netherlands – 131 mg, USA - 106 mg, in Germany - 75 mg, and in Switzerland - 48 mg (Iijima K., et al, 2003).

The average nitrite amount entering the human body is 12-15 mg/day: food - 0.02-0.2 mg, as food additives – 1.2-2.95 mg, endogenously formation from nitrates ions contained in vegetables (10 mg), from water – 0.6 mg and nitrite formation in gastrointestinal system - 0.33 mg.

Following the above, it can be observed that the largest amount of nitrite ions is formed endogenously from nitrate ions, predominantly in oral cavity. According to the literature, nitrite concentration in normal stomach is less than 10 mM (460 mg/l), in the hypo acid stomach > 150 mM (6900 mg/l), and in ulcer disease is < 2 µM/l (0,09 mg/l) (Bogovschi PA et al., 1987).

Ingested nitrates from food are absorbed in the intestine, 30% are transported into the blood, concentrated in the salivary glands and desecrated into the oral cavity (Mirvish S.S., et al., 1995), where they are reduced to nitrite ion by nitrate reducing bacteria. Nitrate reducing bacteria reduce about 30% of nitrate which is then transported into the stomach with swallowed saliva. This process is one of the most important in the accumulation of gastric nitrite. The nitrite concentration in saliva is about 50 µM and can increase up to 200 µM after a portion of salad containing 2 mM of nitrate (Li Honget al., 1997, and Mirvish et al, 2000) reported that in unstimulated saliva contains an amount of nitrate and nitrite ions equal to 38 mg/l (0.61 mM) and 7.9 mg/l (0.17 mM), respectively. Similar values have been also presented by Bodawiet al. (1998). In addition to nitrate ions, the salivary glands also actively take the thiocyanate which is also secreted into saliva. Thiocyanate ions may also be secreted directly into gastric juice (Keszei A.P. et al, 2013).The thiocyanate concentration in saliva and gastric juice is in the range 200 µM- 2 mM, being highest for the smokers.

Thiocyanate ions are important catalysts in the formation of nitrosating agents under acidic conditions (UmeoTakahama et al, 2007).

N- Nitrosamines

Therefore, nitrite ions formed in gastrointestinal system can interact endogenously with amines and amides under acidic or bacterial catalysis to form N-nitrosamines and N-nitrosamines derivatives:

$$NO_2^- + H^+ \underset{}{\overset{k_a}{\rightleftarrows}} HNO_2$$
$$HNO_2 + HNO_2 \overset{k_1}{\rightleftarrows} N_2O_3 + H_2O$$
$$HNO_2 + H^+ \overset{k_2}{\rightleftarrows} H_2ONO^+$$
$$R_1R_2NH + N_2O_3 \overset{k_3}{\rightleftarrows} R_1R_2NNO + HNO_2$$
$$R_1R_2NH + NO^+ \overset{k_4}{\rightleftarrows} R_1R_2NNO + H^+$$

The nitrite ion produced by nitrate reducing bacteria is a determining factor of gastric nitrosation process. Therefore, the exogenous (food, smoke) or endogenous (formed in vivo) N-nitrosocompounds can induce different type of cancer, such as: tongue, stomach, oesophagus, nasal cavity, bladder and oral cavity cancer, and also, leukaemia (Li Hong et al., 1997 and Mirvish S.S et al., 2000).

Nitrosamines can form carbocation active species and shows organotropism for the bladder, stomach, colon, liver. In the presence of enzyme complex P-450, nitrosamines through ά-carbon hydroxylation and oxidative denitrosation process result to the corresponding aldehyde and mono nitro derivatives. Mono alkyl nitrosamines will decompose spontaneously in carbocation, which is considered an alkylation agent for purine, pyrimidine and phosphate DNA residues (Snoz S.V et al., 1995 and Yang C.Set al, 1990).

Nitrosation substrates in the gastrointestinal system can be the peptones, amino acids and amines obtained by protein proteolysis. This process starts in the stomach under the action of pepsin protolithic enzyme which acts mainly on the peptide bond from albumin and globulin chain ends. The pepsin hydrolyses easier the peptide bonds formed by aromatic amino acids (Brain C. Challis et al.,1989).

The protein hydrolysis in the stomach leads to peptones mixture, which is not assimilated in the stomach and arrives with the food bowl into duodenum. From proteolytic enzymes can be cited trypsin, chymotrypsin, carboxypeptidase and other peptidases.

Thus, in the gastrointestinal tract in the presence of proteolytic enzymes, food proteins are transformed into free amino acids (Andra´s P Keszeiet al.,2013). This process is also influenced by intestinal microflora - microorganisms normally present in the intestine. Some amino acids are converted into amines, fatty acids, alcohols, phenols, indole, H_2S and other compounds by enzymatic processes, through the following biotransformation (Guadagni Stefanoet al., 1989):

1) amino acids decarboxylation into amines which can be toxic;
2) amino acids deamination with possible formation of saturated and unsaturated acids, keto- and oxoacids.

Amines obtained from the decarboxylation of amino acids are pharmacologically active compounds, some of them are being toxic. Thus, microbiological deamination of some amino acids can form such amines as putrescine, cadaverine, phenyl ethylamine, indolylethylamine, which are neutralized in the liver (Guadagni Stefano et al., 1998).

The relationship between nitrate, nitrite and N-NDMA concentration and the risk of gastrointestinal cancer has been studied (Knekt Paulet al., 1990). Thus, NNC increases the risk of colorectal cancer. As a result of nitrosation of amines and amides obtained from protein hydrolysis, different endogenous NNC are formed (Bingham S.A et al., 1996). These active compounds can form DNA adducts in colon cells (Bingham Sheila Anne, et al., 2002), thus acting as initiators of carcinogenesis process (Triker A.R et al., 1991). Bacteria at pH 7.5 catalyse the NNC formation, under the anaerobic conditions (Calmels S. et al., 1985, Calmels S. et al., 1988 and Massey R.C. et al., 1991). During the studies of various bioassays it have been proved that colon N-nitrosation process depends on the bacterial flora (Chen Sheng Chonget al., 2001).The cause of colon cancer is nitrite ions formation after the reduction of nitrate ions or taken directly from food (meat, sausages etc.).

RESEARCH AIMS

In order to study the nitrosation process in gastric juice, gastric juice samples (GJS) from various volunteers (Republican Hospital's GastroSurgery Department) with different illnesses such as: chronic pancreatitis, chronic cholecystitis, chronic gastritis, chronic colitis, hepatitis and chronic cirrhosis, were collected. The highest NNC concentrations were determined at pH range 6.0-8.42 and lower at pH range 1.10 -2.99. These values were compared with NNC concentrations found at pH 3.00-5.99. Thus, the obtained NNC concentration depends on nitrite concentration for acidic, neutral or basic media.

RESEARCH METHODS

The samples were stocked in sterile polyethylene tubes. Immediately after collection, the GJ samples were homogenized using a magnetic homogenizer, then their pH values have been determined. It can be observed that pH range is quite wide, from 1.66 to 7.82. For chronic gastritis, the most samples have low pH while in the case of pancreatitis and cholecystitis the pH values are higher. The haemoglobin concentration has also been determined which doesn't vary so much according to the patient's diagnosis. Subsequently, GJS were kept at low temperatures (-30ºC). The total NNC concentration was determined by ATE method (Xu G.P et al., 1993).The homogenized gastric juice aliquot was injected directly into ethyl acetate containing HCl/HBr and then in glacial acetic acid. The NNA concentration was determined from the amount of NO formed in denitrosation process, using equations:

$$R_1R_2NNO \xrightarrow{HBr} R_1R_2NH + NOBr$$
$$2NOBr \longrightarrow Br_2 + 2NO^-$$

Since nitrite ions interfere ATE NNC analysis, they have been removed from GJS by adding the sulfamic acid. The nitrite and nitrate concentration has also been determined in GJS. Subsequently, different amines nitrosation process in SGS induced by nitrites ions have been studied. For this purpose, different amines as DMA, DEA and MORF were chosen. The nitrosation process in GJS has been studied in the absence and in presence of various inhibitors.

Table 1. Nitrate and nitrite concentration collected from different volunteers with various illnesses [NO_3^-], mol/l, [NO_2^-], mol/l

Illnesses	[NO_3^-], mol/l	[NO_2^-], mol/l
Pancreatitis	$(4,07 - 5,95) \cdot 10^{-4}$	$(3,12 - 4,78) \times 10^{-5}$
Cholecystitis	$(1,12 - 1,5) \times 10^{-4}$	$(4,27 - 2,55) \times 10^{-5}$
Gastritis	$(3,81 - 7,01) \times 10^{-4}$	$(2,8 - 2,67) \times 10^{-5}$
Hepatitis	$(7,01 - 7,38) \times 10^{-4}$	$(3,37 - 6,49) \times 10^{-5}$

MAIN FOCUS OF THE CHAPTER

Experimental

Variation of Nitrate and Nitrite Concentration in Gastric Juice on pH

In order to determine the relationship between the intragastric N-nitrosation process, gastric pH and nitrite concentration, different GJS were studied. In these N-nitrosocompounds, the nitrates and nitrites ions concentration has been determined. The nitrite concentration has been found to depend on the stomach pH, at pH < 5 its concentration is very low.

The nitrite concentration in gastric juice is on average 0.12 mg/l (26 µmol), this value is increasing with increasing pH up to 5 (Xu G.P (a) et al., 1993).

An increase in nitrite concentration is observed at patients who have severe anaemia or other pathologies. The nitrite ions in gastric juice is converted in nitric acid during one hour in acidic conditions (Mirvish S.S. et al., 1983). One part of nitric acid is absorbed by the mucosa, and another part is converted to NO_2 and NO, which then is oxidized to nitrite or reacts with the reducing compounds, as amines, forming N-nitroso derivatives (Brambilla G.et al., 2007, Sharma V. et al., 2014 and Duca M. et al.,2003).

From our experimental investigations it was found that the nitrite and nitrate content varies according to the patient's diagnosis (Table 1).

The highest nitrate and nitrite concentration was found in case of chronic hepatitis (Tab.1) and the lowest one - in the case of chronic cholecystitis. The nitrite and nitrate content in the gastric juice is higher for male volunteers.

It has been found that the nitrate and nitrite concentration in GJS depends on the pH value. The nitrite concentration is lower at low pH. In the pH range 1.66-2.96, nitrite concentration range is from 1.41 mg/l (2.04·10-5 mol/l) to 2.98 mg /l (4.32·10-5 mol/l). At pH > 6 – range is from 2.86 mg/l (4.14· 10-5 mol/l) to 5.68 mg/l (8.22·10-5 mol/l).

Under the normal acidic conditions (pH 2.5) bacterial flora doesn't develop and intra-gastric nitrite is only of salivary origin. Increase of gastric pH lead to a growth rate increase of reducing nitrate by microorganisms, the optimal pH for their reducing activity being between 7 and 8. These experimental data are in agreement with the literature (Mirvish S.S. et al., 1983, Duca M. et al., 2003, Stockbrugger R.Wetal, 1998 and Tricker A.R et al., 1992). Increasing of the nitrite concentration with the increase of pH value determined an increase in the enzymatic nitrate reduction rate.

High nitrate concentration could be ingested by polluted water, especially in rural areas where groundwater is used without treatment. By comparing the total nitrate content entering the human body for 12 countries, Mirvish S.S. et al. (1983) found an important correlation between the incidence of gastric cancer (GC) and nitrate concentration. The highest incidence of GC was found in Japan, and the lowest in the US (in Japan being three times higher than in the US). The high nitrate consumption in Japan is related to their high vegetable content, and the low incidence of GC in the US is explained by systematically use of fresh fruits and vegetables.

The nitrate concentration in the analysed GJS is much higher than the nitrite concentration ranging from 1.5·10-4 mole/l (12.8 mg/l) - 1.12·10-3 mole/l (95.2 mg/l). The nitrite concentration increases significantly when incubating GJS at pH 7.82 and temperature of 37°C. The process doesn't occur at pH 2.44. These experimental studies confirm that the nitrite concentration in gastric juice is dependent on the intragastric pH, at pH < 6 the nitrite concentration are lower. This can be explained by instability of nitrite ions at low pH values reacting with amines, amides, peptides, and so on.

Analysis of N-Nitrosamines in Model Systems and the Gastric Juice

Nitrites formed by the nitrate reduction process serve as nitrosating agents for various nitrosatable compounds: amines, amides, amino acids, polypeptides, proteins (Gonța M. et al., 2018).

The highest NNC concentrations were determined at pH range 6.0-8.42 (3.57 ± 0.33 µmol/l) and lower at pH range 1.1-2.99 (1.45 ± 0.17 µmol/l). These values, as compared with the NNC concentrations found at pH 3.00-5.99 (1.02 ± 0.12 µmol/l), are the smallest ones. Thus, the obtained NNC concentration depends on nitrite concentration for acidic, neutral or basic media.

From the total content analysis of volatile N-nitrosamines contained in GJS, it was found that the main NNAs are: N-nitrosodimethylamine, N-nitrosodietylamine, N-nitrosoethylmethylamine, N-nitrosomorpholine, Nitrosopyrrolidine, N-nitrosodipropylamine. The average value of total volatile NNA concentration is 4.84 nmol/l (0-17.7 nmol/l) (Stockbrugger R.Wetal, 1998). Volatile NNA concentration plot versus pH in GJS, shows a positive correlation ranging in pH from 4.0 to 7.5, including total and individual volatile NNA concentration. This is explained by an increase of the growth rate of bacterial culture leading to higher nitrate conversion.

Similarly, the total and individual volatile NNA content is high at low pH (1.0-1.5) this being due to the acid catalysis of nitration process. The secondary amines content in the gastric juice collected from different patients has been reported by Tricker et.al, (1992).The main nitrosable secondary amines in saliva, gastric juice, blood, urine was DMA (0,87 µg/l), pyrrolidine (Pyr 0,18 µg/l), piperidine (PIP 1,35 µg/l) DEA (0.05 µg/l).

Carcinogenic N-nitrosoderivatives, formed in stomach, are potential initiating gastric cancer agents. In this context, the amines nitrosation study in gastric juice, in particular, their inhibition is of great importance. Therefore, the purpose of present work is the study of inhibitor activity of some compounds that allow reducing the nitrite and NNC concentration.

Inhibition Activity of Some Compounds that Allow Reducing the Nitrite and NNC Concentration

As inhibitors, the dihydroxyfumaric acid, carotenoid extract from spirulina (carotenoid pigment, PC), dimethyl esters of tartaric acid and resveratrol have been studied. The inhibition process was monitored by ATE method varying the nitrite and NNC concentration. In the GJS with pH 1.86, the DEA, DMA and MOR nitrosation has been studied (Duca Gh., et al., 2002).

Figure 1. MOR nitrosation reaction yield in the presence of PC in real and model gastric juice system; $[NO_2^-]_0 = 5$ mM, $[MOR]_0 = 5$ mM, $[NaCl]_0 = 0,5$ mM, pH 2,5 (a) and nitrite consumption for morpholine nitrosation in different systems, $[NO_2^-]_0 = 1 \cdot 10^{-4}$M,

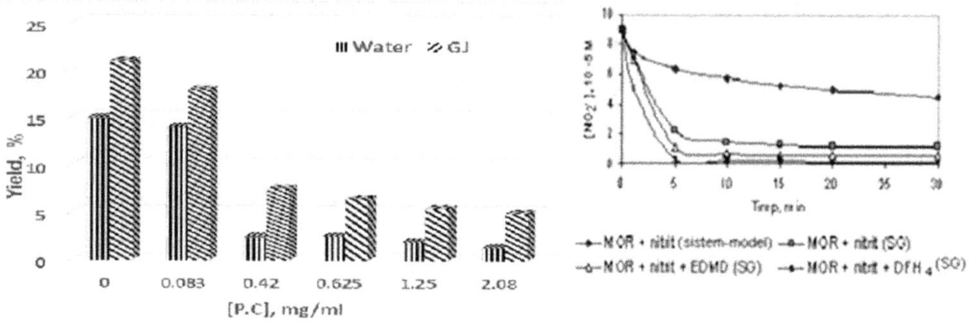

Nitrosation in gastric juice (real system) occurs with higher rate than in the model system. A comparative MOR nitrosation in model and real gastric juice system shows that NMOR concentration is higher in real gastric juice (Cojocaru, C et al., 2013). At the same time, the presence of PC lead to a low NMOR reduction yield with an increase of [PC]o (Fig.1, a). Carotenoid pigment is active not only in the model system but also in the gastric juice, thus inhibiting the N-nitrosomorpholine formation.

Gastric juice is a complex mixture containing predominant chlorhydric acid. However, gastric juice contains also fermentative products as lactic, acetic, valerianic and oleic acid. In gastric juice, different enzymes (pepsin, trypsin), chlorine ions, SCN-, nitrogen based compounds, and so on, are also present. Some of these components serve as catalysts (SCN-) in the nitrosation process (UmeoTakahama et al, 2009).

Experimental results (Fig.1, b) show that inhibitors decrease nitrite concentration and thus NMOR formation.

$[Inhib]_0 = 1 \cdot 10^{-4}$M, $[MOR]_0 = 1 \cdot 10^{-3}$M, pH 3,4, t = 37°C (b).

Based on the results presented in Fig. 1, (b) it was shown that the efficient inhibitor in MOR nitrosation is DFH_4. Thus, at this part we can conclude that NNC formation in stomach depends on catalysts and inhibitors concentration (Dolara P. et al., 2012).

However, the nitrite concentration is the first limiting factor due to an excess of amines. The nitrosation process takes place with nitrosating agents obtained from nitrite ions in the gastric juice.

ENZYMATIC NITROSATION OF PROTEIN METABOLITES WITH NITRITE IONS IN THE PRESENCE OF PEPSIN, TRYPSIN AND CHYMOTRYPSIN

Nitrosation of the Metabolites Formed in Casein Hydrolysis

The main purpose of this research was to study the nitrosation process in conditions that mimic the gastrointestinal system and the inhibitory activity of some compounds on this process. Because of hydrolysis, food proteins are cleaved by pepsin from the gastric juice. This enzyme acts more on the Ala-Ser

Figure 2. Initial casein nitrosation rate in the presence of enzyme; $[NO_2^-]_0 = 1 \cdot 10^{-4} M$, $[cz]_0 = 1\%$, $[enz]_0 = 5$ mg/ml, t = 37ºC (a) and nitrosation rate versus pH for casein proteolysis under trypsin action; $[cz]_0 = 1\%$, $[trypsin]_0 = 2{,}5$ mg/ml, $[NO_2^-]_0 = 1 \cdot 10^{-4} M$, t = 37ºC (b)

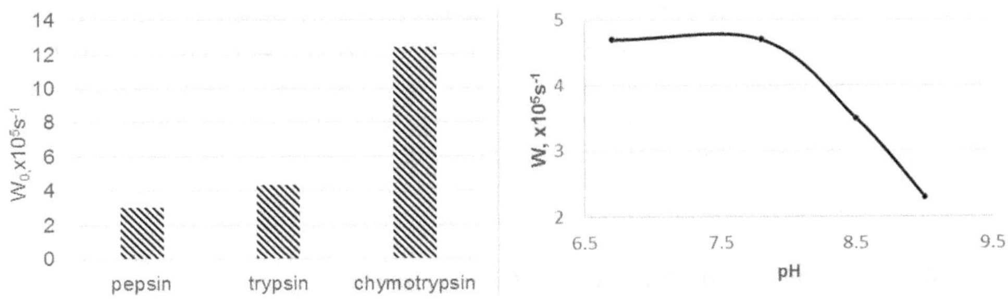

and Ala-Ala bonds from the polypeptide chain ends with peptones formation which are not absorbed in the stomach. In intestine, trypsin, chymotrypsin, carboxypeptidase, catalyse further peptone proteolytic cleavage. Trypsin cleaves especially the amide bonds between arginine and lysine amino acids. Chymotrypsin cleaves the peptide bond of both non-hydrolysed proteins and peptones molecules, especially, linkages that involve aromatic amino acids such as tyrosine, tryptophan and phenylalanine.

Thus, in the gastric juice, metabolites can act as nitrosating agents obtained in result of proteins hydrolytic cleavage:

Experimentally, (Duca Gh. et al., 2002, Duca M. et al., 2003), the NNC formation process was investigated by nitrosation with sodium nitrite of protein (casein and albumin) hydrolysis products varying different parameters: pH, $[NO_2^-]_0$, protein concentration, $[DFH_4]_0$, $[DFH_3Na]_0$, $[(+)\text{-}Ct]_0$and $[AAS]_0$. Analysis of NNC formation has been performed by spectrophotometric-extraction method. Similarly, the NNC concentration was determined by the ATE method (Xu G.P et al., 1993). Therefore, the casein was dissolved in a basic medium on boiling water bath for 10-15 minutes, then cooled to room temperature and acidified with HCl to pH 2.5. Afterwards, citrate-phosphate buffer solution was added and further kept at 37°C.

The proteolytic enzymes are solubilized in the buffer solution with the optimal pH: pepsin is dissolved at pH 2.5, trypsin at pH 7.8 and chymotrypsin at pH 8. During chemical reaction, 5 ml mixture have been taken and extracted first with 4 ml and then with 3 ml of diethyl ether.

The ether solutions obtained were assayed spectrophotometrically at $\lambda 1 = 335$ nm and $\lambda 2 = 235$ nm. The NNC formation process from casein proteolysis depending on the nitrosating agent concentration has been studied. The study was performed for different enzymes at their characteristic pH values. The following concentration ranges were used: casein (2%), pepsin (pH 2.5), trypsin (pH 7.8), chymotrypsin (pH 8) - 5 mg/ml; $NaNO_2$ ($1 \cdot 10^{-4}$, $5 \cdot 10^{-4}$, $1 \cdot 10^{-3}$ M), citrate-phosphate buffer (pH 2.5 and 7.8-8) (Fig. 2).

It was found that NNC concentration is directly proportional with increase of $[NO_2^-]_0$ in the range of 1.10^{-4}-1.10^{-3}M. Comparing the obtained NNC concentration values, it was observed that the highest NNC optical densities are showed by chymotrypsin. This process can be explained by the deeper casein chymotrypsin-catalysed hydrolysis when compared with pepsin and trypsin. It is known that chymotrypsin cleavage predominantly on the peptide bonds of aromatic amino acids. The obtained peptides after nitrosation reaction forms nitroso derivatives.

The initial NNC formation rate increases with increase $[NO_2^-]_0$ concentration being maximal in the presence of chymotrypsin (Fig.2 b). In the case of casein nitrosation, the nitroso- and diazo- compounds ratio is increasing in the presence of chymotrypsin.

The nitrosation of casein metabolites depends on the pH (Duca M. et al., 2003). The NNC formation in the presence of trypsin was studied in the pH range of 6.8 ÷ 9. It has been found that at high pH values of this range NNC formation rate decreases (Fig.2b).

The NNC concentration is maximal in pH range 6.8÷7.8 (Fig.2 b). The NNC initial formation rate decreases as pH increases (pH > 7.8), due to the loses of trypsin activity once the optimum pH interval is exceeded.

Nitrosation of Metabolites from Albumin Proteolysis

During NNC formation, from the albumin hydrolysis catalysed by pepsin, the same regularities have been observed. The nitrosation rate increases with the nitrite concentration increase. Determination of NNC concentration by photometric method at λmax = 235 nm versus on $[NO_2^-]_0$, obtained from albumin pepsin-catalysed hydrolysis, is shown in Fig.3 (1). The reaction order of $[NO_2^-]_0$ is one. Thus, it can be concluded that the nitrosation agent in these conditions is nitrosonium cation.

The nitrosation rate of the metabolites obtained from pepsin-catalysed albumin hydrolysis is decreasing as the pH increase in the range of pH 1.5-4.0 (Fig. 3 (2)). To reduce the NNC concentration obtained from albumin hydrolysis, the following reducers were used: DFH_4, (+)-Ct and AAs. The inhibitory activity has been studied by monitoring NNC and nitrite concentration. If we compare nitrite consumption in albumin hydrolysis for various inhibitors, it can be observed that it is maximal in the presence of DFH_4 and increase with increase $[DFH_4]_0$. The nitrite consumption rate constants depending on reductants are as follows:

$kDFH_4$ > kAAs > k (+) Ct

$0{,}94\cdot10^2$ M-1s-1 $0{,}6\cdot10^2$ M-1s-1 $0{,}4\cdot10^2$ M-1s-1

Nitrosation Process Inhibition of Metabolites from Casein Hydrolysis

This research concerns the process of inhibitor NNC formation from the protein hydrolysis products. Thus, nitrosation process of metabolites from casein hydrolysis in the presence of Bifocal has been studied. From our experimental data, the inhibitory nitrosation activity of DFH_3Na in presence of Bifocal can be observed (Gonţa M. et al., 2003). As shown on Fig.4, the initial rate drops sharply with increasing of ratio $[DFH_3Na]_0$: $[NaNO_2]_0$ The subsequent increase of $[DFH_3Na]_0$does not change the NNC concentration. The total NNC content decreases ~ 1.5 times compared to control sample at $[DFH_3Na]_0 = 5\cdot10^{-4}$M.

In order to ensure a successive substrate hydrolytic cleavage (similar conditions to the natural system) with peptides, amino acids and amino formation has been employed with the following method: in all tubes casein, citrate-phosphate buffer solutions, pH 2.5, sodium nitrite and DFH_3Na in different concentrations, and control test without DFH_3Na was added.

Figure 3. Absorbance plot of NNC at $\lambda = 235$ nm versus $[NO_2^-]_0$ in Alb nitrosation; pH 2,5; $[Alb]_0 = 0,4\%$; $[peps]_0 = 2,5$ mg/l, (1); Nitrate rate consumption in Alb nitrosation versus pH; $[Alb]_0 = 0,4\%$; [peps]0 = 2,5 mg/l; $[NO_2^-]_0 = 1\cdot10^{-4}$M.,(2)

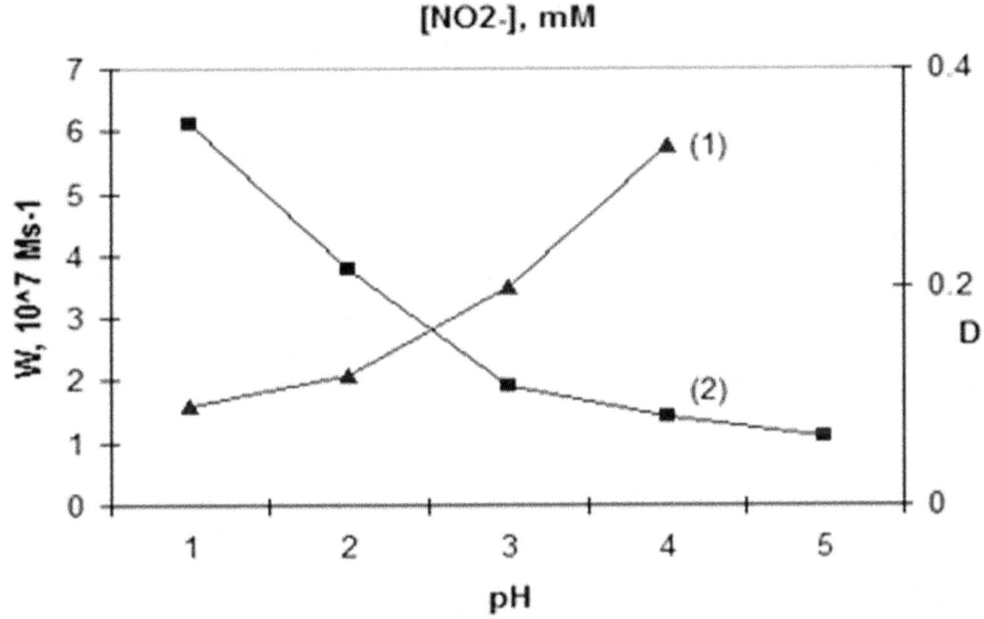

Figure 4. The NNC formation rate versus ($[DFH_3Na]$ o) in presence Bifocal; $[cz]_0 = 2\%$, [Bifocal] = 1 dose, $[NO_2^-]_0 = 1\cdot10^{-4}$ M, t = 37ºC

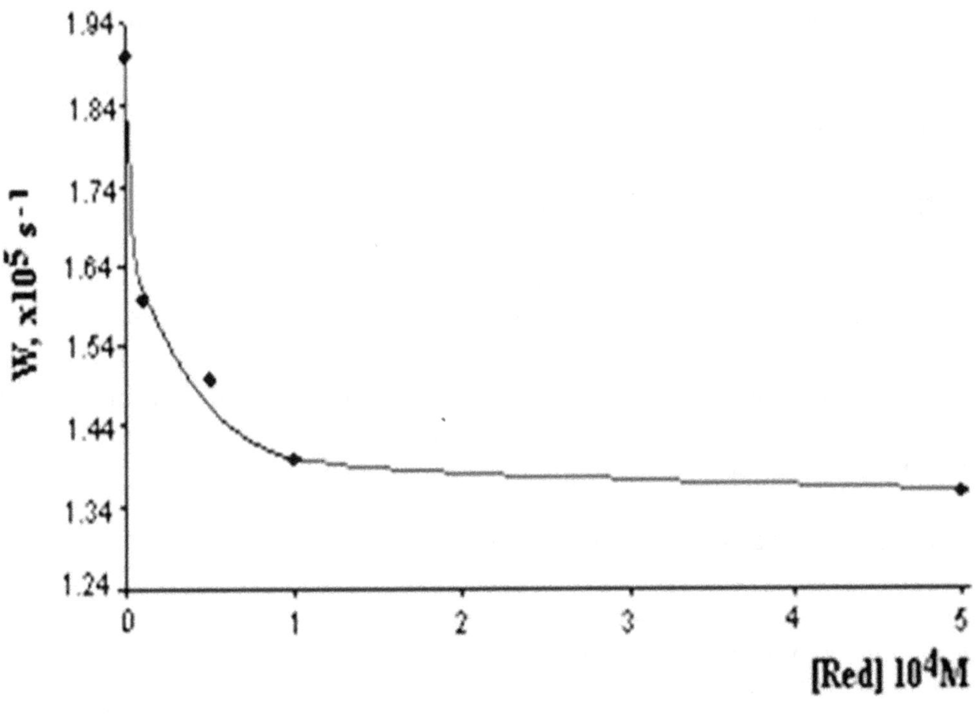

At the temperature 37°C and pH 2.5, pepsin dissolved in citrate-phosphate buffer solution pH = 2.5 in all 6 variants except for control test I was added. In control test the enzyme activity was inhibited using 10% ATA in 1: 1 volume ratio. The substrate pepsin incubation time was 30 minutes. After the first samples were taken for analysis and then the pH has been adjusted from 2.5 to 7.8 with NaOH solution, then trypsin in phosphate-citrate pH 7.8 has been added. In this case, the trypsin incubation time with casein hydrolysate was also 30 minutes. More than 60 minutes after the beginning of the reaction, or 30 minutes after the trypsin addition to the model system, the second samples series are collected. After the pH and temperature monitoring, the chymotrypsin was added for another 30 minutes. Furthermore, samples were collected for analysis and the bacterial culture in reaction mixture was added for another 60 minutes. The incubation time was 150 minutes.

This system which mimics the entire digestive tract, was studied according to $[DFH_3Na]_0$ in the same concentration range. From obtained experimental results, we find that DFH_3Na shows an inhibitor activity in the entire gastrointestinal system. With increasing of DFH_3Na concentration, the NNC content at different pH and enzymes decreases (Gonţa M. et al., 2003). The influence of sodium dihydroxyfumarate on protein and amino acids content obtained from casein hydrolysis was studied.

Using the casein enzymatic digestion method, the summary activity of the proteolytic enzymes was determined. The components of experimental model were: proteases: pepsin (Servos); Trypsin (FlucaBiochemika); Chymotrypsin (Serva); Freeze-dried mass of E. Coli M-17 and B. Bifidum 1 (OOO "Ferment", Moscow); Protein-casein substrate (FlukaBiochemical); Sodium dihydroxyfumarate in four concentrations: N1; N2; N3; N4; sodium nitrite; Phosphate-citrate buffer solution with pH 2.5 and 7.8.

Six samples were studied, four of them based on variation of DFH_3Na concentration and two control tests (control I - enzyme activity was inhibited when hydrolysis start and control II - lack of reductant). The invariable components of incubation medium (40 ml) were taken in the final concentration: 0.44-2% casein, 1.10^{-4} M sodium nitrite, 2.5 mg/ml pepsin, trypsin, and lyophilized bacteria mass (2.5 doses per reaction vessel) dissolved in 3 ml distilled water at 37°C with three hours prior to the start of the experiment. Substrate incubation with enzymes in the presence of sodium nitrite and dihydroxyfumarate was carried out at a constant temperature of 37°C in adapted vessels connected to a thermostat with electric stirrer. The experimental methodology was identical to that described for determination of NNC depending on $[DFH_3Na]$.

Determination of Amino Acid Concentration

After each time interval (30, 60, 90, 150 minutes), 1 ml incubation medium was introduced into Eppendorf tubes with 1 ml of 10% ATA to inhibit the protease activity, protein and large fragments sedimentation from hydrolysis. After 15 minutes, the samples were centrifuged at 10,000 rpm. The supernatant was assayed spectrophotometrically at 280 nm. The amino acids concentration from enzymatic hydrolysis was determined from calibration curve for tyrosine. The intensity of hydrolytic process was estimated based on the aromatic amino acids content.

Determination of Protein Content (Bradford)

The protein amount in incubation medium was determined by photo colorimeter according to Bradford method. The proteins in the experimental sample - 0.2 ml incubation medium of each test - were precipitated with 1/2 volume of 15% ATA at room temperature for about 20 hours. The precipitate, collected after centrifugation (10 minutes, 15,000 rpm), was washed with 5% ATA and then dissolved in equivalent amounts of 0.3N NaOH and 0.3N HCl. From each experimental sample 20 µl protein was taken, then added 80 µl of NaCl (0.15 N) and 1 ml Bradford reagent. After stirring, the samples were kept for 20 minutes more, and then were assayed spectrophotometric at 595 nm. Protein and high molecular weight peptides content (g/ml) insoluble in ATA were calculated.

Proteins SDS-Electrophoresis in Polyacrylamide Gel

A volume of 2 ml medium were taken from all tests and subjected to purification with 5% cooled ATA solution, and three times with cooled acetone, and then centrifuged for 10 minutes on a "MPW-310" centrifuge (Poland) at 10000 rpm. The precipitate obtained was dried in the air and dissolved (10: 1 volumes) in Tris-HCl buffer (pH 6.8) containing 4.24% SDS, sucrose (ca. 20%), β-mercaptoethanol 6% (v/v), and bromophenol blue 0.004%.

After stirring for 10-15 minutes, the samples were boiled in a water bath for 2-3 minutes. Electrophoresis was performed in Laemmle buffers in 1 mm thickness polyacrylamide (GPAA) vertical plates in denaturing conditions and in the presence of SDS. As an eluent, an aqueous solution based on fallowing component has been used: 25 mM Tris, 192 µM glycine, 0.1% SDS. As a stationary gel phase a 12.55% acrylamide + bisacrylamide solution including 4 ml 1.88 M Tris-HCl buffer (pH=8.8), 5 ml acrylamide (48.7%) + Bisacrylamide (1.3%), 2 ml 1% SDS, TEMED, 2 ml distilled water, and 2 ml 0.5% PSA has been used. Concentrate gel (5%) was prepared one hour before protein samples deposition, using 0.638 M Tris-HCl solution with pH 6.8.

The electrophoresis was carried out for three hours at 15 mA and 80 V, until protein samples (in wells were introduced 7.5-10 µl) passed the gel, and at 30 MA, 180 V, until the dye front reached 1 cm from the bottom of gel separation. After detach and matrix disassemble the separated gel has been fixed, coloured, discoloured, shoot, and so on.

Gel fixation after electrophoresis was performed during two hours with a solution containing isopropanol: acetic acid: water in a ratio of 25:10:65.

Post-electrophoretic visualization has been performed using 0.06% CBBG-250 dye in 3.5% HClO4 which was then removed from the polyacrylamide gel by washing with a 7% acetic acid solution.

Molecular relative mass determination of the separate polypeptide fractions was performed in same conditions to those of other samples, electrophoresis of some samples with known molecular weights: β-galactosidases - 116.0 kDa; Phosphorylase b - 97.4 kDa, bovine albumin - 66.0 kDa, ovalbumin - 45.0 kDa, carbonic anhydrase - 29.0 kDa, trypsin inhibitor - 20.1 kDa, α-lactalbumine - 14.2 kDa (Sigma, USA).

From obtained electropherogram in each case a calibration curve Rf and relative molecular masses logarithms has been performed. From these plots the relative molecular masses logarithms of the separated fractions depending on the electrophoretic mobility index has been determined. The difference between experimental samples and the control was assessed employing T-test using the significance threshold of 5% and 1%.

The successive action of proteolytic enzymes on casein in model system: substrate - proteases - nitrosation-reducing agents, lead to the proteins hydrolysis with formation of amino acids and peptides mixture.

Analysing the unhydrolyzed protein content, the same similarities could be observed as in the case of protease activity in all the studied samples (Table 2, Figure 5).

Table 2. Dynamic of insoluble peptide content ATA (mg/ml) in casein hydrolysate in the presence of sodium nitrite and sodium dihydroxyfumarate

Variant $[DFH_3Na]$	Incubation time, min			
	30	60	90	150
Control	5,57±0,47	4,49±0,44	3,64±0,19	3,13±0,30
R1 ($1x10^{-4}$M)	5,44±0,27	4,93±0,52*	3,39±0,25*	3,33±0,29
R2 ($3x10^{-4}$M)	5,67±0,52	4,87±0,35*	3,70±0,25	2,93±0,29
R3 ($5x10^{-4}$M)	5,27±0,63	4,25±0,13	3,23±0,39*	2,95±0,21
R4 ($1x10^{-3}$M)	5,83±0,52	4,50 ±0,25	3,87±0,38	3,55±0,33*

*(Values are statistically valid for *P<0,05; **P<0,01).*

Thus, the evaluation of protein concentration in the hydrolysate during 150 minutes reveals a decrease in dynamics with 84.4% in comparison to the initial casein concentration (Tricker, et al., 1992). The obtained values for this index for all samples containing sodium dihydroxyfumarate show the irrelevant statistical differences.

As a result, in enzymatic digestion in incubation medium, the protein substrate content will be decreased with simultaneously increasing of amino acids concentration.

The quantitative amino acids spectrophotometrically analysis shows, as a result of pepsin action, an increase of concentration for these compounds by a 56-fold in comparison to the control test (0.13 μM enzyme activity was inhibited) and 83 times as a result of the successive hydrolysis of pepsin, trypsin, chymotrypsin (Tab. 3, Fig. 5).

Figure 5. The variation of unhydrolyzed protein substrate content according to $[DFH_3Na]_0$ in casein hydrolysis; $[NO_2^-]_0 = 1\cdot10^{-4}M$, $[enz]_0 = 2{,}5$ mg/ml

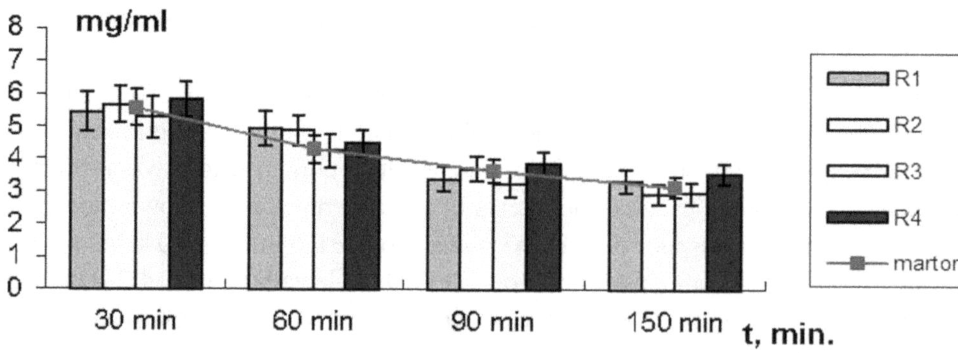

The upward dynamics of this index in the presence of the reductant shows some differences depending on concentration. Thus, in the samples containing DFH_3Na in R2 and R4 concentration, the amino acids content is low when protein hydrolysis is catalysed by pepsin and trypsin about 1.16-1.34 times in comparison with control test. After 90 minutes, the chymotrypsin was added; in the result, amino acids content exceeds the control test with 1.12-1.21 times. In contrast to these two tests, in the incubation medium with DFH_3Na, a higher concentration of protein hydrolysis products with 18.38% after 30

Table 3. Dynamic of amino acids content (μM) from casein hydrolysate in the presence of sodium nitrite and sodium dihydroxyfumarate

Variant $[DFH_3Na]$	Incubation time, minute			
	30	60	90	150
Control	7,29±0,10	10,29±0,60	10,74±0,98	10,80±0,86
R1 (1x10⁻⁴M)	7,47±0,79	8,06±0,55**	11,35±0,95	11,11±0,95
R2 (3x10⁻⁴M)	6,30±0,56**	6,72±0,92**	12,97±0,63**	12,73±1,20**
R3 (5x10⁻⁴M)	8,63±0,82**	9,90±0,5	11,98±0,21**	12,03±1,10*
R4 (1x10⁻³M)	5,43±0,45**	8,27±0,46**	12,05±0,43**	11,95±0,58**

Values are statistically valid for *P<0,05; **P<0,01.

minutes, 11.54% in 90 minutes and 11.39% in 150 minutes was found. In the presence of DFH_3Na with lowest concentration, this index shows a similar dynamic to the control test (Duca M. et al., 2003).

Thus, the analysis of unhydrolyzed substrate content in correlation with protein hydrolysis products shows that the enzymes activity was not reductant-dependent at these concentrations. Different amino acids amount in the presence of the reductant may be explained by the chemical transformations that involve sodium nitrite in water solution (Duca Gh. Et al., 2002). Increasing N-nitrosoamine content implies a decrease of amino acids concentration, which exhibits absorption at 280 nm, and vice versa, the reduction of nitrosating agent's concentration leads to a decrease in nitrosocompounds content. Thus, more free amino acids remain in the reactant mixture compared to control test (Duca Gh. et al., 2002).

Figure 6. The variation of amino acids concentration in casein hydrolysis depending on $[DFH_3Na]_0$; $[NO_2^-]_0 = 1\cdot10^{-4}M$; $[enz]_0$ = 2,5 mg/ml

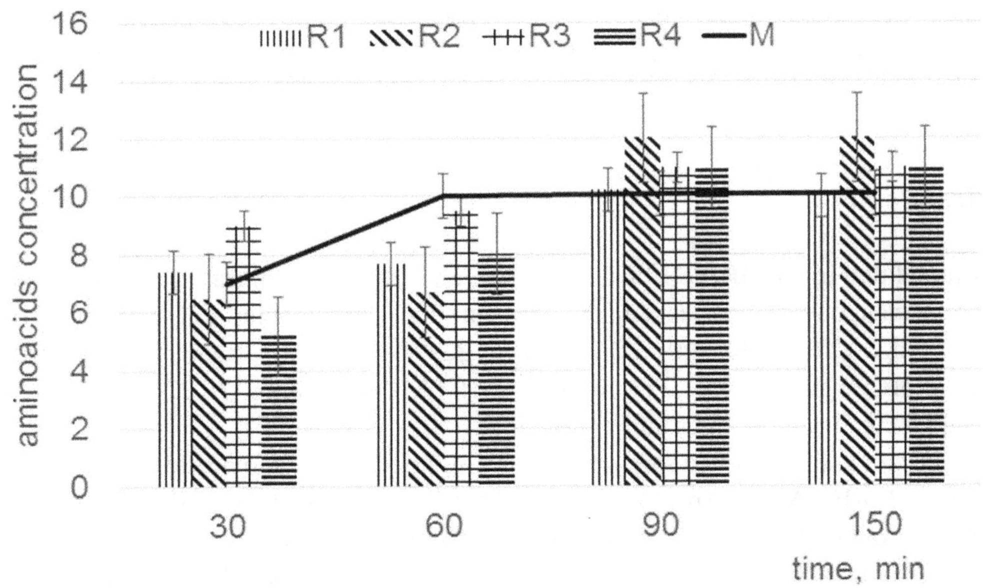

SOLUTIONS AND RECOMMENDATIONS

It was found that some N-nitroso (NOC) compounds generate genotoxic effects and can cause tumours in organisms of the bio tests. NNA can be produced in the gastrointestinal tract by various mechanisms, such as acid catalysis or under the influence of bacterial flora, which increases the risk of colorectal cancer. The formation of nitrite ions after reduction of nitrate ions, through chemical or microbial catalysis, or directly taken from food (meat, meat products etc.).

Therefore, one of the key tasks is to inhibit the nitrosation process of nitrosatable substrates entering the gastrointestinal system. Since it was shown that sodium dihydroxyfumarate, which can be obtained from ecological by products, does not influence the process of protein hydrolysis and lowers the NOC concentration, this compound can be used to inhibit NNA formation. For this purpose, other reducers, such as the spirulina extract containing the carotenoid pigment, can be used.

FUTURE RESEARCH DIRECTIONS

Future research is focused on the study of the processes of drugs nitrosation and the formation of NO. Most of the time, people are not informed that drugs in the gastrointestinal system can form toxic products that could cause adverse effects. Moreover, a considerable number of nitrogen containing drugs (antibiotics, anti-inflammatory drugs, diuretics, etc.), such as amines and amides could undergo nitration reactions. As a result, NOCs can be formed, along with the other reactive species due to the nitration with nitrite-forming ions, penetrated into the gastro-intestinal tract from food.

It is important to study the mechanisms of drugs nitrosation, which have different structures, how the rate of these processes depends on the different physic-chemical parameters. But the most important step is to develop the inhibition methods of the NOC formation process, to reduce the concentration of these carcinogenic and mutagenic compounds. Various natural inhibitors for drug formulation can be used for this purpose.

CONCLUSION

- It was found that in the GJS collected from different patients, the nitrate and nitrite concentration is pH dependent on GJS type. The nitrite concentration is lower at low pH (pH 1.5-2.5) and ranges from 1.41 mg/l (2.04 to 10^{-5} M) to 2.98 mg/l (4 32•10^{-5}). At pH > 6 the nitrite concentration increases in pH range 6-8, from 2.86 mg/l ($4.14 \cdot 10^{-5}$ M) to 5.68 mg/l ($8.22 \cdot 10^{-5}$ M). This can be explained by the bacterial microflora growth.
- The nitrate content in analysed samples is much higher compared to nitrite concentration and varies in range $1.5 \cdot 10^{-4}$ M (12.8 mg/l) - $1.12 \cdot 10^{-2}$ M (95.2 mg/l). As low pH does not favour bacterial microflora growth, the nitrate concentration, compared to nitrite one, is higher at low pH.
- The amines nitrosation rate in gastric juice is higher than in the model system, due to the presence of SCN ions which catalyse this process. As the amines concentration is in excess, the limiting factor is nitrite concentration.

- It was found that nitrosation rate of metabolites obtained from casein hydrolysis is depending on enzymes (pepsin, trypsin, chymotrypsin) being maximal for chymotrypsin. This rate is also pH dependent.
- Our results have demonstrated that DFH_3Na and DFH_4 show the reducing properties and inhibit the NNC formation from casein and albumin hydrolysis in the presence of various enzymes. The NNC concentration decreases 1.5-fold at $[DFH_3Na] = 5 \bullet 10^{-4}$ M compared the control sample.
- The analysis of unhydrolyzed substrate content in correlation with product concentration from protein hydrolysis shows that the enzymes activity is not influenced by DFH_3Na at these concentrations.
- The amino acids concentration obtained in the incubation medium during protein hydrolysis increases over time and does not depend on DFH_3Na concentration. Thus, in the protein-enzyme-nitrite-bacteria-reducing experiment model using different dosing methods of protein hydrolysis (peptides, amino acids, proteins) and N-nitroso components, have observed the inhibitory action of DFH_3Na in NNC formation, with no influence on the proteases activity in digestive tract.
- Comparative study concerning the decrease of nitrite content in albumin hydrolysis in the presence of DFH_4, AAs and (+) - Ct, shows that nitrite consumption rate in the presence of DFH_4 is maximal and the rates constants are follows:

$kDFH_4 > kAAs > k\,(+)\,Ct$

$0{,}94 \cdot 10^2\ M^{-1}\ s^{-1}\ 0{,}6 \cdot 10^2\ M^{-1}\ s^{-1}\ 0{,}4 \cdot 10^2\ M^{-1}\ s^{-1}$

REFERENCES

Anne, Roisin, & Jane. (2002). *Effect of White Versus Red Meat on Endogenous N-Nitrosation in the uman Colon and Further Evidence of a Dose Respons.* American Society for Nutritional Sciences.

Bingham, S. A., Pignatelli, B., Pollock, J. R., Ellul, A., Malaveille, C., Gross, G., ... O'neill, I. K. (1996). Does increased endogenous formation of N-nitroso compounds in the human colon explain the association between red meat and colon cancer? *Carcinogenesis, 17*(3), 515–523. doi:10.1093/carcin/17.3.515 PMID:8631138

Bodawi, A. F., Hosny, G., & Mostafa, M. (1998). Salivary nitrate, nitrite and nitrate reductase activity in relation to risk of oral cancer in Egypt. *Disease Markers, 14*(2), 91–97. doi:10.1155/1998/507653 PMID:9868596

Brain, C. (1989). Challis. Chemistry and biology of nitrosated peptides. *Cancer Surveys, 8*(2), 363–384. PMID:2696585

Brambilla, G., & Martelli, A. (2007). Genotoxic and carcinogenic risk to humans of drug–nitrite interaction products. *Mutation Research, 635*(1), 17–52. doi:10.1016/j.mrrev.2006.09.003 PMID:17157055

Calmels, S., Ohshima, H., & Bartsch, H. (1988). Nitrosamine formation by denitrifying and non-denitrifying bacteria: Implication of nitrite reductase and nitrate reductase in nitrosation catalysis. *Journal of General Microbiology*, *134*, 221–226. PMID:3141563

Calmels, S., Ohshima, H., Vincent, P., Gounot, A. M., & Bartsch, H. (1985). Screening of microorganisms for nitrosation catalysis at pH 7 and kinetic studies on nitrosamine formation from secondary amines by E. coli strains. *Carcinogenesis*, *6*(6), 911–915. doi:10.1093/carcin/6.6.911 PMID:4006079

Chen, S. C., Lin, Z., Ding, X., & Mirvish Sidney, S. (2001). Depentylation of the Rat Esophageal Carcinogen, Methyl-Npentylnitrosamine, By Microsomes from Various Human and Rat Tissues and by Cytochrome P450 2a3. *Drug Metabolism and Disposition: the Biological Fate of Chemicals*, *29*(9), 1221–1228. PMID:11502732

Cojocaru, C., Duca, G.H., & Gonta, M. (2013). Chemical kinetic for methzlurea nitrosation reaction: computer-ided solutions to inverse and direct problems. *Chemical Engineering Journal, 217*, 385-397.

Dolara, P., Bigagli, E., & Collins, A. (2012). Antioxidant vitamins and mineral supplementation, life span expansion and cancer incidence: A critical commentary. *European Journal of Nutrition*, *51*(7), 769–781. doi:10.100700394-012-0389-2 PMID:22684632

Duca, Gh., Gonţa, M., Stepanov, I., Port, A., Bejenaru, I., Porubin, D., & Voloc, N. (2002). Inhibition of carcinogenig N-nitroso compounds formation in simulated gastric juice. *International conference on Ecological Chemistry*, 196-206.

Duca, M., Gonţa, M., Porubin, D., & Voloc, N. (2003). Research on the formation of nitrosocomposites in model systems. *Scientific Annals of the MSU*, 419-425.

Gonţa, M. (2003). The influence of polyphenols on n-nitrosoamines formation during proteins fermentation. *The I-st international conference of the Moldavian chemical society*, 68.

Gonţa, M., & Sîrbu, E. (2018). Kinetics of the processes of nitrosation of amoxicillin with nitrite ions and inhibition of N-nitrozoamoxicillin formation. *Studia Universitatis, 1*(111), 162-169.

Guadagni, S., Pistoia, M. A., Valenti, M., Leocata, P., Coletti, G., Calvisi, G., ... Reed, P. I. (1998). N-Nitroso Compounds, Bacteria, and Carcinoembryonic Antigen in the Gastric Stump, FRACG. *The Journal of Surgical Research*, *80*(985444), 345–351. doi:10.1006/jsre.1998.5444 PMID:9878336

Hong, L., Duncan, C., Townend, J., Killham, K., Smith, L. M., & Johnston, I. (1997). Nitrate- reducing bacteria on rat tongues. *Applied and Environmental Microbiology*, *63*(3), 924–930. doi:10.1128/AEM.63.3.924-930.1997 PMID:9055411

Iijima, K., Eyfe, V., & McColl, K. E. L. (2003). Studies of nitric oxide generation from salivary in human gastric juice. *Scandinavian Journal of Gastroenterology*, *38*(3), 246–252. doi:10.1080/00365520310000708a PMID:12737438

Keszei, A. P., Goldbohm, R. A., Schouten, L. J., Jakszyn, P., & van den Brandt, P. A. (2013). Dietary N-nitroso compounds, endogenous nitrosation, and the risk of esophageal and gastric cancer subtypes in the Netherlands Cohort Study. *The American Journal of Clinical Nutrition*, *97*(1), 135–146. doi:10.3945/ajcn.112.043885 PMID:23193003

Martelli A., Brambilla G. (2012). *Arylamine drugs: genotoxic-carcinogenic activity of NO-derivatives. Frontiers in Bioscience E4*. Department of Internal Medicine, Division of Clinical Pharmacology and Toxicology, University of Genoa, Viale Benedetto.

Massey, R. C., Key, P. E., Mallett, A. K., & Rowland, I. R. (1991). An investigation of the endogenous formation of apparent total N-nitroso compounds in conventional flora and germ-free rats. *Food and Chemical Toxicology*, *26*(7), 595–600. doi:10.1016/0278-6915(88)90230-X PMID:3181835

Mirvish, S. S. (1983). The etiology of gastric cancer. Intra-gastric nitrozamide formation and other theories. *Journal of the National Cancer Institute*, *71*(3), 613–647. PMID:6350677

Mirvish, S. S., Grandjean, A. C., Reimers, K. J., Connelly, B. J., Chen, S.-C., Gallagher, J., & … . (1995). Dosing time with ascorbic acid and nitrate, gum and tobacco chewing, fasting, and other factors affecting N-nitrosoproline formation in healthy subjects taking Proline with a standard meal. *Cancer Epidemiology, Biomarkers & Prevention*, *4*, 775–820. PMID:8672996

Mirvish, S. S., Reimers, K. J., Kutler, B., Chen, S. C., Haorah, J., Morris, C. R., & Lyden, E. R. (2000). Nitrate and nitrite concentrations in human saliva for men and women and different ages and times of the day and their consistency over time. *European Journal of Cancer Prevention*, *9*(5), 335–342. doi:10.1097/00008469-200010000-00008 PMID:11075887

Paul, K. (1990). Risk of Colorectal and Other Gastro-Intestinal Cancers After Exposure to Nitrate, Nitrite and N-Nitroso Compounds: A Follow-Up Study. *International Journal of Cancer*, *80*, 852–856.

Porubin, D., Hecht, S. S., Li, Z., Gonta, M., & Stepanov, I. (2007). Stepanov, I. Endogenous Formation of N'-itrosonornicotine in F344 Rats in the Presence of Some Antioxidants and Grape Seed Extract. *Journal of Agricultural and Food Chemistry*, *55*(17), 7199–7204. doi:10.1021/jf0712191 PMID:17637060

Simonov, Y. A., Fonari, M. S., Duca, G. G., Gonta, M. V., Ganin, E. V., Yavolovskii, A. A., ... Lipkowski, J. (2005). Nitrosation of hydrochlorothiazide and the modes of binding of the N-nitroso derivative with two macrocycles possessing an 18-membered crown ether cavity. *Tetrahedron*, *61*(27), 6596–6601. doi:10.1016/j.tet.2005.04.050

Simonov, Y. A., Gonta, M. V., & Yavolovskii, A. A. (2005). 11-Nitroso-6, 7, 10, 11, 12, 13, 15, 16-octahydro-9H-5, 8, 14, 17-tetraoxa-11-azabenzocyclopentadecene. *Acta Crystallographica. Section E, Structure Reports Online*, *61*(10), o3136–o3139. doi:10.1107/S1600536805027315

Snoz, S. V., & Dmitrenko, N. P. (1995). Activity of the rat cytochrome P-450 system in conditions of the induction immunity and nitric oxide synthesis. *Toxicology Letters*, *78*(Supl.1), 75–76.

Stepanov, I., Hecht, S. S., Mirvish, S. S., & Gonta, M. (2005). Comparative Analysis of Tobacco-Specific Nitrosamines and Total N-Nitroso Compounds in Moldovan Cigarette Tobacco. *Journal of Agricultural and Food Chemistry*, *53*(20), 8082–8086. doi:10.1021/jf050747e PMID:16190674

Stockbrugger, R. W., & Kleinjans, J. C. S. (1998). Volatile N-nitrosamines in gastric juice of patients with various conditions of the gastrointestinal tract determined by gas-cromatography – mass spectrometry and related to intragastric pH and nitrate and nitrite levels. *Cancer Letters*, *124*(2), 119–125. doi:10.1016/S0304-3835(97)00467-9 PMID:9500200

Takahama, U., Hirota, S., & Kawagishi, S. (2009). Effects of pH on nitrite-induced formation of reactive nitrogen oxide species and their scavenging by phenolic antioxidants in human oral cavity. *Free Radical Research*, *43*(3), 250–261. doi:10.1080/10715760802691463 PMID:19169919

Takahama, U., Ryu, K., Oniki, T., & Hirota, S. (2007). Dual-function of thiocyanate on nitrite-induced formation of reactive nitrogen oxide species in human oral cavity: Inhibition under neutral and enhancement under acidic conditions. *Free Radical Research*, *41*(11), 1289–1300. doi:10.1080/10715760701710885 PMID:17957543

Tricher, A.R., Perkins, M.J., Massey, R.C., & McWeeny, D. J. (1984). Synthesis of three N-nitroso-dipeptides N-terminal and a method for determination in food. *Food Additives and Contaminant, 1*, 307-312.

Tricker, A. R., Pfundstein, R., Kalble, T., & Preussmann, R. (1992). Secondary amine precursors to nitrosamines in human saliva, gastric juice, blood, urine and faeces. *Carcinogenesis*, *13*(4), 563–568. doi:10.1093/carcin/13.4.563 PMID:1576707

Triker, A. R., & Preussmann, R. (1991). Carcinogenic N-nitrosamines in the diet: Occurrence, formation, mechanisms and carcinogenic potential. *Mutation Research*, *259*(3-4), 277–289. doi:10.1016/0165-1218(91)90123-4 PMID:2017213

Veena, S., & Rashmi, S. (2014). A review on mechanism of nitrosamine formation, metabolism and toxicity in Vivo. *International Journal of Toxicological and Pharmacological Research*, *6*(4), 86–96.

Xu, G. P., & Reed, P. I. (1993). Method for group determination of total N-nitroso compounds and nitrite in fresh gastric juice by chemical denitrosation and thermal energy analysis. *Analyst (London)*, *118*(7), 877–883. doi:10.1039/an9931800877 PMID:8372977

Xu, G. P., & Reed, P. I. (1993). N-nitroso compounds in fresh gastric juice and their relation in intragastric pH and nitrite employing an improved analytical method. *Carcinogenesis*, *14*(12), 2547–2551. doi:10.1093/carcin/14.12.2547 PMID:8269625

Yang, C. S., Yoo, J.-S. H., Ishizaki, H., & Hong, J. (1990). Cytochrome P-450 IIE1: Roles in nitrosoamine metabolism and mechanisms of regulation. *Drug Metabolism Reviews*, *22*(2-3), 147–159. doi:10.3109/03602539009041082 PMID:2272285

Yoshida, K., & Kasama, K. (1987). Biotransformation of nitric oxide. *Environmental Health Perspectives*, *73*, 201–206. doi:10.1289/ehp.8773201 PMID:3665863

Chapter 24
Moss Biomonitoring in Former Sovet Union Countries: A Review

Inga Zinicovscaia
Joint Institute for Nuclear Research, Russia & Horia Hulubei National Institute for R&D in Physics and Nuclear Engineering, Romania & The Institute of Chemistry, Moldova

Nikita Yushin
Joint Institute for Nuclear Research, Russia

Konstantin Vegel
Joint Institute for Nuclear Research, Russia

Dmitrii Grozdov
Joint Institute for Nuclear Research, Russia

ABSTRACT

Air pollution is a worldwide environmental and health issue. Among environmental pollutants, heavy metals are the most dangerous due to their persistence and bioaccumulation in food chain. Assessment of heavy metal deposition using moss biomonitors is a cheap and effective technique, which was successfully applied in different European counties. The present work revises application of passive biomonitoring in former Soviet Union countries: Belarus, Estonia, Georgia, Kazakhstan, Latvia, Moldova, Russia, and Ukraine. The air pollution sources in each country were identified. The mean concentration of elements considered as environmental pollutants were compared in order to detect the most polluted countries on the post-Soviet space.

DOI: 10.4018/978-1-7998-1241-8.ch024

INTRODUCTION

Air pollution is a major problem in recent decades, which has a significant toxicological impact on human health and the environment (Ghorani-Azam et al., 2016). Various chemicals are emitted into the air from both natural and anthropogenic sources. The natural sources of air pollution include volcanic eruptions, geothermal sources and emissions from land and water, forest fires, sea salt in a coastal area, biological material and radiological decomposition (Pénard-Morand & Annesi-Maesano 2004). It is noteworthy that natural air pollution did not pose a serious problem as it is part of natural environment equilibrium, whereas the anthropogenic pollution sources present a major challenge for the world today. The increase of the pollution level is directly linked to the population growth which is associated with the growing demands in energy and consumer goods. The anthropogenic chemical pollution has no borders and is independent. The pollutants being released into the atmosphere will have an impact over the global environment (Popescu & Ionel 2010). Traditionally, the anthropogenic sources are divided into two groups - stationary and mobile sources (Pénard-Morand&Annesi-Maesano2004). The most important groups of anthropogenic air pollution sources are industrial activity, mining, transport and agricultural systems (Popescu & Ionel 2010; Aksu 2015; Zinicovscaia et al., 2017)

The long-term air pollution causes millions of deaths each year. In addition, it is associated with the number of immediate, medium-term and long-term human health problems such as the respiratory infections and inflammations, cardiovascular dysfunctions, and cancer (Ghorani-Azam et al., 2016; Olmo et al., 2011).

In the most countries, the various regulatory instruments are combined into a coordinated control programme. However, in practice, the air pollution monitoring is a complex problem, which requires identification of emissions sources, evaluation of analytical methods, risk assessment, control of critical emissions and integration of economic aspects (Wolterbeek 2002). In many countries the particular attention is given to determination of gaseous air pollutants such as sulphur dioxide (SO_2), nitrogen oxides (NO_x: NO, NO_2), carbon dioxide (CO_2) and ozone (O_3), etc. At the same time, the serious threat for the environment and human health represent the heavy metals, due to their toxicity, non-degradable nature and ability to be build up in organisms and move up in the food chain.

Dispersion modeling and field measurements of the emissions are used to learn about air pollutants. However, these techniques are characterized by low sensitivity, high costs and require long-term sampling at large numbers of sampling sites (Wolterbeek 2002). The moss biomonitoring technique, introduced in Scandinavian countries around 1970 (Ruhling & Tyler 1971), has proven to be suitable, cheap and efficient technique for the assessment of atmospheric deposition of heavy metals. In 1989/1990 the first moss survey at the European scale was conducted and since then the number of participating countries has greatly expanded (Harmens et al. 2013). Today, the moss biomonitoring studies in the European countries are mainly performed in the framework of the International Cooperative Programme on Effects of Air Pollution on Natural Vegetation and Crops International Cooperative Programme, (UNECE ICP Vegetation) (Harmens et al., 2008; 2009; 2013). Estonia, Lithuania and Latvia take an active part in the European moss survey, while similar information about the Newly Independent States (NIS) is not so well highlighted. The present chapter is the first attempt to collect and present data on moss biomonitoring study in NIS countries. A short description of countries is given as well.

MOSS BIOMONITORING IN FORMER SOVIET UNION COUNTRIES

Following the collapse of the Soviet Union, 15 new independent countries (Republic of Armenia, Republic of Azerbaijan, Republic of Belarus, Republic of Estonia, Republic of Georgia, Republic of Kazakhstan, Kyrgyz Republic, Republic of Latvia, Republic of Lithuania, Republic of Moldova, Russian Federation, Republic of Tajikistan, Republic of Turkmenistan, Ukraine, Republic of Uzbekistan) came into existence. Three of them, Republic of Estonia, Republic of Lithuania and Republic of Latvia, joined UNECE ICP Vegetation programme in 1989, while the Republic of Kazakhstan, Republic of Moldova, Republic of Georgia, Republic of Armenia, Republic of Azerbaijan and Republic of Tajikistan in 2015. In Russian Federation and Ukraine moss surveys were carried out periodically and mainly in polluted regions. On Fig. 1 are highlighted countries for which information about moss biomonitoring studies (article papers, conference papers, atlases) is available.

Figure 1.

Republic of Estonia

Located on the eastern coast of the Baltic Sea, Republic of Estonia (Estonia) is the northern most and also the smallest country in the Baltic region, in terms of both population (1.32 million) and area (45,339 km^2). Estonia's neighbors are Russia to the East, Latvia to the South, Sweden to the West and Finland to the North. Almost half of the Estonia's land is covered by forests (approx. 51%). Estonia's biodiversity is formed due to the differences in the climatic conditions. The local climatic differences are attributed to the neighboring Baltic Sea, which warms up the coastal zone in winter and later has a cooling effect, especially in spring. As a result, the summers are moderately warm and the winters are moderately cold (Kallis et al., 2017).

As it was previously mentioned, Estonia joined the UNECE ICP Vegetation programme of the European Nordic countries in 1989 and national moss study is being carried out in Estonia every five years (1989/90; 1995/96; 2000/01; 2005/06, 2010/11) (Liiv & Kaasik 2004). During 1989-2005, the significant decrease in Cd (by 48%), Cr (by 92%), Fe (by 70%), Ni (by 92%), Pb (by 80%) and V (by 76%) concentrations in Estonia's moss was observed, while the concentrations of Cu and Zn do not exhibit a marked shift. The decrease in the concentrations of Cr, Fe, Ni, Pb and V in Estonian mosses was related to the decrease of emissions of solid particles from the Eesti Power Plant and the cement factory of Kunda Nordic Tsement AS, all of them being located in North-Eastern part of the country. The Pb emission has decreased by 80% due to the introduction of unleaded fuel in Estonia (Kösta & Liiv 2011).

The main air pollutants in Estonia originate from the oil shale-based power engineering and chemical industry, as well as from the building materials industry, the north-eastern part of the country (NE Estonia) is considered particularly polluted (Liiv & Kaasik 2004). For example, in 2002, about 97% of air pollution, 86% of total waste and 23% of water pollution in Estonia came from the power industry, which uses the oil shale as the main input for production process (Raukas 2004). Thus, in 1992, 1997, 2002, and 2007 the moss samples were collected around two large oil shale fired thermal power stations and a cement factory in the North-Eastern part of the country. According to obtained results, Cr, Fe, Ni, Pb and V concentrations in moss samples collected from oil shale fired power plants and the cement factory, were diminished remarkably in NE Estonia between 1992–2008. The changes in Cd, Cu, and Zn concentrations in mosses were smaller (Liiv & Kaasik 2004; Kaasik & Liiv 2007; Kösta & Liiv 2011).

Emissions of heavy metals from thermal power plants in NE Estonia have been fallen remarkably since 1989 due to implementation of environmentally friendly technologies. At the same time, due to the increasing of the vehicles number, the emissions from road transport have become dominating, especially in towns. Although the waiver of leaded gasoline has caused a decline in lead pollution, the emissions from diesel fuel and wear-off tires and machine parts remain a significant sources of air pollution (Kaasik & Liiv 2007).

Republic of Lithuania

Republic of Lithuania (Lithuania) is situated on the Eastern shore of the Baltic Sea. The territory of the Lithuania covers 65 302 km^2. It borders Latvia on the North, Belarus on the Eeast and South, Poland and Russia on the Southwest. Lithuania is the country of lowlands with the highest hills reaching up to 300 m. The agricultural land covers about 52.4% of the total land area. The climate in Lithuania varies from marine to continental with the average temperatures on the coast – 2.5 °C in January and 16 °C in July (Lithuania's seventh National Communication, 2017).

In Lithuania, 133 moss samples were collected in 1995, 138 samples in 2000 and 146 samples in 2005. *Hylocomium splendens* and *Pleurozium schreberi* are the most widespread moss species. There are three major metal pollution sources in Lithuania: two thermal power stations located in Elektrenai and Mazeikiai and a cement factory located in Naujoji Akmene (Ceburnis et al., 2002). During three moss surveys the concentration of As, Cd, Cr, Cu, Fe, Hg, Mn, Ni, Pb, V and Zn in the analyzed moss samples were determined using atomic absorption spectrometry in order to separate the background and anthropogenic contributions to heavy metal concentrations in moss. The decrease of As, Cr, Ni,

Pb, V and Zn content was observed in the Lithuanian mosses during the years 1995–2005, though the concentration of Cu and Hg increased. The decrease in metal concentration can be associated with the decreased of combustion of oil fuel (Šakalys et al., 1999; Šakalys et al., 2009).

Five moss species, namely *Pylocomium splendens*, *Pleurozium schreberi*, *Eurhynchium angustirete*, *Sphagnum* and *Rhytidiadelphus* were used to trace the atmospheric heavy metal depositions, during August–September 1993. Relatively high concentrations of Pb, Ni, Cu, Cr, Cd, Fe were determined near the factory of building materials Naujoji Akmene. The high V concentrations were determined near the Elektrenai and Alytus thermal power stations, indicating that the release of V into the atmosphere is mainly due to the oil refineries and fuels burning (Ceburnis et al., 1997).

Blagnyte & Paliulis (2011) investigated the pollution with heavy metals in mosses along the high intensive traffic in Geležinis Vilkas street, the main rapid traffic street in the city of Vilnius. The moss samples were taken at a different distance (5-45 meters) from the Geležinis Vilkas street. The study results showed that the metals content in all the moss samples tend to decrease with the distance from the Geležinis Vilkas street, the highest metals content being determined in the mosses collected at the distance from 5 to 15 meters from the street. The heavy metals accumulations in the moss samples changed in the following order: Zn > Cu >Pb> Cr, Ni.

The methodological work was carried out by Ceburnis et al. (1999) in order to determine the main sources contributing to heavy metal content in the mosses in Lithuania. The uptake efficiencies of mosses were calculated for Cd, Cr, Cu, Fe, Mn, Ni, V and Zn. It was observed that for Cr, Ni, and Fe the major factor is soil dust influence, so the use of moss to study these elements in atmospheric heavy metal surveys is therefore of limited value, except for the vicinity of local pollution sources. The "Vegetation factor" was dominating for Mn and also significant for Zn, while V is an anthropogenic element.

Republic of Latvia

Republic of Latvia (Latvia) is situated on the edge of the Eastern European Plain near the Baltic Sea. The territory covers an area of 64573 km^{2}in total. It is bordered by Estonia to the North, Lithuania to the South, Russia to the East, and Belarus to the Southeast, and shares a maritime border with Sweden to the West. Latvia is a typical lowland country and its terrain is characterized by flat, low areas and hilly elevations. The air temperature in Latvia has a seasonal nature – February being the coldest month with average air temperature -3.7^{0}C and July being the warmest with +17.4^{0}C (Ratniece et al., 2017)

In Latvia the biomonitoring of atmospheric heavy metal distributions in the framework of UNECE ICP Vegetation programme has been conducted in five national moss surveys: in 1990, using *Hylocomium splendens* collected in 81 plots; and in 1995, 2000, 2005 and 2015, using *Pleurozium schreberi* in 101 plots (Tabors et al., 2017). During the first moss survey conducted in 1990 many industries and thermal power plants were still operating in Latvia, however by the year 2000 the loads from the stationary sources have been decreased by 60% (Nikodemus et al., 2004). The concentrations of seven elements (Cd, Cr, Cu, Pb, Ni, V, Zn) were determined in moss samples. Comparing the result of moss surveys performed between 1990 and 2000 the decrease in Cr, Cd, Ni, Pb and Cu content in moss was noted. The decrease in Pb content was mainly associated with the improved vehicle quality and the waiving of leaded gasoline use. In 2000, the main air pollution sites in Latvia were considered Liepaja, Olaine, Riga, Daugavpils and Broceni (Nikodemus et al., 2004).

In the moss survey conducted in 2015/16 the sampling sites had close locations to those in 1990/91. In Latvia the overall concentrations of heavy metals in mosses in the period from 1990 to 2015 have

Figure 2.

been declined for lead (by 89%), vanadium (by 85%), chromium (by 78%), iron (by 71%), cadmium (by 69%) and nickel (by 66%), followed by zinc (by 21%) and copper (by 14%). In 2000/01, 2005/06 and 2015/16 surveys, the concentrations of Cu and Pb have not decreased significantly. The main sources of heavy metals in Latvia are considered to be transport, thermal power plant, metallurgical industry, pharmaceutical companies, and transboundary transport of pollutants from Poland, Germany and Lithuania (Taborset al., 2017; Nikodemus et al., 2004).

Brumelis et al. (1999) collected moss *Hylocomium splendens* in 43 forests around the metal smelter "Liepajas metalurgs" in order to estimate the distribution of deposition of Zn, Pb and Cu. The results obtained have showed the high concentrations of Zn (>200 µg/g), and elevated concentrations of Pb (38.3 µg/g) and Cu (18.3 µg/g), the metal smelter was identified as the main pollution source in the region.

Russian Federation

Russian Federation (Russia) is the world's largest country (17,098,242 km^2), covering more than 11% of Earth's land areas. Russia is bordered by 14 countries: Azerbaijan, Belarus, China, Estonia, Finland, Georgia, Kazakhstan, North Korea, Latvia, Lithuania, Mongolia, Norway, Poland, and Ukraine. Furthermore, it shares maritime borders with Japan, Sweden, Turkey and the United States. The climate in Russia ranges from steppes in the South through humid continental in much of European Russia; subarctic in Siberia to tundra climate in the polar North; the winters vary from cool along the Black Sea coast to frigid in Siberia; summers vary from warm in the steppes to cool along the Arctic coast (Russia, 2020).

In Russia, the moss survey was performed only in a limited number of regions that can be explained by the big territory and extreme climatic conditions. The regions of Russia, which participated in moss surveys are presented in Fig. 2. In some regions such as Moscow, Tula, Tver region, the biomonitoring studies were conducted for several times, while in the other regions - only once.

Moscow region, one of the most densely populated and industrialized zone in the country, was involved in the two moss survey 2005/06 and 2015/16. Air pollution in Moscow region in both surveys was mainly associated with oil combustion, metallurgy industry and mechanical plants, solid waste disposal, thermal power plants, transport, steel, aerospace, and electronics industry, metallurgical plants, machinery companies, plants for the nuclear fuel production, chemicals, pharmaceutical, fertilizers and pesticides production (Vergel et al., 2009, Vergel et al., 2019). Comparison of the results of performed surveys revealed significant differences for Cl, Ca, K, Co, As, Mo, and Hf between two surveys.

The Tula region is one of the most developed industrial and farming region in Russia. At the same time is considered one of the most polluted region in the Central Russia, more than 450 enterprises are operating within the region. During the first biomonitoring study, summer-autumn of 1998-2000, 83 moss samples were collected throughout the region. The high content of As, Th and U determined in the analyzed moss, was related to the coal mining and refining and coal combustion plants, located in the East of the Tula and in the West of the Tula region. The other anthropogenic sources are ferrovanadium industry (Fe, V), production of chlorine and organic compounds (Cl, I), other industries and general urban activities (Co, Ni, Ba, Cs, Mn, Zn, Cu, Pb,Mo, W) (Ermakova et al., 2004a). Biomonitoring study performed in 2014 showed the decrease of Mn, Zn, and V content due to the reduction of the productive capacity of the "Vanadium-Tula" metallurgical enterprise. An increase of Fe (by 30%), Cr and Co (by 40-43%), As and Cd (by 37%), Sr and Sm (by 22%) was resulted from the ferrous metallurgy enterprises, metal processing and defense industries, ore processing, transport and agricultural activities (Gorelova et al., 2016).

Kaliningrad region participated in three moss surveys moss in 1995/96, 2000/01 and 2005/06. According to the data obtained, the main air pollution sources in Kaliningrad region are engineering, metalworking, fuel and energy complex, agriculture and oil production, transport, peat fires. The transboundary transport is considered an important source of Cr, Cd, Cu and Ag (Korolyova 2010).

During the years 2001-2002, the moss samples were collected in Tver and Yaroslavl regions. These regions have similar geographical and climatic conditions, but for historical reasons Yaroslavl region is more economically developed. Based on the results obtained, the coal combustion, building materials production, industrial activity around Yaroslavl, manufacturing industry in Kashin and steel industry in Cherepovets, the thermal power plant in Konakovo and oil-refinery plant in Yaroslavl were defined as the main pollution sources in the studied regions (Ermakova et al., 2004b). Biomonitoring studies performed in Tver region in 2004 have confirmed that the main pollution sources in the region remain to be the thermal power plant in the town of Konakovo, as well as the factories of the fuel-energy complex allocated in the nearby towns of Kashin and Kalyazin. At the same time, it should be mentioned that the Tver region is considered as a background territory for the Russian Federation (metal concentrations are lower than in other studied regions) (Vergel et al., 2007).

The moss biomonitoring was applied in North-Western region of Russia with the aim to document the environmental impact of the Russian nickel industry near Nikel, Zapoljarnij and Monchegorsk on different compartments of the ecosystem. The high content of Ni, Cr, Fe, Mg, and S was determined in the moss species collected near the nickel roaster in Zapoljarnij, while the high conten of Ag, Ba, Bi, Cd, and Na was associated with the nickel refinery in Monchegorsk. The main source of Al, Ba,

Ca, K, La, Na, P, Rb, Sr and Y was considered to be the large open cast apatite mine in Kirovsk near Apatity (Reimann et al., 1999). The comparison of the accumulation of heavy metals (Ni, Cu) by two dominant moss species *Pleurozium schreberi* and *Hylocomium splendens* collected in the Lapland State Biosphere Reserve located in the impact area of atmospheric emissions from the Severonickel Smelter Combine (Monchegorsk, Murmansk Region, Russia) during 1991 and 2011 was performed by (Barkan and Lyanguzova 2018). In 1991 in the Lapland Reserve, the average nickel concentration was 12 times over its background level and copper content was 23 times over, while in 2011 the content ofNi and Cu was only 4 times higher the background concentrations. Decrease in the Ni and Cu content in mainly associated with the reduction in atmospheric emissions by the Severonickel Combine.

The region of the South Ural Mountains is ranked as the most severely polluted region in Russia and probably is among the most polluted areas in the world (Frontasyeva 2000). This region includes several industrial cities such as Karabash, Chelyabinsk, Ozersk and many others, where the clustering of heavy industry has produced the extremely high levels of air and water pollution. In addition, the substantial emissions of radioactive substances have occurred in the Ural region as the result of the full-scale activities of the radiochemical "Mayak" Production Association (Cherchintsevet al., 1999; Frontasyeva et al., 2004). In 1998 the samples of mosses *Hylocomium splendens* and *Pleurozium schreberi* were collected in Chelyabinsk. The high concentrations of Cu, Zn, As, Ag, Sb and Cd are related to the emissions from Karabash copper smelter. High Ni, Co, Cr, and As concentrations were determined near the Ufalei town, where Ni and Co production plant is located (Frontasyeva et al., 2004). At the same period, the moss samples were collected near Magnitogorsk, the "City of Steel". The values determined for Sb (12-29 ppm) were comparable with the previously reported ones for mosses, and have clearly indicated the strong pollution with this element in the Magnitogorsk region. The high concentrations of Cr, Zn, As, Se, Ag, Cd, Sb and Au have indicated that these elements represent a regional pollution problem. (Frontasyeva et al., 2000).

During the summer-autumn of 2005/06, the moss samples of *Hylocomium splendens* and *Pleurozium schreberi* have been collected in the Southern part of the Udmurt Republic at 79 sites. According to neutron activation analysis data and factor analysis, the results of the main anthropogenic sources of heavy metals are: oil refining, steel and lightindustries, mining and building material industry, chemical weapon destruction plant, pesticide and fertilizer storages (Pankratova et al., 2007).

In the Leningrad region the moss biomonitoring was performed in the area surrounding the Tikhvin town, showing that the main source of Cr, Fe, Ni and V is the ferroalloy plant located in Tikhvin (Vergel et al., 2014). In the Ivanovo region, the monitoring of the air quality performed in 2010 has showed an increased concentrations of heavy metals (V, Fe, Co, Zn) in the town of Rodniki and arsenic in its vicinities (Dunaev et al., 2018).

Ukraine

Ukraine is located in the Central and Eastern Europe and covers 5.7% of the entire territory of Europe. Ukraine is bordered by Belarus, Russia, Moldova, Romania, Hungary, Slovakia, and Poland. The most of the Ukraine's territory is located in the South-West of the East-European Plain and only 5% of the territory is mountainous. The continental climate of the country is greatly affected by the proximity to the Black Sea. The average winter temperature varies from -8 ° to -12 ° C, while the average summer temperature is from +18 ° to 25 ° C (The Sixth National Communication of Ukraine on climate change, 2012).

In Ukraine the moss biomonitoring studies were performed in Polissia zone: Rivne, Zhytomyr, Kyiv and Chernigiv regions. During the spring-summer 1995, a set of 79 moss samples was collected and concentrations of 10 elements (As, Cd, Cr, Cu, Hg, Fe, Ni, Pb, V and Zn) were determined. It was found that the high concentration of Cu as well as Cr, Pb and Zn in the west zone of studied area can be related to copper sandstone. The high zinc content was also determined near the chemical factories in Chernigiv and Belaya Tserkov. High concentration of V around the thermal centrals is due to the burning of the crude oil. The central anomalies in Cr and Fe are related to the engine factories in Kiyv and Brovari. The main source of As can be considered tanning and home chemical making, while Cd and Pb may originate from the old machinery disposed after Chernobil accident (Hervada-Sala et al., 2003).

In 1998 and 1999 the moss samples were collected in Zakarpattia and Chernivtsi regions. The element concentrations in the moss from Zakarpattia were lower than those from Chernivtsi. The distinct air pollution sources were not identified (Blum et al., 2002). The moss biomonitoring studies were also performed in Kiev and Zhitomir Regions. The concentration of 23 elements were determined by ICP-MS in mosses *Pleurozium schreberi* and *Brachythecium oedipodium*. As the main air pollution sources titan ore mining and processing in Zhitomir region and transboundary transfer of pollutants from Belarus were mentioned (Shabatura et al., 2018).

Republic of Belarus

Republic of Belarus (Belarus) occupies the area of 207,600 km^2. The longest distance from West to East is 650 km, from North to South 560 km. The terrain of Belarus is predominantly plain with hills: the average elevation is 160 m above the sea level; the highest elevation is 345 m. Its neighbors are Russia, Latvia, Lithuania, Poland, Ukraine (Aleksiayenak et al., 2010). The South-Eastern part of the country was severely contaminated with fallout from the 1986 accident at the Chernobyl nuclear power plant, situated in Ukraine only 11 km from the Belarus border. About 70% of the total radioactive fallout occurred on Belarus territory (Aleksiayenak et al., 2013).

The moss samples were collected in Belarus during 2005-2008 in Minsk, Gomel, Grodno and Vitebsk regions. The activity of ^{137}Cs and ^{210}Pbactivity was measured in collected moss samples. The maximum activities of ^{137}Cs were observed in the Gomel Region near the town Mazyr (6827 Bq/kg) and the minimum activity in Vitebsk region (4.83 Bq/kg). The "hot spots" near the towns Borisow and Yuratsishki were noticed. The concentrations of ^{210}Pb in moss samples collected over the territory of Belarus varied between 141 and 575 Bq/kg, with a median value of 312 Bq/kg. The low median value of ^{210}Pb obtained for Belarus mosses (312 Bq/kg) is presumably connected with a generally lower uranium abundance in soils in Belarus (Aleksiayenak et al., 2013; Aleksiayenak et al., 2010)

Besides the radionuclides, the concentrations of 36 elements were determined in 156 moss samples collected in the above mentioned regions in the period 2005-2008. The high concentrations of Mg, Co, Se, Mo, Cd, W, Pb, which can be associated with engineering industry were determined near Minsk, Molodechno and Zhodino. The main source of Co, Pb, Fe, Sb, Br, Cr, Cu, Zn, and Sb can be considered metallurgical industry (Aleksiayenak et al., 2015). The results obtained were published in the European atlas of heavy metal atmospheric deposition in 2005/06 and 2010/11.

Figure 3.

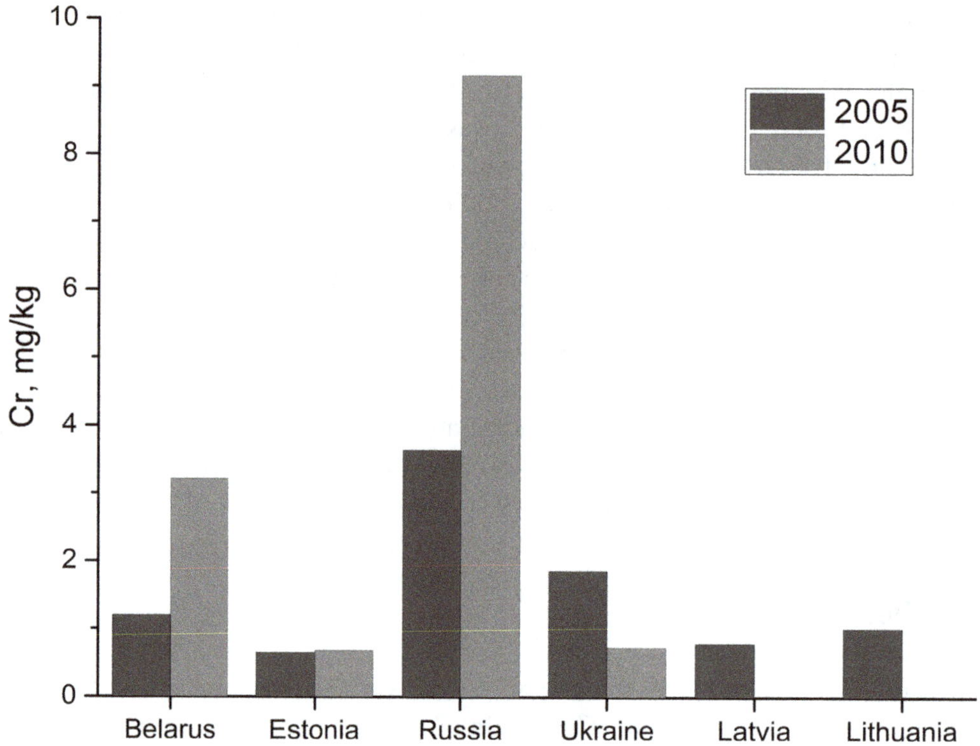

Republic of Moldova

Republic of Moldova (Moldova) is a landlocked country, located in the North-Western Balkans, covering the area of 33,846 km^2. Moldova borders with Ukraine and Romania. The relief of Moldova is represented by hills as well as plains, with uplands mostly in the central part of the country. The climate of Moldova is moderately continental, characterized by the relatively mild winters with little snow, long warm summers, and low humidity. The Republic of Moldova has the unique land resources characterized by the predominant black earth soils (~75 per cent) with high productivity potential. Moldova's rich soil and temperate continental climate have made the country one of the most productive agricultural regions in Southeast Europe since ancient times, and a major supplier of agricultural products (Third National Communication of the Republic of Moldova, 2013).

The first attempt to perform the biomonitoring study in Moldova was done by Romanian scientists in 2001 but covered only a limited area, mostly along the Prut river (Cucu-Man 2006). The concentrations of V, Cr, Fe, Ni, As, Br, La, Ce, Th, U were determined in moss samples using ICP-MS technique. In 2015 neutron activation analysis and atomic absorption spectrometry were used by Zinicovscaia et al. (2017) to determine 41elements (Na, Mg, Al, Cl, K, Ca, Sc, Ti, V, Cr, Mn, Fe, Co, Ni, Cu, Zn, As, Se, Br, Rb, Sr, Zr, Cd, Sb, Cs, Ba, La, Ce, Nd, Sm, Eu, Gd, Tb, Tm, Yb, Hf, Ta, W, Pb, Th, and U) in 33 analysed moss samples. High concentrations of Fe, Cr, As, V and U determined around Chisinau and Balti indicate the influence of the industrial sources, while Pb, Sb and Zn are mainly associated with transport.

The calculated values of Geo-accumulation Index I_{geo} and contamination factor C_F points towards the moderate to moderately severely polluted state for the entire territory, with higher values for U in Chisinau.

The comparison of elements determined in the samples from Moldova collected in 2001 and 2015 (Fig. 3) showed the increase of V, Cr, Ni, Cu and Pb contents within 14 years. This increase can be explained by the use of different analytical techniques as well as the economic revival.

Republic of Georgia

Republic of Georgia (Georgia), is a country of Transcaucasia located at the Eastern end of the Black Sea on the Southern flanks of the main crest of the Greater Caucasus Mountains. The country lies between the latitudes 41° and 44° N, and longitudes 40° and 47° E, with an area of 67,900 km^2. It is bounded on the North and Northeast by Russia, on the East and Southeast by Azerbaijan, on the South by Armenia and Turkey, and on the West by the Black Sea. The capital of Georgia is Tbilisi (Suny et al., 2019).

The Western Georgia's landscape ranges from the low-land marsh-forests, swamps, and temperate rain-forests to eternal snows and glaciers, while the Eastern part of the country even contains a small segment of semi-arid plains. The general landscape of the Eastern Georgia comprises the numerous valleys and gorges separated by mountains. The main part of the Western Georgia lies within the Northern periphery of the humid subtropical zone, while the Eastern Georgia has a transitional climate - from humid subtropical to continental. More than a third of the country is covered by forests and brush (Georgia, 2019).

Georgian industries and agricultural sector provide the considerable anthropogenic impact on the environment of the Caucasus. Georgia is characterized by high-level anthropogenic pollution in some districts, namely Lower Svaneti, Kutaisi, Zestafoni, and others (Shetekauri et al., 2018).

In 2014 for the first time sixteen *Hylocomium splendens*, *Pleurozium schreberi* and *Hypnum cupressiforme* moss samples were collected in Georgia. The samples were collected along the altitudinal gradients in the range of altitudes from 600 m to 2665 m. The concentrations of 25 elements Na, Mg, Al, Cl, K, Ca, Ti, V, Mn, Fe, Zn, As, Br, Rb, Mo, Cd, I, Sb, Ba, La, Sm, W, Au, and U were determined by neutron activation analysis. According to the data obtained, the mining industry, in particular the metallurgical plants and arsenic containing wastes are the main sources of air pollution. The authors have also observed a remarkable increase in concentrations of all elements with the altitude rising, especially for the light and heavy crust elements. The most obvious explanation of this phenomenon is that the content of mineral soil particles trapped by the moss is increasing with altitude due to the gradually disappearing vegetation cover and appearance of the barren areas from which soil particles may be released by wind erosion and physically captured on the moss surface (Shetekauri et al., 2015).

In 2015, 36 moss samples were collected in different regions of Georgia (Meskheti, Khevsureti, Stepantsminda in Dariali gorge, near the Georgian Military Road, Svaneti, Imereti, Rikoti Pass, Kutaisi). The main environmental pollution sources in sampled regions can be considered transport, mining of manganese, metallurgy, machinery, human waste, and burials of old industrial wastes. The results obtained have shown that the Northwest region of Georgia is characterized by the high pollution level and the majority of metallurgical and mining industry enterprises – like the machine-building factory in Kutaisi, Zestaponi Ferroalloy Plant, and old arsenic mining constructed before 1990 (Ailama, Koruldashi) – the Chiatura mine complex and others are located there (Shetekauri et al., 2018).

The comparison of the moss studies performed in 2014 and 2015 shows that the median concentrations of most elements have been decreased in 2015, which may be attributed to the larger number of samples taken from points with different levels of contamination.

Republic of Kazakhstan

Republic of Kazakhstan (Kazakhstan) is the ninth largest country in the world, covering more than 2.7 million km^2 and accounting for 1.8% of the world's land. It is also the largest landlocked country in the world. Kazakhstan shares borders with Russia, China, Kyrgyzstan, Uzbekistan, and Turkmenistan (Kazakhstan, 2019). The landscape of Kazakhstan is diverse. The Northern forest-steppe urns into steppe, half-deserts and deserts in the South. Kazakhstan is mineral rich country, there are nearly 160 oil and gas deposits including one of the largest world deposits, Tenghiz. In Kazakhstan there are world's largest reserves of chromium, vanadium, bismuth and fluorine. Since Kazakhstan has a vast territory, weather can vary substantially in different parts of the country. In winter the temperature may go down to -45°C and in summer up to +45°C (General information about Kazahkstan, 2019)

Kazakhstan is a rapidly growing economy. The enterprises involved in the extraction and processing of coal, oil, gas, non-ferrous and ferrous metals play a leading role in the national economy. These industries have an adverse effect on the environment. The combination of large coal and energy sectors results in high levels of air pollution (Kenessariyev et al., 2013).

As some other NIS countries, Kazahkstan joint the UNECE ICP Vegetation Programme in 2015. Seventy-eight moss samples were collected in summer and autumn 2015, and thirty-five moss samples in summer 2016. A total of 46 elements (Na, Mg, Al, Cl, K, Ca, Sc, Ti, V, Cr, Mn, Fe, Ni, Co, Zn, As, Se, Br, Rb, Sr, Zr, Nb, Mo, Ag, Cd, Sb, Ba, La, Ce, Nd, Sm, Eu, Gd, Tb, Dy, Tm, Hf, Ta, W, Au, Th, and U) were determined in the collected moss samples. As main sources of air pollution in Kazahkstan were defined the metallurgy, mining, engineering and chemical industry (Omarova et al., 2018).

Comparison of the Data Presented in European Atlas of Heavy Metal Atmospheric Deposition

The moss surveys are conducted in European countries every five years and data reported by European countries are presented in the "Spatial and temporal trends in heavy metal accumulation in mosses in Europe (1990-2005)" (Harmens et al., 2008) and "Heavy metals and nitrogen in mosses: spatial patterns in 2010/2011 and long-term temporal trends in Europe" (Harmens et al., 2013). On Fig.3 the comparison of the data from two abovementioned atlases reported by Estonia, Latvia, Lithuania, Belarus, Russia and Ukraine are presented. Since the main part of the NIS countries have only joined the ICP Vegetation Program in 2015, the data for them are not available yet. For comparison values reported for Cr, Fe, Ni, V and Zn were chosen. For Cr maximum values in 2005/06 and 2010/11 moss surveys were reported for Russia. For Fe, Ni, V and Zn maximum value in 2005/06 was reported for Russia, while in 2010/11 for Ukraine. The significant increase of some elements concentrations in Russia and Ukraine in 2010/11 moss survey can be explained by the involvement of different regions in the studies. In the moss survey 2005/06 Russia presented the results for the North-East of Moscow region, Tula, Tver, Udmurt Republic and Ukraine for Volyn and Sumy region. In the 2010/11 moss survey Russia reported data for Ivanovo, Kostromskaya, Tikhvin-Leningradskaya regions, while Ukraine presented data for Donetsk, the most polluted region in the Ukraine.

CONCLUSION

The air pollution is among the top problems faced by the world today. It can be reduced or avoided based on the permanent monitoring of air quality. The conventional techniques of air quality monitoring are often expensive and inefficient. The moss biomonitoring technique could be addressed as a cheap and useful tool for the pollutants estimating in the air. The moss biomonitoring has been actively applied in Estonia, Latvia and Lithuania, while the main part of the NIS countries have only joined the ICP Vegetation program in 2015. Since many pollution sources were identified on the territory of the countries discussed, it is necessary to perform the moss biomonitoring studies more frequently involving more regions.

The new moss survey was announced for 2020/2021. It is very important for national authorities to support scientific groups involved in biomonitoring studies. Assessment of air pollution is important for both citizen and national authorities since it helps to determine the current level of environmental contamination by heavy metals, to depict the major emission sources and source regions, and to estimate the contribution of atmospheric pathway to the contamination of terrestrial and aquatic environment. Collection of new samples and comparison of obtained results with the previous surveys will provide a view on the environmental situation in the participation countries, will enable the identification of a new sources of air pollution as well as will evaluate the contribution of already known pollution sources.

REFERENCES

Aksu, A. (2015). Sources of metal pollution in the urban atmosphere. (A case study: Tuzla, Istabul). *Journal of Environmental Health Science & Engineering, 13*(1), 79. doi:10.118640201-015-0224-9 PMID:26587239

Aleksiayenak, Yu., Frontasyeva, M., Florek, M., Sykora, I., Holy, K., Masarik, J., ... Ramatlhape, K. (2013). Distributions of ^{137}Cs and ^{210}Pb in moss collected from Belarus and Slovakia. *Journal of Environmental Radioactivity, 117*, 19–24. doi:10.1016/j.jenvrad.2012.01.018 PMID:22326019

Aleksiayenak, Yu., Frontasyeva, M., Ostrovnaya, T., & Okina, O. (2015). Moss biomonitoring technique, NAA and AAS in air pollution studies in Belarus. *Problems of Regional Ecology*, 126-134. (in Russian)

Aleksiayenak, Yu., Frontasyeva, M., Steinnes, E., Florek, M., Faanhof, A., & Ramatlhape, K. (2010). Atmospheric deposition of radionuclides in Belarus: 20 years after Chernobyl. *Proceedings of the XVII International Seminar on Interaction of Neutrons with Nuclei (Neutron Spectroscopy, Nuclear Structure, Related Topics).*

Barkan, V., & Lyanguzova, I. (2018). Concentration of Heavy Metals in Dominant Moss Species as an Indicator of Aerial Technogenic Load. *Russian Journal of Ecology, 49*(2), 128–134. doi:10.1134/S1067413618020030

Blagnyte, R., & Paliulis, D. (2011, May). *Determination of heavy metals in mosses (pylaisia polyantha) along the high – intensive traffic flow in geležinisvilkas street (Vilnius, Lithuania).* Paper presented at the conference The 8th International Environmental Engineering Conference, Vilnius, Lithuania.

Blum, O., Culicov, O., & Frontasyeva, M. (2002). *Heavy metal deposition in Ukrainian Carpathians (Zakarpattia and Chernivtsi regions): the regional biomonitoring*. Paper presented at the conference: Urban Air Pollution, Bioindication and Environmental Awareness. EuroBionet, Stuttgart, Germany.

Brumelis, G., Brown, D., Nikodemus, O., & Tjarve, D. (1999). The monitoring and risk assessment of Zn deposition around a metal smelter in Latvia. *Environmental Monitoring and Assessment, 58*(2), 201–212. doi:10.1023/A:1006085220366

Ceburnis, D., Ruhling, A., & Kvietkus, K. (1997). Extended study of atmospheric heavy metal deposition in Lithuania based on moss analysis. *Environmental Monitoring and Assessment, 47*(2), 135–152. doi:10.1023/A:1005779101732

Ceburnis, D., Steinnes, E., & Kvietkus, K. (1999). Estimation of metal uptake efficiencies from precipitation in mosses in Lithuania. *Chemosphere, 38*(2), 445–455. doi:10.1016/S0045-6535(98)00183-0 PMID:10901666

Cherchintsev, V., Frontasyeva, M., Lyapunov, S., & Smirnov, L. (1999). *Biomonitoring air pollution in Chelyabinsk region (Ural Mountains, Russia) through trace-elements andradionuclides: temporal and spatial trends*. Paper presented at the meeting of the International Atomic Energy Agency, Section of Nutritional and Health-Related Environmental Studies, Vienna, Austria.

Cucu-Man, S. (2006). *Studiul depunerilor atmosferice de metale grele pe teritoriul Moldovei cu ajutorul biomonitorilor* (PhD Thesis). University of Iasi, Romania. (in Romanian)

Dunaev, A., Rumyantsev, I., Agapova, I., Grinevich, V., Vergel, K., & Gundorina, S. (2018). Physical and chemical and biological monitoring in Central Russia: investigation of quality of atmospheric air and soil in territory of Rodniki town. *Izvestiya vysshikh uchebnykh zavedenii khimiya khimicheskaya tekhnologiya, 61*(8), 97-104.

Ermakova, E., Frontasyeva, M., Pavlov, S., Povtoreiko, E., Steinnes, E., & Cheremisina, Ye. (2004a). Air Pollution Studies in Central Russia (Tver and Yaroslavl Regions) Using the Moss Biomonitoring Technique and Neutron Activation Analysis. *Journal of Atmospheric Chemistry, 49*(1-3), 549–561. doi:10.100710874-004-1265-0

Ermakova, E., Frontasyeva, M., & Steinnes, E. (2004b). Air pollution studies in Central Russia (Tula Region) using the moss biomonitoring technique, INAA and AAS. *Journal of Radioanalytical and Nuclear Chemistry, 259*(1), 51–58. doi:10.1023/B:JRNC.0000015805.22707.a1

Frontasyeva, M., Smirnov, L., Steinnes, E., Lyapunov, S., & Cherchintsev, V. (2004). Heavy metal atmospheric deposition study in the South Ural Mountains. *Journal of Radioanalytical and Nuclear Chemistry, 259*(1), 19–26. doi:10.1023/B:JRNC.0000015800.41525.28

Frontasyeva, M., Steinnes, E., Lyapunov, S., Cherchintsev, V., & Smirnov, L. (2000). Biomonitoring of Heavy Metal Deposition in the South Ural Region: Some Preliminary Results Obtained by Nuclear and Related Techniques. *Journal of Radioanalytical and Nuclear Chemistry, 245*(2), 415–420. doi:10.1023/A:1006799513736

General information about Kazahkstan. (n.d.). Retrieved from https://www.britishcouncil.kz/sites/default/files/general_information_about_kazakhstan.pdf

Georgia. (2019). Retrieved from https://ro.wikipedia.org/wiki/Georgia

Ghorani-Azam, A., Riahi-Zanjani, B., & Balali-Mood, M. (2016). Effects of air pollution on human health and practical measures for prevention in Iran. *Journal of Research in Medical Sciences*, *21*(1), 65. doi:10.4103/1735-1995.189646 PMID:27904610

Gorelova, S., Babicheva, D., Frontasyeva, M., Vergel, K., & Volkova, E. (2016). Atmospheric Deposition of Trace Elements in Central Russia: Tula Region Case Study. Comparison of Different Moss Species for Biomonitoring. *International Journal of Environmental Science,* 220-229.

Harmens, H., Mills, G., & Hayes, F. (2009). *Air pollution and vegetation: ICP Vegetation annual report 2008/2009 NERC/Centre for Ecology & Hydrology, Bangor, UK.* Retrieved from http://www.unfccc.int/resource/docs/natc/mdanc3.pdf

Harmens, H., & Norris, D. (2008). *Spatial and temporal trends in heavy metal accumulation in mosses in Europe (1990-2005).* Bangor, UK: Centre for Ecology & Hydrology. Retrieved from http://nora.nerc.ac.uk/8672/2/HarmensN008672CP.pdf

Harmens, H., Norris, D., & Mills, G. (2013). *Heavy metals and nitrogen in mosses: spatial patterns in 2010/2011 and long-term temporal trends in Europe.* Bangor, UK: NERC/Centre for Ecology & Hydrology. Retrieved from http://nora.nerc.ac.uk/502676/1/N502676CR.pdf

Hervada-Sala, C., Jarauta-Bragulat, E., Tyutyunnik, Yu., & Blum, O. (2003). *Indirect geostatistical methods to assess environmental pollution by heavy metals. case study: Ukraine.* Paper presented at the Annual Conference of the International Association for Mathematic Geology, Portsmouth, UK.

Informational bulletin "About the state of natural resources on the environment of the Moscow region in 2015". (2016). The Ministry of Ecology and Natural Resources Use of the Moscow Region. Krasnogorsk. (in Russian)

Kaasik, M., & Liiv, S. (2007). Spatial and temporal variability of trace metals in mosses in Estonia. *Forestry Studies: Metsanduslikud Uurimused*, *46*, 35–44.

Kallis, A., Sims, A., Tammik, A., Roose, A., Türkson, C.-T., Kupri, H.-L., . . . Jurtom, T. (2017). *Estonia's Seventh National Communication Under the United Nations Framework Convention on Climate Change*. Tallinn: Ministry of the Environment.

Kazahkstan. (2019). Retrieved from https://en.wikipedia.org/wiki/Kazakhstan

Kenessariyev, U., Golub, A., Brod, M., Dosmukhametov, A., Amrin, M., Erzhanova, A., & Kenessary, D. (2013). Human Health Cost of Air Pollution in Kazakhstan. *Journal of Environmental Protection*, *4*(08), 869–876. doi:10.4236/jep.2013.48101

Korolyova, Yu. (2010) The bioindication of heavy metal precipitation in the Kaliningrad region. *Bulletin of Kant Russian State University, 7,* 39-44.

Kösta, H., & Liiv, S. (n.d.). *Spatial and temporal trends of heavy metal accumulation in mosses in Estonia*. Paper presented at the 8th International Conference on Ecosystems and Sustainable Development, Alicante, Spain. 10.2495/ECO110121

Liiv, S., & Kaasik, M. (2004). Trace metals in mosses in the Estonian oil shale processing region. *Journal of Atmospheric Chemistry*, *49*(1-3), 563–578. doi:10.100710874-004-1266-z

Lithuania's seventh National Communication. (2017). Ministry of Environment of the Republic of Lithuania. Retrieved from https://unfccc.int/files/national_reports/national_communications_and_biennial_reports/application/pdf/142035_lithuania-nc7-1-7th_national_communication.pdf

Nikodemus, O., Brumelis, G., Tabors, G., Lapin¸, L., & Pope, S. (2004). Monitoring of Air Pollution in Latvia Between 1990 and 2000 Using Moss. *Journal of Atmospheric Chemistry*, *49*(1-3), 521–531. doi:10.100710874-004-1263-2

Olmo, N., Saldiva, P., Braga, A., Lin, Ch., Santos, U., & Pereira, L. (2011). A review of low-level air pollution and adverse effects on human health: Implications for epidemiological studies and public policy. *Clinics*, *66*(4), 681–690. doi:10.1590/S1807-59322011000400025 PMID:21655765

Omarova, N., Nurgalieva, D., Nurkasimova, M., Frontasyeva, M., Morzhukhina, S., & Kabdulkarimova, K. (2018). Environmental monitoring in the Republic of Kazakhstan on the content of heavy metals and radionuclides. *International Electronic journal. Sustainable Development Science and Practice, 1*(3), 20-29.

Pankratova, Yu., Frontasyeva, M., Berdnikov, A., Pavlov, S., Granja, C., Leroy, C., & Stekl, I. (2007). Air Pollution Studies in the Republic of Udmurtia, Russian Federation, using Moss Biomonitoring and INAA. *Nuclear Physics Methods and Accelerators in Biology and Medicine*, *958*, 236–237. doi:10.1063/1.2825794

Pénard-Morand, C., & Annesi-Maesano, I. (2004). Air pollution: From sources of emissions to health effects. *Breathe*, *1*(2), 109–119. doi:10.1183/18106838.0102.108

Popescu, F., & Ionel, I. (2010). *Anthropogenic Air Pollution Sources*. Retrieved from https://www.intechopen.com/books/air-quality/anthropogenic-air-pollution-sources

Ratniece, V., Rubene, L., Cakars, I., Siņics, L., Griķe, I., Gaidis, K., . . . Gancone, A. (2017). *Latvia's seventh National Communication and Third Biennial Report under the United Nations Framework Convention on Climate Change*. Retrieved from https://unfccc.int/files/national_reports/national_communications_and_biennial_reports/application/pdf/9308541_latvia-br3-nc7-1-latvia_nc7_29122017.pdf

Raukas, A. (2004). Opening a new decade. *Oil Shale, A Scientific-Technical Journal, 21*(1), 1-2.

Reimann, C., Halleraker, J., Kashulina, G., & Bogatyrev, I. (1999). Comparison of plant and precipitation chemistry in catchments with different levels of pollution in Kola Peninsula, Russia. *The Science of the Total Environment*, *243/244*, 169–191. doi:10.1016/S0048-9697(99)00390-3

Ruhling, A., & Tyler, G. (1971). Regional differences in the deposition of heavy metals over Scandinavia. *Journal of Applied Ecology*, *8*(2), 497–507. doi:10.2307/2402886

Russia. (2019). Retrieved from https://en.wikipedia.org/wiki/Russia

Russia. (2020). Retrieved from https://en.wikipedia.org/wiki/Russia

Šakalys, J., Kvietkus, K., Sucharová, J., Suchara, I., & Valiulis, D. (2009). Changes in total concentrations and assessed background concentrations of heavy metals in moss in Lithuania and the Czech Republic between 1995 and 2005. *Chemosphere*, *76*(1), 91–97. doi:10.1016/j.chemosphere.2009.02.009 PMID:19269004

Shabatura, A., Blum, O., & Tyutyunnik, Yu. (2018). Regional atmogeochemical fields based on briogeochemical indication data in the central part of northern Ukraine. *Biosphere, 10*(1).

Shetekauri, S., Chaligava, O., Shetekauri, T., Kvlividze, A., Kalabegishvili, T., Kirkesali, E., ... Tselmovich, V. (2018). Biomonitoring Air Pollution Using Moss in Georgia. *Polish Journal of Environmental Studies*, *27*(5), 2259–2266. doi:10.15244/pjoes/73798

Shetekauri, S., Shetekauri, T., Kvlividze, A., Chaligava, O., Kalabegishvili, T., Kirkesali, E., ... Chepurchenko, O. (2015). Preliminary results of atmospheric deposition of major and trace elements in the greater and lesser Caucasus mountains studied by the moss technique and neutron activation analysis. *Annals of Botany*, *5*, 89–95.

Suny, R. G., Howe, G. M., Djibladze, M. L., & Marshall Lang, D. (2010). *Georgia*. Retrieved from https://www.britannica.com/place/Georgia

Tabors, G., Nikodemus, O., Dobkeviča, L., Kļaviņa, L., Ajanoviča, A., Viligurs, K., & Krūz, I. (2017). Assessment of atmospheric pollution with heavy metals and nitrogen using *Pleurozium schreberi* mosses as bioindicator in Latvia: Spatial and temporal aspects. *Environmental and Experimental Biology*, *15*, 143–150.

The Sixth National Communication of Ukraine on climate change. (2012). Ministry of Ecology and Natural Resources of Ukraine. Retrieved from http://www.seia.gov.ua/seia/doccatalog/document?id=638134

Third National Communication of the Republic of Moldova. Under the United Nations Framework Convention on Climate Change. (2013). Chisinau: Imprint Plus SRL.

Vergel, K., Frontasyeva, M., Kamanina, I., & Pavlov, S. (2009). Biomonitoring atmosphernukh vupadenii tyazeluh metallov na severo-vostoke Moskovskoi oblasti s pomosh'yu metoda mhov-biomonitorov. *Ecology of urbanized areas III*, 88-95. (in Russian)

Vergel, K., Frontasyeva, M., Pavlov, S., & Povtoreyko, E. (2007). Air Pollution Studies in Tver Region of Russia using Moss-Biomonitoring with Nuclear Analytical Methods. In *Proceedings of AIP Conference*, (vol. 958, p. 240). 10.1063/1.2825797

Vergel, K., Goryainova, Z., Vikhrova, I., & Frontasyeva, M. (2014). Moss Biomonitoring and Employment of the GIS Technology within the Framework of the Assessment of Air Pollution by Industrial Enterprises in the Tikhvin District of the Leningrad Region. Ecology of Urbanized Areas, 2, 92-101.

Vergel, K., Zinicovscaia, I., Yushin, N., & Frontasyeva, M. (2019). Heavy metal atmospheric deposition study in Moscow region, Russia. *Bulletin of Environmental Contamination and Toxicology*, *103*(3), 435–440. doi:10.100700128-019-02672-4 PMID:31267137

Wolterbeek, H. (2002). Biomonitoring of trace element air pollution: Principles, possibilities and perspectives. *Environmental Pollution*, *120*(1), 11–21. doi:10.1016/S0269-7491(02)00124-0 PMID:12199457

Zinicovscaia, I., Hramco, C., Duliu, O., Vergel, K., Culicov, O., Frontasyeva, M., & Duca, G. (2018). Air pollution study in the Republic of Moldova using moss biomonitoring technique. *Bulletin of Environmental Contamination and Toxicology*, *98*(2), 262–269. doi:10.100700128-016-1989-y PMID:27889805

ADDITIONAL READING

Aničić Urošević, M., Vuković, G., & Tomašević, M. (Eds.). (2016). *Biomonitoring of Air Pollution Using Mosses and Lichens, A Passive and Active Approach, State of the Art Research and Perspectives*. Nova Science Publishers.

Di Palma, A. (2016). *Mosses for monitoring air pollution: towards the standardization of moss-bag technique and the set-up of a new biomaterial* (Ph. D. dissertation). Naples, Italy

Fernández, J. A., Aboal, J. R., Real, C., & Carballeira, A. (2007). A new moss biomonitoring method for detecting sources of small scale pollution. *Atmospheric Environment*, *41*(10), 2098–2110. doi:10.1016/j.atmosenv.2006.10.072

Fernández, J. A., Boquete, M. T., Carballeira, A., & Aboal, J. R. (2015). A critical review of protocols for moss biomonitoring of atmospheric deposition: Sampling and sample preparation. *The Science of the Total Environment*, *517*, 132–150. doi:10.1016/j.scitotenv.2015.02.050 PMID:25725198

Frontasyeva, M., Nazarov, M., & Steinnes, E. (1994). Moss as monitor of heavy metal deposition: Comparison of different multi-element analytical techniques. *Journal of Radioanalytical and Nuclear Chemistry*, *181*(2), 363–371. doi:10.1007/BF02037642

Harmens, H., Norris, D. A., Koerber, G. R., Buse, A., Steinnes, E., & Rühling, A. (2008). Temporal trends (1990 - 2000) in the concentration of cadmium, lead and mercury in mosses across Europe. *Environmental Pollution*, *151*(2), 368–376. doi:10.1016/j.envpol.2007.06.043 PMID:17669565

Stankovic, J., Sabovljevic, A., & Sabovljevic, M. S. (2018). Bryophytes and heavy metals: A review. *Acta Botanica Croatica*, *77*(2), 109–118. doi:10.2478/botcro-2018-0014

Zechmeister, H. G., Riss, A., & Hanus-Illnar, A.Biomonitoring of Atmospheric Heavy Metal Deposition by Mosses in the Vicinity of Industrial Sites. (2004, November). Biomonitoring of Atmospheric Heavy Metal Deposition by Mosses in the Vicinity of Industrial Sites. *Journal of Atmospheric Chemistry*, *49*(1-3), 461–477. doi:10.100710874-004-1260-5

KEY TERMS AND DEFINITIONS

Monitoring: The assessment of chemical pollution by measuring the chemicals (metals, organics compounds) in a biological sample.

Moss Survey: The procedure of moss collection in the same sites every five years.

Mosses: Small flowerless plants, without roots, that typically grow in dense green clumps or mats, often in damp or shady locations.

Neutron Activation Analysis: Highly sensitive analytical technique for metal concentration determination in different type of samples.

Pollution Source: Sector of human activity or natural process, which lead to the release of the pollutants in the environment.

Post-Soviet Countries: Countries formed after collapse of the USSR.

Section 5
Educational Methodology

Chapter 25
Ecological Chemistry Through Popular Scientific Articles

Ketevan Kupatadze
Ilia State University, Georgia

ABSTRACT

Popularization of ecological chemistry is the aim of scientific-popular articles, which are published in the online journal for teachers. The articles of this type ecological chemistry are linked with literature and history. Due to this linkage, this module turns into an easily comprehensible one and it becomes fun. In all articles, there is also included very useful and interesting information pertaining to ecological chemistry. The titles of such articles must be underlined because they should not only show the common meaning of article, but they should also attract readers. The utmost interest is generated by the historical papers where chemical issues are connected with history. The period of alchemy is more popular and that's why the alchemical stories are described in the articles. The outcome of the pedagogical experiment has made it clear that such a method of teaching of ecological chemistry with scientific popular articles affects positively school student motivation and changes their attitude towards environmental pollution.

INTRODUCTION

In recent years ecological chemistry has acquired special significance. The same can be said on environmental and green chemistry. Enhanced cataclysms, changes in migration of chemical elements in nature, and climate change are subjects of discussion for consideration. The reasons are many, although a majority agree that one of the most important factors is anthropogenic, i.e. chemical cycles in the environment are disrupted as a result of human action; leading to changed climates, polluted air and water, and increased diseases.

Therefore it is important to provide the next generation with knowledge and insight in order to understand the essence and risks of the problem. Correspondingly, members of society will be as committed as the ecologist to take care of the environment (Kupatadze et al, 2016).

DOI: 10.4018/978-1-7998-1241-8.ch025

In the article we will discuss the issues of ecological chemistry teaching. Generally there are many methods of teaching ecological chemistry and among them is the transfer of information / current issues in scientific-popular form.

In the teaching process, it is essential to maintain student engagement and interest. Approaches differ but applying popular science articles may be helpful. This chapter reviews the popular-science articles on ecological and environmental chemistry, as one of the tools to popularize chemistry and get students (and not only students) interested in ecological and environmental chemistry. The popular-science article we are going to review, are published with the authorship of Prof. K. Kupatadze in an on-line magazine "The Teacher" and are targeted towards at a wide range of readership.

The articles are-in line with the didactic teaching principles. In the framework of didactics and methodology, there are some general didactic principles which deal with a pragmatic approach in teaching (Morais et al, 2015).

For example:

1. The Principle of combining theory and practice-means to combine theoretical knowledge with practical activities. Each popular-science articles contains a protocol of a chemical experiment which can be conducted during a chemistry lesson. Some of the chemical experiments can be conducted at home.
2. The Principle of comprehensibility means-that the learning material should always be adapted to the level of knowledge of the learners. The popular-science articles are written such that a layperson without expertise in chemistry can still understand them.
3. The Principle of using examples means-that article must contain a good practical example which can be used to make the contents easier to understand.
4. The principle of interdisciplinary links - in scope of popular-science articles in Ecological and Environmental Chemistry are connected with other sciences – such as history and literature.
5. The principle of transfer of knowledge- in popular-science articles, the chemical information is linked with everyday life.

Popular Scientific articles has the same goals as "Storytelling". The different research shows that "storytelling" in chemistry inspires school students (and not only them) to overcome their "Chemo phobia" and try to understand this subject (Morais et al, 2015). What can storytelling offer? Stories teach about life, about ourselves and about others. Storytelling is a unique way for students to develop an understanding, respect and appreciation for other culture and can promote a positive attitude to people from different countries (Moen et al, 2006).

It should be underlined that not only school students but also University students (not chemistry specialization) should be involved in chemistry. Mark Green's book about Organic Chemistry realize this purpose and with storytelling students can found related video lectures links (Pruett et al, 2014).

The pedagogical value of storytelling in Science education relies on the fact that its structure allows concepts to be placed in a form that is easy to assimilate and can also provide elements to memorization and re-elaboration. Moreover, if a student is able to place discipline knowledge into a personal and clear storytelling structure, then this knowledge can be put in other contexts that will no longer belong to such discipline, but instead will be integrated to knowledge from other disciplines. In order to do this in such a way that the storytelling becomes meaningful for the writer and readers, a process of adaptation of the language of Science to the language of the writer has to be undertaken (Buckingham et al, 2003; Norris et al, 2003).

According some researches comics have been considered an important pedagogical resource in teaching education programs. If properly explored in the classroom, comics, with its peculiar features, has an enormous power to tell stories and convey messages, contributing to stimulate students' active participation and creativity, along with oral and written language and even the development of technological abilities. They also show potential to capture and maintain students' interest, besides being considered as an intermediate support for learning difficult concepts (Santos et al, 20130.

No less effective and interesting is linking theatre with chemistry, or conducting scientific workshops both for teachers and school students (Djerassi et al, 2012).

Because of the advantages listed above, we have chosen "storytelling" method to promote environmental chemistry. And make it more understandable to a wider audience (Wu et al, 2010). As well as with the help of the articles, it is possible to improve the systemic thinking of school students and give them the opportunity to think not only about the environment but regarding other topics of chemistry (Mahaffy et al, 2019).

There are plenty of methods for active teaching of chemistry, however we will discuss active methods of information dissemination. The dissemination of information to masses is justified in relation to environmental chemistry, because today's school student may not choose chemistry as a specialty, or a student may be studying a completely different field of chemistry, or an adult from humanitarian field, however the society needs environmentally educated individual at all times and of all ages. As in most cases, all environmental disasters, chemical contamination of environment, and diseases conditioned by these reasons take place due to environmentally illiterate people (Shepherd et al, 2015).

LANGUAGE AND FORMAT OF POPULAR-SCIENCE ARTICLES

The popular science articles deal with all applications of chemistry, including ecological chemistry. The article aims to kindle interest in all types of readers, including those who do not have any contact with chemistry, or just believe that chemistry is uninteresting and useless. For that very reason, each article consists of the so called "hook" introduction, which should not have anything in common with chemistry. Its purpose is to allure the reader to a "hook" and engage him/her in reading. The next part of the article directly deals with chemical information, however it should be written in a lightweight, easy and simple language.

In the final part, a light, non-chemistry story can be interpreted as an introductory "hook" or a completely independent story. In addition to engaging the reader, the "hook" is aimed at connecting the chemical information to the history of chemistry, literature, poetry or simply to amusing news stories (the latter is more preferred by students and pupils more).

Furthermore, the title of the article should serve as a "hook" and catch the attention of potential readers. A mini example of articles is composed on each environmental chemistry issue;

"A Story about a Cock, Wanting to cry: Cock-A-Doodle-Doo. …., but had no Voice": The attitude of one of the significant representatives of Islamic alchemy - Avicenna, i.e. Ibn -Sina, towards water is conveyed here. The author's impressions on the most beautiful city Eskişehir, often called little Paris of Turkey, are used as the "hook".

"There are private houses next to the hotel, small gardens, with lots of flowers. There are also many trees. Generally, Eskişehir is a green city. The name means an old town. It seems they've been knowing the price of oxygen since ancient times. In one of these yards there lives a little cock, desperately wishing to utter a crow, but ...doesn't have enough voice. It tries all its best, but in vain… nothing comes out of its mouth except for funny voices. However it does not reconcile and keeps trying every morning. From the balcony Eskişehir "Shatto's" blue towers appear. These are a visitor cards of the city and are located in Sazova park. Here you are, entering and feeling, that you are in a fairy tale. There are lots of adults guests, kids, locals and foreigners, all of them are here. The way to "Shatto" is full of "troubles", first you will have to pass by a moollah Nasredin, who has a big pot in front of him, cooking all the truth and lies. The next stop is a house on stilts. If you want to get on the second floor, you will have to climb on some kind of strapped ladder, and despite the fact that there are long queues of kids willing to climb the house, I am happy to join them.

I am sorry, but one must leave seriousness outside from such a wonderland, as I myself, left it at the entrance gate and now I'm having fun like these kids. There is a trench with some kind of blue water around the "Shatto", where a crocodile is swimming. It isn't alive, it's mechanical, however it jumps like an alive alligator in water. There is a sign board next to the trench: - No swimming.

Why should one wish to jump into it and swim, but as they specifically warn us, it means that someone had already jumped into it and tried to swim. You remember, I wrote about the cock above, every morning trying to say a cock-a-doodle-doo. …. It gets ready, claps its wings, makes its neck longer… Oops, still in vain. This is a really exemplary behavior, you can't do something, but constantly keep trying… and therefore, I am pretty sure, one day this cock will make a real sound in the district. You will find cock on papers of Islamic alchemists' manuscripts…there are different versions about what it might indicate. However, according to the most common opinion, cock means the beginning of life. If the cock drinks water, then in such a manuscript; the discussion is about water, as water is the source of life.

El Kanun Fit-Tibb (Canon Medicine) is the most significant in the history of Islamic science. There are five volumes and the second volume comprising from six chapters is dedicated to medicine. It is written by Ibn Sina, who lived in 980-1037 AD. This book was used as one of the main sources in teaching medicine until the end of the seventeenth century. Avicenna, the same Ibn Sinai, divides water into two categories. The first of them is "good water" and it should be clean by all means, it should be potable and easily absorbed by body…" (Greenberg et al, 2007).

The article continues with the discussion of Avicenna's notes about "good" and "bad" water, and also about the tricks of polluted water treatment. The same article names two monarchs, who took care about water cleanliness. King Vakhtang XI in his book about "Mixing the oils and doing the chemistry", wrote that water treatment would be possible by means of swamp plants.

King of Spain, Philip II used to put rose petals in distilled water and drink. Later, with the help of alchemists, who were invited to the palace, decrypted the method of water distillation written by his compatriot, Ramond Llull and started to make distilled water according to his advice…"

Separate articles are published on Vakhtang VI and Philip II, ("I would like to thank you, my King", "A story written in Terminal"), where their attitudes toward water is discussed in more details (Williams et al, 2005).

The article-"Kocaeli, Izmit, Nicomedia"- describes a modern method of water treatment through constructed wetland.

"Welcome to Turkey, I wish you a happy day, - a customs officer is giving back my passport with an artificial smile, opening the entrance door of the country. – Thanks, likewise" – I am answering with the same smile.

What a happy day - It is raining cats and dogs. I am looking for a person holding a big paper with my name on it, to drive me from Istanbul to Kocaeli. Kocaeli is near Izmit. This is old Nicomedia, located next to Constantinople (now Istanbul) with the shores flaunting in the Sea of Marmara. The city was founded in 712 B.C. and was called Astacos. After being destroyed, it was first rebuilt by King Nicodemus; therefore the city was named after him. Plinius Junior wrote about the landmarks of Nicodemia in his letters, commenting how great it would be to install water pipes and lay the channel between the inner lakes and the gulf of Nicomedia.

So, if we drive along this Gulf, we will get to Kocaeli. It takes half an hour to drive there from Izmit, and about an hour and a half - from Istanbul.

Kocaeli inhabitants are proud of having potable water running directly from taps. Drinking water is taken from the Mount of Chenesuyu (Çenesuyu), which is famous for its quality and purity. They take pride in the whole network of natural treatment systems and take special care of the cleanliness of water.

Today I am going to talk about water for the third time, however - now our discussion will refer to its pollution and treatment" (Sylaios et la, 2014).

The Title of the Article –"What Can be Hard"

What can be meant by the word "hard"? It might mean comma (In Georgian language, "comma" is pronounced the same as the word "hard"). Yes, you are right. It might have such great importance that it can change the whole meaning of the idea if skipped or put in a wrong place. What else can be hard? For instance: life itself. It's a common topic of discussion… I could never understand why some people were awarded by hard life while the others enjoyed the easiest one. Furthermore, the one, who has been indulged with the easiest life, is no better than the one, suffering from hard life. It can be the opposite, but this is the reality and that's all – for some people life is as easy as a water flow, while for others - it is as hard as staying dry in the rain, but can one really feel others suffering?

Impression.. yes, it is also possible to be impressed heavily by something, or make a heavy impression on others. However, the heaviness is perceived differently by various people. Terenti Graneli (a famous Georgian poet) wrote about heavy fate. Although, in the other verse he wrote that neither the fear of death was easy. Sometimes Galaktion's (Galaktion Tabidze, a well-known Georgian poet) heart was as heavy as the gates of Troy, to open for life. What else can be heavy?

Freight, an exam (may be a life test), a mistake…The latter may cause heavy consequences for doctors, teachers and chemists. A doctor does not have the right to play with someone's health and life. A teacher does not have the right to distort someone's spiritual world. Very often a chemist in his/her lab is entrusted with the others' health and life along with his/her own. Someone will ask-whether a worker standing on a scaffold has the right of mistake. No, he/she does not. Likewise, numerous professions can be listed endlessly. It turns out that each of our blunders might be fatal.

However, not all mistakes are grave. (the word grave in Georgian is pronounced the same as words: hard, heavy, severe, comma). For instance, if a housewife mixes sugar with salt in a cake crème, that will not be an error in big science. Furthermore – baking a cake is not a big deal. And even more, like everything, "heavy /hard" is a conditional concept. The life which can seem hard to me might be called by someone easiest, or, for instance, snowy weather is hard for me, while for someone it can be entertaining.

Therefore, let me write about what's really hard.

Heavy metals: mercury, lead, cadmium, chromium, manganese, nickel, cobalt, vanadium, copper, iron, zinc, antimony are considered to be especially dangerous by the UN Commission ". Then article continues describing vanadium.

The article with the title – "Excuse me, are you a Paracelsus?" deals with cadmium.

"Now it is warm outside… the smell of spring subtly touches my olfactory senses. Though born in winter, I am not fond of this season of the year and every time it comes, a strange feeling of sadness seizes my soul. Nor I am a worshiper of wintry Tbilisi. I cannot help agreeing with Tato (nickname of the famous Georgian poet Nikoloz Baratashvili) and admitting that in cold weather "Tbilisi is a city of melancholy and grief". Anyway, you can never judge by my appearance what grievance I am bearing, as nobody is actually interested in what you feel, the world requires you to be only smiling. I paused for a while…It is decided, I will take a walk in such a beautiful weather. I am going home and have no reason to hurry up.

Suddenly my attention is caught by a man in front of me. He should be about 70. He is walking too. Having a habit of going fast, I am leaving him behind in a shortest while, for I want to take him in at a glance. First I slightly overtake him, then turn back and after a while follow him. He has a stocky configuration, not tall, with ruddy cheeks, broad forehead, and back combed hair. He does not even notice that he is being observed. I am burning with desire to approach him and ask:

- Excuse me, are you Paracelsus?

But I do not ask. First of all, he won't understand, second of all, he will refuse by all means, saying that I am mistaken and will start reciting all day long how he was associated with some Para... Wait a minute; let him remember… in short... with some zzel. Or may be celsus? (This is a play of words)... or to an ax (play of words). One can not understand these people…

In order to avoid any accusation or misunderstanding, I will tell you, that Paracelsus was an alchemist. This strange name was taken as a pseudonym, his real name was Philippus Aureolus Theophastus Bombastus von Hohenheim. He tried to compete with an Italian doctor, Avlus Kornelius Cels claiming to be superior to him. That's why he called himself Paracelsus, i.e. superior to Celsus. Although being an alchemist, he used to practice in medicine as well. Or maybe on the contrary, apart from being a doctor, he also practiced in alchemy. He was searching for a philosophers' stone both for obtaining gold and for medical purposes. That time, two major issues were of concern: life prolongation and rejuvenation. In medicine, Paracelsus believed that blood was the most important thing calling it the main liquid. As for alchemy, he thought that Zink or table salt would act as an accelerator facilitating sulfide's (which was made of mercury and sulfur powder) transformation into gold". Despite the fact that the reader might be thinking that the next paragraph of the article will deal with zinc, it continues with describing cadmium, as this is the very element, dispersed during zinc extraction from deposits, harming the environment.

We talk about mercury in the article "**Mercury**". Another article "**Don't play with it, it' s insidious**", is about the same metal, although here we specifically discuss water pollution by mercury and the talk is about well-known "Minamata tragedy"(Sutton et al, 2008)

. Chemical mechanism of pollution is described. The story about "A Crazy Hatter" is applied as a "hook".

"… In my childhood I was particularly fond of one character of fairy tale and it was this crazy hatter. Yes, he was not good, but at the same time, he deserved sympathy. After all, he was laughing about something and did not do anything wrong. In France there really were hatters and crazy hatters by the way. Their profession was blooming during Ludovick XIV. This king was spendthrift, he was wealthy and therefore, each nobleman in his retinue wasn't allowed to wear one and the same hat twice. Because of that, the hatters had continuous orders. The hatters played with paints. Red paint was made of beautiful, stable mineral having various spectrum, Singuri.

Singuri as known in Georgian language may be known elsewhere as "Cinnabar". The Japanese used to compare it with "a red dragon". Its components include mercury sulfides, which is the same as two valence hydrargyrum, the same as mercury sulfide and red pigments. This profession had a bad side as well… the hatters it can be said, were not fit and healthy. They used to cry, or shout, or laugh without reason… They had difficulties with sleeping, and used to walk drunkenly. There were rumors that hatters drank in hiding, although they used to swear and laugh at it, claiming that they were sober. They looked bad, older, easily caught illness and would soon die…

40 years were the maximum of their vital age ... You probably understood why this was happening. For producing the pigment, the mineral cinnabar was made thermally, although they did not know anything about "thermal treatment". They just heated it on a high temperature and that's all. While heating mercury, it evaporated ….. and hatters breathed it.

Imagine what will happen to you, if you breathe mercury all your life. A man is strange, wants more and more. There opened a paint factory in one city of Japan. Along with other paints, they used to make red as well. The raw material was mineral "Cinnabar", the same as "Singuri". The product was sold well. Factory owners got good money. Yes, they knew the ingredients of this mineral and had worked out its treatment technology. Therefore, neither they themselves, nor their workers got poisoned. However, where could they dispose the left over mercury? Decontamination required separate funds, therefore, they found an easy way – to discharge it into Minamata channel. I forgot to mention, the city was also called Minamata and in five years all its population died. How does mercury react in water? In addition to mercury, the complex combination of other poisonous substances also exists in water. Very often, people drink polluted water due to ignorance of elementary rules. This problem is highlighted in two articles: "**Mărţişor**" and "**Money or Health?".**

"There was a forest at the end of the path. In that very turn, the most beautiful spring appeared. It looked here and there and saw snowdrops covered by snow. It decided to help. It shook off snow by hand. Winter was watching from above and got crossed, lifted a hand and created a strong hurricane. The snowdrops leaned on the ground. The Spring covered the flowers. But got caught on a prickly hedge and cut its hand. Blood began to tickle down from the cut, fell on flowers and made them alive. Thus was defeated the winter. Red blood on white snow are Martisor colors. What a nice story – I am smiling and caressing Martisor flower on my breast with my finger.

I forgot to say that I am in the Chisinau airport. This is the minute I went across the border and saw two beautiful girls in the door. The first put the flower on my breast, and the second told me the story. They also advised me to carry the flower for a month and then put it around any tree and make a wish. It is the first of March and Moldova is celebrating the Martisor holiday. Each passenger is met in this way. The holiday dates back from the time of the Roman empire. That time they used to celebrate a new year on the First of March, the month of the God Mars. He was not only the god of wars, but also protected farming and nature revival. The very double approach is reflected in Martisor. In which white and red can be compared to war and peace. Now, we have to go back to the legend and tell the story in a new way. So, Spring came out from the forest, looked around and saw... the land, water and in general, all the environment was polluted by Phthalates...why by phthalates anyway?

Yes, why phthalates, what kind of substances are they, how can they get into water and afterwards in our bodies. This very information is given in details in the main part of the article (Williams et al, 2005). In addition to phthalates and dangerous from plastic vessels, water can be polluted by Arsenic.

The article "Mgelika's source" ("Mgelika" is the word used for petting, in the mountainous areas of Georgia this word also is used for naming the men.) it described the threats existing in mineral water of various sources. Mgelika is a tale character drinking spring water in the forest. "…Mgelika was drinking spring water. I myself drink bottled tap water. Which of them is safer? Spring water is dangerous, it may contain various infectious agents, including - virus -A hepatitis. Therefore it's recommended to first boil and then filter water. Filtered and boiled water can be kept, but it can be used for drinking and cooking only within 12 hours. Spring water may be polluted by harmful substances. One of the pollutants is arsenic...In some countries there are water maps, which, as a rule, are determined by the country's chief physician - Sanitarian. These maps indicate dangerous hazards and possible pollution of surface waters, as well as the types of fish and mollusks".

Clean air is necessary for clean environment. The following articles discuss air pollutants: "A Ladder" and "Fragrances". Both of them review the air existing inside the house, the composition of which is dependent on place, where the house stands, how green is the area and the traffic volume around it. What is the type of furniture furnishing the house (here we mean the used material, glue, etc) the basic pollutants are aldehyde and their derivatives (Rawn et al, 2004). Ibn Sinas' opinion in "Fragrances" is given here. "…Let's recollect that Ibn Sina belonged to the number of those alchemists, who were seeking the philosophic stone only for human health. He believed that the flesh and the soul of man was one whole and that in healthy body there must have been a healthy soul. Or may be on the contrary, it might have been possible to get a healthy body through the healthy soul. Therefore, he paid great attention to the "fragrances". Hereby we should explain that aromatic does not necessarily mean to have a good aroma, it just means, that something has a specific smell."

Amusing stories serve as a good "hook" as well. The author of the article "Smog" begins with the story on how he started to learn Turkish language.

"In these latter days, I attend courses in the Turkish language. We have a teacher, named Esra, a long haired and black eyed young woman. She does not know Georgian, but tries her best to learn it. The class is staffed with students; I am the eldest among them. However; it does not make me uncomfortable, and I am entirely engaged in the learning process. There are only two boys in the class, the rest are the girls and it's not a surprise that there is much noise considering the number of girls assembled together. Esra Hoca (Teacher Esra) knows a couple of Georgian words, that's why sometimes she shouts out with a warning tone:"Gogolar (gogo – means a girl in Georgian), this is a mix of Georgian-Turkish. A girl is pronounced in Turkish as "kız", in plural – "kızlar", however "gogolar" is somehow different, affectionate and strict at the same time.

You can ask her anything ten times, she won't get tired of explaining with the same energy and enthusiasm. All was the same that day, when she was teaching us about the clock, tirelessly drawing various versions of time on the board asking the question "kach Saat? (What time is it?)"; one of the girls joked in Georgian, our Hoca is like fire. And as if someone had asked me I immediately translated to the teacher: "You are the woman like fire". At first she couldn't understand, and then we explained that it meant an energetic person and being like fire was considered to be a compliment in Georgia.

By the way, after early alchemists, fire was first looked at by Jan Baptist van Helmond, followed by Georg Ernst Stahl, and finally by Lavoisier. Each of them saw something different in fire, although all of them agreed that combustion processes took place during fire. And what if combustion process is incomplete? Yes, gases are released and the very smog is formed from these gases.

Today we are going to talk about smog.

In 1661, John Evelyn described London's atmospheric air in his book "Fumifugium". He wrote about strange smoke and smog sometimes formed in London, which made it difficult to see anything and made an impression as if the air was full of ash after volcano eruption (Rawn et al, 2004).

After reviewing smog chemically, the article ends with a funny story.

Once upon a time, a girl and a boy lived in a small borough. They fell in love with each other and met every day. Someone told the girl's father that his daughter was in love with the boy of another man. The girl's father was strict, and he got furious....How she could have dared! He locked up the girl in the house and forbid her to go outside. The girl stopped to eat in protest and declared that she didn't want anyone but him. On the third day of hunger, she started thinking about what her lover was doing while she was sacrificing herself and sent a letter with the help of her sister. (There was no telephone and internet that time). The boy suggested running away from home that night after her sister had unlocked door. And they did so, but the girl's enraged father pursued them armed with lots of bullets.

Esra Hoja paused here ....

1. Then? What happened then?
2. How do I know what happened? You have to think yourself! I told the story for I wanted to put the verbs in the past tense and to make you get used to it. Now we are going to learn how to use verbs in the past tense.

Past tense is necessary of course; however the story must be ended. We can not allow the father to kill the fleeing kids through the double-barreled shotgun, can we? Look, I've got an idea...

The boy and the girl entered the town before the chasing father. There was smog everywhere in the town...So, the father could not see anything, he had never heard about fog either. Finally he gave up and came back to his village. We are going to talk about the family and the number of children the couple will have at our next discussion when explaining the future tense".

In each article, the reader finds a test protocol that is relevant to the topic. This means, that the article can be freely used at Chemistry lesson as a teaching resource. As a result we will get, if not defenders of environment, at least the people valuing clean environment.

RESULTS AND DISCUSSION

We had offered an anonymous questionnaire to the students. Students from two introductory courses and from the biological specialization had participated in this opinion research (about 800 students). The questionnaire contained the following questions (the test is anonymous and student names are unknown, they are requested to be as honest as possible): see Table 1. The answers of students are following: (See in the Table 2)

Table 1. The questionnaire

The questionnaire
1. How often do you read popular-science articles 2. Will it be useful for the student of other specialty to read such articles? 3. Chemical contents outlined with easy language? 4. Which field is more important to link with chemistry in articles like this to make them interesting? 5. What kind of information from ecological chemistry is more interested for you

Table 2. The answers of students

Questions	Answers of students			
	Yes	No	Partially	Other: (If you chose the last answer, please, explain).
How often do you read popular-science articles?	500	20	250	20lack of time; 10-the articles are uninteresting for them;
Will it be useful for the student of other specialty to read such articles?	780	0	20	0
Chemical contents outlined with easy language?	800	0	0	0
Which field is more important to link with chemistry in articles like this to make them interesting?				725-with amusing stories; 40-with history of chemistry;
What kind of information from ecological chemistry is of more interested to you				500-water pollution; 100-heavy metals; 160-Air pollution; 40-everything;

According to the student's survey, most of them often read popular-science articles. They read because it is interesting and the chemical information is conveyed simply and clearly. It is also significant that besides chemistry, it provides other valid information as well. However, the interview revealed those too, who don't read such articles and named the lack of time as an excuse. But the smallest part of them considers such articles to be uninteresting. The majority of students believe that reading such articles will be very useful for students having another specialty, since many articles refer to environmental and the so cold "household chemistry".

The majority of students believe that chemistry should be related with amusing stories in the articles of this type. Other part of students prefer to connect chemistry with history and poetry.

However, the interview revealed those too, who don't read such articles and named the lack of time as an excuse. But the smallest part of them considers such articles to be uninteresting. The majority of students believe that reading such articles will be very useful for students having other specialty, since many articles refer to environmental and the so cold "household chemistry".

The majority of students interested in topics regarding water pollution and treatment. Other parts are interested in air pollution and chemistry of heavy metals.

CONCLUSION

The data of deductive experiment revealed, that most of the students likes popular–science articles and find them interesting and useful at the same time. This kind of articles makes it easy to inform a wide range of readers, giving them information with chemical contents.

After reading the articles, the students discuss and write recommendations on how to implement environmental cleanliness. What a member of society can do to protect the environment. Articles on environmental chemistry enable the future generation (the citizens and the decision-makers of tomorrow) if not becoming a chemist, to get environmentally educated and be aware of the significance of environmental safety.

ACKNOWLEDGMENT

An initial version of this paper has been presented in the 6 International Conference, "ECOLOGICAL & ENVIRONMENTAL CHEMISTRY 2017", Chisinau, Moldova, March 2-3, 2017.

REFERENCES

Buckingham, D. (2003). *Media education: literacy, learning and contemporary culture*. Cambridge: Polity Press.

C. Eamon Masters of Fire. (2010). *Italian Alchemists in the Court of Philip II. In Chymia:Science and Nature in Medieval and Early Modern Europe* (pp. 138–156). Cambridge Scholars Publishing.

Didactic principles. (n.d.). Retrieved from http://vinacel.hcmute.edu.vn/12MODULE/general_stuff/didactic_principles.htm

Greenberg, A. (2007). *From Alchemy to Chemistry in Picture and Story*. Wiley.

Kupatadze, K., & Kiziloz, B. (2016). Natural treatment systems from the point of didactics. *Periódico Tchê Química*, *26*, 69.

Kupatadze, K., & Malazonia, D. (2016). The Overview of Chemical knowledge in Eighteenth Century Georgia according to the Chemistry Book by King Vakhtang VI. Orbital: The Electronic. *Journal of Chemistry*, *8*(3), 204–210.

Mahaffy, P., Matlin, S., Holme, T., & Mackellar, J. (2019). System thinking for education about the molecular basis of sustainability. *Nature Suitability*, *2*, 362.

Moen, T. (2006). Reflections on the Narrative Research Approach. *International Journal of Qualitative Methods*, *5*(4), 56–69. doi:10.1177/160940690600500405

Morais, C. (2015). Storytelling with Chemistry and Related Hands-on Activities: Informal Learning Experiences to Prevent "Chemophobia" and Promote Young Children Scientific Literacy. *Journal of Chemical Education*, *92*(1), 58–65. doi:10.1021/ed5002416

Norris, S., & Phillips, L. (2003). How literacy in its fundamental sense is central to scientific literacy. *Science Education*, *87*(2), 224–240. doi:10.1002ce.10066

Pruett, S. (2014). Review of Organic Chemistry Principles in Context: A Story-Telling Historical Approach. *Journal of Chemical Education*, *91*(5), 624–625. doi:10.1021/ed5001679

Rawn. (2004). General Chemistry. Wm. C. Brown Publishers.

Santos, P., Souza, D., & Saldanha, T.Introducing Comics as an Alternative Scientific Narrative in Chemistry Teaching. (2013). Anadolu University J. *Science Education.*, *8*, 14.

Shepherd, A., Jepson, R., & Watterson, A. (2015). Risk Perceptions of Environmental Hazards and Human Reproduction: A Community-Based survey. International Scholarly Reseach Network. *Public Health*, 9.

Sutton, R., & Rockett, B. (2008). *Chemistry for the Life Sciences*. Taylor & Francis Group.

Sylaios. (2014). *WASTEnet - A Black Sea network promoting integrated natural Wastewater Treatment Systems*. Black Sea Basin 2007-2013 Joint Operational Programme.

Vanloon, G., & Duffy, S. (2011). *Environmental Chemistry: a global perspective*. Oxford University Press.

Williams, I. (2005). *Environmental Chemistry. Wiley.*

Wu, C., & Foos, J. (2010). Making Chemistry Fun to Learn. *Literacy Information and Computer Education Journal*, *1*(1), 3–7. doi:10.20533/licej.2040.2589.2010.0001 PMID:24478915

Compilation of References

Abdullah, E. J. (2013). Quality assessment for Shatt Al-Arab River using heavy metal pollution index and metal index. *Journal of Environment and Earth Science*, *3*(5), 114–120.

Abia, A. (2014). *A bioseparation process for removing heavy metals from waste water using biosorbents*. Academic Press.

Ackley, M., Rege, S., & Saxena, H. (2003). Application of natural zeolites in the purification and separation of gases. *Microporous and Mesoporous Materials*, *61*(1-3), 25–42. doi:10.1016/S1387-1811(03)00353-6

Adams, F., & Rawajfih, Y. (1977). Basaluminite and Alunite: A possible cause of sulfate retention by acid soils. *Soil Science Society of America Journal*, *41*(4), 686–692. doi:10.2136ssaj1977.03615995004100040013x

Adekola, F., Fédoroff, M., Geckeis, H., Kupcik, T., Lefèvre, G., Lützenkirchen, J., ... Schild, D. (2011). Characterization of acid–base properties of two gibbsite samples in the context of literature results. *Journal of Colloid and Interface Science*, *354*(1), 306–317. doi:10.1016/j.jcis.2010.10.014 PMID:21044786

Adeleye, A. S., Conway, J. R., Garner, K., Huang, Y., Su, Y., & Keller, A. A. (2016). Engineered nanomaterials for water treatment and remediation: Costs, benefits, and applicability. *Chemical Engineering Journal*, *286*, 640–662. doi:10.1016/j.cej.2015.10.105

Adeyemo, A. A., Adeoye, I. O., & Bello, O. S. (2017). Adsorption of dyes using different types of clay: A review. *Applied Water Science*, *7*(2), 543–568. doi:10.100713201-015-0322-y

Agyeman, J., Bullard, R. D., & Evans, B. (Eds.). (2003). *Just Sustainabilities: Development in an Unequal World*. London: Earthscan.

Ahmad Khan, N., Ibrahim, S., & Subramaniam, P. (2004). *Elimination of Heavy Metals from Wastewater Using Agricultural Wastes as Adsorbents* (Vol. 23). Academic Press.

Ahmad, M., Ahmed, S., Lal Swami, B., & Ikram, S. (2015). *Adsorption of heavy metal ions: Role of chitosan and cellulose for water treatment* (Vol. 2). Academic Press.

Ahmad, S., Munir, S., Zeb, N., Ullah, A., Khan, B., Ali, J., … Ali, S. (2019). Green nanotechnology: A review on green synthesis of silver nanoparticles — An ecofriendly approach. *International Journal of Nanomedicine*.

Ahmad, M., Ahmed, E., Hong, E., Ahmed, Z. L., Elhissi, A., & Khalid, N. R. (2014). Photocatalytic, sonocatalytic and sonophotocatalytic degradation of rhodamine B using ZnO/CNTs composites photocatalysts. *Ultrasonics Sonochemistry*, *21*(2), 761–773. doi:10.1016/j.ultsonch.2013.08.014 PMID:24055646

Ahrens, S., Peritz, A., & Strassner, T. (2009). Tunable aryl alkyl ionic liquids (TAAILs): The next generation of ionic liquids. *Angewandte Chemie International Edition*, *48*(42), 7908–7910. doi:10.1002/anie.200903399 PMID:19760688

Akhavan, O. (2009). Lasting antibacterial activities of Ag–TiO2/Ag/a-TiO2 nanocomposite thin film photocatalysts under solar light irradiation. *Journal of Colloid and Interface Science, 336*(1), 117–124. doi:10.1016/j.jcis.2009.03.018 PMID:19394952

Akir, S., Barras, A., Coffinier, Y., Bououdina, M., Boukherroub, R., & Omrani, A. D. (2016). Eco-friendly synthesis of ZnO nanoparticles with different morphologies and their visible light photocatalytic performance for the degradation of Rhodamine B. *Ceramics International, 42*(8), 10259–10265. doi:10.1016/j.ceramint.2016.03.153

Aksu, A. (2015). Sources of metal pollution in the urban atmosphere. (A case study: Tuzla, Istabul). *Journal of Environmental Health Science & Engineering, 13*(1), 79. doi:10.118640201-015-0224-9 PMID:26587239

Akunwa, N., Muhammad, M., & Akunna, J. (2014). *Treatment of metal-contaminated wastewater: A comparison of low-cost biosorbents* (Vol. 146). Academic Press.

Alaee, M., Arias, P., Sjödin, A., & Bergman, A. (2003). An overview of commercially used brominated flame retardants, their applications, their use patterns in different countries/regions and possible modes of release. *Environment International, 29*(6), 683–689. doi:10.1016/S0160-4120(03)00121-1 PMID:12850087

Aleksiayenak, Yu., Frontasyeva, M., Ostrovnaya, T., & Okina, O. (2015). Moss biomonitoring technique, NAA and AAS in air pollution studies in Belarus. *Problems of Regional Ecology*, 126-134. (in Russian)

Aleksiayenak, Yu., Frontasyeva, M., Florek, M., Sykora, I., Holy, K., Masarik, J., ... Ramatlhape, K. (2013). Distributions of ^{137}Cs and ^{210}Pb in moss collected from Belarus and Slovakia. *Journal of Environmental Radioactivity, 117*, 19–24. doi:10.1016/j.jenvrad.2012.01.018 PMID:22326019

Aleksiayenak, Yu., Frontasyeva, M., Steinnes, E., Florek, M., Faanhof, A., & Ramatlhape, K. (2010). Atmospheric deposition of radionuclides in Belarus: 20 years after Chernobyl. *Proceedings of the XVII International Seminar on Interaction of Neutrons with Nuclei (Neutron Spectroscopy, Nuclear Structure, Related Topics).*

Ali, I., Asim, M., & Khan, T. A. (2012). Low cost adsorbents for the removal of organic pollutants from wastewater. *Journal of Environmental Management, 113*, 170–183. doi:10.1016/j.jenvman.2012.08.028 PMID:23023039

Ali, L. I., El-Molla, S. A., Ibrahim, M. M., Mahmoud, H. R., & Naghmash, M. A. (2016). Effect of preparation methods and optical band gap of ZnO nanomaterials on photodegradation studies. *Optical Materials, 58*, 484–490. doi:10.1016/j.optmat.2016.05.034

Ali, S. M. (2018). Fabrication of a nanocomposite from an agricultural waste and its application as a biosorbent for organic pollutants. *International Journal of Environmental Science and Technology, 15*(6), 1169–1178. doi:10.100713762-017-1477-x

Alkaff, A. H., Hendri, H., Farozy, I. H., Annisa, M., & Aritonang, R. P. (2018). Article. *Materials Science and Engineering, 299*, 012003.

Al-Saad, K., Amr, M., Hadi, D., Arar, R., Al-Sulaiti, M., Abdulmalik, T., & Alsahamary, N. (2012). *Iron oxide nanoparticles: applicability for heavy metal removal from contaminated water* (Vol. 45). Academic Press.

Alswata, A. A., Ahmad, M. B., Al-Hada, N. M., Kamari, H. M., Hussein, M. Z. B., & Ibrahim, N. A. (2017). Preparation of Zeolite/Zinc Oxide Nanocomposites for toxic metals removal from water. *Results in Physics, 7*, 723–731. doi:10.1016/j.rinp.2017.01.036

Alvarez, M. E., Pennell, R. I., Meijer, P.-J., Ishikawa, A., Dixon, R. A., & Lamb, C.ALVAREZ M.E. (1998). Reactive oxygen intermediates mediate a systemic signal network in the establishment of plant immunity. *Cell, 92*(6), 773–784. doi:10.1016/S0092-8674(00)81405-1 PMID:9529253

Amendola, V., Scaramuzza, S., Carraro, F., & Cattaruzza, E. (2017). Formation of alloy nanoparticles by laser ablation of Au/Fe multilayer films in liquid environment. *Journal of Colloid and Interface Science*, *489*, 18–27. doi:10.1016/j.jcis.2016.10.023 PMID:27770998

Amin, M. T., Alazba, A. A., & Manzoor, U. (2014). A review of removal of pollutants from water/wastewater using different types of nanomaterials. *Advances in Materials Science and Engineering*, *2014*, 2014. doi:10.1155/2014/825910

Amorós, J. A., Pérez-de-los Reyes, C., Francisco, J. G. N., Bravo, S., Chacón, J. L., Martínez, J., & Ballesta, R. J. (2013). Bioaccumulation of mineral elements in grapevine varieties cultivated in "La Mancha". *Journal of Plant Nutrition and Soil Science*, *176*(6), 843–850. doi:10.1002/jpln.201300015

Amrhein, C., Haghnia, G. H., Kim, T. S., Mosher, P. A., Gagajena, R. C., Amanios, T., & de la Torre, L. (1996). Synthesis and Properties of Zeolites from Coal Fly Ash. *Environmental Science & Technology*, *30*(3), 735–742. doi:10.1021/es940482c

Anderson, K., Cortiñas Fernández, S., Hardacre, C., & Marr, P. C. (2004). Preparation of nanoparticulate metal catalysts in porous supports using an ionic liquid route; hydrogenation and C-C coupling. *Inorganic Chemistry Communications*, *7*(1), 73–76. doi:10.1016/j.inoche.2003.10.008

Andrews, J. E., Brimblecombe, P., Jickells, T. D., & Liss, P. S. (1998). *An introduction to environmental chemistry*. Oxford: Blackwell Science Limited, UK.

Angell, C. A., Byrne, N., & Belieres, J. P. (2007). Parallel developments in aprotic and protic ionic liquids: Physical chemistry and applications. *Accounts of Chemical Research*, *40*(11), 1228–1236. doi:10.1021/ar7001842 PMID:17979250

Ani, E., Avramenko, Y., Ktaslawski, A., & Agachi, P. (2011). Identification of pollution sources in the Romanian Somes River using graphical analysis of concentration profiles. *Asia-Pacific Journal of Chemical Engineering*, *6*(5), 801–812. doi:10.1002/apj.522

Anitha, B., Ravidhas, C., Venkatesh, R., Raj, A. M. E., Ravichandran, K., Subramanian, B., & Sanjeeviraja, C. (2017). Self-assembled sulfur induced interconnected nanostructure TiO2 electrode for visible light photoresponse and photocatalytic application. *Physica E, Low-Dimensional Systems and Nanostructures*, *91*, 148–160. doi:10.1016/j.physe.2017.04.017

Anjum, M., Al-Makishah, N. H., & Barakat, M. A. (2016). Wastewater sludge stabilization using pre-treatment methods. *Process Safety and Environmental Protection*, *102*, 615–632. doi:10.1016/j.psep.2016.05.022

Anjum, M., Oves, M., Kumar, R., & Barakat, M. A. (2017). Fabrication of ZnO-ZnS@ polyaniline nanohybrid for enhanced photocatalytic degradation of 2-chlorophenol and microbial contaminants in wastewater. *International Biodeterioration & Biodegradation*, *119*, 66–77. doi:10.1016/j.ibiod.2016.10.018

Anne, Roisin, & Jane. (2002). *Effect of White Versus Red Meat on Endogenous N-Nitrosation in the uman Colon and Further Evidence of a Dose Respons*. American Society for Nutritional Sciences.

Annual report. The surface water quality status according to hydrochemical indices on the territory of the Republic of Moldova in 2015 year. (2016). Chisinau, Republic of Moldova.

Ansari, R., Seyghali, B., Mohammad-khah, A., & Zanjanchi, M. A. (2012). Application of Nano Surfactant Modified Biosorbent as an Efficient Adsorbent for Dye Removal. *Separation Science and Technology*, *47*(12), 1802–1812. doi:10.1080/01496395.2012.658485

Anshits, N., Vereshchagina, T., Fomenko, E., Kruchek, D. M., Bayukov, O., Kyrenskii, L. V., . . . Anshits, A. G. (2005). *Coal fly ash cenospheres and their application for immobilization of liquid radioactive waste* (Vol. 2005). Academic Press.

Appelo, C. A. J., & Postma, D. (2005). *Geochemistry, Groundwater and Pollution* (2nd ed.). Leiden, Netherlands: CRC Press.

Appels, W. M., Bogaart, P. W., & Sjoerd, E. A. T. M. (2011). Influence of spatial variations of microtopography and infiltration on surface runoff and field scale hydrological connectivity. *Advances in Water Resources*, *34*(2), 303–313. doi:10.1016/j.advwatres.2010.12.003

Arcibar-Orozco, J., Delgado, J., Rios-Hurtado, J., & Rangel-Mendez, J. (2014). *Influence of iron content, surface area and charge distribution in the arsenic removal by activated carbons* (Vol. 249). Academic Press.

Arica, Y., Arpa Şahin, Ç., Ergene, A., Bayramoglu, G., & Genç, Ö. (2003). *Ca-alginate as a support for Pb(II) and Zn(II) biosorption with immobilized Phanerochaete chrysosporium* (Vol. 52). Academic Press.

Arslanov, K. A. (1987). Radiocarbon: Geochemistry and geochronology. Publishing house Leningrad State University.

Ashfaq, A., & Kaifiyan, M. (2016). Simultaneous Biosorption of Iron (II) and Zinc (II) onto Flyash from Aqueous Solutions. *International Journal of Current Microbiology and Applied Sciences*, *5*(7), 117–120. doi:10.20546/ijcmas.2016.507.009

Astruc, D. (2008). Transition-metal Nanoparticles in Catalysis: From Historical Background to the State-of-the Art. *Nanoparticles and Catalysis*.

Astruc, D., Lu, F., & Aranzaes, J. R. (2005). Nanoparticles as recyclable catalysts: The frontier between homogeneous and heterogeneous catalysis. *Angewandte Chemie International Edition*, *44*(48), 7852–7872. doi:10.1002/anie.200500766 PMID:16304662

ATSDR. (2016). *Public Health Statement for Trichloroethylene*. Author.

ATSDR. (2017). *Toxicological profile for polybrominated biphenyl and polybrominated diphenyl ethers*. Atlanta, GA: Author.

Austen Angell, C., Ansari, Y., & Zhao, Z. (2012). Ionic Liquids: Past, present and future. *Faraday Discussions*, *154*, 9–27. doi:10.1039/C1FD00112D PMID:22455011

Azimov, O. T., & Bublyas, V. M. (2005). Investigation of geodynamic processes in anomaly transfer zones. *Visnyk KNU. Geology*, *34-35*, 97–102.

Azina, E., Moghimia, H., & Taheri, R. A. (2017). Article. *Desalination and Water Treatment*, *94*, 222–230. doi:10.5004/dwt.2017.21586

Babu, B. V., & Gupta, S. (2008). Removal of Cr(VI) from wastewater using activated tamarind seeds as an adsorbent. *Journal of Environmental Engineering and Science*, *7*(5), 553–557. doi:10.1139/S08-025

Badaluta-Minda, C., & Dunca, A.M. (2016). The analyse of the maximum runoff in Someşul Mic catchment area. *Scientific Bulletin of Politehnica University of Timişoara, Transactions on Hydrotechnics*, *61*(75), 39-42.

Baeissa, E. S. (2016). Photocatalytic degradation of methylene blue dye under visible light irradiation using In/ZnO nanocomposite. *Front Nanosci Nanotech*, *2*(5), 1–5. doi:10.15761/FNN.1000134

Bagasra, O., & Bagasra, A. (2019). Exposure to Toxic Environmental Agents in the Form of Fragrances and Development of Autism Spectrum Disorders (ASD). In P. Papadopoulou, C. Marouli, & A. Misseyanni (Eds.), *Environmental Exposures and Human Health Challenges* (pp. 52–85). Hershey, PA: IGI Global. doi:10.4018/978-1-5225-7635-8.ch003

Bagiyan, G. A., Koroleva, I. K., Ufimtsev, A. V., & .... (2005). Ion-molekulyarnye mekhanizmy kataliticheskogo okisleniya tiol'nyh soedinenij v prisutstvii ionov medi [Ion-molecular mechanisms for the catalytic oxidation of thiol compounds in the presence of copper ions] Himicheskaya fizika. *Chemical Physics*, *24*(6), 51–62.

Bahadir, M., & Duca, G. (Eds.). (2009). *The role of ecological chemistry in pollution research and sustainable development: Proceedings of the NATO Advanced Research Workshop on the role of ecological chemistry in pollution research and sustainable development, Chisinau, Moldova, 8-11 October 2008.* Dordrecht, The Netherlands: Springer. 10.1007/978-90-481-2903-4

Bahadir, A. M., & Duca, Gh. (2008). The role of ecological chemistry în pollution research and sustainable development. In *Proc. of the NATO Advanced Research Workshop*. Chisinau, Moldova: Springer.

Bahadory, M. (2008). Synthesis of noble metal nanoparticles. *Materials Science and Engineering C, 22*(August), 204.

Bahrami, M., & Nezamzadeh-Ejhieh, A. (2014). Effect of supporting and hybridizing of FeO and ZnO semiconductors onto an Iranian clinoptilolite nano-particles and the effect of ZnO/FeO ratio in the solar photodegradation of fish ponds waste water. *Materials Science in Semiconductor Processing, 27*, 833–840. doi:10.1016/j.mssp.2014.08.030

Bai, X., Sun, C., Liu, D., Luo, X., Li, D., Wang, J., ... Zhu, Y. (2017). Photocatalytic degradation of deoxynivalenol using graphene/ZnO hybrids in aqueous suspension. *Applied Catalysis B: Environmental, 204*, 11–20. doi:10.1016/j.apcatb.2016.11.010

Bajkova, I. S., Shtamm, E. V., Vichutinskaya, E. V., & Skurlatov, Yu. I. (2009). Mekhanizm okisleniya nanochastic FeS molekulyarnym kislorodom i peroksidom vodoroda v razbavlennyh vodnyh rastvorah [Mechanism of oxidation of FeS nanoparticles with molecular oxygen and hydrogen peroxide in dilute aqueous solutions] Himicheskaya fizika. *Chemical Physics*, (4): 86–91.

Bakan, G., Boke-Ozkoc, H., Tulek, S., & Cuce, H. (2010). Integrated environmental quality assessment of Kizilirmak River and coastal environment. *Turkish Journal of Fisheries and Aquatic Sciences, 10*(4), 453–462. doi:10.4194/trjfas.2010.0403

Baker, M. G., Venugopal, K., & Howden-Chapman, P. (2011). Household crowding and tuberculosis. In M. Braubach, D. Jacobs, & D. Ormandy (Eds.), *Environmental burden of disease associated with inadequate housing*. Copenhagen: WHO Regional Office for Europe.

Balan, V. (2008). Development of Fruit Growing In the Republic Of Moldova. *Bulletin UASVM. Horticulture, 65*(1), 252–256.

Balasubramanian, A. (2008). *Ecosystem and its Components*. Technical Report. Retrieved from: https://www.researchgate.net/publication/314213426_ECOSYSTEM_AND_ITS_COMPONENTS

Balck, M. (2015). *Phthalates in preschool dust the relation between phthalates and parameters in the preschool environment*. Degree project in biology, Master of Science. Retrieved from https://www.diva-portal.org/smash/get/diva2:852215/FULLTEXT01.pdf

Bali, U., Çatalkaya, E. Ç., & Şengül, F. (2003). Photochemical degradation and mineralization of phenol: A comparative study. *Journal of Environmental Science and Health. Part A, Toxic/Hazardous Substances & Environmental Engineering, 38*(10), 2259–2275. doi:10.1081/ESE-120023373 PMID:14524680

Baltrus, J. P., Wells, A. W., Fauth, D. J., Diehl, J. R., & White, C. M. (2001). Characterization of Carbon Concentrates from Coal-Combustion Fly Ash. *Energy & Fuels, 15*(2), 455–462. doi:10.1021/ef000201o

Barakat, M. A. (2011). New trends in removing heavy metals from industrial wastewater. *Arabian Journal of Chemistry, 4*(4), 361–377. doi:10.1016/j.arabjc.2010.07.019

Barakat, M. A., Al-Hutailah, R. I., Hashim, M. H., Qayyum, E., & Kuhn, J. N. (2013). Titania-supported silver-based bimetallic nanoparticles as photocatalysts. *Environmental Science and Pollution Research International, 20*(6), 3751–3759. doi:10.100711356-012-1306-1 PMID:23161501

Barakat, M. A., Ramadan, M. H., Alghamdi, M. A., Algarny, S. S., Woodcock, H. L., & Kuhn, J. N. (2013). Remediation of Cu (II), Ni (II), and Cr (III) ions from simulated wastewater by dendrimer/titania composites. *Journal of Environmental Management*, *117*, 50–57. doi:10.1016/j.jenvman.2012.12.025 PMID:23353877

Baral, S., Lume-Pereeira, C., Janata, E., & Henglein, A. (1986). Chemistry of colloidal manganese dioxide. III. Formation in the reaction of hydroxyl radicals with Mn2+ ions. *Journal of Physical Chemistry*, *90*(22), 6025–6028. doi:10.1021/j100280a113

Barhoumi, B., Beldean-Galea, M. S., Al-Rawabdeh, A. M., Roba, C., Martonos, I. M., Kahlaoui, M., ... Baciu, C. (2019). Occurrence, distribution and ecological risk of trace metals and organic pollutants in surface sediments from a Southeastern European river (Someşu Mic River, Romania). *The Science of the Total Environment*, *660*(10), 660–676. doi:10.1016/j.scitotenv.2018.12.428 PMID:30641395

Barinova, S., Anissimova, O., Nevo, E., Jarygin, M., & Wasser, S. (2004). Diversity and ecology of algae from the Nahal Qishon River, northern Israel. *Plant Biosystems*, *138*(3), 245–259. doi:10.1080/11263500400006985

Barkan, V., & Lyanguzova, I. (2018). Concentration of Heavy Metals in Dominant Moss Species as an Indicator of Aerial Technogenic Load. *Russian Journal of Ecology*, *49*(2), 128–134. doi:10.1134/S1067413618020030

Barouki, R., Gluckman, P. D., Grandjean, P., Hanson, M., & Heindel, J. J. (2012). Developmental origins of non-communicable disease: Implications for research and public health. *Environmental Health*, *11*(42), 42. doi:10.1186/1476-069X-11-42 PMID:22715989

Barranco, R., Gong, M., Thompson, A., Cloke, M., Hanson, S., Gibb, W., & Lester, E. (2007). *The impact of fly ash resistivity and carbon content on electrostatic precipitator performance* (Vol. 86). Academic Press.

Baruah, S., & Dutta, J. (2009). Hydrothermal growth of ZnO nanostructures. *Science and Technology of Advanced Materials*, *10*(1), 013001. doi:10.1088/1468-6996/10/1/013001 PMID:27877250

Baruah, S., Thanachayanont, C., & Dutta, J. (2008). Growth of ZnO nanowires on nonwoven polyethylene fibers. *Science and Technology of Advanced Materials*, *9*(2), 1–8. doi:10.1088/1468-6996/9/2/025009 PMID:27877984

Basavaraja, S. S., Balaji, S. D., Lagashetty, A. K., Rajasab, A. H., & Venkataraman, A. (2008). Extracellular bio-synthesis of silver nanoparticles using the fungus Fugarium semitectum. *Materials Research Bulletin*, *43*(5), 1164–1170. doi:10.1016/j.materresbull.2007.06.020

Baud, L., & Ardaillou, R. (1986). Reactive oxygen species: Production and role in the kidney. *American Journal of Physiology. Renal, Fluid and Electrolyte Physiology*, *251*(5), F765–F776. doi:10.1152/ajprenal.1986.251.5.F765 PMID:3022602

Becheri, A., Durr, M., Lo Nostro, P., & Baglioni, P. (2008). Synthesis and characterization of zinc oxide nanoparticles: Application to textiles as UV-absorbers. *Journal of Nanoparticle Research*, *10*(4), 679–689. doi:10.100711051-007-9318-3

Beck, U. (1992). *Risk Society: Towards a New Modernity*. New Delhi: Sage.

Bedard, D. L., Bunnell, S. C., & Smullen, L. A. (1996). Stimulation of Microbial para-Dechlorination of Polychlorinated Biphenyls That Have Persisted in Housatonic River Sediment for Decades. *Environmental Science & Technology*, *30*(2), 687–694. doi:10.1021/es950463i

Bedard, D. L., Van Dort, H., & Deweerd, K. A. (1998). Brominated biphenyls prime extensive microbial reductive dehalogenation of Aroclor 1260 in housatonic river sediment. *Applied and Environmental Microbiology*, *64*(5), 1786–1795. doi:10.1128/AEM.64.5.1786-1795.1998 PMID:9572952

Beddu, S., Mohd Zainoodin, M., Kamal, N., Mohamad, D., Sadon, S., & Mohamed Nazri, F. (2018). *Material characterization of coal combustion product (cenosphere) generated from coal – fired power plant in Malaysia*. Academic Press.

Beelen, R., Hoek, R., Vienneau, D., Eeftens, M., Dimakopoulou, K., Pedeli, X., ... de Hoogh, K. (2013). Development of NO_2 and NO_x land use regression models for estimating air pollution exposure in 36 study areas in Europe—The ESCAPE project. *Atmospheric Environment*, *72*, 10–23. doi:10.1016/j.atmosenv.2013.02.037

Begu, A., Lungu, A., & Obuh, P. (2006). Issues of Taxonomy, Ecology, and the Ecobio Indicator Role of Algal Vegetation in the Left Prut of the Middle Prut. *Bulletin ASM*, *2*(299), 172–178.

Behr, A., & Neubert, P. (2012). *Applied homogeneous catalysis*. John Wiley & Sons.

Bekoryukov, V.I., Glazkov, V.N., & Kokin, G.A. (2009). Dolgovremennyye izmeneniya globalnogo ozona okeana [Long-term changes in global ozone of the ocean]. *Izvestiya RAS. Physics of Atmosphere and Ocean, 45*(5), 607–616.

Bellinger, D. C. (2013). Prenatal Exposures to Environmental Chemicals and Children's Neurodevelopment: An Update. *Safety and Health at Work*, *4*(1), 1–11. doi:10.5491/SHAW.2013.4.1.1 PMID:23515885

Belyaev, Y. K., & Hajiyev, A. H. (1979). Statistical Data Analysis [Статистический анализ данных]. Известия АН СССР, Техническая кибернетика, 4, 79-83.

Benitez, F.J., Acero, J.L., Real, F.J., & Maya, C. (2004). Modeling of photooxidation of acetamide herbicides in natural waters by UV radiation and the combinations UV/H2O2 and UV/O3. *Journal of Chemical Technology and Biotechnology, 79*, 987–997. Doi:10.1002/jctb.1073

Beraa, A., Hajjaji, M., Laurent, R., & Caminade, A. M. (2017). Dendrimers-containing organoclays: Characterization and interaction with methylene blue. *Applied Clay Science*, *136*, 142–151. doi:10.1016/j.clay.2016.11.017

Beresnev, E. N. (1992). *Method of residual concentrations*. Moscow: Nauka.

Berkaw, M., Sowers, K. R., & May, H. D. (1996). Anaerobic ortho dechlorination of polychlorinated biphenyls by estuarine sediments from Baltimore Harbor. *Applied and Environmental Microbiology*, *62*(7), 2534–2539. doi:10.1128/AEM.62.7.2534-2539.1996 PMID:16535360

Bethke, C. M. (2007). *Geochemical and biogeochemical reaction modeling*. Cambridge: University Press. doi:10.1017/CBO9780511619670

Bhainsa, K. C., & Souza, S. F. D. (2006). Extracellular biosynthesis of silver nanoparticles using the fungus Aspergillus-fumigatus. *Colloids and Surfaces. B, Biointerfaces*, *47*(2), 160–164. doi:10.1016/j.colsurfb.2005.11.026 PMID:16420977

Bhateria, R., & Singh, R. (2019). A review on nanotechnological application of magnetic iron oxides for heavy metal removal. *Journal of Water Process Engineering*, *31*, 100845. doi:10.1016/j.jwpe.2019.100845

Bhatti, H., Khadim, R., & Hanif, M. (2019). *Biosorption of Pb(II) and Co(II) on red rose waste biomass* (Vol. 30). Academic Press.

Bilen, K., Ozyurt, O., Bakırcı, K., Karslı, S., Erdogan, S., Yılmaz, M., & Comaklı, O. (2018). Energy production, consumption, and environmental pollution for sustainable development: A case study in Turkey. *Renewable & Sustainable Energy Reviews*, *12*(6), 1529–1561. doi:10.1016/j.rser.2007.03.003

Bilgin, A., & Konanc, M. (2016). Evaluation of surface water quality and heavy metal pollution of Coruh River Basin (Turkey) by multivariate statistical methods. *Environmental Earth Sciences*, *75*(12), 1029. doi:10.1007/s12665-016-5821-0

Bingham, S. A., Pignatelli, B., Pollock, J. R., Ellul, A., Malaveille, C., Gross, G., ... O'neill, I. K. (1996). Does increased endogenous formation of N-nitroso compounds in the human colon explain the association between red meat and colon cancer? *Carcinogenesis*, *17*(3), 515–523. doi:10.1093/carcin/17.3.515 PMID:8631138

Birnbaum, L. (2014). Exposure, effects, and public health challenges, Interview. *EMBO Reports*, *15*(4). doi:10.1002/embr.201438564 PMID:24623404

Bi, S., An, S. Q., Tang, W., Yang, M., Qian, H. F., & Wang, C. (2001). Modeling the distribution of aluminium speciation in acid soil solution equilibria with the mineral phase alunite. *Environmental Geology*, *41*(1-2), 25–36. doi:10.1007002540100381

Blagnyte, R., & Paliulis, D. (2011, May). *Determination of heavy metals in mosses (pylaisia polyantha) along the high – intensive traffic flow in geležinisvilkas street (Vilnius, Lithuania).* Paper presented at the conference The 8th International Environmental Engineering Conference, Vilnius, Lithuania.

Bleam, W. (2016). *Soil and Environmental Chemistry* (2nd ed.). New York: Academic Press.

Blissett, R. S., & Rowson, N. A. (2012). A review of the multi-component utilisation of coal fly ash. *Fuel*, *97*, 1–23. doi:10.1016/j.fuel.2012.03.024

Błoński, P., Birczyński, A., Lalowicz, Z., Datka, J., & Łodziana, Z. (2015). *Structure of the Hydroxyl Groups and Adsorbed D 2 O Sites in the DX Zeolite: DFT and Experimental NMR Data* (Vol. 119). Academic Press.

Blowers, A. (2003). Inequality and community and the Challenge to Modernization: Evidence from the Nuclear Oases. In J. Agyeman, R. Evan, & R. D. Bullard (Eds.), *Just Sustainabilities: Development in an Unequal World.* Earthscan Publications Ltd.

Blum, O., Culicov, O., & Frontasyeva, M. (2002). *Heavy metal deposition in Ukrainian Carpathians (Zakarpattia and Chernivtsi regions): the regional biomonitoring.* Paper presented at the conference: Urban Air Pollution, Bioindication and Environmental Awareness. EuroBionet, Stuttgart, Germany.

Bobeica, V., Covaliov, V., Nenno, V., & Covaliova, O. (2015). Phytochemical compounds and plant products containing active principles as stimulators of the methanogenic processes of the anaerobic microorganisms. In *Proceedings of the SIMI 2015 International Symposium "The Environment & The Industry"* (pp. 57-65). INCD ECOIND.

Bodawi, A. F., Hosny, G., & Mostafa, M. (1998). Salivary nitrate, nitrite and nitrate reductase activity in relation to risk of oral cancer in Egypt. *Disease Markers*, *14*(2), 91–97. doi:10.1155/1998/507653 PMID:9868596

Boddu, V. M., Abburi, K., Talbott, J. L., & Smith, E. D. (2003). Removal of Hexavalent Chromium from Wastewater Using a New Composite Chitosan Biosorbent. *Environmental Science & Technology*, *37*(19), 4449–4456. doi:10.1021/es021013a PMID:14572099

Bolin, B. (1972). *The carbon Cycle. Biosphere.* Izd-vo "Mir".

Bolisetty, S., Peydayesh, M., & Mezzenga, R. (2019). Sustainable technologies for water purification from heavy metals: Review and analysis. *Chemical Society Reviews*, *48*(2), 463–487. doi:10.1039/C8CS00493E PMID:30603760

Bollati, V., & Baccarelli, A. (2010). Environmental epigenetics. *Heredity*, *105*(1), 105–112. doi:10.1038/hdy.2010.2 PMID:20179736

Bonnefoi, M. S., Belanger, S. E., Devlin, D. J., Doerrer, N. G., Embry, M. R., Fukushima, S., ... van der Laan, J. W. (2010). Human and environmental health challenges for the next decade (2010–2020). *Critical Reviews in Toxicology*, *40*(10), 893–911. doi:10.3109/10408444.2010.506640 PMID:20854192

Bönnemann, H., & Richards, R. M. (2001). Nanoscopic metal particles– synthetic methods and potential applications. *European Journal of Inorganic Chemistry*, *2001*(10), 2455–2480. doi:10.1002/1099-0682(200109)2001:10<2455::AID-EJIC2455>3.0.CO;2-Z

Bopp, S. K., Kienzler, A., Richarz, A. N., van der Linden, S. C., Paini, A., Parissis, N., & Worth, A. P. (2019, February 7). Regulatory assessment and risk management of chemical mixtures: challenges and ways forward. *Critical Reviews in Toxicology*.

Borglin, S., Wilke, A., Jepsen, R., & Lick, W. (1996). Parameters affecting the desorption of hydrophobic organic chemicals from suspended sediments. *Environmental Toxicology and Chemistry*, *15*(12), 2254–2262. doi:10.1002/etc.5620151219

Borisovich Fenelonov, V., Melgunov, M., & Parmon, V. (2010). *The Properties of Cenospheres and the Mechanism of Their Formation During High-Temperature Coal Combustion at Thermal Power Plans* (Vol. 28). Academic Press.

Borja, J., Taleon, D. M., Auresenia, J., & Gallardo, S. (2005). Polychlorinated biphenyls and their biodegradation. *Process Biochemistry*, *40*(6), 1999–2013. doi:10.1016/j.procbio.2004.08.006

Bose, A., & Shukla, A. (2019). *Processing of cenosphere-cement/asphalt composite materials and evaluation of their mechanical and acoustic properties*. Academic Press.

Botnari, V. (2016). Use of microelements in vegetable growing. In *Microelements in the composition of the biosphere and their application in agriculture and medicine* (pp. 164–167). Chisinau: ASM Publishing House.

Bousquet, J., Anto, J., Auffray, C., Akdis, M., Cambon-Thomsen, A., Keil, T., ... Zuberbier, T. (2011). MeDALL (Mechanisms of the Development of ALLergy): An integrated approach from phenotypes to systems medicine. *Allergy*, *66*(5), 596–604. doi:10.1111/j.1398-9995.2010.02534.x PMID:21261657

Boussouf, I., Medjram, M. S., Ferroudj, N., & Abramson, S. (2019). Unexpected activity of magnetically separable Fenton catalyst in clay slurries. *Environmental Technology*. (accepted)

Bouza, P. J., Ríos, I., Idaszkin, Y. L., & Bortolus, A. (2019). Patagonian salt marsh soils and oxidizable pedogenic pyrite: Solid phases controlling aluminum and iron contents in acidic soil solutions. *Environmental Earth Sciences*, *78*(1), 2. doi:10.100712665-018-7991-4

Bradley, J. S., Hill, E. W., Klein, C., Chaudret, B., & Duteil, A. (1993). Synthesis of monodispersed bimetallic palladium-copper nanoscale colloids. *Chemistry of Materials*, *5*(3), 254–256. doi:10.1021/cm00027a005

Bradman, A., Kogut, K., Eisen, E. A., Jewell, N. P., Quiros-Alcala, L., Castorina, R., ... Eskenazi, B. (2013). Variability of organophosphorous pesticide metabolite levels in spot and 24-hr urine samples collected from young children during 1 week. *Environmental Health Perspectives*, *121*(1), 118–124. doi:10.1289/ehp.1104808 PMID:23052012

Brain, C. (1989). Challis. Chemistry and biology of nitrosated peptides. *Cancer Surveys*, *8*(2), 363–384. PMID:2696585

Brambilla, G., & Martelli, A. (2007). Genotoxic and carcinogenic risk to humans of drug–nitrite interaction products. *Mutation Research*, *635*(1), 17–52. doi:10.1016/j.mrrev.2006.09.003 PMID:17157055

Brame, J., Li, Q., & Alvarez, P. J. (2011). Nanotechnology-enabled water treatment and reuse: Emerging opportunities and challenges for developing countries. *Trends in Food Science & Technology*, *22*(11), 618–624. doi:10.1016/j.tifs.2011.01.004

Braun, J. M., Kalkbrenner, A. E., Calafat, A. M., Bernert, J. T., Ye, X., Silva, M. J., ... Lanphear, B. P. (2011). Variability and predictors of urinary bisphenol A concentrations during pregnancy. *Environmental Health Perspectives*, *119*(1), 131–137. doi:10.1289/ehp.1002366 PMID:21205581

Braun, J. M., Kalkbrenner, A. E., Just, A. C., Yolton, K., Calafat, A. M., Sjödin, A., ... Lanphear, B. P. (2014). Gestational exposure to endocrine-disrupting chemicals and reciprocal social, repetitive, and stereotypic behaviors in 4- and 5-year-old children: The HOME study. *Environmental Health Perspectives*, *122*(5), 513–520. Epub2014Mar12. doi:10.1289/ehp.1307261 PMID:24622245

Breierova, E., Vajczikova, I., Sasinkova, V., Stratilova, E., Fisera, M., Gregor, T., & Sajbidor, J. (2002). Biosorption of cadmium ions by different yeast species. *Z Naturforsch*, *57*(7-8), 634–639. doi:10.1515/znc-2002-7-815 PMID:12240989

Breivik, K., Sweetman, A., Pacyna, J. M., & Jones, K. C. (2002). Towards a global historical emission inventory for selected PCB congeners - A mass balance approach: 1. Global production and consumption. *The Science of the Total Environment*, *290*(1–3), 181–198. doi:10.1016/S0048-9697(01)01075-0 PMID:12083709

Breivik, K., Sweetman, A., Pacyna, J. M., & Jones, K. C. (2007). Towards a global historical emission inventory for selected PCB congeners--a mass balance approach 3. An update. *The Science of the Total Environment*, *377*(2–3), 296–307. doi:10.1016/j.scitotenv.2007.02.026 PMID:17395248

Bridges, T. S., Ells, S., Hayes, D., Mount, D., Nadeau, S. C., Palermo, M. R., … Schroeder, P. (2008). *The Four Rs of Environmental Dredging : Resuspension, Release, Residual, and Risk (Report No. ERDC/EL TR-08-4)*. Washington, DC: US Army Corps of Engineers.

Broeker, W. S., & Walton, A. (1959). Radiocarbon from Nuclear Tests. *Science*, *130*(3371), 309–314. doi:10.1126cience.130.3371.309 PMID:13668560

Brown, P. (1995). Race, class, and environmental health: A review and systematization of the literature. *Environmental Research*, *69*(1), 15–30. doi:10.1006/enrs.1995.1021 PMID:7588491

Brumelis, G., Brown, D., Nikodemus, O., & Tjarve, D. (1999). The monitoring and risk assessment of Zn deposition around a metal smelter in Latvia. *Environmental Monitoring and Assessment*, *58*(2), 201–212. doi:10.1023/A:1006085220366

Bublias, V. N., & Shestopalov, V. M. (2001). Anomalous zones and their role in the redistribution of radionuclides from the soil surface to groundwater. In Water exchange in hydrageological structure and the Chernobyl disaster. Part 1. Distribution of Chernobyl radionuclides in hydrogeological structure. (pp. 251-356) Kiev: Geological Institute.

Buck Louis, G. M., Smarr, M. M., & Patel, C. J. (2017). The Exposome Research Paradigm: An Opportunity to Understand the Environmental Basis for Human Health and Disease. *Current Environmental Health Reports*, *4*(1), 89–98. doi:10.100740572-017-0126-3 PMID:28194614

Buckingham, D. (2003). *Media education: literacy, learning and contemporary culture*. Cambridge: Polity Press.

Bulat, D., Ciornea, V., & Bulat, D. (2018). The state of ichthyofauna of the Bic river in 2018 year. *Studia Universitatis Moldaviae*, *6*(116), 14–19.

Bullard, R., Mohai, P., Saha, R., & Wright, B. (2007). *Toxic wastes and race at twenty: 1987–2007. Grassroots struggles to dismantle environmental racism in the United States*. A report prepared for the United Church of Christ Justice and Witness Ministries: United Church of Christ. Retrieved from https://www.ucc.org/assets/pdfs/toxic20.pdf

Bulut, Y., & Tez, Z. (2003). *Removal of heavy metal ions by modified sawdust of walnut* (Vol. 12). Academic Press.

Bunduchi, E., Gladchi, V., Duca, Gh., & Goreacheva, N. (2012). Kinetic indicators based evaluation of water quality of some tributaries of Dniester river. In *Abstract Book of the V Intern. conf.-simpos. Ecological Chemistry*. Chişinău, Moldova: CEP USM.

Bunduchi, E., Duca, Gh., Gladchi, V., Goreaceva, N., & Mardari, I. (2006). Assessment of kinetic parameters în the waters of the Nistru course în the section Naslavcea – Dubasari. *Chemistry Journal of Moldova. General. Industrial and Ecological Chemistry*, *1*(1), 68–73.

Burgot, J. L. (2012). *Ionic equilibria in analytical chemistry*. Springer Science & Business Media. doi:10.1007/978-1-4419-8382-4

Burkhard, R., Deletic, A., & Craig, A. (2000). Techniques for water and wastewater management: A review of techniques and their integration in planning. *Urban Water*, *2*(3), 197–221. doi:10.1016/S1462-0758(00)00056-X

Burlakova, V. E., & Novikova, A. A. (2017). Extracting of fullerene-like nanoparticles from environmental soots. *Fullerenes, Nanotubes, and Carbon Nanostructures*, *25*(8), 483–487. doi:10.1080/1536383X.2017.1339274

Busset, G., Sangely, M., Montrejaud-Vignoles, M., Thannberger, L., & Sablayrolles, C. (2012). Life cycle assessment of polychlorinated biphenyl contaminated soil remediation processes. *The International Journal of Life Cycle Assessment*, *17*(3), 325–336. doi:10.100711367-011-0366-7

Buszewski, B., & Pomastowski, P. (2015). Influence of heterogeneity biocolloids surface on their electrophoretic separation. *Wiadomości chemiczne*, *69*, 9–10.

Buszewski, B., Rafińska, K., Pomastowski, P., Walczak, J., & Rogowska, A. (2016). Corrigendum to Novel aspects of silver nanoparticles functionalization. *Colloids and Surfaces. A, Physicochemical and Engineering Aspects*, *506*, 170–178. doi:10.1016/j.colsurfa.2016.05.058

Buszewski, B., Railean-Plugarua, V., Pomastowski, P., Rafińska, K., Szultka-Mlynska, M., & Kowalkowski, T. (2017). Antimicrobial Effectiveness of Bioactive Silver Nanoparticles Synthesised by Actinomycetes HGG16n Strain. *Current Pharmaceutical Biotechnology*, *18*(15), 1–9. PMID:28056728

Buszewski, B., Railean-Plugaru, V., Pomastowski, P., Rafinska, K., Szultka-Mlynska, M., Golinska, P., ... Dahm, H. (2018). Antimicrobial activity of biosilver nanoparticles produced by a novel Streptacidiphilus durhamensis strain. *Journal of Microbiology, Immunology, and Infection*, *51*(1), 45–54. doi:10.1016/j.jmii.2016.03.002 PMID:27103501

Butler, J. N. (1998). *Ionic equilibrium: solubility and pH calculations*. New York: Wiley.

Byalko, A. V. (2013). Variations in the concentration and gas exchange atmosphere-ocean. Proc. Russian Acad. of Sciences, 451(1), 28-32.

C. Eamon Masters of Fire. (2010). *Italian Alchemists in the Court of Philip II. In Chymia:Science and Nature in Medieval and Early Modern Europe* (pp. 138–156). Cambridge Scholars Publishing.

Cabras, P., & Angioni, A. (2000). Pesticide Residues in Grapes, Wine, and Their Processing Products. *Journal of Agricultural and Food Chemistry*, *48*(4), 967–973. doi:10.1021/jf990727a PMID:10775335

Cabrera, J. M., Diaz, M. M., Schultz, S., Temporetti, P., & Pedrozo, F. (2016). Iron buffer system in the water column and partitioning in the sediments of the naturally acidic Lake Caviahue, Neuquén, Argentina. *Journal of Volcanology and Geothermal Research*, *318*, 19–26. doi:10.1016/j.jvolgeores.2016.03.001

Cai, Z. X., Liu, C. C., Wu, G. H., Chen, X. M., & Chen, X. (2013). Palladium nanoparticles deposit on multi-walled carbon nanotubes and their catalytic applications for electrooxidation of ethanol and glucose. *Electrochimica Acta*, *112*, 756–762. doi:10.1016/j.electacta.2013.09.040

Calabi-Floody, M., Theng, B., & Mora, M. L. (2009). *Natural nanoclays: Applications and future trends- A chilean perspective* (Vol. 44). Academic Press.

Calmels, S., Ohshima, H., & Bartsch, H. (1988). Nitrosamine formation by denitrifying and non-denitrifying bacteria: Implication of nitrite reductase and nitrate reductase in nitrosation catalysis. *Journal of General Microbiology*, *134*, 221–226. PMID:3141563

Calmels, S., Ohshima, H., Vincent, P., Gounot, A. M., & Bartsch, H. (1985). Screening of microorganisms for nitrosation catalysis at pH 7 and kinetic studies on nitrosamine formation from secondary amines by E. coli strains. *Carcinogenesis*, *6*(6), 911–915. doi:10.1093/carcin/6.6.911 PMID:4006079

Cameron, R. C., Ritchie, G. S. P., & Robson, A. D. (1986). Relative toxicities of inorganic aluminum complexes to barley. *Soil Science Society of America Journal*, *50*(5), 1231–1236. doi:10.2136ssaj1986.03615995005000050029x

Campbell, C. T. (2004). The active site in nanoparticle gold catalysis. *Science*, *306*(5694), 234–235. doi:10.1126cience.1104246 PMID:15472065

Campelo, J. M., Luna, D., Luque, R., Marinas, J. M., & Romero, A. A. (2009). Sustainable preparation of supported metal nanoparticles and their applications in catalysis. *ChemSusChem: Chemistry & Sustainability Energy & Materials*, *2*(1), 18–45. doi:10.1002/cssc.200800227 PMID:19142903

Campisi, S., Chan-Thaw, C. E., Wang, D., Villa, A., & Prati, L. (2016). Metal nanoparticles on carbon based supports: The effect of the protective agent removal. *Catalysis Today*, *278*, 91–96. doi:10.1016/j.cattod.2016.04.026

Campos, J. C., Borges, R. M. H., Oliveira Filho, A. M., Nobrega, R., & Sant'Anna Jr, G. L. (2002). Oilfield wastewater treatment by combined microfiltration and biological processes. *Water Research*, *36*(1), 95–104. doi:10.1016/S0043-1354(01)00203-2 PMID:11767743

Cao, X.-L. (2010). Phthalate Esters in Foods: Sources, Occurrence, and Analytical Methods. *Comprehensive Reviews in Food Science and Food Safety*, *9*(1), 21–43. doi:10.1111/j.1541-4337.2009.00093.x

Cao, Y., Tang, C., Song, X., Liu, C., & Zhang, Y. (2016). Identifying the hydrochemical characteristics of Rivers and groundwater by multivariate statistical analysis in the Sanjiang Plain, China. *Applied Water Science*, *6*(2), 169–178. doi:10.100713201-014-0215-5

Capello, C., Fischer, U., & Hungerbühler, K. (2007). What is a green solvent? A comprehensive framework for the environmental assessment of solvents. *Green Chemistry*, *9*(9), 927–934. doi:10.1039/b617536h

Carabineiro, S. A., Bogdanchikova, N., Avalos-Borja, M., Pestryakov, A., Tavares, P. B., & Figueiredo, J. L. (2011). Gold supported on metal oxides for carbon monoxide oxidation. *Nano Research*, *4*(2), 180–193. doi:10.100712274-010-0068-7

Carlos, L., Cipollone, M., Soria, D., Moreno, M. S., Ogilby, P. R., & Einschlag, G. F. (2012). S a, Mártire D.O. (2012) The effect of humic acid binding to magnetite nanoparticles on the photogeneration of reactive oxygen species. *Separation and Purification Technology*, *91*, 23–29. doi:10.1016/j.seppur.2011.08.028

Carrasco, P. M., García, I., Yate, L., Zaera, R. T., Cabañero, G., Grande, H. J., & Ruiz, V. (2016). Graphene quantum dot membranes as fluorescent sensing platforms for Cr (VI) detection. *Carbon*, *109*, 658–665. doi:10.1016/j.carbon.2016.08.038

Castano, A., Sanchez-Rodriguez, J.E., Canas, A., Esteban, M., Navarro, C., Rodrνguez-Garcνa, A.C., Arribas, M., Dνaz, G., & Jimenez-Guerrero, J.A. (2012). Mercury, lead and cadmium levels in the urine of 170 Spanish adults: A pilot human biomonitoring study. *International Journal of Hygiene and Environmental Health*, *215*(2), 191–195.

Castro, E., Avellaneda, A., & Marco, P. (2014). Combination of advanced oxidation processes and biological treatment for the removal of benzidine-derived dyes. *Environmental Progress & Sustainable Energy*, *33*(3), 873–885. doi:10.1002/ep.11865

Çatalkaya, E. Ç., Bali, U., & Şengül, F. (2003). Photochemical degradation and mineralization of 4-chlorophenol. *Environmental Science and Pollution Research International*, *10*(2), 113–120. doi:10.1065/espr2002.10.135 PMID:12729044

Cavill, R., Kamburov, A., Ellis, J. K., Athersuch, T. J., Blagrove, M. S., Herwig, R., ... Keun, H. C. (2011). Consensus-phenotype integration of transcriptomic and metabolomic data implies a role for metabolism in the chemosensitivity of tumour cells. *PLoS Computational Biology*, *7*(3), e1001113. doi:10.1371/journal.pcbi.1001113 PMID:21483477

Ceburnis, D., Ruhling, A., & Kvietkus, K. (1997). Extended study of atmospheric heavy metal deposition in Lithuania based on moss analysis. *Environmental Monitoring and Assessment*, *47*(2), 135–152. doi:10.1023/A:1005779101732

Ceburnis, D., Steinnes, E., & Kvietkus, K. (1999). Estimation of metal uptake efficiencies from precipitation in mosses in Lithuania. *Chemosphere*, *38*(2), 445–455. doi:10.1016/S0045-6535(98)00183-0 PMID:10901666

Cecil, K. M., Brubaker, C. J., Adler, C. M., Dietrich, K. N., Altaye, M., Egelhoff, J. C., ... Lanphear, B. P. (2008). Brain Volume in Adults with Childhood Lead Exposure. *PLoS Medicine*, *5*(5), e112. doi:10.1371/journal.pmed.0050112 PMID:18507499

Cerofolini, G. F., Meda, L., & Bandosz, T. J. (1999). Adsorption with soft adsorbents or adsorbates. Theory and practice. In A. Dąbrowski (Ed.), *Studies in Surface Science and Catalysis* (Vol. 120, pp. 227–272). Elsevier.

Chaibakhsh, N., Ahmadi, N., & Zanjanchi, M. (2016). Optimization of photo catalytic degradation of neutral red dye using TiO2 nanocatalyst via Box-Behnken design. *Desalination and Water Treatment*, *57*(20), 1–11. doi:10.1080/19443994.2015.1030705

Chakravertty, R. R., & Trivedi, S. S. (1979). *Technology of Bleaching & Dyeing of Textile Fibers*. Ahmedabad, Gujarat, India: Mahajan Brothers.

Chandrakar, R. K., Baghel, R. N., Chandra, V. K., & Chandra, B. P. (2015). Synthesis, characterization and photoluminescence studies of Mn doped ZnS nanoparticles. *Superlattices and Microstructures*, *86*, 256–269. doi:10.1016/j.spmi.2015.07.043

Chang, B.Y.Z., Mehmood, M.S., Pandikumar, A., Huanga, N.M., Limb, H.N., & Marlinda, Ab. R., Yusoff, N., & Chiu, W. (2015). Hydrothermally prepared graphene-titania nanocomposite for the solar photo catalytic degradation of methylene blue. *Desalination and Water Treatment*, *57*, 238–245.

Chang, J. S., Hwang, J. S., Jhung, S. H., Park, S. E., Férey, G., & Cheetham, A. K. (2004). Nanoporous Metal-Containing Nickel Phosphates: A Class of Shape-Selective Catalyst. *Angewandte Chemie International Edition*, *43*(21), 2819–2822. doi:10.1002/anie.200353502 PMID:15150758

Chang, J.-W., Lee, C.-C., Pan, W.-H., Chou, W.-C., Huang, H.-B., Chiang, H.-C., & Huang, P.-C. (2017). Estimated Daily Intake and Cumulative Risk Assessment of Phthalates in the General Taiwanese after the 2011 DEHP Food Scandal. *Scientific Reports*, *7*(1), 45009. doi:10.1038rep45009 PMID:28327585

Chan, Y. L., Pung, S. Y., Hussain, N. S., Sreekantan, S., & Yeoh, F. Y. (2013). Photocatalytic degradation of Rhodamine B using MnO_2 and ZnO nanoparticles. *Materials Science Forum*, *756*, 379–382. doi:10.4028/www.scientific.net/MSF.756.167

Chapman, D. (1996). *Water Quality Assessments - A Guide to Use of Biota, Sediments and Water in Environmental Monitoring* (2nd ed.). University Press, Cambridge.

Charron, D. F. (Ed.). (2012). *Ecohealth Research in Practice: Innovative Applications of an Ecosystem Approach to Health, Insight and Innovation in International Development 1*. Ottawa, Ontario, Canada: International Development Research Centre. doi:10.1007/978-1-4614-0517-7

Charykov, A. K., & Osipov, N. N. (1991). *Carbonic acids and carboxylate complexes in chemical analysis*. Leningrad: Chemistry. (in Russian)

Chatonnet, P., Boutou, S., & Plana, A. (2014). Contamination of wines and spirits by phthalates: Types of contaminants present, contamination sources and means of prevention. *Food Additives & Contaminants. Part A, Chemistry, Analysis, Control, Exposure & Risk Assessment*, *31*(9), 1–11. doi:10.1080/19440049.2014.941947 PMID:25099435

Chatterjee, S., & Hadi, A. S. (2012). *Regression Analysis by Examples*. Springer.

Chaturvedi, S., Dave, P. N., & Shah, N. K. (2012). Applications of nano-catalyst in new era. *Journal of Saudi Chemical Society*, *16*(3), 307–325. doi:10.1016/j.jscs.2011.01.015

Chemist's Handbook of the XXI century. (n.d.). Retrieved from https://chem21.info/page/151172155152186050048232162151237038134048022108/

Chen, C. (2007). Degradation pathways of ethyl violet by photocatalytic reaction with ZnO dispersions. *Journal of Molecular Catalysis A Chemical*, *264*(1-2), 82–92. doi:10.1016/j.molcata.2006.09.013

Chen, C. C., Lin, C. L., & Chen, L. C. (2015). Functionalized carbon nanomaterial supported palladium nano-catalysts for electrocatalytic glucose oxidation reaction. *Electrochimica Acta*, *152*, 408–416. doi:10.1016/j.electacta.2014.11.116

Chen, C. Z., & Zhou, Z. W. (2004). The preparation of nano-ZnO and its middle infrared-ultraviolet-visible light absorption properties. *Journal of Functional Materials*, *35*(1), 97–98.

Chen, D., Huang, J., & Jackson, T. J. (2005). Vegetation water content estimation for corn and soybeans using spectral indices derived from MODIS near- and short- wave infrared bands. *Remote Sensing of Environment*, *98*(2-3), 225–236. doi:10.1016/j.rse.2005.07.008

Chen, F., Ai, H., Wei, M., Qin, C., Feng, Y., Ran, S., ... Chen, L. (2018). Distribution and phytotoxicity of soil labile aluminum fractions and aluminum species in soil water extracts and their effects on tall fescue. *Ecotoxicology and Environmental Safety*, *163*, 180–187. doi:10.1016/j.ecoenv.2018.07.075 PMID:30053588

Cheng, W., Ding, C., Wang, X., Wu, Z., Sun, Y., Yu, S., ... Wang, X. (2016). Competitive sorption of As (V) and Cr (VI) on carbonaceous nanofibers. *Chemical Engineering Journal*, *293*, 311–318. doi:10.1016/j.cej.2016.02.073

Chen, J., Wang, C., Pan, Y., Farzana, S. S., & Tam, N. F.-Y. (2017). Biochar accelerates microbial reductive debromination of 2,2¢,4,4¢-tetrabromodiphenyl ether (BDE-47) in anaerobic mangrove sediments. *Journal of Hazardous Materials*, *341*, 177–186. doi:10.1016/j.jhazmat.2017.07.063 PMID:28777963

Chen, J., Wang, P. F., Wang, C., Miao, H. C., & Wang, X. (2018). How wastewater with different nutrient levels influences microbial degradation of 2,2',4,4'-tetrabromodiphenyl ether (BDE-47) in anaerobic sediments. *Chemosphere*, *211*(1), 128–138. doi:10.1016/j.chemosphere.2018.07.122 PMID:30071424

Chen, J., Zhou, H. C., Wang, C., Zhu, C. Q., & Tam, N. F. Y. (2015). Short-term enhancement effect of nitrogen addition on microbial degradation and plant uptake of polybrominated diphenyl ethers (PBDEs) in contaminated mangrove soil. *Journal of Hazardous Materials*, *300*, 84–92. doi:10.1016/j.jhazmat.2015.06.053 PMID:26156523

Chen, Q., Rao, P., Cheng, Z., Yan, L., Qian, S., Song, R., & Shen, G. (2019). Novel soil remediation technology for simultaneous organic pollutant catalytic degradation and nitrogen supplementation. *Chemical Engineering Journal*, *370*, 27–36. doi:10.1016/j.cej.2019.03.179

Chen, S. C., Lin, Z., Ding, X., & Mirvish Sidney, S. (2001). Depentylation of the Rat Esophageal Carcinogen, Methyl-Npentylnitrosamine, By Microsomes from Various Human and Rat Tissues and by Cytochrome P450 2a3. *Drug Metabolism and Disposition: the Biological Fate of Chemicals*, *29*(9), 1221–1228. PMID:11502732

Chen, W. X., Lee, J. Y., & Liu, Z. (2002). Microwave-assisted synthesis of carbon supported Pt nanoparticles for fuel cell applications. *Chemical Communications*, (21): 2588–2589. doi:10.1039/b208600j

Chen, X., Dai, J., Shi, G., Li, L., Wang, G., & Yang, Y. (2015). Sonocatalytic degradation of Rhodamine B catalyzed by β-$Bi2O_3$ particles under ultrasonic irradiation. *Ultrasonics Sonochemistry*, *27*, 172–177. PMID:26584995

Cherchintsev, V., Frontasyeva, M., Lyapunov, S., & Smirnov, L. (1999). *Biomonitoring air pollution in Chelyabinsk region (Ural Mountains, Russia) through trace-elements andradionuclides: temporal and spatial trends*. Paper presented at the meeting of the International Atomic Energy Agency, Section of Nutritional and Health-Related Environmental Studies, Vienna, Austria.

Chikkanna, A., Mehan, L. P. K. S., & Ghosh, D. (2019). Arsenic Exposures, Poisoning, and Threat to Human Health: Arsenic Affecting Human Health. In P. Papadopoulou, C. Marouli, & A. Misseyanni (Eds.), Environmental Exposures and Human Health Challenges (pp. 86-105). Hershey, PA: IGI Global. doi:10.4018/978-1-5225-7635-8.ch004

Chiu, W. S., Khiew, P. S., Cloke, M., Isa, D., Tan, T. K., Radiman, S., ... Chia, C. H. (2010). Photocatalytic study of two dimensional ZnO nanopellets in the decomposition of methylene blue. *Chemical Engineering Journal, 158*(2), 345–352. doi:10.1016/j.cej.2010.01.052

Choi, J., Chan, S., Yip, G., Joo, H., Yang, H., & Ko, F. K. (2016). Palladium-Zeolite nanofiber as an effective recyclable catalyst membrane for water treatment. *Water Research, 101*, 46–54. doi:10.1016/j.watres.2016.05.051 PMID:27253639

Chong, M. N., Jin, B., Chow, C. W., & Saint, C. (2010). Recent developments in photocatalytic water treatment technology: A review. *Water Research, 44*(10), 2997–3027. doi:10.1016/j.watres.2010.02.039 PMID:20378145

Clevers, J. G. P. W., Kooistra, L., & Schaepman, M. E. (2008). Using spectral information from the NIR water absorption features for the retrieval of canopy water content. *International Journal of Applied Earth Observation and Geoinformation, 10*(3), 388–397. doi:10.1016/j.jag.2008.03.003

Clewell, H. J., Tan, Y. M., Campbell, J. L., & Andersen, M. E. (2008). Quantitative interpretation of human biomonitoring data. *Toxicology and Applied Pharmacology, 231*(1), 122–133. doi:10.1016/j.taap.2008.04.021 PMID:18589468

Cojocaru, C., Duca, G.H., & Gonta, M. (2013). Chemical kinetic for methzlurea nitrosation reaction: computer-ided solutions to inverse and direct problems. *Chemical Engineering Journal, 217*, 385-397.

Colais, P., Faustini, A., Stafoggia, M., Berti, G., Bisanti, L., Cadum, E., Cernigliaro, A., Mallone, S., Pacelli, B., Serinelli, M., Simonato, L., Vigotti, M.A., & Forastiere, F. (2012). Particulate Air Pollution and Hospital Admissions for Cardiac Diseases in Potentially Sensitive Subgroups. *Epidemiology, 23*(3), 473–481.

Cole-Hamilton, D. J. (2003). Homogeneous catalysis--new approaches to catalyst separation, recovery, and recycling. *Science, 299*(5613), 1702–1706. doi:10.1126cience.1081881 PMID:12637737

Cooper, S., & Thorogood, N. (2013). Social construction of health and health promotion. In M. Davies, W. Macdowall, & L. Cragg (Eds.), *Health Promotion Theory* (2nd ed.). Maidenhead, UK: McGraw-Hill Education.

Corain, B., Schmid, G., & Toshima, N. (Eds.). (2011). *Metal nanoclusters in catalysis and materials science: the issue of size control*. Elsevier.

Corburn, J. (2009). *Toward the Healthy City: People, Places, and the Politics of Urban Planning*. Cambridge, MA: The MIT Press. doi:10.7551/mitpress/7583.001.0001

Corrêa, C. M., Bizeto, M. A., & Camilo, F. F. (2016). Direct synthesis of silver nanoparticles in ionic liquid. *Journal of Nanoparticle Research, 18*(5), 1–10. doi:10.100711051-016-3436-8

Covaliov, V., Bobeica, V., Ungureanu, D., Covaliova, O., Duca, Gh., Nenno, V. & Senicovschaia, I. (2012). *Method of anaerobic treatment of liquid organic wastes*. Moldova Patent, No. 4189.

Covaliov, V., Covaliova, O. & Ungureanu, D. (2010). *Process for the obtaining of a volume charge for microflora fixation and anaerobic reactor with its use for biochemical treatment of waste waters*. Moldova Patent, No. 187Y.

Covaliov, V., Covaliova, O. & Ungureanu, D. (2015). *Combined reactor for biogas production*. Moldova Patent, No. 4382.

Covaliov, V., Covaliova, O., Ungureanu, D. & Nenno, V. (2011). *Reactor for biogas production*. Moldova Patent, No. 4130.

Covaliov, V., Duca, Gh., & Covaliova, O. (2017). "Green Energy": Innovative Ecobiotechnologies and Integrated Reactors. Anthology of Inventions. Chisinau, Moldova: CEP USM.

Covaliova, O., Duca, Gh., Covaliov, V., & Bobeica, V. (2018). Stimulation of Biochemical production of Cyanocobalamin (Vitamin B12) with Phyto-Additives and Electrolytic Hydrogen. In Compounds and materials for drugs development and biomedical applications (pp. 159-180). Bucuresti: Ed. Academiei Romane; Braila: Istros Publi. House.

Covaliov, V. (2015). Improving the biogas production systems (Rus.). *Academos*, *4*, 38–44.

Covaliov, V., Nenno, V., Bobeica, V., Senicovscaia, I., Covaliova, O., & Ungureanu, D. (2012). Stimulation of the methanogenic fermentation processes of liquid organic wastes. *Studia Universitatis Moldaviae. Real and Natural Sciences Series.*, *1*(51), 160–168.

Covaliov, V., Senicovscaia, I., Nenno, V., Bobeica, V., & Covaliova, O. (2015). Effects of Amaranths' Seeds on Dehydrogenase Activity and Gases Emission in Methanogenic Bioreactors (Rus.). *Studia Universitatis Moldaviae. Real and Natural Sciences Series*, *1*(81), 230–235.

Covaliov, V., Ungureanu, D., & Covaliova, O. (2012). Theoretical and applied aspects of improving the biogas technology. In *Problems of the Regional Power Engineering* (pp. 3–17). Chisinau, Moldova: Institute of Power Engineering ASM.

Craqmile, P., & Haran, M. (2014). Statistical Methodology. In *Modern Statistical Methods in Ecology*. Elsevier.

Crini, G., & Badot, P. M. (2007). *Traitement et épuration des eaux industrielles polluées: Procédés membranaires, bioadsorption et oxydation chimique*. Presses Univ. Franche-Comté.

Crini, G. (2006). Non-conventional low-cost adsorbents for dye removal: A review. *Bioresource Technology*, *97*(9), 1061–1085. doi:10.1016/j.biortech.2005.05.001 PMID:15993052

Cucu-Man, S. (2006). *Studiul depunerilor atmosferice de metale grele pe teritoriul Moldovei cu ajutorul biomonitorilor* (PhD Thesis). University of Iasi, Romania. (in Romanian)

Cui, Y., Wang, F., Iqbal, M. Z., Wang, Z., Li, Y., & Tu, J. (2015). Synthesis of novel 3D SnO flower-like hierarchical architectures self-assembled by nano-leaves and its photocatalysis. *Materials Research Bulletin*, *70*, 784–788. doi:10.1016/j.materresbull.2015.06.021

Čuš, F., Česnik, H. B., Bolta, Š. V., & Gregorčič, A. (2010). Pesticide residues in grapes and during vinification process. *Food Control*, *21*(11), 1512–1518. doi:10.1016/j.foodcont.2010.04.024

Cushing, B. L., Kolesnichenko, V. L., & O'Connor, C. J. (2004). Recent advances in the liquid-phase syntheses of inorganic nanoparticles. *Chemical Reviews*, *104*(9), 3893–3946. doi:10.1021/cr030027b PMID:15352782

Dahl, M., Liu, Y., & Yin, Y. (2014, October 8). Composite titanium dioxide nanomaterials. *Chemical Reviews*.

Dahl, J. A., Maddux, B. L., & Hutchison, J. E. (2007). Toward greener nanosynthesis. *Chemical Reviews*, *107*(6), 2228–2269. doi:10.1021/cr050943k PMID:17564480

Dai, K., Lv, J., Lu, L., Liu, Q., Zhu, G., & Li, D. (2014). Synthesis of micro-nano heterostructure AgBr/ZnO composite for advanced visible light photocatalysis. *Materials Letters*, *130*, 5–8. doi:10.1016/j.matlet.2014.04.190

Daneshfozoun, S., Abdullah, M., & Abdullah, B. (2017). *Preparation and characterization of magnetic biosorbent based on oil palm empty fruit bunch fibers, cellulose and Ceiba pentandra for heavy metal ions removal* (Vol. 105). Academic Press.

Daneshvar, M., & Raouf Hosseini, M. (2018). Kinetics, isotherm, and optimization of the hexavalent chromium removal from aqueous solution by a magnetic nanobiosorbent. *Environmental Science and Pollution Research International*, *25*(28), 28654–28666. doi:10.100711356-018-2878-1 PMID:30094673

Daniels, L., Sparling, R., & Sprott, G. (1984). The bioenergetics of methanogenesis. *Biochimica et Biophysica Acta*, *6*(2), 113–163. doi:10.1016/0304-4173(84)90002-8 PMID:6236847

Dankovich, T. A., & Smith, J. A. (2014). Incorporation of copper nanoparticles into paper for point-of-use water purification. *Water Research*, *63*, 245–251. doi:10.1016/j.watres.2014.06.022 PMID:25014431

Dar, M. A., Ingle, A., & Rai, M. (2013). Enhanced antimicrobial activity of silver nanoparticles synthesized by Cryphonectria sp. evaluated singly and in combination with antibiotics. *Nanotech. Biology and Medicine*, *9*, 105–110. PMID:22633901

Darweesh, M. (2018). *Adsorption Kinetic Study for Removal of Heavy Metal from Aqueous Solution by Natural Adsorbent* (Vol. 4). Academic Press.

Das, P., Chowdhury, S., Gupta, S., Kumar, I., & Kumar, R. (2010). *Assessment on the Removal of Malachite Green Using Tamarind Fruit Shell as Biosorbent* (Vol. 38). Academic Press.

Das, k., Grewal, A., & Banerjee, M. (2011). A brief review: Heavy metal and their analysis. *International Journal of Pharmaceutical Sciences Review and Research*, *11*, 13–18.

Das, S., Chakraborty, J., Chatterjee, S., & Kumar, H. (2018). Prospects of biosynthesized nanomaterials for the remediation of organic and inorganic environmental contaminants. *Environmental Science. Nano*, *5*(12), 2784–2808. doi:10.1039/C8EN00799C

de Nazelle, A., Seto, E., Donaire-Gonzalez, D., Mendez, M., Matamala, J., Nieuwenhuijsen, M. J., & Jerrett, M. (2013). Improving estimates of air pollution exposure through ubiquitous sensing technologies. *Environ Pollut*, *176*, 2–99. doi:10.1016/j.envpol.2012.12.032 PMID:23416743

de Souza, A. C., Taniguchi, S., Lopes Figueira, R. C., Montone, R. C., Caruso Bícego, M., & Martins, C. C. (2018). Historical records and spatial distribution of high hazard PCBs levels in sediments around a large South American industrial coastal area (Santos Estuary, Brazil). *Journal of Hazardous Materials*, *360*(April), 428–435. doi:10.1016/j.jhazmat.2018.08.041 PMID:30138903

Debnath, B., Chowdhury, R., & Ghosh, S. K. (2018). Sustainability of metal recovery from E-waste. *Frontiers of Environmental Science & Engineering*, *12*(6), 2. doi:10.100711783-018-1044-9

Defarge, N., Spirouxde Vendômois, J., & Séralini, G. E. (2018). Toxicity of formulants and heavy metals in glyphosate-based herbicides and other pesticides. *Toxicology Reports*, *5*, 156–163. doi:10.1016/j.toxrep.2017.12.025 PMID:29321978

Delfino, R. J., Sioutas, C., & Malik, S. (2005). Potential Role of Ultrafine Particles in Associations between Airborne Particle Mass and Cardiovascular Health. *Environmental Health Perspectives*, *113*(8), 934–946. doi:10.1289/ehp.7938 PMID:16079061

Delfosse, T., Elsass, F., & Delvaux, B. (2005). Direct evidence of basic aluminium sulphate minerals in an S-impacted Andosol. *European Journal of Soil Science*, *56*(3), 281–286. doi:10.1111/j.1365-2389.2004.00675.x

Demidenko, E. (1989). *Optimization and Regression [Оптимизация и регрессия]*. Наука.

Demirel, B., Scherer, P., Yenigun, O., & Onay, T. (2010). Production of Methane and Hydrogen from Biomass through Conventional and High-Rate Anaerobic Digestion Processes. *Critical Reviews in Environmental Science and Technology*, *40*(2), 116–146. doi:10.1080/10643380802013415

Demirtepe, H. (2017). *Sustainable Remediation of Aquatic Sediments Contaminated with Polybrominated Diphenyl Ethers and Hexabromocyclododecane*. Ankara, Turkey: Middle East Technical University.

Demirtepe, H., & Imamoglu, I. (2019). Degradation of decabromodiphenyl ether (BDE-209) in microcosms mimicking sediment environment subjected to comparative bioremediation strategies. *Journal of Environmental Management, 233*, 120–130. doi:10.1016/j.jenvman.2018.11.144 PMID:30576959

Demirtepe, H., Kjellerup, B., Sowers, K. R., & Imamoglu, I. (2015). Evaluation of PCB dechlorination pathways in anaerobic sediment microcosms using an anaerobic dechlorination model. *Journal of Hazardous Materials, 296*, 120–127. doi:10.1016/j.jhazmat.2015.04.033 PMID:25913678

Deng, S., & Ting, Y.-P. (2005). *Characterization of PEI-modified biomass and biosorption of Cu(II), Pb(II) and Ni,(II)* (Vol. 39). Academic Press.

Deng, J., Huang, K., Hu, C., Chen, D., Xiao, N., & Ju, S. (2014). Mixed micelle-mediated extraction for simultaneous trace determination of rhodamine B and rhodamine 6Gby spectrophotometry coupled with the partial least squares regression. *ACAIJ, 14*, 150–159.

Denicourt-Nowicki, A., Léger, B., & Roucoux, A. (2011). N-Donor ligands based on bipyridine and ionic liquids: An efficient partnership to stabilize rhodium colloids. Focus on oxygen-containing compounds hydrogenation. *Physical Chemistry Chemical Physics, 13*(30), 13510–13517. doi:10.1039/c1cp20602h PMID:21660329

Dennis, K. K., Marder, E., Balshaw, D. M., Cui, Y., Lynes, M. A., Patti, G. J., ... Barr, D. B. (2017). Biomonitoring in the Era of the Exposome. *Environmental Health Perspectives, 125*(4), 502–510. doi:10.1289/EHP474 PMID:27385067

Desai, M., Jellyman, J. K., & Ross, M. G. (2015). Epigenomics, gestational programming and risk of metabolic syndrome. *International Journal of Obesity, 39*(4), 633–641. doi:10.1038/ijo.2015.13 PMID:25640766

Deshmukh, R. R., Rajagopal, R., & Srinivasan, K. V. (2001). Ultrasound promoted C-C bond formation: Heck reaction at ambient conditions in room temperature ionic liquids. *Chemical Communications, 1*(17), 1544–1545. doi:10.1039/b104532f PMID:12240373

Determination of Migration Rates for Certain Phthalates. (2016). *Survey of chemical substances in consumer products No. 149*. Retrieved from https://www2.mst.dk/Udgiv/publications/2016/08/978-87-93529-01-4.pdf

Devi, P., Kaur, G., Thakur, A., Kaur, N., Grewal, A., & Kumar, P. (2017). Waste derivitized blue luminescent carbon quantum dots for selenite sensing in water. *Talanta, 170*, 49–55. doi:10.1016/j.talanta.2017.03.069 PMID:28501201

DG Health and Food Safety. (2016). Strategic plan 2016-2020. *European Commission*. Retrieved from https://ec.europa.eu/info/sites/info/files/strategic-plan-2016-2020-dg-sante_en_0.pdf

Dhakshinamoorthy, A., Navalon, S., Alvaro, M., & Garcia, H. (2012). Metal Nanoparticles as Heterogeneous Fenton Catalysts. *ChemSusChem, 5*(1), 46–64. doi:10.1002/cssc.201100517 PMID:22250135

Dhanalakshmi, R., Pandikumar, A., Sujatha, K., & Gunasekaran, P. (2011). Photo catalytic and antimicrobial activities of functionalized silicate sol–gel embedded ZnO–TiO_2 nanocomposite materials. *Materials Express, 307*, 291–298.

Dhir, B., & Kumar, R. (2010). *Adsorption of heavy metals by Salvinia biomass and agricultural residues* (Vol. 4). Academic Press.

Di Gregorio, S., Azaizeh, H., & Lorenzi, R. (2013). Biostimulation of the autochthonous microbial community for the depletion of polychlorinated biphenyls (PCBs) in contaminated sediments. *Environmental Science and Pollution Research International, 20*(6), 3989–3999. doi:10.100711356-012-1350-x PMID:23208754

Di Leo, A., Annicchiarico, C., Cardellicchio, N., Cibic, T., Comici, C., Giandomenico, S., & Spada, L. (2016). Mobilization of trace metals and PCBs from contaminated marine sediments of the Mar Piccolo in Taranto during simulated resuspension experiment. *Environmental Science and Pollution Research International*, *23*(13), 12777–12790. doi:10.100711356-015-5472-9 PMID:26438371

Diamond, S. (1986). *Particle morphologies in fly ash* (Vol. 16). Academic Press.

Diamond, M. L., Page, C. A., Campbell, M., McKenna, S., & Lall, R. (1999). Life-cycle framework for assessment of site remediation options: Method and generic survey. *Environmental Toxicology and Chemistry*, *18*(4), 788–800. doi:10.1002/etc.5620180427

Didactic principles. (n.d.). Retrieved from http://vinacel.hcmute.edu.vn/12MODULE/general_stuff/didactic_principles.htm

Ding, K., Miao, Z., Liu, Z., Zhang, Z., Han, B., An, G., ... Xie, Y. (2007). Facile Synthesis of High Quality TiO_2 Nanocrystals in Ionic Liquid via a Microwave-Assisted Process. *Journal of the American Chemical Society*, *129*(20), 6362–6363. doi:10.1021/ja070809c PMID:17469828

Disasa Irge, D. (2016). Ionic Liquids: A Review on Greener Chemistry Applications, Quality Ionic Liquid Synthesis and Economical Viability in a Chemical Processes. *American Journal of Physical Chemistry*, *5*(3), 74. doi:10.11648/j.ajpc.20160503.14

Dmitruk, Y. M., & Stuzhuk, O. V. (2014). Cartographic modeling and soil erosion by method of plastic relief. *Geopolitics and Ecogeodynamics of Regions, 10*(1), 41-43.

Dolara, P., Bigagli, E., & Collins, A. (2012). Antioxidant vitamins and mineral supplementation, life span expansion and cancer incidence: A critical commentary. *European Journal of Nutrition*, *51*(7), 769–781. doi:10.100700394-012-0389-2 PMID:22684632

Domingo, J. L., & Gine Bordonaba, J. (2011). A literature review on the safety assessment of genetically modified plants. *Environment International*, *37*(4), 734–742. doi:10.1016/j.envint.2011.01.003 PMID:21296423

Domínguez-Quintero, O., Martínez, S., Henríquez, Y., D'Ornelas, L., Krentzien, H., & Osuna, J. (2003). Silica-supported palladium nanoparticles show remarkable hydrogenation catalytic activity. *Journal of Molecular Catalysis A Chemical*, *197*(1-2), 185–191. doi:10.1016/S1381-1169(02)00583-6

Dong, K., Liu, X., Dong, H., Zhang, X., & Zhang, S. (2017, May 24). Multiscale Studies on Ionic Liquids. *Chemical Reviews*.

Dostálek, P. (1970). Immobilized Biosorbents for Bioreactors and Commercial Biosorbents. Academic Press.

Dotto, G. L., Cadaval, T. R. S., & Pinto, L. A. A. (2012). Use of Spirulina platensis micro and nanoparticles for the removal synthetic dyes from aqueous solutions by biosorption. *Process Biochemistry*, *47*(9), 1335–1343. doi:10.1016/j.procbio.2012.04.029

Dougherty, D. P., Neta, C. E., McFeeters, R. F., Lubkin, S. R., & Breidt, F. J. (2006). Semi-mechanistic partial buffer approach to modelling pH, the buffer properties, and the distribution of ionic species in complex solutions. *Journal of Agricultural and Food Chemistry*, *54*(16), 6021–6029. doi:10.1021/jf0531508 PMID:16881711

Dragan, E. S., Dinu, M., & Shankar, G. (2015). *Recent developments in composite biosorbents and their applications for wastewater treatment.* Academic Press.

Dreiper, N., & Smith, G. (2016). *Applied Regression Analysis /Прикладной регрессионный анализ.* Диалектика.

Driscoll, C. T., & Schecher, W. D. (1990). The chemistry of Al in the environment. *Environmental Geochemistry and Health, 12*(1-2), 28–50. doi:10.1007/BF01734046 PMID:24202563

Duca, G. (1997). Free radicals in natural water. Free Radicals în Biology and Environment Journal, 27, 475–489.

Duca, G. (2012). Homogeneous catalysis with metal complexes: Fundamentals and applications. Berlin: Springer Series in Chemical Physics. 102, 478 pp. doi:10.1007/978-3-642-24629-6

Duca, G. (2014). Management of water quality in Moldova. Dordrecht: Springer. doi:10.1007/978-3-319-02708-1

Duca, G., Romanciuc, L., & Porubin, D. (2012). Groundwater resources of Moldova and transboundary impact. In: Transboundary aquifers in the eastern borders of the European Union. In Regional cooperation for effective management of water resources. Dordrecht: Springer Series in Environmental Security.

Duca, G., Skurlatov, I., & Miziti, A. (1994). *Introduction in Ecological Chemistry*. Moscow: Vysshaia shcola. (in Russia)

Duca, G., Zanoagă, C., Duca, M., & Gladchi, V. (2001). Redox processes in the environment. Chisinau, Moldova: Moldovan State University.

Duca, M., Gonţa, M., Porubin, D., & Voloc, N. (2003). Research on the formation of nitrosocomposites in model systems. *Scientific Annals of the MSU*, 419-425.

Duca,, G., & Bunduchi,, E., Romanciuc L., & Goreaceva, N. (2008). Estimation of the natural water self-purification capacity from the kinetic standpoint. *Chemistry Journal of Moldova, Series General. Industrial and Ecological Chemistry, 3*(1), 10–21.

Duca, G. (2017). Ecological and environmental chemistry. *Chemistry Journal of Moldova. General. Industrial and Ecological Chemistry., 12*(1), 9–19.

Duca, G., Ciub, L., Skurlatov, Yu., & Sychev, A. (1989). Redox transformation of iron in the natural water. *Journal of Physical Chemistry (Rus), 63*(9), 2366–2371.

Duca, G., Goreaceva, N., & Romanciuc, L. (1996). Investigation of the natural water self-purification capacity under simulated conditions [Rus]. *Water Resources Journal, 23*(5-6), 523–536.

Duca, G., Mereuta, A., Velisco, N., Tanaselia, C., & Mitina, T. (2019). Determination of quality indicators of Prut River water. *Chemistry Journal of Moldova, 14*(1), 61–67. doi:10.19261/cjm.2018.541

Duca, G., & Skurlatov, Y. (2002). *Ecological chemistry*. Chisinau, Moldova: CE USM.

Duca, G., Skurlatov, Y., & Sychev, A. (2002). *Redox catalysis and ecological chemistry*. Chisinau, Moldova: CE USM.

Duca, G., Sturza, R., & Lazacovici, D. (2014, Dec.). Comparative study of the analysis techniques of phthalates in wine products. *Revue Roumaine de Chimie*, 11–15.

Duca, Gh., Gonţa, M., Stepanov, I., Port, A., Bejenaru, I., Porubin, D., & Voloc, N. (2002). Inhibition of carcinogenig N-nitroso compounds formation in simulated gastric juice. *International conference on Ecological Chemistry*, 196-206.

Duca, Gh., Sturza, R., & Siretanu, L. (2012). Estimation of organic pesticide residue content in wines. *CLEAN - Soil, Air, Water, 40*(6), 661–666. doi:10.1002/clen.201100081

Duka, G. G. (1988). Ekologicheskaya himiya stochnyh vod [Ecological chemistry of wastewater]. In *Materialy II Vsesoyuz. shkoly po ekologicheskoj khimii vodnoj sredy* [Materials II All-Union. School on Environmental Chemistry of the Aquatic Environment]. Moscow: Institute of Chemical Physics of the Academy of Sciences of the USSR.

Dunaev, A., Rumyantsev, I., Agapova, I., Grinevich, V., Vergel, K., & Gundorina, S. (2018). Physical and chemical and biological monitoring in Central Russia: investigation of quality of atmospheric air and soil in territory of Rodniki town. *Izvestiya vysshikh uchebnykh zavedenii khimiya khimicheskaya tekhnologiya, 61*(8), 97-104.

Dupont, J. (2004). On the solid, liquid and solution structural organization of imidazolium ionic liquids. *Journal of the Brazilian Chemical Society.*

Dupont, J., & Suarez, P. A. Z. (2006). Physico-chemical processes in imidazolium ionic liquids. *Physical Chemistry Chemical Physics, 8*(21), 2441–2452. doi:10.1039/b602046a PMID:16721427

Dutova, E. M., & Nikitenko, A. N. (2010). Physico-chemical modeling of the behavior of uranium in the system "water-granitoids". [In Russian]. *Bulletin of Tomsk State University, 3*, 202–207.

Dutta, A. K., Maji, S. K., & Adhikary, B. (2014). γ-Fe2O3 nanoparticles: An easily recoverable effective photo-catalyst for the degradation of rose bengal and methylene blue dyes in the waste-water treatment plant. *Materials Research Bulletin, 49*, 28–34. doi:10.1016/j.materresbull.2013.08.024

Dutta, S., & Lodhari, D. R. (2018). *Extraction of nuclear and non-ferrous metals.* Singapore: Springer. doi:10.1007/978-981-10-5172-2

Dwivedi, A., & Jain, M. (2014). Fly ash – waste management and overview. *RE:view, 6.*

Dyatchina, O. V., Semenyak, L. V., & Skurlatov, Yu. I. (1992). Okislenie glutationa kislorodom i peroksidom vodoroda v prisutstvii ionov medi [Oxidation of glutathione with oxygen and hydrogen peroxide in the presence of copper ions] Himicheskaya fizika. *Chemical Physics, 11*(9), 1255–1259.

Ealias, A. M., & Saravanakumar, M. P. (2017). Article. *IOP Conf. Series: Materials Science and Engineering, 263*, 1-14.

Ealias, A. M., & Saravanakumar, M. P. (2017). A review on the classification, characterisation, synthesis of nanoparticles and their application. *IOP Conference Series. Materials Science and Engineering, 263*, 032019. doi:10.1088/1757-899X/263/3/032019

Edelstein, A. S., & Cammaratra, R. C. (1998). *Nanomaterials: synthesis, properties and applications.* CRC Press. doi:10.1201/9781482268591

EFSA. (2009). Cadmium in food. Scientific Opinion of the Panel on Contaminants in the Food Chain (Question No EFSA-Q-2007-138). *EFSA Journal, 980*, 1–139.

EFSA. (2011). *Panel on Contaminants in the Food Chain (CONTAM). Statement on tolerable weekly intake for cadmium. EFSA Journal, 9(2), 1975.*

Eggleton, J., & Thomas, K. V. (2004). A review of factors affecting the release and bioavailability of contaminants during sediment disturbance events. *Environment International, 30*(7), 973–980. doi:10.1016/j.envint.2004.03.001 PMID:15196845

Eisele, T., & Kawatra, S. (2002). *Use of froth flotation to remove unburned carbon from fly ash* (Vol. 23). Academic Press.

El-Gendy, N. S., El-Salamony, R. A., & Younis, S. A. (2016). Green synthesis of fluorapatite from waste animal bones and the photo-catalytic degradation activity of a new ZnO/green biocatalyst nano-composite for removal of chlorophenols. *Journal of Water Process Engineering, 12*, 8–19. doi:10.1016/j.jwpe.2016.05.007

Eliche-Quesada, D., Sandalio-Pérez, J. A., Martínez-Martínez, S., Pérez-Villarejo, L., & Sánchez-Soto, P. (2017). *Investigation of use of coal fly ash in eco-friendly construction materials: Fired clay bricks and silica-calcareous non fired bricks* (Vol. 44). Academic Press.

El-Kemary, M. A., El-mehasseb, I. M., Shoueir, K. R., El-Shafey, S. E., El-Shafey, O. I., Aljohani, H. A., & Fouad, R. R. (2018). Sol-gel TiO_2 decorated on eggshell nanocrystal as engineered adsorbents for removal of acid dye. *Journal of Dispersion Science and Technology*, *39*(6), 911–921. doi:10.1080/01932691.2017.1410829

Elliott, P., Briggs, D., & Morris, S. (2001). Risk of adverse birth outcomes in populations living near landfill sites. *BMJ (Clinical Research Ed.)*, *323*(7309), 363–368. doi:10.1136/bmj.323.7309.363 PMID:11509424

El-Sayed, M. A. (2004). Small is different: Shape-, size-, and composition-dependent properties of some colloidal semiconductor nanocrystals. *Accounts of Chemical Research*, *37*(5), 326–333. doi:10.1021/ar020204f PMID:15147173

Elsey, H., Manandah, S., Sah, D., Khanal, S., MacGuire, F., King, R., ... Baral, S. C. (2016). Public Health Risks in Urban Slums: Findings of the Qualitative 'Healthy Kitchens Healthy Cities' Study in Kathmandu, Nepal. *PLoS One*, *29*(9), 1–17. doi:10.1371/journal.pone.0163798 PMID:27685999

El-Shafey, E.-S. (2009). *Removal of Zn(II) and Hg(II) from Aqueous Solution on a Carbonaceous Sorbent Chemically Prepared from Rice Husk* (Vol. 175). Academic Press.

El-Wakeel, S., Moghazy, R., Labena, A., & Husien, S. (2019). Algal biosorbent as a basic tool for heavy metals removal; the first step for further applications. *Journal of Materials and Environmental Science*, *10*, 75–87.

Ene, A., Popescu, I. V., Stihi, C., Gheboianu, A., Radulescu, C., Tigau, N., & Gosav, S. (2010). Assessment of river water quality in central and eastern parts of Romania using atomic and optical methods. *Journal of Science and Arts. Section C-Physics*, *1*(12), 113–118.

Erickson, M. D. (2001). Introduction: PCB properties, uses, occurrence, and regulatory history. In L. W. Robertson & L. G. Hansen (Eds.), *PCBs: Recent Advances in Environmental Toxicology and Health Effects* (pp. xi–xxx). The University Press of Kentucky.

Ermakova, E., Frontasyeva, M., Pavlov, S., Povtoreiko, E., Steinnes, E., & Cheremisina, Ye. (2004a). Air Pollution Studies in Central Russia (Tver and Yaroslavl Regions) Using the Moss Biomonitoring Technique and Neutron Activation Analysis. *Journal of Atmospheric Chemistry*, *49*(1-3), 549–561. doi:10.100710874-004-1265-0

Ermakova, E., Frontasyeva, M., & Steinnes, E. (2004b). Air pollution studies in Central Russia (Tula Region) using the moss biomonitoring technique, INAA and AAS. *Journal of Radioanalytical and Nuclear Chemistry*, *259*(1), 51–58. doi:10.1023/B:JRNC.0000015805.22707.a1

Ernestova, L. S., & Skurlatov, Yu. I. (1995). Obrazovanie i prevrashchenie svobodnyh radikalov OH i O2- v prirodnyh vodah [Formation and transformation of free radicals OH and O2- in natural waters] Zh. Fiz. himii. *Journal of Physical Chemistry*, *69*(7), 1157–1164.

Eskizeybek, V., Sarı, F., Gülce, H., Gülce, A., & Avcı, A. (2012). Preparation of the new polyaniline/ZnO nanocomposite and its photocatalytic activity for degradation of methylene blue and malachite green dyes under UV and natural sun lights irradiations. *Applied Catalysis B: Environmental*, *119*, 197–206. doi:10.1016/j.apcatb.2012.02.034

Esmaeili, A., & Aghababai Beni, A. (2018). Optimization and design of a continuous biosorption process using brown algae and chitosan/PVA nano-fiber membrane for removal of nickel by a new biosorbent. *International Journal of Environmental Science and Technology*, *15*(4), 765–778. doi:10.100713762-017-1409-9

Esmaeili, A., & Khoshnevisan, N. (2016). Optimization of process parameters for removal of heavy metals by biomass of Cu and Co-doped alginate-coated chitosan nanoparticles. *Bioresource Technology*, *21*, 650–658. doi:10.1016/j.biortech.2016.07.005 PMID:27416515

Espinosa-Ortiz, E. J., Shakya, M., Jain, R., Rene, E. R., Van Hullebusch, E. D., & Lens, P. N. L. (2016). Sorption of zinc onto elemental selenium nanoparticles immobilized in Phanerochaete chrysosporium pellets. *Environmental Science and Pollution Research International, 23*(21), 21619–21630. doi:10.100711356-016-7333-6 PMID:27519902

EURACTIV. (2012). *Pesticides are EU citizens' top food-related health concern.* Retrieved from: https://www.euractiv.com/section/health-consumers/news/pesticides-are-eu-citizens-top-food-related-health-concern/

European Commission. (2004, June 9). Communication of 9 June 2004: "The European Environment & Health Action Plan 2004-2010" [COM(2004) 416 - Official Journal C 49, 28.02.2006].

European Court of Auditors. (2018). *Air pollution: Our health still insufficiently protected.* Special report 23. Retrieved from https://www.eca.europa.eu/Lists/ECADocuments/SR18_23/SR_AIR_QUALITY_EL.pdf

European Environmental Agency (EEA) & Joint Research Center (JRC). (2013). *Environment and human health — Joint EEA-JRC report* (EEA Report No 5/2013), 1/112.

European Environmental Agency (EEA) & Joint Research Center (JRC). (2013). *Environment and human health.* Joint EEA-JRC report (EEA Report No 5/2013), 1/112.

European Environmental Agency (EEA). (2012). *Air quality in Europe — 2012 report.* EEA Report No 4/2012, European Environment Agency, Copenhagen, Denmark.

European Union report on pesticide residues in food. Scientific Report. (2018). EFSA Journal, 16(7). doi:10.2903/j.efsa.2018.5348

Eurostat. (2010). *The use of plant protection products in the European Union.* Retrieved from: https://ec.europa.eu/eurostat/web/products-statistical-books/-/KS-76-06-669

Evans, G. W., & Kim, P. (2007). Childhood Poverty and Health: Cumulative Risk Exposure and Stress Dysregulation. *Psychological Science, 18*(11), 953–957. doi:10.1111/j.1467-9280.2007.02008.x PMID:17958708

Eyckens, D. J., & Henderson, L. C. (2019b, April 18). A review of solvate ionic liquids: Physical parameters and synthetic applications. *Frontiers in Chemistry.*

Eyckens, D. J., & Henderson, L. C. (2019a). *A review of solvate ionic liquids: Physical parameters and synthetic applications. Frontiers in Chemistry .*

Faber, D. R., & Krieg, E. J. (2002). Unequal exposure to ecological hazards: Environmental injustices in the Commonwealth of Massachusetts. *Environmental Health Perspectives, 110*(Suppl 2), 277–288. doi:10.1289/ehp.02110s2277 PMID:11929739

Faber, D. R., & McCarthy, D. (2003). Neo-liberalism, Globalization and the Struggle for Ecological Democracy: Linking Sustainability and Environmental Justice. In J. Agyeman, R. Evan, & R. D. Bullard (Eds.), *Just Sustainabilities: Development in an Unequal World.* Earthscan Publications Ltd.

Fagervold, S. K., Watts, J. E. M., May, H. D., & Sowers, K. R. (2011). Effects of bioaugmentation on indigenous PCB dechlorinating activity in sediment microcosms. *Water Research, 45*(13), 3899–3907. doi:10.1016/j.watres.2011.04.048 PMID:21601905

Fang, J., Fan, H., Ma, Y., Wang, Z., & Chang, Q. (2015). Surface defects control for ZnO nanorods synthesized by quenching and their anti-recombination in photocatalysis. *Applied Surface Science, 332*, 47–54. doi:10.1016/j.apsusc.2015.01.139

Fang, X. S., Zhai, V., Gautam, U. K., Li, L., Wu, L. M., Bando, Y., & Golberg, D. (2011). ZnS nanostructures: From synthesis to applications. *Progress in Materials Science, 56*(2), 175–287. doi:10.1016/j.pmatsci.2010.10.001

Fan, J., & Gao, Y. (2006). Nanoparticle-supported catalysts and catalytic reactions – a mini-review. *Journal of Experimental Nanoscience*, *1*(4), 457–475. doi:10.1080/17458080601067708

Fan, L., Luo, C., Sun, M., Li, X., & Qiu, H. (2013). Highly selective adsorption of lead ions by water-dispersible magnetic chitosan/graphene oxide composites. *Colloids and Surfaces. B, Biointerfaces*, *103*, 523–529. doi:10.1016/j.colsurfb.2012.11.006 PMID:23261576

Fan, L., Zhang, Y., Luo, C., Lu, F., Qiu, H., & Sun, M. (2012). Synthesis and characterization of magnetic β-cyclodextrin–chitosan nanoparticles as nano-adsorbents for removal of methyl blue. *International Journal of Biological Macromolecules*, *50*(2), 444–450. doi:10.1016/j.ijbiomac.2011.12.016 PMID:22227306

Faraday, M. (1857). X. The Bakerian Lecture.—Experimental relations of gold (and other metals) to light. *Philosophical Transactions of the Royal Society of London*, (147): 145–181.

Farcas, N., & Matei, F. (2014). Main factors affecting biogas production - an overview. *Romanian Biotechnological Letters*, *19*(3), 9283–9296.

Farooq, U., Khan, M., & Athar, M. (2007). *Triticum aestivum: A novel biosorbent for lead (II) ions* (Vol. 51). Academic Press.

Feng, J., Sun, J., Liu, X., Zhu, J., Xiong, Y., & Tian, S. (2019). Enhancement and mechanism of nano-BaTiO 3 piezo-catalytic degradation of tricyclazole by co-loading Pt and RuO 2. *Environmental Science: Nano*.

Feng, N.-C., Guo, X.-Y., & Liang, S. (2010). *Enhanced Cu(II) adsorption by orange peel modified with sodium hydroxide* (Vol. 20). Academic Press.

Feng, Y., Yang, F., Wang, Y., Ma, L., Wu, Y., Kerr, P. G., & Yang, L. (2011). Basic dye adsorption onto an agro-based waste material – Sesame hull (Sesamum indicum L.). *Bioresource Technology*, *102*(22), 10280–10285. doi:10.1016/j.biortech.2011.08.090 PMID:21962534

Fernández-Lodeiro, A., Djafari, J., Lopez-Tejedor, D., Perez-Rizquez, C., Rodríguez-González, B., Capelo, J. L., ... Fernández-Lodeiro, J. (2019). Highly accessible aqueous synthesis of well-dispersed dendrimer type platinum nanoparticles and their catalytic applications. *Nano Research*, *12*(5), 1083–1092. doi:10.100712274-019-2350-7

Ferroudj, N., Nzimoto, J., Davidson, A., Talbot, D., Briot, E., Dupuis, V., ... Abramson, S. (2013). Maghemite nanoparticles and maghemite/silica nanocomposite microspheres as magnetic Fenton catalysts for the removal of water pollutants. *Applied Catalysis B: Environmental*, *136*, 9–18. doi:10.1016/j.apcatb.2013.01.046

Fiorentino, A., Gentili, A., Isidori, M., Monaco, P., Nardelli, A., Parrella, A., & Temussi, F. (2003). Environmental Effects Caused by Olive Mill Wastewaters: Toxicity Comparison of Low-Molecular-Weight Phenol Components. *Journal of Agricultural and Food Chemistry*, *51*(4), 1005–1009. doi:10.1021/jf020887d PMID:12568563

Fishtik, I. F., Povar, I. G., & Vataman, I. I. (1986). Thermodynamics of Precipitation and Dissolution of Oxides and Hydroxides in Aqueous Solutions. *Bulletin of Academy of Sciences of MRSS* [In Russian]. *Series of Biological and Chemical Sciences*, *5*, 58–62.

Fishtik, I., & Povar, I. (1987). Side reactions in calculating the thermodynamic functions for the process of formation – dissolution of salt precipitates. [Engl. Transl.]. *Russian Journal of General Chemistry*, *5*, 25–30.

Fishtik, I., & Povar, I. (2006). Buffer capacity in multiple chemical reaction systems involving solid phases. *Canadian Journal of Chemistry*, *84*(8), 1036–1044. doi:10.1139/v06-127

Fishtik, I., Povar, I., & Vataman, I. (1987). Thermodynamics of complex chemical equilibria in the system "precipitate of salt – aqueous solution". The formation of polynuclear species. [In Russian]. *Russian Journal of General Chemistry, 57*, 736–742.

Fleischer, T., & Grunwald, A. (2008). Making nanotechnology developments sustainable. A role for technology assessment? *Journal of Cleaner Production, 16*(8-9), 889–898. doi:10.1016/j.jclepro.2007.04.018

Franus, W., Wdowin, M., & Franus, M. (2014). Synthesis and characterization of zeolites prepared from industrial fly ash. *Environmental Monitoring and Assessment, 186*(9), 5721–5729. doi:10.100710661-014-3815-5 PMID:24838802

Fraser, M. W., Short, J., Kendrick, G., McLean, D., Keesing, J., Byrne, M., ... Wu, P. (2017). Effects of dredging on critical ecological processes for marine invertebrates, seagrasses and macroalgae, and the potential for management with environmental windows using Western Australia as a case study. *Ecological Indicators, 78*, 229–242. doi:10.1016/j.ecolind.2017.03.026

Frattini, A., Pellegri, N., Nicastro, D., & Sanctis, O. (2005). Effect of amine groups in the synthesis of Ag nanoparticles using aminosilanes. *Materials Chemistry and Physics, 94*(1), 148–152. doi:10.1016/j.matchemphys.2005.04.023

Frog, B. N., Skurlatov, Yu. I., Shtamm, E. V., & Vichutinskaya, E. V. (2012) Vliyanie vodorastvorimyh soedinenij sery na toksicheskie svojstva prirodnyh i stochnyh vod [The influence of water-soluble sulfur compounds on the toxic properties of natural and waste waters]. Vestnik MGSU, 6, 105-113.

Frontasyeva, M. V. (2011). Neutron activation analysis for the Life Sciences. A review. *Physics of Particles and Nuclei, 42*(2), 332–378. doi:10.1134/S1063779611020043

Frontasyeva, M., Smirnov, L., Steinnes, E., Lyapunov, S., & Cherchintsev, V. (2004). Heavy metal atmospheric deposition study in the South Ural Mountains. *Journal of Radioanalytical and Nuclear Chemistry, 259*(1), 19–26. doi:10.1023/B:JRNC.0000015800.41525.28

Frontasyeva, M., Steinnes, E., Lyapunov, S., Cherchintsev, V., & Smirnov, L. (2000). Biomonitoring of Heavy Metal Deposition in the South Ural Region: Some Preliminary Results Obtained by Nuclear and Related Techniques. *Journal of Radioanalytical and Nuclear Chemistry, 245*(2), 415–420. doi:10.1023/A:1006799513736

Fu, F., & Wang, Q. (2011). Removal of heavy metal ions from wastewaters: A review. *Journal of Environmental Management, 92*(3), 407–418. doi:10.1016/j.jenvman.2010.11.011 PMID:21138785

Fujishima, A., Zhang, X., & Tryk, D. A. (2008). TiO2 photocatalysis and related surface phenomena. *Surface Science Reports, 63*(12), 515–582. doi:10.1016/j.surfrep.2008.10.001

Fuller, A., Maier, J., Karampinis, E., Kalivodova, J., Grammelis, P., Kakaras, E., & Scheffknecht, G. (2018). Fly Ash Formation and Characteristics from (co-)Combustion of an Herbaceous Biomass and a Greek Lignite (Low-Rank Coal) in a Pulverized Fuel Pilot-Scale Test Facility. *Energies, 11*(6), 1581. doi:10.3390/en11061581

Fu, S.-Y., Sun, Z., Huang, P., Li, Y.-Q., & Hu, N. (2019). *Some basic aspects of polymer nanocomposites: A critical review*. Nano Materials Science; doi:10.1016/j.nanoms.2019.02.006

Gadilohar, B. L., & Shankarling, G. S. (2017, February 1). Choline based ionic liquids and their applications in organic transformation. *Journal of Molecular Liquids*.

Gaffron, P. (2012). Urban transport, environmental justice and human daily activity patterns. *Transport Policy, 20*, 114–127. doi:10.1016/j.tranpol.2012.01.011

Gaikwad, S., Birla, S., Ingle, A., Gade, A., Marcato, P., Rai, M., & Duran, N. (2013). Screening of different *Fusarium* species to select potential species for the synthesis of silver nanoparticles. *Journal of the Brazilian Chemical Society, 24*(2), 1974–1982.

Galani-Nikolakaki, S. M., & Kallithrakas-Kontos, N. G. (2007). Elemental Content of Wines. In P. Szefer & J. O. Nriagu (Eds.), Mineral components in foods (pp. 323–338). Academic Press.

Gallagher, J., Hudgens, E., Williams, A., Inmon, J., Rhoney, S., Andrews, G., ... Hubal, E. C. (2011). Mechanistic indicators of childhood asthma (MICA) study: Piloting an integrative design for evaluating environmental health. *BMC Public Health, 11*(1), 344. doi:10.1186/1471-2458-11-344 PMID:21595901

Ganzenko, O., Huguenot, D., Esposito, G., & Oturan, M. A. (2014). Electrochemical advanced oxidation and biological processes for wastewater treatment: A review of the combined approaches. *Environmental Science and Pollution Research International, 21*(14), 8493–8524. doi:10.100711356-014-2770-6 PMID:24965093

Garrigue, P., Delville, M. H., Labrugère, C., Cloutet, E., Kulesza, P. J., Morand, J. P., & Kuhn, A. (2004). Top– down approach for the preparation of colloidal carbon nanoparticles. *Chemistry of Materials, 16*(16), 2984–2986. doi:10.1021/cm049685i

Gascon, M., Morales, E., Sunyer, J., & Vrijheid, M. (2013a). Effects of persistent organic pollutants on the developing respiratory and immune systems: A systematic review. *Environment International, 52*, 51–65. doi:10.1016/j.envint.2012.11.005 PMID:23291098

Gascon, M., Verner, M. A., Guxens, M., Grimalt, J. O., Forns, J., Ibarluzea, J., ... Vrijheid, M. (2013b). Evaluating the neurotoxic effects of lactational exposure to persistent organic pollutants (POPs) in Spanish children. *Neurotoxicology, 34*, 9–15. doi:10.1016/j.neuro.2012.10.006 PMID:23085522

Gautam, R. K., Rawat, V., Banerjee, S., Sanroman, M. A., Soni, S., Singh, S. K., & Chattopadhyaya, M. C. (2015). Synthesis of bimetallic Fe–Zn nanoparticles and its application towards adsorptive removal of carcinogenic dye malachite green and Congo red in water. *Journal of Molecular Liquids, 212*, 227–236. doi:10.1016/j.molliq.2015.09.006

Geana, I., Iordache, A., Ionete, R., Marinescu, A., Ranca, A., & Culeac, M. (2013). Geographical origin identification of Romanian wines by ICP-MS elemental analysis. *Food Chemistry, 138*(2-3), 1125–1134. doi:10.1016/j.foodchem.2012.11.104 PMID:23411223

Ge, J., Zhang, Q., Zhang, T., & Yin, Y. (2008). Core–satellite nanocomposite catalysts protected by a porous silica shell: Controllable reactivity, high stability, and magnetic recyclability. *Angewandte Chemie International Edition, 47*(46), 8924–8928. doi:10.1002/anie.200803968 PMID:18925600

Gellman, A. J., & Shukla, N. (2009). Nanocatalysis: More than speed. *Nature Materials, 8*(2), 87–88. doi:10.1038/nmat2363 PMID:19165207

Ge, N., Xu, J., Li, F., Peng, B., & Pan, S. (2017). Immobilization of inactivated microbial cells on magnetic Fe3O4@CTS nanoparticles for constructing a new biosorbent for removal of patulin in fruit juice. *Food Control, 82*, 83–90. doi:10.1016/j.foodcont.2017.06.027

General information about Kazahkstan. (n.d.). Retrieved from https://www.britishcouncil.kz/sites/default/files/general_information_about_kazakhstan.pdf

Georgekutty, R., Seery, M. K., & Pillai, S. C. (2008). A highly efficient Ag-ZnO photocatalyst: Synthesis, properties, and mechanism. *Journal of Physical Chemistry, 112*, 13563–13570.

Georgia. (2019). Retrieved from https://ro.wikipedia.org/wiki/Georgia

Ghandi, K. (2014). A Review of Ionic Liquids, Their Limits and Applications. *Green and Sustainable Chemistry*, *04*(01), 44–53. doi:10.4236/gsc.2014.41008

Ghantous, A., Hernandez-Vargas, H., Byrnes, G., Dwyer, T., & Herceg, Z. (2015). Characterising the epigenome as a key component of the fetal exposome in evaluating *in utero* exposures and childhood cancer risk. *Mutagenesis*, *30*(6), 733–742. doi:10.1093/mutage/gev010 PMID:25724893

Ghasemi, M., Ghasemi, N., Zahedi, G., Wan Alwi, S. R., Goodarzi, M., & Javadian, H. (2014). *Kinetic and equilibrium study of Ni(II) sorption from aqueous solutions onto Peganum harmala-L.* Academic Press.

Gherghina, A., Grecu, F., & Molin, P. (2008). Morphometrical Analysis of Microdepressions in the Central Baragan Plain (Romania). *Revista de Geomorfologie, 10*, 31-38.

Ghernaout, D., Ghernaout, B., & Naceur, M. W. (2011, April 15). Embodying the chemical water treatment in the green chemistry-A review. *Desalination*, *271*(1-3), 1–10. doi:10.1016/j.desal.2011.01.032

Gholizadeh, A., Mohammadmehdi, S., Ben-Dor, E., & Borůvka, L. (2018). Monitoring of selected soil contaminants using proximal and remote sensing techniques: Background, state-of-the-art and future perspectives. *Critical Reviews in Environmental Science and Technology*, *48*(3), 1–36. doi:10.1080/10643389.2018.1447717

Ghorai, S., Sarkar, A. K., Panda, A. B., & Pal, S. (2013). Effective removal of Congo red dye from aqueous solution using modified xanthan gum/silica hybrid nanocomposite as adsorbent. *Bioresource Technology*, *144*, 485–491. doi:10.1016/j.biortech.2013.06.108 PMID:23896441

Ghorani-Azam, A., Riahi-Zanjani, B., & Balali-Mood, M. (2016). Effects of air pollution on human health and practical measures for prevention in Iran. *Journal of Research in Medical Sciences*, *21*(1), 65. doi:10.4103/1735-1995.189646 PMID:27904610

Ghoranneviss, M., Soni, A., Talebitaher, A., & Aslan, N. (2015). Nanomaterial synthesis, characterization, and application. *Journal of Nanomaterials*.

Ghosal, S., & Self, S. A. (1995). Particle size-density relation and cenosphere content of coal fly ash. *Fuel*, *74*(4), 522–529. doi:10.1016/0016-2361(95)98354-H

Giddens, A. (1998). *The Third Way: The Renewal of Social Democracy*. Cambridge: Polity.

Gladchi, V., Duca, G., Goreaceva, N., Bunduchi, E., & Lis, A. (2013). Chemical composition from the Dniester river tributaries. *Chemistry Journal of Moldova, 8*(1), 23-32.

Gladchi, V., Goreaceva, N., Duca, Gh., Bunduchi, E., Borodaev, R., Surighina, O., & Lis, A. (2014). Chemical compozition of right bank tributaries of Nistru river and their general impact. In Management of water quality in Moldova. Springer International Publishing.

Gladchi, V., Goreaceva, N., Duca, G., Bunduchi, E., Romanciuc, L., Mardari, I., & Borodaev, R. (2008). The study of redox conditions în the Dniester river. *Chemistry Journal of Moldova. General. Industrial and Ecological Chemistry*, *3*(1), 70–76.

Glansdorf, P., & Prigogine, I. (1973). Termodinamicheskaya teoriya struktury, ustoichivosty I fluktuatsij [Thermodynamic theory of structure, stability and fluctuations]. Moscow: Mir.

Glaspell, G., Hassan, H. M., Elzatahry, A., Abdalsayed, V., & El-Shall, M. S. (2008). Nanocatalysis on supported oxides for CO oxidation. *Topics in Catalysis*, *47*(1-2), 22–31. doi:10.100711244-007-9036-1

Gluckman, P. D., & Hanson, M. A. (2004). Living with the past: Evolution, development, and patterns of disease. *Science, 305*(5691), 1733–1736. doi:10.1126cience.1095292 PMID:15375258

Goerlandt, F., & Reniers, G. (2018). Prediction in a risk analysis context: Implications for selecting a risk perspective in practical applications. *Safety Science, 101*, 344–351. doi:10.1016/j.ssci.2017.09.007

Goetz, L. A., Jalvo, B., Rosal, R., & Mathew, A. P. (2016). Superhydrophilic anti-fouling electrospun cellulose acetate membranes coated with chitin nanocrystals for water filtration. *Journal of Membrane Science, 510*, 238–248. doi:10.1016/j.memsci.2016.02.069

Goher, M., Hassan, A., Abdel-Moniem, I., Fahmy, A., & El-Sayed, S. (2014). Evaluation of surface water quality and heavy metal indices of Ismailia Canal, Nile River, Egypt. *Egyptian Journal of Aquatic Research, 40*(3), 225–233. doi:10.1016/j.ejar.2014.09.001

Gokila, S., Gomathi, T., Sudha, P. N., & Anil, S. (2017). Removal of the heavy metal ion chromiuim(VI) using Chitosan and Alginate nanocomposites. *International Journal of Biological Macromolecules, 104*, 1459–1468. doi:10.1016/j.ijbiomac.2017.05.117 PMID:28551438

Gombotz, W. R., & Wee, S. F. (1998). Protein release from alginate matrices. *Advanced Drug Delivery Reviews, 31*(3), 267–285. doi:10.1016/S0169-409X(97)00124-5 PMID:10837629

Gomes, S., François, M., Abdelmoula, M., Refait, P., Pellissier, C., & Evrard, O. (1999). *Characterization of magnetite in silico-aluminous fly ash by SEM, TEM, XRD, magnetic susceptibility, and Mössbauer spectroscopy* (Vol. 29). Academic Press.

Goncharuk, V. (2012). Chemistry, physics and biology of water. In *Abstract Book of the V Intern. conf.-simpos. Ecological Chemistry, Chisinau, March 2-3, 2012*. Chişinău, Moldova: CEP USM.

Gonţa, M. (2003). The influence of polyphenols on n-nitrosoamines formation during proteins fermentation. *The I-st international conference of the Moldavian chemical society*, 68.

Gonța, M., & Sîrbu, E. (2018). Kinetics of the processes of nitrosation of amoxicillin with nitrite ions and inhibition of N-nitrozoamoxicillin formation. *Studia Universitatis, 1*(111), 162-169.

Gonzálvez, A., Llorens, A., Cervera, M. L., Armenta, S., & Guardia, M. (2009). Elemental fingerprint of wines from the protected designation of origin Valencia. *Food Chemistry, 112*(1), 26–34. doi:10.1016/j.foodchem.2008.05.043

Goodarzi, F., & Sanei, H. (2009). *Plerosphere and its role in reduction of emitted fine fly ash particles from pulverized coal-fired power plants* (Vol. 88). Academic Press.

Goossens, H., & Zwolsman, J. J. G. (1996). Evaluation of the Behaviour of Pollutants During Dredging Activities. *Terra et Aqua, 62*(March), 20–27.

Gordon, C. M. (2001). New developments in catalysis using ionic liquids. *Applied Catalysis A, General, 222*(1–2), 101–117. doi:10.1016/S0926-860X(01)00834-1

Goreacheva, N., Gladchi, V., Bunduchi, E., Borodaev, R., & Romanciuc, L. (2012). On the ecological state of Dniester. In *Abstract Book of the V Intern. conf.-simpos. Ecological Chemistry, Chisinau, March 2-3, 2012*. Chişinău, Moldova: CEP USM.

Gorelova, S., Babicheva, D., Frontasyeva, M., Vergel, K., & Volkova, E. (2016). Atmospheric Deposition of Trace Elements in Central Russia: Tula Region Case Study. Comparison of Different Moss Species for Biomonitoring. *International Journal of Environmental Science*, 220-229.

Gould, I. M. (2007). MRSA bacteraemia. *International Journal of Antimicrobial Agents*, *30*(1), 66–70. doi:10.1016/j.ijantimicag.2007.06.023 PMID:17875386

Goulding, K. W. T. (2016). Soil acidification and the importance of liming agricultural soils with particular reference to the United Kingdom. *Soil Use and Management*, *32*(3), 390–399. doi:10.1111um.12270 PMID:27708478

Gräfe, M., McFarlane, A., & Klauber, C. (2017). Clays and the Minerals Processing Value Chain (MPVC). In A. J. McFarlane, C. Klauber, D. J. Robinson, & M. Gräfe (Eds.), *Clays in the Minerals Processing Value Chain* (pp. 1–80). Cambridge: Cambridge University Press. doi:10.1017/9781316661888.003

Grandjean, P., & Landrigan, P. J. (2014). Neurobehavioural effects of developmental toxicity. *Lancet Neurology*, *13*(3), 330–338. doi:10.1016/S1474-4422(13)70278-3 PMID:24556010

Gray, M., Champagne, K. J., Soong, Y., & Finseth, D. H. (2001). *Parametric study of the column oil agglomeration of fly ash* (Vol. 80). Academic Press.

Grazuleviciene, R., Danileviciute, A., Nadisauskiene, R., & Vencloviene, J. (2009). Maternal smoking, GSTM1 and GSTT1 polymorphism and susceptibility to adverse pregnancy outcomes. *International Journal of Environmental Research and Public Health*, *6*(3), 1282–1297. doi:10.3390/ijerph6031282 PMID:19440446

Greaves, T. L., & Drummond, C. J. (2008, January). Protic ionic liquids: Properties and applications. *Chemical Reviews*, *108*(1), 206–237. doi:10.1021/cr068040u PMID:18095716

Greenberg, A. (2007). *From Alchemy to Chemistry in Picture and Story*. Wiley.

Grey, D., Garrick, D., Blackmore, D., Kelman, J., Muller, M., & Sadoff, C. (2013). Water security in one blue planet: twenty-first-century policy challenges for science. *Philosophical Transactions of the Royal Society A: Mathematical, Physical and Engineering Sciences, 371*(2002), 20120406.

Griffin, S., Masood, M., Nasim, M., Sarfraz, M., Ebokaiwe, A., Schäfer, K.-H., ... Jacob, C. (2017). Natural Nanoparticles: A Particular Matter Inspired by Nature. *Antioxidants*, *7*(1), 3. doi:10.3390/antiox7010003 PMID:29286304

Grinbaum. (2014). *Le point sur les résidus des produits phytosanitaires dans les vins. Œnologies Bordeaux.* Retrieved from: https://www.canal.tv/video/universite_de_bordeaux/

Gruden, K., Hren, M., Herman, A., Blejec, A., Albrecht, T., Selbig, J., ... Jeras, M. (2012). A "crossomics" study analysing variability of different components in peripheral blood of healthy Caucasoid individuals. *PLoS One*, *7*(1), e28761. doi:10.1371/journal.pone.0028761 PMID:22253695

Guadagni, S., Pistoia, M. A., Valenti, M., Leocata, P., Coletti, G., Calvisi, G., ... Reed, P. I. (1998). N-Nitroso Compounds, Bacteria, and Carcinoembryonic Antigen in the Gastric Stump, FRACG. *The Journal of Surgical Research*, *80*(985444), 345–351. doi:10.1006/jsre.1998.5444 PMID:9878336

Guan, S., Hao, L., Lu, Y., Yoshida, H., Pan, F., & Asanuma, H. (2016). Fabrication of oxygen-deficient TiO2 coatings with nano-fiber morphology for visible-light photocatalysis. *Materials Science in Semiconductor Processing*, *41*, 358–363. doi:10.1016/j.mssp.2015.09.018

Gul, K., Sohni, S., Waqar, M., Ahmada, F., Norulaini, N. A. N., & Mohd. Omar, A. K. (2016)... *Carbohydrate Polymers*, *152*, 520–531. doi:10.1016/j.carbpol.2016.06.045 PMID:27516300

Güngen, A. C., Aydemir, Y., Çoban, H., Düzenli, H., & Tasdemir, C. (2016). Lung cancer in patients diagnosed with silicosis should be investigated. *Respiratory Medicine Case Reports*, *18*, 93–95. doi:10.1016/j.rmcr.2016.04.011 PMID:27330963

Guo, J., Wang, R., Weei Tjiu, W., Pan, J., & Liu, T. (2012). *Synthesis of Fe nanoparticles@graphene composites for environmental applications* (Vol. 225-226). Academic Press.

Guo, L., Liang, L. H. B., Xu, H., Simon, P., & Wu, Z. Y. (2002). Regularly shaped, single crystalline ZnO nanorods with wurtzite structure. *Journal of the American Chemical Society, 124*(50), 14864–14865. doi:10.1021/ja027947g PMID:12475325

Guo, M., Weng, X., Wang, T., & Chen, Z. (2017). Biosynthesized iron-based nanoparticles used as a heterogeneous catalyst for the removal of 2, 4-dichlorophenol. *Separation and Purification Technology, 175*, 222–228. doi:10.1016/j.seppur.2016.11.042

Guo, X., Zhu, H., & Li, Q. (2014). Visible-light-driven photo catalytic properties of ZnO/ZnFe2O4 core/shell nanocable-arrays. *Applied Catalysis B: Environmental, 160*, 408–414. doi:10.1016/j.apcatb.2014.05.047

Gupta, V. K., Suhas, Ali, I., & Saini, V. K. (2004). Removal of rhodamine B, fast green, and methylene blue from wastewater using red mud, an aluminum industry waste. *Industrial & Engineering Chemistry Research, 43*(7), 1740–1747. doi:10.1021/ie034218g

Gutha, Y., Munagapati, V. S. (2016). Article. *International Journal of Biological Macromolecules, 93*(A), 408-417.

Güy, N., & Özacar, M. (2016). The influence of noble metals on photocatalytic activity of ZnO for Congo red degradation. *International Journal of Hydrogen Energy, 41*(44), 20100–20112. doi:10.1016/j.ijhydene.2016.07.063

Habibi, B., & Mohammadyari, S. (2015). Facile synthesis of Pd nanoparticles on nano carbon supports and their application as an electrocatalyst for oxidation of ethanol in alkaline media: The effect of support. *International Journal of Hydrogen Energy, 40*(34), 10833–10846. doi:10.1016/j.ijhydene.2015.07.021

Hafshejani, L. D., Nasab, S. B., Gholami, R. M., Moradzadeh, M., Izadpanah, Z., Hafshejani, S. B., & Bhatnagar, A. (2015). Removal of zinc and lead from aqueous solution by nanostructured cedar leaf ash as biosorbent. *Journal of Molecular Liquids, 211*, 448–456. doi:10.1016/j.molliq.2015.07.044

Haider, A. J., Jameel, Z. N., & Al-Hussaini, I. H. M. (2019). Review on: Titanium dioxide applications. *Energy Procedia, 157*, 17–29. doi:10.1016/j.egypro.2018.11.159

Hajiyev, A. H. (2011). Regression Models with Increasing Numbers of Unknown Parameters. In International Encyclopedia. Statist. Sciences. Springer.

Hajiyev, A. H. (2004). *Mathematical Theory of Regression Models with Increasing Number of Unknown Parameters. Nonlinear Models.* Research Report. Department of Computer Science. *Lincoln University*.

Hajiyev, A. H., & Hajiyev, V. S. (2009). Nonlinear Regression Models with Increasing Numbers of Unknown Parameters. *Russian Acad. Sci. Mathem., 79*(3), 339–341.

Hallett, J. P., & Welton, T. (2011, May 11). Room-temperature ionic liquids: Solvents for synthesis and catalysis. *Chemical Reviews*.

Hameed, B. H., Ahmad, A. L., & Latiff, K. N. A. (2007). Adsorption of basic dye (methylene blue) onto activated carbon prepared from rattan sawdust. *Dyes and Pigments, 75*(1), 143–149. doi:10.1016/j.dyepig.2006.05.039

Hammett, L. P. (1970). Physical organic chemistry. Reaction rates, equilibria and mechanisms. New York: McGraw-Hill.

Hanida, S. (2018). High Potency Enviromerntal Physical and Biological Factors of Malaria Transmission Tendency in Regency Working Area of Pandean Health Center Trenggalek. *Jurnal Kesehatan Lingkungan, 10*(1), 82. doi:10.20473/jkl.v10i1.2018.82-91

Hanna, K. E., & Coussens, C.Institute of Medicine. (2001). *Rebuilding the Unity of Health and the Environment: A New Vision of Environmental Health for the 21st Century*. Washington, DC: The National Academies Press. doi:10.17226/10044

Hao, C., Wang, J., Cheng, Q., Bai, Y., Wang, X., & Yang, Y. (2017). Anionic surfactants-assisted solution-phase synthesis of ZnO with improved photocatalytic performance. *Journal of Photochemistry and Photobiology A Chemistry*, *332*, 384–390. doi:10.1016/j.jphotochem.2016.09.013

Harmens, H., Mills, G., & Hayes, F. (2009). *Air pollution and vegetation: ICP Vegetation annual report 2008/2009 NERC/ Centre for Ecology & Hydrology, Bangor, UK*. Retrieved from http://www.unfccc.int/resource/docs/natc/mdanc3.pdf

Harmens, H., Norris, D., & Mills, G. (2013). *Heavy metals and nitrogen in mosses: spatial patterns in 2010/2011 and long-term temporal trends in Europe*. Bangor, UK: NERC/Centre for Ecology & Hydrology. Retrieved from http://nora.nerc.ac.uk/502676/1/N502676CR.pdf

Harmens, H., & Norris, D. (2008). *Spatial and temporal trends in heavy metal accumulation in mosses in Europe (1990-2005)*. Bangor, UK: Centre for Ecology & Hydrology. Retrieved from http://nora.nerc.ac.uk/8672/2/HarmensN008672CP.pdf

Harper, C. A. (2001). *Guide to Ceramic Materials. Handbook of Ceramics, Glasses, and Diamonds*. New York: McGraw-Hill.

Harrel, F. (2001). *Regression Modeling Strategies*. Springer. doi:10.1007/978-1-4757-3462-1

Harrington, D. W., & Elliott, S. J. (2015). Understanding emerging environmental health risks: A framework for responding to the unknown. *The Canadian Geographer. Geographe Canadien*, *59*(3), 283–296. doi:10.1111/cag.12198

Haruta, M. (2003). When gold is not noble: Catalysis by nanoparticles. *Chemical Record (New York, N.Y.)*, *3*(2), 75–87. doi:10.1002/tcr.10053 PMID:12731078

Hasan, M., Khan, M. Z. H., Khan, M., Aktar, S., Rahman, M., & Hossain, F. (2016). Heavy metals distribution and contamination in surface water of the Bay of Bengal coast. Cogent Environmetal Science, *2*(1).

Hayat, K., Gondal, M. A., Khaled, M. M., Ahmed, S., & Shemsi, A. M. (2011). Nano ZnO synthesis by modified sol gel method and its application in heterogeneous photocatalytic removal of phenol from water. *Applied Catalysis A, General*, *393*(1-2), 122–129. doi:10.1016/j.apcata.2010.11.032

Haziev, F. (2005). Methods of soil enzymology. Moscow: Nauka.

HCN. (2004). Nature and Health: The influence of nature on social, psychological and physical well.being, 2004/09E. Health Council of the Netherlands and Dutch Advisory Council for Research on Spatial Planning, Nature and the Environment (RMNO), The Hague, The Netherlands.

He, Z., & Alexandridis, P. (2015, July 28). Nanoparticles in ionic liquids: Interactions and organization. *Physical Chemistry Chemical Physics*.

Healing Cities. (2011). Retrieved from http://healingcities.org/

Health and Environment Alliance. (2010). *The EU Environment and Health Action Plan (EHAP): Assessment and Outlook for future action*. Retrieved from https://www.env-health.org/IMG/pdf/ehap_final_report_final.pdf

Health Effects Institute (HEI). (2004). *Health Effects of Outdoor Air Pollution in Developing Countries of Asia: A Literature Review*. Retrieved from https://www.healtheffects.org/publication/health-effects-outdoor-air-pollution-developing-countries-asia-literature-review

Hedayati, K. (2015). Fabrication and optical characterization of zinc oxide nanoparticles prepared via a simple sol-gel method. *J Nanostruct*, *5*, 395–401.

Hedman, J. E., Tocca, J. S., & Gunnarsson, J. S. (2009). Remobilization of polychlorinated biphenyl from baltic sea sediment: Comparing the roles of bioturbation and physical resuspension. *Environmental Toxicology and Chemistry*, *28*(11), 2241–2249. doi:10.1897/08-576.1 PMID:19499969

Heera, P., & Shanmugam, S. (2015). Review Article Nanoparticle Characterization and Application: An Overview. *Int. J. Curr. Microbiol. App. Sci*, *4*.

Hegde, K., Goswami, R., Sarma, S., Dasu, V., & Brar, S. (2015). *Nano-Ecotoxicology of Natural and Engineered Nanomaterials for Different Ecosystems*. doi:10.1061/9780784414088.ch19

Heindel, J. J. (2019). History of the Obesogen Field: Looking Back to Look Forward. *Frontiers in Endocrinology (Lausanne)*, *10*, 14. doi:10.3389/fendo.2019.00014 PMID:30761083

He, J., Robrock, K. R., & Alvarez-Cohen, L. (2006). Microbial Reductive Debromination of Polybrominated Diphenyl Ethers (PBDEs). *Environmental Science & Technology*, *40*(14), 4429–4434. doi:10.1021/es052508d PMID:16903281

Heleno Alves, R., Vitória da Silva Reis, T., Rovani, S., & Fungaro, D. (2017). *Green Synthesis and Characterization of Biosilica Produced from Sugarcane Waste Ash* (Vol. 2017). Academic Press.

Helming, K., Roth, Ch. H., Wolf, R., & Diestel, H. (1993). Characterization of rainfall-microrelief interactions within runoff using parameters derived from digital elevation models (DEMs). *Soil Technology*, *6*(3), 273–286. doi:10.1016/0933-3630(93)90016-8

Herceg, Z. (2016). Epigenetic Mechanisms as an Interface Between the Environment and Genome. *Advances in Experimental Medicine and Biology*, *903*, 3–15. doi:10.1007/978-1-4899-7678-9_1 PMID:27343085

Herceg, Z., & Vaissière, T. (2011). Epigenetic mechanisms and cancer: An interface between the environment and the genome. *Epigenetics*, *6*(7), 804–819. doi:10.4161/epi.6.7.16262 PMID:21758002

Hernandez, M. D. (2009). Development of alternative photocatalysts to TiO_2: Challenges and opportunities. *Energy & Environmental Science*, *2*(12), 1231–1257. doi:10.1039/b907933e

Hervada-Sala, C., Jarauta-Bragulat, E., Tyutyunnik, Yu., & Blum, O. (2003). *Indirect geostatistical methods to assess environmental pollution by heavy metals. case study: Ukraine*. Paper presented at the Annual Conference of the International Association for Mathematic Geology, Portsmouth, UK.

Hewitt, L. M., Carey, J. H., Dixon, D. G., & Munkittrick, K. R. (1996). Examination of bleached kraft mill effluent fraction for potential inducers of mixed function oxygenase activity in rainbow trout. In M. R. Servos, K. R. Munkittrick, J. H. Carey, & G. J. van der Kraak (Eds.), *Environmental fate and effects of pulp and paper mill effluents* (pp. 79–94). Delray Beach, FL: St. Lucie Press.

He, X., Wang, F., Liu, H., Li, J., & Niu, L. (2017). Synthesis of quartz crystals supporting Ag nanoparticle powder with enhanced antibacterial properties. *Surfaces and Interfaces*, *6*, 122–126. doi:10.1016/j.surfin.2017.01.002

He, Y., Sutton, N. B., Rijnaarts, H. H., & Langenhoff, A. A. (2016). Degradation of pharmaceuticals in wastewater using immobilized TiO2 photocatalysis under simulated solar irradiation. *Applied Catalysis B: Environmental*, *182*, 132–141. doi:10.1016/j.apcatb.2015.09.015

Himmelheber, D. W., & Hughes, J. (2014). In situ biotransformation of contaminants in sediments. In D. D. Reible (Ed.), *Processes, Assessment and Remediation of Contaminated Sediments*. New York: Springer. doi:10.1007/978-1-4614-6726-7_10

Hines, R. N., Sargent, D., Autrup, H., Birnbaum, L. S., Brent, R. L., Doerrer, N. G., ... Slikker, W. (2010). Approaches for assessing risks to sensitive populations: Lessons learned from evaluating risks in the pediatric population. *Toxicological Sciences*, *113*(1), 4–26. doi:10.1093/toxsci/kfp217 PMID:19770482

Hirajima, T., Petrus, H. T. B. M., Oosako, Y., Nonaka, M., Sasaki, K., & Ando, T. (2010). *Recovery of cenospheres from coal fly ash using a dry separation process: Separation estimation and potential application* (Vol. 95). Academic Press.

Hong, H. J., Yu, H., Park, M., & Jeong, H. S. (2019). Recovery of platinum from waste effluent using polyethyleneimine-modified nanocelluloses: Effects of the cellulose source and type. *Carbohydrate Polymers*, *210*, 167–174. doi:10.1016/j.carbpol.2019.01.079 PMID:30732749

Hong, L., Duncan, C., Townend, J., Killham, K., Smith, L. M., & Johnston, I. (1997). Nitrate- reducing bacteria on rat tongues. *Applied and Environmental Microbiology*, *63*(3), 924–930. doi:10.1128/AEM.63.3.924-930.1997 PMID:9055411

Hong, S., Piao, S., Chen, A., Liu, Y., Liu, L., Peng, S., ... Zeng, H. (2018). Afforestation neutralizes soil pH. *Nature Communications*, *9*(1), 520. doi:10.103841467-018-02970-1 PMID:29410472

Hornstein, B. J., & Finke, R. G. (2003). Transition-metal nanocluster catalysts: Scaled-up synthesis, characterization, storage conditions, stability, and catalytic activity before and after storage of polyoxoanion-and tetrabutylammonium-stabilized Ir (0) nanoclusters. *Chemistry of Materials*, *15*(4), 899–909. doi:10.1021/cm020856y

Horsfall, M. Jr, Abia, A., & Spiff, A. (2006). *Kinetic studies on the adsorption of Cd2+, Cu2+ and Zn2+ ions from aqueous solutions by cassava (Manihot sculenta Cranz) tuber bark waste* (Vol. 97). Academic Press.

Horváth, O., & Huszánk, R. (2003). Degradation of surfactants by hydroxyl radicals photogenerated from hydroxoiron(III) complexes. *Photochem. Photobiol. Sci.*, *2003*(2), 960–966. Doi:10.1039/b303697a

Hossain, F., Perales-Perez, O. J., Hwang, S., & Roman, F. (2014). Antimicrobial nanomaterials as water disinfectant: Applications, limitations and future perspectives. *The Science of the Total Environment*, *466*, 1047–1059. doi:10.1016/j.scitotenv.2013.08.009 PMID:23994736

Hou, C., Zhao, D., Zhang, S., & Wang, Y. (2018). Highly selective adsorption of Hg(II) by the monodisperse magnetic functional chitosan nano-biosorbent. *Colloid & Polymer Science*, *296*(3), 547–555. doi:10.100700396-017-4253-z

Hou, L., Wang, D., & Baccarelli, A. (2011). Environmental chemicals and microRNAs. *Mutation Research*, *714*(1–2), 105–112. doi:10.1016/j.mrfmmm.2011.05.004 PMID:21609724

Hou, L., Zhang, X., Wang, D., & Baccarelli, A. (2012). Environmental chemical exposures and human epigenetics. *International Journal of Epidemiology*, *41*(1), 79–105. doi:10.1093/ije/dyr154 PMID:22253299

Hou, X. (2015). ZnO/Ag heterostructured nanoassemblies: Wet-chemical preparation and improved visible-light photocatalytic performance. *Materials Letters*, *139*, 201–204. doi:10.1016/j.matlet.2014.10.053

Hower, J. C., Senior, C. L., Suuberg, E. M., Hurt, R. H., Wilcox, J. L., & Olson, E. S. (2010). Mercury capture by native fly ash carbons in coal-fired power plants. *Progress in Energy and Combustion Science*, *36*(4). doi:10.1016/j.pecs.2009.12.003

Hower, J., Groppo, J., Graham, U., Ward, C., Kostova, I., Maroto-Valer, M., & Dai, S. (2017). Coal-derived unburned carbons in fly ash. *RE:view*, *179*.

Hua, I., & Hoffmann, M. R. (1997). Optimization of ultrasonic irradiation as an advanced oxidation technology. *Environmental Science & Technology*, *31*(8), 2237–2243. doi:10.1021/es960717f

Hua, M., Zhang, S., Pan, B., Zhang, W., Lv, L., & Zhang, Q. (2012). Heavy metal removal from water/wastewater by nanosized metal oxides: A review. *Journal of Hazardous Materials*, *211*, 317–331. doi:10.1016/j.jhazmat.2011.10.016 PMID:22018872

Huamán, G., Maria Souza de Mesquita, L., Torem, M. L., & Adolfo Saavedra Pinto, G. (2006). *Biosorption of cadmium by green coconut shell powder* (Vol. 19). Academic Press.

Huang, X., Li, Z., Yu, Z., Deng, X., Xin, Y., & Jesionowski, T. (2019). Recent Advances in the Synthesis, Properties, and Biological Applications of Platinum Nanoclusters. *Journal of Nanomaterials*.

Huang, H. W., Chang, B. V., & Lee, C. C. (2014). Reductive debromination of decabromodiphenyl ether by anaerobic microbes from river sediment. *International Biodeterioration & Biodegradation*, *87*, 60–65. doi:10.1016/j.ibiod.2013.10.011

Huang, M. H. (2001). Room-temperature ultraviolet nanowire nanolasers. *Science*, *292*(5523), 1897–1899. doi:10.1126cience.1060367 PMID:11397941

Hubbs-Tait, L., Nation, J. R., Krebs, N. F., & Bellinger, D. C. (2005). Neurotoxicants, Micronutrients, and Social Environments: Individual and Combined Effects on Children's Development. *Psychological Science in the Public Interest*, *6*(3), 57–121. doi:10.1111/j.1529-1006.2005.00024.x PMID:26158603

Huck, J. J., Whyatt, J. D., Coulton, P., Davison, B., & Gradinar, A. (2017). Combining physiological, environmental and locational sensors for citizen-oriented health applications. *Environmental Monitoring and Assessment*, *189*(3), 114. doi:10.100710661-017-5817-6 PMID:28210895

Hui, K. S., Chao, C., & Kot, S. (2005). Removal of mixed heavy metal ions in wastewater by zeolite 4A and residual products from recycled coal fly ash. *J Hazard Materrials*, *127*(1-3), 89–101. doi:10.1016/j.jhazmat.2005.06.027 PMID:16076523

Hu, J., Zhang, P., An, W., Liu, L., Liang, Y., & Cui, W. (2019). In-situ Fe-doped g-C3N4 heterogeneous catalyst via photocatalysis-Fenton reaction with enriched photocatalytic performance for removal of complex wastewater. *Applied Catalysis B: Environmental*, *245*, 130–142. doi:10.1016/j.apcatb.2018.12.029

Hussain, S. T., & Khan, M. B. (2013). *U.S. Patent No. 8,487,147*. Washington, DC: U.S. Patent and Trademark Office.

Husseiny, M. I., Ei-Aziz, M. A., Bad, R. Y., & Mahmoud, M. A. (2007). Biosynthesis of gold nanoparticles using Pseudomonas aeruginosa. *Spectrochimica Acta. Part A: Molecular and Biomolecular Spectroscopy*, *673–674*(3-4), 1003–1006. doi:10.1016/j.saa.2006.09.028 PMID:17084659

Hu, X., Zhu, J., & Ding, Q. (2011). Environmental life-cycle comparisons of two polychlorinated biphenyl remediation technologies: Incineration and base catalyzed decomposition. *Journal of Hazardous Materials*, *191*(1–3), 258–268. doi:10.1016/j.jhazmat.2011.04.073 PMID:21571422

IARC. (2014). *IARC monographs on the evaluation of carcinogenic risks to humans: Trichloroethylene, tetrachloroethylene and some other chlorinated agents*. Lyon, France: World Health Organization.

IARC. (2016). *Glyphosate cancer determination challenged by world consensus*. Retrieved from https://geneticliteracyproject.org/glp-facts/iarc-international-agency-research-cancer-glyphosate-determination-world-consensus/

IARC. (2016). IARC Monographs on the Evaluation of Carcinogenic Risks to Humans: Polychlorinated biphenyls and polybrominated biphenyls (Vol. 107). Lyon, France: IARC.

Iijima, K., Eyfe, V., & McColl, K. E. L. (2003). Studies of nitric oxide generation from salivary in human gastric juice. *Scandinavian Journal of Gastroenterology*, *38*(3), 246–252. doi:10.1080/00365520310000708a PMID:12737438

Illés, E., Mizrahi, A., Marks, V., & Meyerstein, D. (2019). Carbonate-radical-anions, and not hydroxyl radicals, are the products of the Fenton reaction in neutral solutions containing bicarbonate. *Free Radical Biology & Medicine*, *131*, 1–6. doi:10.1016/j.freeradbiomed.2018.11.015 PMID:30458276

Informational bulletin "About the state of natural resources on the environment of the Moscow region in 2015". (2016). The Ministry of Ecology and Natural Resources Use of the Moscow Region. Krasnogorsk. (in Russian)

IPBES. (2019). *Assessment-reports*. Retrieved from https://www.ipbes.net/assessment-reports

IPBES. (2019). *Global Assessment*. Retrieved from https://www.ipbes.net/news/Media-Release-Global-Assessment

Islam, N., & Miyazaki, K. (2007, August). Nanotechnology systems of innovation: investigation of scientific disciplines' fusion trend into nanotech. In *PICMET'07-2007 Portland International Conference on Management of Engineering & Technology* (pp. 2922-2931). IEEE.

Ivanova, E., Kravciuk, E., & Kommakova, O. (2007). Ecologist floristic characteristics of phytoplankton of small water bodies in Krasnoyarsk (Russian). *Algol.*, *17*(1), 3–14.

Ivanova, N., Gugleva, V., Dobreva, M., Pehlivanov, I., Stefanov, S., & Andonova, V. (2019). Silver Nanoparticles as Multi-Functional. *Drug Delivery System*. doi:10.5772/intechopen.80238

Jacak, L., Hawrylak, P., & Wojs, A. (1998). Quantum Dots. Springer-Verlag.

Jacoby, M. (2009). *Perfecting solid-catalyst synthesis*. Science & Technology. doi:10.1021/cen-v087n017.p037

Jaffé, R. (1991). Fate of hydrophobic organic pollutants in the aquatic environment: A review. *Environmental Pollution*, *69*(2–3), 237–257. doi:10.1016/0269-7491(91)90147-O PMID:15092165

Jaganath, D., & Mupere, E. (2012). Childhood tuberculosis and malnutrition. *The Journal of Infectious Diseases*, *206*(12), 1809–1815. doi:10.1093/infdis/jis608 PMID:23033147

Jain, R., Sharma, N., & Bhargava, M. (2003). Electrochemical degradation of RhB in textile and paper industry effluents. *Journal of Scientific and Industrial Research (New Delhi, India)*, *62*, 1138–1144.

Jaishankar, M., Tseten, T., Anbalagan, N., Mathew, B. B., & Beeregowda, K. N. (2014). Toxicity, mechanism and health effects of some heavy metals. *Interdisciplinary Toxicology*, *7*(2), 60–72. doi:10.2478/intox-2014-0009 PMID:26109881

Jamal, A., Rahman, M. M., Khan, S. B., Faisal, M., Akhtar, K., Rub, M. A., ... Al-Youbi, A. O. (2012). Cobalt doped antimony oxide nano-particles based chemical sensor and photo-catalyst for environmental pollutants. *Applied Surface Science*, *261*, 52–58. doi:10.1016/j.apsusc.2012.07.066

Janiak, C. (2013). Ionic Liquids for the Synthesis and Stabilization of Metal Nanoparticles. *Z. Naturforsch*, *68*(10), 1059–1089. doi:10.5560/znb.2013-3140

Janiak, C. (2015). Metal nanoparticle synthesis in ionic liquids. *Topics in Organometallic Chemistry*, *51*, 17–54. doi:10.1007/3418_2013_70

Janus, E., Goc-Maciejewska, I., Łozyński, M., & Pernak, J. (2006). Diels-Alder reaction in protic ionic liquids. *Tetrahedron Letters*, *47*(24), 4079–4083. doi:10.1016/j.tetlet.2006.03.172

Jarmuła, A., Obłąk, E., Wawrzycka, D., & Gutowicz, J. (2011). Oporność wielolekowa związana z aktywnym usuwaniem leków z komórek drobnoustrojów. *Postepy Higieny i Medycyny Doswiadczalnej*, *65*, 216–227. doi:10.5604/17322693.937011 PMID:21502698

Javanbakht, V., Amir Alavi, S., & Zilouei, H. (2014). *Mechanisms of heavy metal removal using microorganisms as biosorbent* (Vol. 69). Academic Press.

Jeevanandam, J., Barhoum, A., Chan, Y. S., Dufresne, A., & Danquah, M. K. (2018, April 3). Review on nanoparticles and nanostructured materials: History, sources, toxicity and regulations. *Beilstein Journal of Nanotechnology.*

Jeevanandam, J., Barhoum, A., Chan, Y. S., Dufresne, A., & Danquah, M. K. (2018). Review on nanoparticles and nanostructured materials: History, sources, toxicity and regulations. *Beilstein Journal of Nanotechnology*, *9*, 1050–1074. doi:10.3762/bjnano.9.98 PMID:29719757

Jerold, M., Vasantharaj, K., Joseph, D., & Sivasubramanian, V. (2017). Fabrication of hybrid biosorbent nanoscale zero-valent iron- *Sargassum swartzii* biocomposite for the removal of crystal violet from aqueous solution. *International Journal of Phytoremediation*, *19*(3), 214–224. doi:10.1080/15226514.2016.1207607 PMID:27420340

Jiao, Y., Zhu, M., Feng, C., & Zhang, J. (2013). Hydrothermal preparation of lanthanum-doped titanium dioxide nano-crystals with high photocatalytic activity (English). *Chinese Journal of Catalysis*, (3): 585–592. doi:10.1016/S1872-2067(11)60483-X

Jin, L., Li, W., Xu, Q., & Sun, Q. (2015). Amino-functionalized nanocrystalline cellulose as an adsorbent for anionic dyes. *Cellulose (London, England)*, *22*(4), 2443–2456. doi:10.100710570-015-0649-4

Jin, L., Sun, Q., Xu, Q., & Xu, Y. (2015). Adsorptive removal of anionic dyes from aqueous solutions using microgel based on nanocellulose and polyvinylamine. *Bioresource Technology*, *197*, 348–355. doi:10.1016/j.biortech.2015.08.093 PMID:26344242

Johnson, C. H., Athersuch, T. J., Collman, G. W., Dhungana, S., Grant, D. F., Jones, D. P., & Vasiliou, V. (2017). Yale school of public health symposium on lifetime exposures and human health: The exposome; summary and future reflections. *Human Genomics*, *11*(1), 32. doi:10.118640246-017-0128-0 PMID:29221465

Johnson, K. E., Pagni, R. M., & Bartmess, J. (2007). Brønsted acids in ionic liquids: Fundamentals, organic reactions, and comparisons. *Monatshefte für Chemie*, *138*(11), 1077–1101. doi:10.100700706-007-0755-6

Jones, A. M., Collins, R. N., & Waite, T. D. (2011). Mineral species control of aluminum solubility in sulfate-rich acidic waters. *Geochimica et Cosmochimica Acta*, *75*(4), 965–977. doi:10.1016/j.gca.2010.12.001

Jose, J. C., Debs, K. B., Carvalh, J. T. T., Labuto, G., & Carrilho, E. N. V. M. (2019). Article. *Chemical Engineering Communications.*

JRC. (2008). *European Union Risk Assessment Report: Bis(2-ethylhexyl) phthalate (DEHP), Existing Substances Second Priority List* (Vol. 80). Brussels: European Commission Joint Research Centre.

Kaasik, M., & Liiv, S. (2007). Spatial and temporal variability of trace metals in mosses in Estonia. *Forestry Studies: Metsanduslikud Uurimused*, *46*, 35–44.

Kalarikkal, N., Thomas, S., & Koshy, O. (Eds.). (2018). Nanomaterials. Oakville, Canada: Apple Academic Press.

Kalidindi, S. B., & Jagirdar, B. R. (2012). Nanocatalysis and Prospects of Green Chemistry. *ChemSusChem*, *5*(1), 65–75. doi:10.1002/cssc.201100377 PMID:22190344

Kaliuzhny, S. (2004). High-intensive anaerobic industrial wastewaters treatment technologies. *Cataysis in Industry*, *6*, 42-50.

Kalliopi, Petros-Achillefs, Anne-Katrin, Jan, Sandra, & Franz (2016). *Delivering on EU Food Safety and Nutrition in 2050 – Future challenges and policy preparedness.* Publications Office of the European Union.

Kallis, A., Sims, A., Tammik, A., Roose, A., Türkson, C.-T., Kupri, H.-L., . . . Jurtom, T. (2017). *Estonia's Seventh National Communication Under the United Nations Framework Convention on Climate Change*. Tallinn: Ministry of the Environment.

Kamarudin, M. Z. F., Kamarudin, S. K., Masdar, M. S., & Daud, W. R. W. (2013). Direct ethanol fuel cells. *International Journal of Hydrogen Energy*, *38*(22), 9438–9453. doi:10.1016/j.ijhydene.2012.07.059

Kamiya, T., & Hirotsuji, J. (1998). New combined system of biological process and intermittent ozonation for advanced wastewater treatment. *Water Science and Technology*, *38*(8-9), 145–153, 147–153. doi:10.2166/wst.1998.0801

Kannan, R., Lakshmi, S., Aparna, N., Prabhakar, S., Thilagaraj, W. R. (2016). Article. *International Journal of Industrial Chemistry*. DOI doi:10.100740090-016-0088-6

Kansal, S. K., Singh, M., & Sud, D. (2007). Studies on photodegradation of two commercial dyes in aqueous phase using different photocatalysts. *Journal of Hazardous Materials*, *141*(3), 581–590. doi:10.1016/j.jhazmat.2006.07.035 PMID:16919871

Kapoor, S., Lawless, D., Kennepohl, P., Meisel, D., & Serpone, N. (1994). Reduction and Aggregation of Silver Ions in Aqueous Gelatin Solutions. *Langmuir*, *10*(9), 3018–3022. doi:10.1021/la00021a026

Karak, T., Paul, R. K., Das, D. K., & Boruah, R. K. (2016). Complexation of DTPA and EDTA with Cd 2+: Stability constants and thermodynamic parameters at the soil-water interface. *Environmental Monitoring and Assessment*, *188*(12), 670. doi:10.100710661-016-5685-5 PMID:27848112

Karami, H., Moosakhani, Z., & Goli, F. (2017). Walnut skin nanoparticles as a novel and fast nano-biosorbent for the fast removal of lead ions from polluted water. *Desalination and Water Treatment*, *60*, 291–300. doi:10.5004/dwt.2017.0088

Karami, H., Yaghoobi, F., & Goli, F. (2017). Application of sawdust nanoparticles as a cheap and nontoxic biosorbent for the removal of lead ions from polluted waters. *Desalination and Water Treatment*, *58*, 120–128. doi:10.5004/dwt.2017.1754

Karnib, M., Kabbani, A., Holail, H., & Olama, Z. (2014). *Heavy Metals Removal Using Activated Carbon* (Vol. 50). Silica and Silica Activated Carbon Composite.

Kashkin, V.B., Rubleva, T.V., & Hlebopros, R.G. (2015). *Stratosfernyi ozone: vid s kosmicheskoi orbity* [Stratospheric ozone: a view from orbit]. Krasnoyarsk: SFU.

Katiyar, N., & Balasubramanian, K. (2014). Thermal modelling of hybrid composites of nano cenosphere and polycarbonate for a thermal protection system. *RSC Advances*, *4*(88), 47529–47535. doi:10.1039/C4RA07973F

Katunina, E. (2015). Kinetic parameters of sea water quality in assessing the ecological state of the Lakspi Bay water area in the Black Sea. In Abstract Book on Modern problems of hydrochemistry and monitoring of surface water quality. Rostov on Don, Russia: Academic Press.

Kaur, N., & Singh, V. (2017). Current status and future challenges in ionic liquids, functionalized ionic liquids and deep eutectic solvent-mediated synthesis of nanostructured TiO2: A review. *New Journal of Chemistry*, *41*(8), 2844–2868. doi:10.1039/C6NJ04073J

Kaur, P., Shorey, L. E., Ho, E., Dashwood, R. H., & Williams, D. E. (2013). The epigenome as a potential mediator of cancer and disease prevention in prenatal development. *Nutrition Reviews*, *71*(7), 441–457. doi:10.1111/nure.12030 PMID:23815143

Kay, J. W., Steven, R. J., McGuigan, J. A. S., & Elder, H. Y. (2008). Automatic determination of ligand purity and apparent dissociation constant (Kapp) in Ca^{2+}/Mg^{2+} buffer solutions and the Kapp for Ca^{2+}/Mg^{2+} anion binding in physiological solutions from Ca^{2+}/Mg^{2+} - macroelectrode measurements. *Computers in Biology and Medicine*, *38*, 101–110. doi:10.1016/j.compbiomed.2007.07.008 PMID:17854794

Kazahkstan. (2019). Retrieved from https://en.wikipedia.org/wiki/Kazakhstan

Kazemi, M., Jahanshahi, M., & Peyravi, M. (2018). Chitosan-sodium alginate multilayer membrane developed by Fe0@WO3 nanoparticles: Photocatalytic removal of hexavalent chromium. *Carbohydrate Polymers*, *198*, 164–174. doi:10.1016/j.carbpol.2018.06.069 PMID:30092986

Kearney, G., & Kiros, G. E. (2009). A spatial evaluation of socio demographics surrounding National Priorities List sites in Florida using a distance-based approach. *International Journal of Health Geographics*, *8*(1), 33. doi:10.1186/1476-072X-8-33 PMID:19531266

Kędziora, A., & Sobik, K. (2013). Oporność bakterii na srebro- problem stary czy nowy? *Kosmos (Stockholm)*, *62*(4), 557–570.

Keesstra, S., Geissen, V., Mosse, K., Piiranen, S., Scudiero, E., Leistra, M., & van Schaik, L. (2012). Soil as a filter for groundwater quality. *Current Opinion in Environmental Sustainability*, *4*(5), 507–516. doi:10.1016/j.cosust.2012.10.007

Kenessariyev, U., Golub, A., Brod, M., Dosmukhametov, A., Amrin, M., Erzhanova, A., & Kenessary, D. (2013). Human Health Cost of Air Pollution in Kazakhstan. *Journal of Environmental Protection*, *4*(08), 869–876. doi:10.4236/jep.2013.48101

Kenworthy & Laube. (2002). Urban Transport Patterns in a Global Sample of Citiesand their Linkages to Transport Infrastructure, Land Use, Economics and Environment. *World Transport Policy and Practice*, *3*(3).

Keshvardoostchokami, M., Babaei, L., Zamani, A. A., Parizanganeh, A. H., & Piri, F. (2017). Global. *Journal of Environmental Science and Management*, *3*, 267–278.

Keszei, A. P., Goldbohm, R. A., Schouten, L. J., Jakszyn, P., & van den Brandt, P. A. (2013). Dietary N-nitroso compounds, endogenous nitrosation, and the risk of esophageal and gastric cancer subtypes in the Netherlands Cohort Study. *The American Journal of Clinical Nutrition*, *97*(1), 135–146. doi:10.3945/ajcn.112.043885 PMID:23193003

Keum, Y.-S., & Li, Q. X. (2005). Reductive debromination of polybrominated diphenyl ethers by zerovalent iron. *Environmental Science & Technology*, *39*(7), 2280–2286. doi:10.1021/es048846g PMID:15871265

Khalid, A., Arshad, M., Anjum, M., Mahmood, T., & Dawson, L. (2011). The anaerobic digestion of solid organic waste. *Waste Management (New York, N.Y.)*, *31*(8), 1737–1744. doi:10.1016/j.wasman.2011.03.021 PMID:21530224

Khallaf, H., Oladeji, I. O., & Chow, L. (2008). Optimization of chemical bath deposited CdS thin films using nitrilotriacetic acid as a complexing agent. *Thin Solid Films*, *516*(18), 5967–5973. doi:10.1016/j.tsf.2007.10.079

Khan, I., Saeed, K., & Khan, I. (2017). Nanoparticles: Properties, applications and toxicities. *Arabian Journal of Chemistry*. doi:10.1016/j.arabjc.2017.05.011

Kharytonov, M. M., Mitsik, O., Stankevich, S. A., & Titarenko, O. V. (2013). Geomining Sate ecological assessment and reclamation along costal line of the Kerch peninsula. In A. Veziroglu & Tsitskishvili (Ed.), Black Sea Energy Resource Development and Hydrogen Energy Problems (pp. 325-336). Dordrecht: Springer.

Khatri, N., & Tyagi, S. (2015). Influences of natural and anthropogenic factors on surface and groundwater quality in rural and urban areas. *Frontiers in Life Science*, *8*(1), 23–39. doi:10.1080/21553769.2014.933716

Khoramzadeh, E., Nasernejad, B., & Halladj, R. (2013). *Mercury biosorption from aqueous solutions by Sugarcane Bagasse* (Vol. 44). Academic Press.

Kilic, M., Kırbıyık, Ç., Çepelioğullar Mutlu, Ö., & Pütün, A. (2013). *Adsorption of heavy metal ions from aqueous solutions by bio-char, a by-product of pyrolysis* (Vol. 283). Academic Press.

Kim, K. J., Sung, W. S., Moon, S. K., Choi, J. S., Kim, J. G., & Lee, D. G. (2008). Antifungal effect of silver nanoparticles on dermatophytes. *Journal of Microbiology and Biotechnology, 18*, 1482–1484. PMID:18756112

Kim, K.-S., Demberelnyamba, D., & Lee, H. (2004). Size-Selective Synthesis of Gold and Platinum Nanoparticles Using Novel Thiol-Functionalized Ionic Liquids. *Langmuir, 20*(3), 556–560. doi:10.1021/la0355848 PMID:15773074

Kleinman, L., Lee, Y.-N., Springston, S. R., Nunnermacker, L., Zhou, X., Brown, R., ... Newman, L. (1994). Ozone formation at a rural site in the southeastern United States. *Journal of Geophysical Research, 99*(D2), 3469–3482. doi:10.1029/93JD02991

Kloster, G. A., Mosiewicki, M. A., & Marcovich, N. E. (2019). Chitosan/iron oxide nanocomposite films: Effect of the composition and preparation methods on the adsorption of congo red. *Carbohydrate Polymers, 221*, 186–194. doi:10.1016/j.carbpol.2019.05.089 PMID:31227157

Koch, H., & Angerer, J. (2012). Phthalates: Biomarkers and Human Biomonitoring. In Biomarkers and Human Biomonitoring, Volume 1: Ongoing Programs and Exposures. Royal Society of Chemistry.

Kolay, P., & Singh, D. (2001). *Physical, chemical, mineralogical, and thermal properties of cenospheres from an ash lagoon* (Vol. 31). Academic Press.

Komari, N. P. (1984). *Chemical metrology. Ionic heterogeneous equilibria*. Kharkov: Vishcha Shkola. (in Russian)

Kong, X., Yang, B., Xiong, H., Zhou, Y., Xue, S., Xu, B.-Q., & Wang, S.-X. (2014). *Selective removal of heavy metal ions from aqueous solutions with surface functionalized silica nanoparticles by different functional groups* (Vol. 21). Academic Press.

Korolyova, Yu. (2010) The bioindication of heavy metal precipitation in the Kaliningrad region. *Bulletin of Kant Russian State University, 7,* 39-44.

Korte, F., Bahadir, M., & Klein, W. (1997). Ecological Chemistry. In Basics and Concepts. Mir.

Kösta, H., & Liiv, S. (n.d.). *Spatial and temporal trends of heavy metal accumulation in mosses in Estonia*. Paper presented at the 8th International Conference on Ecosystems and Sustainable Development, Alicante, Spain. 10.2495/ECO110121

Kostuychanko, Y. V., Bilous, Yu., Movchan, D., Laszlo, M., & Kopachevsky, I. (2013). Toward methodology of satellite observation utilization for agricultural production risk assessment. IERI Procedia, 1, 1-6.

Kostyuchenko, Y. V., Movchan, D., Artemenko, I., & Kopachevsky, I. (2016). Stochastic Approach to Uncertainty Control in Multiphysics Systems: Modeling of Carbon Balance and Analysis of GHG Emission Using Satellite Tools. In R. Mangey & J. Paulo Davim (Ed.), Mathematical Contepts and Applications in Mechanical Engineering and Mechatronics (pp. 350-378). IGI Global. DOI: 10.4018/978-1-5225-1639-2.ch017

Kotrba, P., Mackova, M., & Macek, T. (2011). *Microbial Biosorption of Metals*. Academic Press.

Kouassi, A. M., Zika, R. G., & Plane, J. M. C. (1990). Light-induced alteration of the photophysical properties of dissolved organic matter in seawater Part II. Estimates of the environmental rates of the natural water fluorescence. *Netherlands Journal of Sea Research, 27*(1), 33–41. doi:10.1016/0077-7579(90)90032-C

Kovac, L., Luptacik, P., Miklisova, D., & Mati, R. (2001). Soil Oribatida and Collembola communitias across a land depression in an arable field. *European Journal of Soil Biology*, *37*(4), 285–289. doi:10.1016/S1164-5563(01)01106-2

Kowshik, M., Ashtaputre, S., Kharrazi, S., Vogel, W., Urban, J., Kulkarni, S. K., & Paknikar, K. M. (2003). Extracellular synthesis of silver nanoparticles by a silver-tolerant yeast strain MKY3. *Nanotechnology*, *14*(1), 95–100. doi:10.1088/0957-4484/14/1/321

Krämer, J., Redel, E., Thomann, R., & Janiak, C. (2008). Use of Ionic Liquids for the Synthesis of Iron, Ruthenium, and Osmium Nanoparticles from Their Metal Carbonyl Precursors. *Organometallics*, *27*(9), 1976–1978. doi:10.1021/om800056z

Krimsky, S. (2019). Glyphosate Toxicology: What We Can Learn From the Current Controversy. In P. Papadopoulou, C. Marouli, & A. Misseyanni (Eds.), *Environmental Exposures and Human Health Challenges* (pp. 343–356). Hershey, PA: IGI Global. doi:10.4018/978-1-5225-7635-8.ch015

Krimsky, S. (2019a). *GMOs Decoded: A Skeptic's View of Genetically Modified Foods*. Cambridge, MA: The MIT Press. doi:10.7551/mitpress/10824.001.0001

Krishna, R., Kumar Chintalpudi, V., & Muddada, S. (2018). *Application of Biosorption for Removal of Heavy Metals from Wastewater*. Academic Press.

Krishnani, K., Meng, X., Christodoulatos, C., & Boddu, V. (2008). *Biosorption Mechanism of Nine Different Heavy Metals onto Biomatrix from Rice Husk* (Vol. 153). Academic Press.

Kristie, L. E., Semenza, J. C., & Rocklöv, J. (2016). Current medical research funding and frameworks are insufficient to address the health risks of global environmental change. *Environmental Health*, *15*(1), 108. doi:10.118612940-016-0183-3 PMID:27835959

Kroto, H. (1993). *The Birth of C60: Buckminsterfullerene*. Paper presented at the Electronic Properties of Fullerenes, Berlin, Germany.

Krumins, V., Park, J. W., Son, E. K., Rodenburg, L. A., Kerkhof, L. J., Häggblom, M. M., & Fennell, D. E. (2009). PCB dechlorination enhancement in Anacostia River sediment microcosms. *Water Research*, *43*(18), 4549–4558. doi:10.1016/j.watres.2009.08.003 PMID:19744693

Kuang, Y., Du, J., Zhou, R., Chen, Z., Megharaj, M., & Naidu, R. (2015). Calcium alginate encapsulated Ni/Fe nanoparticles beads for simultaneous removal of Cu (II) and monochlorobenzene. *Journal of Colloid and Interface Science*, *447*, 85–91. doi:10.1016/j.jcis.2015.01.080 PMID:25700214

Kulkova, M. A. (2011). Radiocarbon (14C) in the environment and the radiocarbon Dating method. In *Training Handbook*. Izdatel'stvo RGPU im.

Kumar, A., & Ghosh, A. K. (2019). Arsenic and Cancer. In P. Papadopoulou, C. Marouli, & A. Misseyanni (Eds.), *Environmental Exposures and Human Health Challenges* (pp. 106–132). Hershey, PA: IGI Global. doi:10.4018/978-1-5225-7635-8.ch005

Kumar, A., & Schweizer, H. (2005). Bacterial resistance to antibiotics: Active efflux and reduced uptake. *Advanced Drug Delivery Reviews*, *57*(10), 1486–1513. doi:10.1016/j.addr.2005.04.004 PMID:15939505

Kumar, A., Subash, B., Krishnakumar, B., Sobral, A. J. F. N., & Sankaran, K. R. (2016). Synthesis, characterization and excellent catalytic activity of modified ZnO photocatalyst for RR 120 dye degradation under UV-A and solar light illumination. *Journal of Water Process Engineering*, *13*, 6–15. doi:10.1016/j.jwpe.2016.07.009

Kumar, S. S., Venkateswarlu, P., Ranga Rao, V., & Nageswara Rao, G. (2013). Synthesis, characterization and optical properties of zinc oxide nanoparticles. *International Nano Letters*, *30*, 1–6.

Kumwenda, S., Samanyika, Y., Chingaipe, E., Mamba, K., Lungu, K., Mwendera, C., & Kalulu, K. (2014). The Emerging Environmental Health Risks and Challenges for Tomorrow: Prospects for Malawi. *Malawi Journal of Applied Sciences and Innovation*, *1*(1), 21–26.

Kuo-Chuan, H., Lin, C.-Y., Lai, Y.-H., & Chen, H.-W. (2011). Highly efficient dye-sensitized solar cell with a ZnO nanosheet-based photoanode. *Energy & Environment*, *4*(9), 3448–3455. doi:10.1039/c0ee00587h

Kupatadze, K., & Kiziloz, B. (2016). Natural treatment systems from the point of didactics. *Periódico Tchê Química*, *26*, 69.

Kupatadze, K., & Malazonia, D. (2016). The Overview of Chemical knowledge in Eighteenth Century Georgia according to the Chemistry Book by King Vakhtang VI. Orbital: The Electronic. *Journal of Chemistry*, *8*(3), 204–210.

Kurian, M., & Nair, D. S. (2015). Heterogeneous Fenton behavior of nano nickel zinc ferrite catalysts in the degradation of 4-chlorophenol from water under neutral conditions. *Journal of Water Process Engineering*, *8*, e37–e49. doi:10.1016/j.jwpe.2014.10.011

Kurian, M., & Nair, D. S. (2016). Effect of preparation conditions on nickel zinc ferrite nanoparticles: A comparison between sol–gel auto combustion and co-precipitation methods. *Journal of Saudi Chemical Society*, *20*, S517–S522. doi:10.1016/j.jscs.2013.03.003

Kurian, M., Nair, D. S., & Rahnamol, A. M. (2014). Influence of the synthesis conditions on the catalytic efficiency of NiFe 2 O 4 and ZnFe 2 O 4 nanoparticles towards the wet peroxide oxidation of 4-chlorophenol. *Reaction Kinetics, Mechanisms and Catalysis*, *111*(2), 591–604. doi:10.100711144-013-0667-x

Kwok, K. C. M., Koong, L. F., Ansari, T. A., & McKay, G. (2018). Adsorption/desorption of arsenite and arsenate on chitosan and nanochitosan. *Environmental Science and Pollution Research International*, *25*(15), 14734–14742. doi:10.100711356-018-1501-9 PMID:29536422

Kyzas, G. Z., & Matis, K. A. (2015). Nanoadsorbents for pollutants removal: A review. *Journal of Molecular Liquids*, *203*, 159–168. doi:10.1016/j.molliq.2015.01.004

La Guardia, M. J., Hale, R. C., & Harvey, E. (2006). Detailed polybrominated diphenyl ether (PBDE) congener composition of the widely used penta-, octa-, and deca-PBDE technical flame-retardant mixtures. *Environmental Science & Technology*, *40*(20), 6247–6254. doi:10.1021/es060630m PMID:17120549

La Merrill, M., & Birnbaum, L. S. (2011). Childhood obesity and environmental chemicals. *The Mount Sinai Journal of Medicine, New York*, *78*(1), 22–48. doi:10.1002/msj.20229 PMID:21259261

Lagadic, I., Mitchell, M., & Payne, B. (2001). *Highly Effective Adsorption of Heavy Metal Ions by a Thiol-Functionalized Magnesium Phyllosilicate Clay* (Vol. 35). Academic Press.

Lamon, L., Aschberger, K., Asturiol, D., Richarz, A., & Worth, A. (2019, January 2). Grouping of nanomaterials to read-across hazard endpoints: a review. *Nanotoxicology*.

Lam, S. M., Sin, J. C., Abdullah, A. Z., & Mohamed, A. R. (2012). Degradation of wastewaters containing organic dyes photocatalysed by zinc oxide: A review. *Desalination and Water Treatment*, *41*(1-3), 131–169. doi:10.1080/19443994.2012.664698

Lancelot, A., Clavería-Gimeno, R., Velázquez-Campoy, A., Abian, O., Serrano, J. L., & Sierra, T. (2017). Nanostructures based on ammonium-terminated amphiphilic Janus dendrimers as camptothecin carriers with antiviral activity. *European Polymer Journal*, *90*, 136–149. doi:10.1016/j.eurpolymj.2017.03.012

Lara-Vásquez, E. J., Solache-Ríos, M., & Gutiérrez-Segura, E. (2016). Malachite green dye behaviors in the presence of biosorbents from maize (Zea mays L.), their Fe-Cu nanoparticles composites and Fe-Cu nanoparticles. *Journal of Environmental Chemical Engineering*, *4*(2), 1594–1603. doi:10.1016/j.jece.2016.02.008

Larkin, A., & Hystad, P. (2017). Towards Personal Exposures: How Technology Is Changing Air Pollution and Health Research. *Current Environmental Health Reports*, *4*(4), 463–471. doi:10.100740572-017-0163-y PMID:28983874

Larsen, P. S., Kamper-Jørgensen, M., Adamson, A., Barros, H., Bonde, J. P., Brescianini, S., ... Andersen, A.-M. N. (2013). Pregnancy and birth cohort resources in Europe: A large opportunity for aetiological child health research. *Paediatric and Perinatal Epidemiology*, *27*(4), 393–414. doi:10.1111/ppe.12060 PMID:23772942

Larssen, T., Jiling, X., Vogt, R. D., Seip, H. M., Bohan, L., & Dianwu, Z. (1998). Studies of soils, soil water and stream water at a small catchment near Guiyang, China. *Water, Air, and Soil Pollution*, *101*(1-4), 137–162. doi:10.1023/A:1004985209931

Lastochkin, A. N. (2016a). *Foundations of the general theory of geosystems: a textbook in 2 parts. Part 1*. Russia, St. Petersburg: SPbGU.

Lastochkin, A. N. (2016b). *Foundations of the general theory of geosystems: a textbook in 2 parts. Part 2*. Russia, St. Petersburg: SPbGU.

Lata, S., & Samadder, S. R. (2016). Removal of arsenic from water using nano adsorbents and challenges: A review. *Journal of Environmental Management*, *166*, 387–406. doi:10.1016/j.jenvman.2015.10.039 PMID:26546885

Laurian, L. (2008). Environmental Injustice in France. *Journal of Environmental Planning and Management*, *51*(1), 55–79. doi:10.1080/09640560701712267

Lavric, M., & Ceban, E. (2011). Ecological problems of Bic river. *Mediul Ambiant*, *2*(56), 21–23.

Lebeau, T. (2011). Bioaugmentation for in situ soil remediation: How to ensure the success of such a process. In A. Singh, N. Parmar, & R. C. Kuhad (Eds.), *Bioaugmentation, Biostimulation, and Biocontrol* (pp. 129–186). Heidelberg, Germany: Springer Berlin Heidelberg. doi:10.1007/978-3-642-19769-7_7

Lee, L. K., & He, J. (2010). Reductive debromination of polybrominated diphenyl ethers by anaerobic bacteria from soils and sediments. *Applied and Environmental Microbiology*, *76*(3), 794–802. doi:10.1128/AEM.01872-09 PMID:20008168

Lee, P. C., & Meisel, D. (1982). Adsorption and surface-enhanced Raman of dyes on silver and gold sols. *The Journal of Chemical Physics*, *86*(17), 3391–3395. doi:10.1021/j100214a025

Léger, B., Denicourt-Nowicki, A., Olivier-Bourbigou, H., & Roucoux, A. (2008). Rhodium nanocatalysts stabilized by various bipyridine ligands in nonaqueous ionic liquids: Influence of the bipyridine coordination modes in arene catalytic hydrogenation. *Inorganic Chemistry*, *47*(19), 9090–9096. doi:10.1021/ic8010713 PMID:18781733

Lei, Z., Chen, B., Koo, Y.-M., & MacFarlane, D. R. (2017). Introduction: Ionic Liquids. *Chemical Reviews*, *117*(10), 6633–6635. doi:10.1021/acs.chemrev.7b00246 PMID:28535681

Lemming, G., Hauschild, M. Z., & Bjerg, P. L. (2010). Life cycle assessment of soil and groundwater remediation technologies: Literature review. *The International Journal of Life Cycle Assessment*, *15*(1), 115–127. doi:10.100711367-009-0129-x

Leung, C. C., Lam, T. H., Ho, K. S., Yew, W. W., Tam, C. M., Chan, W. M., ... Au, K. F. (2010). Passive smoking and tuberculosis. *Archives of Internal Medicine*, *170*(3), 287–292. doi:10.1001/archinternmed.2009.506 PMID:20142576

Leung, W.-H., Lo, W.-H., & Chan, P.-H. (2015). Amyloid fibrils as rapid and efficient nano-biosorbents for removal of dye pollutants. *RSC Advances*, *5*(109), 90022–90030. doi:10.1039/C5RA17182B

Levei, E., Senila, M., Miclean, M., Roman, C., Abraham, B., & Cordos, E. (2008, May). Surface water pollution with heavy metals in Baia Mare mining basin. *Proceedings of the 15 th International Soils Conservation Organization (ISCO) Congress.*

Levei, E., Senila, M., Miclean, M., Abraham, B., Roman, C., Stefanescu, L., & Moldovan, O. T. (2011). Influence of Rosia Poieni and Rosia Montana mining areas on the water quality of the Aries River. *Environmental Engineering and Management Journal*, *10*(1), 23–29. doi:10.30638/eemj.2011.004

Lewis, R. W., Bertsch, P. M., & McNear, D. H. (2019, March 16). Nanotoxicity of engineered nanomaterials (ENMs) to environmentally relevant beneficial soil bacteria–a critical review. *Nanotoxicology.*

Li, C., Wang, X., Meng, D., Zhou, L. (2018). Article. *International Journal of Biological Macromolecules, 107*(Part B), 1871-1878.

Li, J., Gan, J., Wu, L., & Yi, J. (2016). *Preparation of Fly ash Based Adsorbents for Removal Active Red X-3B from Dying Wastewater* (Vol. 67). Academic Press.

Liang, D., Siegert, M., Liu, X., Lu, Q., Xu, G., Qiu, L., ... Yu, L. (2018). Electron transport chains in organohalide-respiring bacteria and bioremediation implications. *Biotechnology Advances*, *36*(4), 1194–1206. doi:10.1016/j.biotechadv.2018.03.018 PMID:29631017

Lian, J. J., Huang, Y. G., Chen, B., Wang, S. S., Wang, P., Niu, S. P., & Liu, Z. L. (2017). Article. *Water Science and Technology*, *3*, 859–868.

Li, G., Deng, L., Liu, J., Cao, Y., Zhang, H., & Ran, J. (2015). A New Technique for Removing Unburned Carbon from Coal Fly Ash at an Industrial Scale. *International Journal of Coal Preparation and Utilization*, *35*(5), 273–279. doi:10.1080/19392699.2015.1008098

Li, H., Qiu, Y. F., Wang, X. L., Yang, J., Yu, Y. J., Chen, Y. Q., & Liu, Y. (2017). Biochar supported Ni/Fe bimetallic nanoparticles to remove 1, 1, 1-trichloroethane under various reaction conditions. *Chemosphere*, *169*, 534–541. doi:10.1016/j.chemosphere.2016.11.117 PMID:27898326

Liiv, S., & Kaasik, M. (2004). Trace metals in mosses in the Estonian oil shale processing region. *Journal of Atmospheric Chemistry*, *49*(1-3), 563–578. doi:10.100710874-004-1266-z

Li, J., Jin, P., Dai, W., Wang, C., Li, R., Wu, T., & Tang, C. (2017). Excellent performance for water purification achieved by activated porous boron nitride nanosheets. *Materials Chemistry and Physics*, *196*, 186–193. doi:10.1016/j.matchemphys.2017.02.049

Lillesand, T. M., Kiefer, R. W., & Chipman, J. W. (2015). *Remote sensing and image interpretation* (7th ed.). Hoboken, NJ: John Wiley & Sons, Inc.

Lin, D., Liu, J., Hu, Z., Cottrell, J. E., & Kass, I. S. (2018). Neonatal anesthesia exposure impacts brain microRNAs and their associated neurodevelopmental processes. *Scientific Reports*, *8*(1), 10656. doi:10.103841598-018-28874-0 PMID:30006558

Lin, L., Wang, H., Luo, H., & Xu, P. (2015). Enhanced photocatalysis using side-glowing optical fibers coated with Fe-doped TiO2 nanocomposite thin films. *Journal of Photochemistry and Photobiology A Chemistry*, *307*, 88–98. doi:10.1016/j.jphotochem.2015.04.010

Lin, S. T., Thirumavalavan, M., Jiang, T. Y., & Lee, J. F. (2014). Synthesis of ZnO/Zn nano photocatalyst using modified polysaccharides for photodegradation of dyes. *Carbohydrate Polymers*, *105*, 1–9. doi:10.1016/j.carbpol.2014.01.017 PMID:24708945

Lin, S., Huang, R., Cheng, Y., Liu, J., Lau, B. L., & Wiesner, M. R. (2013). Silver nanoparticle-alginate composite beads for point-of-use drinking water disinfection. *Water Research, 47*(12), 3959–3965. doi:10.1016/j.watres.2012.09.005 PMID:23036278

Lioy, P. J., & Rappaport, S. M. (2011). Exposure science and the exposome: An opportunity for coherence in the environmental health sciences [Editorial]. *Environmental Health Perspectives, 119*(11), A466–A467. doi:10.1289/ehp.1104387 PMID:22171373

Li, Q., Mahendra, S., Lyon, D. Y., Brunet, L., Liga, M. V., Li, D., & Alvarez, P. J. (2008). Antimicrobial nanomaterials for water disinfection and microbial control: Potential applications and implications. *Water Research, 42*(18), 4591–4602. doi:10.1016/j.watres.2008.08.015 PMID:18804836

Lis, A., Duca, Gh., Bunduchi, E., Gladchi, V., & Goreaceva, N. (2010). The study of the buffering capacity of several water objects in the Republic of Moldova. *Chemistry Journal of Moldova. Series General. Industrial and Ecological Chemistry.*, *5*(2), 30–36.

Litchfield, M. H. (1989). *Toxicology evaluation of pesticides in drinking water*. Brussels: GIFAP.

Lithuania's seventh National Communication. (2017). Ministry of Environment of the Republic of Lithuania. Retrieved from https://unfccc.int/files/national_reports/national_communications_and_biennial_reports/application/pdf/142035_lithuania-nc7-1-7th_national_communication.pdf

Liu, C., Omer, A. M., & Ouyang, X.-K. (2018). Adsorptive removal of cationic methylene blue dye using carboxymethyl cellulose/k-carrageenan/activated montmorillonite composite beads: Isotherm and kinetic studies. *International Journal of Biological Macromolecules, 106*, 823–833. doi:10.1016/j.ijbiomac.2017.08.084 PMID:28834705

Liu, G., Li, G., Qiu, X., & Li, L. (2009). Synthesis of ZnO/titanate nanocomposites with highly photocatalytic activity under visible light irradiation. *Journal of Alloys and Compounds, 481*(1-2), 492–497. doi:10.1016/j.jallcom.2009.03.021

Liu, H., Gong, C., Wang, J., Liu, X., Liu, H., Cheng, F., ... Wen, S. (2016). Chitosan/silica coated carbon nanotubes composite proton exchange membranes for fuel cell applications. *Carbohydrate Polymers, 136*, 1379–1385. doi:10.1016/j.carbpol.2015.09.085 PMID:26572483

Liu, W., Xu, Y., Zhou, W., Zhang, X., Cheng, X., Zhao, H., ... Huo, L. (2017). A Facile Synthesis of Hierarchically Porous TiO_2 Microspheres with Carbonaceous Species for Visible-light Photocatalysis. *Journal of Materials Science and Technology, 33*(1), 39–46. doi:10.1016/j.jmst.2016.04.007

Liu, Y., Deng, J., Xie, S., Wang, Z., & Dai, H. (2016). Catalytic removal of volatile organic compounds using ordered porous transition metal oxide and supported noble metal catalysts. *Chinese Journal of Catalysis, 37*(8), 1193–1205. doi:10.1016/S1872-2067(16)62457-9

Liu, Z., Gan, L. M., Hong, L., Chen, W., & Lee, J. Y. (2005). Carbon-supported Pt nanoparticles as catalysts for proton exchange membrane fuel cells. *Journal of Power Sources, 139*(1-2), 73–78. doi:10.1016/j.jpowsour.2004.07.012

Li, W., Ju, B., & Zhang, S. (2019). Preparation of cysteamine-modified cellulose nanocrystal adsorbent for removal of mercury ions from aqueous solutions. *Cellulose (London, England), 26*(8), 4971–4985. doi:10.100710570-019-02420-1

Li, W., Zhang, Y., Tian, G., Xie, S., Xu, Q., Wang, L., ... Bu, Y. (2016). Fabrication of graphene-modified nano-sized red phosphorus for enhanced photocatalytic performance. *Journal of Molecular Catalysis A Chemical, 423*, 356–364. doi:10.1016/j.molcata.2016.07.039

Li, X., Yang, L., Zhou, Q., Qi, T., Liu, G., & Peng, Z. (2020). A split-combination method for estimating the thermodynamic properties (Gfo and Hfo) of multicomponent minerals. *Applied Clay Science, 185*, 105406. doi:10.1016/j.clay.2019.105406

Li, X., Zhou, H., Wu, W., Wei, S., Xu, Y., & Kuang, Y. (2015). Studies of heavy metal ion adsorption on Chitosan/Sulfydryl-functionalized graphene oxide composites. *Journal of Colloid and Interface Science, 448*, 389–397. doi:10.1016/j.jcis.2015.02.039 PMID:25746192

Li, Y. F., & Liu, Z. P. (2011). Particle size, shape and activity for photocatalysis on titania anatase nanoparticles in aqueous surroundings. *Journal of the American Chemical Society, 133*(39), 15743–15752. doi:10.1021/ja206153v PMID:21879719

Li, Y., Guo, W., Wu, Z., Jin, L., Ke, Y., Guo, Q., & Hu, S. (2016). Determination of ultra-trace rare earth elements in high-salt groundwater using aerosol dilution inductively coupled plasma-mass spectrometry (ICP-MS) after iron hydroxide co-precipitation. *Microchemical Journal, 126*, 194–199. doi:10.1016/j.microc.2015.12.006

Li, Y., & Ju, D. (2017). The Application, Neurotoxicity, and Related Mechanism of Cationic Polymers. In X. Jiang & H. Gao (Eds.), *Neurotoxicity of Nanomaterials and Nanomedicine* (pp. 285–329). Academic Press. doi:10.1016/B978-0-12-804598-5.00012-X

Loh, M., Sarigiannis, D., Gotti, A., Karakitsios, S., Pronk, A., Kuijpers, E., ... Cherrie, J. W. (2017). How Sensors Might Help Define the External Exposome. *International Journal of Environmental Research and Public Health, 14*(4), 434. doi:10.3390/ijerph14040434 PMID:28420222

Long, H., Li, X., Wang, H., & Jia, J. (2013). Biomass resources and their bioenergy potential estimation: A review. *Renewable & Sustainable Energy Reviews, 26*, 344–352. doi:10.1016/j.rser.2013.05.035

Long, W., & Li, F. (2011). Hydrothermal synthesis of ZnO flower-reduced graphene oxide composite for electrochemical determination of ascorbic acid. *Energy & Environment, 4*, 3448–3455.

López Alonso, M., Prieto Montaña, F., Miranda, M., Castillo, C., Hernández, J., & Luis Benedito, J. (2004). Interactions between toxic (As, Cd, Hg and Pb) and nutritional essential (Ca, Co, Cr, Cu, Fe, Mn, Mo, Ni, Se, Zn) elements in the tissues of cattle from NW Spain. *Biometals, 17*(4), 389–397. doi:10.1023/B:BIOM.0000029434.89679.a2 PMID:15259359

Lotfy, S. (2009). Controlling degradation of low-molecular-weight natural polymer "dextrin" using gamma irradiation. *International Journal of Biological Macromolecules, 44*(1), 57–63. doi:10.1016/j.ijbiomac.2008.10.002 PMID:18996409

Lozano, V. L., Defarge, N., Rocque, L. M., Mesnage, R., Hennequin, D., Cassier, R., ... Amiel, C. (2018). Sex-dependent impact of Roundup on the rat gut microbiome. *Toxicology Reports, 5*, 96–107. doi:10.1016/j.toxrep.2017.12.005 PMID:29854581

Lubick, N., & Betts, K. (2008). *Silver socks have cloudy lining| Court bans widely used flame retardant*. Academic Press.

Lu, J., Li, L., Wang, Z., Wen, B., & Cao, J. (2013). Synthesis of visible-light-active TiO_2-based photo-catalysts by a modified sol–gel method. *Materials Letters, 94*, 147–149. doi:10.1016/j.matlet.2012.11.137

Lu, K., Yang, X., Gielen, G., Bolan, N., Ok, Y. S., Niazi, N. K., ... Wang, H. (2017). Effect of bamboo and rice straw biochars on the mobility and redistribution of heavy metals (Cd, Cu, Pb and Zn) in contaminated soil. *Journal of Environmental Management, 186*(part 2), 285–292. doi:10.1016/j.jenvman.2016.05.068 PMID:27264699

Lu, T., Xiang, T., Huang, X.-L., Li, C., Zhao, W.-F., Zhang, Q., & Zhao, C.-S. (2015). Post-crosslinking towards stimuli-responsive sodium alginate beads for the removal of dye and heavy metals. *Carbohydrate Polymers, 133*, 587–595. doi:10.1016/j.carbpol.2015.07.048 PMID:26344317

Lu, Y. C., Wang, L. L., Wang, D. J., Xie, T. F., Chen, L. P., & Lin, Y. H. (2011). A comparative study on plate-like and flower-like ZnO nanocrystals surface photovoltage property and photocatalytic activity. *Materials Chemistry and Physics, 129*(1-2), 281–287. doi:10.1016/j.matchemphys.2011.04.004

Lv, J.-L., Zhai, S.-R., Wang, Z.-Z., Lei, Z.-M., & An, Q.-D. (2016). Carbon–silica composite bio-sorbents with a high density of oxygen-containing sites for efficient methylene blue adsorption. *Research on Chemical Intermediates, 42*(2), 839–854. doi:10.100711164-015-2058-6

Lyalko, V. I., & Sakhatsky, A. I. (2013). The Use of multispectral satellite images for the hydrogeological model certain to evaluate the water balance and groundwater resources. In A. G. Zagorodny & Yu. M. Yermoliev (Ed.), Integrated modeling of food, energy and water security management for sustainable social, economic and environmental developments (pp. 308–319). Akademperiodyka.

Lyalko, V. I., Sakhatsky, A. I., Zholobak, G. M., & Apostolov, A. A. (2016). The research of the dynamics of the annual average of gross primary productivity in Ukraine from 2000-2010. *Ukrainian Journal of Remote Sensing, 9*, 30–35. https://ujrs.org.ua/ujrs/article/view/77/96

Lytras, M. D., & Papadopoulou, P. (2018). *Applying Big Data Analytics in Bioinformatics and Medicine*. Hershey, PA: IGI Global. doi:10.4018/978-1-5225-2607-0

Mackay, D., Shiu, W. Y., Ma, K. C., & Lee, S. C. (2006). Handbook of Physical- Chemical Properties and Environmental Fate for Organic Chemicals.: Vol. II. *Halogenated Hydrocarbons*. Boca Raton, FL: CRC Press Taylor & Francis Group.

Maeda, K., Kondo, A., & Muir Wood, D. (2015). Micro and Macro Modeling of Internal Erosion. In *Geomechanics from Micro to Macro* (pp. 871–882). London: Taylor & Francis Group.

Ma, H., Wang, H., & Na, C. (2015). Microwave-assisted optimization of platinum-nickel nanoalloys for catalytic water treatment. *Applied Catalysis B: Environmental, 163*, 198–204. doi:10.1016/j.apcatb.2014.07.062

Mahaffy, P., Matlin, S., Holme, T., & Mackellar, J. (2019). System thinking for education about the molecular basis of sustainability. *Nature Suitability, 2*, 362.

Mahajan, D., Gütlich, P., Ensling, J., Pandya, K., Stumm, U., & Vijayaraghavan, P. (2003). Evaluation of nanosized iron in slurry-phase Fischer– Tropsch synthesis. *Energy & Fuels, 17*(5), 1210–1221. doi:10.1021/ef0300343

Mahasin, F., Al-Kadhemy, Israa, F., Alsharue, Ali Abid, D., & Al-Zuky. (2011). Analysis of the effect of the concentration of rhodamine B in ethanol on the fluorescence spectrum using the "gauss mod" function. *Journal of Physical Chemistry, 22*(2), 77–86.

Maksimenko, P. Y., Skurlatov, Y. I., & Shtamm, E. V. (2000) Veroyatnaya rol' serosoderzhashchih soedinenij v formirovanii toksicheskih svojstv stochnyh vod cellyulozno-bumazhnogo proizvodstva [The probable role of sulfur-containing compounds in the formation of the toxic properties of wastewater from pulp and paper production]. Vestnik RUDN, Ser. Ekologiya i bezopasnost' zhiznedeyatel'nosti, 4, 63-70.

Ma, L., Peng, Y., Wu, B., Lei, D., & Xu, H. (2013). Article. *Chemical Engineering Journal (Amsterdam, Netherlands), 225*, 59–67.

Malato, S., Fernandez-Ibanez, P., Maldonado, M. I., Blanco, J., & Gernjak, W. (2009). Decontamination and disinfection of water by solar photocatalysis: Recent overview and trends. *Catalysis Today, 147*(1), 1–59. doi:10.1016/j.cattod.2009.06.018

Malik, R., Dahiya, S., & Lata, S. (2017). *An Experimental and Quantum Chemical Study of Removal of Utmostly Quantified Heavy Metals in Wastewater Using Coconut Husk: A Novel Approach to Mechanism* (Vol. 98). Academic Press.

Mallakpour, S., Abdolmaleki, A., & Tabesh, F. (2018). Ultrasonic-assisted manufacturing of new hydrogel nanocomposite biosorbent containing calcium carbonate nanoparticles and tragacanth gum for removal of heavy metal. *Ultrasonics Sonochemistry*, *41*, 572–581. doi:10.1016/j.ultsonch.2017.10.022 PMID:29137788

Mallakpour, S., & Tabesh, F. (2019). Tragacanth gum based hydrogel nanocomposites for the adsorption of methylene blue: Comparison of linear and non-linear forms of different adsorption isotherm and kinetics models. *International Journal of Biological Macromolecules*, *133*, 754–766. doi:10.1016/j.ijbiomac.2019.04.129 PMID:31004645

Mallampati, R., Xuanjun, L., Adin, A., & Valiyaveettil, S. (2015). Fruit Peels as Efficient Renewable Adsorbents for Removal of Dissolved Heavy Metals and Dyes from Water. *ACS Sustainable Chemistry & Engineering*, *3*(6), 1117–1124. doi:10.1021/acssuschemeng.5b00207

Ma, M., Wang, H., Zhen, Q., Zhang, M., & Du, X. (2017). Development of nitrogen-enriched carbonaceous material coated titania nanotubes array as a fiber coating for solid-phase microextraction of ultraviolet filters in environmental water. *Talanta*, *167*, 118–125. doi:10.1016/j.talanta.2017.02.002 PMID:28340701

Mandai, T., Yoshida, K., Ueno, K., Dokko, K., & Watanabe, M. (2014, May 21). Criteria for solvate ionic liquids. *Physical Chemistry Chemical Physics*.

Mandavgane, S., Tajane, S., & Jatav, P. (2018). *A Process of Carbon Enrichment of Bottom Slag Ash for Value-Added Applications*. Academic Press.

Manikkam, M., Tracey, R., Guerrero-Bosagna, C., & Skinner, M. K. (2013). Plastics Derived Endocrine Disruptors (BPA, DEHP and DBP) Induce Epigenetic Transgenerational Inheritance of Obesity, Reproductive Disease and Sperm Epimutations. *PLoS One*, *8*(1), e55387. doi:10.1371/journal.pone.0055387 PMID:23359474

Manilov, A. I., Kozinetz, A. V., Litvinenko, S. V., Skryshevsky, V. A., Al Araimi, M., & Rozhin, A. (2019). Effect of nano-carbon dispersions on signal in silicon-based sensor structure with photoelectrical transducer principle. *Current Applied Physics*, *19*(3), 308–313. doi:10.1016/j.cap.2018.12.012

Manocha, L. M., Ram, K. A., & Manocha, S. (2011). *Separation of Cenospheres from Fly Ashes by Floatation Method* (Vol. 13). Academic Press.

Manousakas, M., Papaefthymiou, H., Eleftheriadis, K., & Katsanou, K. (2014). Determination of water-soluble and insoluble elements in PM2.5 by ICP-MS. *The Science of the Total Environment*, *493*, 694–700. doi:10.1016/j.scitotenv.2014.06.043 PMID:24992462

Manrai, A. K., Cui, Y., Bushel, P. R., Hall, M., Karakitsios, S., Mattingly, C., ... Patel, C. J. (2017). Informatics and Data Analytics to Support Exposome-Based Discovery for Public Health. *Annual Review of Public Health*, *38*(1), 279–294. doi:10.1146/annurev-publhealth-082516-012737 PMID:28068484

Manzoor, Q., Nadeem, R., Iqbal, M., Saeed, R., & Ansari, T. (2013). *Organic acids pretreatment effect on Rosa bourbonia phyto-biomass for removal of Pb(II) and Cu(II) from aqueous media* (Vol. 132). Academic Press.

Mao, C., Feng, Y., Wang, X., & Ren, G. (2015). Review on research achievements of biogas from anaerobic digestion. *Renewable & Sustainable Energy Reviews*, *45*, 540–555. doi:10.1016/j.rser.2015.02.032

Marcelo, C. R., Lopes, R. P., Cruz, J. C., Nascimento, M. A., Silva, A. A., & Lima, C. F. (2016). Evaluation of different parameters on the acetamiprid degradation by bimetallic Fe/Ni nanoparticles. *Separation and Purification Technology*, *171*, 256–262. doi:10.1016/j.seppur.2016.07.032

Marchisio, A., Minella, M., Maurino, V., Minero, C., & Vione, D. (2015). Photogeneration of reactive transient species upon irradiation of natural water samples: Formation quantum yields in different spectral intervals, and implications for the photochemistry of surface waters. *Water Research, 73*, 145–156. doi:10.1016/j.watres.2015.01.016 PMID:25655321

Mariño, M., Lopes da Silva, L., Durán, N., & Tasic, L. (2015). Enhanced materials from nature: Nanocellulose from citrus waste. *Molecules (Basel, Switzerland), 20*(4), 5908–5923. doi:10.3390/molecules20045908 PMID:25854755

Marinoni, N., & Broekmans, M. A. T. M. (2013). Microstructure of selected aggregate quartz by XRD, and a critical review of the crystallinity index. *Cement and Concrete Research, 54*, 215–225. doi:10.1016/j.cemconres.2013.08.007

Marković, M., Radmanović, S., Đurović-Pejčev, R., Milinovic, J., & Kljajić, P. (2009). Assessment of Heavy Metal and Pesticide Levels in Soil and Plant Products from Agricultural Area of Belgrade, Serbia. *Archives of Environmental Contamination and Toxicology, 58*(2), 341–351. doi:10.100700244-009-9359-y PMID:19603130

Marlinda, A., Pandikumar, A., Yusoff, N., Huang, N., & Lim, H. N. (2015). Electrochemical sensing of nitrite using a glassy carbon electrode modified with reduced functionalized graphene oxide decorated with flower-like zinc oxide. *Mikrochimica Acta, 182*(5-6), 1113–1123. doi:10.100700604-014-1436-x

Marouli, C. (1995). Women Resisting (in) the City: Struggles, Gender, Class and Space in Athens. *International Journal of Urban and Regional Research, 19*(4), 534–548. doi:10.1111/j.1468-2427.1995.tb00527.x

Marquardt, D., Barthel, J., Braun, M., Ganter, C., & Janiak, C. (2012). Weakly-coordinated stable platinum nanocrystals. *CrystEngComm, 14*(22), 7607–7615. doi:10.1039/c2ce25904d

Marques, G., & Pitarma, R. (2016). An Indoor Monitoring System for Ambient Assisted Living Based on Internet of Things Architecture. *International Journal of Environmental Research and Public Health, 13*(11), 1152. doi:10.3390/ijerph13111152 PMID:27869682

Martelli A., Brambilla G. (2012). *Arylamine drugs: genotoxic-carcinogenic activity of NO-derivatives. Frontiers in Bioscience E4.* Department of Internal Medicine, Division of Clinical Pharmacology and Toxicology, University of Genoa, Viale Benedetto.

Martínez-Castañòn, G. A., Niño-Martínez, N., Martínez-Gutierrez, F., Martínez-Mendoza, J. R., & Ruiz, F. J. (2008). Synthesis and antibacterial activity of silver nanoparticles with different sizes. *Journal of Nanoparticle Research, 10*(8), 1343–1348. doi:10.100711051-008-9428-6

Martins, A., & Mata, T. (2015). Nanotechnology and Sustainability— Current Status and Future Challenges. Academic Press.

Martuzzi, M., Mitis, F., Bianchi, F., Minichilli, F., Comba, P., & Fazzo, L. (2009). Cancer mortality and congenital anomalies in a region of Italy with intense environmental pressure due to waste. *Occupational and Environmental Medicine, 66*(11), 725–732. doi:10.1136/oem.2008.044115 PMID:19416805

Martuzzi, M., Mitis, F., & Fiorastere, F. (2010). Inequalities, inequities, environmental justice in waste management and health. *European Journal of Public Health, 20*(1), 21–26. doi:10.1093/eurpub/ckp216 PMID:20061348

Marvin, C., Waltho, J., Jia, J., & Burniston, D. (2013). Spatial distributions and temporal trends in polybrominated diphenyl ethers in Detroit River suspended sediments. *Chemosphere, 91*(6), 778–783. doi:10.1016/j.chemosphere.2013.02.009 PMID:23478126

Ma, S. S., Li, R., Lv, C. P., Xu, W., & Gou, X. L. (2011). Facile synthesis of ZnO nanorod arrays and hierarchical nanostructures for photocatalysis and gas sensor applications. *Journal of Hazardous Materials, 192*(2), 730–740. doi:10.1016/j.jhazmat.2011.05.082 PMID:21684076

Massey, R. C., Key, P. E., Mallett, A. K., & Rowland, I. R. (1991). An investigation of the endogenous formation of apparent total N-nitroso compounds in conventional flora and germ-free rats. *Food and Chemical Toxicology*, *26*(7), 595–600. doi:10.1016/0278-6915(88)90230-X PMID:3181835

Masuda, J. R., Poland, B., & Baxter, J. (2010). Reaching for environmental health justice: Canadian experiences for a comprehensive research, policy and advocacy agenda in health promotion. *Health Promotion International*, *25*(4), 453–463. doi:10.1093/heapro/daq041 PMID:20615911

Matache, M. L., David, I. G., Dinu, C., & Radu, L. G. (2018). Trace metals in water and sediments of the Prut River, Romania. *Environmental Engineering and Management Journal*, *17*(6), 1363–1371. doi:10.30638/eemj.2018.135

Mateen, F. J. and Brook, R. D. (2011). Air pollution as an emerging global risk factor for stroke. *JAMA: the Journal of the American Medical Association*, *305*(12), 1 240–1 241.

Mathew, D. S., & Juang, R.-S. (2007). An overview of the structure and magnetism of spinel ferrite nanoparticles and their synthesis in microemulsions. *Chemical Engineering Journal*, *129*(1-3), 51–65. doi:10.1016/j.cej.2006.11.001

Matilainen, A., & Sillanpää, M. (2010). Removal of natural organic matter from drinking water by advanced oxidation processes. *Chemosphere*, *80*(4), 351–365. doi:10.1016/j.chemosphere.2010.04.067 PMID:20494399

Matturro, B., Ubaldi, C., Grenni, P., Caracciolo, A. B., & Rossetti, S. (2016). Polychlorinated biphenyl (PCB) anaerobic degradation in marine sediments: Microcosm study and role of autochthonous microbial communities. *Environmental Science and Pollution Research International*, *23*(13), 12613–12623. doi:10.100711356-015-4960-2 PMID:26162439

Ma, Y.-X., Shao, W.-J., Sun, W., Kou, Y.-L., Li, X., & Yang, H.-P. (2018). One-step fabrication of β-cyclodextrin modified magnetic graphene oxide nanohybrids for adsorption of Pb(II), Cu(II) and methylene blue in aqueous solutions. *Applied Surface Science*, *459*, 544–553. doi:10.1016/j.apsusc.2018.08.025

Ma, Z., & Dai, S. (2011). Development of novel supported gold catalysts: A materials perspective. *Nano Research*, *4*(1), 3–32. doi:10.100712274-010-0025-5

McBride, M. B., & Blasiak, J. J. (1979). Zinc and Copper Solubility as a Function of pH in an Acid Soil. *Soil Science Society of America Journal*, *43*(5), 866–870. doi:10.2136ssaj1979.03615995004300050009x

McCarthy, G., Solem, J., Manz, O., & Hassett, D. (1989). *Use of a Database of Chemical, Mineralogical and Physical Properties of North American Fly Ash to Study the Nature of Fly Ashand Its Utilization as a Mineral Admixture in Concrete* (Vol. 178). Academic Press.

McCarthy, P. L. (2001). The development of anaerobic treatment and its future. *Water Science and Technology*, *44*(8), 149–156. doi:10.2166/wst.2001.0487 PMID:11730130

McGoldrick, D. J., Pelletier, M., de Solla, S. R., Marvin, C. H., & Martin, P. A. (2018). Legacy of legacies: Chlorinated naphthalenes in Lake Trout, Walleye, Herring Gull eggs and sediments from the Laurentian Great Lakes indicate possible resuspension during contaminated sediment remediation. *The Science of the Total Environment*, *634*, 1424–1434. doi:10.1016/j.scitotenv.2018.04.077 PMID:29710642

McLean, M. I., Evers, L., Bowman, A. W., Bonte, M., & Jones, W. R. (2019). Statistical modelling of groundwater contamination monitoring data: A comparison of spatial and spatiotemporal methods. *The Science of the Total Environment*, *652*, 1339–1346. doi:10.1016/j.scitotenv.2018.10.231 PMID:30586819

Medical Advisory, S. (2006). Nanotechnology: An evidence-based analysis. *Ontario Health Technology Assessment Series*, *6*(19), 1–43. PMID:23074489

Mehdizadeh, S., Sadjadi, S., Ahmadi, S. J., & Outokesh, M. (2014). Removal of heavy metals from aqueous solution using platinum nanopartcles/Zeolite-4A. *Journal of Environmental Health Science & Engineering*, *12*(1), 7–7. doi:10.1186/2052-336X-12-7 PMID:24397886

Meng, Q., Chen, H., Lin, J., Lin, Z., & Sun, J. (2017). Zeolite A synthesized from alkaline assisted pre-activated halloysite for efficient heavy metal removal in polluted river water and industrial wastewater. *Journal of Environmental Sciences (China)*, *56*, 254–262. doi:10.1016/j.jes.2016.10.010 PMID:28571862

Mesnage, R., Defarge, N., Spiroux de Vendomois, J., & Seralini, G. E. (2015). Potential toxic effects of glyphosate and its commercial formulations below regulatory limits. *Food and Chemical Toxicology*, *84*, 133–153. doi:10.1016/j.fct.2015.08.012 PMID:26282372

Message in a Bottle. (2008). *Supporting Information.* Retrieved from: https://www.pan-europe.info/old/ Resources/Briefings/Message_in_a_bottle_Background.pdf

Mévellec, V., Roucoux, A., Ramirez, E., Philippot, K., & Chaudret, B. (2004). Surfactant-stabilized aqueous iridium (0) colloidal suspension: An efficient reusable catalyst for hydrogenation of arenes in biphasic media. *Advanced Synthesis & Catalysis*, *346*(1), 72–76. doi:10.1002/adsc.200303157

Meyts, E. R.-D. (2006). Developmental model for the pathogenesis of testicular carcinoma in situ: Genetic and environmental aspects. *Human Reproduction Update*, *12*(3), 303–323. doi:10.1093/humupd/dmk006 PMID:16540528

Michael, A., Kohanski, D. J., Dwyer, J., & Collins, J. (2010). How antibiotics kill bacteria: From targets to networks. *Nature Reviews. Microbiology*, *8*(6), 423–433. doi:10.1038/nrmicro2333 PMID:20440275

Michelet, X., Dyck, L., Hogan, A., Loftus, R. M., Duquette, D., Wei, K., ... Lynch, L. (2018). Metabolic reprogramming of natural killer cells in obesity limits antitumor responses. *Nature Immunology*, *19*(12), 1330–1340. doi:10.103841590-018-0251-7 PMID:30420624

Miller, G. C., & Zepp, R. G. (1979). Effects of dissolved pollutants. *Water (Basel)*, 453.

Mills, C. W. (1959). *The Sociological Imagination.* Oxford University Press.

MINTEQA2/PRODEFA2, A Geochemical Assessment Model for Environmental Systems: User Manual Supplement for Version 4.0. (1998). = Allison Geoscience Consultants. Inc. Flowery Branch.

Miretzky, P., & Cirelli, A. (2009). Hg(II) Removal from Water by Chitosan and Chitosan Derivatives. *RE:view*, *167*. PMID:19232467

Mirvish, S. S. (1983). The etiology of gastric cancer. Intra-gastric nitrozamide formation and other theories. *Journal of the National Cancer Institute*, *71*(3), 613–647. PMID:6350677

Mirvish, S. S., Grandjean, A. C., Reimers, K. J., Connelly, B. J., Chen, S.-C., Gallagher, J., & ... . (1995). Dosing time with ascorbic acid and nitrate, gum and tobacco chewing, fasting, and other factors affecting N-nitrosoproline formation in healthy subjects taking Proline with a standard meal. *Cancer Epidemiology, Biomarkers & Prevention*, *4*, 775–820. PMID:8672996

Mirvish, S. S., Reimers, K. J., Kutler, B., Chen, S. C., Haorah, J., Morris, C. R., & Lyden, E. R. (2000). Nitrate and nitrite concentrations in human saliva for men and women and different ages and times of the day and their consistency over time. *European Journal of Cancer Prevention*, *9*(5), 335–342. doi:10.1097/00008469-200010000-00008 PMID:11075887

Mirzajani, R., Pourreza, N., & Burromandpiroze, J. (2017). Synthesis of chitosan-SnO2 nanoparticles biocomposite for dispersive micro-solid phase extraction of iron in environmental and wastewater samples and its determination by microvolume spectrophotometic detection. *Desalination and Water Treatment*, *97*, 213–224. doi:10.5004/dwt.2017.21536

Mittal, M., Sharma, C., & Singh, R. (2012). *Estimates of Emissions from Coal Fired Thermal Power Plants in India*. Academic Press.

Mittal, A. K., Bhaumik, J., Kumar, S., & Banerjee, U. C. (2014). Biosynthesis of silver nanoparticles: Elucidation of prospective mechanism and therapeutic potential. *Journal of Colloid and Interface Science*, *415*, 39–47. doi:10.1016/j.jcis.2013.10.018 PMID:24267328

Modirshahla, N., Hassani, A., & Mohammad, A. (2011). Effect of operational parameters on decolorization of acid yellow 23 from wastewater by UV irradiation using ZnO and ZnO/SnO_2 photocatalysts. *Desalination*, *271*(1-3), 187–192. doi:10.1016/j.desal.2010.12.027

Moen, T. (2006). Reflections on the Narrative Research Approach. *International Journal of Qualitative Methods*, *5*(4), 56–69. doi:10.1177/160940690600500405

Mohaghegh, N., Tasviri, M., Rahimi, E., & Gholami, M. R. (2015). Comparative studies on Ag3PO4/BiPO4–metal-organic framework–graphene-based nanocomposites for photocatalysis application. *Applied Surface Science*, *351*, 216–224. doi:10.1016/j.apsusc.2015.05.135

Mohajerani, M. S., Lak, A., & Simchi, A. (2009). Effect of morphology on the solar photocatalytic behavior of ZnO nanostructures. *Journal of Alloys and Compounds*, *485*(1-2), 616–620. doi:10.1016/j.jallcom.2009.06.054

Mohanpuria, P., Rana, N. K., & Yadav, S. K. (2007). Biosynthesis of nanoparticles: Technological concepts and future applications. *Journal of Nanoparticle Research*, *10*(3), 507–517. doi:10.100711051-007-9275-x

Mold, A., & Berridge, V. (2013). The history of health promotion. In Health Promotion Theory (2nd ed.). Maidenhead, UK: McGraw-Hill Education.

Moldovan, Z., Chira, R., & Alder, A. C. (2009). Environmental exposure of pharmaceuticals and musk fragrances in the Somes River before and after upgrading the municipal wastewater treatment plant Cluj-Napoca, Romania. *Environmental Science and Pollution Research International*, *16*(46), 46–54. doi:10.100711356-008-0047-7 PMID:18972147

Molea, A., Popescu, V., Rowson, N. A., & Dinescu, A. M. (2014). Influence of pH on the formulation of TiO_2 nanocrystalline powders with high photocatalytic activity. *Powder Technology*, *253*, 22–28. doi:10.1016/j.powtec.2013.10.040

Moletta, R., Verrier, D., & Albagnac, G. (1986). Dynamic modeling of anaerobic digestion. *Water Research*, *20*(4), 427–434. doi:10.1016/0043-1354(86)90189-2

Molitor, J., Papathomas, M., Jerrett, M., & Richardson, S. (2010). Bayesian profile regression with an application to the National Survey of Children's Health. *Biostatistics (Oxford, England)*, *11*(3), 484–498. doi:10.1093/biostatistics/kxq013 PMID:20350957

Mopoung, S., Sriprang, N., & Namahoot, J. (2014). Sintered filter materials with controlled porosity for water purification prepared from mixtures with optimal ratio of zeolite, bentonite, kaolinite, and charcoal. *Applied Clay Science*, *88*, 123–128. doi:10.1016/j.clay.2013.11.035

Morais, C. (2015). Storytelling with Chemistry and Related Hands-on Activities: Informal Learning Experiences to Prevent "Chemophobia" and Promote Young Children Scientific Literacy. *Journal of Chemical Education*, *92*(1), 58–65. doi:10.1021/ed5002416

Morales, R. (2018). *Nanomaterials – Do the benefits outweigh their risks*. Academic Press.

Morsi, R. E., Alsabagh, A. M., Nasr, S. A., & Zaki, M. M. (2017). Multifunctional nanocomposites of chitosan, silver nanoparticles, copper nanoparticles and carbon nanotubes for water treatment: Antimicrobial characteristics. *International Journal of Biological Macromolecules*, *97*, 264–269. doi:10.1016/j.ijbiomac.2017.01.032 PMID:28082228

Mosse, K. P., Verheyen, T. V., Cruickshank, A. J., Patti, A. F., & Cavagnaro, T. R. (2013). Soluble organic components of winery wastewater and implications for reuse. *Agricultural Water Management, 120*, 5–10. doi:10.1016/j.agwat.2012.05.011

Mostaquimur Rahman, A. S. M., Rosolem, R., Kollet, S. J., & Wagener, T. (2019). Towards a computationally efficient free-surface groundwater flow boundary condition for large-scale hydrological modelling. *Advances in Water Resources, 123*, 225–233. doi:10.1016/j.advwatres.2018.11.015

Mousavi, S. M., Hashemi, S. A., Esmaeili, H., Amani, A. M., & Mojoudi, F. (2018). Synthesis of Fe3O4 Nanoparticles Modified by Oak Shell for Treatment of Wastewater Containing Ni(II). *Acta Chimica Slovenica, 65*(3), 750–756. doi:10.17344/acsi.2018.4536

Mukhtar, N. H., & See, H. H. (2016). Carbonaceous nanomaterials immobilised mixed matrix membrane microextraction for the determination of polycyclic aromatic hydrocarbons in sewage pond water samples. *Analytica Chimica Acta, 931*, 57–63. doi:10.1016/j.aca.2016.04.032 PMID:27282751

Mukit, M. S. H., Rahman, M., & Roy, D. (2016). Investigation on the relative degradation of methylene blue (MB) and rhodamine-B (RB) dyes under UV-visible light using thermally treated commercial and doped TiO_2/ZnO photocatalysts. *Int. J. Integ. Sci. Tech, 2*, 14–18.

Muñoz-Fernandez, L., Sierra-Fernandez, A., Milošević, O., & Rabanal, M. E. (2016). Solvothermal synthesis of Ag/ZnO and Pt/ZnO nanocomposites and comparison of their photocatalytic behaviors on dyes degradation. *Advanced Powder Technology, 27*(3), 983–993. doi:10.1016/j.apt.2016.03.021

Murtić, S., Brkovic, D., Djuric, M., & Vujinovic, I. (2014). Heavy metal dynamics in the soil–leaf–fruit system under intensive apple cultivation. *Acta Agriculturae Serbica, 19*(38), 123–132. doi:10.5937/AASer1438123M

Murti, K. C. R., & Nag, D. (1991). *Human health impact of pesticide in the environment*. Cambridge.

Murugesan, S., Suresh, S., Pandikumar, A., Ramaraj, R., & Paul Raj, S. (2011). Metal-free low-cost organic dye-sensitized zno-nanorod photoanode for solid-state solar cell. *Materials Express, 1*(4), 307–314. doi:10.1166/mex.2011.1035

Mushtaq, F., Chen, X., Hoop, M., Torlakcik, H., Pellicer, E., Sort, J., ... Pané, S. (2018). Piezoelectrically Enhanced Photocatalysis with BiFeO3 Nanostructures for Efficient Water Remediation. *iScience, 4*, 236-246.

Mustafic, H. (2012). Main Air Pollutants andMyocardial Infarction: A Systematic Review and Meta-analysis. *Journal of the American Medical Association, 307*(7), 713. doi:10.1001/jama.2012.126 PMID:22337682

Mustea, M., & Holban, V. (2002). The main causes of pollution of water resources. *Management Resources Water*, 57-59.

Myers, P. J., Antoniou, M. N., Blumberg, B., Carroll, L., Colborn, T., Everett, L. G., ... Benbrook, C. M. (2016). Concerns over use of glyphosate-based herbicides and risks associated with exposures: A consensus statement. *Environmental Health, 15*, 19. doi:10.118612940-016-0117-0

Nadagouda, M. N., & Varma, R. S. (2006). Green and controlled synthesis of gold and platinum nanomaterials using vitamin B2: Density-assisted self-assembly of nanospheres, wires and rods. *Green Chemistry, 8*(6), 516–518. doi:10.1039/b601271j

Nagy, A., & Mestl, G. (1999). High temperature partial oxidation reactions over silver catalysts. *Applied Catalysis A, 188*(1-2), 337–353. doi:10.1016/S0926-860X(99)00246-X

Naja, G. M., Murphy, V., & Volesky, B. (2010). Biosorption, Metals. Encyclopedia of Industrial Biotechnology. doi:10.1002/9780470054581.eib166

Namvari, M., & Namazi, H. (2016). Magnetic sweet graphene nanosheets: Preparation, characterization and application in removal of methylene blue. *International Journal of Environmental Science and Technology*, *13*(2), 599–606. doi:10.100713762-015-0885-z

Narwade, V. N., Khairnar, R. S., & Kokol, V. (2018). In Situ Synthesized Hydroxyapatite—Cellulose Nanofibrils as Biosorbents for Heavy Metal Ions Removal. *Journal of Polymers and the Environment*, *26*(5), 2130–2141. doi:10.100710924-017-1101-7

NASA Ozone Watch. (2020). Retrieved from: https://ozonewatch.gsfc.nasa.gov/

Nasir, M. I., Mohd Ghazi, T. I., & Omar, R. (2012). Production of biogas from solid organic wastes through anaerobic digestion: A review. *Applied Microbiology and Biotechnology*, *95*(2), 321–329. doi:10.100700253-012-4152-7 PMID:22622840

Nastasiuc, L., Bogdevici, O., Overcenco, A., Culighin, E., Sidorenko, A., & Vaseashta, A. (2016). Monitoring Water Contaminants: A Case Study for the Republic of Moldova. *Polish Journal of Environmental Studies*, *25*(1), 1–11. doi:10.15244/pjoes/58888

Natarajan, M. R., Nye, J., Wu, W., Wang, H., Jain, M. K., & International, M. B. I. (1997). Reductive Dechlorination of PCB- Contaminated Raisin River Sediments by Anaerobic Microbial Granules. *Biotechnology and Bioengineering*, *55*(1), 182–190. doi:10.1002/(SICI)1097-0290(19970705)55:1<182::AID-BIT19>3.0.CO;2-O PMID:18636456

Nath, B. K., Chaliha, C., Kalita, E., & Kalita, M. C. (2016). Synthesis and characterization of ZnO:CeO2:nanocellulose:PANI bionanocomposite. A bimodal agent for arsenic adsorption and antibacterial action. *Carbohydrate Polymers*, *148*, 397–405. doi:10.1016/j.carbpol.2016.03.091 PMID:27185154

National Research Council. (2007). *Sediment Dredging at Superfund Megasites*. Washington, DC: Author.

National Research Council. (2012). *Exposure Science in the 21st Century: A Vision and Strategy*. Washington, DC: National Academies Press.

Nazim, S., Kousar, T., Shahid, M., Khan, M. A., Nasar, G., Sher, M., & Warsi, M. F. (2016). New graphene-CoxZn1−xFe2O4 nano-heterostructures: Magnetically separable visible light photocatalytic materials. *Ceramics International*, *42*(6), 7647–7654. doi:10.1016/j.ceramint.2016.01.177

Nazir, S., Wani, I. A., & Masoodi, F. A. (2017). Extraction optimization of mucilage from Basil (Ocimum basilicum L.) seeds using response surface methodology. *Journal of Advanced Research*, *8*(3), 235–244. doi:10.1016/j.jare.2017.01.003 PMID:28239494

Negi, R., Satpathy, G., Tyagi, Y. K., & Gupta, R. K. (2012). Biosorption of heavy metals by utilising onion and garlic wastes. *International Journal of Environment and Pollution*, *49*(3/4), 179–196. doi:10.1504/IJEP.2012.050898

Nelson, L. H., Saulsbery, A. I., & Lenz, K. M. (2018). Small cells with big implications: Microglia and sex differences in brain development, plasticity and behavioral health. *Progress in Neurobiology*. doi:10.1016/j.pneurobio.2018.09.002 PMID:30193820

Nematidil, N., & Sadeghi, M. (2018). Article. *Polymer Bulletin.*

Neyens, E., & Baeyens, J. (2003). A review of classic Fenton's peroxidation as an advanced oxidation technique. *Journal of Hazardous Materials*, *98*(1-3), 33–50. doi:10.1016/S0304-3894(02)00282-0 PMID:12628776

Ngu, L., Wu, H., & Zhang, D. (2007). Characterization of Ash Cenospheres in Fly Ash from Australian Power Stations. *Energy & Fuels*, *21*(6), 3437–3445. doi:10.1021/ef700340k

Nhandeyara do Carmo Ramos, S., Xavier, A. L., Teodoro, F., Frédéric Gil, L., & Gurgel, L. (2016). *Removal of cobalt(II), copper(II), and nickel(II) ions from aqueous solutions using phthalate-functionalized sugarcane bagasse: Mono- and multicomponent adsorption in batch mode* (Vol. 79). Academic Press.

Nicolis, G., & Prigogine, I. (1979). Samoorganizatsiya v neravnovesnykh sistemakh. In Dissipativnych struktur k uporyadochennosti cherez fluktuatsii [Self-organization in nonequilibrium systems: From dissipative structures to order through fluctuations]. Moscow: Mir.

Nienhuis, P., Buijse, A., Leuven, R., Smits, A. J. M., de Nooij, R. J. W., & Samborska, E. M. (2002). Ecological rehabilitation of the lowland basin of the River Rhine (NW Europe). *Hydrobiologia*, *478*(1/3), 53–72. doi:10.1023/A:1021070428566

Nieuwenhuijsen, M. J., Donaire-Gonzalez, D., Foraster, M., Martinez, D., & Cisneros, A. (2014). Using Personal Sensors to Assess the Exposome and Acute Health Effects. *International Journal of Environmental Research and Public Health*, *11*(8), 7805–7819. doi:10.3390/ijerph110807805 PMID:25101766

Nieuwenhuijsen, M. J., Donaire-Gonzalez, D., Rivas, I., de Castro, M., Cirach, M., Hoek, G., ... Sunyer, J. (2015). Variability in and agreement between modeled and personal continuously measured black carbon levels using novel smartphone and sensor technologies. *Environmental Science & Technology*, *49*(5), 2977–2982. doi:10.1021/es505362x PMID:25621420

Nieuwenhuijsen, M. J., Smith, R., Golfinopoulos, S., Best, N., Bennett, J., Aggazzotti, G., ... Kogevinas, M. (2009). Health impacts of long-term exposure to disinfection by-products in drinking water in Europe: HIWATE. *Journal of Water and Health*, *7*(2), 185–207. doi:10.2166/wh.2009.073 PMID:19240347

Nigiz, F. U., Veli, S., & Hilmioglu, N. D. (2017). Deep purification of seawater using a novel zeolite 3A incorporated polyether-block-amide composite membrane. *Separation and Purification Technology*, *188*, 90–97. doi:10.1016/j.seppur.2017.07.017

Nikodemus, O., Brumelis, G., Tabors, G., Lapin¸, L., & Pope, S. (2004). Monitoring of Air Pollution in Latvia Between 1990 and 2000 Using Moss. *Journal of Atmospheric Chemistry*, *49*(1-3), 521–531. doi:10.100710874-004-1263-2

Ni, M., Leung, D., Leung, M., & Sumathy, K. (2006). An overview of hydrogen production from biomass. *Fuel Processing Technology*, *87*(5), 461–472. doi:10.1016/j.fuproc.2005.11.003

Ni, M., Leung, M. K., Leung, D. Y., & Sumathy, K. (2007). A review and recent developments in photocatalytic water-splitting using TiO2 for hydrogen production. *Renewable & Sustainable Energy Reviews*, *11*(3), 401–425. doi:10.1016/j.rser.2005.01.009

Nithya, R., Gomathi, T., Sudha, P. N., Venkatesan, J., Anil, S., & Kim, S. (2016). Removal of Cr(VI) from aqueous solution using chitosan-g-poly(butyl acrylate)/silica gel nanocomposite. *International Journal of Biological Macromolecules*, *87*, 545–554. doi:10.1016/j.ijbiomac.2016.02.076 PMID:26952703

NOAA/ESRL. (n.d.). Retrieved from https://genomicscience.energy.gov/

Noor Azwen, N., Ahmad Ashrif, B., Norhana, A., Sarada, I., & Abdul Rahman, A. (2017). Enhancement of ZnO-rGO nanocomposite thin films by gamma radiation for E.coli sensor. *Applied Surface Science*, *392*, 1134–1143. doi:10.1016/j.apsusc.2016.09.144

Nordstrom, D. K. (1982). The effect of sulfate on aluminum concentrations in natural waters: Some stability relations in the system Al_2O_3-SO_3-H_2O at 298 K. *Geochimica et Cosmochimica Acta*, *46*(4), 681–692. doi:10.1016/0016-7037(82)90168-5

Norris, S., & Phillips, L. (2003). How literacy in its fundamental sense is central to scientific literacy. *Science Education*, *87*(2), 224–240. doi:10.1002ce.10066

Norton, G. A., Markuszewski, R., & Shanks, H. R. (1986). Morphological and chemical characterization of iron-rich fly ash fractions. *Environmental Science & Technology*, *20*(4), 409–413. doi:10.1021/es00146a016 PMID:22300215

Novi, V., Kumar, V., Charles, C., Vishnu, D., Ponnusamy, S. K., . . . Kumar Vaidyanathan, V. (2016). *Review on Nano Adsorbents: A Solution For Heavy Metal Removal From Waste Water* (Vol. 11). Academic Press.

Ntim, S. A., & Mitra, S. (2012). Adsorption of arsenic on multiwall carbon nanotube–zirconia nanohybrid for potential drinking water purification. *Journal of Colloid and Interface Science*, *375*(1), 154–159. doi:10.1016/j.jcis.2012.01.063 PMID:22424815

Nuengmatcha, P., Chanthai, S., Mahachai, R., & Oh, W. C. (2016). Visible light-driven photocatalytic degradation of rhodamine B and industrial dyes (texbrite BAC-L and texbrite NFW-L) by ZnO-graphene-TiO_2 composite. *Journal of Environmental Chemical Engineering*, *4*(2), 2170–2177. doi:10.1016/j.jece.2016.03.045

OECD. (2012). *OECD Environmental Outlook to 2050*. Paris, France: Organisation for Economic Co-operation and Development.

Ohio State University. (2018). How childhood stress contributes to anxiety, depression: Research in animals illuminates important role of mast cells. *ScienceDaily*. Retrieved November 18, 2018 from www.sciencedaily.com/releases/2018/11/181105160833.htm

Okoh, A. I., Sibanda, T., & Gusha, S. S. (2010). Inadequately treated wastewater as a source of human enteric viruses in the environment. *International Journal of Environmental Research and Public Health*, *7*(6), 2620–2637. doi:10.3390/ijerph7062620 PMID:20644692

Olajire, A. A., Abidemi, J. J., Lateef, A., & Benson, N. U. (2017). Adsorptive desulphurization of model oil by Ag nanoparticles-modified activated carbon prepared from brewer's spent grains. *Journal of Environmental Chemical Engineering*, *5*(1), 147–159. doi:10.1016/j.jece.2016.11.033

Oller, I., Malato, S., & Sánchez-Pérez, J. A. (2011). Combination of Advanced Oxidation Processes and biological treatments for wastewater decontamination - A review. *The Science of the Total Environment*, *409*(20), 4141–4166. doi:10.1016/j.scitotenv.2010.08.061 PMID:20956012

Olmo, N., Saldiva, P., Braga, A., Lin, Ch., Santos, U., & Pereira, L. (2011). A review of low-level air pollution and adverse effects on human health: Implications for epidemiological studies and public policy. *Clinics*, *66*(4), 681–690. doi:10.1590/S1807-59322011000400025 PMID:21655765

Olveira, S., Forster, S. P., & Seeger, S. (2014). *Nanocatalysis: Academic Discipline and Industrial Realities*. Academic Press.

Olveira, S., Forster, S. P., & Seeger, S. (2014). Nanocatalysis: Academic discipline and industrial realities. *Journal of Nanotechnology*.

Omarova, N., Nurgalieva, D., Nurkasimova, M., Frontasyeva, M., Morzhukhina, S., & Kabdulkarimova, K. (2018). Environmental monitoring in the Republic of Kazakhstan on the content of heavy metals and radionuclides. *International Electronic journal. Sustainable Development Science and Practice, 1*(3), 20-29.

Opopol, N. (2006). Water management in Republic of Moldova. Principal Threats. In Management of Intentional and Accidental Water Pollution (pp. 167-176). Springer.

Ortega-Liebana, M. C., Hueso, J. L., Ferdousi, S., Yeung, K. L., & Santamaria, J. (2016). Nitrogen-doped luminescent carbon nanodots for optimal photo-generation of hydroxyl radicals and visible-light expanded photo-catalysis. *Diamond and Related Materials*, *65*, 176–182. doi:10.1016/j.diamond.2016.03.021

Ouchi, Y., Yanagisawa, H., & Fujimaki, S. (2019). Evaluating Phthalate Contaminant Migration Using Thermal Desorption–Gas Chromatography–Mass Spectrometry (TD–GC–MS). *Polymers*, *11*(4), 683. doi:10.3390/polym11040683 PMID:30991697

Ozbek, N., & Akman, S. (2016). Method development for the determination of calcium, copper, magnesium, manganese, iron, potassium, phosphorus and zinc in different types of breads by microwave induced plasma-atomic emission spectrometry. *Food Chemistry, 200*, 245–248. doi:10.1016/j.foodchem.2016.01.043 PMID:26830585

Ozdemir, S., Mohamedsaid, S. A., Kilinc, E., Yildirim, A., & Soylak, M. (2018). Application of magnetized fungal solid phase extractor with Fe2O3 nanoparticle for determination and preconcentration of Co(II) and Hg(II) from natural water samples. *Microchemical Journal, 143*, 198–204. doi:10.1016/j.microc.2018.07.032

Özer, M., Basha, O. M., & Morsi, B. (2017). Coal-Agglomeration Processes: A Review. *International Journal of Coal Preparation and Utilization, 37*(3), 131–167. doi:10.1080/19392699.2016.1142443

Pacchioni, G. (2009). Nanocatalysis: Staying put. *Nature Materials, 8*(3), 167–168. doi:10.1038/nmat2394 PMID:19229260

Pagnanelli, F., Mainelli, S., Veglio, F., & Toro, L. (2003). *Heavy metal removal by olive pomace: Biosorbent characterisation and equilibrium modelling* (Vol. 58). Academic Press.

Palermo, M. R., Schroeder, P. R., Estes, T. J., & Francingues, N. R. (2008). *Technical Guidelines for Environmental Dredging of Contaminated Sediments (Report No. ERDC/EL TR-08-29). US Army Corps of Engineers*. Vicksburg, MS: US Army Corps of Engineers Engineer Research and Development Center.

Palermo, M., & Hays, D. F. (2014). Sediment Dredging, Treatment and Disposal. In D. D. Reible (Ed.), *Processes, Assessment and Remediation of Contaminated Sediments*. New York: Springer. doi:10.1007/978-1-4614-6726-7_13

Palm, A., Cousins, I. T., Mackay, D., Tysklind, M., Metcalfe, C., & Alaee, M. (2002). Assessing the environmental fate of chemicals of emerging concern: a case study of the polybrominated diphenyl ethers. *Environmental Pollution, 117*(2), 195–213. Retrieved from https://www.ncbi.nlm.nih.gov/pubmed/11916035

Pal, S. L., Jana, U., Manna, P. K., Mohanta, G. P., & Manavalan, R. (2011). Article. *Journal of Applied Pharmaceutical Science*, *01*(06), 228–234.

Palumbo, A., Daughney, C., Slade, A., & Glover, C. (2013). *Influence of pH and natural organic matter on zinc biosorption in a model lignocellulosic biofuel biorefinery effluent* (Vol. 146C). Academic Press.

Panayotova, M. (2016). Mining and mineral processing as heavy metals pollution source. Academic Press.

Pandian, N., Rajasekhar, C., & Sridharan, A. (1998). *Studies of the Specific Gravity of Some Indian Coal Ashes* (Vol. 26). Academic Press.

Pandikumar, A., Saranya, K. M., & Ramaraj, R. (2012). Sheaf-like-ZnO@Ag nanocomposite materials modified photoanode for low-cost metal-free organic dye-sensitized solid-state solar cell. *Applied Physics Letters, 101*(9), 187–192. doi:10.1063/1.4748287

Pangi, Z., Beletsi, A., & Evangelatos, K. (2003). Article. *Advanced Drug Delivery Reviews, 24*, 403–419.

Pang, Y. L., Abdullah, A. Z., & Bhatia, S. (2010). Effect of annealing temperature on the characteristics, sonocatalytic activity and reusability of nanotubes TiO_2 in the degradation of Rhodamine B. *Applied Catalysis B: Environmental, 100*(1-2), 393–402. doi:10.1016/j.apcatb.2010.08.016

Pankratova, Yu., Frontasyeva, M., Berdnikov, A., Pavlov, S., Granja, C., Leroy, C., & Stekl, I. (2007). Air Pollution Studies in the Republic of Udmurtia, Russian Federation, using Moss Biomonitoring and INAA. *Nuclear Physics Methods and Accelerators in Biology and Medicine*, *958*, 236–237. doi:10.1063/1.2825794

Pan, Y., Chen, J., Zhou, H., & Tam, N. F. Y. (2018). Changes in microbial community during removal of BDE-153 in four types of aquatic sediments. *The Science of the Total Environment*, *613–614*, 644–652. doi:10.1016/j.scitotenv.2017.09.130 PMID:28934686

Papadopoulou, P., Lytras, M., & Marouli, C. (2018). Bioinformatics as Applied to Medicine: Challenges Faced Moving from Big Data to Smart Data to Wise Data. In M. Lytras & P. Papadopoulou (Eds.), *Applying Big Data Analytics in Bioinformatics and Medicine* (pp. 1–25). Hershey, PA: IGI Global. doi:10.4018/978-1-5225-2607-0.ch001

Papageorgiou, S., Kouvelos, E. P., & Katsaros, F. (2008). *Calcium alginate beads from Laminaria digitata for the removal of Cu+2 and Cd+2 from dilute aqueous metal solutions* (Vol. 224). Academic Press.

Parawira, W. (2011). Enzyme research and applications in biotechnological intensification of biogas production. *Critical Reviews in Biotechnology*, *32*(2), 172–186. doi:10.3109/07388551.2011.595384 PMID:21851320

Pare, B., Jonnalagadda, S. B., Tomar, H., Singh, P., & Bhagwat, V. M. (2008). ZnO assisted photocatalytic degradation of acridineorange in aqueous solution using visible irradiation. *Desalination*, *232*(1-3), 80–90. doi:10.1016/j.desal.2008.01.007

Park, D., Yun, Y.-S., & Park, J. (2010). *The Past* (Vol. 15). Present, and Future Trends of Biosorption.

Parsons, S. A., & Jefferson, B. (2006). *Introduction to potable water treatment processes*. Blackwell Publishing. doi:10.1002/9781444305470

Patel, A., Sv, A., Shah, N., & Verma, D. (2018). *Lactic Acid Bacteria As Metal Quenchers To Improve Food Safety And Quality* (Vol. 6). Academic Press.

Patel, C. J., Bhattacharya, J., & Butte, A. J. (2010). An environment-wide association study (EWAS) on type 2 diabetes mellitus. *PLoS One*, *5*(5), e10746. doi:10.1371/journal.pone.0010746 PMID:20505766

Patel, C. J., Chen, R., Kodama, K., Ioannidis, J. P., & Butte, A. J. (2013). Systematic identification of interaction effects between genome- and environment-wide associations in type 2 diabetes mellitus. *Human Genetics*, *132*(5), 495–508. doi:10.100700439-012-1258-z PMID:23334806

Paterson, E., Goodman, B. A., & Farmer, V. C. (1991). The Chemistry of Aluminium, Iron and Manganese Oxides in Acid Soils. In B. Ulrich & M. E. Sumner (Eds.), *Soil Acidity*. Berlin: Springer. doi:10.1007/978-3-642-74442-6_5

Pathak, P., Mandavgane, S., & Kulkarni, B. (2015). *Fruit peel waste as a novel low-cost bio adsorbent* (Vol. 31). Academic Press.

Patmont, C., LaRosa, P., Narayanan, R., & Forrest, C. (2018). Environmental dredging residual generation and management. *Integrated Environmental Assessment and Management*, *14*(3), 335–343. doi:10.1002/ieam.4032 PMID:29400910

Patra, J. K., & Baek, K.-H. (2014). Green Nanobiotechnology: Factors Affecting Synthesis and Characterization Techniques. *Journal of Nanomaterials*, *12*. doi:10.1155/2014/417305

Patterson, T. F. (2007). Treatment and prevention of fungal infections. Focus on candidemia. New York. *Applied Clinical Education*, *2007*, VII–VIII.

Patzke, G. R., Zhou, Y., Kontic, R., & Conrad, F. (2011). Oxide nanomaterials: Synthetic developments, mechanistic studies, and technological innovations. *Angewandte Chemie International Edition*, *50*(4), 826–859. doi:10.1002/anie.201000235 PMID:21246678

Paul, B. (2012). Heavy Metals Toxicity and the Environment. *EXS*, *101*, 133–164. doi:10.1007/978-3-7643-8340-4_6 PMID:22945569

Paul, K. (1990). Risk of Colorectal and Other Gastro-Intestinal Cancers After Exposure to Nitrate, Nitrite and N-Nitroso Compounds: A Follow-Up Study. *International Journal of Cancer*, *80*, 852–856.

Payne, R. B., Fagervold, S. K., May, H. D., & Sowers, K. R. (2013). Remediation of polychlorinated biphenyl impacted sediment by concurrent bioaugmentation with anaerobic halorespiring and aerobic degrading bacteria. *Environmental Science & Technology*, *47*(8), 3807–3815. doi:10.1021/es304372t PMID:23463900

Payne, R. B., Ghosh, U., May, H. D., Marshall, C. W., & Sowers, K. R. (2019). A Pilot-Scale Field Study: In Situ Treatment of PCB-Impacted Sediments with Bioamended Activated Carbon. *Environmental Science & Technology*, *53*(5), 2626–2634. doi:10.1021/acs.est.8b05019

Payne, R. B., May, H. D., & Sowers, K. R. (2011). Enhanced Reductive Dechlorination of Polychlorinated Biphenyl Impacted Sediment by Bioaugmentation with a Dehalorespiring Bacterium. *Environmental Science & Technology*, *45*(20), 8772–8779. doi:10.1021/es201553c PMID:21902247

Pazynych, N., & Romanciuc, I. (2018). Characteristic of the soil moisture state based on cartographic methods and satellite survey data. Odessa National University, Geography and Geology, 23(33), 43-54. doi:10.18524/2303-9914.2018.2(33).146631

Pearce, M.S., Glinianaia, S.V., Ghosh, R., Rankin, J., Rushton, S., Charlton, M., Parker, L., & Pless-Mulloli, T. (2012). Particulate matter exposure during pregnancy is associated with birth weight, but not gestational age, 1962–1992: a cohort study. *Environmental Health*, *11*(1), 13.

Peden, M. M. & World Health Organization. (2004). *World report on road traffic injury prevention*. Geneva: World Health Organization.

Pei, A., Butchosa, N., Berglund, L. A., & Zhou, Q. (2013). Surface quaternized cellulose nanofibrils with high water absorbency and adsorption capacity for anionic dyes. *Soft Matter*, *9*(6), 2047–2055. doi:10.1039/c2sm27344f

Pelizzetti, E., & Calza, P. (2002). Photochemical processes in the euphotic zone of sea water: progress and problems. In *Chemistry of Marine Water and Sediments. Springer-Verlag*. doi:10.1007/978-3-662-04935-8_3

Pénard-Morand, C., & Annesi-Maesano, I. (2004). Air pollution: From sources of emissions to health effects. *Breathe*, *1*(2), 109–119. doi:10.1183/18106838.0102.108

Penilla, R. P., Guerrero Bustos, A., & Goñi Elizalde, S. (2006). *Immobilization of Cs, Cd, Pb and Cr by synthetic zeolites from Spanish low-calcium coal fly ash* (Vol. 85). Academic Press.

Perelo, L. W. (2010). Review: In situ and bioremediation of organic pollutants in aquatic sediments. *Journal of Hazardous Materials*, *177*(1–3), 81–89. doi:10.1016/j.jhazmat.2009.12.090 PMID:20138425

Perrin, D. D. (2012). *Buffers for pH and metal ion control*. Springer Science & Business Media.

Pertusatti, J., & Prado, A. G. (2007). Buffer capacity of humic acid: Thermodynamic approach. *Journal of Colloid and Interface Science*, *314*(2), 484–489. doi:10.1016/j.jcis.2007.06.006 PMID:17603069

Pfendt, L. B. (1995). Two-phase liquid and solid-liquid pH buffers based on solubility equilibria: Theoretical considerations. *Analyst (London)*, *120*(8), 2129–2144. doi:10.1039/AN9952002129

Philippat, C., Wolff, M. S., Calafat, A. M., Ye, X., Bausell, R., Meadows, M., ... Engel, S. M. (2013). Prenatal exposure to environmental phenols: Concentrations in amniotic fluid and variability in urinary concentrations during pregnancy. *Environmental Health Perspectives*, *121*(10), 1225–1231. doi:10.1289/ehp.1206335 PMID:23942273

Phthalates and their alternatives health and environmental concerns. (2011). University of Massachusetts Lowell.

Phuruangrat, A., Kongpet, W., Yayapao, O., Kuntalue, B., Thongtem, S., & Thongtem, T. (2014). Ultrasonic-assisted synthesis, characterization, and optical properties of Sb doped ZnO and their photocatalytic activities. *Journal of Nanomaterials*, 1–10.

Pi, S., Li, A., Wei, W., Feng, L., Zhang, G., Chen, T., ... Ma, F. (2017). Synthesis of a novel magnetic nano-scale biosorbent using extracellular polymeric substances from Klebsiella sp. J1 for tetracycline adsorption. *Bioresource Technology*, *24*, 471–476. doi:10.1016/j.biortech.2017.08.190 PMID:28898846

Pischulov, S. A. (2013). Sors: An example of arid landforms. *Geomorfologiya*, *3*, 88–98. doi:10.15356/0435-4281-2013-3-

Pitchaimuthu, S., Lakshmi, G., & Velusamy, P. (2015). Enhanced photo catalytic activity of TiO_2 using β-cyclodextrin on solar light assisted decoloration of azocarmine G dye. *J. Adv. Chem. Sci*, *1*, 9–14.

Plechkova, N. V., & Seddon, K. R. (2008). Applications of ionic liquids in the chemical industry. *Chemical Society Reviews*, *37*(1), 123–150. doi:10.1039/B006677J PMID:18197338

Plohl, O., Ajdnik, U., Ban, I., Gyergyek, S., Vesel, A., Glaser, T. K., & Zemljic, L. (2019). Superior stability and high biosorbent efficiency of carboxymethylchitosan covalently linked to silica-coated core-shell magnetic nanoparticles for application in copper removal. *Journal of Environmental Chemical Engineering*, *7*(1), 102913. doi:10.1016/j.jece.2019.102913

Polshettiwar, V., & Varma, R. S. (2010). Green chemistry by nano-catalysis. *Green Chemistry*, *12*(5), 743–754. doi:10.1039/b921171c

Pomastowski, P., Szultka-Młyńska, M., Kupczyk, W., Jackowski, M., & Buszewski, B. (2015). Evaluation of Intact Cell Matrix-Assisted Laser Desorption/Ionization Time-of-Flight Mass Spectrometry for Capillary Electrophoresis Detection of Controlled Bacterial Clumping. *Journal of Analytical & Bioanalytical Techniques, S13,008,* 1–7.

Pomastowski, P., Sprynskyy, M., Žuvela, P., Rafińska, K., Milanowski, M., Liu, J. J., ... Buszewski, B. (2016). Silver-Lactoferrin Nanocomplexes as a Potent Antimicrobial Agent. *Journal of the American Chemical Society*, *138*(25), 7899–7909. doi:10.1021/jacs.6b02699 PMID:27263865

Popescu, F., & Ionel, I. (2010). *Anthropogenic Air Pollution Sources*. Retrieved from https://www.intechopen.com/books/air-quality/anthropogenic-air-pollution-sources

Porter, F. (1991). *Zinc dust and compounds. Zinc handbook properties, processing, and use in design*. New York: Marcel Dekker. doi:10.1201/9781482276947

Porubin, D., Hecht, S. S., Li, Z., Gonta, M., & Stepanov, I. (2007). Stepanov, I. Endogenous Formation of N'-itrosonornicotine in F344 Rats in the Presence of Some Antioxidants and Grape Seed Extract. *Journal of Agricultural and Food Chemistry*, *55*(17), 7199–7204. doi:10.1021/jf0712191 PMID:17637060

Povar, I. (1994). Thermodynamic calculation of the minimum solubility pH of slightly soluble oxides and hydroxides in polynuclear hydrolysis of metal ion. [In Russian]. *Ukrainian Chemistry Journal*, *60*, 371–378.

Povar, I. (1996). Buffer properties of heterogeneous systems containing sparingly soluble acids and metal hydroxides. *Russian Journal of Inorganic Chemistry*, *41*, 1167–1171.

Povar, I. (1997). Method for graphic representation of heterogeneous chemical equilibria in systems sparingly soluble compound-complexing agent-aqueous solution. [Engl. Transl.]. *Russian Journal of Inorganic Chemistry*, *42*, 607–612.

Povar, I. (2000 b). On Homogeneous Ion Buffer Systems. *Russian Journal of Inorganic Chemistry*, *45*, 1628–1631.

Povar, I. (2000). Potentiometric determination of solubility products and equilibrium ion concentrations for sparingly soluble salts. *Russian Journal of General Chemistry*, *70*, 501–507.

Povar, I. (2000a). Buffer properties of heterogeneous water-salt systems in relation to the components of a low-solubility precipitate. *Russian Journal of Inorganic Chemistry*, *45*, 1632–1636.

Povar, I. (2001). Determination of the stability of slightly soluble complexonate from pH metric data. *Canadian Journal of Chemistry*, *79*, 1166–1172. doi:10.1139/v01-065

Povar, I., Fishtik, I., & Vataman, I. (1989). Calculation of optimal conditions of underground leaching of uranium ores. *Bulletin of Academy of Sciences of MRSS* [In Russian]. *Series of Biological and Chemical Sciences*, *5*, 57–61.

Povar, I., & Luca, C. (2003). Considerations regarding ionic-molecular buffer solutions. *Revista de Chimie*, *54*, 312–316.

Povar, I., & Rusu, V. (2012). Aluminium heterogeneous speciation in natural waters. *Canadian Journal of Chemistry*, *290*(4), 326–332. doi:10.1139/v2012-003

Povar, I., & Rusu, V. (2012). Buffer capacity of heterogeneous chemical equilibria in natural waters. *Canadian Journal of Chemistry*, *90*(4), 395–402. doi:10.1139/v2012-008

Povar, I., & Spinu, O. (2014a). The role of hydroxy aluminium sulfate minerals in controlling Al^{3+} concentration and speciation in acidic soils. *Open Chemistry*, *12*(8), 877–885. doi:10.247811532-014-0540-4

Povar, I., & Spinu, O. (2014b). *Thermodynamics of Complex Chemical Equilibria in Heterogeneous Multicomponent Systems*. Chisinau: Printing House of the Academy of Sciences of Moldova. (in Romanian)

Povar, I., & Spinu, O. (2015). Application of the buffer theory for evaluating attenuation and natural remediation of ionic pollutants in aquatic ecosystems. *Ecological Processes*, *4*(1), 17. doi:10.118613717-015-0043-z

Pöykiö, R., Nurmesniemi, H., & Keiski, R. (2007). Environmental risk assessment of heavy metal extractability in a biosludge from the biological wastewater treatment plant of a pulp and paper mill. *Environmental Monitoring and Assessment*, *128*(1-3), 153–164. doi:10.100710661-006-9301-y PMID:17057998

Pradeep, T., & Anshup. (2009). Noble metal nanoparticles for water purification: A critical review. *Thin Solid Films*, *517*(24), 6441–6478. doi:10.1016/j.tsf.2009.03.195

Pranjali, G., Deepa, M., & Nair, A. B. (2013). Nanotechnology in waste water treatment: A review. *International Journal of Chemtech Research*, *5*(5), 2303–2308.

Preau, J. L. Jr, Wong, L. Y., Silva, M. J., Needham, L. L., & Calafat, A. M. (2010). Variability over 1 week in the urinary concentrations of metabolites of diethyl phthalate and di(2-ethylhexyl) phthalate among eight adults: An observational study. *Environmental Health Perspectives*, *118*(12), 1748–1754. doi:10.1289/ehp.1002231 PMID:20797930

Přech, J., Pizarro, P., Serrano, D. P., & Čejka, J. (2018). From 3D to 2D zeolite catalytic materials. *Chemical Society Reviews*, *47*(22), 8263–8306. doi:10.1039/C8CS00370J PMID:30167621

Prechtl, M. H. G. (2016). *Nanocatalysis in Ionic Liquids. Nanocatalysis in Ionic Liquids*. Wiley.

Prechtl, M. H. G., Scholten, J. D., & Dupont, J. (2010). Carbon-carbon cross coupling reactions in ionic liquids catalysed by palladium metal nanoparticles. *Molecules (Basel, Switzerland)*, *15*(5), 3441–3461. doi:10.3390/molecules15053441 PMID:20657493

Prigogine, I. (1985). *Ot suschestvuyuschego k voznikayuschemu, Vremya I slozhnost v phizicheskikh naukakh* [From Being to Becoming: Time and Complexity in the Physical Sciences]. Moscow: Nauka.

Prigogine, I., & Stengers, I. (1986). *Poryadok iz khaosa. Noviy dialog cheloveka s prirodoy* [Order out of chaos. Man's new dialogue with nature]. Moscow: Progress.

Prince, W., Perami, R., Espagne, M., Hollman, G. G., Steenbruggen, G., & Janssen-Jurkovicova, M. (1999). *A Two-step Process for the Synthesis of Zeolites from Coal Fly Ash* (Vol. 78). Academic Press.

Pristaer, B. S., Kliuchnykov, A. A., Baryakhtar, V. G., Shestopalov, V. M., & Kuhar, V. P. (Eds.). (2016). The safety problems of the nuclear power the lessons of chernobyl. NAS of Ukraine.

Pruett, S. (2014). Review of Organic Chemistry Principles in Context: A Story-Telling Historical Approach. *Journal of Chemical Education*, *91*(5), 624–625. doi:10.1021/ed5001679

Pruss-Usten, A., Wolf, J., Corvalán, C., Bos, R., & Neira, M. (2016). *Preventing disease through healthy environments. A global assessment of the burden of disease from environmental risks.* World Health Organization (WHO).

Prüss-Ustün, A., Bartram, J., Clasen, T., Colford, J. M. Jr, Cumming, O., Curtis, V., ... Cairncross, S. (2014). Burden of disease from inadequate water, sanitation and hygiene in low- and middle-income settings: A retrospective analysis of data from 145 countries. *Tropical medicine & international health. TM & IH*, *19*(8), 894–905. doi:10.1111/tmi.12329 PMID:24779548

Pruss-Ustun, A., Wolf, J., Corvalán, C., Bos, R., & Neira, M. (2016). *Preventing disease through healthy environments: A global assessment of the burden of disease from environmental risks.* Geneva: WHO.

Qian, H., Pretzer, L. A., Velazquez, J. C., Zhao, Z., & Wong, M. S. (2013). Gold nanoparticles for cleaning contaminated water. *Journal of Chemical Technology and Biotechnology (Oxford, Oxfordshire)*, *88*(5), 735–741. doi:10.1002/jctb.4030

Qiao, H., Zhou, Y., Yu, F., Wang, E., Min, Y., Huang, Q., ... Ma, T. (2015). Effective removal of cationic dyes using carboxylate-functionalized cellulose nanocrystals. *Chemosphere*, *141*, 297–303. doi:10.1016/j.chemosphere.2015.07.078 PMID:26298027

Qiu, L., Peng, Y., Liu, B., Lin, B., Peng, Y., Malik, M. J., & Yan, F. (2012). Polypyrrole nanotube-supported gold nanoparticles: An efficient electrocatalyst for oxygen reduction and catalytic reduction of 4-nitrophenol. *Applied Catalysis A, General*, *413*, 230–237. doi:10.1016/j.apcata.2011.11.013

Qiu, M., Chen, X., Deng, D., Guo, J., Sun, G., Mai, B., & Xu, M. (2012). Effects of electron donors on anaerobic microbial debromination of polybrominated diphenyl ethers (PBDEs). *Biodegradation*, *23*(3), 351–361. doi:10.100710532-011-9514-9 PMID:21910024

Qiu, X., Li, G., Sun, X., Li, L., & Fu, X. (2008). Doping effects of Co^{2+} ions on ZnO nanorods and their photo catalytic properties. *Journal of Nanotechnology*, *19*(21), 215703. doi:10.1088/0957-4484/19/21/215703 PMID:21730582

Qureshi, M. (2015). *Effect of Heavy Metals from Tannery Effluent on the Soil and Groundwater using Multivariate Analysis in District Sheikhupura*. Academic Press.

Qu, X., Alvarez, P. J., & Li, Q. (2013). Applications of nanotechnology in water and wastewater treatment. *Water Research*, *47*(12), 3931–3946. doi:10.1016/j.watres.2012.09.058 PMID:23571110

Qu, X., Brame, J., Li, Q., & Alvarez, P. J. (2012). Nanotechnology for a safe and sustainable water supply: Enabling integrated water treatment and reuse. *Accounts of Chemical Research*, *46*(3), 834–843. doi:10.1021/ar300029v PMID:22738389

Rabbani, D., Mahmoudkashi, N., Mehdizad, F., & Shaterian, M. (2016).. . *Journal of Environmental Science and Technology*, *9*(1), 121–130. doi:10.3923/jest.2016.121.130

Rabotnova, I. L. (1996). *Die Bedentung physikalish-chemischer Factoren (pH und rH_2) fur die Leben-Statigkeit der Microorganismen*. Jena: Fischer-Verlag.

Rahim, M., & Haris, M. R. H. M. (2015). Application of biopolymer composites in arsenic removal from aqueous medium: A review. *Journal of Radiation Research and Applied Sciences*, *8*(2), 255–263. doi:10.1016/j.jrras.2015.03.001

Rahman, F., Langford, K. H., Scrimshaw, M. D., & Lester, J. N. (2001). Polybrominated diphenyl ether (PBDE) flame retardants. *The Science of the Total Environment*, *275*(1-3), 1–17. doi:10.1016/S0048-9697(01)00852-X PMID:11482396

Rahman, N. M., Roy, D., & Mukit, M. S. H. (2016). Investigation on the relative degradation of Methylene Blue (MB) and Rhodamine-B (RB) dyes under UV-visible light using thermally treated commercial and doped TiO2/ZnO photocatalysts. *J Mat Let*, *2*, 14–18.

Rahman, Q. I., Ahmad, M., Misra, S., & Lohani, M. (2013). Effective photocatalytic degradation of rhodamine B dye by ZnO nanoparticles. *J Mat Let*, *91*, 170–174. doi:10.1016/j.matlet.2012.09.044

Railean-Plugaru, V., Pomastowski, P., Kowalkowski, T., Sprynskyy, M., & Buszewski, B. (2018). Physicochemical study of natural fractionated biocolloid by asymmetric flow field-flow fractionation in tandem with various complementary techniques using biologically synthesized silver nanocomposites. *Analytical and Bioanalytical Chemistry*, *410*(11), 2837–2847. doi:10.100700216-018-0967-0 PMID:29616293

Railean-Plugaru, V., Pomastowski, P., Meller, K., Zloch, M., Rafinska, K., & Buszewski, B. (2017). Lactococcus lactis as a safe and inexpensive source of bioactive silver composites. *Applied Microbiology and Biotechnology*, *101*(19), 7141–7153. doi:10.100700253-017-8443-x PMID:28842740

Railean-Plugaru, V., Pomastowski, P., Rafińska, K., Kupczyk, W., Jackowski, M., & Buszewski, B. (2016). Antimicrobial properties of biosynthesized silver nanoparticles studied by flow cytometry and related techniques. *Electrophoresis*, *37*(5–6), 752–761C. doi:10.1002/elps.201500507 PMID:26763104

Railean-Plugaru, V., Pomastowski, P., Szultka, M., & Buszewski, B. (2014). Bionanokompleksy srebra w walce z patogenami. *Analityka*, *3*, 8–17.

Railean-Plugaru, V., Pomastowski, P., Wypij, M., Szultka-Młyńska, M., Rafińska, K., Golińska, P., ... Buszewski, B. (2016). Study of silver nanoparticles synthesized by acidophilic strain of Actinobacteria isolated from the of Picea sitchensis forest soil. *Journal of Applied Microbiology*, *120*(5), 1250–1263. doi:10.1111/jam.13093 PMID:26864807

Rajalakshmi, S., Kannan, N., & Velusamy, P. (2017). Enhanced photocatalytic activity of metal oxidesβ-cyclodextrin nanocomposites for decoloration of Rhodamine B dye under solar light irradiation. *Applied Water Science*, *7*(1), 115–1274. doi:10.100713201-014-0223-5

Rajeswari, A., Vismaiya, S., & Pius, A. (2017). Preparation, characterization of nano ZnO-blended cellulose acetate-polyurethane membrane for photocatalytic degradation of dyes from water. *Chemical Engineering Journal*, *313*, 928–937. doi:10.1016/j.cej.2016.10.124

Ramachandra, T. V., N, A., & Rd, K. (2005). *Biosorption: Techniques and Mechanisms*. Academic Press.

Ramgopal, G., Vidya, Y. S., Anantharaju, K. S., Prasad, B. D., Sharma, S. C., Prashantha, S. C., ... Nagabhushana, H. (2015). Bio-inspired synthesis of Y2O3: Eu3+ red nanophosphor for eco-friendly photocatalysis. *Spectrochimica Acta. Part A: Molecular and Biomolecular Spectroscopy*, *141*, 149–160. doi:10.1016/j.saa.2015.01.055

Rani, A. (2009). Cytotoxicity and genotoxicity of silver nanoparticles in human cells. *ACS Nano*, *3*(2), 279–290. doi:10.1021/nn800596w PMID:19236062

Ranjbar, N., & Künzel, C. (2017). Cenospheres. *RE:view*, *207*.

Rappaport, S. M. (2011). Implications of the exposome for exposure science. *Journal of Exposure Science & Environmental Epidemiology*, *21*(1), 5–9. doi:10.1038/jes.2010.50 PMID:21081972

Rappaport, S. M., & Smith, M. T. (2010). Environment and disease risks. *Science*, *330*(6003), 460–461. doi:10.1126cience.1192603 PMID:20966241

Rathinam, K., & Sankaran, M. (2014). Article. *Chemical Engineering Journal.*

Ratniece, V., Rubene, L., Cakars, I., Siņics, L., Griķe, I., Gaidis, K., . . . Gancone, A. (2017). *Latvia's seventh National Communication and Third Biennial Report under the United Nations Framework Convention on Climate Change.* Retrieved from https://unfccc.int/files/national_reports/national_communications_and_biennial_reports/application/pdf/9308541_latvia-br3-nc7-1-latvia_nc7_29122017.pdf

Ratti, R. (2014). Ionic Liquids: Synthesis and Applications in Catalysis. *Advances in Chemistry*, *2014*, 1–16. doi:10.1155/2014/729842

Raukas, A. (2004). Opening a new decade. *Oil Shale, A Scientific-Technical Journal, 21*(1), 1-2.

Raveendran, P., Fu, J., & Wallen, S. L. (2003). Completely "Green" Synthesis and Stabilization of Metal Nanoparticles. *Journal of the American Chemical Society*, *125*(46), 13940–13941. doi:10.1021/ja029267j PMID:14611213

Rawn. (2004). General Chemistry. Wm. C. Brown Publishers.

Reddy, S. R., Saritha, V., Karnena, M. K., & Dwarapureddi, B. K. (2017). Combined SBR and RO pilot scale treatment for pharmaceutical wastewater. *Desalination and Water Treatment*, *98*, 45–51. doi:10.5004/dwt.2017.21579

Rehman, S., Ullah, R., Butt, A. M., & Gohar, N. D. (2009). Strategies of making TiO_2 and ZnO visible light active. *Journal of Hazardous Materials*, *170*(2-3), 560–569. doi:10.1016/j.jhazmat.2009.05.064 PMID:19540666

Reimann, C., Halleraker, J., Kashulina, G., & Bogatyrev, I. (1999). Comparison of plant and precipitation chemistry in catchments with different levels of pollution in Kola Peninsula, Russia. *The Science of the Total Environment*, *243/244*, 169–191. doi:10.1016/S0048-9697(99)00390-3

Renna, L. A., Boyle, C. J., Gehan, T. S., & Venkataraman, D. (2015). Polymer Nanoparticle Assemblies: A Versatile Route to Functional Mesostructures. *Macromolecules*, *48*(18), 6353–6368. doi:10.1021/acs.macromol.5b00375

Ren, Z. F., Lao, J. Y., Huang, J. Y., & Wang, D. Z. (2003). ZnO nanobridges and nanonails. *Nano Letters*, *3*(2), 235–238. doi:10.1021/nl025884u

Rice, C. P., & White, D. S. (1987). PCB availability assessment of river dredging using caged clams and fish. *Environmental Toxicology and Chemistry*, *6*(4), 259–274. doi:10.1002/etc.5620060404

Ritchie, J. D., & Perdue, E. M. (2003). Proton-binding study of standard and reference fulvic acids, humic acids, and natural organic matter. *Geochimica et Cosmochimica Acta*, *67*(1), 85–96. doi:10.1016/S0016-7037(02)01044-X

Rixecher, S. S., & Tipene-Matua, B. (2003). Maori Kaupapa and the Inseparability of Social and Environmental Justice: An Analysis of Bioprospecting and a People's Resistance to (Bio)cultural Assimilation. In J. Agyeman, R. Evan, & R. D. Bullard (Eds.), *Just Sustainabilities: Development in an Unequal World.* Earthscan Publications Ltd.

Robinson, L., & Miller, R. (2015). The Impact of Bisphenol A and Phthalates on Allergy, Asthma, and Immune Function: A Review of Latest Findings. *Current Environmental Health Reports*, *2*(4), 379–387. doi:10.100740572-015-0066-8 PMID:26337065

Robinson, O., & Vrijheid, M. (2015). The Pregnancy Exposome. *Current Environmental Health Reports*, *2*(2), 204–213. doi:10.100740572-015-0043-2 PMID:26231368

Robrock, K. R., Korytár, P., & Alvarez-Cohen, L. (2008). Pathways for the anaerobic microbial debromination of polybrominated diphenyl ethers. *Environmental Science & Technology*, *42*(8), 2845–2852. doi:10.1021/es0720917 PMID:18497133

Rodica, S. (2016). Microelements in food. In Microelements in the composition of the biosphere and their application in agriculture and medicine, (pp.174-194). Chisinau: ASM Publishing House.

Rodica, S., Boris, G., Roxana, I. E., & Diana, C. (2017). The authenticity and harmlessness of uvological products. Chişinău, "MS Logo" Publishing House.

Rodriguez-Liviano, S., Nuñez, N. O., Rivera-Fernández, S., de la Fuente, J. M., & Ocaña, M. (2013). Ionic Liquid Mediated Synthesis and Surface Modification of Multifunctional Mesoporous Eu:GdF_3 Nanoparticles for Biomedical Applications. *Langmuir*, *29*(10), 3411–3418. doi:10.1021/la4001076 PMID:23402647

Rojas, D., Rager, J. E., Smeester, L., Bailey, K. A., Drobná, Z., Rubio-Andrade, M., ... Fryet, R. C. (2015). Prenatal arsenic exposure and the epigenome: Identifying sites of 5-methylcytosine alterations that predict functional changes in gene expression in newborn cord blood and subsequent birth outcomes. *Toxicological Sciences*, *143*(1), 97–106. doi:10.1093/toxsci/kfu210 PMID:25304211

Rokesh, K., Pandikumar, A., & Jothivenkatachalam, K. (2014). ZincOxide nanopillar: Preparation, characterization and its photoelectron catalytic activity. *Mater. Focus*, *3*(5), 1–5. doi:10.1166/mat.2014.1172

Roldan Cuenya, B., & Behafarid, F. (2015). *Nanocatalysis: Size- and shape-dependent chemisorption and catalytic reactivity. Surface Science Reports.* Elsevier B.V.

Romanciuc, I. F., Sakhatsky A. I., & Apostolov A. A. (2018). The estimation of soil humidity by the satellite Sentinel-2 imageries (object of study is the Baryshevs'kyi district of the Kiev region. Dopov. Nac. akad. nauk Ukr, 1, 60-66. doi:10.15407/dopovidi2018.01.060

Romanciuc, I. F. (2018). Investigation the influence of soils moisture regime on their degradation using the remote sensing and ground field verification. *Ukrainian Journal of Remote Sensing*, *17*(17), 26–30. doi:10.36023/ujrs.2018.17.126

Rossi, L. M., Dupont, J., Machado, G., Fichtner, P. F. P., Radtke, C., Baumvol, I. J. R., & Teixeira, S. R. (2004). Ruthenium dioxide nanoparticles in ionic liquids: Synthesis, characterization and catalytic properties in hydrogenation of olefins and arenes. *Journal of the Brazilian Chemical Society*, *15*(6), 904–910. doi:10.1590/S0103-50532004000600018

Roucoux, A., Schulz, J., & Patin, H. (2002). Reduced transition metal colloids: A novel family of reusable catalysts? *Chemical Reviews*, *102*(10), 3757–3778. doi:10.1021/cr010350j PMID:12371901

Rowley, J. A., Madlambayan, G., & Mooney, D. J. (1999). Alginate hydrogels as synthetic extracellular matrix materials. *Biomaterials*, *20*(1), 45–53. doi:10.1016/S0142-9612(98)00107-0 PMID:9916770

Roy, L. N., Roy, R. N., Wollen, J. T., Harmon, M. A., Stegner, J. M., Shah, A. A., & Henson, I. B. (2011b). Buffer standards for the biological pH of the amino acid n-[2-hydroxyethyl] piperazine-n¢-[3-propanesulfonic acid], Hepps, from (278.15 to 328.15) K. *Journal of Chemical & Engineering Data*, *56*(11), 4126–4132. doi:10.1021/je200629y PMID:22096257

Roy, R. N., Roy, L. N., Stegner, J. M., Sechler, S. A., Jenkins, A. L., Krishchenko, R., & Henson, I. B. (2011a). Buffer standards for the physiological pH of N, N - bis (2- hydroxyethyl) - 2-aminoethanesulfonic acid (BES) from (278.15 to 328.15) K. *Electroanalytical Chemistry*, *663*(1), 8–13. doi:10.1016/j.jelechem.2011.09.019 PMID:22247703

Rozich, A., & Bordacs, K. (2002). Use of thermophilic biological aerobic technology for industrial wastewater treatment. *Water Science and Technology*, *46*(4-5), 83–89. doi:10.2166/wst.2002.0557 PMID:12361053

Ruhling, A., & Tyler, G. (1971). Regional differences in the deposition of heavy metals over Scandinavia. *Journal of Applied Ecology*, *8*(2), 497–507. doi:10.2307/2402886

Rusiecki, J. A., Baccarelli, A., Bollati, V., Tarantini, L., Moore, L. E., & Bonefeld-Jorgensen, E. C. (2008). Global DNA hypomethylation is associated with high serum-persistent organic pollutants in Greenlandic Inuit. *Environmental Health Perspectives*, *116*(11), 1547–1552. doi:10.1289/ehp.11338 PMID:19057709

Russia. (2019). Retrieved from https://en.wikipedia.org/wiki/Russia

Russia. (2020). Retrieved from https://en.wikipedia.org/wiki/Russia

Ruta, M., Laurenczy, G., Dyson, P. J., & Kiwi-Minsker, L. (2008). Pd Nanoparticles in a Supported Ionic Liquid Phase: Highly Stable Catalysts for Selective Acetylene Hydrogenation under Continuous-Flow Conditions. *The Journal of Physical Chemistry C*, *112*(46), 17814–17819. doi:10.1021/jp806603f

Ryabenkii, V., & Tsynkov, S. (2019). *A Theoretical Introduction to Numerical Analysis*. Chapman and Hall/CRC.

Rytwo, G. (2012). The use of clay-polymer nanocomposites in wastewater pretreatment. *TheScientificWorldJournal*, *2012*, 498503–498503. doi:10.1100/2012/498503 PMID:22454607

S., A., & M., S. (2016). [EMIM] BF4 ionic liquid-mediated synthesis of TiO2 nanoparticles using Vitex negundo Linn extract and its antibacterial activity. *Journal of Molecular Liquids, 221*, 986–992.

Sabbaghan, M., Anaraki Firooz, A., & Jan Ahmadi, V. (2012). The effect of template on morphology, optical and photocatalytic properties of ZnO nanostructures. *Journal of Molecular Liquids*, *175*, 135–140. doi:10.1016/j.molliq.2012.08.019

Sadhasivam, S., Shanmugam, P., & Yun, K.-S. (2010). Biosynthesis of silver nanoparticles by Streptpmyces hygroscopicus and antimicrobial activity against medically important pathogenic microorganism. *Colloids and Surfaces. B, Biointerfaces*, *81*(1), 358–362. doi:10.1016/j.colsurfb.2010.07.036 PMID:20705438

Safari, E., Rahemi, N., Kahforoushan, D., & Allahyari, S. (2019). Article. *Journal of Environmental Chemical Engineering*.

Safe, S. H. (1994). Polychlorinated biphenyls (PCBs): Environmental impact, biochemical and toxic responses, and implications for risk assessment. *Critical Reviews in Toxicology*, *24*(2), 87–149. doi:10.3109/10408449409049308 PMID:8037844

Sahoo, P., Kim, K., Powell, M., & Equeenuddin, S. (2016). *Recovery of metals and other beneficial products from coal fly ash: a sustainable approach for fly ash management* (Vol. 3). Academic Press.

Sahu, P. (2019). *Characterization of Coal Combustion By-Products (CCBs) for their Effective Management and Utilization*. Academic Press.

Šakalys, J., Kvietkus, K., Sucharová, J., Suchara, I., & Valiulis, D. (2009). Changes in total concentrations and assessed background concentrations of heavy metals in moss in Lithuania and the Czech Republic between 1995 and 2005. *Chemosphere*, *76*(1), 91–97. doi:10.1016/j.chemosphere.2009.02.009 PMID:19269004

Sakharov, A. D. (1958). Radioactive carbon in nuclear explosions and biological non threshold effects. Atomic Energy, 4(6).

Sakhatsky, A. I. (2007). Improved hydrological forecasts using water index multispectral satellite filming. *Geological Journal*, *4*, 42–47.

Sakthivel, S., Neppolian, B., Shankar, M. V., Arabindoo, B., Palanichamy, M., & Murugesan, V. (2003). Solar photocatalytic degradationof azo dye: Comparison of photocatalytic efficiency of ZnO and TiO_2. *Solar Energy Materials and Solar Cells*, *77*(1), 65–82. doi:10.1016/S0927-0248(02)00255-6

Sakulpitakphon, T., Hower, J., S. Trimble, A., H. Schram, W., & Thomas, G. (2000). *Mercury Capture by Fly Ash: Study of the Combustion of a High-Mercury Coal at a Utility Boiler* (Vol. 14). Academic Press.

Salah, N., Habib, S. S., Khan, Z. H., Alshahrie, A., Memic, A., & Al-ghamdi, A. A. (2016). Carbon rich fly ash and their nanostructures. *Carbon Letters, 19*, 23-31. doi:10.5714/cl.2016.19.023

Saleh, R., & Djaja, N. F. (2014). Transition-metal-doped ZnO nanoparticles: Synthesis, characterization and photocatalytic activity under UV light. *Spectrochimica Acta. Part A: Molecular and Biomolecular Spectroscopy*, *130*, 581–590. doi:10.1016/j.saa.2014.03.089 PMID:24813289

Salvador, P. (2007). On the nature of photogenerated radical species active in the oxidative degradation of dissolved pollutants with TiO2 aqueous suspensions: A revision in the light of the electronic structure of adsorbed water. *The Journal of Physical Chemistry C*, *2007*(111), 17038–17043. doi:10.1021/jp074451i

Samiee, S., Goharshadi, E. K., & Nancarrow, P. (2016). Successful degradation of Reactive Black 5 by engineered Fe/Pd nanoparticles: Mechanism and kinetics aspects. *Journal of the Taiwan Institute of Chemical Engineers*, *67*, 406–417. doi:10.1016/j.jtice.2016.07.012

Samsudin, E. M., Goh, S. N., Wu, T. Y., Ling, T. T., Hamid, S. B. A., & Juan, J. C. (2015). Evaluation on the photocatalytic degradation activity of reactive blue 4 using pure anatase nano-TiO2. *Sains Malaysiana*, *44*(7), 1011–1019. doi:10.17576/jsm-2015-4407-13

Sánchez-España, J., Yusta, I., & Diez-Ercilla, M. (2011). Schwertmannite and hydrobasaluminite: A re-evaluation of their solubility and control on the iron and aluminium concentration in acidic pit lakes. *Applied Geochemistry*, *26*(9-10), 1752–1774. doi:10.1016/j.apgeochem.2011.06.020

Sandu, M., Tarita, A., Mosanu, E., & Turcan, S. (2017). *Surface water pollution index. Case study - water from the forest bypass Hârjauca. Scientific-practical guide.* Chisinau, Republic of Moldova: Primex - Com.

Sandu, M., Moşanu, E., Gladchi, V., Tarita, A., Duca, Gh., Spataru, P., ... Turcan, S. (2010). Study of springs water quality as sources of potable water and for irrigation în Rezina district. *Chemistry Journal of Moldova: General. Industrial and Ecological Chemistry*, *5*(1), 84–89.

Sandu, M., Tarita, A., Lozan, R., Gladchi, V., Duca, G. H., Turcan, S., ... Prepelita, A. (2013). Water of springs – sources for water supply and irrigation in the Nistru river basin. *Chemistry Journal of Moldova. Series General. Industrial and Ecological Chemistry.*, *8*(2), 42–50.

Sandvik, S. L. H., Bilski, P., Pakulski, J. D., Chignell, C. F., & Coffin, R. B. (2000). Photogeneration of singlet oxygen and free radicals in dissolved organic matter isolated from the Mississippi and Atchafalaya River plumes. *Marine Chemistry*, *69*(1-2), 139–152. doi:10.1016/S0304-4203(99)00101-2

Sangave, P. C., & Pandit, A. B. (2004). Ultrasound pre-treatment for enhanced biodegradability of the distillery wastewater. *Ultrasonics Sonochemistry*, *11*(3-4), 197–203. doi:10.1016/j.ultsonch.2004.01.026 PMID:15081981

Santhosh, C., Velmurugan, V., Jacob, G., Jeong, S. K., Grace, A. N., & Bhatnagar, A. (2016). Role of nanomaterials in water treatment applications: A review. *Chemical Engineering Journal*, *306*, 1116–1137. doi:10.1016/j.cej.2016.08.053

Santos, P., Souza, D., & Saldanha, T.Introducing Comics as an Alternative Scientific Narrative in Chemistry Teaching. (2013). Anadolu University J. *Science Education.*, *8*, 14.

Saravanan, R., Karthikeyan, S., Gupta, V. K., Sekaran, G., Narayanan, V., & Stephen, A. (2013). Enhanced photocatalytic activity of ZnO/CuO nanocomposite for the degradation of textile dye on visible light illumination. *Materials Science and Engineering C*, *33*(1), 91–98. doi:10.1016/j.msec.2012.08.011 PMID:25428048

Sarigiannis, D. A. (2017). Assessing the impact of hazardous waste on children's health: The exposome paradigm. *Environmental Research*, *158*, 531–541. Epub2017Jul14. doi:10.1016/j.envres.2017.06.031 PMID:28711809

Sarigiannis, D. A. (2019). The Exposome Paradigm in Environmental Health. In P. Papadopoulou, C. Marouli, & A. Misseyanni (Eds.), *Environmental Exposures and Human Health Challenges* (pp. 1–29). Hershey, PA: IGI Global. doi:10.4018/978-1-5225-7635-8.ch001

Sarigiannis, D. A., Gotti, A., Handakas, E., & Karakitsios, S. P. (2018). Informatics and Data Analytics to Support Exposome-Based Discovery: Part 2 - Computational Exposure Biology. In M. Lytras & P. Papadopoulou (Eds.), *Applying Big Data Analytics in Bioinformatics and Medicine* (pp. 145–187). Hershey, PA: IGI Global. doi:10.4018/978-1-5225-2607-0.ch007

Sarigiannis, D. A., Karakitsios, S. P., Handakas, E., Papadaki, K., Chapizanis, D., & Gotti, A. (2018). Informatics and Data Analytics to Support Exposome-Based Discovery: Part 1 - Assessment of External and Internal Exposure. In M. Lytras & P. Papadopoulou (Eds.), *Applying Big Data Analytics in Bioinformatics and Medicine* (pp. 115–144). Hershey, PA: IGI Global. doi:10.4018/978-1-5225-2607-0.ch006

Saritha, V., Srinivas, N., & Srikanth Vuppala, N. V. (2017). Analysis and optimization of coagulation and flocculation process. *Applied Water Science*, *7*(1), 451–460. doi:10.100713201-014-0262-y

Sartore, L., & Dey, K. (2019). Preparation and Heavy Metal Ions Chelating Properties of Multifunctional Polymer-Grafted Silica Hybrid Materials. *Advances in Materials Science and Engineering*, *11*, 1–11. doi:10.1155/2019/7260851

Sartz, L., Bäckström, M., Karlsson, S., & Allard, B. (2016). Mixing of acid rock drainage with alkaline ash leachates: Formation of solid precipitates and pH-buffering. *Mine Water and the Environment*, *35*(1), 64–76. doi:10.100710230-015-0347-3

Sathishkumar, P., Sweena, R., Wu, J. J., & Anandan, S. (2011). Synthesis of CuO-ZnO nanophotocatalyst for visible light assisted degradation of a textile dye in aqueous solution. *Chemical Engineering Journal*, *171*(1), 136–140. doi:10.1016/j.cej.2011.03.074

Sathyamurthy, R., Dharmaraj, V., & Thadhani J, P. (2015). *Thermal and wear analysis on the effect of different compositions of flyash in Mg/SiC composite*. Academic Press.

Sato, T., Shimosato, T., & Klinman, D. M. (2018). Silicosis and lung cancer: Current perspectives. *Lung Cancer (Auckland, N.Z.)*, *9*, 91–101. doi:10.2147/LCTT.S156376 PMID:30498384

Savadogo, O., Lee, K., Oishi, K., Mitsushima, S., Kamiya, N., & Ota, K. I. (2004). New palladium alloys catalyst for the oxygen reduction reaction in an acid medium. *Electrochemistry Communications*, *6*(2), 105–109. doi:10.1016/j.elecom.2003.10.020

Saxena, R. C., Adhikari, D. K., & Goyal, H. B. (2009). Biomass-based energy fuel through biochemical routes: A review. *Renewable & Sustainable Energy Reviews*, *13*(1), 167–178. doi:10.1016/j.rser.2007.07.011

Schätz, A., Long, T. R., Grass, R. N., Stark, W. J., Hanson, P. R., & Reiser, O. (2010). Immobilization on a nanomagnetic Co/C surface using ROM polymerization: Generation of a hybrid material as support for a recyclable palladium catalyst. *Advanced Functional Materials*, *20*(24), 4323–4328. doi:10.1002/adfm.201000959 PMID:22308111

Schecter, A., Lorber, M., Guo, Y., Wu, Q., Yun, S. H., Kannan, K., ... Birnbaum, L. S. (2013). Phthalate Concentrations and Dietary Exposure from Food Purchased in New York State. *Environmental Health Perspectives*, *121*(4), 473–479. doi:10.1289/ehp.1206367 PMID:23461894

Schiewer, S., & Iqbal, M. (2010). *The Role of Pectin in Cd Binding by Orange Peel Biosorbents: A Comparison of Peels, Depectinated Peels and Pectic Acid* (Vol. 177). Academic Press.

Schmidt, F. (2001). New catalyst preparation technologies—Observed from an industrial viewpoint. *Applied Catalysis A, General*, *221*(1-2), 15–21. doi:10.1016/S0926-860X(01)00802-X

Schneider, S., Halbig, P., Grau, H., & Nickel, U. (1994). Reproducible preparation of silver sols with uniform particle size for application in surface-enhanced raman spectroscopy. *Photochemistry and Photobiology*, *60*(6), 605–610. doi:10.1111/j.1751-1097.1994.tb05156.x

Schnitzer, M., & Gupta, U. C. (1965). Determination of Acidity in Soil Organic Matter. *Soil Science Society of America Journal*, *29*(3), 274–277. doi:10.2136ssaj1965.03615995002900030016x

Scholten, J. D., Ebeling, G., & Dupont, J. (2006). On the involvement of NHC carbenes in catalytic reactions by iridium complexes, nanoparticle and bulk metal dispersed in imidazolium ionic liquids. *Journal of the Chemical Society, Dalton Transactions: Inorganic Chemistry*, (47): 5554–5560. PMID:18043817

Scholten, J. D., Leal, B. C., & Dupont, J. (2012). Transition Metal Nanoparticle Catalysis in Ionic Liquids. *ACS Catalysis*, *2*(1), 184–200. doi:10.1021/cs200525e

Schrödinger E. (1944). *What is life? The physical aspect of the living cell.* Based on lectures delivered under the auspices of the Dublin Institute for Advanced Studies at Trinity College, Dublin, Ireland.

Schwarzenbach, R. P., Escher, B. I., Fenner, K., Hofstetter, T. B., Johnson, C. A., Von Gunten, U., & Wehrli, B. (2006). The challenge of micropollutants in aquatic systems. *Science*, *313*(5790), 1072–1077. doi:10.1126cience.1127291 PMID:16931750

Scierbak, V., & Kuzminciuk, V. (2006). The occurrence of species and the diversity of plankton algae of the Teterev River and its reservoirs (Ucraine). *Algol.*, *6*(1), 336–344.

Scott, J., Guang, D., Naeramitmarnsuk, K., Thabuot, M., & Amal, R. (2002). *Zeolite synthesis from coal fly ash for the removal of lead ions from aqueous solution* (Vol. 77). Academic Press.

Seelye, J. G., Hesselberg, R. J., & Mac, M. J. (1982). Accumulation by Fish of Contaminants Released from Dredged Sediments. *Environmental Science & Technology*, *16*(8), 459–464. doi:10.1021/es00102a006 PMID:22236268

Seinfeld, J. H. (1989). Urban air pollution: State of the science. *Science*, *243*(4892), 745–752. doi:10.1126cience.243.4892.745 PMID:17820421

Seinfeld, J. H., & Pandis, S. N. (2006). *Atmospheric chemistry and physics. From air pollution to climate change* (2nd ed.). John Wiley & Sons, Inc.

Selvam, N. C. S., Narayanan, S., Kennedy, L. J., & Vijaya, J. J. (2013). Pure and Mg-doped self-assembled ZnO nanoparticles for the enhanced photocatalytic degradation of 4-chlorophenol. *Journal of Environmental Sciences (China), 25*(10), 2157–2167. doi:10.1016/S1001-0742(12)60277-0 PMID:24494504

Séralini, G.-E., Clair, E., Mesnage, R., Gress, S., Defarge, N., Malatesta, M., ... Spiroux de Vendômois, J. (2014). Republished study: Long-term toxicity of a Roundup herbicide and a Roundup-tolerantgenetically modified maize. *Environmental Sciences Europe*, *26*(1), 14. doi:10.118612302-014-0014-5

Seralini, G.-E., Mesnage, R., Clair, E., Gress, S., de Vendomois, J., & Cellier, D. (2011). Genetically modified crops safety assessments: Present limits and possibleimprovements. *Environmental Sciences Europe*, *23*(1), 10. doi:10.1186/2190-4715-23-10

Seravkin, I. B., Kosarev, A. M., & Puchkov, V. N. (2017). Geodynamic conditions of formation of massive sulfide deposits in the Magnitogorsk Megazone, Southern Urals, and prospection criteria. *Geology of Ore Deposits*, *59*(3), 227–243. doi:10.1134/S1075701517030059

Serdyutskaya, L. F., & Kameneva, I. P. (2004). Modeli vodnyh ecosystem i ih phazovye portrety (na primere modeli krugovorota azota v Kuibyshevskom vodohranilische) [Models of aquatic ecosystems and their phase portraits (on the example of the nitrogen cycle in the Kuibyshev reservoir)]. *Ibid*, *1*(Appendix), 408.

Serrano, S. E. (2014, June 2). Phthalates and diet: A review of the food monitoring and epidemiology data. *Environmental Health*, 13. PMID:24894065

Shabatura, A., Blum, O., & Tyutyunnik, Yu. (2018). Regional atmogeochemical fields based on briogeochemical indication data in the central part of northern Ukraine. *Biosphere, 10*(1).

Shaffer, R. M., Smith, M. N., & Faustman, E. M. (2017). Developing the Regulatory Utility of the Exposome: Mapping Exposures for Risk Assessment through Lifestage Exposome Snapshots (LEnS). *Environmental Health Perspectives*, *125*(8), 085003. doi:10.1289/EHP1250 PMID:28796633

Shahverdi, A. R., Minaian, S., Shahverdi, H. R., Jamalifar, H., & Nohi, A. A. (2007). Rapid synthesis of silver nanoparticles using culture supernatatnt of *Enterobacteria*: A novel biological approach. *Process Biochemistry (Barking, London, England), 42*(5), 919–923. doi:10.1016/j.procbio.2007.02.005

Shakoor, S., & Nasar, A. (2016). *Removal of methylene blue dye from artificially contaminated water using citrus limetta peel waste as a very low cost adsorbent* (Vol. 66). Academic Press.

Shaligram, N. S., Bule, M., Bhambure, R., Singhal, R. S., Singh, S. K., Zakacs, G., & Pandey, A. (2009). Biosynthesis of silver nanoparticles using aqueous extract from the compactin producing fungal strain. *Process Biochemistry (Barking, London, England), 44*(8), 939–943. doi:10.1016/j.procbio.2009.04.009

Shamsizadeh, A., Ghaedi, M., Ansari, A., Azizian, S., & Purkait, M. K. (2014). Tin oxide nanoparticle loaded on activated carbon as new adsorbent for efficient removal of malachite green-oxalate: Non-linear kinetics and isotherm study. *Journal of Molecular Liquids*, *195*, 212–218. doi:10.1016/j.molliq.2014.02.035

Shanmugarajah, B., Chew, I. M., Mubarak, N. M., Choong, T. S. Y., Yoo, C. K., & Tan, K. W. (2019). Valorization of palm oil agro-waste into cellulose biosorbents for highly effective textile effluent remediation. *Journal of Cleaner Production*, *210*, 697–709. doi:10.1016/j.jclepro.2018.10.342

Shao, Y., Jia, D., & Liu, B. (2009). *Characterization of porous silicon nitride ceramics by pressureless sintering using fly ash cenosphere as a pore-forming agent* (Vol. 29). Academic Press.

Shao, M., Tao, P., Wang, M., Jia, H., & Li, Y. F. (2016). Trophic magnification of polybrominated diphenyl ethers in the marine food web from coastal area of Bohai Bay, North China. *Environmental Pollution, 213*, 379–385. doi:10.1016/j.envpol.2016.02.034 PMID:26942685

Shao, Z., Huang, X., Yang, F., Zhao, W., Zhou, X., & Zhao, C. (2018). Engineering sodium alginate-based cross-linked beads with high removal ability of toxic metal ions and cationic dyes. *Carbohydrate Polymers, 187*, 85–93. doi:10.1016/j.carbpol.2018.01.092 PMID:29486848

Shapiro, M., & Galperin, V. (2005). Air classification of solid particles. *RE:view, 44*.

Sharma, D., Kanchi, S., & Bisetty, K. (2019, December 1). Biogenic synthesis of nanoparticles: A review. *Arabian Journal of Chemistry*.

Sharma, S., Hasan, A., Kumar, N., & Pandey, L. M. (2018). Removal of methylene blue dye from aqueous solution using immobilized Agrobacterium fabrum biomass along with iron oxide nanoparticles as biosorbent. *Environmental Science and Pollution Research International, 25*(22), 21605–21615. doi:10.100711356-018-2280-z PMID:29785597

Sharonova, O. M., Anshits, N. N., Fedorchak, M. A., Zhizhaev, A. M., & Anshits, A. G. (2015). Characterization of Ferrospheres Recovered from High-Calcium Fly Ash. *Energy & Fuels, 29*(8), 5404–5414. doi:10.1021/acs.energyfuels.5b01618

Shavisi, Y., Sharifnia, S., & Mohamadi, Z. (2016). Solar-light-harvesting degradation of aqueous ammonia by CuO/ZnO immobilized on pottery plate: Linear kinetic modeling for adsorption and photocatalysis process. *Journal of Environmental Chemical Engineering, 4*(3), 2736–2744. doi:10.1016/j.jece.2016.04.035

Shekari, Z., Younesi, H., Heydari, A., Tajbakhsh, M., Chaichi, M., Shahbazi, A., & Saberi, D. (2017). Fluorescence chemosensory determination of Cu^{2+} using a new rhodamine–morpholine conjugate. *Chemosensors, 5*(3), 26–40. doi:10.3390/chemosensors5030026

Sheldon, R. A. (2012, February 21). Fundamentals of green chemistry: Efficiency in reaction design. *Chemical Society Reviews, 41*(4), 1437–1451. doi:10.1039/C1CS15219J PMID:22033698

Shenhar, R., Norsten, T. B., & Rotello, V. M. (2005). Polymer-Mediated Nanoparticle Assembly: Structural Control and Applications. *Advanced Materials, 17*(6), 657–669. doi:10.1002/adma.200401291

Shepherd, A., Jepson, R., & Watterson, A. (2015). Risk Perceptions of Environmental Hazards and Human Reproduction: A Community-Based survey. International Scholarly Reseach Network. *Public Health*, 9.

Sheshmani, S., & Mashhadi, S. (2017). Article. *Polymer Composites*. doi:10.1002/pc.24608

Shestopalov, V. M., & Bublyas, V. N. (2016). On formation of depression-channel migration structures. *Geological Journal, 3*(356), 73–88. doi:10.30836/igs.1025-6814.2016.3.97248

Shetekauri, S., Chaligava, O., Shetekauri, T., Kvlividze, A., Kalabegishvili, T., Kirkesali, E., ... Tselmovich, V. (2018). Biomonitoring Air Pollution Using Moss in Georgia. *Polish Journal of Environmental Studies, 27*(5), 2259–2266. doi:10.15244/pjoes/73798

Shetekauri, S., Shetekauri, T., Kvlividze, A., Chaligava, O., Kalabegishvili, T., Kirkesali, E., ... Chepurchenko, O. (2015). Preliminary results of atmospheric deposition of major and trace elements in the greater and lesser Caucasus mountains studied by the moss technique and neutron activation analysis. *Annals of Botany, 5*, 89–95.

Shima, S. (2014). Enzyme chemistry of methanogenesis and anaerobic oxidation of methane. *Proceedings of Max Plank Institute for Terrestrial Microbiology, 52*, 307–312.

Shima, S., Warkentin, E., Thauer, R., & Ermler, U. (2002). Structure and function of enzymes involved in the methanogenic pathway utilizing carbon dioxide and molecular hydrogen. *Journal of Bioscience and Bioengineering*, *93*(6), 519–530. doi:10.1016/S1389-1723(02)80232-8 PMID:16233244

Shirkhanloo, H., Osanloo, M., Ghazaghi, M., & Hassani, H. (2017). Validation of a new and cost-effective method for mercury vapor removal based on silver nanoparticles coating on micro glassy balls. *Atmospheric Pollution Research*, *8*(2), 359–365. doi:10.1016/j.apr.2016.10.004

Shitikov, V. K., Rozenberg, G. S., & Zinchenko, T. D. (2003). *Kolichestvennaya gidroekologiya: metody sistemnoy identifikatsii* [Quantitative Hydroecology: methods of system identification]. Tolyatti: IEVB RAS.

Shivaji, S., Madhu, S., & Singh, S. (2011). Extracellular synthesis of antibacterial silver nanoparticles using psychrophilic bacteria. *Process Biochemistry (Barking, London, England)*, *46*(9), 1800–1807. doi:10.1016/j.procbio.2011.06.008

Shostak, S. (2013). *Exposed Science: Genes, the Environment, and the Politics of Population Health*. University of California Press. doi:10.1525/california/9780520275171.001.0001

Show, K. Y., Tay, J. H., & Hung, Y. T. (2010). Global Perspective of Anaerobic Treatment of Industrial Wastewater. In L. Wang, J. H. Tay, S. Tay, & Y. T. Hung (Eds.), *Environmental Bioengineering. Handbook of Environmental Engineering* (Vol. 11, pp. 773–807). Totowa, NJ: Humana Press. doi:10.1007/978-1-60327-031-1_23

Shtamm, E. V. (1988). Redoks-sostoyanie vodnoj sredy i voprosy vosproizvodstva rybnyh resursov [Redox state of the aquatic environment and issues of reproduction of fish resources]. In Digest Ekologicheskaya himiya vodnoj sredy. Materialy 1 Vsesoyuznoy shkoly [Ecological chemistry of water environment]. Materials of the I All-Union School.

Shtamm, E. V., Purmal, A. P., & Skurlatov, Y. I. (1991). Rol' peroksida vodoroda v prirodnoj vodnoj srede [The role of hydrogen peroxide in a natural water environment]. *Uspekhi Khimii*, *60*(11), 2373–2411.

Shtamm, E. V., Skurlatov, Yu. I., & Shvydkij, V. O. (2015). Priroda toksicheskogo vozdejstviya stochnyh vod predpriyatij cellyulozno-bumazhnogo proizvodstva na vodnye ekosistemy [The nature of the toxic effects of wastewater from pulp and paper production enterprises on aquatic ecosystems] Himicheskaya fizika. *Chemical Physics*, *34*(6), 22–29.

Siegel, J., Lyutakov, O., Polívková, M., Staszek, M., Hubáček, T., & Švorčík, V. (2017). Laser-assisted immobilization of colloid silver nanoparticles on polyethyleneterephthalate. *Applied Surface Science*, *420*, 661–668. doi:10.1016/j.apsusc.2017.05.151

Silva, I. M. P., Byzynski, G., Ribeiro, C., & Longo, E. (2016). Different dye degradation mechanisms for ZnO and ZnO doped with N (ZnO:N). *Journal of Molecular Catalysis A Chemical*, *417*, 89–100. doi:10.1016/j.molcata.2016.02.027

Simonov, Y. A., Fonari, M. S., Duca, G. G., Gonta, M. V., Ganin, E. V., Yavolovskii, A. A., ... Lipkowski, J. (2005). Nitrosation of hydrochlorothiazide and the modes of binding of the N-nitroso derivative with two macrocycles possessing an 18-membered crown ether cavity. *Tetrahedron*, *61*(27), 6596–6601. doi:10.1016/j.tet.2005.04.050

Simonov, Y. A., Gonta, M. V., & Yavolovskii, A. A. (2005). 11-Nitroso-6, 7, 10, 11, 12, 13, 15, 16-octahydro-9H-5, 8, 14, 17-tetraoxa-11-azabenzocyclopentadecene. *Acta Crystallographica. Section E, Structure Reports Online*, *61*(10), o3136–o3139. doi:10.1107/S1600536805027315

Singh, A. V., Laux, P., Luch, A., Sudrik, C., Wiehr, S., Wild, A. M., … Sitti, M. (2019b, June 13). Review of emerging concepts in nanotoxicology: opportunities and challenges for safer nanomaterial design. *Toxicology Mechanisms and Methods*.

Singh, A. V., Laux, P., Luch, A., Sudrik, C., Wiehr, S., Wild, A.-M., ... Sitti, M. (2019a). Review of emerging concepts in nanotoxicology: Opportunities and challenges for safer nanomaterial design. *Toxicology Mechanisms and Methods, 29*(5), 378–387. doi:10.1080/15376516.2019.1566425 PMID:30636497

Singh, A., Parmar, N., Kuhad, R. C., & Ward, O. P. (2011). Bioaugmentation, Biostimulation, and Biocontrol in Soil Biology. In A. Singh, N. Parmar, & R. C. Kuhad (Eds.), *Bioaugmentation, Biostimulation, and Biocontrol* (pp. 1–23). Berlin: Springer. doi:10.1007/978-3-642-19769-7_1

Singh, P., Singh, S. K., Bajpai, J., Bajpai, A. K., & Shrivastava, R. B. (2014). Iron crosslinked alginate as novel nanosorbents for removal of arsenic ions and bacteriological contamination from water. *Journal of Materials Research and Technology, 3*(3), 195–202. doi:10.1016/j.jmrt.2014.03.005

Singh, R., Gautam, N., Mishra, A., & Gupta, R. (2011). Heavy metals and living systems: An overview. *Indian Journal of Pharmacology, 43*(3), 246–253. doi:10.4103/0253-7613.81505 PMID:21713085

Singh, S., Tripathi, D. K., Singh, S., Sharma, S., Dubey, N. K., Chauhan, D. K., & Vaculík, M. (2017). Toxicity of aluminium on various levels of plant cells and organism: A review. *Environmental and Experimental Botany, 137*, 177–193. doi:10.1016/j.envexpbot.2017.01.005

Sitorus, B., Sukandar, & Panjaitan, S. D. (2013). Biogas Recovery from Anaerobic Digestion Process of Mixed Fruit -Vegetable Wastes. *Energy Procedia, 32*, 176–182. doi:10.1016/j.egypro.2013.05.023

Sivasubramanian, V., & Jayaraj, S. (2014). Biogas Generation through Anaerobic Digestion Process. *Research Journal of Chemistry and Environment, 18*(5), 80–93.

Skurlatov, Y. I., & Purmal, A. P. (1984). Ekologicheskaya khimiya vodnoj sredy [Ecological chemistry of the aquatic environment]. Priroda, 10, 94-103.

Skurlatov, Y., & Shtamm, E. (2012). Redox and free-radical processes balance in natural aquatic environment as a factor for the stability of the aerobic aquatic ecosystems under anthropogenic impacts. In *Abstract Book of the V Intern. conf.-simpos. Ecological Chemistry, Chisinau, March 2-3, 2012*. Chişinău, Moldova: CEP USM.

Skurlatov, Y., Duca, G., Batir, D., & Travin, S. (1989). Coordination compounds of the partial charge transfer complexes in the redox processes. *Coordination Chemistry Journal, 15*(3), 291–307.

Skurlatov, Y., Ernestova, L.S., & Travin, S.O. (1987). Oxidation processes in natural waters. Water Resources, 5, 66-72.

Skurlatov, Y., Stamm, E., & Shvydkii, V. (2017). Role of oxidation-reduction processes in formation of toxic properties of a natural aqueous environment. In Abstract Book on Ecological and Environmental Chemistry. Chisinau, Republic of Moldova: Academy of Sciences of Moldova.

Skurlatov, Y.I., Duka, G.G., & Mizitty, A. (1994). *Vvedenie v ekologicheskuyu khimiyu* [Introduction to ecological chemistry]. Moscow: Vyshaya shkola.

Skurlatov, Yu. I., Sirenko, L. A., & Shtamm, E. V. (1992) Svobodno-radikal'nye processy v vode Dneprovskih vodohranilishch posle avarii na ChAES [Free radical processes in the water of the Dnieper reservoirs after the Chernobyl accident]. In Radioaktivnoe i himicheskoe zagryaznenie Dnepra i ego vodohranilishch posle avarii na Chernobyl'skoj AES [Radioactive and chemical pollution of the Dnieper and its reservoirs after the accident at the Chernobyl nuclear power plant, Kiev]. Kiev: Naukova Dumka.

Skurlatov, Y. I., Shtamm, E. V., Kozlova, N. B., Zajceva, N. I., & Aleksandrova, E. V. (2011). Biotestirovanie v ocenke effektivnosti tekhnologij ochistki stochnyh vod [Biotesting in assessing the effectiveness of wastewater treatment technologies]. *Водные ресурсы, 38*(2), 232–238.

Skurlatov, Y., Duca, G., & Ernestova, L. (1983). The processes of toxication and mechanisms of natural water self-purification in the condition of anthropogenic impact. *News of the Academy of Sciences of Moldova Series of Biological and Chemical Sciences*, *5*, 3–20.

Skurlatov, Y., Duca, G., & Misiti, A. (1994). *Introduction in ecological chemistry: Manual for chem. and chem.-tech.* VyshaiaShkola.

Skurlatov, Yu. I., Ernestova, L. S., & Shtamm, E. V. (1984). Redoks-sostoyanie i sezonnaya toksichnost' prirodnoj vody [Redox State and Seasonal Toxicity of Natural Water] [Proceedings of Academy of Sciences of the USSR]. *Dokl. AN SSSR*, *276*(4), 1014–1016.

Slattery, J. M., Daguenet, C., Dyson, P. J., Schubert, T. J. S., & Krossing, I. (2007). How to predict the physical properties of ionic liquids: A volume-based approach. *Angewandte Chemie International Edition*, *46*(28), 5384–5388. doi:10.1002/anie.200700941 PMID:17568467

Slowing, I. I., Vivero-Escoto, J. L., Trewyn, B. G., & Lin, V. S. Y. (2010). Mesoporous silica nanoparticles: Structural design and applications. *Journal of Materials Chemistry*, *20*(37), 7924–7937. doi:10.1039/c0jm00554a

Smical, I., Muntean, A., & Nour, E. (2015). Research on the Surface Water Quality in Mining Influenced Area in North-Western part of Romania. *Geographica Pannonica*, *19*(1), 20–30. doi:10.5937/GeoPan1501020S

Smith, S. C., & Rodrigues, D. F. (2015). Carbon-based nanomaterials for removal of chemical and biological contaminants from water: A review of mechanisms and applications. *Carbon*, *91*, 122–143. doi:10.1016/j.carbon.2015.04.043

Smuleac, L., Imbrea, F., Corodan, I., Ienciu, A., Smuleac, A., & Manea, D. (2017). The influence of anthropic activity on Somes river water quality. *AgroLife Scientific Journal*, *6*(2), 178–182.

Snehal, Y., Chandra, M., & Prakash, M. (2016). Biosynthesis of zinc oxide nanoparticles using ixora coccinea leaf extract-a green approach. *Open Journal of Synthesis Theory and Applications*, *5*(01), 1–14. doi:10.4236/ojsta.2016.51001

Snoz, S. V., & Dmitrenko, N. P. (1995). Activity of the rat cytochrome P-450 system in conditions of the induction immunity and nitric oxide synthesis. *Toxicology Letters*, *78*(Supl.1), 75–76.

Sobus, J. R., Tan, Y. M., Pleil, J. D., & Sheldon, L. S. (2011). A biomonitoring framework to support exposure and risk assessments. *The Science of the Total Environment*, *409*(22), 4875–4884. doi:10.1016/j.scitotenv.2011.07.046 PMID:21906784

Somerset, V., Petrik, L. A. White, R., Klink, M. J., Key, D., & Iwuoha, E. (2005). *Alkaline hydrothermal zeolites synthesized from high SiO2 and Al2O3 co-disposal fly ash filtrates* (Vol. 84). Academic Press.

Somorjai, G. A., & Borodko, Y. G. (2001). Research in nanosciences–great opportunity for catalysis science. *Catalysis Letters*, *76*(1-2), 1–5. doi:10.1023/A:1016711323302

Søndergaard, G. L., & Owsianiak, M. (2018). LCA of soil and groundwater remediation. In M. Z. Hauschild, S. I. Olsen, & R. K. Rosenbaum (Eds.), *Life Cycle Assessment Theory and Practice*. Springer. doi:10.1007/978-3-319-56475-3_36

Song, M., Luo, C., Li, F., Jiang, L., Wang, Y., Zhang, D., & Zhang, G. (2015). Anaerobic degradation of Polychlorinated Biphenyls (PCBs) and Polychlorinated Biphenyls Ethers (PBDEs), and microbial community dynamics of electronic waste-contaminated soil. *The Science of the Total Environment*, *502*, 426–433. doi:10.1016/j.scitotenv.2014.09.045 PMID:25268572

Sowers, K. R., & May, H. D. (2013). In situ treatment of PCBs by anaerobic microbial dechlorination in aquatic sediment: Are we there yet? *Current Opinion in Biotechnology*, *24*(3), 482–488. doi:10.1016/j.copbio.2012.10.004 PMID:23102490

Sowmiah, S., & Cheng, I, C., & Chu, Y.-H. (2012). Ionic Liquids for Green Organic Synthesis. *Current Organic Synthesis, 9*(1), 74–95. doi:10.2174/157017912798889116

Sowmiah, S., Srinivasadesikan, V., Tseng, M. C., & Chu, Y. H. (2009, September). On the chemical stabilities of ionic liquids. *Molecules (Basel, Switzerland), 14*(9), 3780–3813. doi:10.3390/molecules14093780 PMID:19783957

Spencer, T. (2016). *Risk Perception: Theories and Approaches*. Hauppauge, NY: Nova Science Publishers, Inc.

Spinu, O., & Povar, I. (2015). Buffer capacity in heterogeneous multicomponent systems. [Review]. *Chemistry Journal of Moldova, 10*(2), 8–25. doi:10.19261/cjm.2015.10(2).01

Spruit, M., & Lammertink, M. (2018). Effective and Efficient Business Intelligence Dashboard Design: Gestalt Theory in Dutch Long-Term and Chronic Healthcare. In M. Lytras & P. Papadopoulou (Eds.), *Applying Big Data Analytics in Bioinformatics and Medicine* (pp. 243–271). Hershey, PA: IGI Global. doi:10.4018/978-1-5225-2607-0.ch010

Sreeram, A., Hadi, P., Hui, C. W., Ansari, T. A., & McKay, G. (2017). Optimisation of the removal of arsenate from water using nanochitosan. *Desalination and Water Treatment, 70*, 235–243. doi:10.5004/dwt.2017.20561

Staodubtsev, V. M., & Basarab, R. M. (2019). Ecologic and economic estimation of land productivity spatial heterogeneity in forest-steppe zone. *Resour Environ Econ., 1*(1), 39–42. doi:10.25082/REE.2019.01.005

Stasiev, G., & Toma, S. (2016). Ecological content and evaluation of microelements in soil (2016). In *Microelements in the composition of the biosphere and their application in agriculture and medicine* (pp. 27–56). Chisinau: ASM Publishing House.

State Register of Pesticides Approved in the Republic of Moldova. (2016). Retrieved from http://www.pesticide.md/registrul-de-stat/

Steckling, N., Gotti, A., Bose-O'Reilly, S., Chapizanis, D., Costopoulou, D., De Vocht, F., ... Sarigiannis, D. A. (2018). Biomarkers of exposure in environment-wide association studies - Opportunities to decode the exposome using human biomonitoring data. *Environmental Research, 164*, 597–624. Epub2018. doi:10.1016/j.envres.2018.02.041 PMID:29626821

Stepanov, I. N. (2006). Theory of relief plasticity and new thematic maps. Moscow: Nauka.

Stepanov, I., Hecht, S. S., Mirvish, S. S., & Gonta, M. (2005). Comparative Analysis of Tobacco-Specific Nitrosamines and Total N-Nitroso Compounds in Moldovan Cigarette Tobacco. *Journal of Agricultural and Food Chemistry, 53*(20), 8082–8086. doi:10.1021/jf050747e PMID:16190674

Stewart, P. S. (2002). Mechanisms of antibiotic resistance in bacterial biofilms. *International Journal of Medical Microbiology, 292*(2), 110–113. doi:10.1078/1438-4221-00196 PMID:12195733

Stingone, J. A., Buck Louis, G. M., Nakayama, S. F., Vermeulen, R. C., Kwok, R. K., Cui, Y., ... Teitelbaum, S. L. (2017). Toward Greater Implementation of the Exposome Research Paradigm within Environmental Epidemiology. *Annual Review of Public Health, 20*(38), 315–327. Epub2017Jan6. doi:10.1146/annurev-publhealth-082516-012750 PMID:28125387

Stockbrugger, R. W., & Kleinjans, J. C. S. (1998). Volatile N-nitrosamines in gastric juice of patients with various conditions of the gastrointestinal tract determined by gas-cromatography – mass spectrometry and related to intragastric pH and nitrate and nitrite levels. *Cancer Letters, 124*(2), 119–125. doi:10.1016/S0304-3835(97)00467-9 PMID:9500200

Stockholm Convention. (2019). *Stockholm Convention Website*. Retrieved from http://pops.int/

Stoffregen, R. E., & Cydan, G. L. (1990). An experimental study of Na-K exchange between alunite and aqueous sulfate solution. *The American Mineralogist, 75*, 209–220.

Stowell, C. A., & Korgel, B. A. (2005). Iridium nanocrystal synthesis and surface coating-dependent catalytic activity. *Nano Letters*, *5*(7), 1203–1207. doi:10.1021/nl050648f PMID:16178211

Stumm, W., & Morgan, J. J. (1996). *Aquatic Chemistry: Chemical Equilibria and Rates in Natural Waters*. New York: John Wiley & Sons.

Sturza & Lazacovich. (2017). *Monitoring the overall phthalate burden – a public health problem*. RIVEmed no.2, "Grigore T. Popa" Publishing House Iasi.

Sturza, R., Bilici, C., Zinicovscaia, I., Culicov, O. A., Gundorina, S., & Duca, G. (2015). Moldavian wine analysis by ICP-AES and NAA techniques: Comparison study. *Revue Roumaine de Chimie*, *60*(11-12), 1065–1071.

Stutter, M. I., & Richards, S. (2012). Relationships between soil physicochemical, microbiological properties, and nutrient release in buffer soils compared to field soils. *Journal of Environmental Quality*, *41*(2), 400–409. doi:10.2134/jeq2010.0456 PMID:22370402

Suárez-Ruiz, I., & Parra, J. B. (2007). Relationship between Textural Properties, Fly Ash Carbons, and Hg Capture in Fly Ashes Derived from the Combustion of Anthracitic Pulverized Feed Blends. *Energy & Fuels*, *21*(4), 1915–1923. doi:10.1021/ef060656z

Sudarsanam, P., Peeters, E., Makshina, E. V., Parvulescu, V. I., & Sels, B. F. (2019). Advances in porous and nanoscale catalysts for viable biomass conversion. *Chemical Society Reviews*, *48*(8), 2366–2421. doi:10.1039/C8CS00452H PMID:30785143

Suèr, P., Nilsson-Påledal, S., & Norrman, J. (2004). LCA for site remediation: A literature review. *Soil & Sediment Contamination*, *13*(4), 415–425. doi:10.1080/10588330490471304

Sulaymon, A. (2014). Biosorption of Heavy Metals. *RE:view*, *3*.

Sumpter, C., & Chandramohan, D. (2013). Systematic review and metaanalysis of the associations between indoor air pollution and tuberculosis. *Tropical Medicine & International Health*, *18*(1), 101–108. doi:10.1111/tmi.12013 PMID:23130953

Sun, J., & Tan, H. (2013). Alginate-Based Biomaterials for Regenerative Medicine Applications. *Materials (Basel)*, *6*(4), 1285–1309. doi:10.3390/ma6041285 PMID:28809210

Suny, R. G., Howe, G. M., Djibladze, M. L., & Marshall Lang, D. (2010). *Georgia*. Retrieved from https://www.britannica.com/place/Georgia

Suresh, S., Pandikumar, A., Murugesan, S., Ramaraj, R., & Paul Raj, S. (2011). Photovoltaic performance of solid-state solar cells based on ZnO nanosheets sensitized with low-cost metal-free organic dye. *Solar Energy*, *85*(9), 1787–1793. doi:10.1016/j.solener.2011.04.016

Sutaryo, S., Ward, A. J., & Moller, H. B. (2014). The effect of mixed-enzyme addition in anaerobic digestion on methane yield of dairy cattle manure. *Environmental Technology*, *35*(17-20), 2476–2482. doi:10.1080/09593330.2014.911356 PMID:25145202

Sutton, R., & Rockett, B. (2008). *Chemistry for the Life Sciences*. Taylor & Francis Group.

Suvilampi, J., & Rintala, J. (2003). Thermophilic aerobic wastewater treatment, process performance, biomass characteristics, and effluent quality. *Reviews in Environmental Science and Biotechnology*, *2*(1), 35–51. doi:10.1023/B:RESB.0000022959.46025.9a

Swarnalatha, P., & Sevugan, P. (Eds.). (2018). *Big Data Analytics for Satellite Image Processing and Remote Sensing*. IGI Global. doi:10.4018/978-1-5225-3643-7

Swenson, P., Tanchuk, B., Bastida, E., An, W., & Kuznicki, S. M. (2012). Water desalination and de-oiling with natural zeolite membranes—Potential application for purification of SAGD process water. *Desalination*, *286*, 442–446. doi:10.1016/j.desal.2011.12.008

Sychev, A., Travin, S., Duca, G., & Skurlatov, Y. (1983). Catalitical reactions and environmental protection. Chisinau, Moldova: Stiinta.

Sychev, A. Y., Travin, S. O., Duka, G. G., & Skurlatov, Y. I. (1983). *Katalitícheskie reakcii i ohrana okruzhayushchej sredy* [Catalytic reactions and environmental protection]. Chisinau: Shtiintsa.

Sycheva, S., Khokhlova, O., Pushkina, P., & Ukrainsky, P. (2019). Interrelations of the Bryansk paleosol (end of MIS 3) with the Holocene surface soils in microdepressions of the central forest-steppe within the Russian Upland. *Catena*, *172*, 619–633. doi:10.1016/j.catena.2018.09.021

Sylaios. (2014). *WASTEnet - A Black Sea network promoting integrated natural Wastewater Treatment Systems*. Black Sea Basin 2007-2013 Joint Operational Programme.

Syme, S., & Wang, C. (2012). The social Environment and Health. In *The praeger handbook of environmental health*. ABC-CLIO, LLC. Retrieved from http://ebookcentral.proquest.com

Tabors, G., Nikodemus, O., Dobkeviča, L., Kļaviņa, L., Ajanoviča, A., Viligurs, K., & Krūz, I. (2017). Assessment of atmospheric pollution with heavy metals and nitrogen using *Pleurozium schreberi* mosses as bioindicator in Latvia: Spatial and temporal aspects. *Environmental and Experimental Biology*, *15*, 143–150.

Taggart, R., Hower, J., S Dwyer, G., & Hsu-Kim, H. (2016). *Trends in the Rare Earth Element Content of U.S.-Based Coal Combustion Fly Ashes* (Vol. 50). Academic Press.

Takahama, U., Hirota, S., & Kawagishi, S. (2009). Effects of pH on nitrite-induced formation of reactive nitrogen oxide species and their scavenging by phenolic antioxidants in human oral cavity. *Free Radical Research*, *43*(3), 250–261. doi:10.1080/10715760802691463 PMID:19169919

Takahama, U., Ryu, K., Oniki, T., & Hirota, S. (2007). Dual-function of thiocyanate on nitrite-induced formation of reactive nitrogen oxide species in human oral cavity: Inhibition under neutral and enhancement under acidic conditions. *Free Radical Research*, *41*(11), 1289–1300. doi:10.1080/10715760701710885 PMID:17957543

Takano, T., Nakamura, K., & Watanabe, M. (2002). Urban residential environments and senior citizens' longevity in megacity areas: the importance of walkable green spaces. *Journal of Epidemiology and Community Health*, *56*(12), 913–918.

Tamasi, G., & Cini, R. (2004). Heavy metals in drinking waters from Mount Amiata (Tuscany, Italy). Possible risk from arsenic for public health in the Province of Siena. *The Science of the Total Environment*, *327*(1-3), 41–51. doi:10.1016/j.scitotenv.2003.10.011 PMID:15172570

Tang, H., Yu, C. H., Oduoro, W., He, H., & Tsang, S. C. (2008). Engineering of a monodisperse core− shell magnetic Ti− O− Si oxidation nanocatalyst. *Langmuir*, *24*(5), 1587–1590. doi:10.1021/la702489s PMID:18225927

Tang, J., Song, Y., Zhao, F., Spinney, S., Bernardes, J., & Tam, K. C. (2019). Compressible cellulose nanofibril (CNF) based aerogels produced via a bio-inspired strategy for heavy metal ion and dye removal. *Carbohydrate Polymers*, *208*, 404–412. doi:10.1016/j.carbpol.2018.12.079 PMID:30658817

Tang, S. C., & Lo, I. M. (2013). Magnetic nanoparticles: Essential factors for sustainable environmental applications. *Water Research*, *47*(8), 2613–2632. doi:10.1016/j.watres.2013.02.039 PMID:23515106

Tang, X., Zhang, Q., Liu, Z., Pan, K., Dong, Y., & Li, Y. (2014). Removal of Cu (II) by loofah fibers as a natural and low-cost adsorbent from aqueous solutions. *Journal of Molecular Liquids*, *199*, 401–407. doi:10.1016/j.molliq.2014.09.033

Tano, T., Esumi, K., & Meguro, K. (1989). Preparation of organopalladium sols by thermal decomposition of palladium acetate. *Journal of Colloid and Interface Science*, *133*(2), 530–533. doi:10.1016/S0021-9797(89)80069-4

Tao, A., Sinsermsuksaku, P., & Yang, P. (2006). Polyhedral Silver Nanocrystals with Distinct Scattering Signatures. *Angewandte Chemie International Edition*, *45*(28), 4597–4601. doi:10.1002/anie.200601277 PMID:16791902

Teunissen, W., Bol, A. A., & Geus, J. W. (1999). Magnetic catalyst bodies. *Catalysis Today*, *48*(1-4), 329–336. doi:10.1016/S0920-5861(98)00389-7

Thakur, S., Pandey, S., & Arotiba, O. A. (2016). Development of a sodium alginate-based organic/inorganic superabsorbent composite hydrogel for adsorption of methylene blue. *Carbohydrate Polymers*, *153*, 34–46. doi:10.1016/j.carbpol.2016.06.104 PMID:27561469

Thannickal, V. J., & Fanburg, B. F. (2000). Reactive oxygen species in cell signaling. *American Journal of Physiology. Lung Cellular and Molecular Physiology*, *279*(6), L1005–L1028. doi:10.1152/ajplung.2000.279.6.L1005 PMID:11076791

The Norms on Classification of Surface Water Quality in order to determine the ecological status of water bodies. Order of the Romanian Ministry nr. 161/2006.

The Regulation on Environmental Requirements for Surface Waters. GD Nr.890/2013. MO 262-267 (22.11.2013), art. Nr.1006.

The Sixth National Communication of Ukraine on climate change. (2012). Ministry of Ecology and Natural Resources of Ukraine. Retrieved from http://www.seia.gov.ua/seia/doccatalog/document?id=638134

The state of the environment in the Republic of Moldova 2004. (National report). (2005). Chisinau, Republic of Moldova.

The state of the environment in the Republic of Moldova 2006. (National report). (2007). Chisinau, Republic of Moldova.

Theoharides, T. C., Sant, J., & Giota, M. (2019). Autism Spectrum Disorder, Fear Response, and Environmental Exposures. In P. Papadopoulou, C. Marouli, & A. Misseyanni (Eds.), *Environmental Exposures and Human Health Challenges* (pp. 30–51). Hershey, PA: IGI Global. doi:10.4018/978-1-5225-7635-8.ch002

Theoharides, T. C., Tsilioni, I., Patel, A. B., & Doyle, R. (2016). Atopic diseases and inflammation of the brain in the pathogenesis of autism spectrum disorders. *Translational Psychiatry*, *6*(6), e844. doi:10.1038/tp.2016.77 PMID:27351598

Thines, R. K., Mubarak, N. M., Nizamuddin, S., Sahu, J. N., Abdullah, E. C., & Ganesan, P. (2017). Application potential of carbon nanomaterials in water and wastewater treatment: A review. *Journal of the Taiwan Institute of Chemical Engineers*, *72*, 116–133. doi:10.1016/j.jtice.2017.01.018

Third National Communication of the Republic of Moldova. Under the United Nations Framework Convention on Climate Change. (2013). Chisinau: Imprint Plus SRL.

Tian, L., Li, Y., Ren, T., Tong, Y., Yang, B., & Li, Y. (2017). Novel bimetallic gold– silver nanoclusters with "Synergy"-enhanced fluorescence for cyanide sensing, cell imaging and temperature sensing. *Talanta*, *170*, 530–539. doi:10.1016/j.talanta.2017.03.107 PMID:28501207

Tian, W., Li, H., Zhou, J., & Guo, Y. (2017). Preparation, characterization and the adsorption characteristics of lignin/silica nanocomposites from cellulosic ethanol residue. *RSC Advances*, *7*(65), 41176–41181. doi:10.1039/C7RA06322A

Tojo, G., & Fernandez, M. (2006). *Oxidation of Alcohols to Aldehydes and Ketones: A Guide to Current Common Practice*. Academic Press.

Tokarz, J. A. III, Ahn, M.-Y., Leng, J., Filley, T. R., & Nies, L. (2008). Reductive debromination of polybrominated diphenyl ethers in anaerobic sediment and a biomimetic system. *Environmental Science & Technology*, *42*(4), 1157–1164. doi:10.1021/es071989t PMID:18351087

Tokuda, H., Tsuzuki, S., Susan, M. A. B. H., Hayamizu, K., & Watanabe, M. (2006). How ionic are room-temperature ionic liquids? An indicator of the physicochemical properties. *The Journal of Physical Chemistry B*, *110*(39), 19593–19600. doi:10.1021/jp064159v PMID:17004825

Torresi, M., Fornarelli, F., Fortunato, B., Camporeale, M. S., & Saponaro, A. (2017). Assessment against Experiments of Devolatilization and Char Burnout Models for the Simulation of an Aerodynamically Staged Swirled Low-NOx Pulverized Coal Burner. *Energies*, *10*(1), 66. doi:10.3390/en10010066

Toshima, N., & Yonezawa, T. (1998). Bimetallic nanoparticles—Novel materials for chemical and physical applications. *New Journal of Chemistry*, *22*(11), 1179–1201. doi:10.1039/a805753b

Totsche, O., Fyson, A., Kalin, M., & Steinberg, C. E. W. (2006). Titration Curves: A useful instrument for assessing the buffer systems of acidic mining waters. *Environmental Science and Pollution Research International*, *13*(4), 215–224. doi:10.1065/espr2005.09.284 PMID:16910118

Townsend, M., Peck, C., Meng, W., Heaton, M., Robison, R., & O'Neill, K. (2017). Evaluation of various glyphosate concentrations on DNA damage in human Raji cells and its impact on cytotoxicity. *Regulatory Toxicology and Pharmacology*, *85*, 79–85. doi:10.1016/j.yrtph.2017.02.002

Tran, C. D., Duri, S., & Harkins, A. L. (2013). Recyclable synthesis, characterization, and antimicrobial activity of chitosan-based polysaccharide composite materials. *Journal of Biomedical Materials Research. Part A*, *101*(8), 2248–2257. doi:10.1002/jbm.a.34520 PMID:23349116

Trasande, L., Cronk, C., Durkin, M., Weiss, M., Schoeller, D. A., Gall, E. A., ... Gillman, M. W. (2009). Environment and obesity in the National Children's Study. *Environmental Health Perspectives*, *117*(2), 159–166. doi:10.1289/ehp.11839 PMID:19270782

Travin, S., Duca, G., Skurlatov, Y., & Sychev, A. (1986). Kinetics and mechanisms of intermediary complexes formation in the redox system. *Journal of Physical Chemistry (Rus)*, *60*(6), 1394–1398.

Tricher, A.R., Perkins, M.J., Massey, R.C., & McWeeny, D. J. (1984). Synthesis of three N-nitroso-dipeptides N-terminal and a method for determination in food. *Food Additives and Contaminant, 1*, 307-312.

Tricker, A. R., Pfundstein, R., Kalble, T., & Preussmann, R. (1992). Secondary amine precursors to nitrosamines in human saliva, gastric juice, blood, urine and faeces. *Carcinogenesis*, *13*(4), 563–568. doi:10.1093/carcin/13.4.563 PMID:1576707

Triker, A. R., & Preussmann, R. (1991). Carcinogenic N-nitrosamines in the diet: Occurrence, formation, mechanisms and carcinogenic potential. *Mutation Research*, *259*(3-4), 277–289. doi:10.1016/0165-1218(91)90123-4 PMID:2017213

Tristao, J. C., Magalhães, F., Corio, P., & Sansiviero, M. T. C. (2006). Electronic characterization and photocatalytic properties of CdS/TiO2 semiconductor composite. *Journal of Photochemistry and Photobiology A Chemistry*, *181*(2-3), 152–157. doi:10.1016/j.jphotochem.2005.11.018

Tsuda, T., Yoshii, K., Torimoto, T., & Kuwabata, S. (2010). Oxygen reduction catalytic ability of platinum nanoparticles prepared by room-temperature ionic liquid-sputtering method. *Journal of Power Sources*, *195*(18), 5980–5985. doi:10.1016/j.jpowsour.2009.11.027

Tural, B., Ertas, E., Enez, B., Fincan, S. A., & Tural, S. (2017). Preparation and characterization of a novel magnetic biosorbent functionalized with biomass of Bacillus Subtilis: Kinetic and isotherm studies of biosorption processes in the removal of Methylene Blue. *Journal of Environmental Chemical Engineering*, *5*(5), 4795–4802. doi:10.1016/j.jece.2017.09.019

Tural, B., Ertas, E., & Tural, S. (2016). Removal of phenolic pollutants from aqueous solutions by a simple magnetic separation. *Desalination and Water Treatment*, *54*(54), 26153–26164. doi:10.1080/19443994.2016.1162202

Turner, M. C., Nieuwenhuijsen, M., Anderson, K., Balshaw, D., Cui, Y., Dunton, G., ... Jerrett, M. (2017). Assessing the Exposome with External Measures: Commentary on the State of the Science and Research Recommendations. *Annual Review of Public Health*, *20*(38), 215–239. doi:10.1146/annurev-publhealth-082516-012802 PMID:28384083

Turro, N. J., Ramamurthy, V., Cherry, W., & Farneth, W. (1978). The effect of wavelength on organic photoreactions in solution reaction from upper excited states. Chem. Rew., 78, 125. doi:10.1021/cr60312a003

Uçurum, M., Toraman, O., Depci, T., & Yoğurtcuoğlu, E. (2011). *A Study on Characterization and Use of Flotation to Separate Unburned Carbon in Bottom Ash from Çayirhan Power Plant* (Vol. 33). Academic Press.

Udoh, F. V., Udoh, P. B., & Umoh, E. E. (2005). Activity of Alkaloid Extract of Carica papaya. Seeds on Reproductive Functions in Male Wistar Rats. *Pharmaceutical Biology*, *43*(6), 563–567. doi:10.1080/13880200500220961

Ueno, K., Tokuda, H., & Watanabe, M. (2010). Ionicity in ionic liquids: Correlation with ionic structure and physicochemical properties. *Physical Chemistry Chemical Physics*, *12*(8), 1649–1658. doi:10.1039/b921462n PMID:20145829

Uhlmann, W., Büttcher, H., Totsche, O., & Steinberg, C. E. W. (2004). Buffering of acidic mine lakes: The relevance of surface exchange and solid-bound sulphate. *Mine Water and the Environment*, *23*(1), 20–27. doi:10.100710230-004-0032-4

Ulaszewska, M. M., Ciffroy, P., Tahraoui, F., Zeman, F. A., Capri, E., & Brochot, C. (2012). Interpreting PCB levels in breast milk using a physiologically based pharmacokinetic model to reconstruct the dynamic exposure of Italian women. *Journal of Exposure Science & Environmental Epidemiology*, *22*(6), 601–609. doi:10.1038/jes.2012.36 PMID:22760444

Umar, A. (2009). Growth of comb-like ZnO nanostructures for dyesensitized solar cells applications. *Nanoscale Research Letters*, *4*(9), 1004–1008. doi:10.100711671-009-9353-3 PMID:20596445

Umar, A., Chauhan, M. S., Chauhan, S., Kumar, R., Kumar, G., Al-Sayari, S. A., ... Al-Hajry, A. (2011). Large-scale synthesis of ZnO balls made of fluffy thin nanosheets by simple solution process: Structural, optical and photocatalytic properties. *Journal of Colloid and Interface Science*, *363*(2), 521–528. doi:10.1016/j.jcis.2011.07.058 PMID:21862032

Umpierre, A. P., Machado, G., Fecher, G. H., Morais, J., & Dupont, J. (2005). Selective Hydrogenation of 1,3-Butadiene to 1-Butene by Pd(0) Nanoparticles Embedded in Imidazolium Ionic Liquids. *Advanced Synthesis & Catalysis*, *347*(10), 1404–1412. doi:10.1002/adsc.200404313

UN. (2019). *Office for Distaster Risk Reduction*. Retrieved from https://www.unisdr.org/we/inform/publications

UNEP. (2010). *Technical Review of the Implications of Recycling Commercial Pentabromodiphenyl Ether and Commercial Octabromodiphenyl Ether*. UNEP.

UNEP. (2013). *Minamata Convention Agreed by Nations*. Retrieved from https://www.unenvironment.org/news-and-stories/press-release/minamata-convention-agreed-nations

United Nations. (2015). *Transforming our world: the 2030 Agenda for Sustainable Development*. Retrieved from https://www.un.org/ga/search/view_doc.asp?symbol=A/RES/70/1&Lang=E

University of California - San Francisco. (2018). Brain signature of depressed mood unveiled in new study: Direct recordings of human brain activity link memory, emotion, and anxiety during bouts of low mood. *ScienceDaily*. Retrieved from www.sciencedaily.com/releases/2018/11/181108142410.htm

Upadhyay, P., & Srivastava, V. (2019). Synthesis of Ionic Liquid Mediated Nanoparticle Synthesis. In *Nanomaterials* (pp. 67–89). Apple Academic Press.

Upadhyay, N., Majestic, B. J., & Herckes, P. (2011). Solubility and speciation of atmospheric iron in buffer systems simulating cloud conditions. *Atmospheric Environment*, *45*(10), 1858–1866. doi:10.1016/j.atmosenv.2011.01.010

Upadhyay, P. R., & Srivastava, V. (2016a). Heterogeneous Silica Tethered Ruthenium Catalysts for Carbon Sequestration Reaction. *Catalysis Letters*, *146*(8), 1478–1486. doi:10.100710562-016-1772-z

Upadhyay, P. R., & Srivastava, V. (2016b). Recyclable graphene-supported palladium nanocomposites for Suzuki coupling reaction. *Green Processing and Synthesis*, *5*(2), 123–129. doi:10.1515/gps-2015-0112

Upadhyay, P. R., & Srivastava, V. (2016c). Selective Hydrogenation of CO2 Using Ruthenium Nanoparticles Intercalated Montmorillonite Clay. *Letters in Organic Chemistry*, *13*(6), 459–465. doi:10.2174/1570178613666160815095733

Upadhyay, P. R., & Srivastava, V. (2016d). Silica Tethered Ruthenium Catalyst for the Hydrogenation of CO2 Gas. *Letters in Organic Chemistry*, *13*(5), 380–387. doi:10.2174/1570178613666160303002129

Upadhyay, P. R., & Srivastava, V. (2017). Ionic Liquid Mediated In Situ Synthesis of Ru Nanoparticles for CO2 Hydrogenation Reaction. *Catalysis Letters*, *147*(4), 1051–1060. doi:10.100710562-017-1995-7

Upadhyay, P., & Srivastava, V. (2015). Synthesis of supported Ru-nanoparticles for selective hydrogenation of carbonyl compounds. *Letters in Organic Chemistry*, *12*(8), 528–533. doi:10.2174/1570178612081508261l2807

Upadhyay, P., & Srivastava, V. (2016). Synthesis of Monometallic Ru/TiO2 Catalysts and Selective Hydrogenation of CO2 to Formic Acid in Ionic Liquid. *Catalysis Letters*, *146*(1), 12–21. doi:10.100710562-015-1654-9

Urunkar, Y., Pandit, A., Bhargava, P., Joshi, J. B., Mathpati, C. S., Sudarshan, V., . . . More, V. (2018). *Light-weight thermal insulating fly ash cenosphere ceramics*. Academic Press.

US EPA. (1996). *Assessment and remediation of contaminated sediments (ARCS) program*. Author.

US EPA. (2005). *Contaminated sediment remediation guidance for hazardous waste sites (Report No. EPA-540-R-05-012)*. Retrieved from https://nepis.epa.gov/

USEPA. (2010). *An Exposure Assessment of Polybrominated Diphenyl Ethers (Report No. EPA/600/R-08-086F)*. Washington, DC: National Center for Environmental Assessment Office of Research and Development.

Vaghetti, J., Lima, E., Royer, B., Cardoso, N., Martins, B., & Calvete, T. (2009). *Pecan Nutshell as Biosorbent to Remove Toxic Metals from Aqueous Solution* (Vol. 44). Academic Press.

Vajtai, R. (2013). *Springer handbook of nanomaterials*. Springer Science & Business Media. doi:10.1007/978-3-642-20595-8

Vallet-Regí, M., & Balas, F. (2008). Silica materials for medical applications. *The Open Biomedical Engineering Journal*, *2*(1), 1–9. doi:10.2174/1874120700802010001 PMID:19662110

Van den Berg, M., Birnbaum, L. S., Denison, M., De Vito, M., Farland, W., Feeley, M., ... Peterson, R. E. (2006). The 2005 World Health Organization reevaluation of human and Mammalian toxic equivalency factors for dioxins and dioxin-like compounds. *Toxicological Sciences*, *93*(2), 223–241. doi:10.1093/toxsci/kfl055 PMID:16829543

Van Dijk, S. J., Molloy, P. L., Varinli, H., Morrison, J. L., & Muhlhausler, B. S. (2015). Recent developments on the role of epigenetics in obesity and metabolic disease. *Clinical Epigenetics*, *7*(66), 66. doi:10.118613148-015-0101-5 PMID:27408648

van Gelder, M. M. H. J., van Rooij, I. A. L. M., Miller, R. K., Zielhuis, G. A., de Jong-van den Berg, L. T. W., & Roeleveld, N. (2010). Teratogenic mechanisms of medical drugs. *Human Reproduction Update*, *16*(4), 378–394. doi:10.1093/humupd/dmp052 PMID:20061329

Van Kempen, E., Fischer, P., Janssen, N., Houthuijs, D., Van Kamp, I., Stansfeld, S., & Cassee, F. (2012). Neurobehavioral effects of exposure to traffic-related air pollution and transportation noise in primary schoolchildren. *Environmental Research*, *115*, 18–25. doi:10.1016/j.envres.2012.03.002 PMID:22483436

Van Leeuwen, P. W. (2006). *Homogeneous catalysis: understanding the art*. Springer Science & Business Media.

Vanloon, G., & Duffy, S. (2011). *Environmental Chemistry: a global perspective*. Oxford University Press.

Vanrolleghem, P., & Lee, D. (2003). *On-Line Monitoring Equipment for Wastewater Treatment Processes: State of the Art* (Vol. 47). Academic Press.

Varga, C., Kiss, I., & Ember, I. (2002). The lack of environmental justice in Central and Eastern Europe. *Environmental Health Perspectives*, *110*(11), A662–A1. doi:10.1289/ehp.110-a662 PMID:12417491

Vasilenko, I. Y., Osipov, V. A., & Rublevsky, V. P. (1992). Radioactive carbon. *Nature, 12*, 59-65.

Vassilev, S., Menendez, R., Díaz-Somoano, M., & Martínez-Tarazona, M. (2004). *Phase-mineral and chemical composition of coal fly ashes as a basis for their multicomponent utilization. 2. Characterization of ceramic cenosphere and salt concentrates* (Vol. 83). Academic Press.

Veena, S., & Rashmi, S. (2014). A review on mechanism of nitrosamine formation, metabolism and toxicity in Vivo. *International Journal of Toxicological and Pharmacological Research*, *6*(4), 86–96.

Vekariya, R. L. (2017, February 1). A review of ionic liquids: Applications towards catalytic organic transformations. *Journal of Molecular Liquids*.

Velarde, M.D., Fry, G., & Tveit, M. (2007). Health effects of viewing landscapes — Landscape types in environmental psychology. *Urban Forestry & Urban Greening*, *6*(4), 199–212.

Velev, P. M., Kalambur, O. D., Rabolt, A. T., Lenhoff, J. F., Kaler, A. M., & Kaler, E. W. (2000). Assembly of Gold Nanostructured Films Templated by Colloidal Crystals and Use in Surface-Enhanced Raman Spectroscopy. *Journal of the American Chemical Society*, *122*(39), 9554–9555. doi:10.1021/ja0022831

Veliksar, S. (2016). Microelements in achieving the adaptive productivity potential and resistance in horticulture. In *Microelements in the composition of the biosphere and their application in agriculture and medicine* (pp. 142–163). Chisinau: ASM Publishing House.

Velmurugan, S., Deepanraj, B., & Jayaraj, S. (2014). Biogas Generation through Anaerobic Digestion Process – An Overview. *Research Journal of Chemistry and Environment*, *18*(5), 80–94.

Velusamy, P., Rajalakshmi, S., Pitchaimuthu, S., & Kannan, N. (2014). Modification of the photocatalytic activity of TiO_2 by β-Cyclodextrin in decoloration of ethyl violet dye. *Journal of Advanced Research*, *5*(1), 19–25. doi:10.1016/j.jare.2012.10.001 PMID:25685468

Venkatesan, R., Prechtl, M. H. G., Scholten, J. D., Pezzi, R. P., MacHado, G., & Dupont, J. (2011). Palladium nanoparticle catalysts in ionic liquids: Synthesis, characterisation and selective partial hydrogenation of alkynes to Z-alkenes. *Journal of Materials Chemistry, 21*(9), 3030–3036. doi:10.1039/c0jm03557b

Venkateswaran, P., Vellaichamy, S., & Palanivelu, K. (2007). Speciation of heavy metals in electroplating industry sludge and wastewater residue using inductively coupled plasma. *International Journal of Environment Science and Technology, 4*(4). doi:10.1007/BF03325986

Venkateswarlu, P., Vijaya Durga, G., Chitti Babu, N., & Venkateswara Rao, M. (2008). *Biosorption of Zn(II) from an Aqueous Solution by Erythrina variegata Orientalis Leaf Powder* (Vol. 3). Academic Press.

Verbyla, M. E., Cairns, M. R., Gonzalez, P. A., Whiteford, L. M., & Mihelcic, J. R. (2015). Emerging challenges and synergies for pathogen control and resource recovery in natural wastewater treatment systems. *WIREs. Water, 2*(6), 701–714. doi:10.1002/wat2.1101

Vergel, K., Frontasyeva, M., Kamanina, I., & Pavlov, S. (2009). Biomonitoring atmosphernukh vupadenii tyazeluh metallov na severo-vostoke Moskovskoi oblasti s pomosh'yu metoda mhov-biomonitorov. *Ecology of urbanized areas III*, 88-95. (in Russian)

Vergel, K., Goryainova, Z., Vikhrova, I., & Frontasyeva, M. (2014). Moss Biomonitoring and Employment of the GIS Technology within the Framework of the Assessment of Air Pollution by Industrial Enterprises in the Tikhvin District of the Leningrad Region. Ecology of Urbanized Areas, 2, 92-101.

Vergel, K., Frontasyeva, M., Pavlov, S., & Povtoreyko, E. (2007). Air Pollution Studies in Tver Region of Russia using Moss-Biomonitoring with Nuclear Analytical Methods. In *Proceedings of AIP Conference*, (vol. 958, p. 240). 10.1063/1.2825797

Vergel, K., Zinicovscaia, I., Yushin, N., & Frontasyeva, M. (2019). Heavy metal atmospheric deposition study in Moscow region, Russia. *Bulletin of Environmental Contamination and Toxicology, 103*(3), 435–440. doi:10.100700128-019-02672-4 PMID:31267137

Verma, C., Ebenso, E. E., & Quraishi, M. A. (2019, February 15). Transition metal nanoparticles in ionic liquids: Synthesis and stabilization. *Journal of Molecular Liquids.*

Vijayalakshmi, K., Devi, B. M., Latha, S., Gomathi, T., Sudha, P. N., Venkatesan, J., & Anil, S. (2017). Batch adsorption and desorption studies on the removal of lead (II) from aqueous solution using nanochitosan/sodium alginate/microcrystalline cellulose beads. *International Journal of Biological Macromolecules B., 104*, 1483–1494. doi:10.1016/j.ijbiomac.2017.04.120

Villa, R., Alvarez, E., Porcar, R., Garcia-Verdugo, E., Luis, S. V., & Lozano, P. (2019, December 10). Ionic liquids as an enabling tool to integrate reaction and separation processes. *Green Chemistry.*

Vilor, N. V., & Kuzmin, L. A. (2007). Petrology, geochemistry and mineralogy, the use of physical and chemical modeling in the study of sulfoarsenid complexes in hydrothermal solutions. *Geology and Geophysics, 48*, 588. (in Russian)

Vione, D., Falletti, G., Maurino, V., Minero, C., Pelizzetti, E., Landrino, M., ... Lariu, R.-G. (2006). And Arsene C. (2006) Sources and Sinks of Hydroxyl Radicals upon Irradiation of Natural Water Samples Environ. *Sci. Technol., 40*(12), 3775–3781. doi:10.1021/es052206b PMID:16830541

Visentin, C., da Silva Trentin, A. W., Braun, A. B., & Thomé, A. (2019). Application of life cycle assessment as a tool for evaluating the sustainability of contaminated sites remediation: A systematic and bibliographic analysis. *The Science of the Total Environment, 672*, 893–905. doi:10.1016/j.scitotenv.2019.04.034 PMID:30978551

Voica, C., Kovacs, M. H., Dehelean, A., Ristoiu, D., & Iordache, A. (2012). ICP-MS determinations of heavy metals in surface waters from Transylvania. *Romanian Journal of Physics*, *57*(7-8), 1184–1193.

Voicu, R., & Banaduc, D. (2016). Suggestions of scenarios for restoring longitudinal connectivity to sustain fish fauna migration upstream and downstream of Apahida bottom sill. *Lakes, Reservoirs and Ponds*, *10*(2), 126–138.

Voloshin, V., Korol, P., & Rudyk, O. (2013). Mathematical and cartographic support for monitoring studies of erosion-prone land using a method of relief plastics. *Lviv Polytechnic National University*, *1*(25), 96-99. http://ena.lp.edu.ua/handle/ntb/18863

Voloshin, V. I., & Korchinsky, V. M. (Eds.). (2007). *Distant sounding of the Earth Information Technology, collection, and the maintenance of the Earth's Earth Watch*. Prospectus.

Vrijheid, M. (2014). The exposome: A new paradigm to study the impact of environment on health. *Thorax*, *69*(9), 876–878. doi:10.1136/thoraxjnl-2013-204949 PMID:24906490

Vrijheid, M., Casas, M., Bergstrom, A., Carmichael, A., Cordier, S., Eggesbo, M., & Nieuwenhuijsen, M. (2012). European birth cohorts for environmental health research. *Environmental Health Perspectives*, *120*(1), 29–37. doi:10.1289/ehp.1103823 PMID:21878421

Vrijheid, M., Slama, R., Robinson, O., Chatzi, L., Coen, M., van den Hazel, P., ... Nieuwenhuijsen, M. J. (2014). The Human Early-Life Exposome (HELIX): Project rationale and design. *Environmental Health Perspectives*, *122*(6), 535–544. doi:10.1289/ehp.1307204 PMID:24610234

Waeger, P.A., Schluep, M., Móller, E., & Gloor, R. (2012). RoHS regulated Substances in Mixed Plastics from Waste Electrical and Electronic Equipment. *Environmental Science & Technology*, *46*(2), 628–635.

Waite, T. D. (1990). Photochemistry of colloids and surfaces in natural waters and water treatment. In R. Beckett (Ed.), Surface and Colloid Chemistry in Natural Waters and Water Treatment. Springer Science+Business Media.

Wakeel, A., Gul, M., & Sanaullah, M. (2013). *Potassium dynamics in three alluvial soils differing in clay contents* (Vol. 25). Academic Press.

Walker, A., & Wheelock, T. D. (2006). Separation of Carbon from Fly Ash Using Froth Flotation. *Coal Preparation*, *26*(4), 235–250. doi:10.1080/07349340601104883

Walker, G. (2012). *Environmental Justice: Concepts, Evidence and Politics*. London: Routledge. doi:10.4324/9780203610671

Walters, C., Pool, E., & Somerset, V. (2016). Nanotoxicology: A Review. In Toxicology - New Aspects to This Scientific Conundrum. InTech. doi:10.5772/64754

Wang, B., Wan, Y., Zheng, Y. P., Lee, X., Liu, T., Yu, Z., . . . Gao, B. (2018). *Alginate-based composites for environmental applications: a critical review*. Academic Press.

Wang, J., & Chen, C. (2009). *Biosorbents for Heavy Metal Removal and Their Future* (Vol. 27). Academic Press.

Wang, K. (2007). Chemical Precipitation (vol. 37). Academic Press.

Wang, C., Liu, H., Liu, Z., Gao, Y., Wu, B., & Xursos, H. (2018). Fe_3O_4 nanoparticle-coated mushroom source biomaterial for Cr(VI) polluted liquid treatment and mechanism research. *Royal Society Open Science*, *5*(5), 171776–171790. doi:10.1098/rsos.171776 PMID:29892371

Wang, D. K., Elma, M., Motuzas, J., Hou, W. C., Schmeda-Lopez, D. R., Zhang, T., & Zhang, X. (2016). Physicochemical and photocatalytic properties of carbonaceous char and titania composite hollow fibers for wastewater treatment. *Carbon*, *109*, 182–191. doi:10.1016/j.carbon.2016.08.001

Wang, H., Xie, C., Zhang, W., Cai, S., Yang, Z., & Gui, Y. (2007). Comparison of dye degradation efficiency using ZnO powders with various size scales. *Journal of Hazardous Materials, 141*(3), 645–652. doi:10.1016/j.jhazmat.2006.07.021 PMID:16930825

Wang, L., & Wang, A. (2007). Adsorption characteristics of Congo Red onto the chitosan/montmorillonite nanocomposite. *Journal of Hazardous Materials, 147*(3), 979–985. doi:10.1016/j.jhazmat.2007.01.145 PMID:17349744

Wang, L., Wu, Y., Chen, F., & Yang, X. (2014). Photocatalytic enhancement of Mg-doped znO nanocrystals hybridized with reduced graphene oxide sheets. *Progress in Natural Science: Materials International, 24*(1), 6–12. doi:10.1016/j.pnsc.2014.01.002

Wang, W., Leung, A. O. W., Chu, L. H., & Wong, M. H. (2018). Phthalates Contamination in China: Status, Trends and Human Exposure-with an Emphasis on Oral Intake. *Environmental Pollution, 238*, 771–782. doi:10.1016/j.envpol.2018.02.088 PMID:29625301

Wang, Y., Wang, R., Xu, D., Sun, C., Ni, L., Fu, W., ... Qiu, S. (2016). Synthesis and properties of MFI zeolites with microporous, mesoporous and macroporous hierarchical structures by a gel-casting technique. *New Journal of Chemistry, 40*(5), 4398–4405. doi:10.1039/C5NJ03387J

Wang, Y., Yang, J., Li, Y., Jiang, T., Chen, J., & Wang, J. (2015). Controllable preparation of ZnO nanostructure using hydrothermal-electrodeposited method and its properties. *Materials Chemistry and Physics, 153*, 266–273. doi:10.1016/j.matchemphys.2015.01.013

Wang, Y., Yang, Y., Xi, L., Zhang, X., Jia, M., Xu, H., & Wu, H. (2016). A simple hydrothermal synthesis of flower-like ZnO microspheres and their improved photocatalytic activity. *Materials Letters, 180*, 55–58. doi:10.1016/j.matlet.2016.05.107

Wang, Y., Yan, L., Zhang, Z., Prado, E., Fu, L., Xu, X., & Du, L. (2018). Epigenetic regulation and its therapeutic potential in pulmonary hypertension. *Frontiers in Pharmacology, 9*, 241. doi:10.3389/fphar.2018.00241 PMID:29615911

Wang, Y., Zhang, Y., Hou, C., & Liu, M. (2016). Mussel-inspired synthesis of magnetic polydopamine–chitosan nanoparticles as biosorbent for dyes and metals removal. *Journal of the Taiwan Institute of Chemical Engineers, 61*, 292–298. doi:10.1016/j.jtice.2016.01.008

Wang, Y., Zhao, S., Li, M., Li, W., Zhao, Y., Qi, J., & Cui, X. (2017). Graphene quantum dots decorated graphene as an enhanced sensing platform for sensitive and selective detection of copper (II). *Journal of Electroanalytical Chemistry (Lausanne, Switzerland), 797*, 113–120. doi:10.1016/j.jelechem.2017.05.031

Wang, Z. L. (2004). Zinc oxide nanostructures: Growth, properties and applications. *Journal of Physics Condensed Matter, 16*(25), R829–R858. doi:10.1088/0953-8984/16/25/R01

Wan, Q., Wang, T. H., & Zhao, J. C. (2005). Enhanced photocatalytic activity of ZnO nanotetrapods. *Applied Physics Letters, 87*(8), 083105–3. doi:10.1063/1.2034092

Wardhono, A. (2018). *Comparison Study of Class F and Class C Fly Ashes as Cement Replacement Material on Strength Development of Non-Cement Mortar* (Vol. 288). Academic Press.

Wariki, W. M., Ota, E., Mori, R., Koyanagi, A., Hori, N., & Shibuya, K. (2012). Behavioral interventions to reduce the transmission of HIV infection among sex workers and their clients in low- and middle-income countries. *Cochrane Database of Systematic Reviews (Online), 2*. doi:10.1002/14651858.CD005272.pub3 PMID:22336811

Warule, S. S., Chaudhari, N. S., Kale, M. A., & More, M. A. (2009). Novel sonochemical assisted hydrothermal approach towards the controllable synthesis of ZnO nanorods, nanocups and nanoneedles and their photocatalytic study. *CrystEngComm, 11*(12), 2776–2783. doi:10.1039/b909916f

Wasserscheid, P., & Keim, W. (2000). Ionic Liquids—New "Solutions" for Transition Metal Catalysis. *Angewandte Chemie, 39*(21), 3772–3789. doi:10.1002/1521-3773(20001103)39:21<3772::AID-ANIE3772>3.0.CO;2-5 PMID:11091453

Wegner, S., & Janiak, C. (n.d.). Metal Nanoparticles in Ionic Liquids. *Topics in Current Chemistry*. PMID:28589266

Weiss, J. M., Lignell, S., Darnerud, P. O., & Kotova, N. (2016). Human Biomonitoring of Environmental Contaminants - Examples Offering Tools Towards Safe Food in Sweden. *European Journal of Nutrition & Food Safety, 6*(3), 132–147. doi:10.9734/EJNFS/2016/20212

Welderfael, T., Pattabi, M., Pattabi, R. M., & G, A. K. T. (2016). Photocatalytic activity of Ag-N co-doped ZnO nanorods under visible and solar light irradiations for MB degradation. *Journal of Water Process Engineering, 14*, 117–123. doi:10.1016/j.jwpe.2016.11.001

Welton, T. (2018, June 1). Ionic liquids: a brief history. *Biophysical Reviews*.

Welton, T. (1999). Room-Temperature Ionic Liquids. Solvents for Synthesis and Catalysis. *Chemical Reviews, 99*(8), 2071–2083. doi:10.1021/cr980032t PMID:11849019

Weng, X., Huang, L., Chen, Z., Megharaj, M., & Naidu, R. (2013). Synthesis of iron-based nanoparticles by green tea extract and their degradation of malachite. *Industrial Crops and Products, 51*, 342–347. doi:10.1016/j.indcrop.2013.09.024

West, J. E., O'Neill, S. M., & Ylitalo, G. M. (2017). Time Trends of Persistent Organic Pollutants in Benthic and Pelagic Indicator Fishes from Puget Sound, Washington, USA. *Archives of Environmental Contamination and Toxicology, 73*(2), 207–229. doi:10.100700244-017-0383-z PMID:28528416

Wheelock, T. D., & Markuszewski, R. (1984). Coal Preparation and Cleaning. In B. R. Cooper & W. A. Ellingson (Eds.), *The Science and Technology of Coal and Coal Utilization* (pp. 47–123). Boston, MA: Springer US. doi:10.1007/978-1-4684-4580-0_3

Whigham, D. F., & Jordan, T. E. (2003). Isolated wetlands and water quality. *Wetlands. Springer, 23*(3), 541–549. doi:10.1672/0277-5212(2003)023[0541:IWAWQ]2.0.CO;2

WHO. (1986). *The Ottawa Charter for Health Promotion*. Retrieved from: https://www.who.int/healthpromotion/conferences/previous/ottawa/en/

WHO. (2005). *Trichloroethylene in drinking-water: Background document for development of WHO guidelines for drinking-water quality (Report No. WHO/SDE/WSH/05.08/22)*. WHO.

WHO. (2006). *Concise international chemical assessment document: Tetrachloroethene*. Geneva: World Health Organization.

WHO. (2006a). *Air quality guidelines. Global update 2005. Particulate matter, ozone, nitrogen dioxide and sulfur dioxide*. Copenhagen, Denmark: World Health Organization, Regional Office for Europe.

WHO. (2006b). *Health risks of particulate matter from long-range transboundary air pollution*. Copenhagen, Denmark: World Health Organization, Regional Office for Europe.

WHO. (2007). *Health risks of heavy metals from long-range transboundary air pollution. Joint WHO/Convention Task Force on the Health Aspects of Air Pollution*. Copenhagen, Denmark: World Health Organization, Regional Office for Europe.

WHO. (2011). Guidelines for drinking-water quality. *WHO Chronicle, 38*, 104–108. PMID:6485306

WHO. (2012). *Soil-transmitted helminthiases: eliminating soil-transmitted helminthiases as a public health problem in children: progress report 2001-2010 and strategic plan 2011-2020*. Geneva: World Health Organization.

WHO. (2012b). *Endocrine disrupters and child health - Possible developmental early effects of endocrine disrupters on child health.* Geneva, Switzerland: World Health Organization.

WHO. (2013a). *Review of evidence on health aspects of air pollution — REVIHAAP*. Copenhagen, Denmark: World Health Organization, Regional Office for Europe.

WHO. (2014). *Global Tuberculosis Report 2014*. Geneva: World Health Organization.

WHO. (2015). *Drinking-water: Fact sheet No. 391*. Retrieved from https://www.who.int/mediacentre/factsheets/fs391/en/

WHO/JRC. (2011). Burden of disease from environmental noise. In L. Fritschi, A. L. Brown, R. Kim, D. Schwela, & S. Kephalopoulos (Eds.), *World Health Organization, Regional Office for Europe*. Copenhagen, Denmark: Author.

Wibowo, E., Rokhmat, M., & Abdullah, M. (2017). Reduction of seawater salinity by natural zeolite (Clinoptilolite): Adsorption isotherms, thermodynamics and kinetics. *Desalination*, *409*, 146–156. doi:10.1016/j.desal.2017.01.026

Widegren, J. A., & Finke, R. G. (2003). A review of the problem of distinguishing true homogeneous catalysis from soluble or other metal-particle heterogeneous catalysis under reducing conditions. *Journal of Molecular Catalysis A Chemical*, *198*(1-2), 317–341. doi:10.1016/S1381-1169(02)00728-8

Wiegel, J., & Wu, Q. (2000). Microbial reductive dehalogenation of polychlorinated biphenyls. *FEMS Microbiology Ecology*, *32*(1), 1–15. doi:10.1111/j.1574-6941.2000.tb00693.x PMID:10779614

Wigger, H., & Nowack, B. (2019). Material-specific properties applied to an environmental risk assessment of engineered nanomaterials–implications on grouping and read-across concepts. *Nanotoxicology*, *13*(5), 623–643. doi:10.1080/17435390.2019.1568604 PMID:30727799

Wigle, D. T. (1977). Cancer mortality in relation to asbestos in municipal water supplies. *Archives of Environmental Health*, *32*(July/August), 185–190. doi:10.1080/00039896.1977.10667278 PMID:889357

Wild, C. P. (2005). Complementing the genome with an "exposome": The outstanding challenge of environmental exposure measurement in molecular epidemiology. *Cancer Epidemiology, Biomarkers & Prevention*, *14*(8), 1847–1850. doi:10.1158/1055-9965.EPI-05-0456 PMID:16103423

Wild, C. P., Scalbert, A., & Herceg, Z. (2013). Measuring the exposome: A powerful basis for evaluating environmental exposures and cancer risk. *Environmental and Molecular Mutagenesis*, *54*(7), 480–499. doi:10.1002/em.21777 PMID:23681765

Wiley, B., Sun, Y., Mayers, B., & Xi, Y. (2005). Shape-Controlled Synthesis of Metal Nanostructures: The Case of Silver. *Chemistry (Weinheim an der Bergstrasse, Germany)*, *11*(2), 454–463. doi:10.1002/chem.200400927 PMID:15565727

Williams, I. (2005). *Environmental Chemistry. Wiley.*

Willner, I., & Willner, B. (2010). Biomolecule-Based Nanomaterials and Nanostructures. *Nano Letters*, *10*(10), 3805–3815. doi:10.1021/nl102083j PMID:20843088

Winchell, L. J., & Novak, P. J. (2008). Enhancing polychlorinated biphenyl dechlorination in fresh water sediment with biostimulation and bioaugmentation. *Chemosphere*, *71*(1), 176–182. doi:10.1016/j.chemosphere.2007.10.021 PMID:18037157

Wolterbeek, H. (2002). Biomonitoring of trace element air pollution: Principles, possibilities and perspectives. *Environmental Pollution*, *120*(1), 11–21. doi:10.1016/S0269-7491(02)00124-0 PMID:12199457

Woo, H., & Park, K. H. (2016). Recent developments in hybrid iron oxide–noble metal nanocatalysts for organic reactions. *Catalysis Today*, *278*, 209–226. doi:10.1016/j.cattod.2016.01.030

World Health Organization (WHO). (1946). *Constitution of the World Health Organization.* Retrieved from http://apps.who.int/gb/bd/PDF/bd47/EN/constitution-en.pdf?ua=1

World Health Orgranization. (2003). *Guidelines for Drinking Water Quality. Iron in Drinking Water.* Retrieved from https://www.who.int/water_sanitation_health/dwq/chemicals/iron.pdf

Wu Jacknife, C. F. (1986). Bootstrap and other Resampling Methods in Regression Analysis. *Annals of Statistics, 14*(4), 1261–1350. doi:10.1214/aos/1176350142

Wu, S.-P., Dai, X.-Z., Kan, J.-R., Shilong, F.-D., & Zhu, M.-Y. (2016). *Fabrication of carboxymethyl chitosan–hemicellulose resin for adsorptive removal of heavy metals from wastewater* (Vol. 28). Academic Press.

Wu, Y., Bianco, A., Brigante, M., Dong, W., Sainte-Claire, D. P., Hanna, K., & Mailhot, G. (2015). Sulfate Radical Photogeneration using Fe-EDDS: Influence of Critical Parameters and Naturally Occurring Scavengers. *Environ. Sci. Technol.* Doi:10.1021/acs.est.5b03316

Wu, C., & Foos, J. (2010). Making Chemistry Fun to Learn. *Literacy Information and Computer Education Journal, 1*(1), 3–7. doi:10.20533/licej.2040.2589.2010.0001 PMID:24478915

Wuebbles, D. (2020). Ozone Depletion. *Encyclopedia Britannica.* Retrieved from: https://www.britannica.com/science/ozone-depletion

Wu, H., & Feng, Q. (2017). Fabrication of bimetallic Ag/Fe immobilized on modified biochar for removal of carbon tetrachloride. *Journal of Environmental Sciences (China), 54*, 346–357. doi:10.1016/j.jes.2016.11.017 PMID:28391946

Wu, H., Feng, Q., Yang, H., Alam, E., Gao, B., & Gu, D. (2017). Modified biochar supported Ag/Fe nanoparticles used for removal of cephalexin in solution: Characterization, kinetics and mechanisms. *Colloids and Surfaces. A, Physicochemical and Engineering Aspects, 517*, 63–71. doi:10.1016/j.colsurfa.2017.01.005

Xia, M., Chen, C., Long, M., Chen, C., Cai, W., & Zhou, B. (2011). Magnetically separable mesoporous silica nanocomposite and its application in Fenton catalysis. *Microporous and Mesoporous Materials, 145*(1-3), 217–223. doi:10.1016/j.micromeso.2011.05.017

Xiao, S., Hu, S., Zhang, Y., Zhao, X., & Pan, W. (2018). Influence of sewage treatment plant effluent discharge into multipurpose river on its water quality: A quantitative health risk assessment of Cryptosporidium and Giardia. *Environmental Pollution, 233*, 797–805. doi:10.1016/j.envpol.2017.11.010 PMID:29141237

Xiao, Z., Zhang, H., Xu, Y., Yuan, M., Jing, X., Huang, J., ... Sun, D. (2017). Ultra-efficient removal of chromium from aqueous medium by biogenic iron based nanoparticles. *Separation and Purification Technology, 174*, 466–473. doi:10.1016/j.seppur.2016.10.047

Xia, T., Zhu, F., Cui, Y., Yang, Y., Wang, Z., & Qian, G. (2017). Highly selective luminescent sensing of picric acid based on a water-stable europium metal-organic framework. *Journal of Solid State Chemistry, 245*, 127–131. doi:10.1016/j.jssc.2016.10.016

Xia, Y., Xu, L., Peng, J., Han, J., Guo, S., Zhang, L., ... Komarneni, S. (2019). TiO2@ g-C3N4 core/shell spheres with uniform mesoporous structures for high performance visible-light photocatalytic application. *Ceramics International, 45*(15), 18844–18851. doi:10.1016/j.ceramint.2019.06.118

Xie, J., Li, Y. T., Zhao, W., Bian, L., & Wei, Y. (2011). Simple fabrication and photocatalytic activity of ZnO particles with different morphologies. *Powder Technology, 207*(1-3), 140–144. doi:10.1016/j.powtec.2010.10.019

Xie, K., Zhao, W., & He, X. (2011). Adsorption properties of nano-cellulose hybrid containing polyhedral oligomeric silsesquioxane and removal of reactive dyes from aqueous solution. *Carbohydrate Polymers*, *83*(4), 1516–1520. doi:10.1016/j.carbpol.2010.09.064

Xie, Y., Li, X., Wang, H., & Li, W. (2013). Spatio-temporal variation analysis of hydrochemical characteristics in the Luanhe River Basin, China. *Water Science and Technology*, *67*(6), 1332–1338. doi:10.2166/wst.2013.007 PMID:23508159

Xin, S., Zeng, Z., Zhou, X., Luo, W., Shi, X., Wang, Q., Deng, H., Du, Y. (2017). Article. *Journal of Hazardous Materials, 324*(B), 365-372.

Xu, A. N., Liu, Y. N., Zhou, X., Wang, X., Liang, K., Yang, Z. K., ... Sahar, S. (2017). Hydrogenation/oxidation induced efficient reversible color switching between methylene blue and leuco-methylene blue. *RSC Advances*, *7*(48), 30080–30085. doi:10.1039/C7RA04498D

Xue, Q., & Lu, S. (2008). Microstructure of ferrospheres in fly ashes: SEM, EDX and ESEM analysis. *Journal of Zhejiang University. Science A*, *9*(11), 1595–1600. doi:10.1631/jzus.A0820051

Xu, G. P., & Reed, P. I. (1993). Method for group determination of total N-nitroso compounds and nitrite in fresh gastric juice by chemical denitrosation and thermal energy analysis. *Analyst (London)*, *118*(7), 877–883. doi:10.1039/an9931800877 PMID:8372977

Xu, G. P., & Reed, P. I. (1993). N-nitroso compounds in fresh gastric juice and their relation in intragastric pH and nitrite employing an improved analytical method. *Carcinogenesis*, *14*(12), 2547–2551. doi:10.1093/carcin/14.12.2547 PMID:8269625

Xu, G., Lu, Q., Yu, L., & Wang, S. (2019). Tetrachloroethene primes reductive dechlorination of polychlorinated biphenyls in a river sediment microcosm. *Water Research*, *152*, 87–95. doi:10.1016/j.watres.2018.12.061 PMID:30665163

Xu, J., Han, X., Liu, H., & Hu, Y. (2006). Synthesis of silver nanoparticles in revers micelles stabilized by natural biosurfactant. *Colloids and Surfaces. A, Physicochemical and Engineering Aspects*, *273*, 179–183. doi:10.1016/j.colsurfa.2005.08.019

Yadav, V. K., & Fulekar, M. H. (2018). The current scenario of thermal power plants and fly ash: Production and utilization with a focus in India. *International Journal of Advance Engineering and Research Development*, *5*(4), 768–777.

Yakovlev, S., & Kariukhina, T. (1980). Biochemical processes in the biochemical treatment of wastewaters. Moscow: Academic Press.

Yakymchuk, C. (2017). Applying phase equilibria modelling to metamorphic and geological processes: Recent developments and future potential. *Geoscience Canada: Journal of the Geological Association of Canada*, *44*(1), 27–45. doi:10.12789/geocanj.2017.44.114

Yang, C. S., Yoo, J.-S. H., Ishizaki, H., & Hong, J. (1990). Cytochrome P-450 IIE1: Roles in nitrosoamine metabolism and mechanisms of regulation. *Drug Metabolism Reviews*, *22*(2-3), 147–159. doi:10.3109/03602539009041082 PMID:2272285

Yang, J., Qiu, L., Liu, B., Peng, Y., Yan, F., & Shang, S. (2011). Synthesis of polymeric ionic liquid microsphere/Pt nanoparticle hybrids for electrocatalytic oxidation of methanol and catalytic oxidation of benzyl alcohol. *Journal of Polymer Science. Part A, Polymer Chemistry*, *49*(21), 4531–4538. doi:10.1002/pola.24895

Yang, Y., Li, H., Hou, F., Hu, J., Zhang, X., & Wang, Y. (2016). Facile synthesis of ZnO/Ag nanocomposites with enhanced photocatalytic properties under visible light. *Materials Letters*, *180*, 97–100. doi:10.1016/j.matlet.2016.05.117

Yan, N., Yuan, Y., & Dyson, P. J. (2011). Rhodium nanoparticle catalysts stabilized with a polymer that enhances stability without compromising activity. *Chemical Communications*, *47*(9), 2529–2531. doi:10.1039/c0cc04641h PMID:21225042

Yannas, I. V., & Burke, J. F. (1980). Design of an artificial skin. I. Basic design principles. *Journal of Biomedical Materials Research*, *14*(1), 65–81. doi:10.1002/jbm.820140108 PMID:6987234

Yantasee, W., Rutledge, R., Chouyyok, W., Sukwarotwat, V., Orr, G., Warner, C., . . . Shane Addleman, R. (2010). *Functionalized Nanoporous Silica for the Removal of Heavy Metals from Biological Systems: Adsorption and Application* (Vol. 2). Academic Press.

Yan, W. H., Hou, J. B., Fu, Z. P., Yang, B. F., Yang, P. H., Liu, K. P., ... Chen, Y. J. (2009). Growth and photocatalytic properties of one-dimensional ZnO nanostructures prepared by thermal evaporation. *Materials Research Bulletin*, *44*(10), 1954–1958. doi:10.1016/j.materresbull.2009.06.014

Yao, Y., Lu, F., Zhu, Y., Wei, F., Liu, X., Lian, C., & Wang, S. (2015). Magnetic core–shell CuFe2O4@ C3N4 hybrids for visible light photocatalysis of Orange II. *Journal of Hazardous Materials*, *297*, 224–233. doi:10.1016/j.jhazmat.2015.04.046 PMID:25974659

Yao, Z. T., Xia, M. S., Sarker, P. K., & Chen, T. (2014). A review of the alumina recovery from coal fly ash, with a focus in China. *Fuel*, *120*, 74–85. doi:10.1016/j.fuel.2013.12.003

Ye, C. H., Bando, Y., Shen, G. Z., & Golberg, D. (2006). Thickness-dependent photocatalytic performance of ZnO nanoplatel. *The Journal of Physical Chemistry B*, *110*(31), 15146–15151. doi:10.1021/jp061874w PMID:16884228

Yen, C. H., Lien, H. L., Chung, J. S., & Yeh, H. D. (2017). Adsorption of precious metals in water by dendrimer modified magnetic nanoparticles. *Journal of Hazardous Materials*, *322*, 215–222. doi:10.1016/j.jhazmat.2016.02.029 PMID:26921983

Yilmaz, B., & Müller, U. (2009). Catalytic applications of zeolites in chemical industry. *Topics in Catalysis*, *52*(6-7), 888–895. doi:10.100711244-009-9226-0

Ying, Y., Ying, W., Li, Q., Meng, D., Ren, G., Yan, R., & Peng, X. (2017). Recent advances of nanomaterial-based membrane for water purification. *Applied Materials Today*, *7*, 144–158. doi:10.1016/j.apmt.2017.02.010

Yoon, T. J., Lee, W., Oh, Y. S., & Lee, J. K. (2003). Magnetic nanoparticles as a catalyst vehicle for simple and easy recycling. *New Journal of Chemistry*, *27*(2), 227–229. doi:10.1039/b209391j

Yoshida, K., & Kasama, K. (1987). Biotransformation of nitric oxide. *Environmental Health Perspectives*, *73*, 201–206. doi:10.1289/ehp.8773201 PMID:3665863

Yoshii, K., Tsuda, T., Arimura, T., Imanishi, A., Torimoto, T., & Kuwabata, S. (2012). Platinum nanoparticle immobilization onto carbon nanotubes using Pt-sputtered room-temperature ionic liquid. *RSC Advances*, *2*(22), 8262. doi:10.1039/c2ra21243a

Yuan, X., Yan, N., Katsyuba, S. A., Zvereva, E. E., Kou, Y., & Dyson, P. J. (2012). A remarkable anion effect on palladium nanoparticle formation and stabilization in hydroxyl-functionalized ionic liquids. *Physical Chemistry Chemical Physics*, *14*(17), 6026–6033. doi:10.1039/c2cp23931k PMID:22446618

Yu, F., Sun, L., Zhou, Y., Gao, B., Gao, W., Bao, C., ... Li, Y. (2016). Biosorbents based on agricultural wastes for ionic liquid removal: An approach to agricultural wastes management. *Chemosphere*, *165*, 94–99. doi:10.1016/j.chemosphere.2016.08.133 PMID:27639078

Yu, S., & Gh, D. (1985). *Catalitical reactions in the natural waters/* Chisinau, Moldova: News of the Academy of Sciences of Moldova. *Series of Biological and Chemical Sciences (Rus)*, *3*, 74–76.

Yusuf, A., Olasehinde, A., Mboringong, M. N., Tabale, R. P., & Daniel, E. P. (2018). Evaluation of heavy metals concentration in groundwater around Kashere and its environs, upper Benue trough, northeastern Nigeria. *Global Journal of Geological Sciences*, *16*(1), 25–36. doi:10.4314/gjgs.v16i1.4

Zäch, M., Hägglund, C., Chakarov, D., & Kasemo, B. (2006). Nanoscience and nanotechnology for advanced energy systems. *Current Opinion in Solid State and Materials Science*, *10*(3-4), 132–143. doi:10.1016/j.cossms.2007.04.004

Zafiriou, O. C., Joussot-Dubien, J., Zepp, R. G., & Zika, R. G. (1984). Photochemistry in natural waters. *Environmental Science & Technology*, *18*(12), 1984. doi:10.1021/es00130a711

Zaleska-Medynska, A., Marchelek, M., Diak, M., & Grabowska, E. (2016). Noble metal-based bimetallic nanoparticles: The effect of the structure on the optical, catalytic and photocatalytic properties. *Advances in Colloid and Interface Science*, *229*, 80–107. doi:10.1016/j.cis.2015.12.008 PMID:26805520

Zănoagă, C. (1995). Role of the redox state in informing the biological quality of natural waters. Self-Purification Processes in Natural waters, 218-245.

Zare, E. N., Lakouraj, M. M., & Kasirian, N. (2018). Development of effective nano-biosorbent based on poly m-phenylenediamine grafted dextrin for removal of Pb (II) and methylene blue from water. *Carbohydrate Polymers*, *201*, 539–548. doi:10.1016/j.carbpol.2018.08.091 PMID:30241851

Żelazowska, R., & Pasternak, K. (2007). Matale szlachetne: Srebro (Au), złoto (Au), platyna w biologii i medycynie. *Bromatologia i Chemia Toksykologiczna*, *2*, 205–209.

Zelmanov, G., & Semiat, R. (2008). Phenol oxidation kinetics in water solution using iron (3)-oxide-based nano-catalysts. *Water Research*, *42*(14), 3848-3856.

Zeng, J. H., Jin, B., & Wang, F. (2009). Facet enhanced photocatalytic effect with uniform single-crystalline zinc oxide nanodisks. *Chemical Physics Letters*, *472*(1-3), 90–95. doi:10.1016/j.cplett.2009.02.082

Zepp, R. (1980). Assessing the photochemistry of organic pollutants in aquatic environments in dynamics, eposure and hazard assessment of toxic chemical. Arbor Science Publishers.

Zepp, R. G., & Cline, D. M. (1977). Rates of direct photolysis in aquatic environments. *Environmental Science & Technology*, *11*(4), 359–366. doi:10.1021/es60127a013

Zepp, R. G., Skurlatov, Yu. I., & Rutmiller, L. F. (1988). Effects of aquatic humic substances on analysis for hydrogen peroxide using peroxidase catalyzed oxidation of triarylmethanes or p-hydroxyphenylacetic acid. *Environmental Technology Letters*, *9*(4), 287–298. doi:10.1080/09593338809384569

Zgu, R., Alivov, Y. Z., Liu, C., Teke, A., Reshchikov, C. A., Dogan, S., ... Morkoc, H¸. A. (2005). Comprehensive review of ZnO materials and devices. *Journal of Applied Physics*, *98*, 1–103.

Zhang, B., & Yan, N. (2013). Towards Rational Design of Nanoparticle Catalysis in Ionic Liquids. *Catalysts*, *3*(2), 543–562. doi:10.3390/catal3020543

Zhang, C., & Diaz Anadon, L. (2013). Life Cycle Water Use of Energy Production and Its Environmental Impacts in China. *Environmental Science & Technology*, *47*(24), 14459–14467. doi:10.1021/es402556x PMID:24125477

Zhang, F., Chen, X., Wu, F., & Ji, Y. (2016). High adsorption capability and selectivity of ZnO nanoparticles for dye removal. *Colloids and Surfaces. A, Physicochemical and Engineering Aspects*, *509*, 474–483. doi:10.1016/j.colsurfa.2016.09.059

Zhang, H., & Cui, H. (2009). Synthesis and Characterization of Functionalized Ionic Liquid-Stabilized Metal (Gold and Platinum) Nanoparticles and Metal Nanoparticle/Carbon Nanotube Hybrids. *Langmuir*, *25*(5), 2604–2612. doi:10.1021/la803347h PMID:19437685

Zhang, J., Li, L., Li, Y., & Yang, C. (2018). Adsorptive filtration of low-concentration lead with wet-laid filter media made from nano-TiO_2 decorated cellulose fibers. *Journal of Chemical Technology and Biotechnology (Oxford, Oxfordshire)*, *93*(2), 432–439. doi:10.1002/jctb.5373

Zhang, J., Worley, J., Dénommée, S., Kingston, C., Jakubek, Z. J., Deslandes, Y., ... Botton, G. A. (2003). Synthesis of metal alloy nanoparticles in solution by laser irradiation of a metal powder suspension. *The Journal of Physical Chemistry B*, *107*(29), 6920–6923. doi:10.1021/jp027269k

Zhang, L., Liu, Z., Wang, Y., Xie, R., Ju, X. J., Wang, W., ... Chu, L. Y. (2017). Facile immobilization of Ag nanoparticles on microchannel walls in microreactors for catalytic applications. *Chemical Engineering Journal*, *309*, 691–699. doi:10.1016/j.cej.2016.10.038

Zhang, L., Xia, W., Teng, B., Liu, X., & Zhang, W. (2013). Zirconium cross-linked chitosan composite: Preparation, characterization and application in adsorption of Cr(VI). *Chemical Engineering Journal*, *229*, 1–8. doi:10.1016/j.cej.2013.05.102

Zhang, L., Yang, H. Q., Ma, J. H., Li, L., Wang, X. W., Zhang, L. H., ... Wang, X. Y. (2010). Controllable synthesis and shape-dependent photocatalytic activity of ZnO nanorods with a cone and different aspect ratios and of short-and-fat ZnO microrods by varying the reaction temperature and time. *Appl. Phys. Astronomy*, *100*, 106–1067.

Zhang, Q. F., & Cao, G. Z. (2011). Nanostructured photoelectrodes for dye-sensitized solar cells. *Nano Today*, *6*(1), 91–109. doi:10.1016/j.nantod.2010.12.007

Zhang, Q., Xu, R., Xu, P., Chen, R., He, Q., Zhong, J., & Gu, X. (2014). a. Performance study of ZrO2 ceramic microfiltration membranes used in pretreatment of DMF wastewater. *Desalination*, *346*, 1–8. doi:10.1016/j.desal.2014.05.006

Zhang, W., Qiao, H., & Chen, J. (2007b). Synthesis of silver nanoparticles—Effects of concerned parameters in water/oil microemulsion. *Materials Science and Engineering B*, *142*(1), 1–15. doi:10.1016/j.mseb.2007.06.014

Zhang, W., Qiao, X., & Chen, J. (2007a). Synthesis of Nanosilver Colloidal Particles in Water/Oil Microemulsion. *Colloids and Surfaces. A, Physicochemical and Engineering Aspects*, *299*(1-3), 22–28. doi:10.1016/j.colsurfa.2006.11.012

Zhang, W., Zhou, C., Zhou, W., Lei, A., Zhang, Q., Wan, Q., & Zou, B. (2011). Fast and considerable adsorption of methylene blue dye onto graphene. *Bulletin of Environmental Contamination and Toxicology*, *87*(1), 86–90. doi:10.100700128-011-0304-1 PMID:21567134

Zhang, X. (2014). A study on the water environmental quality assessment of fenjiang River in Yaan city of Sichuan Province in China. *ScienceDirect IERI Procedia*, *9*, 102–109. doi:10.1016/j.ieri.2014.09.048

Zhang, X., Zhao, J., Cheng, L., Lu, C., Wang, Y., He, X., & Zhang, W. (2014). Acrylic acid grafted and acrylic acid/sodium humate grafted bamboo cellulose nanofibers for Cu^{2+} adsorption. *RSC Advances*, *4*(98), 55195–55201. doi:10.1039/C4RA08307E

Zhang, Y., Wu, B., Xu, H., Liu, H., Wang, M., He, Y., & Pan, B. (2016). Nanomaterials-enabled water and wastewater treatment. *NanoImpact*, *3*, 22–39. doi:10.1016/j.impact.2016.09.004

Zhang, Y., Yan, L., Xu, W., Guo, X., Cui, L., Gao, L., ... Du, B. (2014). b. Adsorption of Pb (II) and Hg (II) from aqueous solution using magnetic CoFe2O4-reduced graphene oxide. *Journal of Molecular Liquids*, *191*, 177–182. doi:10.1016/j.molliq.2013.12.015

Zhang, Y., Zhu, C., Liu, F., Yuan, Y., Wu, H., & Li, A. (2019). Effects of ionic strength on removal of toxic pollutants from aqueous media with multifarious adsorbents: A review. *The Science of the Total Environment*, *646*, 265–279. doi:10.1016/j.scitotenv.2018.07.279 PMID:30055489

Zhao, M., Zhong, N., & Ji, Y. (2017). Ultra-stable colloidal porous Pt-Au-Ag nanoparticles. *Materials Letters*, *191*, 38–41. doi:10.1016/j.matlet.2017.01.003

Zha, Y., Li, C., Liu, X., Gu, F., Du, H. L., & Shi, L. (2008). Zn-doped TiO_2 nanoparticles with high photocatalytic activity synthesized by hydrogen–oxygen diffusion flame. *Applied Catalysis B: Environmental*, *79*(3), 208–215. doi:10.1016/j.apcatb.2007.09.044

Zheng, W., Liu, X., Yan, Z., & Zhu, L. (2009). Ionic liquid-assisted synthesis of large-scale TiO2 nanoparticles with controllable phase by hydrolysis of TiCl4. *ACS Nano*, *3*(1), 115–122. doi:10.1021/nn800713w PMID:19206257

Zhou, B., Balee, R., & Groenendaal, R. (2005). Nanoparticle and nanostructure catalysts: Technologies and markets. *Nanotech. L. & Bus.*, *2*, 222.

Zhou, C., Wu, Q., Lei, T., & Negulescu, I. I. (2014). Adsorption kinetic and equilibrium studies for methylene blue dye by partially hydrolyzed polyacrylamide/cellulose nanocrystal nanocomposite hydrogels. *Chemical Engineering Journal*, *251*, 17–24. doi:10.1016/j.cej.2014.04.034

Zhou, J., Liu, Y., Zhou, X., Ren, J., & Zhong, C. (2017). Removal of mercury ions from aqueous solution by thiourea-functionalized magnetic biosorbent: Preparation and mechanism study. *Journal of Colloid and Interface Science*, *507*, 107–118. doi:10.1016/j.jcis.2017.07.110 PMID:28780330

Zhou, W., Zhou, Z., Song, S., Li, W., Sun, G., Tsiakaras, P., & Xin, Q. (2003). Pt based anode catalysts for direct ethanol fuel cells. *Applied Catalysis B: Environmental*, *46*(2), 273–285. doi:10.1016/S0926-3373(03)00218-2

Zhou, X., Jing, G., Lv, B., Zhou, Z., & Zhu, R. (2016). Highly efficient removal of chromium (VI) by Fe/Ni bimetallic nanoparticles in an ultrasound-assisted system. *Chemosphere*, *160*, 332–341. doi:10.1016/j.chemosphere.2016.06.103 PMID:27393969

Zhu, Y., & Hosmane, N. S. (2015, June 15). Nanocatalysis: Recent advances and applications in boron chemistry. *Coordination Chemistry Reviews*.

Zhuang, J., Dai, W., Tian, Q., Li, Z., Xie, J., Wang, J., ... Wang, D. (2010). Photocatalytic degradation of RhB over TiO_2 bilayer films: Effect of defects and their location. *Langmuir*, *26*(12), 9686–9694. doi:10.1021/la100302m PMID:20334418

Zhu, H., Jiang, R., Xiao, L., Chang, Y., Guan, Y., Li, X., & Zeng, G. (2009). Photocatalytic decolorization and degradation of Congo Red on innovative crosslinked chitosan/nano-CdS composite catalyst under visible light irradiation. *Journal of Hazardous Materials*, *169*(1-3), 933–940. doi:10.1016/j.jhazmat.2009.04.037 PMID:19477069

Zhu, H., & Njuguna, J. (2014). 7 - Nanolayered silicates/clay minerals: uses and effects on health. In J. Njuguna, K. Pielichowski, & H. Zhu (Eds.), *Health and Environmental Safety of Nanomaterials* (pp. 133–146). Woodhead Publishing. doi:10.1533/9780857096678.3.133

Zhu, H., Wang, Y., Wang, X., Luan, T., & Tam, N. F. Y. (2014). Intrinsic debromination potential of polybrominated diphenyl ethers in different sediment slurries. *Environmental Science & Technology*, *48*(9), 4724–4731. doi:10.1021/es4053818 PMID:24679249

Zhu, Q. Y., Chen, J., Zhu, Q., Cui, Y. M., Liu, L., Li, B., & Zhou, X. F. (2010). Monodispersed hollow microsphere of ZnO mesoporous nanopieces: Preparation, growth mechanism and photocatalytic performance. *Materials Research Bulletin*, *45*(12), 2024–2030. doi:10.1016/j.materresbull.2010.06.016

Zinicovscaia, Duliu, Culicov, Frontasyeva, & Sturza. (2018). Major and trace elements distribution in Moldavian soils. Rom. Rep. Phys., 70(2).

Zinicovscaia, I., Duliu, O. G., Culicov, O. A., Sturza, R., Bilici, C., & Gundorina, S. (2017). Geographical Origin Identification of Moldavian Wines by Neutron Activation Analysis. *Food Analytical Methods*, *10*(11), 3523–3530. doi:10.100712161-017-0913-3

Zinicovscaia, I., Hramco, C., Duliu, O., Vergel, K., Culicov, O., Frontasyeva, M., & Duca, G. (2018). Air pollution study in the Republic of Moldova using moss biomonitoring technique. *Bulletin of Environmental Contamination and Toxicology*, *98*(2), 262–269. doi:10.100700128-016-1989-y PMID:27889805

Zinicovscaia, I., Sturza, R., Gurmeza, I., Vergel, K., Gundorina, S., & Duca, G. (2018). Metal bioaccumulation in the soil-leaf-fruit system determined by neutron activation analysis. *Journal of Food Measurement and Characterization*. doi:10.100711694-018-9972-4

Zou, B., Wang, X., Zhang, Q., Wan, Q., Dai, G., & Zhou, C. (2011). Controllable ZnO architectures by ethanolamine-assisted hydrothermal reaction for enhanced photocatalytic activity. *The Journal of Physical Chemistry C*, *115*(6), 2769–2775. doi:10.1021/jp1096822

Zveagintsev, D. (1991). Methods of soil microbiology and biochemistry. Moscow: MGU.

Zvereva, E. E., Katsyuba, S. A., Dyson, P. J., & Aleksandrov, A. V. (2016). Solvation of Palladium Clusters in an Ionic Liquid: A QM/MM Molecular Dynamics Study. *The Journal of Physical Chemistry C*, *120*(8), 4596–4604. doi:10.1021/acs.jpcc.5b11229

Zvyagintsev, A. M., & Kuznetsova, I. N. (2013). Formirovanie urovney ozona v prizemnom vozdukhe [The formation of ozone levels in surface air]. In Collection «Vtoroye mezhdunarodnoe soveschanie-seminar: Problemy monitoringa prizemnogo ozona i puti neytralizatsii ego vrednogo vliyaniya, Tarusa 6-7 June 2012 [Second international meeting-seminar on Problems of monitoring ground-level ozone and ways to neutralize its harmful influence." Tarusa 6-7 June 2012]. Moscow: Academic Press.

Żyrkowski, M., Neto, R. C., Santos, L. F., & Witkowski, K. (2016). Characterization of fly-ash cenospheres from coal-fired power plant unit. *Fuel*, *174*, 49–53. doi:10.1016/j.fuel.2016.01.061

About the Contributors

Gheorghe Duca is the author of 1760 edited works, including 418 scientific papers, 136 editorial works, 165 patents, 83 monographs and textbooks, 513 articles on science popularization and more than 445 conference presentations and materials, including the first textbook "Ecological Chemistry," published in three editions and translated into three languages: Romanian , Russian, English. Prof. Gh. Duca developed and taught courses on Physical chemistry, Chemical technology, Ecological chemistry, Processes and equipment, Physico-chemical research methods, Chemical risk assessment, Kinetics and thermodynamics of ecological systems. He held the lectures at the State University of Moldova, State University "Dmitrie Cantemir", Universities of Rome, Turin (Italy), Texas, California (USA), Zurich (Switzerland), Budapest (Hungary), Prague (Czechia), Iasi, Timisoara, Cluj, Bucharest (Romania). Under his supervision 30 PhD were granted, including thesis in cooperation with INSA, Lyon, France (1998) and 5 Habilitate Doctors. He founded Department of Industrial and Ecological Chemistry, SUM (1991), Research Center of Ecological and Applied Chemistry, SUM (1992), Faculty of Ecology, International University of Moldova (1992), Research Center of Physical and Inorganically Chemistry (2006), High School for gifted children of the ASM (2007), University of the Academy of Sciences (2007), Nongovernmental organization for environmental protection "Terra Nostra" (1992), Moldovan Research and Development iAssociation, MRDA (2000), and the journals "Environment" (2002), "Chemistry Journal of Moldova" (2005), Journal of Science, Innovation, Culture and Art "Akademos" (2005). Acad. Gh. Duca is a member of Board of Directors of BSCSF-International Fund for Cooperation and Partnership of the Black Sea and the Caspian Sea (2010), JRC-European Union Joint Research Center (2013), ALLEA-All European Academy (2018) and was delegated by the Government of Moldova to sign the Memorandum of Understanding for participation of the Republic of Moldova in the FP7, EU Framework Programme for Research, Technological Development and Demonstration Activities (2007-2013), Association Agreement for participation of the Republic of Moldova in the H2020, EU Framework Programme for Research and Innovation (2014-2020) and supervise the cooperation of the Republic of Moldova with the U.S. Civilian Research and Development Foundation, CRDF (2000-2010). Besides, he is Honorary and Full Member of the Balkan Academy of Science, Bulgarian Academy of Science and Art (2018), Culture New and Sustainable Development, Bulgaria (2004), Academy of Agricultural and Forestry Sciences "Gheorghe Ionescu-Siseşti" Romania (2006), Academy of Sciences and Arts of Montenegro (2006), Honorary Member of the Romanian Academy (2007), National Academy of Sciences of Ukraine (2009), Russian Academy of Sciences (2012), Research Network for Advance Studies of River System DANUBIUS (2013), APOLLONIA University, Iasi, Romania (2018), BABES-BOLYAI University, Cluj-Napoca University, Craiova University and Oradea University, Romania (2018). He was appreciated with a lots of awards and distinctions such as: Knight of the Order for Inventions of the Kingdom of Belgium (2003), the Cross of Commander of the Order of Honour, Poland (2004), the Gold Medal for Eminent Services to the cause of progress, Brussels (2005), Order "Cultural Merit" of Romania (2006), Gold Medal "N. S. Kurnakov ", Russian Federation (2007), International Socrates Award, Oxford, UK (2009), Silver Medal of the Association of Innovators of China (2010), Gold Medal "EUROINVENT", Brussels (2011), Order of the Republic of Moldova (2011), Award for outstanding achievements in the field of European Integration of the Economic Cooperation Organization of the Black Sea (2013), Medal of the National Commission of the Republic of Moldova for UNESCO (2014),the Order "The crown of Romania" with the rank of Commander (2015), Award of the Parliamentary Assembly of the Black Sea Economic Cooperation Organization (2016), Runners up Prize for Scientific Papers, Japan ICMSEM-2017, European Innovation Award with Diploma of Commander, European Commission, Belgium (2018), etc. Acad. Gh. Duca is not only a scientist, professor and prominent specialist in its field, but also a great organizer of research, skillful leader of the most important and prestigious institution of Moldova.

Ashok Vaseashta currently serves as the Chief Research Officer, Executive Director of Research and Chair of Institution Review Board and Office of Research Integrity at NJCU. Concurrently, he serves as Chaired Professor of Nanotechnology at the Academy of Sciences, Academician at Euro-Mediterranean Academy of Arts and Sciences and Senior Strategic Research Advisor for government and non-government organizations.. He is a scholar and thought leader who provides strategic leadership to promote and advance research initiatives and priorities in support of international defense, safety and security. Prior to joining NJCU, he served as Vice Provost for Research at the Molecular Science Research Center in Orangeburg, South Carolina. He also served as the Director of Research at the Institute for Advanced Sciences Convergence (IASC) and International Clean Water Institute (ICWI) for Norwich University's Applied Research Institutes and currently serves as strategic advisor. Prior to his appointment at NUARI, he served as a Professor of Physics and Physical Sciences and Director of Research at the Nanomaterials Processing and Characterization Laboratories, Graduate Program in Physical Sciences at Marshall University. Concurrently, he held Visiting Professorship at the 3 Nano-SAE Research Centre, University of Bucharest, Romania and visiting scientist at the Helen and Martin Kimmel Center of Nanoscale Science at the Weizmann Institute of Science, Israel. In 2007-08, he was detailed as a William C. Foster fellow to the Bureau of International Security and Nonproliferation at the US Department of State working with the Office of Weapons of Mass Destruction and Terrorism and Foreign Consequence Management program. He also served (2009-13) as a Franklin Fellow and strategic S&T advisor in the office of Verification and Transparency Technologies/Arms Verification and Control at the U.S. Department of State. His research interests span nanotechnology, environmental/ecological science and safety and security. Much of his work in nanotechnology has been on improving the understanding, design, and performance of nanofibers and sensors/detectors, mainly for applications such as wearable electronics, target drug delivery, detection of biomarkers and toxicity of nano and xenobiotic materials. In the security arena, he has worked on counterterrorism, countering unconventional warfare and hybrid threats, critical-Infrastructure protection, biosecurity, dual-use research concerns, and mitigating hybrid threats including fake-news. In addition, he has made numerous contributions in environmental/ecological science including directing research for International Clean Water Institute, pollution monitoring, contamination detection and remediation, and sustainability through green nanotechnology. He has authored over 250 research publications, edited/authored seven books on nanotechnology, and presented many keynote and invited lectures worldwide. Two books and a special edition of an international journal are currently under contract. He serves on the editorial board of Polish Journal of Environmental Studies, Digest

Journal of Nano-bio structures, International Journal of Green Nanotechnology. He received PhD from the Virginia Polytechnic Institute and State University, Blacksburg, VA in 1990 followed by Kobe post-doctoral fellowship. Since 2003. he served as the Director of several NATO Advanced Study Institutes (ASI)/Advanced Research workshops (ARW) supported by Emerging Security Challenges Division of the Science for Peace and Security. In 2008, he served as co-chair of an International Symposium on Nanotechnology and Environmental Pollution Prevention (ISNEPP). In 2009, he led the U.S. position on Nanotechnology in High Technology Coordination Group (HTCG) to joint US and India delegation. In addition, he served as a member of the U.S. Department of Commerce, NIST, and ANSI delegation to the U.K. representing the U.S. position on Standards in Nanotechnologies in the ISO/TAG to TC-229. He is a member of NATO-SET-040, an exploratory team panel investigating security and surveillance applications of nanotechnology. He serves as an expert counsel to the UNESCO, ObservatoryNANO, COSENT – south-east consortium on Nanotechnologies on NANO-Science and Technologies, and S&T advisor to NATO. He is an active member of several national and international professional organizations. He is a fellow of the American Physical Society (FAPS), Institute of Nanotechnology (FIoN), and the New York Academy of Sciences (FNYAS). He was awarded a Gold medal by the State Engineering University of Armenia for his contribution to Nanotechnology. In addition, he has earned several other fellowships and awards for his meritorious service including 2004/2005 Distinguished Artist and Scholar award.

* * *

Aliona Mereuta is associate professor at Dimitrie Cantemir State University. She obtained a master's degree in environmental science in 2019 and she got her PhD in chemistry at the State University of Moldova, specialty - Technology of special products (2004). A. Mereuta has managerial experience and implementation of waste treatment technologies. In the last five years, it has carried out three projects with technology transfer and a bilateral Romania-RM project. Among the results obtained: analysis of water quality in the area of the Prut river; section for the treatment of infectious medical waste; technology for the processing of plastics waste; methods for obtaining valuable substances from the winery wastes. She co-authored more than 20 scientific articles, and is the author of 10 patented inventions, and co-author of one monograph. She successfully participated in 8 national projects as coordinator and in 4 international projects as member of the research group.

Bogusław Buszewski is the head of the Chair of Environmental Chemistry & Bioanalytics at the Faculty of Chemistry, Nicolaus Copernicus University. He is also the president of the Central European Group for Separation Sciences and the chairmen of the Committee of Analytical Chemistry of Polish Academy of Sciences. Former, He was the president of the Polish Chemical Society and European Society for Separation Science. He was awarded by numerous national and international organizations. His main scientific interests are concerned with separation science including theoretical and practical aspects in chromatography and related techniques, adsorption, sample preparation, bioanalytic and bioanalysis (metabolomics, proteomics, biomarkers), nanotechnology and chemometrics. He is authored or co-authored 15 books, patents and more than 560 scientific papers (over 10.000 citations, h= 46) and member of the editorial boards of 26 national and international journals in field of analytical chemistry and separation sciences.

Victor Covaliov works as a Coordinating researcher and Moldova State University, Institute of Research and Innovation, Chisinau, Moldova. He is author of over 400 research publications, including 18 books and monographs on industrial ecology, as well as over 350 inventions and patents in the fields of plating, environmentally friendly technologies of wastes and waste waters treatment, gases separation and purification, biochemical processes, agro-industrial wastes management with the obtaining of value-added products. Dr. V. Covaliov is one of the leading scientists in the Republic of Moldova in elaboration and implementation of intensive and efficient water treatment technologies, was granted over 150 honorary diploma and medals at the international exhibitions on inventors and technology transfer, has an honorary title of Outstanding inventor (MD). Has taught lectures to university students on ecological audit, environmental engineering, general chemical technology, plating. His research interests involve: Intensified water treatment technologies pollution prevention, biogas processes, environmental engineering, physical-chemical and biochemical treatment of wastewaters and solid wastes containing refractory organics, heavy metal ions.

Olga Covaliova, Hab. in Chemistry, is a coordinating researcher at the Institute of Chemistry, Chisinau, Moldova (2015 till now), and a coordinating researcher at Moldova State University, Chisinau, Moldova (1982-till now, part-time). Synergistic Activities: Advisory Board Member of Water JPI (Joint Programming Initiative) (www.waterjpi.eu), since 2018; Member of Organizing Committee of International Conferences on Environmental and Ecological Chemistry, Chisinau, Moldova (1995, 2012, 2017); Grants Assistance Program Manager at Moldovan Research and Development Association (MRDA) (2004-2009), Chisinau, Moldova; Project Coordinator (2012-2015) of "WASTEnet – A Black Sea network promoting integrated natural Wastewater treatment systems" Project, Black Sea Joint Operational Program 2007-2013. Research interests: Water pollution prevention, biogas processes, environmental engineering, physical-chemical and biochemical treatment of wastewaters and solid wastes containing refractory organics, heavy metal ions. Is author and co-author of over 200 research publications, including 8 monographs, and about 100 patents for inventions (MD).

Hale Demirtepe received her Ph.D. degree in Environmental Engineering from Middle East Technical University in 2017. She then worked as a Research Specialist in the Research Center for Toxic Compounds in the Environment (RECETOX) at Masaryk University. She currently works as an Assistant Professor at the Department of Environmental Engineering, Izmir Institute of Technology. Her research interests include fate of pollutants in the environment, bioremediation of organic pollutants and indoor pollution.

Mintu Maan Dutta was born in Nazira, Sibsagar, Assam, India in May 10,1992. He obtained his B.Sc. degree from Cotton College in 2012 and M.Sc (Physical Chemistry) degree in 2014, from Gauhati University, Assam, India. He is carrying out his Ph.D. work in the Department of Chemistry, Gauhati University. His current research interest includes the use of magnetic nanoparticles for various organic transformations.

Viorica Gladchi, PhD, Associate Professor, Department of Industrial and Ecological Chemistry, Moldova State University. Dean of the Chemistry and Chemical Technology Faculty, Moldova State University. Research interests are focused on the studying the kinetics and mechanisms of free radical or photochemical redox catalytic processes responsible for transformation of various chemical pollutants in the model and natural aquatic systems; evaluation of surface waters self-purification capacity applying different methods of redox sate and kinetic parameters analysis, elaboration of new indicators of water quality estimation, prognosis and its ability for auto purification, as well as recommendations for their application.

Maria Vasile Gonța graduated Moldova State University, Faculty of Chemistry in 1973. In 1986 she graduated PhD, Physical Chemistry field. Since 2004 from to 2006 she has been studied Post-Doc. Maria Gonta has been head of the Environmental and Industrial Chemistry Department, SUM since 1998 from to 2016. Also, since 1995 – till present she has leaded Scientific Research laboratory. She is professor and has taught undergraduate level: The chemical management of water, The chemical management of air, Environmental Chemistry, Chemistry and Quality of food products, Graduate level (MSc): Ecological Chemistry of nitrate, nitrite and N-nitrosamine, organics dyes, Biopharmacy and pharmacodynamics. She is supervisor of the MSc and PhD thesis. Maria Gonta has elaborated more 280 scientific publications in the field of ecological chemistry, physical chemistry and environmental protection of with 92 articles in peer reviewed journals, 8 patents, 2 monographs, 5 textbooks.

Asaf Hajiyev is an internationally recognized scholar, a graduate of renowned Kolmogorov's probability school at Lomonosov Moscow State University (MSU). He has spent more than fifteen years at the MSU, in the Laboratory of Statistical Methods created and led by A. N. Kolmogorov, where he went from being a student to becoming a professor. He created control theory for queues with recurrent service, which can be applied for transportation systems. This theory allows to predict traffic jams and calculate some characteristics of traffic systems. Some fragments of this theory were unexpected and have been refereed as the paradoxe of theory of probability (see, K. Szekely, Paradoxes of theory of probability and mathematical statistics. Academia, Budapest, 1980). In 1980 for contribution to Queuing Theory, A. Hajiyev was awarded the Lenin Komsomol Prize in Science and Engineering. Another field of scientific interest is regression models. He constructed new type of regression models with increasing numbers of unknown parameters, where variances of observation errors unknown and different. Such models are typical for applications, but their investigation is challenging. For (linear and non-linear) such type of regression models, A. Hajiyev found necessary and sufficient conditions when various estimators (least square, L and M estimators) are consistent and proposed the methods of estimation of the elements of a covariance matric of a deviation vector. These results helped construct a confident band for unknown functions in regression models. A. Hajiyev also created the new method for constructing an optimal placement of the letters for different alphabets on a computer keyboard. This approach has been used for an optimal placement of the letters of Azerbaijani alphabet on a computer keyboard and has been adopted as a State Standard Placement by Azerbaijan State Agency on Standardization, Metrology and Patents and is used today in the Windows 8.1 and later versions. A. Hajiyev is a Chair of the Department of Applied Probability and Statistics in the Institute of Control Systems of the Azerbaijan National Academy of Sciences, where under his leadership more than 50 scholars successfully are working. A. Hajiyev is the President of the Scientific Council on Problems of Mathematics in Azerbaijan. His research have been published in the prestigious world leading journals (Scand. J. Statistics., Theory of Probability and its Applications; Stochastic Processes and their Applications; Russian Academy of Sciences, Doklady, Mathematics; Mathematical Notes and others) and he was invited as speaker for different universities of USA (Berkeley, USC (University of Southern California), G. Washington University, Lincoln University), Russia (Lomonosov Moscow State University, Bauman Moscow State Technical University), Sweden (Royal Institute of Technology, Chalmers University, Stockholm University), Germany (Munich University,

Frey University, Fern University), Portugal (Lisbon University), China (Chengdu University), Turkey (Bogazici University, Middle East Technical University) and others. A. Hajiyev created and introduced new probabilistic and statistical terminologies in Azerbaijanian language and published the Dictionary of Probabilistic and Statistical Terminologies in four languages (Azerbaijanian-English-Russian-Turkish). He is also co-author of Probability and Statistics Encyclopaedia in Azerbaijani language (1300p.) with translation of all terminologies and conceptions to English and Russian languages, which is used nowadays by scholars in the different fields. He is organizer and editor of the Proceedings of several International conferences, published by Springer and board member of various international scientific journals (Applied and Computational Mathematics, Management Science and Engineering Management, Transaction of the Azerbaijan National Academy of Sciences, Azerbaijan Mathematical Journal and others). A. Hajiyev is academician of the Azerbaijan National Academy of Sciences, Fellow of TWAS (The World Academy of Sciences), Foreign member of Mongolian National Academy of Sciences, Honorary Academician of Academy of Sciences of Moldova, Honorary Professor of Chengdu University (China). During 2005-2011 he was senior associated of the International Centre for Theoretical Physics (ICTP). For twenty years (1995-2015) A. Hajiyev was member of the Azerbaijan Parliament, where as a member of the Committee of Science and Education he has brought contribution for development of science and education in independent Azerbaijan. In 2015 he was elected Secretary General of Parliamentary Assembly of the Black Sea Economic Cooperation (PABSEC) and in 2019 re-elected for the next 5 years.

Manoj Kumar Karnena is a Researcher in the Department of Environmental Science. He has 3 years of research experience working on various lab-scale projects. He has publications in reputed international and national journals. His research interests include Water and Wastewater Treatment, Natural Coagulants, Nanocatalysis, Storage Technologies and membrane technology.

Ketevan Kupatadze graduated from S.S. Orbeliani State Pedagogical University, faculty of Natural Sciences, Chemistry and Biology specialization (With honors). In 1997, she graduated from the same University postgraduate studies. In 1998, she defended a Candidate (PhD) and in 2003, she defended her Doctoral dissertations. She is the author of more than 70 published scientific works, monograph, manual as well as the author of three educational–virtual programs in Chemistry and Biochemistry. She authored over 200 scientific-popular articles in chemistry. She graduated from the secondary faculty Journalism specialization and published in many newspapers. From 2017 till now every Autumn and Spring, she has active collaboration with the Science Popularization Center of Eskisehir (Turkey), where she conducts special workshops for chemistry teachers and students. She is technical editor of "Southern Brazilian Journal of Chemistry" and editorial member of "Tchê Química" and "SCIREA Journal of Chemistry". 1998-2003 _ Docent (Senior teacher) of the Faculty of Natural Sciences at S.S. Orbeliani State Pedagogical University. 2003-2005 _ Professor at the same University. She was a member of the Scientific Council of the mentioned University and in 2000-2005 she served as the Head of the Department of Postgraduate Studies. At Ilia State University she works from 2006, first at the faculty of life sciences, then the faculty of Science and Art and nowadays she is Professor Faculty of Natural Sciences and Medicine. She manages courses in Chemistry and Biochemistry direction (Bachelor and Master level). The main research area-chemistry, didactic of chemistry, History of chemistry. She is expert of chemistry and author in the internet journal "Maststsavlebeli" (The Teacher) Scientific interests / research interests Chemistry, Ecological Chemistry, Didactic of Chemistry, History of Chemistry.

Erika Andrea Levei Originally from Cluj-Napoca, Romania, Erika has 15 years of experience in environmental and analytical chemistry. Erika currently works as a senior researcher at INCDO-INOE 2000 Research Institute for Analytical Instrumentation in Cluj-Napoca and has more than 70 refereed publications and 50 presentations in international conferences. Her research interests cover water and soil chemistry, environment quality assessment, source apportionment in contaminated areas, metals mobility and bioavailability.

Christina Marouli studied biochemistry (B.Sc. from Brandeis University), urban and environmental policy (M.A. from Tufts University) and sociology (M.A. and PhD from the University of California, Santa Cruz) in the USA – with a specialization on the environment, health and social inequalities. Her present research interests include education for sustainability; sustainable, healthy, smart cities; environmental behaviors and social change. She is teaching at the Environmental Studies Program of Deree – American College of Greece (ACG) and she is the founder and ex-Director of the Center of Excellence for Sustainability at ACG. She is also a recipient of a Fulbright award in the context of which she did research on multicultural environmental education programmes in the USA in collaboration with a Turkish professor. She has extensive consulting experience on environmental and occupational health issues. She has also worked in NGOs on women's and children's issues as Director or expert and she was a co-founder of the Emergency Research Center. She recently co-edited a book on "Environmental Exposures and Human Health Challenges".

Anastasia Misseyanni is Associate Professor of Environmental Studies at Deree – The American College of Greece (ACG). She holds a Chemistry degree from the University of Athens and graduate degrees in Biochemistry from the University of Paris 7 (DEA) and the University of Marburg (PhD). She has a long career in teaching during which she has developed and taught a significant number of undergraduate natural science and environmental studies courses with emphasis on chemistry, biological chemistry, environmental science, ecology, water resources and methodology for environmental studies. She has conducted research in the fields of protein-DNA interactions and hormone-regulated gene expression. She has also been active in the scholarship of teaching. Her present research interests include Mediterranean biodiversity and green roofs, environmental health issues, sustainability initiatives in higher education, innovative teaching and learning strategies in STEM disciplines, active and experiential learning. She coordinated the development of the Deree Environmental Studies program and has served as its program coordinator since 2010. Since fall 2014, she serves as Head of the Department of Science and Mathematics at Deree-ACG. She has contributed to efforts and initiatives to promote sustainability at ACG.

Paraskevi Papadopoulou is Professor of Biology at Deree-The American College of Greece. She holds Biology degrees from the University of California, Los Angeles and from the University of Athens and a PhD in Biology from the University of Athens. She has served as Head of the department of Science, Technology and Mathematics at Deree-ACG. Her research interests are focused in the fields of Structural Biology/Molecular Biophysics and Bioinformatics, on structural and self-assembly studies of fibrous proteins. She has also contributed to the development of genetic testing protocols for Tuberous Sclerosis and of Neurofibromatosis type 1. Her current research engagement is on environmental health issues, sexually transmitted diseases, Mediterranean biodiversity and green roofs, and innovative ways of teaching and learning in STEM disciplines plus on big data analytics in bioinformatics and medicine.

She teaches biology, molecular biology, human genetics and cell and molecular neurobiology and the natural sciences research methods course for the environmental studies program. She had a major contribution in the development of the Biomedical Sciences program at Deree-ACG.

Igor Povar is a head of the department "Physicochemical Methods of Research and Analysis" at Institute of Chemistry. He is a well-known specialist in the area of modeling and theoretical research, mainly, involving chemical reaction equilibrium in natural systems. The theory of the chemical equilibrium, properties, mechanism of action and buffer capacity of the two-phase buffers has been developed by our team. Our developed approach can be used for research of the buffer action in the case of formation of one or several other solid phases in natural waters. The use of results of this research allows predicting the variations of the composition and response of the environment at the technogenic loads increasing. The proposed approach is intended to determine the dominant processes that are responsible for observed changes in contaminant concentrations at a site and provide researchers with the knowledge needed to set reliable limits of ion (metal) levels in the environment. During the 2001-2009 years he worked at the University of Western Ontario, London, Canada, where he was actively involved in research in chemical kinetics and thermodynamics of reactions in the organic chemistry using various ab-initio theoretical techniques, modeling and theoretical research.

Viorica Railean-Plugaru is involved in projects related in biosynthesis of metal and its oxides nanocomposites (AgNCs, ZnONCs), as a new antiseptic agent. To her topics belongs synthesis and physicochemical characterization of nanomaterials by using multi instrumental approaches such as spectroscopy, optical and electron microscopy, mass spectrometry and separation techniques. In 2010 has been graduated the master's degree in chemical technology and Biotechnologies, specialization: Technology of Drugs and Cosmetic Products at State University of Moldova. In 2014 initiated the PhD study on "Biosilver nanocomposites as a safe antimicrobial agent" at the Faculty of Chemistry, Nicolaus Copernicus University in Torun (Poland). She defended in 2018. The research approaches into the field of new antimicrobial agent have been published in the form of monographs (2) and scientific publications (20), 19 of this from Journal Citation Reports. Additionally, has participated at national (14) and international conferences (8) presented the research approaches as a form of poster and oral presentations. Dr. Viorica Railean-Plugaru is co-author of 2 patent applications and 2 patent deposits.

Inna Romanciuc received her Master of Science in cartography in Taras Shevchenko National University of Kyiv in 2014. From 2014 till now working in State institution "Scientific Centre for Aerospace Research of the Earth Institute of Geological Science National Academy of Sciences of Ukraine" as lead engineer. Research interests include estimation of energy and mass exchange in geosystems, soil moisture content, drought disaster, ecological pollution via remote sensing data.

Lidia Romanciuc has a scientific background in the physical chemistry: kinetics, catalysis and ecological chemistry. The main point of scientific interest is focused on the natural water self-purification mechanisms research, chemical and photochemical transformation of pollutants in natural water, redox-processes in aquatic medium, water quality monitoring and management, hydrochemistry etc. She is graduated from the State University of Moldova. Passed several scientific fellowships in the Europe, Japan, Russia, USA. Has up to 100 scientific publications and patents. Contributed to the organization of series of international scientific conferences on Ecological & Environmental Chemistry with participation of experienced and young scientists from NIS, Europe, USA, etc.

Anatolie Sidorenko completed his doctoral degree in 1979 at the Institute for Low Temperatures of Ukrainian Academy of Sciences. Working in the field of superconductivity of low dimensional and layered systems, received his Habilitation degree in 1991. From 1993 to 1995 he was Humboldt Fellow at the Technical University of Karlsruhe, in 2000-2003 - visiting professor in University of Augsburg, in 2003-2004 - visiting professor in Institute of Nanotechnology, Karlsruhe, in 2004-2007 - head of Laboratory of Superconductivity at the Institute of Applied Physics of Moldavian Academy of Sciences. Since 2008 he is Director of Institute of Electronic Engineering and Nanotechnologies, Chisinau, Moldova. Anatolie Sidorenko was the editor of 4 books published in "Springer", he is member of Deutsche Physikalische Gesellschaft since 2001 and member of Moldavian Academy of Sciences since 2012.

Rohini Singh is assistant professor in Sitarambhai Naranji Patel Institute of Technology & Research Centre, Department of Chemical Engineering in Gujrat, India.

K. Sivakumar is a Professor and Head of the Department of Chemistry at Sri Chandrasekharendra Saraswathi Viswa Mahavidyalaya deemed to be university, Kanchipuram, Tamilnadu, India. His research areas are supramolecular chemistry and computational chemistry. He has around 60 publications in high impact journals. His h-index is 12. He has published three books for undergraduate students. He is Guiding Ph.D and M.Phil scholars in supramolecular chemistry. He is a recipient of a research fellowship awarded by the Indian Academy of Sciences. He is functioning as Associate editor of "Chemical Papers" journal published by Springer Nature, editor of "Open Chemistry" journal published by De Gruyter, Germany and functioning as editorial board member of various international and national journals. He is functioning as reviewer for many journals published by American Chemical Society, Elsevier, Springer, and Taylor & Francis. He has received fundings from DST, UGC, Dept. of Atomic Energy, Indian academy of sciences and Tamilnadu state council for science and technology for conducting and organizing various projects and programmes.

Yuri Ivanovich Skurlatov, Semenov Institute of Chemical Physics. Born in Pochinki village of Pochinkovsky district of Gorky region on December 17, 1941. Graduated Moscow Institute of physics and technology, (1966). Academic degree - Doctor of chemical Sciences (1981), Professor (1991). Specialist in physical and environmental chemistry, kinetics and catalysis of chemical reactions. Research interests: Environmental chemistry, problems of chemical safety of environmental objects, assessment of impact on water resources, soil cover, the environment in the event of emergencies, including the assessment of measures to minimize the risk of possible accidents, environmental monitoring, industrial environmental control.

Oxana Spinu is a scientific researcher of the department "Physicochemical Methods of Research and Analysis" at Institute of Chemistry. The main focus of her research activity is to analyze the way by which heavy metals move and transform within the environment, the distribution of metals in ecosystems, their deposition and cycling in the terrestrial environment, such kind of research helps to predict long-term effects in natural decrease as a remediation alternative and the obtained results are indented to provide researchers with a tool needed to help them to set reliable limits of ion (metal) levels in the environment. The use of results of this research allows predicting the variations of the composition and response of the environment at the technogenic loads increasing. Her developed approach is intended to determine the dominant processes that are responsible for observed changes in contaminant concentrations at a site and provide researchers with the knowledge needed to set reliable limits of ion (metal) levels in the environment.

Rodica Sturza is the Head of Oenology and Chemistry Department, Technical University of Moldova PhD in Chemistry, Institute of Chemistry and Chemical Technology of the Academy of Science of Lithuania. Vilnius (Lithuania), 1989. Habilitatus in Technical Science, Technical University of Moldova, 2008. Coordinator of the State Program of the Republic of Moldova "Food Security and Safety in the Period of Demographic Transition" (2018-2021). Area of scientific interest: • The impact of environmental factors on food safety, migration of POPs, phthalates from the soil and water in vegetal raw materials and food; • New functional nutritive elements from traditional raw materials (horticultural products and wastes from processing industry); • Processes of ecological recovery of plant extracts rich in antioxidants and colorants from horticultural sources and wastes; • The evolution of microbiological, physicochemical properties, antioxidant and antiradical capacity and shelf life of the products with added plant extracts. Management of research projects (11). Publications - Over 250 published scientific papers. Laureate of the "Nicolae Garbalau" Prize in Chemistry of Moldavian Academy of Sciences on 2017. Doctor Honoris Causa of the "Ştefan cel Mare University" from Suceava, Romania (2012).

Claudiu Tănăselia Originally from Cluj-Napoca, Romania, Claudiu works at INCDO-INOE 2000 Research Institute for Analytical Instrumentation in Cluj-Napoca for more than 13 years (currently a senior researcher) and has more than 12 years experience in inductively coupled plasma mass spectrometry method development and instrument operation. His work was published in more than 50 impact-factor journals and at over 30 international conferences. Research area involves elemental analysis of environmental samples (water, soil, sediments, and meteorites) and he's a member of The Meteoritical Society for the last 11 years.

Sergey Travin received his degrees from Moscow Institute of Physics and Technology (MIPT), faculty of molecular and chemical physics 1976, diploma with honors, specialty engineer-physicist, followed by his PhD in Chemistry 1979 Doctor of Sciences in Economics, Professor 2001 and Full member of the Russian Academy of Natural Sciences in the section "Education and sustainable development" 2002. He is co-author of three books and more than a hundred articles, including on chemical kinetics, environmental chemistry, Bank information technologies, corporate data warehouses. He is currently working with the Institute of Chemical Physics, USSR Academy of Sciences since 1973. Experience 43 years. Senior Research Scientist, with the scope of interest: Science, education; Chemistry, physics, mathematical modeling, and Computer science, Information systems. Chemical kinetics. Study of the mechanisms of complex multistage reactions. Automation of the experiment, creation and maintenance of computer systems for automated processing of experimental data. Mathematical modeling of the behavior of multicomponent systems. Application of mathematical methods of forecasting. Probability research in non-Gaussian cases with the distribution of Zipf-Mandelbrot with "long tails.

Natalia Velisco is a head of department Chemistry, Mathematics and Informatics of State University ''Dimitrie Cantemir'' (Chisinau, Republic of Moldova). Also, she is currently an assistant professor at the same university. She received her M.S. (2005) in environment protection and Ph.D. (2013) in technology of special products (chemistry science) from State University of Moldova. Her general area of research is treatment of wastes, especially winery wastes, and technological transfer. Current research emphases include assessment of water quality.

V. K. Yadav is working as an assistant professor, Department of Microbiology, Sankalchand Patel University, Visnagar, Mehsana, India. He has a PhD in Nanosciences, from Central university of Gujarat, India. He has his master's and bachelor's in science in microbiology from HNBGU and Delhi University. He is also a lifetime member of "Electron Microscope Society of India" India. The author has also one patent on the topic "A method of separation of ferrous, alumina and silica from fly ash" with application number. The author has published several papers in national and international journals. His expertise is in the field of microbial synthesis of nanoparticles, nanoparticles characterization, waste management, bio and Nano remediation of heavy metals and wastewater treatment.

Inga Zinicovscaia, PhD in Chemistry, Researcher in the Horia Hulubei National Institute for Physics and Nuclear Engineering, Romania seconded to the Sector of Neutron Activation Analysis and Applied Research, Frank Laboratory of Neutron Physics, Joint Institute for Nuclear Research., Russia. Her main area of interest is application of neutron activation analysis in the environmental studies, development of the technologies for wastewater treatment, air pollution studies, statistical analysis. She is the author and co-author of more than 60 scientific publications. Her achievements have been marked with numerous prizes and awards, including two Gold medals.

Index

O

P

R

S

T

U

V

W

www.ingramcontent.com/pod-product-compliance
Lightning Source LLC
LaVergne TN
LVHW080357011026
849161LV00009B/2329